Routledge Handbook of Landscape and Food

Since the turn of the millennium, there has been a burgeoning interest in, and literature of, both landscape studies and food studies. Landscape describes places as relationships and processes. Landscapes create people's identities and guide their actions and their preferences, while at the same time are shaped by the actions and forces of people. Food, as currency, medium, and sustenance, is a fundamental part of those landscape relationships.

This volume brings together over fifty contributors from around the world in forty profoundly interdisciplinary chapters. Chapter authors represent an astonishing range of disciplines, from agronomy, anthropology, archaeology, conservation, countryside management, cultural studies, ecology, ethics, geography, heritage studies, landscape architecture, landscape management and planning, literature, urban design, and architecture. Both food studies and landscape studies defy comprehension from the perspective of a single discipline, and thus such a range is both necessary and enriching.

The *Routledge Handbook of Landscape and Food* is intended as a first port of call for scholars and researchers seeking to undertake new work at the many intersections of landscape and food. Each chapter provides an authoritative overview, a broad range of pertinent readings and references, and seeks to identify areas where new research is needed – though these may also be identified in the many fertile areas in which subjects and chapters overlap within the book.

Joshua Zeunert is Senior Lecturer at the University of New South Wales in Sydney, Australia.

Tim Waterman is Senior Lecturer and Landscape Architecture Theory Coordinator at the University of Greenwich, London, UK.

Routledge Handbook of Landscape and Food

Edited by Joshua Zeunert and Tim Waterman

Routledge
Taylor & Francis Group
LONDON AND NEW YORK

First published 2018 by Routledge

2 Park Square, Milton Park, Abingdon, Oxfordshire OX14 4RN
52 Vanderbilt Avenue, New York, NY 10017

Routledge is an imprint of the Taylor & Francis Group, an informa business

First issued in paperback 2020

British Library Cataloguing-in-Publication Data
A catalogue record for this book is available from the British Library

Library of Congress Cataloging-in-Publication Data
A catalog record for this book has been requested.

ISBN: 978-1-138-12515-5 (hbk)
ISBN: 978-0-367-50212-6 (pbk)

Typeset in Bembo
by Apex CoVantage, LLC

Contents

Illustrations

Figures

Tables

Boxes

Contributors

Jacques Abelman MLA is an Assistant Professor of Landscape Architecture at the University of Oregon. He has also taught at the University of Edinburgh and for the European Master's in Landscape Architecture Program. He completed his MLA at the Amsterdam Academy of Architecture and spent over ten years in professional practice with projects in the Netherlands, Belgium, France, and the UK. His research focuses on multifunctional green infrastructure, food systems, and social justice.

Tal Alon-Mozes PhD is a Landscape Architect and an Associate Professor of Architecture and Town Planning at the Technion. She has an MLA degree from the University of California, Berkeley, and a PhD from the Technion. Her interests include history and theory of gardens and landscape architecture, landscape and culture, and especially the cultural dimensions of landscape production in Palestine and Israel. Among the topics of her papers are the history of gardens of pre-state Israel and its current landscapes, planning and design of Israel's national parks, memorial parks, and the culture of urban agriculture in contemporary Israel.

Gary Austin is a Professor of Landscape Architecture at the University of Idaho. He has experience in public and private landscape architecture practice, being a licensed landscape architect in the state of Washington, USA. He is the author of two recent books: *Constructed Wetlands and Sustainable Development* (with Kongjian Yu, 2016) and *Green Infrastructure for Landscape Planning: Integrating Natural and Human Systems* (2014). Professor Austin is currently at work on a book of sustainable development lessons derived from his study of several European Green Capital cities and another book of planning and design methods to increase urban biological diversity.

Stephan Barthel PhD is an Associate Professor in Urban Planning at the University of Gävle, Sweden, and he holds a research leader position at Stockholm Resilience Centre. His research concerns what role nature environments in cities play for social-ecological resilience by integrating systems theories with anthropology and psychology. He has been part of developing highly cited concepts and theories such as 'social-ecological memory', 'urban green commons', and 'biocultural refugia'.

Jody Beck PhD is an Assistant Professor in the Department of Landscape Architecture at the University of Colorado, Denver. He completed his PhD at the University of Pennsylvania. His book, titled *John Nolen and the Metropolitan Environment*, is based on his dissertation in which he examined the political underpinnings of John Nolen's work. Dr. Beck's work begins with the concept that landscapes are our foundational political medium and his interest in food systems grows from this interest in grounded politics. Dr. Beck grew up on a cattle ranch in rural Montana and his family eats (mostly) from their city's foodshed.

Beau B. Beza PhD is Associate Head of School in Teaching and Learning at the School of Architecture and Built Environment at Deakin University. He has years of professional and academic experience working on a variety of projects in Australia, Bosnia-Herzegovina, the United States, Norway, Mexico, Colombia, Nepal, and Hashemite Kingdom of Jordan. In consultancy he uses an asset-based and co-production model to realise places in urban and rural settings. His research revolves around developing an understanding in how people derive meaning from the natural and built environment.

Megan Blake PhD is a Senior Lecturer in Human Geography and the Director of an MA in Food Security and Food Justice at the University of Sheffield in the United Kingdom. She has worked at universities in the United Kingdom, Hong Kong, and the United States. Her interests focus on how the physical and institutional design of cities shapes how people are enabled or dis-enabled to provision, cook, and eat the food that they want. In addition to publishing in the academic press, Megan actively engages with a number of organisations seeking to make a fairer food system for all.

Andrea Bosio PhD is an Architect currently based in Milan. He graduated both BArch and MArch from the Polytechnic of Milan, and later was awarded a PhD by Griffith University. While his main expertise in the practice involves projects on social housing and urban renewal, his research activity includes investigations on the relationship between the built environment and the food system; the historical connection of city and countryside; and the implications of architecture as a discipline upon societal patterns of inclusion/exclusion, freedom/subjection, and homologation/distinction.

Isis Brook PhD is Director of Crossfields Institute International, which is based in Stroud, UK. She has been a University Lecturer for over twenty years mainly teaching environmental philosophy and aesthetics and is the author of numerous papers in these areas. Currently Brook is teaching on the MA Philosophy of Social Innovation: Researching Holistic Agroecology. Her research interests include place, landscape aesthetics, education, and phenomenological research methods.

Sandra Caquimbo-Salazar MSc is Assistant Professor of the School of Architecture and Design and Editorial Coordinator of the journal *Cuadernos de Vivienda y Urbanismo* at the Pontificia Universidad Javeriana in Bogotá, Colombia. She has worked at universities in Colombia and Chile. Her research interests include landscape architecture, informal settlements/housing, and sustainable urban development. She is member of the Research Group: *Calidad y Habitabilidad de la Vivienda*. She is co-author of *Espacio Público, Periferia Urbana y Derecho a la ciudad, Intervención Parque Caracolí, Ciudad Bolívar* (2017), and *Desarrollo: Prácticas y Discursos Emergentes en América Latina* (2016).

Joy Carey is an independent Sustainable Food Systems Consultant and Filmmaker. She is Director of f3 local food consultants, from 2010–2014 was lead researcher on impact assessment within the Big Lottery Local Food Fund, and has worked for the Soil Association and the Supurbfood project. She is a member of Bristol Food Policy Council, a Director of Bristol Food Network, and an elected board member of Bristol Green Capital Partnership.

Rachel Carey PhD is a Research Fellow in the Faculty of Veterinary and Agricultural Sciences at the University of Melbourne. Her research focuses on sustainable food systems, particularly the resilience of city region food systems in the face of challenges from urban sprawl, climate change, and declining supplies of natural resources. She leads the Foodprint Melbourne project, which is investigating the significance of Melbourne's foodbowl for the city's food security. She has a background in food policy and has been involved in the development of food policies for the City of Melbourne and City of Greater Geelong.

Philip A. Clarke PhD is a consultant Anthropologist specialising in Indigenous relationships with the land, in particular Aboriginal use and perception of plants and animals. For much of his career he was based in the South Australian Museum, and he has curated exhibitions on Aboriginal culture and published over a hundred papers and chapters, along with five commercial books. He is currently working on several native title projects across Australia.

Jonathan Code MEd is the Deputy Director and a member of the faculty of Crossfields Institute International (CII), UK. He co-leads on CII's MA in Researching Holistic Agroecology while also lecturing and running workshops internationally on interdisciplinary approaches to environmental education. His interests focus on agroecology, integrative education, and human development. He is the author of *Muck and Mind: Encountering Biodynamic Agriculture* (Lindisfarne, 2014).

David Crouch Professor of Cultural Geography, University of Derby, has chapters in over eighty books and articles in thirty academic journals. He has given numerous conference papers across Europe and the US, and has been Visiting Professor in Swedish universities and Guest Scholar at Berkeley. He works with qualitative research and writing on art practice and theory; community art projects; landscape studies; everyday living and its culture in leisure and tourism; developing theory through a focus on human life and relationally with the 'other-than human'. His most recent books are *Flirting with Space: Journeys and Creativity* (2010; 2016) and an edited cross-disciplinary collection, *The Question of Space* (2017).

Pay Drechsel PhD in Environmental Sciences, is a Principal Researcher at the International Water Management Institute (IWMI), based in Colombo, Sri Lanka. Drechsel has twenty-five years of experience in the rural-urban interface of developing countries, coordinating projects in the domain of natural resource management. He has authored about three hundred publications, edited several books, and worked extensively in West and East Africa and South and South-East Asia. In 2015, Drechsel received the IWA Water and Development Award for Research.

Efrat Eizenberg PhD is an Assistant Professor at the Faculty of Architecture and Town Planning at the Technion. She completed her PhD in the Department of Environmental Psychology at the Graduate Center of the City University of New York in 2008 and her postdoc at the Planning for the Environment with Communities research lab (PECLAB) in the Department of Geography and Human Environment, Tel Aviv University, in 2011. Her research topics include urban nature and landscape perception, planning with communities, urban regeneration, urban struggles, and the politics of space. She is the author of *From the Ground Up: Community Gardens in New York City and the Politics of Spatial Transformation* (Routledge, 2016).

Alexandra Florea is a PhD student at Deakin University in Australia. She is focusing on the study of villages and rural landscape in Romania, following their transformation through the twentieth century until contemporaneity. She has worked in the architectural industry, tutored at Deakin University, and presented papers at international conferences.

Nina Mukerjee Furstenau MA, journalist and author, is Director of Food Systems Communication at the University of Missouri Science and Agricultural Journalism program in the US. She is the author of the MFK Fisher award-winning food memoir *Biting Through the Skin: An Indian Kitchen in America's Heartland* and *Savor Missouri: River Hills Country Food and Wine*, a book on the flavours of regions and the people who produce them. Her interests include food and identity, traditional food cultures, and the tie between heritage foods and health. She is currently researching low-protein diets in rural Mozambique.

Afshin Ghahramani PhD is an Agro-Ecosystem Modeller and Analyst at the Institute for Agriculture and Environment at the University of Southern Queensland, Australia. Dr. Ghahramani has worked in universities in Australia and Japan and at research organisations, as well as in industry with leading agricultural and environmental projects. He has more than seventeen years of experience in agriculture, water resources, and environmental innovation. Dr. Ghahramani is doing research on the trade-offs between biological, social, economic, and human components of the systems, and how they interact to contribute to sustainability. He has led highly innovative, cross-organisational and interdisciplinary research teams at national scales.

Munir A. Hanjra PhD is an Economist in the International Water Management Institute (IWMI) in Colombo, Sri Lanka. Previously he worked as Senior Research Fellow at Charles Sturt University (CSU); Institute for Land, Water and Society (ILWS); and CSIRO Land and Water, Australia. Hanjra has over twenty years of professional experience on sustainable water management, climate change, food security, wastewater economics, and public policy. He has authored about one hundred peer-reviewed publications and worked in South and South-East Asia as well as East, West, and Southern Africa.

Jaime Hernández-García PhD is an Associate Professor of the School of Architecture and Design at the Pontificia Universidad Javeriana in Bogotá, Colombia, where he currently serves as Director of the Aesthetics Department. His doctoral thesis "*El Parque de mi Barrio*: Production and Consumption of Open Spaces in Popular Settlements in Bogotá" won the Iberoamerican Award for research theses in México in 2011. His research interests include informal settlements/housing, open/public space, social/everyday aesthetics, and community participation. He is the author of *Public Space in Informal Settlements, the Barrios of Bogotá* (2013) and *Arquitectura, Participación y Hábitat Popular* (2008).

Peter R. Hobson PhD is a Professor in Biodiversity Conservation and Sustainability within the School of Sustainable Environments, Writtle University College, UK. He is also currently acting co-director for the Centre of Econics and Ecosystem Management in partnership with Eberswalde University for Sustainable Development, Germany. His research interests include the application of ecosystem theory propositions, specifically thermodynamics, to natural ecosystems and cultural landscapes. For the last ten years Hobson's research has focused on thermodynamic measures of ecosystem function in forested landscapes and how this research may contribute to a new paradigm of sustainable land use practice referred to as 'econics'.

Christian Isendahl PhD is a Senior Lecturer in Archaeology at the Department of Historical Studies, University of Gothenburg, Sweden. His research concerns issues of sustainability, vulnerability, and resilience, and he applies a historical ecological lens to study urbanism, farming systems, water management, and sociopolitical organisation in the Neotropics, including the Maya lowlands, the Andes, and the Amazon Basin. He is the co-editor, with Daryl Stump, of *Oxford Handbook of Historical Ecology and Applied Archaeology* (Oxford University Press).

Ruth Jackson PhD is a Visiting Research Fellow at Deakin University in Australia. She is currently co-editing a book by women who have conducted research in Africa, and researching the intersections between people, the Montane rainforest in southwest Ethiopia, and food production during one of the Ethiopian Orthodox feasts. These research interests build on earlier work that focused on international development and maternal health, and how the goal of reducing maternal mortality fits into Ethiopia's development agenda.

Sarah James PhD is a Visiting Fellow at the School of Regulation and Global Governance (REGNET) at the Australian National University. She is a human geographer whose research focuses on sustainable food systems for population health in a changing climate. Her work across Australia and the Pacific involves cross-sector collaborations with government, industry, and non-government organisations. Recent research themes include healthy and sustainable diets in Australia and food security in the Pacific. She is the author of *Farming on the Fringe: Peri-Urban Agriculture, Cultural Diversity and Sustainability in Sydney* (2016).

David S. Jones PhD, Foundation Professor of Environmental Planning and Landscape Architecture at Deakin University, has been instrumental in developing Deakin's postgraduate planning and landscape architecture courses that include strong learning experiences in Indigenous Knowledge Systems as part of their distinctiveness. His multidisciplinary research interests rotate around the meaning and crafting of places through both Western and Indigenous lens; national/world heritage recognition of designed and Indigenous landscapes; and include inquiries in the fields of landscape architecture, urban and regional planning, urban design, cultural heritage landscapes, and 'Country' in Australia as well as Indigenous landscapes nationally and internationally.

Valerie Kantelberg is a Graduate Forester and Research Associate for Environmental Monitoring, Forest Functions, and Recreation at the Bavarian State Institute of Forestry (an agency within the Bavarian Forestry Administration); contract teacher of api-biology at the Technical University of Munich, Division of Forest Science and Resource Management; and beekeeper with a focus on hollow log tree beekeeping. Kantelberg's research interests include recreation ecology and regional marketing of timber and non-timber forest products.

Max Kelly PhD lectures in International and Community Development Studies at Deakin University, specialising in food security, food systems, and sustainability from local to international. Kelly is an agricultural scientist turned social scientist, with a PhD in Sustainable Livelihoods in Malawi. She researches in areas from local food systems and land use to global development policy, currently working on the role of civil society in post-conflict development, animals and ethics in development, and food systems, and has recently co-edited a book on the impact of foreign aid on poverty and social justice. She has researched, consulted, and worked from farms to offices across the globe.

Ann Light DPhil is Professor of Design and Creative Technology at the University of Sussex and a qualitative researcher specialising in design for social well-being, participatory design, and social innovation. Her interests span culture change, sustainable living, ageing, and human-computer interaction, and she brings professional experience from the design sector, drawing on qualifications in humanities, arts, artificial intelligence, and computer science and work with arts and grass-roots organisations and marginalised groups on five continents, in local, transnational and international development settings. She has also written on gardening practice in understanding future design challenges and incorporating mindfulness and attentiveness.

Mirjana Lozanovska PhD is Senior Lecturer at Deakin University. Her research deploys multidisciplinary theories of space to examine migration, identity, and architecture, and the reinvention of the city. Recent publications include *Performing Equality: The Exceptional Story of Mimoza Nestorova-Tomić in Post 1963 Earthquake Reconstruction of Skopje* (Routledge, 2016) and *House Behaviour in the Australian Suburb* (Consuming Architecture, 2014). She is editor for the anthology *Ethno-Architecture and the Politics of Migration* (Routledge, 2016), and serves on the editorial boards for the journals *Interstices* and *Space and Culture*.

Gerd Lupp PhD freelancer was employed as the Chair for Strategic Landscape Change and Management at the Technical University of Munich until May 2017. He has worked for several research and practice institutions in the field of forestry, nature conservation, recreation ecology, and communication of sustainable land uses. His broad interests cover forest ecology, management, outdoor recreation, renewable energy, marketing of timber and non-timber forest products, as well as human-nature interactions.

Mary Lydecker MLA is a Principal with Hargreaves Associates, where she has worked on a range of projects from downtown parks and campus landscapes to urban riverfronts and shared-use trails. She is committed to the power of landscape to transform cities, with a focus on the reclamation of post-industrial landscapes that integrate infrastructural and ecological systems with the recreational, cultural, and economic goals of urban communities. Lydecker holds a BA from Williams College and an MLA from Harvard's Graduate School of Design. She is a registered landscape architect and a LEED AP BD+C accredited professional.

Damian Maye PhD is Reader in Agri-Food Studies at the Countryside and Community Research Institute in the University of Gloucestershire, England. He has a long-standing research interest in the geography and sociology of agri-food restructuring, including alternative networks, ethical foodscapes, geographies of food security, agricultural biosecurity, and sustainability transitions. He co-edited a special issue on 'food security' for *Journal of Rural Studies* in 2013 and another on 'sustainable food system transitions' for *Sociologia Ruralis* in 2017. Recent research examines multidimensional assessments of food chain performance, short food chains in city regions, and new institutional arrangements for agriculture and fisheries.

Saruhan Mosler PhD is a Senior Lecturer in Landscape Architecture at Writtle University College, UK. She has been leading the postgraduate landscape architecture course and supervising PhD and MA student dissertations for the last five years. Her interests focus on historic urban city developments and their contemporary design and planning including post-industrial sites, sustainable heritage strategies, and landscape theory. She has published several articles and book chapters in peer-reviewed journals and books.

Claire Nettle PhD is a Community Gardener and Researcher. She currently holds research positions at Flinders University and The University of South Australia. She is author of *Community Gardening as Social Action* (2014).

Zachary Nowak is a Doctoral Candidate in American Studies at Harvard University. He is also the Associate Director of the Food and Sustainability Studies Program at the Umbra Institute. He is interested in environmental and spatial history, and is currently working on a research project provisionally titled "The American Train Station: A History." He is the author of *Truffle: A Global History*, the translator of *Why Architects Still Draw*, and the editor and translator of *Inventing The Pizzeria: A History of Pizza Making in Naples*. He is also a co-editor of a volume on Italian food, *Representing Italy Through Food*.

E. Gunilla A. Olsson PhD is Professor in Human Ecology at the University of Gothenburg, Sweden. Her interdisciplinary work in research and teaching is on use and management of natural resources and biodiversity involving knowledge integration in social-ecological systems in Europe and Africa. She has also been working on those issues for the Swedish government and the United Nations Environment Program. Current research is on food security and urban resilience and the linking of urban and rural regions within the context of sustainable development. She is a co-author to global assessments for the UN-Intergovernmental Panel for Biodiversity and Ecosystem Services (IPBES).

Kenneth R. Olwig PhD is a Professor Emeritus of Landscape Planning at the Swedish University of Agricultural Sciences (SLU) in Sweden. He has also worked in the geography departments of five other universities internationally, and as a research fellow at universities in Denmark, Norway, and California. His interests focus on a combination of aesthetic, legal, literary, and cultural geographical approaches to landscape and the relationship between society and nature. These issues are the topics of two monographs: *Nature's Ideological Landscape* (1984) and *Landscape, Nature and the Body Politic* (2002). He has also published many articles and edited books.

Hannes Palang PhD is Professor of Human Geography and Head of the Centre of Landscape and Culture at Tallinn University, Estonia. He has been leading many research projects focusing on landscape, practice, and heritage. The most recent is titled *Culturescapes in Transformation: Towards an Integrated Theory of Meaning Making*. He was also President of the PECSRL (Permanent European Conference for the Study of the Rural Landscape).

Susan Parham PhD works at the University of Hertfordshire where as part of her role as Head of Urbanism, she is Academic Director of the International Garden Cities Institute in Letchworth. Parham has undertaken pioneering research since 1990 into the design and planning interplay between food and place in traditional cities, planned settlements, and contemporary conurbations. Her most recent book, *Food and Urbanism* (2015) delves into food and place design from the scale of the table to the critical region.

Johannes Paul PhD is a researcher at the International Water Management Institute (IWMI) in Colombo, Sri Lanka. During his thirty years of professional work he has gained substantial experience in various academic and development assignments with focus on water and waste management in Germany, EU countries, and South/South-East Asia. His special areas of interest are the recycling of organic waste from municipal and agricultural waste streams, the informal waste sector, and the development of appropriate technologies that are suitable to support waste management in low-income countries.

Anu Printsmann MSc is Researcher in Landscape Studies at the Centre for Landscape and Culture, Tallinn University, Estonia. She has been involved both in scientific and applied projects for historical, valuable, transboundary, urban, protected, seasonal, mire heritage, agricultural, coastal, and everyday landscapes. Her interests include historical, social, cultural, and moral geographies within landscape perception, combining GIS and oral histories in landscape reading and collaborative planning.

Angela E. Raffle FFPH is a Consultant in Public Health and Honorary Senior Lecturer at the University of Bristol. She qualified in Medicine in 1980, working in England and overseas before specialising in Public Health. She is a national figure in UK screening programmes and, since 2010, has devoted most of her professional practice to food, farming, social justice, and ecological matters. She helped found Bristol's Food Policy Council of which she is vice chair. She helped found, and Chairs, the Management Board of 'The Community Farm'.

Maggie Roe MDesS is Reader in Landscape Planning Research & Policy Engagement, Newcastle University, UK; a Director of the UK-based international charity, Landscape Research Group Ltd (LRG); and an editor of *Landscape Research*. Her research examines issues of landscape planning and sustainability with a special focus on participatory landscape planning, new cultural landscapes and landscape change. She has given numerous lectures and examined in many countries. Funded by national and international government agencies, research councils, and environmental bodies, she has published widely and has extensive reviewing experience including the UK REF2014. Recent research includes the transdisciplinary AHRC-funded 'Hydrocitizenship' project.

Jo Russell-Clarke PhD is a Registered Landscape Architect and Senior Lecturer at the University of Adelaide in Australia. Following graduation and practice in award-winning firms in Melbourne, Australia, she focused on teaching and contribution to design commentary through publication with various journals and on radio. Her interests span an earlier career as an orchestral musician, an ongoing love of productive gardening, and a fascination with the ways in which cultures are made and manifest through practices located in specific spatial locales. She currently co-hosts a weekly radio broadcast on food and dining with Radio Adelaide.

Saman Seneweera PhD is an Associate Professor at University of Southern Queensland, Australia. Dr. Seneweera is also an Adjunct Professor in the Department of Biological Sciences at the University of Ruhuna, Sri Lanka, and also Honorary Associate Professor at the University of Melbourne. Dr. Seneweera leads the 'Plant Stress Biology' research group at USQ to elucidate the mechanism of how plants adapt to climate stress, such as low water or low nutrients, high temperature, and high CO_2 concentrations. Internationally, Dr. Seneweera is recognised as one of the leading researchers in plant ecophysiology and climate change biology.

Helen Sooväli-Sepping is Professor of Environmental Management and Senior Researcher in Cultural Geography at Tallinn University, Estonia. She is Head of two Master's Programmes: Urban Governance and Environmental Governance. Her research focus is on the practice perception and representation of landscape and heritage, spatial contestations in urban and rural environments.

Suzanne Speak PhD is Senior Lecturer in International Spatial Planning as Newcastle University, England. She teaches also at universities in India and South Africa and has been involved in consultancy work for a number of governments and UN Habitat. Her research focuses on the everyday lives of the urban poor in the Global South. In particular, she researches and teaches on issues of homelessness, housing, informality, and livelihood practices. Her work on homelessness and informal settlements drove an interest in food security for the urban poor.

Jeremy Strong PhD is Professor of Literature and Film at the University of West London. Chair of the Association of Adaptation Studies (2010–2016), he is widely published in the field of literature-on-screen, but also writes on food and culture, including his book *Educated Tastes: Food, Drink and Connoisseur Culture* (2011). He is currently editing *James Bond Uncovered* for Palgrave-Macmillan.

Simon Tangerding Dipl. ing. silv. is a Forester currently employed at the Bavarian State Forest. He has studied forestry science at the Technical University of Munich in Germany and the State University of Petrozavodsk (Russia). He gained work experience during assignments in Germany, Zambia (Africa), Petrozavodsk (Russia), and Bishkek (Kyrgyzstan) as well as working for national and international consultancies in Western Europe.

Anders Wästfelt PhD and Associate Professor, works at the Department of Human Geography at Stockholm University in Sweden. His interests are geographical theory and the transformation of agriculture in global integration processes, and social and environmental aspects of agriculture. His recent research includes studies of agriculture leasehold, land use transformations, and developing methods for satellite image interpretation of land use. He has published articles in leading international journals within the field of critical landscape studies, rural geography, and remote sensing. Among his most recent publications is *Reclaiming Localisation for Revitalising Agriculture: A Case Study of Peri-Urban Agricultural Change in Gothenburg, Sweden* (with Zhang Qian, *Journal of Rural Studies*, 2016).

Tim Waterman MLA is a former restaurateur and Senior Lecturer and Landscape Architecture Theory Coordinator at the University of Greenwich in London. He is also a thesis tutor at the Bartlett School of Architecture, University College London, and a Non-Executive Director at the arts organisation Furtherfield. Until recently he was the honorary editor of *Landscape*, the journal of the Landscape Institute, and he also writes regularly for *Landscape Architecture Magazine* (LAM). His research interests include practice theory, everyday life, and processes of the propositional imagination. This forms the basis for other explorations of power and democracy and their shaping of public space and public life; taste, etiquette, belief, and ritual; and foodways in community and civic life and landscape. He is also co-editor, with Ed Wall, of *Landscape and Agency: Critical Essays* (2017).

Christina Welch PhD lectures in Religious Studies at the University of Winchester where she is a Knowledge Exchange fellow. Her research is largely in the area of religion and material/visual culture with two specific strands: one concerns issues around the socio-religious appropriation of North American Indian signifiers, and the other relates to death imagery, notably late-medieval cadaver sculptures. She is also interested in religion and the natural world, and especially how diasporic peoples keep and/or adapt their traditional worldviews about nature to their new surroundings; food and food growing with its myths and rituals being an essential part of this.

Dennis Wichelns PhD joined Stockholm Environment Institute (SEI) Asia as a Senior Research Fellow in October 2015. He is an agricultural and natural resource economist. Dr. Wichelns has served for many years at the University of Rhode Island, California State University Fresno, Hanover College, and Indiana University Northwest. He has served also as a researcher with the International Water Management Institute (IWMI), based at IWMI's global headquarters in Sri Lanka and engaged in interdisciplinary research in Asia and Africa. Dr. Wichelns serves currently as one of the editors of *Agricultural Water Management* and as editor of *Water Resources & Rural Development*.

Joshua Zeunert is Senior Lecturer at the University of New South Wales in Sydney, Australia, and a Registered Landscape Architect in Australia (AILA) and the USA (ASLA). He has worked at five universities internationally, as well as in industry with leading landscape architecture and urban design practices for multiple award-winning projects. His research interests focus on the nexus of academia and professional practice in environmental design, planning, strategy, and food systems. He is the author of *Landscape Architecture and Environmental Sustainability* (Bloomsbury, 2017) and numerous book chapters and journal and professional articles. His PhD explores multidimensional sustainability and public urban agriculture in landscape architecture.

Introduction

Tim Waterman and Joshua Zeunert

When this book was conceived, we, its editors, were sharing an office and teaching landscape architecture in a school of art and design unusually located in a small agricultural college in a sunny corner of south-eastern England within London's gravitational pull. We had worked hard with sympathetic colleagues across the institution's disciplines to capitalise upon that fertile association. Design and a landscape perspective, we knew, could help to reimagine processes that linked cities and countrysides inseparably together, with food and foodways being the medium of exchange in so many realms of human interaction: economic, social, cultural, historical; work, play, friendship, conviviality, and so on. We wanted to reach into landscape and food practice and history to plan, as an institution, for a sustainable, resilient, and regenerative future. We worked to put together two programmes, which, added to the college's existing offers in agriculture, horticulture, global ecosystem management, equine and other animal science and management, fine art, landscape architecture, urban design, and garden design, would help to complete a suite of courses that would have given the institution a very particular meaning, and very current and future relevance. One was a programme in urban agriculture (in the broadest sense, as outlined in Chapter 12) and the other in food studies. We also had slightly more distant plans to create a partnership with a culinary institute to complete a necessary virtuous loop. Alas, this was not to be. While we were trying to build the institution of the future, we were racing against a rampaging governmental ideology of austerity and the bite of draconian budget cuts.

The history of both food and landscape are reflective, not just of periods of scarcity and austerity but to our collective hope and striving for prosperity, fulfilment, and good lives. This book was born out of a desire to provide a framework to explain the intricate interconnections between food and landscape, between urban and rural life, and humanity, human ambition, and the natural world that sustains us all. We fervently hope that it will help other institutions to begin to move towards the same vision we had at what was then Writtle College.

Landscape and food

Landscape and food are both the most vital and necessary topics, as the first comprises the total environment that sustains us physically, culturally, and emotionally, while the second is both the necessary fuel for our existence and the symbolic centre of our metaphoric lives. Or is it the other way around? Or both? Thinking across food and landscape requires a relational frame of mind, in which ecologies of place and biota are inextricably embedded in ecologies of culture, affection, and ambition. As Felipe Fernández-Armesto avers, working with landscape for cultivation is itself a form of cookery: "*terram excoquere*, as Vergil called it – exposing clods to the baking sun, turning the earth into an oven for seeds" (2002: 3). It engages us all.

At the turn of the millennium, both landscape studies and particularly food studies were relatively unappreciated, with many of the best presentations in various media coming not from academia, but from popular commentators. Though this has changed radically since this time due to an explosion of interest in both landscape and food in academia, it still remains abundantly true that the interest in the subject remains shared quite disproportionately with an immense number of amateurs. Growing, cooking, and eating are experiences and endeavours common to all. For every trained chef, how many avid home cooks? For every farmer or specialty horticulturist, how many green-fingered weekend gardeners? For every landscape architect, how many designers in the vernacular? What this stresses, for us, is not the need to professionalise or theorise every realm of food and landscape, and to separate it from the quotidian, but that everyday practices are of central importance to everyone, and as such are worthy of study at every level. While it might seem apropos to issue a call for the firming up of methodologies in food and landscape studies, this book perhaps militates against that universalising possibility because of its immensely broad inter- and transdisciplinarity.

For this volume, which has the intention of introducing a range of key topics at the intersections of food and landscape, we have not for a moment tried to define either a singular approach or a specific canon, as this would simply be impossible. It might, indeed, also be undesirable, as the scope of both food and landscape allow for improbable and delightful connections to be made, and thus we have surrendered ourselves to diversity. What is most important is that writers in all areas of landscape and food studies must strive for clarity, and to explain their terms. Thus the reader will find, for example, Zachary Nowak's approach to *terroir* (Chapter 36), grounded in sociology, geography, and comparative studies, in the same volume with Afshin Ghahramani's and Saman Seneweera's (Chapter 18) very methodical agronomic approach of adapting wheat varieties to increased CO_2 levels in the atmosphere. Each of these chapters attempts to be legible to readers from outside the core discipline represented.

Then there are some peculiarities to this book that should be pointed out. Many chapters, for example, talk about 'landscape scale', and it might seem odd to ascribe dimensions to what is conceived as an interplay of human and natural processes and forces in spaces with often very fuzzy borders. It is, however, useful to the study of landscape to ask its size. The landscape scale is that to which landscape management practices can be applied, for various objectives and benefits (see Chapter 21). In a sense, 'landscape scale' also actually means 'multi-scalar' and is particularly constructed in opposition to conventional site-based approaches. As such, the 'landscape scale' is precisely the right focus for the study of food – a focus that can zoom in and out, and take in multiple objects and processes simultaneously, from dinner table and garden to watershed and continent. Accordingly, this volume incorporates landscape contexts from the sprawling fields of vast global agribusiness (Chapters 9 and 16) and remnant ecologies that agriculture has not yet commandeered (Chapter 17) to inner- and intra-urban (Chapters 12 and 27) and peri-urban contexts (Chapters 14 and 15). Similarly, chapters move between vast scales, such as entire continents (such as Africa in Chapter 21 – which also focuses on two countries), 'developed' countries (Chapter 12), 'developing' countries (Part 4 – see the following section on the structure of this book), focus on the characteristics of a settlement (Chapter 7), or individual village (Chapter 8).

Many chapters discuss pressing issues of rapid population growth and urbanisation (Chapters 4, 12, 15, 16, 19, 20, 21, 23, 29, and 40), noting that these strongly relate to industrialisation of agriculture and its displacement of rural inhabitants to urban areas, often to live in urban poverty (Chapter 20) and reside in informal settlements (Chapter 22). While many chapters necessarily base their conclusions upon projections of continued global population growth and continuing processes of urbanisation, the editors intend this volume as a whole to point towards the idea that *such a future might be averted*; that global population might be planned into decline, and that a focus on improving the quality of life and labour in rural areas might also begin to reverse processes of urbanisation, bringing people back into meaningful engagement with, and stewardship of, landscapes. Accordingly, this book deliberately provides a balance between rural and urban contexts as a counterpoint to the contemporary obsession with the urban in much design and wider discourse.

The structure of this book

It is entirely possible that any reader approaching this book from a narrow disciplinary perspective will find its contents and variety of approaches perplexingly wide-ranging, and this is intentional. Some chapters arising from disciplines in the humanities use the first person to elaborate narratives about the sensual and subjective experience of landscape and food (Chapters 34, 35, 36, 37, and 38), while others are scientific and are peppered with graphs, tables, and statistics (Chapters 15, 17, 18, and 19). It was always going to be the case that there would be a certain amount of dissonance resulting from this approach. This dissonance is, in fact, something to be embraced. Both landscape and food benefit from study that brings together science, art, affect, culture, and environment. Indeed, both landscape and food reflect the full range of human experience and human thought. Landscape and food also make palpable human connections and utter reliance upon the landscape and the integrity of its natural systems for survival (Chapters 10, 16, 17, 24) and pleasure (Chapters 34, 35, 36, and 37). Also apparent is how other species and their ecosystems (Chapter 17) are adversely impacted by the vast effects resulting from what we grow and how we eat (Chapters 10 and 16).

As much as this book holds differences in approach together, it also very usefully includes many overlapping subject areas, which are illuminated by that overlap. This book's original intention was to provide the broadest and most comprehensive possible overview of all the subject areas that might be included in food and landscape studies. This intention proved overambitious, and there were some thirty or so topics among those originally identified that have escaped capture here. What has become apparent as the chapters have fallen into place is that rather than the broad, smooth coverage we had wished for, the end result is a mix of overview and what might be called 'core samples'. We have mixed the 'big picture' chapters with those with more specific focus so that they are complementary, and roughly chronological through the book's seven parts. If the reader comes to the book with this understanding, then they may very fruitfully find vital new areas of research to explore in the connections between them as well as from those areas specifically identified within them.

In addition to overviews and excavations of subject areas, each chapter offers extensive and vital references which will help researchers using this book to identify key and seminal readings and current discourses. Many of the chapter's authors are authorities in their area, or are conducting research presently that requires of them a significant range of reading. Then, there are a handful of chapters interspersed throughout the book that are case studies – in a way, another kind of core sample. These case studies, such as Chapters 7 and 8's narratives revealing intergenerational dissolving of traditional growing and cultural practices, most often reveal wider patterns and trends; which are often mirrored in chapters providing overviews of their broad topic area (Chapters 9, 10, 11, 12, 13, 16, 17, 21, 24, 31, 35, 36). The book's seven sections are as follows:

> Part I *From Hunting and Gathering to Agriculture* undertakes two related projects: first, it seeks to trace human, food, and landscape trajectories from prehistory to the present, showing that these traces have relevance to today's lived landscape; and second, it demonstrates how landscapes may be understood from the standpoint of cultural evolution, and that the food relationships which are constitutive of landscapes may be read in very deep time. The first five chapters of the section do a lot of explanatory heavy lifting, including, in Chapters 2 (Olsson) and 3 (Jones and Clarke), a discussion of traditional ecological knowledge (TEK) and its current and future relevance for food systems and sustainable landscape management. The last three chapters are very useful case studies which show how practices persist, coexist, and evolve.

Part II *Agricultures* examines scales and modes of agriculture situated in the dominant drivers and practices of particular times. Emphasis is on the tremendous change in agriculture in the past century, and crucially, what that means now. This commences with Wästfelt's useful overview of shifts in rural agricultural practice (Chapter 9) and incorporates varying contexts of production from entire regions (such as the Mediterranean in Chapter 13 by Alon-Mozes and Eizenberg) seascapes (Chapter 11, Roe) to increasingly vast urban and metropolitan regions (Chapter 12, Zeunert) as well as focus on crucial peri-urban zones (Chapters 14 and 15, Olsson, and Carey and James). Chapters 10 (Code) and 12 (Zeunert) provide important taxonomies of agricultural practices.

Part III *Ecology, Resources, Sustainability, and Climate Change* contains four chapters that set out powerful frameworks for understanding how food systems and landscape practices might be remade for the future. All four chapters articulate many of the pressing challenges facing landscape and food, with Chapters 16 (Zeunert) and 17 (Austin) reminding us of the global magnitude of impact. Despite the scale and extent of challenges, these chapters nonetheless outline solutions and paths forward to shift the planet towards a state of ecological equilibrium. Yet they present and represent varying perspectives on these vast topics and thus provide diverse perspectives on where to go from here.

Part IV is titled *Developing Worlds*. We have approached the term with some reservations, as the notion of a 'developing world' includes the insinuation that places counted as such will someday become 'developed'. The persistence of structural inequality makes such a transformative process extremely unlikely, if not impossible. The first three chapters in this part focus on such places and the strategies that have been employed to improve lives and landscapes. The last (Chapter 23, Hanjra, Lydecker, Drechsel, and Paul) sees hope within the existing global paradigm and serves as a useful counterpoint for those seeking to navigate between the visions driving global development and the realities of third world poverty, exploitation, and expropriation.

Part V *Intellectual, Political, and Economic Realms* and Part VI *Social Practices and Meanings* can be seen as interdependent parts which begin with Isis Brook's (Chapter 24) grounding in foundational values through ethical frameworks that effectively call into question many of the assumptions of contemporary agriculture, which are reflected in Kelly's overview and positioning of the crucial topic of food sovereignty (Chapter 31). Chapters in these parts discuss enduring milieus of sacrifice and the commons (Chapter 28, Olwig), food markets (Chapter 29, Parham), and allotments and community gardens (Chapter 30, Nettle and Crouch). Giving recent focus are interrelated explanations of food pricing (Chapter 25, Maye), food justice, and spatial justice (Chapter 32, Blake), the tactics of urban planning and design (Chapter 26, Jones and Beza, and Chapter 27, Parham and Abelman), and an account of how a strong local food movement can make and remake a city's foodways (Chapter 33, Raffle and Carey).

Part VII, the final part, titled *Food Cultures and Foodways*, expands well beyond the standard phenomenological discussion in landscape and food theory to encompass the emerging field of practice theory (Chapter 34, Waterman) and the many layers of understanding and representation (and marketing) presented in such realms as food tourism (Chapter 35, Russell-Clarke) and the ever-expanding discussion of terroir (Nowak, Chapter 36). Finally, Jody Beck (Chapter 40) delivers a history of the future of landscape and food, reminding us that our futures are always rooted in our pasts, and in the ideas formed around our collective hopes and dreams.

A distant campfire: from the dawn of time to the emerging future

Somewhere about two million years ago, smoke rose from a distant campfire upon which our prehistoric ancestors, who had slowly evolved from the australopithecines, first cooked their food. Some commentators have argued that the added nutrition released by cooking food allowed the great accelerations in human evolution and cultural evolution. Food anthropologist Richard Wrangham, for example, writes that:

> Humans are adapted to eating cooked food in the same essential way as cows are adapted to eating grass, or fleas to sucking blood, or any other animal to its signature diet. We are tied to our adapted diet of cooked food, and the results pervade our lives, from our bodies to our minds.
>
> *(2009: 14)*

That food, first 'cooked' in the earth, as in Vergil's *terram excoquere*, and 'cooked' with the energy of the sun, was further cooked by flame in what was to become a complex cultural, social, meaningful relationship between people and the places they lived, places which have evolved into what we now see as landscape. Landscape, in a sense, is something people have cooked up together, but people are equally its product. This book begins at a decisive point in human development in which sedentist agriculture emerges; the point at which the next great acceleration of human cultural evolution and landscape evolution occurs – and the point, too, when people's environments become visibly, tangibly *built* – or in other terms, *cultural* rather than *natural* landscapes; a fact we can see in the archaeological record. This book ends, then, in a view towards our mutual and planetary future, tying together the flame of a distant campfire; the warmth and conviviality of growing, cooking, and eating; the emergence of the flamethrowers of industry and industrial agriculture; the great physical and intellectual struggles over land, sustenance, and survival; the stories of landscape and food.

References

Fernández-Armesto, F. (2002) *Near a thousand tables: A history of food*. New York and London: The Free Press.

Wrangham, R. (2009) *Catching fire: How cooking made us human*. London: Profile Books.

Part I

From hunting and gathering to agriculture

1

Transformation of the landscape

The relationships between food and land use in prehistoric British and European societies

Saruhan Mosler and Peter R. Hobson

Introduction

The following account describes the changes in land use patterns and practices that occurred across Europe between the late Mesolithic period, beginning around 10,000 BCE through the Neolithic times and the Bronze Age, ca. 6000 to 800 BCE, and up to the end of the Iron Age in 43 CE with the Roman invasion of the British Isles. The unique relationship between humans and nature, driven by the quest for food and the compulsion to establish territorial boundaries, is charted through the chronological changes to the European landscape. However rudimentary it may appear to our common understanding of land use practices, the use of burning by post-Ice Age settlers in order to clear vegetation signalled a shift in the forces dominating landscapes in north-western Europe and marked the beginnings of what was eventually to become the agricultural revolution. The switch from wilderness to 'culture-scape' was irreversible and uniquely defining for Europe. It was the genesis of wood pastures, heathlands, and grasslands that most Europeans identify as the vernacular landscape (Blackford, 2003).

Early human exploitation of natural resources for food, and the evolution of food cultures across the world have transformed the landscapes we inhabit today. Kareiva et al. (2007) state that the domestication of plants and animals may be the single most defining characteristic of human domination of the planet. During the late Mesolithic period, between 13,000 and 30,000 years ago, apart from the domestication of wolves there is little evidence that European cultures had begun the process of systematically cultivating plants or the husbandry of livestock (Driscoll et al., 2009). It was not until around 11,000 BCE that formalised practices of breeding other species of animal for consumption and the growing and harvesting of plants became widespread and defined as the 'Neolithic' (or agricultural) revolution.

Ethno-botanical studies reveal an estimated 7,000 species of plant have contributed, at some point in history, to the diet of humans (Grivetti and Ogle, 2000). Wild plants were the first harvested plants for food and medicine. The upper Palaeolithic site of Ohalo II, near the Sea of Galilee in Israel's Rift Valley, dated ca. 23,000 BCE, offers good evidence for the use of wild grasses both for food and for a grass bedding on the floor in a dwelling hut which consists of bunches of *Puccinellia* (alkali grass) stems and leaves, covered by a thin layer of clay arranged in a repeated pattern (Zapata, 2014: 15, cited in Nadel et al., 2004: 6821). For prehistoric hunter-gatherers cereals were seasonal plant foods, but agriculture made them permanent staples dating back to 8000 BCE in the Middle East region.

The substantial transformation of landscapes in relation to food and humans' way of living starts with human ability to control fire and use it in cooking (Wrangham and Carmody, 2010). Fernández-Armesto

(2003) argues that the origin of cooking sets humans apart from other species. By the Mesolithic period the procurement and consumption of food had developed a level of sophistication that could possibly be described as a food culture (Milner, 2009), and is where we start the story of people, food, and landscape.

Hunter-gatherers of the Holocene-Atlantic period: the early 'shape-shifters' of Europe

About 11,000 BCE, at the tail end of the last ice age, small bands of hunter-gathers moved quickly into more northern parts of Europe, using fire to clear vegetation and open up grazing lands for species of wild herbivore frequently hunted for food. For many communities life consisted of seasonal movements between coasts and inland valleys and uplands to ensure a ready supply of food (Tolan-Smith, 1998: 146). We now know there were also semi-permanent settlements around lakes, rivers, and coastal plains where food was more accessible (Flandrin et al., 1999; Fischer, 2007).

The picture of European Mesolithic landscapes is not entirely clear although it is generally accepted there was a gradual transformation from steppe to woodland as the climate warmed. As the steppe retreated so too its 'megafauna', a vital source of food for an essentially Mesolithic 'steppe peoples' (Kunz, 2016). Forests and wetlands replaced much of the tundra and cold steppes across central and northern Europe, providing new and varied sources of plant and animal foods (Whittle, 1996; Flandrin et al., 1999). It is not clear how extensive the spread of forest across Europe was during this period and whether fire in all its forms, together with wild grazing, kept in check the complete transformation to *Urwald* (closed forest landscape). Recent advances in palaeoecological science have made it possible to attempt a pictorial reconstruction of early forest landscapes and to move the baseline of understanding much further back from the earliest written historical accounts (Hodder et al., 2009). With the evidence from these new discoveries, long-held views of a primeval Europe clothed in closed forest have been challenged and an alternative picture offered of a forested landscape interrupted by variable-sized patches of mixed grassland, scrub and fen, and in some cases extensive wood pasture (Vera, 2000; Kreuz, 2008; Whitehouse and Smith, 2010). All the evidence suggests that the characterisation of early European forested landscapes within the context of the 'Urwald-culture-steppe' has still to be resolved. Instead, it invites a more sophisticated interpretation of the growing scientific evidence and a considered debate about the theory of a landscape subject to complex patterns of diachronic phases of early forest succession, complicated by the erratic colonisation by early human settlements (Kreuz, 2008). The impacts from the movements and behaviour of large fauna cannot be ignored in the argument. At present there is not enough understanding of primeval European forests to construct a detailed modern analogue of old growth forests for the region: a 'one model fits all'.

In the UK there is evidence from pollen records that Dartmoor in south-west England was heathland and transitioned to woodland around 10,000 BCE. By 7000 BCE most of Dartmoor was wooded with hazel, oak, and elm, and it remained relatively undisturbed for the next 2,000 years (White, 2000).

The activities of Mesolithic cultures played an important part in the evolutionary development of early-Holocene Atlantic wooded landscapes (Aaby, 1986; Latalowa, 1992; Latalowa et al., 2004; Simmons, 2003). Fire had been mastered as a means of sculpting landscapes to favour conditions for hunting, and what started to emerge out of the original forest and scrub was a more complex mosaic of wood pasture, often referred to as the 'culture-steppe' – a landform pattern described by Edwards in 1975 and Simmons in 1996 (Whitehouse and Smith, 2004). Hunters would make clearings in woodland around sources of drinking water to attract large herbivores. Such practices around 'watering holes', that is to say, ponds, lakes, and rivers where animals regularly drank, would have encouraged certain communities to establish more permanent settlements, as suggested by Pryor (2010: 29). He describes sedentary lifestyles involving the building of substantial dwellings, mostly for seasonal usage in carefully selected landscape settings, including at Star Carr in North Yorkshire and other strategic positions on river banks and the shores of lakes (Conneller, 2003; Conneller et al., 2011). A number of the seasonally used houses excavated at Star Carr were occupied around 8700 BCE for 300 years and were made from planks that had been split or hewn from a much larger piece of wood, such as a tree trunk or a large branch, suggesting skills and tools capable of felling trees and opening up the forest to create clearings. At Howick, in Northumberland, evidence of a large circular building dating to around 7600 BCE suggests not merely a seasonal but a permanent dwelling (Warren, 2017; Tolan-Smith, 2002). In north Germany, at Duven Lake, Mesolithic seasonal settlements – special 'task' camps – were discovered, dating back to the period between 11,000 BCE and 8500 BCE. The sites were used as food processing and storage camps with structures constructed specifically to process hazelnuts, including nut cracking implements, roasting hearths, and oil pressing and grinding facilities (Holst, 2010).

The picture painted here of a 'proto-sophisticated' culture runs contrary to earlier perceptions of Mesolithic life. A study by Childe (1925), describes the early Asturians as poverty-stricken and miserable populations of food gatherers who lived largely on shellfish, and this view persisted till more recent work by Clark (1975), in which he describes hunter-gatherers as maximising their resources to the extent that fishing was carried out from boats and included the hunting of seal, whale, eels, and crabs (see also Fischer, 2007; Milner, 2009). Small communities on coastal shores and along floodplains would in summer switch from shellfish camps to deer-hunting camps stationed inland (Clark, 1972; Pitts, 1979). The main meat hunted was red and roe deer, elk, auroch, and wild boar. Other fur animals were also trapped (Milner, 2009). Notwithstanding, despite the lower calorific value of aquatic animals, extensive fish traps were deployed to catch eels, and large quantities of shellfish were consumed (Fischer, 2007).

Advances made in tool technology gave hunter-gatherer communities greater access to 'nature's larder', and also improved techniques for processing harvested food. At Star Carr archaeological evidence revealed barbed antler points for use as harpoons for hunting beaver, fish, deer, or elk, and flints shaped for use as the tips of spears, arrows, or javelins, which in turn implies the manufacture of wooden handles and hafts. Ample bone evidence testifies to the range of food species consumed at the site: large mammals, such as brown bear, red and roe deer, elk, wild pig and cattle, as well as smaller species, including a variety of birds, badger, and hare (Conneller, 2003).

The commonly held view of early land use change across Europe maintains the major transition from wild lands to culture-scape occurred once hunter-gatherers abandoned their practices and became cultivators. At a site in the UK, pollen, charcoal, and fungal spore analysis of a peat profile at North Gill, North York Moors in north-east England, provides detailed evidence for an episode of fire-disturbance of woodland around 6300 BCE, which marks a turning point in late Mesolithic – early Neolithic culture (Innes and Blackford, 2003).

A profound shift in lifestyle patterns would have triggered social, technological, and economic changes, thus providing sling-shot shifts towards a new paradigm and relationship with the natural world (Bar-Yosef, 1998: 141–143). According to Rowley-Conwy (2011) full sedentary agricultural lifestyles emerged rapidly, marking the beginning of the Neolithic revolution. There remains very little evidence of transformational communities during the latter period of the Mesolithic.

Early signs of the harvesting of wild cereals and other plant species by hunter-gatherers during the late Mesolithic period have been observed in studies of the Natufian culture of the Levant, which flourished in the Near East between 12,000 and 10,000 BCE (Flandrin et al., 1999). As the practice became more sophisticated it started to spread throughout Europe at an estimated average rate of 0.6–1.3 km/year.

Hayden (2001) presents the case for Mesolithic domestication of the first animals and plants, and the eventual transition to a farming lifestyle as an emergent property of ritualised 'competitive feasting'. The significance of 'feasting' on the livelihoods of late Mesolithic communities would have presented challenges in the procurement, storage, and distribution of food for festivities. Special structures and implements were created to meet demands for the occasions (Hayden, 1990; Hayden, 2006).

Taming the land: the Neolithic revolution

The Neolithic revolution was the beginning of irreversible change in Europe's landscapes, from 'Urwald' to the new 'culture-steppe' that provided the footprint to modern-day agriculture in this part of the world.

Landscapes came to represent a place of permanency and a place to possess. It marked a paradigm shift in the human–nature relationship, from one of 'working with the grain of nature' – synchronising behaviour with the structures and dynamics of the natural world – to one of carving out new forms and of controlling processes where possible in order to maintain the status quo. A field system was created with forms of order and symmetry not seen before. Fire–fallow cultivation enabled early farmers to colonise new ground whilst allowing spent soils to recover under fallow and pasture.

The Neolithic revolution marks the profound transition from a hunting-foraging way of life to a more sedentary lifestyle centred on a culture of agriculture that took place across Europe about 10,000 BCE. The new wave of 'maritime immigrants' into Europe island-hopped their way across the Mediterranean from the Near East about 9,000 years ago (Paschou et al., 2014), and brought with them fresh, innovative ideas and technology, leading to a radical transformation of society and landscape. The term 'revolution' implies a radical and abrupt change, but a more correct interpretation of the transition period was of a gradual process over time, in some cases involving co-existence between communities (Bocquet-Appel, 2011; Bellwood, 2004; Barker, 2006; Brown et al., 2009), built on the reciprocal trading of goods and food. Early forms of growing and harvesting plants are believed to originate from practices of seeding and growing (Figure 1.1), a carefully selected group of species that were naturally less bitter and toxic (Fuller, 2007), and with the advancement of farming techniques, growers learned the process of 'genetic selection', favouring plants that produced higher yields and also showed greater resistance to pests and extreme weather conditions (ibid.).

A more formalised and ordered relationship with nature required the investment of time and resources to carefully plan and manage the necessary growing conditions for crops. Sites with fertile soil and protected from bouts of bad weather were favoured as were areas with less dense forest cover, possibly old wooded pastures established earlier through the slash-and-burn practices of the hunter-gatherer communities. The evidence from Europe for settled lands and the absence of shifting cultivation can be found in the fossil record for 'weed' species (see McClatchie et al., 2014). Plants associated with disturbed ground, including *Galium aparine* and *Plantago lanceolata*, as well as a dominance of annual species, indicates farmers were intensively managing small holdings for relatively long periods of time. In contrast, a shifting cultivation landscape would be indicated by the presence of perennial weeds, particularly woodland perennials (McClatchie et al., 2014). It is very probable Neolithic settlements were supplementing their diets with locally available wild sources of food, such as fruits, tubers, nuts, and leafy greens (Kirleis et al., 2012).

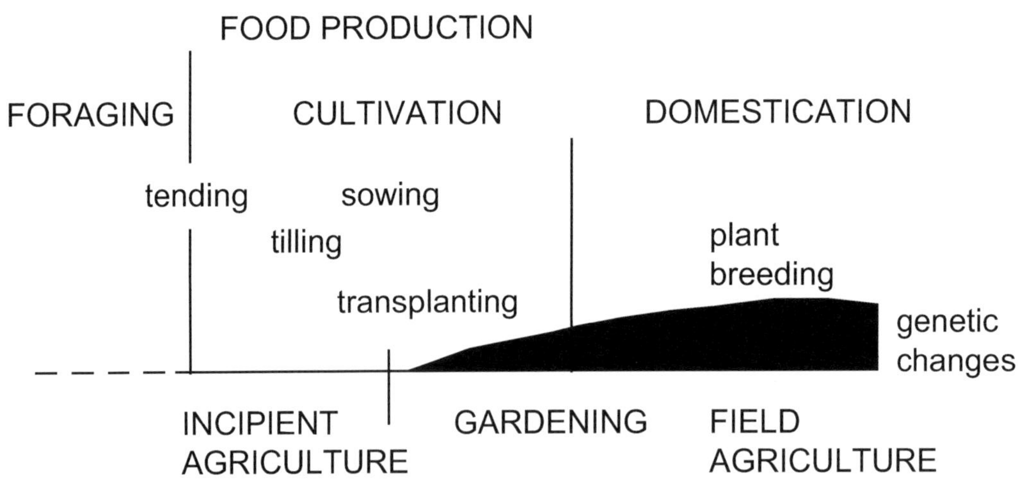

Figure 1.1 Ford's stages and methods of plant food production.

Source: Redrawn from Ford, 1985 (Smith, 2001)

In the fields, farmers, particularly those in the north-western region of Europe, were mainly growing cereals. Evidence from a study by Jones and Rowley-Conwy (2007) indicates cereals were present at 82% of recorded sites from the early Neolithic period, and included both wheat and barley. For both cultural and environmental reasons the situation in southern Europe, around the Mediterranean region, was rather different from that of the north in that a greater range of cereals and pulses were grown (Colledge et al., 2005; also see Chapter 13).

The Neolithic settlements introduced a new landscape character to Europe, the 'culture-scape'. As more land was brought under pasture and plough, settlements coalesced into communities, often marked out by small fields set aside from a cluster of houses within a protective "barrow" (Whittle, 1996: 68).

During this period landscapes were used more intensively to produce, prepare, collect, store, process, and exchange food and related resources across the territories. Further revolutions and developments in food systems led to the reorganisation of landscapes, creating efficient modes of transportation and movement of goods – the beginnings of infrastructure development that also reflected changes in political and economic corridors of power (Bélanger, 2016: 190). Local topography changed in many parts of Europe where there were settlements with the construction of causewayed enclosures, long barrows, long enclosures, and cursuses (Stuiver and Reimer, 1993).

In the early stages of development, most Neolithic settlements across Europe practiced a subsistence lifestyle with limited capacity to create trading links with other communities (Whittle, 1996). This position changed once a degree of permanency and security had set in to the new agronomic lifestyle. Networks of tracks linking settlements were established, providing the means for trading and sharing knowledge (Whittle, 1996). An important trading commodity at the time was salt. The emergence of salt in Neolithic diets signalled a change in eating habits, and possibly an increasing sophistication in the preparation and preservation of food – meat in particular. Brunwasser (2013) maintains salt was widely used by Neolithic communities in the preparation of food, and the importance attached to this commodity gave rise to the establishment of the 'salt-ways' that cut across large swathes of European landscapes.

As food became more abundant by the middle to late European Neolithic period it took on further symbolic status and featured at festivals, funerals, and ritual ceremonies, which were collective events staged in barrows and causewayed enclosures (Marciniak, 2005: 230). Archaeological evidence from Durrington Walls, near Stonehenge and Windmill Hill, Avebury, both in Wiltshire, England, reveals the full scale of some of the feasting ceremonies, which often involved the slaughtering and preparation of substantial numbers of pigs and cattle (see Albarella and Serjeantson, 2002; Fairbairn, 1999; van der Veen, 2003: 417). People from communities scattered widely across a region would travel to feasting sites bearing ceremonial gifts and food in decorated ceramic pots. It was an opportunity to forge new relationships, exchange knowledge, and acquire fresh livestock (Whittle, 1996: 40). There is good archaeological evidence for the processing and discarding of animal carcasses at specific locations to indicate there were many occasions where the ritualised slaughter of livestock took place as part of an elaborate, ritualised ceremony (Burkert, 1986). More generally, in Neolithic society, the increasing use of ceramic tableware signalled not only a growing occasion-related commensality amongst many communities but also an emerging divide between the more prosperous and the less fortunate (Halstead, 2012).

A settled, agronomic lifestyle provided the necessary security for an expanding population across Europe, which had an impact on the landscape and natural resources. Large tracts of forest were cleared for cultivation and where there remained intact forest, large areas were converted to woodland with clearings used to graze pigs, cattle, sheep, and goats. The larger, more difficult to fell trees would have been left after the removal of underwood, initiating a conversion of the forest to wood pasture (Kirby and Watkins, 2015).

A new form of 'countryside' was beginning to emerge: a human-altered landscape made up of large swathes of 'sere-stage' such as meadow, scrub, and heathland (ibid.). The constraints on land conversion imposed by a technology still tightly bound to stone and wood, and the lack of a system-based agriculture created a complex mosaic landscape of shifting, semi-permanent and permanent patches. Woodlands remained an important source of food, material, and fuel and so were retained in proximity to settlements.

Just as the Mesolithic people had prepared the conditions for the Neolithic revolution, so too the stone-age farmers tamed many regions of the European landscape, enough to pave the way for the Bronze Age.

Bronze age and the emergence of settlements

Around 4500 BCE Britain witnessed the arrival from the continent of a cultured people not seen before. The 'Bell Beaker' culture brought with them metal tools and weapons, as well as fine pottery utensils, new technology, and knowledge that was to have a profound impact on Neolithic people and the landscape they lived in. The newcomers were used to systematic planning on a rather larger scale to the clan-like settlements of Neolithic communities. Agriculture was seen as a means of producing enough food to feed a network of settlements within newly established sociopolitical territories. Borders were defended and field boundaries defined by markers such as dry stone walls. A system of land division known as 'reaves' were developed to divide up territories and demark the field systems within them.

A fine example of a British Bronze Age settlement is an excavated site in Essex near Chelmsford, England. The river systems provided important routes linking farmlands on the heavy clays of the Essex hinterland with the Blackwater estuary (Brown and Medlycott, 2013). The 'Chelmsford' enclosure was constructed on a small spur, defined by the shallow valleys of two small streams (now no more than spring-fed field ditches), and is defined by a ditch with post holes that were once part of an internal rampart (Figure 1.2). At the east end of the enclosure was a gate structure leading to a large, centrally positioned roundhouse, surrounded by several smaller buildings (ibid.: 19). Remnants of plant material excavated at the same site give us an idea of the crop types they were growing for food, and they include late spelt (*Triticum spelta*), emmer (*Triticum dicoccum*), free-threshing hexaploid wheat (*Triticum aestivum* s. l.), six-row hulled barley (*Hordeum vulgare*), horse beans (*Vicia faba* var. minor), peas (*Pisum sativum*), flax/linseed (*Linum usitatissimum*) and opium poppy (*Papaver somniferum*). Seeds of marsh and wet grassland plants such as *Ranunculus flammula* were also recorded at low frequencies, as were the seeds of fruit and nut trees such as hazel (*Corylus avellana*), sloe (*Prunus spinosa*), hawthorn (*Crataegus monogyna*) and elder (*Sambucus nigra*) (Brown and Medlycott, 2013).

Figure 1.2 Remnants of the ditch at the Springfield Late Bronze Age site in Chelmsford, Essex.
Source: Mosler, 2016

There is clear evidence of extensive later Bronze Age pastoral and arable activities from excavations in other areas, for instance, at Rectory Farm, West Deeping, Linconshire. Here, macrofossils of dogwood (*Cornus sanguinea*), hawthorn (*Crataegus monogyna*), bullace (*Prunus domestica* subsp. *insititia* var. *nigra* sp.), sloe (*Prunus spinosa*), bramble (*Rubus fruticosus agg.*), elder (*Sambucus nigra*) and three-nerved sandwort (*Moehringia trinervia*) have been unearthed from a Late Bronze Age field ditch, suggesting it was possibly bounded by a hedge (Brown and Murphy, 2000). Large quantities of oak (*Quercus* sp.) charcoal was also recorded, which could be related to the clearance of scrub on the site or charcoal burning for cooking (Brown and Medlycott, 2013: 135).

By the late Bronze Age from around 1000 to 800 BCE, European communities had greatly improved food security through the introduction of large storage structures designed to contain surplus food to carry them through the lean months of the year (Haselgrove, 2009: 157). Greater efficiency measures in food procurement and processing became part of the advancements made in agricultural land use and spatial organisation of settlements. With the transition to the Iron Age around 1200 BCE, goods and services were more tightly controlled, and settlements that were industrially and commercially developed were heavily fortified (Gibson et al., 1991). Improvements made during the Iron Age in agricultural practices and organisation of farmsteads, such as the establishment of the long house (barn house) was carried forward in to the Roman Period (de Hingh, 2000).

By 2400 BCE swathes of Europe were under Celtic influence and they brought with them a more intensive and ordered form of agriculture assisted by the use of the iron ploughshare, which was believed to be a superior version of the plough used later by the Romans. Oats, millet, and rye were grown under a system of rectangular fields, many of which were demarcated by stone walls or early forms of hedging similar to those seen today in parts of Wales and Cornwall (Rackham, 2000; Reynolds, 1995). In terms of what was being grown, there are very few accounts of crops other than cereals being eaten amongst Celtic people. Reynolds (1995) identifies four wheat types, four barley types, oats, rye, and millet as the staple plant food although diets were supplemented with leguminous crops, including beans (*Fabaceae* family), peas (*Pisum* spp.), and vetch (*Vicia* spp.). In addition, a primitive form of parsnip (*Pastinaca* spp.) was eaten along with fat hen (*Chenopodium album*), and with the latter, the young leaves were eaten like spinach, the mature plant cropped as winter animal fodder, and the seeds ground up into flour to be made into bread (Rackham, 2000; Reynolds, 1995). Similarly, black bindweed (*Polygonum bilderdykia*) was dried and ground into flour to make bread. Common orache (*Atriplex patula*), another typical weed of prehistoric fields, was also grown and prepared in much the same way as fat hen.

Cattle were an important part of Celtic farming culture, and the Celtic shorthorn (*Bos taurus brachycephalus celticus*), a relatively small animal, is believed to be the ancestor to the modern day Dexter (*Bos taurus*) (Reynolds, 1995). Wild animals such as red and roe deer and wild boar continued to be hunted for sport as well as for feasting (see Chapter 6). Wild fowl, primarily ducks and geese, were also consumed, as were fish, including eel. Agricultural intensification brought with it problems of over-exploitation of land and the depletion of soils (Koch, 2005). In parts of Britain there is evidence of disused chalk pits that were part of the Celtic solution to soil depletion. A form of manuring was carried out in many farmed areas that included use of mast, loam, and chalk (Pryor, 2010; Chapters 2 and 9).

Around 325 BCE a Greek geographer, Pytheas, on a visit to Britain, described (and later quoted in *Geographica* by Strabo) a landscape occupied by Celtic settlements as one of fields of 'corn' (a collective term used for cultivated cereals). His depiction of Britain's landscape made clear a distinction between the lowlands of the warmer south-east and the colder and wetter hills and mountains of the north-west, where he noticed a gradual disappearance of the more orderly farmed landscape. The contemporary 'Celtic heartlands' of Britain have retained many of their original legacies – rain-drenched, nutrient-poor hills and mountains covered in sheep-grazed grass and heath with small copses and spinneys. Some centuries after Pytheas's account of ancient Britain, Strabo in his comprehensive publication, *Geographica* (book IV, chapter 5; see Jones, 1917–1932), describes the dearth of comfort available to the inhabitants or visitors of the cold and wet territories of Britain, which he described as being "destitute of the cultivated fruits and almost deprived of the domestic animals". Strabo goes on to depict a local lifestyle as one that could best be described as 'neo-Mesolithic';"the food of the inhabitants consisted of millet, herbs, fruits, and roots. When they had corn and honey they made drink of them" (see Jones, 1917–1932).

Overcrowding of the more fertile lowlands forced settlements to move into less productive uplands, where there may have been a preference for pastoralism over arable cropping. Certainly, cattle were an important part of Celtic culture and could have accounted for 60% or more of meat consumption (Koch, 2005). Drinking and feasting were an integral part of life in Iron Age Europe and the British Isles. The distribution of food and especially drink in prescribed fashion played a key role in establishing and maintaining social relationships. Alcoholic beverages were important consumable status items in prehistoric Europe, serving as a social lubricant as well as a social barrier (Arnold, 2009).

In western and northern regions of Britain evidence remains of small hill enclosures cleared of trees and rocks to provide pasture for livestock. Rocks and boulders were moved onto unproductive land, often where there were mires or surface springs covered in woodland scrub (Figures 1.3 and 1.4), or were used to construct "Celtic hedges" (see Rackham, 2000). Although cattle were favoured by the Celts (see Koch, 2005), other livestock included sheep and pigs, the latter making up a small part of the diet, possibly as little as a tenth of the total meat consumed (Koch, 2005). Today's traditional breeds of Soay, Manx, and Hebridean sheep (*Ovis aries*) are believed to be direct descendants of Iron Age stock. While geese, chickens, and hares were considered to be part of the Celtic farmed landscape, an account by Caesar (see Crawfurd, 1867) suggests they were kept more as pets, and slaughtering of 'fowl and hare' was considered taboo.

What does this tell us about the relationship between people and the landscape during the Iron Age period? Archaeological evidence offers a picture of early European cultures, more specifically Celtic and others influenced by them, almost entirely dependent on agricultural landscapes for food and fibre. The transformation of the original wilderness to a prototype of the modern 'culture-scape', or culture-steppe (Vera, 2009), was well underway. What remained of the original old growth forest that was once more prominent across Europe continued to be heavily exploited for timber and fuel. To a lesser extent, wild food and medicinal plants were also harvested from the forests. For instance, Celtic royalty put great store in hunting of wild game, in particular boar in the forest (Ó hÓgáin, 2007).

The introduction of agricultural systems and planned farming practices strengthened the decoupling of nature and culture. By the time Claudius landed in ancient Britain in 43 BCE, he described the south-east

Figure 1.3 Three images show examples of old Celtic hedges, constructed from dry stone walls, topped with turves and then planted with either oak or hazel saplings, Llyn peninsula, North Wales, UK.

Source: Hobson, 2016

Figure 1.4 Ancient spring-fed marginal land littered with boulders cast out from adjoining enclosed meadows, Llyn peninsula, North Wales, UK.

Source: Hobson, 2016

of England as a well-ordered and productive agricultural landscape, inhabited by communities able to produce enough surplus food to export and for trade (Frere, 1987).

Conclusion

Europe is a geographical kaleidoscope of mountains, rivers, lakes, floodplains, coasts, grasslands, and forests, free of many of the diseases that plague humans in tropical regions of the world. Once the ice retreated 12,000 years ago it was a naturally rich area for small bands of hunter-gatherers to tap into yet unexploited natural resources. Humans were as much a part of the evolving ecosystem in Europe as the large fauna species and trees. So, we cannot talk of original European wilderness without considering the influences on the environment of Mesolithic and Neolithic cultures. There is no clear picture of what the landscape looked like at the time of hunter-gatherers. Pollen records give us some clue of the species of plant around at the time but we know little of the conditions or dynamics driving trends in species abundance and distribution. We can conjecture with a certain degree of evidence, that human use of fire for hunting and clearing the landscape of woody vegetation would have had profound effects at various locations across Europe. We can also assume from the large amounts of shell deposits left by Mesolithic peoples across the continent that aquatic ecosystems were pivotal in the survival of communities, and that coastal as well as riverside environments would have felt the deep footprint of humankind during the early stages of post-ice age recovery.

The talk of abrupt transition, a regime shift, from Mesolithic to Neolithic periods is not universally accepted by all scholars studying this period in history (see Milner, 2009). Traditionally held views of a wretched existence eked out by hunter-gatherers, grubbing around in shallow waters for shellfish, and impatience to drop the spear for the 'plough' hinges on earlier research by Childe (1925) and others who believed this to be the case. In contrast, Milner (2009) describes a much more sophisticated culture of well-nourished and robustly built peoples, most likely in a better state of health than the Neolithic people. She goes on to propose an alternative hypothesis of transition that could have taken as long as 2,000 years (see Milner, 2009). This would seem to make sense given the artefacts and fossil items during the early Neolithic period that included a range of wild food. However, the picture is not consistent across Europe, as we see from the evidence presented by Richards and Hedges (1999). They suggest Neolithic settlements, even at sites close to the coast, were eating a diet based mainly on livestock and crops. Little, if any, wild food was being harvested from watery environments. Then again, Boyle (2006) describes hunting sites across Spain, Hungary, and Alpine foreland where wild animal bones make up a large proportion of the finds at excavation sites.

In the Neolithic period a diet heavy in cattle, sheep, and pig (see Shulting, 2008), encourages us to think about the shaping of the landscape beyond the small field systems representative of early crop lands. Forest clearance for fresh pasture would have been widespread but patchy, depending on local conditions. It is very possible that by mid to late Neolithic period substantial tracts of Europe had undergone enough change to be considered culture-scape – a patchwork of field, open grazing lands, woodlands, and relic forest and wetlands. Food was being extracted from all the quarters of the wider landscape and the demographic changes triggered by food-producing economies were major agents of the mid-Holocene shifts in land cover (see Lechterbeck et al., 2014). Primary vegetation communities are believed to have peaked by early Neolithic period (Lechterbeck et al., 2014), but by mid Neolithic secondary woodland with increases in hazel (*Corylus avellana*), suggests a slash and burn practice of agriculture was an important driver of change (Lechterbeck, 2013).

During the Bronze and Iron Ages the more intensive mastering of landscapes with much improved technology and husbandry practices amounted to a scaling up of activities, which in turn, greatly increased conversion of primary vegetation to secondary stage communities. Values for secondary vegetation cover are by this stage comparable with those of early Middle Ages (see Lechterbeck et al., 2014). The farming styles of the earlier Neolithic cultures that provided a diverse array of herbs and spices used in cooking were gradually phased out in favour of larger scale, monocrop production, primarily of wheat and barley. Although millet was also grown to feed the large population of horses used in battle and in the fields (see Küster, 2000). Large, well established fields often growing a single crop were to provide the template for modern agricultural production and land use practices across Europe.

Further reading

Aaby, B. (1986) Trees as anthropogenic indicators in regional pollen diagrams from Eastern Denmark. In: Behre, K.E. (ed.), *Anthropogenic indicators in polen diagrams*. Balkema: Rotterdam, pp. 73–93.

Albarella, U. & Serjeantson, D. (2002) A passion for pork: meat consumption at the British late Neolithic site of Durrington Walls. In P.Miracle & N.Milner (eds). Consuming Passions and Patterns of Consumption, pp.33-49. Cambridge: Monographs of the McDonald InstituteAnschuetz, K., Wilshusen, R.H., & Scheick, C. (2001) An archaeology of landscapes: Perspectives and directions. *Journal of Archaeological Research*, 9(2), 157–211.

Applebaum, S. (1958) Agriculture in Roman Britain. *The Agricultural History Review*, 6(2), 66–86.

Arnold, B. (2009) 'Drinking the Feast': Alcohol and the legitimation of power in celtic Europe. *Cambridge Archaeological Journal*, Cambridge University Press, 9(1), 71–93. Available from: doi.org/10.1017/S0959774300015213 [Accessed: 3rd May 2017].

Ashmore, W., & Knapp, B. (1999) *Archaeologies of landscape: Contemporary perspectives*. London: Blackwell Publishing.

Atkins, P., Simmons, I., & Roberts, B. (2014) *People, land and time: An historical introduction to the relations between landscape, culture and environment*, 2nd ed. London: Routledge. Available from: DOI: 10.1006/jasc.2002.0832 [Accessed: 8th May 2017].

Bailey, G., & Spikins, P. (2008) *Mesolithic Europe*. Cambridge: Cambridge University Press.

Barker, G. (2006) *The agricultural revolution in prehistory: Why did foragers become farmers?* Oxford: Oxford University Press.

Barrett, J. (1999) Chronologies of landscape. In: Ucko, P., & Layton, R. (eds.), *The archaeology and anthropology of landscape: shaping your landscape*. London: Routledge, pp. 21–30.

Bar-Yosef, O. (1998) On the nature of transitions: The middle to upper palaeolithic and the neolithic revolution. *Cambridge Archaeological Journal*, 8(2), 141.

Beagon, M. (1996) Nature and landscapes in Pliny the Elder. In: Schipley, G., & Salmon, J. (eds.), *Human landscapes in classical antiquity: Environment and culture*. Oxon: Routledge, pp. 284–310.

Bélanger, P. (2016) Is landscape infrastructure? In: Waldheim, C., & Doherty, G. (eds.), *Is landscape . . . ? Essays on the identity of landscape*. London: Routledge, pp. 190–228.

Bellwood, P. (2004) *The first farmers: Origins of agricultural societies*. Oxford: Wiley-Blackwell.

Blackford, J.I. (2003) The ecology of late Mesolithic woodland disturbances: Model testing with fungal spore assemblage data. *Journal of Archaeological Science*, 30, 185–194.

Bocquet-Appel, J.P. (2011) When the world's population took off: The springboard of the neolithic demographic transition. *Science*, 333(6042), 560–561. Available from doi:10.1126/science.1208880 [Accessed 3 February 2017].

Bocquet-Appel, J.P., & Bar-Yosef, O. (2008) *The neolithic demographic transition and its consequences*. London: Springer, Science-Business Media B.V.

Boyle, K.V. (2006) Neolithic wild animals in Western Europe: The question of hunting. In: Serjeantson, D., & Field, D. (eds.), *Animals in the neolithic of Britain and Europe*. Oxford: Oxbow, pp. 10–23.

Brown, N., & Medlycott, M. (2013) The Neolithic and Bronze Age enclosures at Springfield Lyons, Essex: Excavations 1981–1991. *East Anglian Archaeology Report No. 149*. Place Services: Essex County Council: Chelmsford.

Brown, T.A., Jones, M.K., Powell, W., & Allaby, R.G. (2009) The complex origins of domesticated crops in the Fertile Crescent. *Trends in Ecology & Evolution*, 24(2), 103–109. Available from: doi.org/10.1016/j.tree.2008.09.008 [Accessed: 9th March 2017].

Brown, N. and Murphy, P. (2000) Neolthic and Bronze Age. In N. Brown and J. Glazebrook (eds.). research and Archaeology: A Framework for the eastern counties 2. Research Agenda and strategy East Anglian Archaeology Occasional paper. 8,9-13.

Brunwasser, M. (2013) Salt and the city. *Archaeology*, 66(4), 35–39.

Burkert, P. (1986) *Homo Necans: The anthropology of ancient Greek sacrificial ritual and myth*. London: University of California Press, Ltd.

Champion, T. (2009) The later Bronze Age. In: Hunter, J., & Ralston, I. (eds.), *The archaeology of Britain: An introduction from earliest times to the twenty-first century*, 2nd ed. London: Routledge, pp. 126–149.

Chessa, B., Pereira, F., Arnaud, A., Amorim, F., Goyache, I., Mainland, R., Kao, R., Pemberton, J.M., Beraldi, D., Stear, M., Alberti, A., Pittau, M., Iannuzzi, L., Banabazi, M.H., Kazwala, R., Zhang, Y.P., Arranz, J.J., Ali, B.A., Wang, Z., Uzun, M., Dione, M., Olsaker, I., Holm, L.E., Saarma, U., Ahmad, S., Marzanov, N., Eythorsdottir, E., Holland, M.J., Ajmone-Marsan, P., Bruford, M.W., Spencer, T.E., & Palmarini, M. (2009) Revealing the history of sheep domestication using retrovirus integrations. *Science*, 324(5926), 532–536. Available from: DOI: 10.1126/science.1170587 [Accessed: 14th March 2017].

Chevalier, A., Marinova, E., & Pena-Chocarro, L. (2014) *Plants and people: Choices and diversity through time*, Earth Vol.I. Oxford: Oxbow Books.

Childe, V.G. (1925) *The dawn of European civilization*. London: Routledge, Trench, Trubner.

Clark, J.G.D. (1972) *Star Carr: A case study in bioarchaeology*. Publications in Anthropology No. 10. Reading, MA: Addison-Wesley Publishing Co.

Clark, J.G.D. (1975) *The earlier stone Age settlement of Scandinavia*. Cambridge: Cambridge University Press.

Cleary, S.E. (2009) Roman Britain: Civil and rural society. In: Hunter, J., & Ralston, I. (eds.), *The archaeology of Britain: An introduction from earliest times to the twenty-first century*, 2nd ed. London: Routledge, pp. 198–219.

Colledge, S., Conolly, J., & Shennan, S. (2005) The evolution of Neolithic farming from SW Asian Origins to NW European Limits. Journal European Journal of Archaeology Volume 8, - Issue 2.

Conneller, C. (2003) Star Carr recontextualized. In: Moore, J., & Bevan, L. (eds.), *Peopling the mesolithic in a northern environment*. Oxford: Archaeopress, pp. 81–86; Oxford: BAR.

Conneller, C., Milner, N., & Taylor, B. (2011) Recent excavations at Star Carr. *Yorkshire Archaeological Society Bulletin*, 48.

Coones, P. (1992) The unity of landscapes. In: Macinnes, L., & Wickham-Jones, C.R. (eds.), *All natural things: Archaeology and the green debate*. London: Oxbow Books, pp. 22–40.

Crawfurd, J. (1867) On Caesar's account of Britain and its inhabitants in reference to ethnology. *Transactions of the Ethnological Society of London*, 5, 202–215.

Cunliffe, B. (1995) *Iron Age Britain*. London: B.T. Batsford Ltd.
Dáithí Ó hÓgáin (2007) *The celts: A history*. New Edition. Martlesham, Boydell & Brewer.
Davidson, A. (1999) *The Oxford companion to food*, 2nd ed. Oxford: Oxford University Press.
de Hingh, AE (2000) *Archaeology of the Bronze Age and Iron Age society*. https://openaccess.leidenuniv.nl/bitstream/handle/1887/13513/ASLU07_02.pdf;jsessionid=56A3DEEDE1BD351F3581F9099B0F06EA?sequence=7 (accessed 3 May 2017).
Delwen, S. (2006) Archaeology of food. In: Davidson, A. (ed.), *The Oxford companion to food*, 2nd ed. Oxford: Oxford University Press, pp. 33–34.
Dietler, M., & Hayden, B. (2006) *Archaeological and ethnographic perspectives on food, politics and power*. Tuscaloosa, AL: The University of Alabama Press.
Donatella Magri, D. (2008) Patterns of post-glacial spread and the extent of glacial refugia of European beech (Fagus sylvatica). *Journal of Biogeography*, 35(3), 450–463.
Driscoll, C.A., Macdonald, D.W., & O'Brien, S.J. (2009) From wild animals to domestic pets, an evolutionary view of domestication. *PNAS*, 106(1), 9971–9978. Available from doi: 10.1073/pnas.0901586106 [Accessed 13 May 2017].
Driscoll, C.A., Macdonald, D.W., & O'Brien, S.J. (2009) From wild animals to domestic pets, an evolutionary view of domestication. *PNAS*, 106(1). Available from www.pnas.org/content/106/Supplement_1/9971.full [Accessed 3 May 2017].
Fairbairn, A. 1999. Charred plant remains. In A. Whittle, J. Pollard and C. Grigson The harmony of symbols, the Windmill Hill causewayed enclosure, 139-56. Oxford: Oxbow Books.
Fernández-Armesto, F. (2003) *Near a thousand tables: A history of food*. London: Free Press.
Fischer, A. (2007) Coastal fishing in Stone Age Denmark: Evidence from below and above present sea level and from human bones. In: Milner, N., Craig, O.E., & Bailey, G.N. (eds.), *Shell middens in Atlantic Europe*. Oxford: Oxbow Books, pp. 54–69.
Flandrin, J., Montanari, M., & Sonnenfeld, A. (1999) *Food. A culinary history antiquity to the present from European perspectives: A series in social thought and cultural criticism*. Colombia: Colombia University Press.
Frere, S. (1987) *Britania: A history of Roman Britain*, 3rd revised ed. London: Routledge and Kegan Paul.
Fuller, D. Q. (2007). Contrasting patterns in crop domestication and domestication rates. *Annals of Botany* 100, 903-924.
Fuller, D.Q., Allaby, R.G., & Stevens, C. (2010) Domestication as innovation: The entanglement of techniques, technology and chance in the domestication of cereal crops. *World Archaeology*, 42(1), 13–18.
Gerrard, S. (1997) *Book of Dartmoor: Landscapes through time*. London: B.T. Batsford & English Heritage.
Gibson, S., Ibbs, B. and Kloner, A. (1991) The Sataf Project of landscape archaeology in the Judeaen HIklls; A preliminary report on four seasons of survey and excavation (1987 89). Levant 23: 29-54.
Gosden, C., & Lock, G. (1998) Prehistoric histories. *World Archaeology*, 30(1), 2–12.
Grivetti, L.E., & Ogle, B.M. (2000) Value of traditional foods in meeting macro- and micronutrient needs: The wild plant connection. *Nutrition Research Reviews*, 13, 31–46.
Jones, H.L. (1917–1932) *The geography of Strabo*, Vols. 1–8, containing Books 1–17. Boston, MA: Harvard University Press and Heinemann.
Halstead, P. (2012) Feast, food and fodder in Neolithic-Bronze Age Greece: Commensality and construction of value. *eTopoi Journal for Ancient Studies*, 2, 21–51.
Harris, D., & Hillman, G.C. (1989) *Foraging and farming: The evolution of plant exploitation*. London: Unwin.
Haselgrove, C. (2009) The Iron Age. In: Hunter, J., & Ralston, I. (eds.), *The archaeology of Britain: An introduction from earliest times to the twenty-first century*, 2nd ed. London: Routledge, pp. 149–175.
Hayden, B. (1990) Nimrods, piscators, pluckers and planters: The emergence of food production. *Journal of Anthropologiocal Archaeology*, 9: 31-69.
Hayden, B. (2001) Fabulous feasts: A prolegomenon to the importance of feasting. In: Dietler, M., & Hayden, B. (eds.), *Archaeological and ethnographic perspectives on food, politics and power*. Tuscaloosa, AL: The University of Alabama Press. pp. 23–65.
Hirsch, E. (1995) Landscape: Between place and space. In: Hirsch, E., & O'Hanlon, M. (eds.), *The anthropology of landscape: Perspectives on place and space*. Oxford: Clarendon Press, pp. 1–30.
Hirsch, E., & O'Hanlon, M. (1995) *The anthropology of landscape: Perspectives on place and space*. Oxford: Clarendon Press.
Hodder, K.H., Buckland, P.C., Kirby, K.J. and Bullock, J.M. (2009) Can the pre-Neolthic provide suitable models for re-wilding the landscape in Britain? *British Wildlife*, 20(5), Special supplement.

Holst, D. (2010) Hazelnut economy of early Holocene hunter-gatherers: A case study from Mesolithic Duvensee, northern Germany. *Journal of Archaeological Science*, 37(11), 2871–2880.
Holst, D. (2010) Hazelnut economy of early holocene hunter-gatherers: A case study from mesolithic Duvensee, Northern Germany. *Journal of Archaeological Science*, 37(11), 2871–2880.
Hooke, D. (2001) *The landscape of Anglo-Saxon England*. London: Leicester University Press.
Hunter, J., & Ralston, I. (2009) *The archaeology of Britain: An introduction from earliest times to the twenty-first century*, 2nd ed. London: Routledge.
Innes, J.B., & Blackford, J.J. (2003) The ecology of Late Mesolithic woodland disturbances: Model testing with fungal spore assemblage data. *Journal of Archaeological Science*, 30(2), 185–194. Available from: DOI: 10.1006/jasc.2002.0832 [Accessed 12 March 2017].
Jefferies, R., Thompson, V., & Milner, G.R. (2005) Archaic hunter-gatherer landscape use in West-Central Kentucky. *Journal of Field Archaeology*, 30(1), 3–23.
Jones, G., & Rowley-Conwy, P. (2007) On the importance of cereal cultivation in the British Neolithic. In: Colledge, S., & Conolly, J. (eds.), *The origins and spread of domestic plants in southwest Asia and Europe*. Walnut Creek, CA: Left Coast Press, pp. 391–419.
Kaniewski, D., Van Campo, E., Guiot, J., Le Burel, S., Otto, T., & Baeteman, C. (2013) Environmental roots of the late Bronze Age crisis. *PLoS One*, 8(8), e71004. Available from: http://dx.doi.org/10.1371/journal.pone.0071004 [Accessed 22 June 2016].
Kaniewski, D., Paulissen, E., Van Campo, E., Weiss, H., Otto, T., Bretschneider, J., & Van Lerberghe, K. (2010) Late second – early first millennium B.C.E. abrupt climate changes in coastal Syria and their possible significance for the history of the Eastern Mediterranean. *Quaternary Research*, 74, 207–215.
Kareiva, P., Watts, S., McDonald, R., & Boucher, T. (2007) Domesticated nature: Shaping landscapes and ecosystems for human welfare. *Science*, 316, 1866–1869.
Kirby, K.J., & Watkins, C. (2015) *Europe's changing woods and forests: From wildwood to managed landscapes*. Wallingford, UK: CABI.
Kirch, P. (2005) Archaeology and global change: The Holocene record. *Annual Review of Environment and Resources*, 30(1), 409–440.
Kirleis, W., Klooß, S., Kroll, H., & Müller, J. (2012) Crop growing and gathering in the northern German Neolithic: A review supplemented by new results. *Vegetation History and Archaeobotany*, 21(3), 221–242.
Knapp, B., & Manning, S. (2016) Crisis in context: The end of the late Bronze Age in the Mediterranean. *American Journal of Archaeology*, 120(1), 99–149.
Koch, J.T. (2005) *Celtic culture: A historical encyclopedia*. London: ABC-CLIO Ltd.
Kreuz, A. (2008) Closed forest or open woodland as natural vegetation in the surroundings of Linearbandkeramik settlements? *Vegetation History and Archaeobotany*, 17, 51–64.
Kunz, W. (2016) *Species conservation in managed habitats: The myth of a pristine nature*. Verlag, Weinheim: Wiley-VCH.
KüsterÜster, H. (2000) VC.5., Northern Europe – Germany and surrounding regions: The history and culture of food and drink in Europe. In Kiple, K.F., & Ornelas, K.C. (eds.), *The Cambridge world history of food*, Vol. 2. Cambridge: Cambridge University Press.
Latalowa, M. (1992) Forest changes at the Mesolithic/early Neolithic transition and in the migration period on Wolin Island (northwest Poland). *Palaeoklimaforschung*, 8, 139–155.
Latalowa, M., Ralska-Jasiewiczowa, M., Miotk-Szpiganowicz, G., & Zachowicz, J. (2004) Fagus sylvatica L. – beech. Late-glacial and Holocene history of vegetation in Poland based on isopollen maps. Krakow, W. Szafer Institute of Botany. *Polish Academy of Sciences*, 95–104.
Lechterbeck, J. (2013) Evaluation of economic activity through palynological data: Modelling agricultural pressure on landscape (REVEALS and LOVE). In: Kerig, T., & Zimmermann, A. (eds.), *Economic archaeology: From structure to performance in European archaeology* (Universitätsforschungen zur prähistorischen Archäologie 237). Bonn: Habelt, pp. 129–141.
Lechterbeck, J., Edinborough, K., Kerig, T., Fyfe, R., Roberts, N., & Shennan, S. (2014) Is Neolithic land use correlated with demography? An evaluation of pollenderived land cover and radiocarbon-inferred demographic change from Central Europe. *The Holocene*, 24(10), 1297–1307.
Marciniak, A. (2005) *Placing animals in the neolithic: Social zooarcheaology of prehistoric farming communities*. London: UCL Press.

Marston, J. (2012) Agricultural strategies and political economy in ancient Anatolia. *American Journal of Archeology*, 116(3), 377–403.

McClatchie, M., Bogaard, A., Colledge, S., Whitehouse, N.J., Schulting, R.J., Barratt, P., & McLaughlin, T.R. (2014) Neolithic farming in north-western Europe: Archaeobotanical evidence from Ireland. *Journal of Archaeological Science*, 51, 206–215.

Milner, N. (2009) Mesolithic consumption practices: Food for thought. *Journal of Nordic Archaeological Science*, 16, 49–63.

Mosler, S. (2014) Springfield Lyons Late Bronze Age heritage Site Development report. A Study on Interpretation and Presentation Models. Chelmsford City Council Report.

Morphy, H. (1995) Landscape and the reproduction of the ancestral past. In: E. Hirsch & M. O'Hanlon (eds.), *The anthropology of landscape: Perspectives on place and space*. Oxford: Clarendon Press, pp. 184–209.

Murray, M.L., & Schoeninger, M.J. (1988) Diet, status and complex social structure in Iron Age Central Europe: Some contributions of bone chemistry. In: Gibson, D.B., & Geselowitz, M.N. (eds.), *Tribe and polity in late prehistoric Europe*. New York: Springer Science and Business Media, pp. 155–176.

Nadel, D., Weiss, E., Simchoni, O., Tsatskin, A., Danin, A. and Kislev, M. (2004) Stone Hut in Israel yields world's oldest evidence of bedding. PNAS (proceedings of the National Academy of Sciences of the United States of America) 101 (17), 6821-6826.

Nelson, S. (2009) *God and the land: The metaphysics of farming in Hesiod and Vergil*. New York: Oxford University Press.

Paschou, P., Drineas, P., Yannaki, E., Razou, A., Kanaki, K., & Tsetsos, F. (2014) Maritime route of colonization of Europe. *Proceedings of the National Academy of Science U S A*, 111, 9211–9216. Available from: DOI: 10.1073/pnas.1320811111 [Accessed 24 November 2016].

Pearson, M.P., Pollard, J., Richards, C., Thomas, J.J., Welham, K., Albarella, U., Chan, B., Marshall, P., & Viner, S. (2011) Feeding Stonehenge: Feasting in late neolithic Britain. In: Jimenez, G.A., Monton-Subias, S., & Sanchez, M.R. (eds.), *Guess who's coming to dinner: Feasting rituals in the prehistoric societies of Europe and the Near East*. Oxford: Oxbow Books, pp. 73–90.

Pinhasi, R., & Von Cramon-Taubadel, N. (2012) A craniometrics perspective on the transition to agriculture in Europe. *Human Biology*, 84(1), 45–66.

Pinhasi, R., Fort, J.M., & Ammerman, A.J. (2005) Tracing the origin and spread of agriculture in Europe. *PloSBiol*, 3(12). Available from: https://doi.org/10.1371/journal.pbio.0030410 [Accessed 15 October 2016].

Pitts, M. (1979) Hides and Antlers: A new look at gathering-hunter site of Star Carr, North Yorkshire, England. *World Archaeology II*, 32–42.

Pollard, P. (2008) *Prehistoric Britain*. London: Wiley.

Pryor, F. (2010) *The making of the British landscape*. London: Penguin Book Ltd.

Rackham, O. (2000) *The history of the countryside*. London: Orion Publishing Co.

Reynolds, P.J. (1995) *The food of the prehistoric celts*. In: Wilkins, J., Harvey, D., & Dobson, M. (eds.), *Food in antiquity*. Exeter: University of Exeter Press. Available from: http://butser.org.uk/Food%20of%20the%20Prehistoric%20Celts.pdf [Accessed 14 March 2017].

Richards, M.P., & Hedges, R.E.M. (1999) A Neolithic revolution? New evidence of diet in the British Neolithic. *Antiquity*, 73, 891–897.

Rippon, S. (2004) *Historic landscape analysis. Deciphering the countryside*. Ripponden: Pennine Printing Services.

Rowley-Conwy, P. (2011) Westward ho! The spread of agriculture from Central Europe to the Atlantic. *Current Anthropology*, 52(4), S431–S451.

Runnels, C., & Van Andel, T.H. (2003) The early stone age of the Nomos of Preveza: Landscape and settlement. In: Wiseman, J., & Zachos, K. (eds.), *Landscape archaeology in Southern Epirus, Greece I*. Athens: The American School of Classical Studies at Athens, pp. 47–135.

Samuel, D. (1996) Approaches to the archaeology of food. *Petits Propos Culinaires*, 54, 12–21.

Simmons, I.G. (2003) *Moorlands of England and Wales – an environmental history 8,000 BC–AD 2,000*. Edinburgh: Edinburg University Press.

Shipley, G. (1996) *Ancient history and landscape histories*. In: Shipley, G., & Salmon, J. (eds.), *Human landscapes in classical antiquity: Environment and culture*. Oxon: Routledge, pp. 1–16.

Shipley, G., & Salmon, J (1996) *Human landscapes in classical antiquity. Environment and Culture*. Oxon: Routledge.

Shulting, R.J. (2008) Foodways and social ecologies from the early Mesolithic to the early Bronze Age. In: Pollard, J. (ed.), *Prehistoric Britain*. London: Blackwell, pp. 90–120.

Smith, B. (2001) Low-level food production. *Journal of Archaeological Research*, 9(1). Available from: http://anthropology.si.edu/archaeobio/images/smith2001_llfp_jar.pdf [Accessed 24 November 2016].

Smith, M. (2006) The archaeology of food preference. *American Anthropologist*, 108(3), 480–493.

Stuiver, M., & Reimer, P.J. (1993) Extended C14 data base and revised calib 3.0 C14 calibration SW Asian origins to NW European limits. *European Journal of Archaeology*, 8, 137–156.

Tannahill, R. (1989) *Food in history*. London: Crown Publishers.

Thommen, L. (2012) *An environmental history of ancient Greece and Rome*. Cambridge: Cambridge University Press.

Tolan-Smith, C. (1998) Radiocarbon chronology and the late-glacial and post-glacial resettlement of the British Isles. In: Eriksen, B.V., & Straus, L.G. (eds.), *As the world warmed: Human adaptation across the pleistocene/holocene boundary. Quaternary International*, 49/50, 21–27.

Tolan-Smith, C. (2002) Colonization – event or process. In: Larsson, L., Kindgren, H., Knutsson, K., Leoffler, D., & Akerlund, A. (eds.), *Mesolithic on the move: Papers presented at the sixth international conference on the mesolithic in Europe*. Stockholm and Oxford: Oxbow, pp. 52–56.

Toussaint-Samat, M. (2009) *History of food*. London: Wiley-Blackwell.

Ucko, P., & Layton, R. (eds.) (1999) *The archaeology and anthropology of landscape: Shaping your landscape*. London: Routledge.

Van der Veen, M. (2003) When is food a luxury? *World Archaeology, 34*(3), 405-427.

Vera, F.W.M. (2000) *Grazing ecology and forest history*. Oxfordshire: Cabi Publishing.

Vera, F.W.M. (2009) Large-scale nature development – the Oostvaardersplassen. *Br Wildl*, 20(5), 28–36.

Waldheim, C., & Doherty, G. (2016) *Is landscape . . . ? Essays on the identity of landscape*. London: Routledge.

Warren, G. (2017) Making the familiar past; northwest European hunter-gatherers, analogoes and comparisons. In B. Finlayson and G. Warren (eds.) *The Diversity of Hunter Gatherer Pasts*. Oxford: Oxbow Books

White, P. (2000) *Ancient Dartmoor – an introduction*. London: Bossiney Books.

Whitehouse, N.J., & Smith, D.N. (2004) Islands´ in Holocene forests: Implications for forest openness, landscape clearance and Culture-Steppe´ species. *Environmental Archaeology*, 9, 203–212.

Whitehouse, N.J. and Smith, D.N. (2010) How fragmented was the British Holocene wildwood? Perspectives on the "Vera" grazing debate. *Quaternary Science Reviews 29*(3-4), 539-553.

Whittle, A. (1996) *Europe from the fossil beetle record in the neolithic: The creation of new worlds*. Cambridge: The Cambridge University Press.

Wiseman, J., & Zachos, K. (2003) *Landscape archaeology in Southern Epirus, Greece I. The Nikopolis project*. Hesperia Supplement 32. Athens, American School of Classical Studies.

Wrangham, R.W., & Carmody, R.N. (2010) Human adaptation to the control of fire. *Evolutionary Anthropology*, 19(5), 187–199.

Zapata, L. (2014) Exploring diversity in the past: An introduction. In: Chevalier, A., Marinova, E., & Pena-Chocarro, L. (eds.), *Plants and people: Choices and diversity through time*, Earth Vol. I. Oxford: Oxbow Books, pp. 15–20.

2

The shaping of food landscapes from the Neolithic to Industrial period

Changing agro-ecosystems between three agrarian revolutions

E. Gunilla A. Olsson

Introduction

More than 25 years ago, I wrote a paper with the title 'Agro-ecosystems from Neolithic time to the present' (Olsson, 1991). It was based on results obtained in the large interdisciplinary research project *The cultural landscape during 6000 years in Southern Sweden* (Berglund, 1991). The approach of that project was pioneering and inspired a number of following landscape projects in the 1990s and onwards (Tress et al., 2005, 2007). The empirical data was extracted from field data and historical sources for a specific region, the Ystad area, in southern Sweden. The previous paper and findings from the interdisciplinary project lay the background for this chapter; although the view on the foundations for agro-ecosystems has not changed, indeed our interpretation and understanding of key factors for landscape changes along spatial and temporal scales have widened as we shall see in the following. The present chapter focuses on European agricultural landscapes.

The transition from foraging to farming

When did humans become subsistence farmers? The *Neolithic revolution* is used as an expression for the assumedly dramatic step in human resource economy from nomadic gatherers and hunters to arable farmers. The onset of food cultivation comprises two important processes: (1) transition from foraging (Chapter 5) in a nomadic lifestyle to a sedentary agricultural economy and (2) domestication of plants and animals. The earliest evidence of cultivation of grain crops, wild predecessors of barley and, later, emmer wheat (Gross, 2013) were found in the Fertile Crescent in the Middle East, some 12,000 years ago; however, archaeologists have found evidence of independently arising grain crop cultivation in different parts of the world (ibid.). Today's interpretation of the Neolithic transition is less dramatic since the process is seen as a co-evolution or mutual domestication between humans and the domesticated and cultivated crops and domesticated animals (ibid.; Chapter 1). The transition from a hunter-gatherer economy to food production in an agro-ecosystem was both gradual and sporadic in different places, depending on combinations of environmental factors, population density, social organisation and mobility (Ullah et al., 2015). Today, it seems likely that the transition was preceded by landscape changes brought about by the late Pleistocene humans in their foraging activities (ibid.). The pre-Neolithic landscape was strongly imprinted by the grazing activities of wild herbivores such as wild horses and aurochs, and was possibly like a semi-open parkland with grasslands interspersed with woodlands (Vera, 2000: 87 cont.). Detailed evidence and data on the Neolithic transition are still scarce, and there are ongoing intensive scientific discussions on the steps in this process (Fuller et al., 2014; de Saulieu and Testart, 2015; Lancelotti et al., 2016).

The agro-ecosystem concept and the challenge to maintain the agricultural nutrient balance

Agro-ecosystems are subject to the same processes as natural ecosystems (ecosystems that include humans but have not been modified by sedentary agriculture) and they are powered by solar energy for production of plant biomass. Biomass production of both wild and cultivated plants is maintained by availability of solar energy, carbon dioxide, water and plant nutrients (particularly nitrogen, potassium and phosphorus). In natural ecosystems, the circulation of nutrients is relatively balanced. Most of the nutrients in the biomass produced are circulated via the decomposer chain and via additions from stored sources by weathering processes. In agro-ecosystems, the continuous harvest of biomass and exposed soil surfaces from recurrent planting and seeding encourages increased nutrient leaching, which leads to degradation of the soil nutrient pool unless this is balanced by a continuous supply of additional nutrients. The major difference between natural ecosystems and agro-ecosystems is that the latter are manipulated in various degrees to maximise the production of one aspect of plant biomass – the cultivated crops. Cultivation seeks to reduce competition for limited resources for crop plants at the expense of biological diversity. Generally, water and plant nutrients can be considered the most important limiting factors for crop production. This is the general understanding of agricultural crop production. However, our knowledge on crop diversity and general knowledge on agro-biodiversity has widened; thus, now we know that polycultures have been widespread both in time and space, from the Neolithic period onwards – and still exist in many parts of the world (Angus Fraser et al., 2015). Such polycultures contain not only different crop species but also different life forms, from grasses to trees and other woody species, facilitating harvest of biomass at different periods of time. Also, species that are today labelled as 'weeds' were included in pre-industrial cultivation. Some of those species were nitrogen-fixing which contributed to maintaining the nutrient balance in the cultivated spots (Small and Raizada, 2017). The most significant factor involved in the transition from the hunter-gatherer economy to agriculture and the use and maintenance of agro-ecosystems is the transport and use of plant nutrients (Emanuelsson, 1988; Olsson, 1988a, b; Olsson, 1991; Emanuelsson, 2009: 25; Backéus et al., 2016; Figure 2.1).

The development of agricultural periods and the agricultural revolutions

With respect to the plant-nutrient challenge in agriculture, the period from the Neolithic period to the Industrial can be divided into three sub-periods according to the main agricultural practices applied (Figure 2.2). The time notations in the following sections are related to the findings from southern Sweden (Berglund, 1991; Olsson, 1991), but the general processes are relevant for most European landscapes – except for the prehistoric periods – some 500–800 years earlier in the Eastern Mediterranean region.

Shifting cultivation system: Neolithic to Early Bronze Age (4000–800 BCE)*

For the Neolithic farmers, the nutrient challenge was solved by: moving arable fields; creating new patches for cultivation when the harvest yield declined; allowing for regeneration of wild vegetation; and successively restoring the soil nutrient pool before the same patch could be cultivated again. This form of agriculture, shifting cultivation with the variant slash and burn agriculture or swidden agriculture, allowed for the use of the stored nutrient pool in soil and vegetation for short-term cultivation, followed by fallow

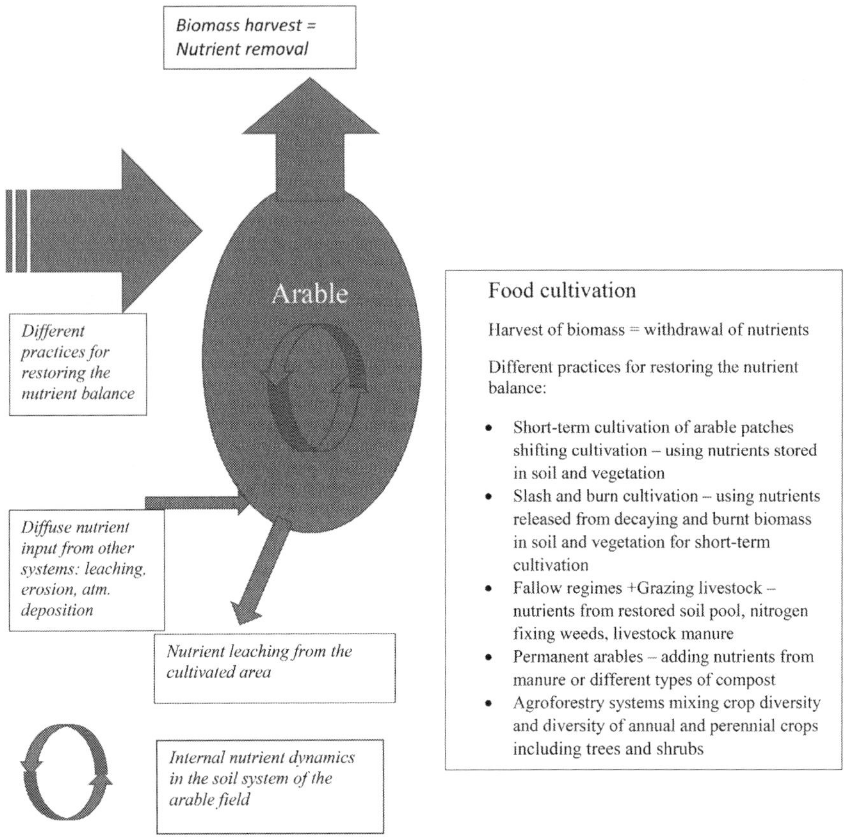

Figure 2.1 The generalised nutrient challenge in arable agriculture.

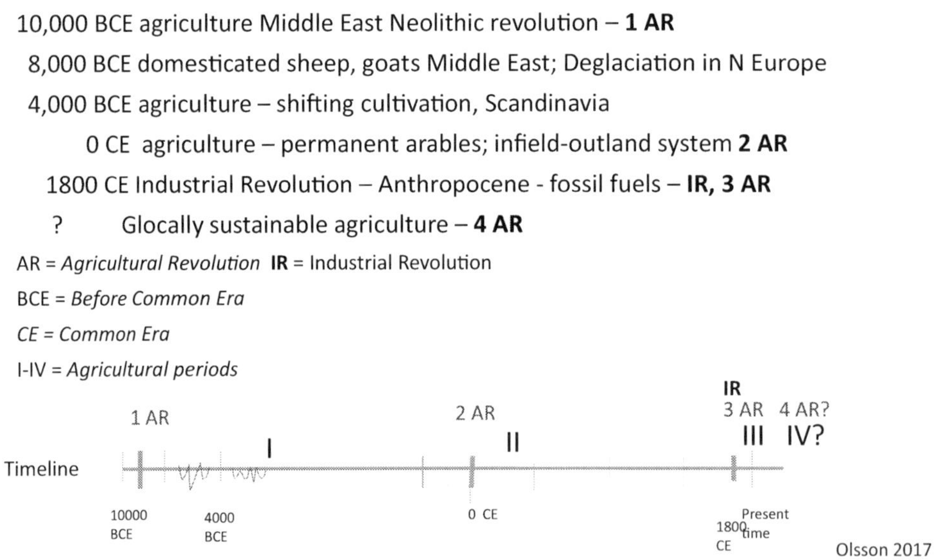

Figure 2.2 Time dimensions of the agricultural periods. See main text for elaboration.

periods, where the soil nutrients were able to recover through the secondary succession processes with developments into shrub and woodland seres. The shifting cultivation system needed large territories for the moving arable fields and for livestock grazing. The length of the fallow periods was dependent on available area, human population size, and environmental conditions, as elaborated in influential texts by Ester Boserup (1965, 1981). The length of fallow periods also influenced the degree and structure of re-growth, shrub lands, or wooded habitats. It is possible that this system also included the management of coppices dispersed among and in the cultivated patches, thereby constituting the early agroforestry systems. Such systems, with nutrients and soil-water captured by tree roots and nutrients released from decaying tree roots related to the coppicing cycles, could probably allow for cultivation at the same spot for long periods, reducing the need for long fallows.

The earliest indication of domesticated livestock is of sheep, goats, cattle and pigs from the Fertile Crescent in the Middle East (Syria, Iraq and neighbouring regions) some 11,000–10,000 years BP (Zeder, 2008). The climate during the Neolithic and Bronze Ages was warmer than today, approximately 2°C warmer on average in the summer (Berglund et al., 1991). This implies that the livestock in northern Europe could also have been nourished by year-round grazing and were thus involved in the creation of semi-natural grasslands and in the shaping of a mosaic landscape with woodlands, small cultivated patches, horticultural activities, and numerous successional stages with shrubs and woody species. The agricultural practices used in shifting cultivation and coppicing were performed with the use of simple tools made of wood, stone, bone and horn, as well as the management of grasslands by livestock grazing. For almost 3,000 years, these were the main agricultural activities, and they relied upon resources from local ecosystems. Parallel to the farming activities, food resources for the human population were obtained from hunting of wild animals, honey collection, and collecting important resources from aquatic ecosystems, both marine and fresh water. The human foraging and agricultural activities in the terrestrial and aquatic landscapes had an impact on different dimensions of the ecosystems, including biological diversity. Hunting decimated the species diversity of wild, large herbivores, but the increasing number of domestic livestock, that is, grazers such as sheep and cattle and browsers like goats and some cattle varieties, might partly have replaced the ecological role of the wild herbivores, in terms of the impact of grazing and browsing (cf. Vera, 2000). The north European prehistoric landscape is interpreted as woodlands, intermingled with open areas of grasslands and arable patches, increasing towards end of the Bronze Age (Berglund et al., 1991; Vera, 2000: 'parkland'). This landscape was very heterogeneous and had considerably high biodiversity due to the manifold of successional stages, small-scale cultivation patches, coppices and grasslands. The human activities were instrumental for changing and increasing the biodiversity by shaping different habitats. The long-lived myth of wilderness untouched by humans as a treasure for biodiversity was dismissed already in 1992 as a result of joint work by ecologists and anthropologists (Gómez-Pompa and Kaus, 1992). They presented several observations from tropical forests, previously presumably untouched by humans, but in reality those forests, such as the Amazon, had been used by human communities for different forms of shifting cultivations and slash and burn agriculture for long periods of time, thereby positively influencing the biological diversity. Similar results and interpretations have been presented later by several researchers (Padoch and Pinedo-Vasquez, 2010).

Transition period to permanent arable: from shifting cultivation to infield-outland system, early Bronze to pre-Roman Iron Age (800 BCE–0 CE)

The introduction of permanent arable fields based on the use of manure can be called the Second Agricultural Revolution. This system may have been introduced in southern Scandinavia around 2800 BP according to interpretation of archaeological findings from the Ystad area (Berglund, 1991). According to this interpretation, the habit of using permanent arable fields gradually became more important, however, used side by side with the shifting cultivation/swidden agriculture. The crop plant diversity was larger than during the Neolithic period; in the Ystad region, southern Scandinavia, cultivation included naked and hulled barley, einkorn and emmer wheat, millet and oats (Olsson, 1991). The large number of macro fossils of the plants *Camelina sativa* and *Bromus secalinus*, both classified today as weeds, indicates they were cultivated as crop plants in this area (Engelmark and Viklund, 1990). Evidence from palaeoarcheological data expressed in pollen diagrams confirms the extended agrarian imprints in the landscape by increasing the share of anthropochores (plants related to human settlements and agriculture) and declining arboreal (tree species) pollen. Overall, the increase in plant biodiversity that started in the Neolithic period continued through this period (Berglund et al., 1991). Another source for confirming increased agrarian activity and/or larger areas of arable field cultivation during this time period is the increased rate of lake sedimentation, resulting from increased topsoil erosion (Dearing, 1991). The time dimension of the different human food acquisition systems is seen in Figure 2.2, and the principles of the infield-outland system are shown in Figure 2.3.

Permanent arable system: infield-outland system and the diversity of agro-pastoral systems (0 CE–1800 CE)

The infield-outland system – a generalised principle

From around the birth of Christ, the *infield-outland* system seems to have been common in the Ystad area as well as in large parts of Europe. This system included cultivation of arable lands and livestock keeping, and the vital principle was the utilisation of certain land-use categories as nutrient sources and the permanent arable land as nutrient sinks. The livestock acted as converters of the nutrients from the outland to the infield arable lands. The generalised understanding of this system in temperate environments is that the fence keeping the livestock out of the infields maintained the border between the 'infields' and 'outlands'. Infields comprised arable fields and infield meadows harvested for winter fodder. Outlands comprised the main grazing areas but could also harbour fenced-in meadows for hay harvest and arable patches. In temperate seasonal climates, livestock had to be fed indoors during winter months. By grazing the outlands, and by consuming hay and tree leaves harvested as winter fodder, livestock transferred biomass nutrients via manure to the arable lands. The transition of manure nutrients was accomplished by night folding of the livestock in the infields, and/or by stalling the livestock at night and in the winter. The system could work as long as there was balance between the size of the arable land (nutrient sink area) and the nutrient-supporting infield meadows and outlands (nutrient source areas; Figures 2.3 and 2.4). Such nutrient balance can be generalised so that the number of livestock was a function of the acreage and resource availability of the infield meadows together with the outlands. From this basic principle evolved and diverged a multitude of pre-industrial agro-ecosystems in Europe; furthermore, since there was a volatile boundary for/around the infield-outland systems in terms of livestock allowance in the different land use categories, it was more consequent to use the broader term *agro-pastoral systems* that became modified and adapted after regional and local environments, in combination with the development and application of traditional and local ecological knowledge of farming communities (Emanuelsson, 2009; Agnoletti and Emanueli, 2016).

This is not the place to review the multitude of pre-industrial agro-ecosystems that developed in Europe as well as globally. However, for explanation, two examples from two agro-pastoral systems are useful: *agro-forestry* and *transhumance* and one vegetation type, or rather ecosystem, shaped by land uses in the outlands within the pre-industrial agro-ecosystems, *the Atlantic heathlands*; further elaboration follows. As is seen in the following text the land use categories are overlapping and thus the term 'infield-outland system' becomes a generalised and abridged concept, although the nutrient transfer principle is valid through the different variations displayed.

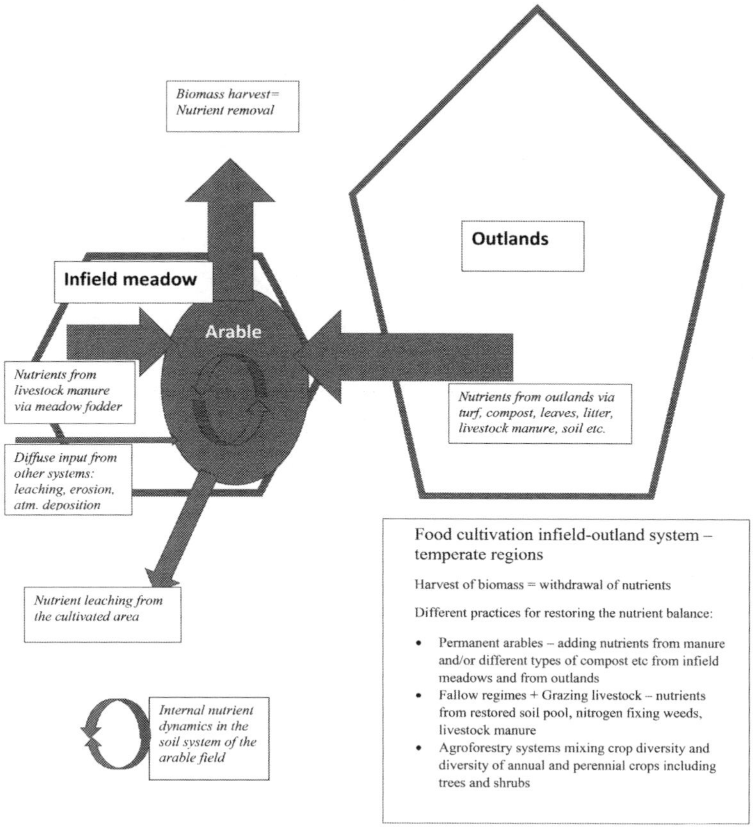

Figure 2.3 Infield-outland system.

Agroforestry systems

Integrating trees and shrubs in and among the cultivated areas, often in combination with livestock, is the basis of agroforestry systems (Mosquera-Losada et al., 2008; FAO, 2015); thus, it has prehistoric roots in the Neolithic time as seen earlier. Characteristic for agroforestry systems is high agro-biodiversity including different life forms, annuals and perennials (trees and woody species), and the use of different cropping, harvesting and pastoral practices. The harvest period is prolonged and productivity of such systems is generally higher than in agro-ecosystems based on monocultures. Agroforestry systems in Scandinavia and several other countries in Europe include the management of fruit and nut bearing trees for human food sources and deciduous tree species for leaf-fodder for livestock in wooded meadows and along borders of arable fields. This practice resulted in diversity in the harvest, in terms of human food and feed for livestock from arable crops, meadows, and from woody species. The harvest of leaves, twigs, and branches from trees and shrubs was accomplished with different techniques adapted after the environment and necessity, for example coppicing, pollarding, shredding and lopping (Rackham, 1980). Those practices provided both leaves and twigs for fodder and also wood from branches and stems for multiple utility needs in the pre-industrial agricultural systems (ibid.). The repeated cutting of leaves and branches implied continuous decay and re-growth of root biomass of the woody species that contributed nutrients to the soil and thus functioned as a nutrient source for the agroforestry system (Olsson, 2004). There is a large diversity of

Figure 2.4 The infield-outland system illustrated on the island Herrön, 1980. In the center is seen cultivated arable fields, mown infields and surrounding extant outlands, here composed of granite rocks with small ponds, fens and woody groves in shallow depressions with soil accumulation. Bohuslän, Western Sweden.

Source: E.G.A. Olsson, 1980

agroforestry systems both globally and in Europe (Mosquera-Losada et al., 2008) with variation in techniques, practices, frequency of cutting and adaptations after local environments. The *wooded meadows/grass land landscapes* are examples of agroforesty in eastern Scandinavia favouring the tree species ash (*Fraxinus excelsior*), lime (*Tilia cordata*) and sometimes birch (*Betula* sp.) and willows (*Salix* spp.) – see for example studies from Åland, Finland (Haeggström, 1998) and from Sweden (Slotte, 2000). The highest grassland plant species diversity in Europe is reported from wooded meadows in Estonia (Kull and Zobel, 1991). Ancient agroforestry systems are found in Romania where there are waste areas of wooded grasslands, with long management history and very high diversity of plant species (Dahlström et al., 2013). Another form of agroforestry involving cutting of tree branches and stems as coppice is seen in Mediterranean regions using species like prickly oak (*Quercus coccifera*), different species of ash (*Fraxinus* sp.) or chestnut (*Castanea sativa*) (Grove and Rackham, 2003).

The variation and endless combinations of practices, scale of cultivated arable fields, labour intensity, and livestock management – all related and adapted to the preconditions in the local environment – imply that agro-pastoral systems can result in *poly-cultural landscapes, agro-silvo-pastoral landscapes* and in the Mediterranean classified as *garden landscapes* (Barbera and Culotta, 2016). This is very evident in the Mediterranean region with the longest agricultural history (see preceding), displaying a large diversity of habitats with a mixture of semi-natural and cultivated patches and a very high biodiversity of wild and cultivated species (also see Chapter 13).

Another form of agro-pastoral system and agroforestry is the well-known and specific *dehesa* in Spain and Portugal (Joffre et al., 1999; Blondel, 2006; Olea and San Miguel-Ayanz, 2006). This system involves grazing livestock in a woodland landscape, often Mediterranean oak species, and sometimes the land amongst the trees is cultivated with annual crops. The *dehesa* is famous for its cork-producing oaks (*Quercus suber*). An example of *dehesa* from Andalucia, Spain (Asha and Newing, 2015) was based on family managed subsistence production of vegetables, honey, grain crops, coal, cork, pigs and goats. The landscape was savannah-like with relatively low oak tree density and understory, managed for goat and pig farming and small patches of grain crops. The acorn produce from the oaks was important for the livestock, in combination with grazing and browsing (Grove and Rackham, 2003).

The agroforestry system unambiguously characterising the Mediterranean region is *olive cultivation*. This system has prehistoric origins and has been in continuous development to increase olive production but within the borders of local and regional agro-ecosystems. A study of this system at the end of pre-industrial time, in Andalusia in 1750, shows the complexity of links and loops of biomass and nutrients directed by agrarian practices where local knowledge was applied to maintain productivity, as described by Infante-Amate et al. (2016). The combination of olive trees, livestock grazing, browsing and recycling of all organic by-products – in particular pomace as manure and fertiliser to the system – was crucial. This system yielded above all olive products – olive fruits, oil for food and lighting, soap – but also meat, milk, wool from livestock, and wood, fuels, as well as arable crops like grains and vegetables when cultivated areas were included. The olive agroforestry system was often developed in combination with cultivation terraces (see Grove and Rackham, 2003).

Transhumance systems

The biological production in seasonal landscapes is governed by seasonal changes in light, precipitation and temperature, which pose challenges for farming communities. The challenges of seasonality are magnified in mountains, with shorter vegetation periods along the altitudinal gradient. An efficient way of utilising biological resources in seasonal landscapes has been the development of different forms of *transhumance*. The transhumance system is an agro-pastoral system and implies moving parts of the human community and their livestock to temporary seasonal settlements in regions where grazing grounds for the flock are available. Such systems are found globally. Transhumance and seasonal resource use in mountain areas in Europe is described in the classical text from the Alps by Netting (1981). The importance of the mountain habitats offering livestock fodder, both summer grazing as well as winter fodder from the meadows and grasslands is demonstrated from several European mountains, including Scandinavia (Dodgshon and Olsson, 2007). The fine-tuning of human practices in time and space is illustrated by the sequential use of temporal settlements and habitats along the altitudinal gradient. There is also an interesting example of the traditional ecological knowledge (TEK) developed by mountain farmers in Norway, for example, by using lichen as winter fodder for their livestock when the environmental conditions of the local ecosystem led to scarcity of grassland habitats (ibid.). The intensive use of the mountains by a variety of practices, of which the livestock grazing was most important, had direct implications for the shaping of plant communities, development of bio-cultural diversity (Austrheim et al., 1999; Olsson et al., 2000; Tømmervik et al., 2010) and even changing position of the treeline (Wehn et al., 2012).

Transhumance in southern Europe is well documented, for example from the Pyrenées (Fernandez-Gimenez and Fillat Estaque, 2012), where the pastoralist's detailed knowledge of palatable grazing plants as well as their ability to monitor and assess the grassland quality for their herds is demonstrated. Another form of transhumance took place along the famous Consequense Drove Road in central Spain (Oteros-Rosaz et al., 2013). Herders drove their flocks in a seasonal schedule. Winter grazing in the southern lowlands, a typical Mediterranean dehesa landscape and an agro-silvopastoral ecosystem aimed mainly at extensive livestock grazing (Oteros-Rozaz et al., 2013). The summer grazing lands in the highland slopes in central Spain is a landscape characterised by mixture of forest groves and agricultural patches (ibid.). The transhumance along the Consequense Drove Road is still alive, although declining (Emanuelsson, 2009: 152); however, at the same time, it is put forward as an example of sustainable resource use with bearing on sustainable food production for the future (Intergovermental Panel for Biodiversity and Ecosystem Services (IPBES), ECA, 2018). The value of transhumance practices for sustainable food production also in the future is also proposed by Dominguez (2016) in mountain transhumance in the Pyrenees and the Atlas mountains in northern Morocco.

The shaping of the outlands and the development of landscape forming vegetation and ecosystems: the example of Atlantic heathlands

As we have seen earlier, the numerous agro-pastoral systems have a very large variation in their land uses adjusted to local, social-environmental conditions. In those systems, although with large variation, the outlands undoubtedly had a key function as important resource cradles for maintenance of the agro-ecosystem. The outlands were the main grazing areas for livestock, but what could also take place was cultivation of arable patches as the first step in land reclamation or as a kind of shifting cultivation. A number of resource extraction activities occurred, for example collection of: peat as fuel for cooking and heating; heather and grass as fodder for livestock in byres; medicinal plants for humans and livestock; young trees and saplings as wood for housings and implements; and not least, litter, compost and sometimes even soil to improve the nutrient status of the agricultural fields (Grove and Rackham, 2003; Dodgshon and Olsson, 2006; Wehn et al., 2012). In the Mediterranean regions, the intensive use of the outlands also involved frequent use of fire for management of the grazing lands. The long-term human use of the Mediterranean outlands has shaped the specific landscape forming vegetation types, namely, *garigue* and *macchia* (Grove and Rackham, 2003; Emanuelsson, 2009: 147).

The western part of Europe that is influenced by the Atlantic climate with relatively cool and rainy summers and mild and wet winters offers year-round grazing for livestock; thus, the need for pasturing areas was significantly larger than in continental regions in the pre-industrial agro-ecosystems (Mosqutera-Losada et al., 2008). This land use, in combination with biophysical conditions, has formed the characteristic landscape-forming vegetation type or rather ecosystem – the *Atlantic heathlands* that exist along the European west coasts, from northern Portugal along the Norwegian coast beyond the Polar circle, including the British Isles and Ireland (Webb, 2002). It is an open tree-less landscape with a vegetation of evergreen dwarf shrubs dominated by heather (*Calluna vulgaris*). Although there is variation in the composition of the plant communities across the heathland region, heather is always the dominant species and has a key role for human use (Gimingham, 1972). The heathland vegetation develops on acidic soils with low fertility and the combination of high precipitation and heathland use – see below – amplifies podsolisation of the soil profile and promotes the life cycle of *Calluna* (Gimingham, 1972; Webb, 2002). From palaeoecological evidence, its origin is traced back to prehistory, some 5,000 years ago (Kaland, 1986; Webb, 2002; Kvamme et al., 2004). The *Calluna* heathland is a genuine human-shaped landscape, starting with deforestation and demanding human influence for its maintenance. The use of the heathlands includes continuous livestock grazing and periodic burning to promote the quality of the grazing vegetation. This management with regulated burning frequency and livestock grazing pressure, adjusted after the local environmental conditions, stimulated cyclic regeneration of nutritious young shoots of the *Calluna* and contributed to development of habitat mosaics with patches of grassland intermingled with the heather vegetation and fostered a rich biological diversity (Kvamme et al., 2004; Rosa Garcia et al., 2013). Besides offering basic resources for livestock grazing, the heathland was a source for collecting a number of resources (see above). *Calluna* shrubs were mown and collected for livestock feed in byres (Webb, 2002; Kvamme et al., 2004; Dodgshon and Olsson, 2006). This use prevented forest re-colonisation; however, today when the traditional use has decreased or ceased, the greatest threat against this ecosystem is forest re-colonisation and forest plantations. Remains of the Atlantic heathlands exist today in some protected areas as Natura 2000 sites in western European countries. Conservation measures applied here are combinations of livestock grazing and regulated burning (Kvamme et al., 2004; EC, 2008; Rosa Garcia et al., 2013).

The nutrient status in the agricultural landscapes at the end of pre-industrial time in Europe

During the existence of the infield-outland system, several improvements in agricultural techniques took place. The use of cropping systems with periods of fallow was able to improve nutrient recovery in the arable areas. During the 800 years when this system was in operation, changes took place gradually, but few were of dramatic enough importance to change the fundamental principles of the system (Figure 2.3). The need for larger harvests to feed increasing human populations was met by expanding arable fields at the expense of the grasslands, although only on the dry soils, and also by increased land reclamation in the outlands. The general result was decreasing crop harvests due to insufficient nutrient availability. The agricultural ecosystems in the framework of the infield-outland system had little external input of energy and nutrients, other than from local and regional systems.

For the Ystad area in southern Scandinavia, the balance between nutrient source and sink areas was already disrupted by the late seventeenth century since most agricultural units (farms, villages) in this area appeared with considerably smaller outlands than arable areas (Olsson, 1991). It is estimated that the total area of meadows and pastures in this region of Europe should, at a minimum, attain twice the size of arable areas in the boulder clay areas and three times or more in the sandy plain areas in order to sustain winter fodder for livestock at a sufficient quantity to produce manure to maintain fertility of arable fields (Olsson, 1991). Dependence of the regional system increased, and during the eighteenth century became an obligate relationship for the three-course villages[1] on the boulder clay areas. The parishes in the northern part of the Ystad area with greater availability of pasture lands became the nutrient sources for the villages in the boulder clay areas, where the grassland had been transferred to arable land (ibid.). Leaching from the arable lands increased due to the large areas used for tillage and due to recovery times too short for the fallows that were exposed to high grazing pressure from domestic pigs and small livestock. The productivity of arable fields generally decreased in the three-course villages towards the end of the eighteenth century (Olsson, 1991). The increasing pressure on the infield-outland system for maintaining arable yields demanded other ways of ensuring plant nutrients. During a very short period in the nineteenth century (ca. 1840–1900), water meadows were used in the Ystad area to supply plant nutrients and thereby raise the productivity of the hay meadows and were widespread in the agro-pastoral systems in Europe (Emanuelsson, 2009: 281). Improved crop rotation and cultivation of nitrogen-fixing crops on the fallows were other methods of raising the agricultural productivity within the framework of the local agro-ecosystem.

From the existence of detailed fiscal maps in Sweden produced during the eighteenth century, detailed landscape, ecological and land use information is offered. For the Ystad area, 63 such large-scale maps were compiled to produce a regional overview during the mid eighteenth century (Linusson and Olsson, 1991). Two features in this landscape are prominent – large numbers of wetlands and a high diversity of habitats. The grasslands on dry and mesic ground had been transformed to arable fields, but it was not possible to cultivate the wetlands at this time and they provided valuable resources for the farming communities. The many different habitats were partly a product of the landscape topography and hydrology, but also of the diverse local resource extraction practices maintaining coppice areas, mowed wetlands, grazed fens, peat cutting areas, and many more. It seems obvious and logical that this landscape could harbour a large biological diversity. It was a landscape structured by land ownership and manual labour in combination with traditional ecological knowledge. The diversity of habitats produced the diversity of land uses built by manual labour with local knowledge – labour landscapes around the peasant villages. It was also a recreation landscape with hunting grounds as forest parks in the vicinity of the manors owned by nobility (Germundsson et al., 1991). In this respect, the landscape in the Ystad region was representative of much of the pre-industrial European countryside.

The third agrarian revolution – the emergence of the Industrial period and Anthropocene era – permanent arable: external nutrient import

The beginning of the Industrial period is often dated to the introduction of the steam engine, fuelled by fossil coal, in Britain in 1764 (Crutzen, 2002). The Anthropocene era, according to Crutzen (2002) and Steffen et al. (2011), is named after the global impact of increasing use of fossil fuels in the western world on the planet's ecosystems and global climate (also see Chapter 16). The Anthropocene era is closely linked to the Industrial period and an attendant inequality of cause – by a global wealthy minority – and effects endured by a global majority of poor people (Malm and Hornborg, 2014). The invention of the steam engine marked the development of an exponential increase in energy demanding activities and processes in all sectors of western societies as well as in agriculture, where manual and animal labour could be replaced by engines driven by fossil fuels, subsequently including not just coal, but oil and gas. This is *the third revolution in agriculture* (see Figure 2.2), which meant independence from transferring nutrients from the local ecosystem to the arable fields and decoupling of the combination of livestock and arable crop production. Among the different activities exercised in the agro-ecosystems were two processes that have particularly affected the modification of European agricultural landscapes. One such process is the *draining of wetlands* on a larger scale by motorised labour for ditching works as well as through the pumping technology used to remove water.

The other influential process is the *industrial manufacturing of fertilisers* based on atmospheric nitrogen (Chapters 10 and 16). This process made nitrogen fertilisers available in large scale, leading to a substantial increase in yield of arable crops. The fodder producing areas for livestock also increased yield by fertilisation and by growing grass on 'improved' grasslands or growing leys[2] as an arable crop. This led subsequently to the abandonment of grazing on semi-natural, non-ploughed grasslands. The use of pesticides[3] in agricultural production facilitated the transition to monocultures and had impact on all landscape elements and habitats in the agricultural landscape. The outcomes of industrial agricultural processes on agro-ecosystems are well documented, such as loss of biodiversity for large numbers of organism groups, soil erosion, decreased water quality, and emissions of greenhouse gases contributing to climate change (MA, 2005; De Chazal and Rounsevell, 2009; Foley et al., 2011; Kuyper and Struik, 2014; Chapter 16). The price of the raised yields from crops and livestock products in the industrialised agricultural production has meant an ongoing decline in biodiversity and ecosystem services (ibid.; Chapter 17). Linked to those processes is also a rapid growth of global trade of input products (such as phosphorus and soy products as livestock feed) for agriculture as well as food items, thus, literally moving food production to other parts of the world (Kastner et al., 2014).

Implicit in those processes is the use of nutrient and energy sources from remote ecosystems, which also implies food production, linked to the global food system (McMichael, 2011), with growing distance between producers and consumers (Chapter 16). This has led to polarisation of landscape patterns with crop production in monocultures and elsewhere abandoned agricultural areas and semi-natural pastures, in processes of secondary forest succession or large-scale forest plantations – or protected areas (Olsson et al., 2016). Despite progress in implementing European policies (such the Birds and Habitats Directives and the Water Framework Directive) as well as environmental measures within the framework of Common Agricultural Policy, EU-CAP, Europe has experienced major decline in biodiversity associated with agro-ecosystems and grasslands. A recent reporting process on the conservation status of species and habitats under the EU-Habitats Directive, showed that 63% of the assessments of habitats linked to grasslands are unfavourable, which has been linked to increased specialisation and intensification in agriculture (EEA, 2015; Chapter 17).

The co-evolution of local knowledge and biological resources for the shaping of the pre-industrial agro-ecosystems

Today, there is an understanding and agreement among most researchers in the international scientific community of the positive influence of local knowledge for the use of natural resources and continuous shaping of habitats, biodiversity and ecosystem services in the pre-industrial agro-ecosystems (Berkes, 1999; Aikenhead and Ogawa, 2007; Molnar, 2012; Roué and Molnar, 2016; Chapters 3 and 17). This is a rather new insight among scientists, and there is an ongoing debate on how to incorporate traditional and local knowledge in scientific accounts since little of local knowledge is published. In the ongoing efforts within the United Nations' IPBES on global and regional assessment of biodiversity and ecosystem services, and to find pathways towards future sustainable resource use in human communities, the issue of traditional and local knowledge integration is a core challenge (Diaz et al., 2015a). With the creation of IPBES' framework, there was an attempt to link scientific and local knowledge and different worldviews (Diaz et al., 2015b). This is neither easy nor clear cut, and both the framework and the concept of local knowledge are evolving. An example of the ambiguity in the concept of *Local Knowledge* with sensitive and political connotations is that it has changed from the term *Traditional Ecological Knowledge, TEK,* suggested by Berkes (1999) and with wide use among ecologists and anthropologists to *Indigenous and Local Knowledge and Practices, ILKP,* which is now recommended for use in the IPBES documents (IPBES, 2014). However, awareness of the crucial role of local knowledge for shaping European agricultural landscapes is growing and can be seen as a valuable adaptive knowledge for handling unexpected situations and disturbances and responding to global environmental change. It can provide insights for the management of biodiversity and habitats, to maintain ecosystem services like food production, and for management of human-shaped landscapes in general. A recent survey of research on local knowledge in European agricultural landscapes (Hernandez-Morcillo et al., 2014) reported on the rapid changes in maintenance of the TEK and decline in intergenerational turnover of this knowledge. Similar trends have been observed from subtropical agro-ecosystems in Bolivia (Fernandez-Llamazares et al., 2015) and in west Africa (Angus Fraser et al., 2015).

In this survey of the evolution of agricultural landscapes with a long-term perspective, it is noteworthy to realise the persistence of previous land uses in the human shaping of biodiversity and landscapes. This was demonstrated by studies on diversity of plant species in Swedish agricultural landscapes (Cousins, 2009). By using ancient maps (>300 years old) for landscape reconstruction, Cousins found that past and present grassland distribution was a major factor in determining recent patterns in grassland plant species (ibid.). This can be called the legacy effects of human land use (Bürgi et al., 2017) and has important lessons for future management and food production from a sustainability perspective.

Food-producing landscapes for the future

Food production in the future must happen within a framework of global equity in order to live up to global agreements of sustainable development (UN-SDGs, 2015; Chapter 31). The possibilities for sustainable food production converge towards one bundle of pathways, guided by the combination of environmental conditions and applied local knowledge. The numerous variants of agro-pastoral systems, agroforestry, polycultures, etc., previously mentioned, were all based upon such combinations. When fossil energy sources became cheaply available, many agro-pastoral systems were abandoned or considered irrelevant and backwards. Today, the situation is very different under the threats from global environmental change, increasing uncertainties, economic and political instability and insight on peaking non-renewable resources such as fossil fuels, phosphorus and certain minerals. Pathways of sustainable food production for the future will resume work on the challenges of local environmental limitations and apply the local knowledge that was developed within the farming communities in Europe as well as in all parts of the world – within a 10,000 year timeframe (Chapter 4). Instead of striving for monocultures, embracing diversity can be seen as an agricultural strategy (Barthel et al., 2013) and involves crop and habitat diversity as important resilience dimensions in facing and handling different disturbances and uncertainties – not just the effects of climate change. In scenarios of changing pressures and disturbances, the major challenge will be to maintain basic ecosystem functions and services such as food production for humans. The traditional agro-ecosystems that developed in pre-industrial time over thousands of years offer today a treasure for development in combination with innovations and transition towards sustainable communities. Remains of such systems still existing today have been called 'bio-cultural refugia' (Barthel et al., 2013); in essence, they are the seeds of food systems for a sustainable future. Interesting examples of this vision are provided from the work on transhumance agro-pastoral systems in the Pyrenées (Fernandez-Gimenez and Fillat Estaque, 2012) as well as the Consequense Drove Road in south-central Spain (Oteros-Rozas et al., 2013). Although both cases report conflicts with industrial food systems and other threats, visions elaborated among stakeholders including regional government authorities, herders and locals show the possibilities of the systems for maintaining food production (meat), together with a number of other ecosystem services and a high biodiversity in the grasslands, wooded pastures and other semi-natural habitats grazed by the transhumant livestock (Oteros-Rozas et al., 2013). The intergenerational turnover of specific local knowledge involved in transhumance practices would be maintained by targeted education and developing professional competence related both to food production and landscape qualities. In this way the maintenance of the system also has repercussions on the development of viable rural regions by job opportunities (ibid.).

Other examples of visions for sustainable food production from wetland ecosystems are presented by Palomo et al. (2011). Here, food production is based on terrestrial and aquatic resources, and a whole watershed is the focus for use and management. This creates and maintains habitats and biodiversity and allows the ecosystem to produce a number of ecosystem services. It offers a diversity of food products that allows local branding, helping this landscape attract visitors for recreation and nature tourism (ibid.).

The numerous examples of urban agriculture (UA) spontaneously arising in cities can be seen as expressions of people's desire to influence their own food access (Chapters 12, 13 and 22). Food production activities in UA are often based on agro-ecological principles, local participation, and traditional knowledge (Opitz et al., 2015). A promising effort towards increasing resilience is that several cities and municipalities have started work on local food strategies involving peri-urban areas and agro-ecological methods as parts of their sustainability work and strive to reduce urban vulnerability (Olsson et al., 2016; Chapter 12).

Towards the post-industrial and sustainable future, there is a persistent call for approaches of food production that involve transformation towards global shifts in agricultural practices, allowing human well-being to be more equally distributed than at present (Garibaldi et al., 2017). Luckily, we are not short on knowledge for the application and results of such practices since there exist a number of studies on the yield outcomes from different parts of the world (Altieri, 2004; Altieri et al., 2012; Björklund et al., 2012). Those studies report on food production adapted to local ecological conditions, using locally adapted and genetically diverse crop cultivars and approaching the nutrient demand in agriculture by using practices reducing leakage and traditional composting techniques (Altieri, 2004; Altieri et al., 2010; Björklund et al., 2012). It is rather a question of whether the transition time will be swift enough to cope with the ongoing global environmental changes and the fast disappearance of local knowledge for food production.

Notes

* BCE = Before Common Era

1 Three-course village: village where the arable fields had expanded at the expense of fodder areas, meadows and outlands. The cropping system for the arable fields was three-course rotation: a fallow year each third year. There was a general shortage of manure and the system was applied on the most fertile soils such as the boulder-clay areas (Olsson, 1991).

2 Ley: cultivation of grass for livestock feed on arable fields by ploughing, re-seeding and fertilisation.

3 Pesticides: organic chemical compounds used to combat attacks on crops from insects, weeds or fungi. The production of pesticides demands energy and organic chemical sources such as fossil oil.

Further reading

Agnoletti, M., & Emanueli, F. (eds.) (2016) *Biocultural diversity in Europe*. Switzerland: Springer International Publ.

Aikenhead, G.S., & Ogawa, M. (2007) Indigenous knowledge and science revisited. *Cultural Studies of Science Education*, 2, 539–520.

Altieri, M.A. (2004) Linking ecologists and traditional farmers in the search for sustainable agriculture. *Frontiers in Ecology & Environment*, 2(1), 35–42.

Altieri, M.A., Fuenes-Monzote, F.R., & Petersen, P. (2012) Agroecologically efficient agricultural systems for smallholder farmers: Contributions to food sovereignty. *Agronomy and Sustainable Development*, 32, 1–13.

Angus Fraser, J., Frausin, V., & Jarvis, A. (2015) An intergenerational transmission of sustainability? Ancestral habitus and food production in a traditional agro-ecosystem of the Upper Guinea Forest, West Africa. *Global Environmental Change*, 31, 226–238.

Asha, A., & Newing, H.S. (2015) Cork Oak landscapes, promised or compromised lands? A case study of a traditional cultural landscape in Southern Spain. *Human Ecology*, 43, 601–611.

Austrheim, G., Olsson, E.G.A., & Grøntvedt, E. (1999) Land-use impact on plant communities in semi- natural subalpine grasslands of Budalen, central Norway. *Biological Conservation*, 87, 369–379.

Backéus, I., Emanuelsson, U., & Petersson, M. (2016) *The rural landscapes of northeast Asia*. FriTanke förlag. Acta Phytogeographica Suecica 89.

Barbera, G., & Culotta, S. (2016) The traditional Mediterranean polycultural landscape as cultural heritage. In: Agnoletti, M., & Emanueli, F. (eds.), *Biocultural diversity in Europe*. Switzerland: Springer International Publ., pp. 21–48.

Barthel, S., Crumbley, C., & Svedin, U. (2013) Bio-cultural refugia – safeguarding diversity of practices for food security and biodiversity. *Global Environmental Change*, 23, 1142–1152.

Berglund, B.E. (Ed.). (1991) The cultural landscape during 6000 years. *Ecological Bulletins*, 41.

Berglund, B.E., Larsson, L., Lewan, N., Olsson, E.G.A., & Skansjö, S. (1991) Ecological and social factors behind the landscape changes. In: Berglund, B.E. (ed.), *The cultural landscape during 6000 years. Ecological Bulletins*, 41, 425–448.

Berglund, B.E., Malmer, N., & Persson, T. (1991) Landscape-ecological aspects of long-term changes in the Ystad area. In: Berglund, B.E. (ed.), *The cultural landscape during 6000 years. Ecological Bulletins*, 41, 405–424.

Berkes, F. (1999) *Sacred ecology: Traditional ecological knowledge and resource management*. Philadelphia, PA: Taylor & Francis.

Björklund, J., Araya, H., Edwards, S., Goncalves, A., Höök, K., Lundberg, J., & Medina, C. (2012) Ecosystem-based agriculture combining production and conservation – a viable way to feed the world in the long term? *Journal of Sustainable Agriculture*, 36, 824–855.

Blondel, J. (2006) The 'Design' of Mediterranean landscapes: A millennial story of humans and ecological systems during the historic period. *Human Ecology*, 34, 713–729.

Boserup, E. (1965) *Conditions of agricultural growth*. London: Allen and Unwin.

Boserup, E. (1981) *Population and technology*. Oxford: Basil Blackwell.

Bürgi, M., Östlund, L., & Mladenoff, D.J. (2017) Legacy effects of human land use: Ecosystems as time-lagged systems. *Ecosystems*, 20, 94–103.

Cousins, S. (2009) Landscape history and soil properties affect grassland decline and plant species richness in rural landscapes. *Biological Conservation*, 142(11), 2752–2758.

Crutzen, P.J. (2002) Geology of mankind. *Nature*, 415, 23.

Dahlström, A., Iuga, A-M., & Lennartsson, T. (2013) Managing biodiversity rich hay meadows in the EU: A comparison of Swedish and Romanian grasslands. *Environmental Conservation*, 40(2), 194–205.

De Chazal, J., & Rounsevell, M.D.A. (2009) Land-use and climate change within assessments of biodiversity change: A review. *Global Environmental Change*, 19, 306–315.

De Saulieu, G., & Testart, A. (2015) Innovations, food storage and origins of agriculture. *Environmental Archaeology*, 20, 314–320.

Dearing, J. (1991) Erosion and land use. In: Berglund, B.E. (ed.), *The cultural landscape during 6000 years. Ecological Bulletins*, 41, 283–292.

Diaz, S., Demissew, S., Joly, C., Lonsdale, W.M., & Larigauderie, A. (2015a). A rosetta stone for nature's benefit to people. *PLOS Biology*, 13(1), 1–8.

Diaz, S., et al. (2015b). The IPBES conceptual framework – connecting nature and people. *Current Opinion in Environmental Sustainability*, 14, 1–16.

Dodgshon, R.A., & Olsson, E.G.A. (2006) Heather moorland in the Scottish Highlands: The history of a cultural landscape, 1600–1880. *Journal of Historical Geography*, 32, 21–37.

Dodgshon, R.A., & Olsson, E.G.A. (2007) Seasonality in mountain areas: A study in human ecology. In Palang, H., & Fry, G. (eds.), *Landscape interfaces*. Dordrecht; London: Kluwer Academic, pp. 85–102.

Dominguez, P. (2016) A comparative study of two Mediterranean transhumant systems and the biocultural diversity associated with them. In: Agnoletti, M., & Emanueli, F. (eds.), *Biocultural diversity in Europe*. Switzerland: Springer International Publ., pp. 105–122.

EC. (2008) Management of Natura2000 wet heathlands. *Technical report* 2008 08/24. European Commission.

EEA. (2015) *State of environment report 2015, SOER 2015*. European Environment Agency. www.eea.europa.eu/soer-2015/europe/agriculture (accessed 20 February 2017).

Emanuelsson, U. (1988) A model for describing the development of the cultural landscape. In: Birks, H.H., et al. (eds.), *The cultural landscape: Past, present and future*. Cambridge: Cambridge University Press, pp. 111–121.

Emanuelsson, U. (2009) *The rural landscapes of Europe. How man has shaped European nature*. Stockholm: Formas.

Engelmark, R., & Viklund, K. (1990) Makrofossilanalys av växtrester – kunskap om odlandets karaktär och historia. *Bebyggelsehistorisk Tidskrift*, 19, 33–42.

FAO (2015) *United Nations Food and Agriculture Organisation. Agroforestry definition*. www.fao.org/forestry/agroforestry/80338/en/ (accessed 15 February 2017).

Fernandez-Gimenez, M., & Fillat Estaque, F. (2012) Pyrenean pastoralists' ecological knowledge: Documentation and application to natural resource management and adaptation. *Human Ecology*, 40, 287–300.

Fernandez-Llamazares, A., Diaz-Reviriego, I., Luz, A.C., Cabeza, M., Pyhälä, A., & Reyes-Garcıa, V. (2015) Rapid ecosystem change challenges the adaptive capacity of local environmental knowledge. *Global Environmental Change*, 31, 272–284.

Foley, J., et al. (2011) Solutions for a cultivated planet. *Nature*, 478: 337–342.

Fuller, D.Q. et al. (2014) Convergent evolution and parallelism in plant domestication revealed by an expanding archaeological record. *PNAS*, 111, 6147–6152.

Garibaldi, L.A., Gemmill-Herren, B., D'Annolfo, R., Graeub, B.E., Sunningham, S.A., & Breeze, T.D. (2017) Farming approaches for greater biodiversity, livelihoods, and food security. *Trends in Ecology & Evolution*, 32, 68–80.

Germundsson, T. & Olsson, E.G.A. (1991) Landscape and landholding changes in Bjäresjö since 1800. In: Berglund, B.E. (ed.), *The cultural landscape during 6000 years. Ecological Bulletins*, 41, ch. 4.2.8, pp. 205–214.

Gimingham, C.H. (1972) *The ecology of heathlands*. London: Chapman & Hall.

Gómez-Pompa, A., & Kaus, A. (1992) Taming the wilderness myth. *BioScience*, 42(4), 271–279.

Gross, M. (2013) The paradoxical evolution of agriculture. *Current Biology*, 23(16), 667–670.

Grove, A.T., & Rackham, O. (2003) *The nature of Mediterranean Europe: An ecological history*. New Haven, CT: Yale University press.

Haeggström, C.-A. (1998) Pollard meadows: Multiple use of human-made nature. In: Kirby, K.J., & Watkins, C. (eds.), *The ecological history of European forests*. Wallingford: CAB International, pp. 33–41.

Hernández-Morcillo, M., Hoberg, J., Oteros-Rozas, E., Plieninger, T., Gómez-Baggethun, E., & Reyes-García, V. (2014) Traditional ecological knowledge in Europe: Status Quo and insights for the environmental policy agenda. *Environment: Science and Policy for Sustainable Development*, 56(1), 3–17.

Hobbs, R.J., & Gimingham, C.H. (1987) Vegetation, fire and herbivore interactions in heathland. *Advances in Ecological Research*, 16, 87–173.

Infante-Amate, J. et al. (2016) The making of olive landscapes in the South of Spain. In: Agnoletti, M., & Emanueli, F. (eds.), *Biocultural diversity in Europe*. Switzerland: Springer International Publ., pp. 157–180.

Infante-Amate, J., & Gonzalez de Molina, M. (2013) The socio-ecological transition on a crop scale: The case of olive orchards in Southern Spain (1750–2000). *Human Ecology*, 41(6), 961–969.

IPBES (2014) *Procedures for bringing indigenous and local knowledge into the Platform's assessments*. Annex II to decision IPBES-4/3. www.ipbes.net (accessed 20 February 2017).

IPBES (2018) IPBES Europe and Central Asia regional assessment report on biodiversity and ecosystem services. Germany: Secretariat of the Intergovernmental Science-Policy Platform on Biodiversity and Ecosystem Services.

Joffre, R., Rambal, S., & Ratte, J.P. (1999) The dehesa system of southern Spain and Portugal as a natural ecosystem mimic. *Agroforestry Systems*, 45, 57–79.

Kaland, P.E. (1986) The origin and management of Norwegian coastal heaths, as reflected by pollen analysis. In: Behre, K.E. (ed.), *Anthropogenic indicators in pollen diagrams*. Balkema: Rotterdam, pp. 19–32.

Kastner, T., Erb, K-H., & Haberl, H. (2014) Rapid growth in agricultural trade: Effects on global area efficiency and the role of management. *Environmental Research Letters*, 9, 1–10.

Kull, K., & Zobel, M. (1991) High species diversity in an Estonian wooded meadow. *Journal of Vegetation Science*, 2, 715–718.

Kuyper, T.W., & Struik, P.C. (2014) Epilogue: Global food security, rhetoric, and the sustainable intensification debate. *Current Opinion in Environmental Sustainability*, 8, 71–79.

Kvamme, M., Kaland, P.E., & Brekke, N.G. (2004) Conservation and management of North European coastal heathlands: Case study: The Heathland Centre, Lygra, Western Norway. HEATHGUARD. Heathland Centre, Lygra.

Lancelotti, C., et al. (2016) Resilience of small-scale societies' livelihoods: a framework for studying the transition from food gathering to food production. *Ecology and Society*, 21(4), 8.

Linusson, A.C., & Olsson, E.G.A. (1991) A detailed map of the 18th century landscape in the Ystad area. In: Berglund, B.E. (ed.), *The cultural landscape during 6000 years. Ecological Bulletins*, 41, Appendix.

MA (2005) *Millennium ecosystem assessment. Ecosystems and human wellbeing. General synthesis.* Washington, DC: Island Press. Available from: www.MAweb.org.

Malm, Hornborg, A. (2014) A geology of mankind? A critique of the Anthropocene narrative. *The Anthropocene Review*, 1, 62–69.

McMichael, P. (2011) Food system sustainability: Questions of environmental governance in the new world (dis)order. *Global Environmental Change*, 21, 804–812.

Molnar, Z. (2012) Classification of pasture habitats by Hungarian herders in a steppe landscape. *Journal of Ethnobiology and Ethnomedicine*, 8, 28.

Mosquera-Losada, M.R., McAdam, J.H., Romero-Franco, R., Santiago-Freijanes, J.J., & Rigueiro-Rodríguez, A. (2008) Definitions and components of agroforestry practices in Europe. In: Rigueiro-Rodríguez, A., et al. (eds.), *Agroforestry in Europe: Current status and future prospects*. Dortrecht, Netherlands: Springer, pp. 3–19.

Netting, R.M.C. (1981) *Balancing on an alp: ecological change and continuity in a Swiss mountain community*. New York: Cambridge University Press.

Olea, L., & San Miguel-Ayanz, A. (2006) The Spanish dehesa: A traditional Mediterranean silvopastoral system linking production and nature conservation. *21st General Meeting of the European Grassland Federation*, Badajoz, Spain, April.

Olsson, E.G.A. (1988a). Nutrient use and productivity for different cropping systems in South Sweden during the 18th century. In: Birks, H.H., et al. (eds.), *The cultural landscape: Past, present and future*. Cambridge: Cambridge University Press, pp. 123–137.

Olsson, E.G.A. (1988b). Nutrient use efficiency as a tool for evaluating function of preindustrial agroecosystems. *Geografiska Annaler*, 70B, 69–73.

Olsson, E.G.A. (1991) Agro-ecosystems from Neolithic time to the present. In: Berglund, B.E. (ed.), *The cultural landscape during 6000 years. Ecological Bulletins*, 41, pp. 293–314.

Olsson, E.G.A. (2004) Forests and grasslands as cradles for agriculture. In: Squires, V.R. (ed.), *The role of food, agriculture, forestry and fisheries in human nutrition*. Encyclopedia of Life Support Systems (EOLSS), UNESCO, EOLSS Publishers. Available from: www.eolss.net.

Olsson, E.G.A. (2005) The use and management of Norwegian mountains reflected in biodiversity values and cultural heritage – what are the options for future food production? In: Thompson, D.B.A., Price, M.F., & Galbraith, C.A. (eds.), *Mountains of Northern Europe: Conservation, management, people and nature*. Edinburgh, Scotland: TSO, pp. 151–162.

Olsson, E.G.A., Austrheim, G., & Grenne, S.N. (2000) Landscape change patterns in mountains, land use and environmental diversity, Mid-Norway, 1960–1993. *Landscape Ecology*, 15, 155–170.

Olsson, E.G.A., Kerselaers, E., Søderkvist Kristensen, L., Primdahl, J., Rogge, E., & Wästfelt, A. (2016) Peri-urban food production and its relation to urban resilience. *Sustainability*, 8, 1340–1361.

Opitz, I., Berges, R., Piorr, A., & Krikser, T. (2015) Contributing to food security in urban areas: Differences between urban agriculture and peri-urban agriculture in the global North. *Agriculture and Human Values*, 33, 341–358.

Oteros-Rozas, E., Ontillera-Sánchez, R., Sanosa, P., Gómez-Baggethun, E., Reyes-García, V., & González, J.A. (2013) Traditional ecological knowledge among transhumant pastoralists in Mediterranean Spain. *Ecology and Society*, 18(3), 33.

Padoch, C., & Pinedo-Vasquez, M. (2010) Saving slash-and-burn to save biodiversity. *Biotropica*, 42(5), 550–552.

Palomo, I., Martin-Lopez, B., Lopez-Santiago, C., & Montes, C. (2011) Participatory scenario planning for protected areas management under the ecosystem services framework: The Doñana social-ecological system in Southwestern Spain. *Ecology and Society*, 16(1), 23.

Plieninger, T., & Bieling, C. (eds.) (2012) *Resilience and the cultural landscape: Understanding and managing change in human-shaped environments*. Cambridge: Cambridge University Press.

Rackham, O. (1980) *Ancient woodland: Its history, ecology and vegetation in England*. London: Edward Arnold.

Rosa García, R., Fraser, M.D., Celaya, R., et al. (2013) Grazing land management and biodiversity in the Atlantic European heathlands: A review. *Agroforestry Systems*, 87, 19–43.

Roué, M., & Molnar, Z. (eds.) (2016) *Indigenous and local knowledge of biodiversity and ecosystem services in Europe and Central Asia*. Knowledges of Nature 9. UNESCO.

Slotte, H. (2000) *Lövtäkt i Sverige och på Öland. Metoder och påverkan på landskapet*. Swedish Agricultural University. Agraria 236.

Small, F.A.A., & Raizada, M.N. (2017) Mitigating dry season food insecurity in the subtropics by prospecting drought-tolerant, nitrogen-fixing weeds. *Agriculture & Food Security*, 6, 23.

Tømmervik, H., Dunfjeld, S., & Olsson, E.G.A. (2010) Detection of ancient reindeer pens, cultural remains and anthropogenic influenced vegetation in Byrkije (Børgefjell) mountains, Fennoscandia. *Landscape and Urban Planning*, 98, 56–71.

Tress, B., Tress, G., & Fry, G. (2005) Researchers experiences, positive and negative, in integrative landscape projects. *Environmental Management*, 36(6), 792–807

Tress, B., Tress, G., & Fry, G. (2007) Analysis of the barriers to integration in landscape research projects. *Land Use Policy*, 24(2), 374–385.

Ullah, I.I.T., Kuijt, I., & Freeman, J. (2015) Toward a theory of punctuated subsistence change. *PNAS*, 112(31), 9579–9584.

UN SDGs (2015) United Nations Sustainable Development Goals, http://un.org/sustainabledevelopment/sustainable-development-goals/

Vera, F.W.M. (2000) *Grazing ecology and forest history*. Wallingford: CABI Publishing.

Webb, N. R. (2002) Atlantic heathlands. In: Perrow, M.R., & Davy, A.J. (eds.), *Handbook of ecological restoration*. Cambridge: Cambridge University Press, pp. 401–418.

Wehn, S., Olsson, E.G.A., & Hanssen, S.K. (2012) Forest line changes after 1960 in a Norwegian mountain region – implications for the future? *Norwegian Journal of Geography*, 66, 2–10.

Zeder, M.A. (2008) Domestication and early agriculture in the Mediterranean Basin: Origins, diffusion, and impact. *PNAS*, 105(33), 11597–11604.

3

Aboriginal culture and food-landscape relationships in Australia

Indigenous knowledge for *Country* and landscape

David S. Jones and Philip A. Clarke

Introduction

The ancestors of the contemporary Aboriginal people first arrived in Australia over 50,000 years ago (Clarke, 2003; Malaspinas et al., 2016), and during their long occupancy as semi-sedentary hunter-gatherers on this environmentally diverse continent they developed the means to survive several major shifts in both geology and climate. This landscape has successfully sustained many culturally and linguistically diverse Aboriginal communities, who possessed strict protocols for foraging and maintaining food harvesting limits (Clarke, 2007; Mulvaney and Kamminga, 1999; Tindale, 1981). The subsistence strategies employed were those that were capable of supporting Indigenous communities without the need for food importation (Pascoe, 2014; Pryor, 2014). In good years when there was a sufficient surplus, neighbouring groups were invited 'on *Country*' to feast and to hold joint ceremonies (Clarke, 2007; Jones, 1997; Wettenhall et al., 2010). While many foraging practices have been decimated since European colonisation of the Australian landscape (especially outside of northern and central Australia), contemporary Aboriginal people believe that the conservation of knowledge concerning their former food sources is important for maintaining their identity (Butler et al., 2012; BMECBMRG, 2010; MAC, 2017; Woodward et al., 2012; YRNTBC, 2011).

Today, the emergent wild bush food market being developed by commercial grower groups (with government support) has selected only a few of the many species that Aboriginal foragers once used (Altyerre-ipenhe et al., 2011; Cherikoff, 2000; Clarke, 2012), the core species of which are listed in Table 3.1. There is also interest in cookbooks in this realm (Bruneteau, 1996; Cherikoff, 1994; Fielke, 2017; Mayall, 2014; Olive, 2006; Outback Chef Store, 2017; Robins, 1997; Weatherhead, 2016).

Scholars have argued that the Aboriginal impact on the Australian environment over the tens of thousands of years of dwelling is such that there were few, if any, true wilderness areas when Europeans arrived (Flannery, 1994). There is an ongoing academic debate within Australia concerning the role of Aboriginal people in managing the Australian environment. Historian Gammage (2011) has controversially called Australia the 'biggest estate on earth', arguing that Aboriginal people culturally manipulated the landscape through regulated vegetation burning practices in order to create a sophisticated, successful, and sensitive

Table 3.1 Plant species being developed for the wild bush food market.

Species	*Origin*	*Parts Used*
Lemon myrtle (*Backhousia citriodora*)	subtropical rainforests of Queensland	leaf and oil
Anise myrtle (*Backhousia anisata*)	subtropical rainforests of northern New South Wales	leaf and oil
Elegant wattle seed (*Acacia victoriae*)	southern arid regions	seed
Bush tomato (*Solanum centrale*)	Central Australia	fruit
Davidson plum (*Davidsonia* species)	subtropical/tropical east coast of New South Wales and Queensland	fruit
Riberry (*Syzygium leuhmanii*)	subtropical rainforests of northern New South Wales and coastal Queensland	fruit
Kakadu plum (*Terminalia ferdinandiana*)	tropical northern Australia	fruit
Muntries (*Kunzea pomifera*)	southeast South Australia and western Victoria	fruit
Lemon aspen (*Acronychia acidula* and *A. oblongifolia*)	eastern ranges of Australia	leaf and oil
Desert limes (*Citrus glauca*)	arid inland regions of eastern Australia	fruit
Finger limes (*Citrus australasica*)	subtropical/tropical east coast of New South Wales and Queensland	fruit
Quandong (*Santalum acuminatum*)	southern arid regions	fruit
Mountain pepper (*Tasmannia lanceolata*)	wet forests of south-eastern Australia	leaf and berry

Source: Clarke, 2012; Ryder and Latham, 2005; Ryder et al., 2009

farming regime that was quickly erased with colonialism, questioning why we cannot learn from this regime and appropriate some of these lessons.

Archaeologists, such as Hiscock (2014), have been critical of Gammage's model of Aboriginal relationships with the environment, claiming that it is too static and ignores the evidence of constant change across space and time. As a part of his critique, Hiscock has argued that since the beginning of the Holocene, major economic reorganisations are known to have occurred in certain parts of Australia. For instance, over the previous 1000 to 3000 years, the intensive earth mound building activities in the western Victorian wetlands appears to represent both an expanded emphasis for the exploitation of resources such as short-finned eels (*Anguilla australis*), underground plant foods, marine plant foods, and shellfish, as well as an increased sedentism. In the same period elsewhere on the continent, Hiscock has concluded that Aboriginal foragers along the northern coastline forced greater mobility because their intensive exploitation of the mollusc beds was being challenged by mangroves colonising what had earlier been open beaches. For scholars, merit and argument lie in acceptance of either Gammage's description of Aboriginal people as 'farmers' who were locked into a particular human/environment relationship, or Hiscock's model of hunter-gatherers maintaining their flexibility in order to respond to environmental change, both seasonal in terms of annual cycles and long-term in the case of climate change.

This chapter surveys Australian Aboriginal culture and its unique food-security paradigm, explains hunting and gathering protocols, examines how and why several of the food-sharing events occurred and then gives examples of the key food types that were harvested. It offers an alternate insight as to how this generational Indigenous knowledge is being and could be further utilised today in Australian landscape architecture, planning, and landscape management practice.

International context of Indigenous traditional food resource access

Indigenous peoples' terrestrial and aquatic wild food fisheries are gaining increased recognition and acknowledgement globally arising from the drafting and signing of Indigenous-relevant international agreements, relating to the environment, fisheries and human rights, and evolving fisheries management approaches. For example, in terms of freshwater and saline fisheries, the Food and Agriculture Organisation's (FAO) *Code of Conduct for Responsible Fisheries* (CCRF) encourages parties:

> When deciding on the use, conservation and management of fisheries resources, due recognition should be given, as appropriate, in accordance with national laws and regulations, to the traditional practices, needs and interests of Indigenous people and local fishing communities which are highly dependent on fishery resources for their livelihood.
>
> *(FAO, 1995, 2015)*

To aid this policy, the FAO published an operational guide for the CCRF designed for Indigenous fishing communities and other people working with those communities to guide all stakeholders in interpreting the CCRF in the light of associated UNESCO human rights instruments on Indigenous peoples' rights (FAO, 2009).

A core UNESCO document is the 2007 *United Nations Declaration on the Rights of Indigenous Peoples* that articulates that Indigenous peoples have a right to use, develop, and control resources associated with their traditional territories; see in particular Articles 25–29 (UN General Assembly, 2007).

Another example is the 1992 *Convention on Biological Diversity* that recognises that a special relationship exists between Indigenous peoples and biodiversity and therefore encourages all signatory parties to conserve and respect components and values of that relationship pertaining to traditional knowledge and the customary use of biological resources (Secretariat of the Convention on Biological Diversity, 2005).

Increased recognition of the interests and rights of Indigenous peoples, land and fisheries management and deliberations have expanded from simple Western science-informed consideration of biological and technological information to include in such plans the needs and aspirations of the people who rely upon terrestrial and aquatic bush foods and/or fishing for their livelihoods and wellbeing (Rettig et al., 1989).

A key aspect of such a consideration is the provision of traditional terrestrial and aerial bush foods and saltwater and freshwater aquatic bush foods and fishing access for Indigenous peoples (Davis and Jentoft, 2001; Pinkerton, 1989). Understanding the characteristics of traditional terrestrial and aquatic bush foods and fisheries – in particular the species harvested – is fundamental to ensuring that policy and management initiatives are aligned to address desired customary and livelihood outcomes for Indigenous 'fishers' and other stakeholders (see for example Hamilton and Walter, 1999; Menzies and Butler, 2007).

As an example, through the lens of fisheries as a source of traditional foods and customary practices in Australia, Aboriginal and Torres Strait Islander freshwater and saline fisheries (Indigenous fisheries) are acknowledged as one of three distinct sectors, along with the commercial and recreational sectors of industry. Yet legal recognition of this fact in Australia commenced some 30 years ago, and is intertwined with the developing recognition of Indigenous fishing rights and provisions under the Australian *Native Title Act 1993* (Peterson and Rigsby, 2014; Smyth et al., 2010). In the absence of a general national or state-level legislative framework for establishing Indigenous fishing rights in Australia, unlike in countries (such as Canada, USA, and New Zealand) where hunting and fishing rights of Indigenous people are specified in regional treaties, legal and policy determinations about the rights of Indigenous Australians to access traditional hunting and fishing resources were and are often subject to legal arguments regarding common law and native title in the context of parochial, jurisdictional legislation (Australian Law Reform Commission, 1986; Sweeney, 1993). Aboriginal Corporations, together with Aboriginals and Torres Strait Islander peoples have successfully demonstrated the need to recognise Indigenous traditional saltwater and freshwater fishing rights, that these are directly integral to customary rights, as validated in landmark Mabo (1992) High Court of Australia decision led by *Mabo*, *Passi*, and *Rice* of the *Meriam* people (from Mer/Murray Island) in the Torres Strait, that recognising native title in Australia for the first time. While this has led to the granting of fishing rights to Indigenous traditional owners in some parts of the country (e.g., the 'Blue

Mud Bay case' (2008) in *Northern Territory of Australia v Arnhem Land Aboriginal Land Trust* [2008] HCA 29 (30 July 2008) (Cwth)), the status and nature of Indigenous fishing rights in many parts of Australia remains uncertain and contested (Cane, 1998; Jackson et al., 2005; *Mabo and Others v Queensland (No. 2)* [1992] HCA 23; (1992) 175 CLR 1; *Northern Territory of Australia v Arnhem Land Aboriginal Land Trust* [2008] HCA 29 (30 July 2008)).

In terms of freshwater and saline fisheries the assumption is that this realm pertains to fish, eel (Hall et al., 1990; McNiven and Bell, 2010), and shellfish resources. There is an increasing awareness that seaweed also offers bush food and medicinal properties (Clarke, 2008a, 2008b; Donovan and Van Oploo, 2015; Lee, 2010; Sanderson and Di Benedetto, 1998; Thurstan et al., 2018; Werner, 1992).

Another example is the historical use and harvesting of Bogong Moths (*Agrotis infusa*) (Flood, 1996, 1980).

At the state level, the Australian leader in advancing Indigenous rights of access and policy guidance has been New South Wales (NSW). In 2002, the NSW State government launched the first Indigenous Fisheries Strategy (IFS) in Australia, to encourage the "practical recognition of the traditional and cultural fishing heritage of Aboriginal people and communities, including their access to and use of the fisheries resource" (NSW Department Primary of Industries, 2002: 2). The National Indigenous Fishing Technical Working Group in 2004 proposed the following set of principles on Indigenous fisheries for application across all jurisdictions in Australia to guide the development of Indigenous fishing strategies:

- Indigenous people were the first custodians of Australia's marine and freshwater environments and that Australia's fisheries and aquatic environment management strategies should respect and accommodate this.
- Customary fishing is to be defined and incorporated by Governments into fisheries management regimes, so as to afford it protection.
- Customary fishing is fishing in accordance with relevant Indigenous laws and customs for the purpose of satisfying personal, domestic or non-commercial communal needs. Specific frameworks for customary fishing may vary throughout Australia by reference, for example, to marine zones, fish species, Indigenous community locations and traditions, or their access to land and water.
- Recognition of customary fishing will translate, wherever possible, into a share in the overall allocation of sustainably managed fisheries.
- In the allocation of marine and freshwater resources, the customary sector should be recognised as a sector in its own right, alongside recreational and commercial sectors, ideally within the context of future integrated fisheries management strategies.

The NSW government amended the *Fisheries Management Act 1994* (FMA) in 2009 to formally acknowledge Aboriginal cultural fishing by inserting a new objective in the Act that states:

- To recognise the spiritual, social and customary significance to Aboriginal persons of fisheries resources and to protect, and promote the continuation of, Aboriginal cultural fishing.

Other changes to the FMA included provisions to vary species bag limits in order to meet Aboriginal cultural needs (Smyth et al., 2010).

Aboriginal Australia

In the mythology of Aboriginal Australia, it was tradition that spirit ancestors in the Creation made all the plants, animals, and people, and then gave the country meaning by shaping it and establishing the law

(Clarke, 2003). Most archaeologists today believe that the ancestors of modern Aboriginal people first arrived in Australia between 45,000 and 60,000 years ago during a period of low sea level (Mulvaney and Kamminga, 1999; Hiscock, 2007). There are difficulties when attempting to establish the population size of Aboriginal Australia when Europeans first arrived, but it is estimated that it may have been in the order of about 500,000 in the late 1700s (Taylor, 2005).

Across the continent in excess of 250 distinct languages (and up to 700 dialects) were spoken, and due to the impact of European colonisation most of these are now endangered, threatened, or extinct (McConvell and Thieberger, 2001). An interactive map of Indigenous Aboriginal language, tribal or nation groups is accessible at www.aiatsis.gov.au/explore/articles/aiatsis-map-indigenous-australia, which assists in communicating language and group habitation patterns across the Australian continent.

While the Australian landscape was seen through the eyes of the early European explorers as 'wilderness', it was historically humanised by its Indigenous inhabitants (Clarke, 2003). The Euro-centric definition of 'wilderness' parallels the colonising British assertion, that was subsequently enshrined in Australian colonial and post-Commonwealth legislation and legal judgements up until the *Mabo and Others v Queensland (No. 2)* (1989) judgement by the Australian High Court, that the Australian continent was *terra nullius;* a 'land of no-one'. Both 'wilderness' and *terra nullius* when applied to Australia are now redundant and inaccurate across all disciplines and jurisdictions. The High Court concluded that the doctrine of *terra nullius*, which imported all laws of England to a new land, did not apply in circumstances where there were already inhabitants present – even if those inhabitants had been regarded at the time as 'uncivilised'. Thus, English law was applicable where the land was barren and uninhabited, and could not be applied where an existing people were settled. The Court's conclusion was that existing customary laws, that were present at the time of United Kingdom settlement, and thus that any Indigenous land rights which had not been extinguished by subsequent Crown grants continued to exist in Australia's lands and waters.

Additionally, the Australian landscape, at the time of European settlement, was the oldest continuing *cultural landscape*. Under UNESCO's definition of 'cultural landscapes', a large portion of which still demonstrates the "combined works of nature and humankind, they express a long and intimate relationship between peoples and their natural environment", native foods are and often an integral, still respected, and harvested as part of this continuity of occupancy, warranting World Heritage registration (UNESCO, 2017a). Such evidence is demonstrated in the registrations of Uluṟu-Kata Tjuṯa National Park (UNESCO, 2017b) and Kakadu (UNESCO, 2017c) regions in Northern Territory, and what may be forthcoming for the Budj Bim region in Victoria (Laurie, 2017; Wright, 2017).

For Aboriginal Australians, their land was layered with cultural and social meanings developed through their long occupation of their respective *Country's* (lands and waters and skies and its air). They are united by their consciousness that saw (and continues to see and perceive today) their *Country* as being full of the places and objects that their spiritual ancestors had created (Clarke, 2003; Rose, 1996). The physical manifestation of that *Country* is the Australian landscape; night sky, day sky, land surface, below land, its waters, and the air within. Thus, *Country* is/was four dimensional, and it was not limited to the terrestrial landmass or 'landscape' of the Australian continent to which the Western definition of 'land' implies, nor to the immediate area physically inhabited by Aboriginals groups at the time of European colonisation. Thus, to an Aboriginal, everything in their *Country* is alive and everything is embodied in relationships, whereby the past, present, and future are one, and where both spiritual and physical worlds of *Country* interact (Clarke, 2014b, 2015a; Rose, 1996). The *Dreaming* is an ongoing celebration and reverence for past events: the creation of the land, the creation of law, and the creation of people. Stories have been given to and vested in Aboriginal peoples from the *Dreaming*, everything comes into being through the expression or journey through story, and the *Dreaming* is the ancestors. All things exist eternally in the *Dreaming*; the *Dreaming* is alive. The individual is born to *Country*, not just in *Country* but from *Country*, and his or her identity is inextricably and eternally linked to the *Dreaming* (Clarke, 2003; Rose, 1996).

Rose (1996: 7) suggests that:

> In Aboriginal English, the word 'Country' is both a common noun and a proper noun. People talk about Country in the same way that they would talk about a person: they speak to Country, sing to Country, visit Country, worry about Country, grieve for Country and long for Country. People say that Country knows, hears, smells, takes notice, takes care, and feels sorry or happy. Country is a living entity with a yesterday, a today and tomorrow, with consciousness, action, and a will toward life. Because of this richness of meaning, Country is home and peace: nourishment for body, mind and spirit; and heart's ease.

Seasonal cycles were integral to this understanding of place, and as hunter-gatherers, Aboriginals knew their *Country*'s seasonal cycles and responded to them (Clarke, 2009; Heyes, 2010).

Much has changed for Australia's Indigenous peoples since British colonisation began in 1788, resulting in an invasion of their respective *Countries* militarily, economically, environmentally, by disease, and through wildlife and human assimilation strategies and actions. Temperate and coastal regions have received the heaviest impacts and disturbances from colonisation. Aboriginal population levels have now recovered from the sharp decline during the colonial period, but there are increasingly questions about percentage of 'Aboriginality' given inter-breeding and inter-marriage that implicated all populations on the continent.

In 2011, the estimated resident Aboriginal and Torres Strait Islander population in Australia was 669,900 people, representing 3% of the total Australian population. Between 2001 and 2011 the Aboriginal and Torres Strait Islander population increased by 2.3% (from 534,700 people) per year on average, compared with 1.5% for the total Australian population. The population of Aboriginal and Torres Strait Islander Australians is projected to increase to between 907,800 and 945,600 people in 2026, at an average growth rate of between 2.0% and 2.3% per year. (ABS, 2014). Increasingly, Indigenous Australians are living in urban and nearby rural areas (Biddle and Yap, 2010). Today, over 33% of Australia's land surface is owned by Aboriginal people with up to 75% having declared Indigenous interests through legal processes, such as land rights and native title (National Native Title Tribunal, 2016). Much of Aboriginal-controlled land is in areas marginal to European-style agriculture, which favours temperate regions.

Foraging practices

In Aboriginal Australia there were two main categories of food recognised by foragers: animal and vegetable (Clarke, 2003; Pascoe, 2014; Pryor, 2014). The job of collecting food for the small foraging band was divided by gender (also see Chapter 5). In general, women focussed on gathering plant foods, shellfish, seaweed, ground insects, and small burrowing animals, while the men fished and hunted more mobile animals, such as wallabies (*Macropus* sp., *Petrogale* sp., *Lagostrophus* sp., *Lagorchestes sp.*, *Dorcopsis* sp., *Onychogalea* sp., *Wallabia* sp., and *Thylogale* sp.), kangaroos (*Macropus* sp.), and emus (*Dromaius novaehollandiae*). Protein meat sources included insects, crustaceans, shellfish, fish, eels, reptiles, birds, and mammals. Aboriginal people utilised a wide variety of terrestrial and aquatic plant parts: fruits, seeds, nuts, sporocarps (seed-like growths on ferns), foliage, stems, galls, gums, leaf exudates, roots, rhizomes, tubers, and fungi (Gott and Conran, 1991; Zola and Gott, 1992). The foraging for both animal and plant foods was highly seasonal. In some areas the proportion of animal food in the overall diet varied through the year from being a major source to a minor one. When available, animal food was generally highly desirable, over more reliable terrestrial and aquatic plant food sources, which were either less palatable or more time consuming to prepare (Clarke, 2008a, 2015b, 2017; Ens et al., 2014a, 2014b, 2017; Gott, 1983, 2008; Gott and Conran, 1991; Jones, 1993; Latz, 1995, 1999; Rose, 1996, 2005; Schnierer and Egan, 2016; Tindale, 1977, 1981; Zola and Gott, 1992).

The diversity of terrestrial and aquatic wild food available to Aboriginal foragers across Australia was enormous when compared to the restricted range of horticultural foods brought out by the first European colonists (Clarke, 2008a). For instance, ethnobotanical research in the temperate region of the South East of South Australia has found historical evidence that local Aboriginal people utilised at least 32 plant species as food, with the analysis of studies from similar nearby areas suggesting a further 38 species were also used (Clarke, 2015b). The true figure for the number of plant foods would have been higher still, with many other plant species used medicinally and for making artefacts. In parts of Australia that have a higher diversity of plant species, such as the Tiwi Islands in the tropics of the Northern Territory, the total number of available plant foods is even greater (Puruntatameri et al., 2001).

The seasonal abundance of certain food sources allowed for a surplus to be produced for trade and provided an opportunity for visiting kin from neighbouring areas to share in a feast. Due to their economic and political importance, Aboriginal people actively managed and protected these resources. For instance, individual Golden Wattle (*Acacia pycnantha*) trees that exuded large quantities of edible gum during the warmer seasons were regarded as the property of particular Aboriginal clans across western Victoria among people such as the *Gunditjmara* (Dawson, 1881; Dearnaley, 2014). In the Coorong area of the Lower Murray in South Australia, fruit gathered from Muntries (*Kunzea pomifera*) by *Ngarrindjeri* peoples was made into flour-based cakes for eating later during the southern winter and for trading in exchange for weapons and toolmaking stone (Tindale, 1981). In south eastern Queensland and adjacent north eastern New South Wales, the seasonal availability of the nut from the Bunya Bunya Pine (*Araucaria bidwillii*) brought in Aboriginal groups from a wide area to hold ceremonies (Maiden, 1889; Petrie and Campbell Petrie, 1932; Smyth, 1878; Swan, 2017). Plant foods had a social importance that went beyond subsistence.

Aboriginal hunters and gatherers needed to know where in the landscape they needed to be in order to make most of the food when it came into season. Aboriginal people possessed calendars that were specific to their *Country*, ranging from four to seven seasons in some temperate and desert areas, to as many as nine in parts of the Australian tropics (Altyerre-ipenhe et al., 2011; Clarke, 2009; FAT and WMAC, 2004; Heyes, 1999; Jones, 2002; Jones et al., 1998; PV and GMTOAC, 2015; Woodward et al., 2012). This variability in seasonal calendars was determined by a number of factors, which included such things as wind direction, star movements, and the flowering of certain terrestrial or aquatic plants. For example, at Marion Bay on Yorke Peninsula in South Australia, the prolific flowering of the Black Teatree (*Melaleuca lanceolata*) was a sign for the *Narangga* people that the Yelloweye Mullet fish (*Aldrichetta forsteri*) were soon to come to the coast in large numbers (Davies, 1952, cited in Clarke, 2007). Here, initiation ceremonies were held at this time to take advantage of a seasonally abundant food source, which was required to support a large number of visitors.

Some of the food gathered was eaten while on the move, while the bulk was taken back to the main camp at the end of the day for cooking and eating by the rest of the band. Older people controlled the food distribution, which was determined by age, gender, and kinship obligations. For instance, on the Eyre Peninsula of South Australia, a colonist observed that game was divided up by the band so that men ate the adult male animals, the women ate adult female animals, and children ate small animals (Wilhelmi, 1861). In this way, foraging success was shared within the group. Apart from periods of hardship, such as brought on by prolonged drought and disease epidemics, Aboriginal people generally ate well due to the diversity of seasonally available foods and maintained a rich diet of plant and animal foods free of artificial sugars, alcohols, preservatives that are common in today's diet and also prevalent in Western diets at the time of Australian colonisation (Dawson, 1881; Gott, 1983, 2008).

Managing resources

Australian foragers were able to minimise their impact upon the landscape by dispersing themselves thinly and by maintaining a nomadic lifestyle, whereby they constantly moved according to season (Clarke, 2003; Dawson, 1881; Gammage, 2011; Jones, 1969). Aboriginal people actively manipulated the environment to extract long-term benefits for food production. For example, in south eastern Queensland, foragers collected edible marine mollusc borers from the waterlogged timber, and these were then thrown back into the sea for harvesting again next season (McCarthy, 1957). Aboriginal people also routinely replanted undersized tubers and other plant fragments to encourage the food growth (Pascoe, 2014). For instance, a European explorer on the central coast of Western Australia recorded in the late nineteenth century that Aboriginal people gathering the Warran (*Dioscorea hastifolia*) roots would "invariably re-insert the head of the yams so as to be sure of a future crop" (Gregory, 1887: 131).

As an example, Colonial Protector of Aborigines in Victoria, James Dawson (1881) wrote that:

> Of roots and vegetables they have plenty. The muurang (Murnong) [*Microseris lanceolata*] which somewhat resembles a small parsnip, with a flower like a buttercup, grows chiefly on the open [Western District] plains. It is much esteemed on account of its sweetness, and is dug up by the women with the muurang pole. The roots are washed and pit into a rush basket made on purposes and placed in the oven in the evening to be ready for the next morning's breakfast. When several families live near each other and cooks their roots together, sometimes the baskets form a pile three feet high.
>
> *(Dawson, 1881 in Zola and Gott, 1992: 6–7)*

On the western plains near Melbourne in Victoria there was once an extensive system of mounds that *Wadawurrung* and *Wurundjeri* peoples had created by their 'accidental gardening' when digging up Yam Daisy (or Murnong) (*Microseris lanceolata*) tubers (Batey, 1909–10, cited in Frankel, 1982; Gott, 1983). The gatherers would have realised the effect they were having on the numbers of edible roots in the ground. The act of digging would have assisted in the dispersal and replanting of undersized tubers, which could then be collected in the same season the following year. It also helped prepare the soil for plant growth. Aboriginal people were aware of the impact of their foraging upon the local environment, with their experiences and understandings being the basis of their traditional ecological knowledge (TEK) that was handed down through the generations (also see Chapter 2).

Early explorers or 'Aboriginal Protectors' Mitchell (1839), Stapylton (1971), Robinson (Clark, 1990), and Dawson (1881) wrote extensively of their period observations of these practices being applied by Aboriginals across the Western District landscape that Clark (1988, 1990), Clark et al. (1987), Jones (1993) and Kiddle (1962) have anthropologically or ethno-ecologically synthesised and narrated, and Rolls (1969) applied in the Pilliga of NSW. In contemporary writings, Gammage (2011), Gott (1983, 2008), Gott and Conran (1991), Zola and Gott (1992), Clarke (2003, 2007, 2008a, 2008b), Bindon (1996), Bonney (2007, 2010), Isaacs (1987, 1996), Latz (1995, 1999), Pascoe (2014), Werner (1992), and Kenneally et al. (1996) have all recorded the fabric, medicinal, edible, and material uses that Australian terrestrial and aquatic plant species were historically put to by Australian Aboriginals, drawing upon both contemporary and period sources.

This is the realm of ethnobotany that involves the applied science of the role and potential of plants through the lens of Indigenous communities. While plants were integral to feeding communities, their harvesting was dependent upon the fire management of the Australian landscape. While plants are often first thought of as a food source, for Aboriginal communities plants also offered signals as to changes in the season, medicinal aids and cures, materials for implements and artefacts, materials for clothing and basketry, and – depending on their properties and attributes – could be traded between communities and shifted around Australia. Given the longevity and formality of some agricultural practices – such as sowing/harvesting and baking grain 15,000 years before the Egyptians – there is particular significance in the history and practices of Aboriginal activities (Pascoe, 2014; Pryor, 2014).

Apart from the gathering of plants, deep ecological understandings were also fundamental for hunting and fishing. Knowledge of the food consumed by game species made their capture easier. For instance, at Borroloola in the southwest Gulf of Carpentaria region, *Karrwa* hunters constructed bough shelters in large Emu Apple-trees (*Owenia acidula*) to enable them to spear from above any unsuspecting Emus (*Dromaius novaehollandiae*) that came to eat the fallen fruit (Bandaluka et al., 2003). At Daly River, south of Darwin, Aboriginal fishers stood under Cluster Fig trees (*Ficus racemosa*) growing over water, in order to spear fish attracted by the fallen fruit (Marrfurra, 1995).

Across Australia, Aboriginal people constructed large fish-traps, or aquaculture systems, using wood and stone, and built weirs from the earth to help them catch fish. For instance, at *Tae Rak* (Lake Condah) in *Gunditjmara Country* in western Victoria, there were large and extensive trench constructions that were used to draw the Short-finned Eels (*Anguilla australis*) out from the swamps (Flood, 1983; EHA, 2011; Hall et al., 1990; McNiven and Bell, 2010; Wettenhall et al., 2010). By altering the water flow, foragers controlled fish movement and made it easier to catch them in large wicker baskets. Fish-traps were typically placed strategically in mangrove inlets or the mouths of tidal creeks to catch fish returning to the sea with the outgoing tide. Fish poisons based on plant materials were also widely used in lagoons and other relatively small bodies of water (Hamlyn-Harris and Smith, 1916; Webb, 1959).

In terms of freshwater and saline aquatic food resources, traditional fishing practices still occur in a parts of New South Wales for many Aboriginal communities, particularly the *Wiradjuri* peoples (Schnierer and Egan, 2016). Such practices continue both customary and livelihood practices – and involve fish – and thus impact upon NSW fishery policies and governance regimes. The species harvested include more than 150 species of fish and invertebrates, over 90% of which are also harvested by commercial and recreational fishers (Schnierer and Egan, 2016).

In the subtropics of northern Australia, Aboriginal people had labour-intensive practices for the preparation of cycad (*Cycas* sp.) nuts, which involved grinding them into a pulp that was then thoroughly washed and leached before it was cooked into damper (Clarke, 2008a; Levitt, 1981). While having the status of a staple food, it is highly toxic without laborious preparation, although it was considered good to eat as damper when travelling and was often stored in this form for use during ceremonial events when large groups converged at one camp. In Central Australia, strips of kangaroo flesh were dried on hot stones for putting aside (Tonkinson, 1978). The skins of several species of *Solanum* sp., which Europeans call 'desert raisins' or 'bush tomatoes', were dried and skewered on sticks for later use, at which time they were soaked in water or ground into a paste. Across the arid zone there was a widespread practice of storing excess grass seed and Nardoo (*Marsilea hirsuta*) fern sporocarps (Clarke, 2003; Tindale, 1977), and very careful management of potable water resources (Bayly, 1999; Magarey, 1895). Extra food, when properly prepared and contained, not only prolonged the period when particular food types would be used, but also provided a surplus for trade. Aboriginal people used their ceremonies to help maintain the abundance of economically and culturally important plants and animals (Howitt, 1904; Strehlow, 1947).

Historical records from the early colonial period indicate that Aboriginal people often burnt the vegetation in order to drive animals for hunting (Clarke, 2003; Gammage, 2011; Gott, 2005; Hallam, 1975; Jones, 1969). Fire was also used to open up the landscape by removing the understorey, which allowed easier travelling and promoted the future growth of grass for game species, such as kangaroos and wallabies. Aboriginal people also preferred not to walk through dense prickly thickets if they could avoid it. Many of the plant foods eaten by Aboriginal people benefited from the opening of understorey and the build-up of ash. Although some Aboriginal burnings were random and directed at flushing out game, others were more structured. In the southern Gulf of Carpentaria region, *Yanyuwa* people actively managed their hunting territories with fires, the first starting in the early dry-season (Baker, 1999). Before starting a fire the permission of the traditional owner was required. Here, a ban was imposed on hunting in the *Country* of a person recently dead, which would have enabled game to increase in numbers (Baker, 1999). It has also been suggested that hunting prohibitions in *Country* around sacred sites also serves as a refuge for game species (Newsome, 1980).

An archaeologist has described the Aboriginal manipulation of the environment through the use of fire as 'fire-stick farming' (Jones, 1969). A form of 'proto-agriculture' was attributed to Aboriginal people who used their ecological knowledge in this manner to deliberately bring about changes favourable to their hunting and gathering practices. The shell mounds of northern Queensland, created by the build-up of cooking fire remains from Aboriginal camps, are places that favour the establishment of vine forests (Cribb et al., 1988; Mulvaney and Kamminga, 1999). To local Aboriginal people, the plant resources associated with the shell mounds include foods, medicines, and raw materials for artefact making and the construction of shelters.

Commercialising native bush food

Today, the Australian native bush food industry occupies a niche market that is expanding both nationally and globally (Clarke, 2012; Graham and Hart, 1997; Phelps, 1997), and is associated with terrestrial plant foods rather than aquatic plant foods (Lee, 2010; Sanderson and Di Benedetto, 1988; Werner, 1992). It is based upon foods from a small range of Australian indigenous plant species, most of which formerly sustained Aboriginal hunter-gatherers and were used as emergency foods by European explorers and settlers (Clarke, 2008a). There are 13 plant species which are the mainstay of the contemporary Australian native bush food industry, and these have been targeted for further research and development (Clarke, 2012; Graham and Hart, 1997; Ryder and Latham, 2005; Ryder et al., 2009). These are all species new to horticulture, and out of a vast number of species that were previously utilised by Aboriginal people only these are being considered for development because they already possess definable farm-gate or wild-harvested values (see Table 3.1). They are predominantly fruits and condiments, and their elevation and acceptance has largely been hindered by Western predilection and cultural entrenchment of conventional Northern hemisphere foodstuffs.

The native bush food network, as an industry, involves wild harvesting, growing, processing, value adding, catering, and spin-off guiding/tour businesses. Most of the raw material is processed in south-eastern Australia and then distributed through outlets in the capital cities. While other plant species have potential for future development, it considered that the market is not yet mature enough to support more variety. Market success with the 13 main native bush foods may well encourage a second wave of research and development involving other species. The preceding list excludes the macadamia nut (*Macadamia integrifolia*), which due to its early horticultural development is not considered part of the contemporary Australian native bush food industry. Perhaps the only commercially successful terrestrial species mass cultivated to date in Australia, macadamia nuts have been grown in plantations since the late nineteenth century, including extensively overseas, and in Australia alone is the basis of an industry estimated to be worth AU$14 million annually (Robins, 2007).

The full economic impact of the Australian terrestrial and aquatic native bush food industry upon Indigenous communities is difficult to quantify, as it is still a developing industry. While the wild harvest provides an opportunity for Aboriginal people with access to *Country* to benefit economically, there are major challenges for the development of economically and environmentally robust Indigenous businesses (Cunningham et al., 2009a, 2009b). These include the need for joint ventures between Aboriginal communities and commercial operators at various points in the food chain. Also important is the branding and certification of the foods, in order to protect intellectual property rights of TEK. Such highlights the ongoing challenge of Western economic models and their unsuitability to Indigenous landscape practices and traditions, that include questions like 'Is it more important to get people on country than an economic return?'

While the shift from the wild-harvest sourcing of terrestrial and aquatic native bush foods to cultivated supplies decreases the risk of variable supply for the native bush food chain, it is claimed that it has preferentially benefited non-Aboriginal producers. There is a risk of Aboriginal involvement being displaced as species move from wild harvest to cultivation, which is of concern to the whole industry because Aboriginal participation brings authenticity and integrity to its product (Clarke, 2012). Native animals, such as crocodiles (*Crocodylus* sp.), short-finned eels (*Anguilla australis*), kangaroos, emus, and muttonbirds (*Ardenna* sp.), have also been commercially hunted for food, but as with Australian species of fish and shellfish, they are part of a market that is already well established, and with some Aboriginal involvement (Wilson et al., 1992). Among the sources of wild animal foods are exotic species, such as rabbits (*Oryctolagus cuniculus*), camel (*Camelus dromedaries*), goat (*Capra aegagrus*), horse (*Equus* sp.), water buffalo (*Bubalus sp.*) and pigs (*Sus* sp.) that are primarily hunted in tropical northern Australia.

Present-day climate change, which is much faster than ever before, presents a range of problems for Aboriginal communities who are trying to become more involved in the emerging native bush food industry (Low Choy et al., 2013). A major impediment is the lack of financial support and access to professional advice to develop businesses, while also adapting to climate change. A solid business plan is essential when acquiring and maintaining legal control of unique varieties of plant foods growing on Aboriginal managed land. In many cases, Aboriginal communities are having their land assets deteriorate environmentally due to the effects of climate change (especially land erosion, invasive plant, and animal species). Interfering with their plans are the restrictive controls by government land management authorities, such as being denied the right to regularly burn *Country* to mitigate the threat of wild fires and encourage the growth of plant foods. Changes in timing of cycles for plant flowering and animal breeding, coupled with rising temperatures and decreasing rainfall, plus the more favourable conditions for the spread of plant diseases, weeds, and pests would all impact negatively on the indigenous food network (see Indigenous Food Systems Network, 2017).

Aboriginal people operating at the supply end of the native bush food network are potentially the most exposed to the impact of climate change, through requiring much more economic support to adapt. Aboriginal-run businesses at the restaurant and catering end in cities are likely to fare better, as they are less exposed to environmental conditions and already source many of their food ingredients from non-Indigenous businesses within the industry.

Applying traditional ecological knowledge (TEK)

Traditional ecological knowledge (TEK), or 'Indigenous biocultural knowledge' as used by several by natural resource management researchers (Berkes et al., 2000; Ens et al., 2014a, 2014b, 2015, 2017; Woodward et al., 2012), or 'phenology' by environmental scientists (Keatley et al., 2013; Noormets, 2009; Schwartz, 2013), has an earlier platform in landscape architecture research (Jones, 2002; McHarg, 1966, 1969; Seddon, 1979). While phenology involves the study of periodic plant and animal life cycle events and how these are influenced by seasonal and inter-annual variations in climate, as well as habitat factors (such as elevation), TEK research engages Aboriginal knowledge to service natural resource management objectives. The former lacks Indigenous cultural perspectives and acceptance of their knowledge whereas Indigenous biocultural knowledge research has incorporated this cultural perspective including recognising movements, land management, and harvesting activities, and an explanation of being. TEK research offers benefits to the disciplines of architecture, planning, and landscape management/architecture in Australia in framing a design and or planning strategy that is informed by the attributes of that landscape, that *Country*. This framing is underpinned by a vision through a lens of often tens of thousands of years of occupancy, use, management and respect, rather than the Western constraints of a Gregorian calendar and quadripartite seasonal structure and conventional 5–20 year length management planning and design aims, objectives, and policy making protocols.

While TEK research is increasingly commonplace in land use planning, land management planning, landscape architecture, natural resource management, and interpretative research and practice applications across northern Australia, because of the ease of accessing Aboriginal knowledge and engagement, Aboriginal seasons and TEK have been little researched in Australia's south-eastern landscapes. Yet, this information offers much potential to enrich the disciplines of architecture, planning, and landscape management/architecture in south-eastern Australia for both conventional and Indigenous consultancy engagements and projects.

Aboriginal seasonal calendars represent significant Indigenous biocultural knowledge translations. Aboriginal seasonal calendars provide mental structuring devices, fluid and oral in their form, that contain multiple announcements, signals, and cultural obligations as to land management, stewardship, and occupancy traditions and relationships. They are unwritten management plans in their own right that can have direct applicability into Western land management regimes. In Kakadu and Uluru Kata-Tjuta National Parks, the respective *Yolngu* (or Yolŋu) and *Arrernte* seasonal calendars structure Commonwealth-approved national park management plans and each apportioned season determines what set of landscape management actions to prepare for and action, and it foreshadows regular post-action activities (Australia, 2010, 2016). The logic of this approach has also informed more recent uniquely structured and drafted management plans for the *Gunditjmara* of south-western Victoria (Parks Victoria et al., 2015), the *Bunya* people of the Bunya Bunya Mountains in south-eastern Queensland (BMECBMRG, 2010), the *Yawuru* peoples (YRNTBC, 2011) of the Broome region, and the *Murujuga* people (Murujuga Aboriginal Corporation, 2016) of the Pilbara in Western Australia.

To Aboriginal people, seasons may comprise five to seven temporal segments, each with a different time band, and fluid commencement dates whereby both time and date are linked to environmental variables that do not adhere to the Western-adopted mathematically precise Gregorian calendar (Summer, Winter, Spring, and Autumn). In addition, dates and/or times of commencement and cessation are determined by variable natural events or agencies, and have little to do with human acts; time is thus determined by the environment in contrast to being determined by fixed temporal measurable Western parameters. The Commonwealth Scientific and Industrial Research Organisation (CSIRO) has devised, in conjunction with respective communities, various seasonal calendars (e.g. *Gulmoerrgin* (*Larrakia*), *Walmajarri*, *Ngadju*, *Ngan'gi*, *Wagiman*, *Gooniyandi*, *MalakMalak*, *Matngala*) across the Arnhem Land region, in their northern Australian research that influence their conclusions and recommendations to the ecosystem sciences (including Indigenous engagement, fire and climate science, and fisheries and wetlands management) by this knowledge (Bauman, 2006; Ens et al., 2014a, 2014b, 2015, 2017; Rose et al., 2002). In some remote regions, attempts are being made to reinstate the human control of the fire regime, by utilising some Aboriginal burning techniques (Yibarbuk et al., 2001). Ecologists and government land managers are increasingly recognising the potential of TEK in the management of country (Ens et al., 2014a, 2014b, 2017). The future of the Australian environment will rely heavily upon its Indigenous environmental managers (Altman et al., 2007; Ens et al., 2014b).

Comprising 'Indigenous biocultural knowledge' in the ecosystem science and management discipline, these seasonal models incorporate plant, aquatic and terrestrial animals, fire, climate, drought, flood, and astronomical information (that could be termed a 'spectrum of knowledge') drawn from Aboriginal sources and corroborated by contemporary Western science (Rose, 2005). While scientific acceptance of this knowledge for northern Australia is now academically documented (Butler et al., 2012; Ens et al., 2014a, 2014b, 2015; Keatley et al., 2013; Prober et al., 2011; Rose, 2005; Woodward et al., 2012), little research on the topic has been entertained in south-eastern Australia.

In *Wurundjeri Country*, naturalist Reid (1995) suggested that six seasons are prevalent in the middle Yarra Valley of Melbourne, naturalist Jameson (1996) proposed an analogous six seasons hypothesis, and botanist Entwisle (2014) has proposed a five-season hypothesis for Melbourne, of which all three scenarios are based upon cyclical ecological patterns with little direct *Wurundjeri* engagement nor the full scope of the Indigenous biocultural knowledge spectrum used in the CSIRO investigations. The latter, however, was explored by Jones et al. (1997) for the upper Yarra Valley (Jones, 2002), informed by *Wurundjeri* elders, concluding that there were seven seasons and two additional variables, and this thesis underpins the cultural design of Museum Victoria's multi-peer award-winning Forest Gallery (Jones et al., 1998).

For western Victoria, the *Gunditjmara* have conceptualised their seasons (Wettenhall & The Gunditjmara People, 2010) which have been directly incorporated into the *Ngootyoong Gunditj Ngootyoong Mara South West Management Plan* (Parks Victoria et al., 2015) being the first use of 'Indigenous biocultural knowledge' for a Victorian state-approved land management plan. Western science and conventional land use planning regimes in Australia have been highly influenced by the theory and practice of northern hemisphere land management science, which take no account of the generational knowledge held by Australia's Aboriginal peoples. This knowledge is and can be well summarised in respective region-specific seasonal calendars. Gammage (2011) and Rose (2005) have argued for a translation of Aboriginal seasons as being essential for re-awakening of Australian land management. This overarching argument is evidenced in Attenbrow (2002, 2010) who has argued for strengthening *Gadigal* (Sydney) *Country* understanding, *D'harawal* elder Aunty Fran Bodkin in articulating the *D'harawal: seasons and climatic cycles* (2008) as a land management platform, and Clarke (2009) in calling for a greater appreciation of the *Ngarrindjeri Country* of The Coorong. This argument is also expressed by Walsh and Mitchell (2002) in 'Planning for *Country*'. The magic of this 'science' has been applied in the *Yawuru* cultural management plan for the Broome region (YRNTBC, 2011), in the *Ngootyoong Gunditj Ngootyoong Mara South West Management Plan* for the Gunditjmara (PV et al., 2015), in the *Murujuga Cultural Management Plan* (MAC/MLSU, 2017) in the Pilbara, as well as in the Bunya Bunya Mountains in the *Bonye Bu'ru Booburrgan Ngmmunge* (2010), that as Indigenous-devised and expressed integrated land/sea/water management plans have been accepted as valid Western-styled management plan tools by their respective Indigenous and state government entities. Research by Huntington-McBurney (2000) for *Narangga Country* of the Yorke Peninsula, Heyes (1999, 2010) and Jones (1999, 2002, 2005, 2010) for *Kaurna Country* of the Adelaide Plains (with direct support from *Kaurna* Elders and representatives) established an innovative landscape architecture-based research focus in Adelaide; this new knowledge was applied in the award-winning Australian Aboriginal Gallery of the South Australian Museum (Clarke, 2009, 2014a; Ens et al., 2015).

Food and the Australian Indigenous perspective

What is increasingly clear is that Australian Aboriginal knowledge of plant foods is intimately interwoven in their traditional ecological knowledge. Removing parts of this knowledge means that the larger picture and tapestry deteriorates. Re-invigorating, re-thinking, and refreshing that knowledge is an integral part of both healing past landscape despoliation actions as well as rekindling parts of knowledge integral to custodial responsibilities. Respecting, embracing, and harnessing this knowledge offers new insights as to how to alternatively 'harvest' the Australian landscape responsibly, but also to bring to the 'table' alternate food options, food management regimes, and a new way of appreciating and engaging with the Australian landscape and its psyche.

The salient themes are that respecting a *Country*-specific Aboriginal lens, through comprehending traditional ecological knowledge, *Country* nomenclature and place names, seasonal calendars, and a deeper mapping and research appraisal of the ethnobotany of a *Country* (or *Country's*) and its peoples, offers a remarkable botanical encyclopaedic resource that can better inform conventional Western practices.

Acknowledgement

We would like to acknowledge the Aboriginal and Torres Strait Islander peoples of Australia's lands and waters who are the Traditional Owners of these lands and *Country's*. We would also like to pay our respect to the Elders both past and present of these Nations and extend that respect to other Aboriginals and Torres Strait Islanders.

Further reading

Altman, J.C., Buchanan, G.J., & Larsen, L. (2007) *The environmental significance of the Indigenous estate: Natural resource management as economic development in remote Australia*. Canberra: Australian National University.

Merne Altyerre-ipenhe (Food from the Creation time) Reference Group, Douglas, J., & Walsh, F. (2011) *Aboriginal people, bush foods knowledge and products from central Australia: Ethical guidelines for commercial bush food research, industry and enterprises*, DKCRC Report 71. Alice Springs: Ninti One Limited.

Attenbrow, V.J. (2002) *Sydney's Aboriginal past: Investigating the archaeological and historical records*. Sydney: UNSW Press.

Attenbrow, V.J. (2010) *Sydney's Aboriginal past: Investigating the archaeological and historical records*, 2nd edition. Sydney: UNSW Press.

Australia (2010) *Uluru-Kata Tjuta National Park Management Plan 2010–2020: Tjukurpa Katuta Ngarantja*. Canberra: Director of National Parks.

Australia (2016) *Kadadu National Park management plan 2016–2026: A living cultural landscape*. Canberra: Director of National Parks.

Australian Bureau of Statistics. [ABS] (2014) *Estimates and projections, Aboriginal and Torres Strait Islander Australians*, www.abs.gov.au/ausstats/abs@.nsf/Products/3238.0~2001+to+2026~Chapter~Main+Features?OpenDocument (accessed 1 January 2017).

Australian Institute of Aboriginal and Torres Strait Islander Studies. (nd) *AIATSIS map of indigenous Australia*, http://aiatsis.gov.au/explore/articles/aiatsis-map-indigenous-australia (accessed 1 January 2017).

Australian Law Reform Commission. (1986) Hunting, fishing and gathering rights: Legislation or common law? Recognition of aboriginal Customary laws, *ALRC Report* 31.

Baker, R.M. (1999) *Land is life. From bush to town. The story of the Yanyuwa people*. Sydney: Allen and Unwin.

Bandaluka, K.B., Kanjiburanya, T.D., Dulbarrimarra, M.K., & Yarrburrngalina, P.M. (2003) *Karrwa bush foods and medicines. Kunjba wardan baki kunjba jingkal ngambalangi munji munji*. Batchelor, NT: Curriculum and Resource Unit, Batchelor Institute of Indigenous Tertiary Education.

Bauman, T. (2006) *Aboriginal Darwin: A guide to exploring important sites of the past and present*. Canberra: Aboriginal Studies Press.

Bayly, L.A.E. (1999) Review of how indigenous people managed for water in desert regions of Australia. *Journal of the Royal Society of Western Australia*, 82(part 1), 17–25.

Berkes, F., Colding, J., & Folke, C. (2000) Rediscovery of traditional ecological knowledge as adaptive management. *Ecological Applications*, 10, 1251–1262.

Biddle, N., & Yap, M. (2010) *Demographic and socioeconomic outcomes across the Indigenous Australian lifecourse: Evidence from the 2006 Census*. Canberra: ANU Press.

Bindon, P. (1996) *Useful bush plants*. Perth: Western Australian Museum.

Bodkin, F. (2008) *D'harawal: Seasons and climatic cycles*. Sydney: F Bodkin and L Robertson.

Bonney, N. (2007) *Adnyamathanha and beyond: Useful plants of an ancient land*. Unley, SA: Australian Plants Society (South Australian Region).

Bonney, N. (2010) *Knowing, growing, eating . . . edible wild native plants for southern Australia*. Tantanoola, SA: N. Bonney.

Bruneteau, J-P. (1996) *Tukka: Real Australian food*. Sydney: HarperCollinsPublishers.

Bunya Mountains Elders Council and Burnett Mary Regional Group (BMECBMRG) (2010) *Bonye Bu'ru Booburrgan Ngmmunge*. Bundaberg, Qld.: BMECandBMRG.

Butler, J.R.A., Tawake, A., Skewes, T., Tawake, L., & McGrath, V. (2012) Integrating traditional ecological knowledge and fisheries management in the Torres Strait, Australia. *Ecology and Society*, 17(4), 34–52.

Cane, S. (1998) Aboriginal fishing rights on the New South Wales South Coast: A court case. In: Petersen, N., & Rigsby, B. (eds.), *Customary marine tenure in Australia*. Sydney: Sydney University Press, pp. 111–146.

Cherikoff, V. (1994) *Uniquely Australian: A wild food cookbook: The beginnings of an Australian bushfood cuisine*. Boronia Park, NSW: Bush Tucker Supply Australia.

Cherikoff, V. (2000) *Marketing the Australian native food industry*. Canberra: Rural Industries Research and Development Corporation, RIRDC Publication No 00/61.

Clark, I.D. (ed.) (1988) *The Port Phillip journals of George Augustus Robinson: 8 March–7 April 1842 and 18 March–29 April 1843*. Clayton: Department of Geography, Monash University (Monash Publications in Geography No. 34).

Clark, I.D. (1990) *Aboriginal languages and clans: An historical atlas of Western and Central Victoria, 1800–1900*. Clayton: Department of Geography and Environmental Science, Monash University (Monash Publications in Geography No. 37).

Clark, M., Clark, I.D., Gott, B., Gunn, B., Critchett, J., & Poynter, J.R. (1987) *Australia felix: The Chap Wurrung and Major Mitchell*. Dunkeld: Dunkeld and District Historical Museum.

Clarke, M. (2012) *Australian native food industry stocktake*. Canberra: Rural Industries Research and Development Corporation.

Clarke, P.A. (2003) *Where the ancestors walked: Australia as an Aboriginal landscape*. Sydney: Allen and Unwin.

Clarke, P.A. (2007) *Aboriginal people and their plants*. Dural Delivery Centre: Rosenberg Publishing.

Clarke, P.A. (2008a) *Aboriginal plant collectors: Botanists and Australian Aboriginal people in the nineteenth century*. Dural Delivery Centre: Rosenberg Publishing.

Clarke, P.A. (2008b) Aboriginal healing practices and Australian bush medicine. *Journal of the Anthropological Society of South Australia*, 33, 3–38.

Clarke, P.A. (2009) Australian aboriginal ethnometeorology and seasonal calendars. *History and Anthropology*, 20(2), 79–106.

Clarke, P.A. (2014a) Australian aboriginal astronomy and cosmology. In: Ruggles. C.L.N. (ed.), *Handbook of archaeoastronomy and ethnoastronomy*, pp. 2223–2230.

Clarke, P.A. (2014b) The aboriginal Australian cosmic landscape. Part 1: The ethnobotany of the Skyworld. *Journal of Astronomical History and Heritage*, 17(3), 307–335.

Clarke, P.A. (2015a) The aboriginal Australian cosmic landscape. Part 2: Plant connections with the Skyworld. *Journal of Astronomical History and Heritage*, 18(1), 23–37.

Clarke, P.A. (2015b) The aboriginal ethnobotany of the South East of South Australia region. Part 2: Foods, medicines and narcotics. *Transactions of the Royal Society of South Australia*, 139(2), 247–272.

Clarke, P.A. (2017) Early indigenous practices of bord foraging in the Lower Murray, South Australia. *Transactions of the Royal Society of South Australia*, 141, 26–47.

Cribb, R., Walmbeng, R., Wolmby, R., & Taisman, C. (1988) Landscape as cultural artefact: Shell mounds and plants in Aurukun, Cape York Peninsula. *Australian Aboriginal Studies*, 2, 60–73.

Cunningham, A., Garnett, S., & Gorman, J. (2009a) Policy lessons from practice: Australian bush products for commercial markets. *Geo Journal*, 74(5), 429–440.

Cunningham, A., Garnett, S., Gorman, J., Courtenay, K., & Boehme, D. (2009b) Eco-enterprises and *Terminalia ferdinandiana*: 'best laid plans' and Australian policy lessons. *Economic Botany*, 63(1), 16–28.

Davis, A. and Jentoft, S. (2001) The challenge and the promise of Indigenous peoples' fishing rights – from dependency to agency. *Maritime Policy*, 25(3), 223–237.

Dawson, J. (1881) *Australian Aborigines: The languages and customs of several tribes of Aborigines in the Western District of Victoria, Australia*. Melbourne: G. Robertson; reissued by Cambridge University Press, 2009.

Dearnaley, J. (2014) *Wathaurong medicinal plant walk*, unpublished MLArch thesis, School of Architecture and Built Environment, Deakin University, Geelong, VIC.

Donovan, C., & Van Oploo, B. (2015) *The great Australian cookbook*. Richmond: PQ Blackwell with The Five Mile Press.

Engineering Heritage Australia (Newcastle). (2011) *Engineering works of the Gunditjmara at Lake Condah (Tae Rak) and Tyrendarra nomination under Heritage Recognition Program of Engineering heritage Australia*. Newcastle, NSW: Engineering Heritage Australia (Newcastle). www.engineersaustralia.org.au/portal/system/files/engineering-heritage-australia/nomination-title/Budj%20Bim%20Lake%20Condah%20Nomination.pdf (accessed 1 Januray 2017).

Ens, E.J., Burns, E., Russell-Smith, J., Sparrow, B., & Wardle, G.M. (2014a) The cultural imperative: Broadening the vision of long-term ecological monitoring to enhance environmental policy and management outcomes. In: Lindenmayer, D.B., Burns, E., Thurgate, N., & Lowe, A. (eds.), *Biodiversity and environmental change: Monitoring, challenges and direction*. Collingwood: CSIRO Publishing, pp. 85–110.

Ens, E.J., Pert, P., Clarke, P.A., Budden, M., Clubb, L., Doran, B., Douras, C., Gaikwad, J., Gott, B., Leonard, S., Locke, J., Packer, J., Turpin, G., & Wason, S. (2014b) Indigenous biocultural knowledge in ecosystem science and management: Review and insight from Australia. Biological Conservation, 181, 133–149.

Ens, E.J., Walsh, F., & Clarke, P. (2017) Aboriginal people and Australia's vegetation: Past and current interactions. In: Keith, D.A. (ed.), Australian Vegetation, 3rd ed. Cambridge: Cambridge University Press, pp. 89-112.

Entwisle, T. (2014) *Sprinter and sprummer*. Melbourne: CSIRO Publishing.

Fielke, A. (2017) *Creative native: Food service solution*. www.andrewfielke.com/ (accessed 1 February 2017).

Flannery, T.F. (1994) *The future eaters: An ecological history of the Australasian lands and people*. Sydney: Reed Books.

Flood, J.M. (1966) *Moth hunters of the Australian Capital Territory: Aboriginal traditional life in the Canberra region*. Canberra: J.M. Flood.

Flood, J.M. (1980) *The moth hunters: Aboriginal prehistory of the Australian Alps*. Canberra: Australian Institute of Aboriginal Studies.

Flood, J.M. (1983) *Archaeology of the dreamtime*. Sydney: Collins.

Food and Agriculture Organization (FAO) (1995) *Code of conduct for responsible fisheries*. Rome: Food and Agriculture Organization of the United Nations.

Food and Agriculture Organization (FAO) (2009) *Code of conduct for responsible fisheries: An operational guide*. Rome: Food and Agriculture Organization of the United Nations.

Food and Agriculture Organization (FAO) (2015) *Voluntary guidelines for securing sustainable small-scale fisheries in the context of food security and poverty eradication*. Rome: Food and Agriculture Organization of the United Nations.

Framlingham Aboriginal Trust (FAT) and Winda Mara Aboriginal Corporation (WMAC) (2004) *Kooyang sea country plan*. Purnim/Heywood: FAT and WMAC.

Frankel, D. (1982) An account of aboriginal use of the yam-daisy. *The Artefact*, 7, 43–45.

Gammage, B. (2011) *The biggest estate on Earth: How Aborigines made Australia*. Sydney: Allen and Unwin.

Gorman, J.T., Whitehead, P.J., Griffiths, A.D., & Petheram, L. (2008) Production from marginal lands: Indigenous commercial use of wild animals in northern Australia. *International Journal of Sustainable Development and World Ecology*, 15(3), 240–250.

Gott, B. (1983) Murnong – *Microseris scapigera*, *Australian Aboriginal Studies*, part 2, 2–18.

Gott, B. (2005) Aboriginal fire management in south-eastern Australia: Aims and frequency. *Journal of Biogeography*, 32(part 7), 1203–1208.

Gott, B. (2008) Root use by aborigines of Southern Australia. *Archaeology in Oceania*, 17(1), 59–67.

Gott, B., & Conran, J. (1991) *Victorian Koorie plants: Some plants used by Victorian Koories for food, fibre, medicines and implements*. Hamilton, VIC: Yangennanock Women's Group, Aboriginal Keeping Place.

Graham, C., & Hart, D. (1997) *Prospects for the Australian native bushfood industry*. Canberra: Rural Industries Research and Development Corporation.

Gregory, A.C. (1887) Memoranda on the aborigines of Australia. *Journal of the Anthropological Institute*, 16, 131–133.

Hall, D.N., Harrington, D.J., & Fairbrother, P.S. (1990) *The commercial eel fishery in Victoria*. Heidelberg, VIC: Dept. of Conservation and Environment, Arthur Rylah Institute for Environmental Research.

Hallam, S.J. (1975) *Fire and hearth: A study of Aboriginal usage and European usurpation in South-western Australia*. Canberra: Australian Institute of Aboriginal Studies.

Hamilton, R., & Walter, R. (1999) Indigenous ecological knowledge and its role in fisheries research design: A case study from Roviana Lagoon, Western Province, Solomon Islands. *SPC Traditional Marine Resource Management* and *Knowledge Information* Bulletin, 11, 13–25.

Hamlyn-Harris, R., & Smith, F. (1916) On fish poisoning and poisons employed among the Aborigines of Queensland. *Memoirs of the Queensland Museum*, 5, 1–9.

Heyes, S. (1999) *The Kaurna calendar: Seasons of the adelaide plains*. Unpublished MLArch honours paper, School of Architecture, Landscape Architecture, and Urban Design, University of Adelaide, Adelaide, South Australia.

Heyes, S. (2010) Water: The lifeblood of a landscape, the heart of the seasons. In: Daniels, C.B. (ed.), *Adelaide: Water of a city*. Adelaide: Wakefield Press, pp. 146–150.

Hiscock, P. (2007) *Archaeology of ancient Australia*. London: Routledge.

Hiscock, P. (2014) Creators or destroyers? The burning questions of human impact in ancient Aboriginal Australia. *Humanities Australia*, 5, 40–52.

Howitt, A.W. (1904) *Native tribes of south-east Australia*. London: Macmillan.

Huntington-McBurney, V. (2000). *The Narangga calendar*. Unpublished BLArch honours paper, School of Architecture, Landscape Architecture, and Urban Design, University of Adelaide, Adelaide, South Australia.

Indigenous Food Systems Network. (2017) *Indigenous Food Systems Network*. www.indigenousfoodsystems.org/ (accessed 1 February 2017).

Isaacs, J. (1987) *Bush food: Aboriginal food and herbal medicine*. McMahons Point, NSW: Weldons.

Isaacs, J. (1996) *A companion guide to bush food*. Sydney: Lansdowne.

Jackson, S., Storrs, M., & Morrison, J. (2005) Recognition of aboriginal rights, interests and values in river research and management: Perspectives from northern Australia. *Ecological Management and Restoration*, 6, 105–110.

Jameson, G. (1996–97). Middle Yarra timelines. *Victorian Naturalist*, 113, 26–27, 122, 269, 313; 114, 45, 116.

Jones, D.S. (1993) Traces in the country of the White Cockatoo (*Chinna junnak cha knæk grugidj*): A quest for landscape meaning in the Western District, Victoria, Australia. Unpublished PhD, University of Pennsylvania, Philadelphia PA.

Jones, D.S. (1999) A traditional seasonal calendar. *Journal of the Science Teachers' Association of Victoria for Primary Teachers*, 16(3), 19–21.

Jones, D.S. (2002) Time, seasonality and design: Reconsidering temporal dimension and patterns of the Australian landscape *Australian Institute of Landscape Architects National Conference: People + Place – the Spirit of Design, Darwin, 22nd–24th August 2002*. www.aila.org.au/conference/papers/David-Jones/davidjones-intro.htm.

Jones, D.S., Mackay, S., Paton, I., & Pisani, A-M. (1998) *Seasons of the Upper Yarra Valley*. Adelaide, SA: The University of Adelaide.

Jones, D.S., Mackay, S., & Pisani, A-M. (1997) Patterns in the valley of the Christmas Bush. *Victorian Naturalist*, 114(5), 242–245.

Jones, D.S. (2005) The ecological history of Adelaide 3: The historical evolution of the present landscape. In: Daniels, C.B., & Tait, C.J. (eds.), *Adelaide – Nature of a City: The Ecology of a Dynamic City from 1836 to 2036*. Adelaide: BioCity, pp. 56–74.

Jones, D.S. (2010) Kaurna cultural heritage. In: Daniels, C.B. (ed.), *Adelaide – Water of a City*. Adelaide: Wakefield Press & Barbara Hardy Centre for Sustainable Urban Environments, pp. 131–153.

Jones, R. (1969) Fire-stick farming. *Australian Natural History*, 16, 224–228.

Keatley, M.R., Chambers, L.E., & Phillips, R. (2013) Australia and New Zealand. In: Schwartz, M.D. (ed.), *Phenology*. Dordrecht, The Netherlands: Springer, pp. 23–52.

Kenneally, K., Edinger, F., Choules, D., & Willing, T. (1996) *Broome and beyond: Plants and people of the Dampier Peninsula, Kimberley, Western Australia*. Como, WA: Dept of Conservation and Land Management.

Kiddle, M. (1962) *Men of yesterday: A social history of the Western district of Victoria 1834–1890*. Melbourne: Melbourne University Press.

Latz, P. (1995) *Bushfires and bushtucker: Aboriginal plant use in Central Australia*. Alice Springs: IAD Press, 400 pp.

Latz, P. (1999) *Pocket bushtucker: A field guide to the plants of Central Australia and their traditional uses*. Alice Springs: IAD Press.

Laurie, V. (2017) Indigenous hopes for heritage status. *The Weekend Australian*, January 21, p. 3.

Lee, B. (2010) *Cultivated Seaweed and Seaweed industry development in Australia, Publication No. 10/164, Project No. PRJ-004681*. Canberra: Rural Industries Research and Development Corporation.

Levitt, D. (1981) *Plants and people: Aboriginal uses in plants on Groote Eylandt*. Canberra: Aboriginal Studies Press.

Low Choy, D., Clarke, P.A., Jones, D.S., Serrao-Neumann, S., Hales, R., & Koschade, O. (2013) *Aboriginal reconnections: Understanding coastal urban and peri-urban Indigenous people's vulnerability and adaptive capacity to climate change*. Gold Coast, Qld.: National Climate Change Adaptation Research Facility.

Mabo and Others v Queensland (No. 2) [1992] HCA 23; (1992) 175 CLR 1.

Mabo v Queensland (1989) 166 CLR 186; see www.austlii.edu.au/au/cases/cth/HCA/1988/69.html (accessed 1 January 2017).

Magarey, A.T. (1895) Aboriginal water quest. *Proceedings of the Royal Geographical Society of Australasia. South Australia Branch*. Printed as a booklet, p. 15.

Maiden, J.H. (1889) *The useful native plants of Australia*. London: Trubner.

Malaspinas, A-S., et al. (2016) A genomic history of aboriginal Australia. *Nature*, 538, 21 October, 207–214.

Marrfurra, P. (1995) *Ngan'gikurunggurr and Ngangiwumirri ethnobotany: Aboriginal plant use from the Daly River area, northern Australia*. Northern Territory Botanical Bulletin No. 22. Darwin, NT: Conservation Commission of the Northern Territory.

Mayall, J. (2014) *The outback chef: Cooking with native Australian ingredients*. Sydney: New Holland Publishers.

McCarthy, F.D. (1957) *Australia's Aborigines: Their life and culture*. Melbourne: Colorgravure.

McConvell, P., & Thieberger, N. (2001) *State of indigenous languages in Australia – 2001*. Canberra: Department of the Environment and Heritage.

McHarg, I.L. (1966) Ecological determinism. In Darling, F.F., & Milton, J.P. (eds.), *Future environments of North America*. Garden City, NY: Natural History Press.

McHarg, I.L. (1969) *Design with nature*. Garden City, NY: Natural History Press.

McNiven, I.J., & Bell, D. (2010) Fishers and farmers: Historicising the Gunditjmara freshwater fishery, western Victoria. *The La Trobe Journal*, (85), 83–91.

Menzies, C.R., & Butler, C.F. (2007) Returning to selective fishing through indigenous fisheries knowledge: the example of K'moda, Gitxaala Territory. *American Indian Quarterly*, 31, 441–464.

Mitchell, T.L. (1839) *Three expeditions into the interior of Eastern Australia, with descriptions of the recently explored region of Australia felix, and the present colony of New South Wales*. London: T. & W. Boone, 1839; facsimile ed., Adelaide: Libraries Board of South Australia, 1965.

Mulvaney, D.J., & Kamminga, J. (1999) *Prehistory of Australia*. Sydney: Allen and Unwin.

Murujuga Aboriginal Corporation [MAC], Murjuga Land and Sea Unit [MLSU] and Urban Design Landscape Architects [UDLA] (2017) *Murujuga cultural management plan*. Karratha, WA: Murujuga Aboriginal Corporation [MAC].

National Native Title Tribunal (2016) Determinations of native title and claimant applications map. *National Native Title Tribunal*. www.nntt.gov.au/Maps/Schedule_and_Determinations_A0_map.pdf (accessed 22 February 2016) and Indigenous Australian Lands Mapped Against Remoteness. *National Native Title Tribunal*. www.humanrights.gov.au/sites/default/files/content/about/media/speeches/social_justice/2007/gif/slide1.jpg (accessed 22 February 2016).

Newsome, A.E. (1980) The eco-mythology of the red kangaroo in Central Australia. *Mankind*, 12(4), 327–333.

Newton, J. (2016) *The oldest foods on Earth: A history of Australian native foods, with recipes*. Sydney: NewSouth Publishing.

Noormets, A. (ed.) (2009) *Phenology of ecosystem processes*. Dordrecht, The Netherlands: Springer.

Northern Territory of Australia v Arnhem Land Aboriginal Land Trust [2008] HCA 29 (30 July 2008) (Cwth) [legally referred to as the 'Blue Mud Bay case'].

NSW Department of Primary Industries (2002) *Indigenous fisheries strategy and implementation plan*, December 2002. www.dpi.nsw.gov.au/fisheries/aboriginal-fishing/strategy/nsw-ifs/nsw-ifs (accessed 1 February 2017).

NSW Department of Primary Industries (2014) *Status of fisheries resources in NSW 2011–2012 summary*. Sydney: Department of Primary Industries.

NSW Department of Primary Industries (2016) Commercial fishing – our fisheries. www.dpi.nsw.gov.au/fishing/commercial/fisheries (accessed 1 February 2017).

Olive, M. (2006) *Mark Olive's outback café: A taste of Australia*. North Melbourne, VIC: Black Olive Catering.

Outback Chef Store. (2017) www.outbackchef.com.au/ (accessed 1 February 2017).

Parks Victoria (PV), Gunditj Mirring Traditional Owners Aboriginal Corporation (GMTOAC) and Winda Mara Aboriginal Corporation (WMAC) (2015) *Ngootyoong Gunditj Ngootyoong Mara South West management plan*. Melbourne: PV.

Pascoe, B. (2014) *Dark emu: Black seeds agriculture or accident*? Broome, WA: Magabala Books.

Peterson, N., & Rigsby, B. (eds.) (2014) *Customary marine tenure in Australia*. Sydney: Sydney University Press.

Petrie, T., & Campbell Petrie, C. (1932) *Tom Petrie's reminiscences of early Queensland (dating from 1837)*. Brisbane: Queensland Book Depot.

Phelps, D.G. (1997) *Feasibility of a sustainable bush food industry in western Queensland*. Canberra: Rural Industries Research and Development Corporation.

Pinkerton, E. (1989) Introduction: Attaining better fisheries management through co-management – prospects, problems and propositions. In: Pinkerton, E. (ed.), *Co-operative management of local fisheries: New directions for improved management and community development*. Vancouver: University of British Columbia Press, pp. 3–32.

Prober, S.M., O'Connor, M., & Walsh, F. (2011) Australian Aboriginal peoples' seasonal knowledge: A potential basis for shared understanding in environmental management. *Ecology and Society*, 16(2), 12. https://www.ecologyandsociety.org/vol16/iss2/art12/ (accessed 1 Januray 2017).

Pryor, C. (2014) Rethinking indigenous Australia's agricultural past. *ABC Bush Telegraph*. www.abc.net.au/radionational/programs/bushtelegraph/rethinking-indigenous-australias-agricultural-past/5452454 (accessed 1 January 2017).

Puruntatameri, J., Puruntatameri, R., Pangiraminni, A., Burak, L., Tipuamantymirri, C., Tipakalippa, M., Puruntatameri, J., Puruntatameri, P., Pupangamirri, J.P., Kerinaiua, R., Tipiloura, D., Orsto, M., Kantilla, B., Kurrupuwu, M., Puruntatameri, P.F., Puruntatameri, T.D., Puruntatameri, L., Kantilla, K., Wilson, J., Cusack, J., Jackson, D., & Wightman, G. (2001) *Tiwi plants and animals: Aboriginal flora and fauna knowledge from Bathurst and Melville Islands, northern Australia*. Darwin: Parks and Wildlife Commission of the Northern Territory and Tiwi Land Council.

Reid, A.J. (1995) *Banksias and bilbies: Seasons of Australia*. Melbourne: Gould League of Victoria.

Rettig, R.B., Berkes, F., & Pinkerton, E. (1989) The future of fisheries co-management: A multidisciplinary assessment. In: Pinkerton, E. (ed.), *Co-operative management of local fisheries: New directions for improved management and community development*. Vancouver: University of British Columbia Press, pp. 273–289.

Robins, J. (1997) *Wild lime – cooking from the Bushfood Garden*. Sydney: Allen and Unwin.

Robins, L. (2007) *Outback Spirit Bush foods: A learning model in marketing and supply chain management*. Canberra: Rural Industries Research and Development Corporation.

Rolls, E.C. (1969) *They all ran wild: The story of pests on the land in Australia*. Sydney: Angus and Robertson.

Rose, D.B. (1996) *Nourishing terrains: Australian Aboriginal view of landscape and wilderness*. Canberra: Australian Heritage Commission. http://www.environment.gov.au/heritage/ahc/publications/nourishing-terrains (accessed 1 January 2017).

Rose, D.B. (2005) Rhythms, patterns and connectivities: Indigenous concepts of seasons and change. In: Sherratt, T., Griffiths, T., & Robin, L. (eds.), *A change in the weather*. Canberra: National Museum of Australia, pp. 32–41.

Rose, D.B., Daiyi, N., Deveraux, K., Daiyi, M., Ford, L., & Bright, A. (2002) *Country of the heart: An indigenous Australian homeland*. Canberra: Aboriginal Studies Press.

Ryan, J.C. (2013) Toward a phen(omen)ology of the seasons. *Environment, Space, Place*, 5(1), 103–131.

Ryder, M., & Latham, Y. (2005) *Cultivation of native food plants in south-eastern Australia*. Canberra: Rural Industries Research and Development Corporation.

Ryder, M., Walsh, F., Douglas, J., Waycott, M., Robson, H., Singh, Z., Sousa Majer, Md., Collins, T., White, J., & Cheers, B. (2009) *Sustainable bush produce systems: Progress report 2004–2006*. Alice Springs: Desert Knowledge CRC.

Sanderson, J.C., & Di Benedetto, R. (1988) *Tasmanian seaweeds for the edible market*, Technical Report. Hobart, Tasmania: Marine Laboratories, Department of Sea Fisheries.

Schnierer, S., & Egan, H. (2016) Composition of the aboriginal harvest of fisheries resources in coastal New South Wales, Australia. *Reviews in Fish Biology and Fisheries*, 26, 693–709.

Schwartz, M.D. (ed.) (2013) *Phenology: An integrative environmental science*. Dordrecht, The Netherlands: Springer.

Secretariat of the Convention on Biological Diversity (2005) *Handbook of the convention on biological diversity including its Cartagena protocol on biosafety*, 3rd ed. Montreal, Canada: Secretariat of the Convention on Biological Diversity.

Seddon, G. (1979) The *genius loci* and Australian landscape. *Landscape Australia*, 2/1979, 66–73.

Seddon, G. (1997) The *genius loci* and Australian landscape. In: Seddon, G. (ed.), *Landprints reflections on place and landscape*. Cambridge: Cambridge University Press, pp. 109–118, 136–142.

Smyth, D., Isherwood, M., & Schnierer, S. (2010) *Right to use country: Towards a freestanding statutory right for Traditional Owners to non-commercial access to and use of natural resources*, in Victoria Final Report, to the Native Title Unit, Victorian Department of Justice.

Smyth, R.B. (1878) *The Aborigines of Victoria*. Melbourne: Victorian Government Printer.

Stapylton, G.W.C. (1971) The journal of Granville Wilson Chetwynd Stapylton – kept while accompanying Major Sir Thomas Livingstone Mitchell on his Australia Felix expedition, 1836. In: Douglas, M.H., & O'Brien, L. (eds.), *The natural history of Western Victoria: Proceedings of the symposium*. Horsham: Australian Institute of Agricultural Science.

Strehlow, T.G.H. (1947) *Aranda traditions*. Melbourne: Melbourne University Press.

Swan, D.K. (2017) *Bunya Tukka tracks: Investigating traditional travelling routes of Eastern Australia*. Unpublished MArch thesis, Institute of Koorie Education and School of Architecture and Built Environment, Deakin University, Geelong.

Sweeney, D. (1993) Fishing, hunting and gathering rights of Aboriginal peoples in Australia. *University of NSW Law Journal*, 16, 97–160.

Taylor, J. (ed.) (2005) Population and patterns of residence. In: Arthur, B., & Morphy, F. (eds.), *Macquarie atlas of indigenous Australia: Culture and society through space and time*. Sydney: Macquarie and Macmillan, pp. 66–77.

Thurstan, R., Brittain, Z., Jones, D.S., Bellgrove, A., & Dearnaley, J. (2018) Archival evidence of indigenous uses of seaweeds in temperate Australia. *Journal of Applied Phycology* (in press).

Tindale, N.B. (1977) Adaptive significance of the Panara or grass seed culture of Australia. In: Wright, R.V.S. (ed.), *Stone tools as cultural markers*. Canberra: Australian Institute of Aboriginal Studies, pp. 345–349.

Tindale, N.B. (1981) Desert Aborigines and the southern coastal peoples: Some comparisons. in Keast, A. (ed.), *Ecological biogeography of Australia*. The Hague: Junk, pp. 1855–1884.

Tonkinson, R. (1978) *The Mardudjara Aborigines: Living the dream in Australia's desert*. New York: Holt, Rinehart and Winston.

UN General Assembly. (2007) United Nations Declaration on the Rights of Indigenous Peoples: Resolution/adopted by the General Assembly, 2 October 2007, A/RES/61/295. www.refworld.org/docid/471355a82.html (accessed 1 February 2017).

UNESCO. (2017a) *Cultural landscapes*. http://whc.unesco.org/en/culturallandscape/ (accessed 1 January 2017).

UNESCO. (2017b) *Uluru-Kata Tjuta National Park*. http://whc.unesco.org/en/list/447 (accessed 1 January 2017).

UNESCO. (2017c) *Kakadu National Park*. http://whc.unesco.org/en/list/147 (accessed 1 January 2017).

Walsh, F., & Mitchell, P. (eds.). (2002) *Planning for country*. Alice Springs, NT: Jukurrpa Books.

Weatherhead, J. (2016) *Australian native food harvest: A guide for the passionate cook and gardener*. Tynong North, VIC: Peppermint Ridge Farm.

Webb, L.J. (1959) The use of plant medicines and poisons by Australian aborigines. *Mankind*, 7, 137–146.

Werner, G.F. (1992) *Edible seaweeds of South-Eastern Australia*. Calton: Department of Ecology and Evolutionary Biology, Monash University.

Wettenhall, G., & The Gunditjmara People (2010) *People of Budj Bim: Engineers of aquaculture, builders of stone house settlements and warriors defending country*. Ballarat: VicEm PRESS.

Wilhelmi, J.F.C. (1861) Manners and customs of the Australian natives in particular of the Port Lincoln district. *Transactions of the Royal Society of Victoria*, 5, 164–203.

Wilson, G., McNee, A., & Platts, P. (1992) *Wild animal resources: Their use by aboriginal communities*. Canberra: Bureau of Rural Resources.

Woodward, E., Jackson, S., Finn, M., & McTaggart, P.M. (2012) Utilising indigenous seasonal knowledge to understand aquatic resource use and information water resource management in northern Australia. *Ecological Management and Restoration*, 13(1), 58–64.

Wright, T. (2017) A country of stones is held up to the world. *The Age*, January 21, pp. 8–9.

Yawuru Registered Native Title Body Corporate (YRNTBC) (2011) *Walyala-jala buru jayida jarringgun buru Nyamba Yawuru ngan-ga mirli mirli*. Broome, WA: YRNTBC.

Yibarbuk, D., Whitehead, P.J., Russell-Smith, J., Jackson, D., Godjuwa, C., Fisher, A., Cooke, P., Choquenot, D., & Bowman, D.M.J.S. (2001) Fire ecology and aboriginal land management in central Arnhem Land, northern Australia: A tradition of ecosystem management. *Journal of Biogeography*, 28, 325–343.

Zola, N., & Gott, B. (1992) *Koorie plants, Koorie people: Traditional Aboriginal food, fibre and healing plants of Victoria*. Melbourne: Melbourne Koorie Heritage Trust.

4

Archaeology, history, and urban food security

Integrating cross-cultural and long-term perspectives

Christian Isendahl and Stephan Barthel

Food security in cities – a timeless challenge

Since the evolution of behaviourally modern *Homo sapiens* some 75,000 years ago human ingenuity has excelled in finding ways to satisfy the most fundamental need of any species: to procure food from the landscape and seascape. The historical sciences – archaeology and cognate disciplines such as palaeobotany and palaeozoology in particular – are advancing to approximate the full range of global food procurement diversity in the past and to understand the sustainability and resilience of these social practices. One of the key insights that archaeology brings to the global change discourse is that evaluating sustainability and detailing the resilience of social practices necessitate a long-term analytical perspective and an integrated comparative frame of reference of both conventional and alternative models, historical trajectories, and potential solutions. Urban farming is a good example of a social practice that warrants a longer-term cross-cultural analytical lens, because if chrono- and occidental-centric outlooks on urban metabolic systems in the recent past are maintained, food production in cities will appear a quaint anomalistic human practice of little value for understanding cities in general, urban problems, and the solutions to these challenges (see Chapter 12). Exploring the archaeological and historical records of past cities in different cultural contexts can offer insights on timeless urban issues, that is, issues that were addressed in the past, remain relevant at present, and are likely to form challenges in the future. One such permanent global challenge is to maintain food security for all in cities. Today, over half of the Earth's population lives in cities, and it has been estimated that this share will increase to two thirds by 2050, some 6.3 billion people, with cities in developing countries growing the fastest (UN, 2014: 11). Such projections are based on only a partial selection of the full range of interrelated decisive variables, however, and assume that there are no incentives to maintain rural practices or re-establish lost rural communities, and that current government policy and attitudes remain unchanged (Tim Waterman, pers. comm., 2016). Indeed, if the market continues to freely roam, global food production may have to increase by 70% to meet shifts in food preferences following broad-based transitions to urban lifestyles (IAASTD, 2009; Chapters 16, 19 and 25). Such projections, uncertain as they may be, entail massive challenges to provide food security equitably (Chapters 20, 21 and 32).

According to the Food and Agriculture Organization of the United Nations (FAO, 1996), food security can be understood as the situation when people have physical and economic access to sufficient, safe, and nutritious food to meet their dietary needs (Barthel and Isendahl, 2013: 225; Chapters 20 and 25). People continue to address urban food security in many different ways (and probably have conceptualised it in just as many), developing a series of technological systems, infrastructures, and organisational principles of governance to feed and be fed in the city. Food systems can be defined as "the chain of activities connecting food production, processing, distribution, consumption, and waste management, as well as all the associated regulatory institutions and activities" (Pothukuchi and Kaufman, 2000: 113). A food system can also be understood as a set of solutions to address food security that is dependent on changing cultural values, historical ecological dynamics, and social relations acting at several different scales, from the household to global socio-economic structures. The efficacy of food systems to provide long-term food security varies, and each introduces susceptibility to different kinds of vulnerabilities (Chapter 16). Although urban scholars and food scientists necessarily reference the past, the scope of their referential frame is often limited in relation to the amount of available data. Isendahl and Smith (2013: 132) suggest that there is a need to draw from as wide a range of cases and models as possible when considering options to inform the planning of sustainable and resilient future cities. In this view, archaeology and cognate sciences contribute two main bodies of knowledge to the understanding of food security in cities: (1) to generate *case studies* of food system performance in different cultural, social, and spatio-temporal contexts, and (2) to apply archaeology's unique temporal analytical and evaluative perspective on social change to study food security as a *long-term process*, particularly addressing which factors build resilience and which introduce vulnerabilities to a food system's efficacy to build food security. These kinds of studies are particularly productive to generate important insights if addressed from a cross-cultural comparative frame of analysis, contrasting different cases to identify trends and patterns of diversity and similarity.

This chapter aims to demonstrate the potential of integrating archaeological and historical research with investigations of contemporary conditions, ultimately to inform planning for the future. It discusses the timeless and permanent challenge of food security drawing on four distinct cases from different global regions and epochs of the last 2,000 years: (1) the cities of the pre-Columbian Maya civilisation in Central America (that is, pre-1500 CE), (2) medieval Constantinople in the eastern Mediterranean (being about 500–1500 CE), (3) London during the first half of the 20th century, and (4) the rapidly growing contemporary mega region of the Ganges-Brahmaputra Delta in Bangladesh and Indian West Bengal. Despite fundamental cultural, historical, economic, governance, and city-planning differences, urban and peri-urban food production have played and continue to play a key role for food security and city resilience.

Modernist urban essentialism limits options

There is a need to reflect on why the food systems of past cities are more than simply idiosyncratic cases. Are they of any greater relevance in our globalised world of unprecedented urbanisation, or only of special interest to the small scholarly guilds of, for instance, Mayanists or Byzantinologists? Since the beginning of the 1900s, the city has largely been associated with innovation and progress and is often framed in essentialist terms as a uniform type of social formation *inter alia* defined by being distinct from rural zones of agricultural production (Barthel and Isendahl, 2013; Isendahl, 2012a; Reader, 2004; also see Wirth, 1938; Woolston, 1912). The modernist understanding, including that of the 20th-century design movement, seems to view cities as separate entities detached from their broader life-support systems, a view which ultimately builds on the basic elements – or 'haunts' – of the emergence of the modern world, such as colonialism, Eurocentrism, racialisation, and capitalism (Orser, 2014). Such a modernist view is associated with industrialisation's innovations and social changes, particularly in infrastructure and transport, global trade in foodstuffs, and asymmetrical economic growth facilitated by cheap fossil fuels (that is, high energy gain – at least until energy prices rise due to increasing scarcity (see Tainter and Allen, 2015; Chapter 16). These processes provided the impulse for a first wave of time-space compression: the acceleration of velocity and reduction of the significance of distance (Harvey, 1990). Hence, modernist ideologies underpinning the emergence of urban planning during the early decades of the 1900s distinctly separated near-residential agriculture as obsolete in normative understandings of cities as autonomous social systems (Eisenstadt, 1966; Chapters 12, 14 and 15). This perception is, however, untenable when taking a longer-term perspective on global urban history. Archaeological research demonstrates that urbanism is a global phenomenon with great time-depth, and that different historical ecological trajectories have resulted in considerable spatial diversity and temporal variation, suggesting that the integration of cross-cultural urban archaeology has a distinct role to play by broadening the frame of reference of cities (Isendahl and Smith, 2013). Orthodox urban essentialism conceptualises urban metabolism (Wolman, 1965) as a one-way flow pattern of foodstuffs from external sources of production into cities as centres of consumption. But this is based on a recent phenomenon that has destructive effects as a universal model by losing sight of the range of alternative options available in the human experience to build food security (Barthel and Isendahl, 2013; Isendahl, 2012a).

Today, urbanisation is in a second wave of time-space compression driven by the internet, jet travel, and globalised economies (Warf, 2008; Chapters 9, 20, 21 and 25). The accelerating pace at which urban life proceeds and the decreasing importance of distance are qualitatively different in terms of their intensity and scope compared to even 50 years ago (Harvey, 1990; Sassen, 1991; Chapter 16). Large cities mainly feed themselves through access to global food production and transport systems relying on fossil fuels (Curtis and Ehrenfield, 2012; Fraser and Rimas, 2010; McMichael, 2011) to appropriate foodstuffs from every corner of the world (Deutsch, 2004; Folke et al., 1997). The global food system is highly effective, at least in the short-term (the long-term productive efficiency and full range of tradeoffs of the global agribusiness food system are concealed because insufficient time has elapsed to fully evaluate these processes). But the environmental impacts are as daunting (Fraser and Rimas, 2010; Steffen et al., 2011; Chapter 16) as they are complex (since to a significant extent geographically displaced from global centres of consumption they are asymmetrically distributed (Hornborg, 2001). High global connectivity between cities and remote food supplies can decrease cities' vulnerability to food shortages and build resilience during medium-severe crises (Ernstson et al., 2010). However, the global food system supporting urban metabolism is a highly complex system, and as in any complex system, variable change (such as a reorientation of the energy support system) may precipitate unexpected system dynamics and emergent properties (Folke et al., 2004), potentially posing major threats to both urban and rural food security (Barthel et al., 2015; Newman et al., 2009; Chapter 16). While many contemporary cities of developed countries can rely almost exclusively on food imports to meet daily subsistence needs (Porter et al., 2014), other strategies for food security exist in the archaeological, historical, and contemporary records.

Four cases of urban–exurban food systems addressing food security

In many tropical countries, urban populations are projected to increase dramatically over the next 35 years (UN, 2014: Table I). Thus, it makes sense to begin our sample of cases with the cities of the tropical lowland Maya in present-day south-eastern Mexico, Guatemala, Belize, and Honduras. Maya civilisation, including state-level polities and cities, emerged in the first millennium BCE and its long-term history suggests cycles of growth, decline, and re-organisation during the course of the Preclassic (2000 BCE–250 CE), Classic (250–1000 CE), and Postclassic periods (1000–1550 CE) (Chase and Scarborough, 2014). The Maya constructed hundreds of cities and towns of varying sizes and longevity, some lasting more than a millennium, others a few centuries, but most – to put these urban histories (and the notion of historic urban resilience and sustainability) in comparative light – functioned longer than any contemporary North American city has so far. For instance, the pre-Columbian Maya city of Caracol, today located in the rain forest of Belize, was occupied between 600 BCE and 900 CE, covered about 200 square kilometres, and had a population of about 100,000 in 650 CE (Chase and Chase, 2016). The Maya traded a variety of resources and commodities both short- and long-distance, including particular foodstuffs such as cacao and salt (see for example McKillop, 2002; Staller and Carrasco, 2010). However, the Maya had no beasts of burden and they lacked wheeled carriers, making overland transport costly in terms of energy and time compared to goods transportation in Eurasia and Africa (Barthel and Isendahl, 2013; Isendahl, 2012b; Isendahl and Smith, 2013). The Maya transported goods by sea and rivers, at least seasonally, but many inland regions lacked permanent navigable streams. The scarcity of transport energy put a relatively high cost on trade, inhibited exchange of bulk food goods that largely could be produced anywhere (such as the staple maize), and was a contributing factor to stimulate urban farming. Maya cities have been characterised as 'low-density cities' in a global comparison of urban traditions pre-1500 CE (Fletcher, 2009); domestic household residential units were dispersed in the landscape around central civic-ceremonial and elite residential building sectors providing economic, political, social, and religious urban functions. Cities interlaced built areas with spaces used for food production, including household gardens, infields at various stages of succession, orchards, and other green areas. The spatial patterning of urban land use has led scholars to describe Maya cities as 'green cities' (Graham, 1999), 'garden cities' (Dunning and Beach, 2010), 'forest gardens' (Ford and Nigh, 2015), and 'agro-urban landscapes' (Isendahl, 2012b). Although land, labour, and produce was controlled in a hierarchic system of socio-economic obligations, the household-based urban farmstead, its labour force and land, formed the basic production unit and building block of Maya agro-urban landscapes. Autonomous Maya farmsteads were functional production units, essentially able to maintain food security independently, but also flexible to adapt to and function within complex forms of organisation in large urban settlements. The farmstead garden concentrated agricultural knowledge and practice – owing to residential proximity it was most carefully tended, fertilised by the organic waste concentrated by city dwellers (possibly including human waste), and used for plant breeding and experimentation, seed storage and buffering, and for managing plant diversity (Isendahl, 2012b; Ford and Nigh, 2015). However, to produce the surplus needed to finance and maintain city-states, local food production depended on high investments in water management infrastructure such as reservoirs, which increased vulnerability to macro-scale economic fluctuations (see for example Isendahl et al., 2014; Isendahl and Heckbert, 2017; Isendahl et al., forthcoming). Nevertheless, the farmstead institution survived the political turmoil and widespread city-state collapse that took place at the end of the Classic period, owing to its ability to flexibly self-organise in a number of ways.

Constantinople, capital of the Roman and subsequently Byzantine Empires from the 4th century CE until conquered by the Ottomans in 1453, was very different from Maya cities in many respects. And yet there are certain commonalities in terms of how food security was addressed during crisis events. In contrast to many other European cities of the period (as well as the major pre-Columbian Maya cities), Constantinople did not lose its key position as a global city, subsequently continuing – as Istanbul – as the capital of the Ottoman Empire (Ljungkvist et al., 2010). Istanbul maintains its position as a major global city today; with a population of 14 million in 2014 it ranks as the largest European city and the 15th largest in the world (UN, 2014: 26). Early in Constantinople's imperial history food security in the capital relied on a global food system with distant production zones and an efficient transport system. The main source of food was bulk imports of grain staples produced in the Nile Valley brought in by trading vessels averaging 40–50 tons in capacity (Barthel and Isendahl, 2013). Although these supply lines were subject to the difficult winds of the eastern Mediterranean and the fluctuations of the Nile, the most severe threat to food security in Constantinople was regional conflict (Ljungkvist et al., 2010; Barthel, 2014). Access to the food supplies of the Nile was lost to Constantinople in the 7th and 8th centuries CE following the Moslem expansion, forcing the global food system to down-scale and focus on production zones closer to the capital region (Haldon, 1990). In the centuries that followed until 1453, the city of Constantinople was repeatedly involved in military conflict with city sieges and blockades of trade routes occurring on average every 65 years, effectively cutting food imports (Barthel and Isendahl, 2013). The most severe blockade on supply lines, at the end of the 14th century CE, lasted eight years but did not succeed in getting the city's population to surrender (Ljungkvist et al., 2010). The key to Constantinople's resilience was how the city and its population addressed food security and adapted, with a social memory of repeated sieges and blockades, to food shortages by developing strategies of horticulture and agriculture in the city and the immediately surrounding area. This adaptation seems to have had elements of both top-down governance and self-organising principles. For instance, an outer city wall (the Theodosian wall) was erected 1.5 kilometres westwards from the old wall, motivated not by a need to expand settlement for a growing population, but by the necessity to protect a water storage system and agricultural space. The city invested in a system of large water reservoirs in the area between the inner and outer city walls with the capacity to store more than a million cubic metres of water. In addition, the 3 square kilometre area between the walls was allocated for food production (Ljungkvist et al., 2010). A 2 kilometre wide buffer zone of farm fields immediately outside the Theodosian wall was included in the local food infrastructure, allowing the city to use 15 square kilometres of agricultural lands in direct proximity to the urban core. In combination with household gardens located throughout the city and exceptionally rich fishing at the Golden Horn, urban food production had the potential to feed hundreds of thousands of people during recurrent crisis events when the main, geographically expansive food system network failed (Barthel et al., 2010b).

Modern western history shows that urban gardens and exurban agriculture have provided relief for urban people during periods of temporal cuts in global food trade, much similar to medieval Constantinople. Disturbances in food systems supplying food to Europe have been from various causes, but economic crises and wars are the most common (Barthel, 2014). From the late 1800s onwards, the rise and demise of food shortages have been ubiquitous across Europe, most dramatically during the two World Wars. In European cities, governments and city dwellers often responded to food crises by increasing food production within the cities. In the UK, for instance, there were two major peaks of urban farming during the 20th century (Barthel, 2014). The first occurred during WWI, when supply lines to outside food sources were severed. Facing mass starvation, the UK government instigated the *Every Man a Gardener* campaign, permitting food production in urban lots, parks, and sports fields. The number of urban gardens increased dramatically from about 0.6 million to 1.5 million, providing almost 2 million tons of fresh vegetables by 1918 (Crouch and Ward, 1997; Select Committee, 1998). WWII generated another explosion of urban gardening; the number of allotment gardens rose from about 0.8 million prior to the war to 1.4 million during its peak, providing British citizens with almost 1.3 million tons of food (Select Committee, 1998; Chapter 30). Such widespread cultivation in cities provided relief, increasing food security by decreasing the gap between the nutritional needs of concentrated urban populations and the food available to them. This pattern of a growing share of city produce to the total food resource framework was similar across the increasingly agro-urban landscapes of Europe. For instance, although Sweden was not directly involved as a combatant in either war, firm food rationing during WWII resulted in a sharp increase of urban gardening and agriculture, rising from about 30,000 urban gardens prior to the war to about 130,000 during its peak (Barthel et al., 2010a; Barthel et al., 2015). In Germany, the number of urban gardens rose from a few hundred in the 19th century to about 0.45 million during the financial crises of the 1930s, to about 0.8 million at the end of WWII (Barthel et al., 2013; Barthel et al., 2015). Since the emergence of the world as we know it following the great acceleration of economic development post-1950, modernisation and the second wave of time-space compression have diminished local urban food production infrastructure in Europe and developed countries (Chapter 30). Indeed, from the perspective of global agribusiness, local urban food systems are considered inconsequential in cities like London, Stockholm, Paris, Berlin, or New York. In many of the cities of developing countries, however, local and regional food infrastructures are still providing both food security and livelihood security to the less affluent in the urban–exurban interphase, but perhaps not for much longer (Chapters 20 and 21).

The proportion, rate, and spatial extent of current urbanisation in developing countries are unprecedented in human history (UN, 2014). One of the main challenges is how to address food security in rapidly growing informal settlements (see Chapters 20 and 22). The mega region of the Ganges-Brahmaputra Delta (GBD) is a case in point. The GBD in Bangladesh and Indian West Bengal is the largest delta in the world with an estimated total population of almost 150 million people. The GBD is home to more than 40 million urban dwellers (of whom about half live in informal settlements), including Dhaka (17 million inhabitants; ranking 11 in world city size), Kolkata (15 million inhabitants; world rank 14), and a number of other cities (UN, 2014: 26). In 2007–2008 climate-related hazards pushed staple food prices on the global market to double (Chapter 25) at the same time as the GBD experienced severe flooding and a cyclone that destroyed paddy fields in the north of the delta. Increasing food prices sparked food riots in many parts of the world, for instance in the Middle East, with unforeseeable social and political consequences, but this did not happen in the cities of the GBD. Instead, the incomes of the urban poor in the GBD increased during this period (Keck and Etzold, 2013).

Self-organised local food systems form alternatives that provide food security to the urban poor. The GBD cities and surrounding wetlands form an integrated resource recycling and food producing complex: urban waste is transported to food producing areas at the rim of the cities to be applied as fertilisers, and food for urban consumption is transported back to markets for sale. Canals connecting urban markets with marshes, wetlands, and paddy fields, as well as with major rivers and the sea, form part of an informal infrastructure for local food systems. Boats, vehicles, and other equipment are used to carry waste and sewage out of the city, which serve as fertilisers in a great diversity of peri-urban horticulture, rice paddy fields, and aquaculture (with freshwater carp particularly important), and products are brought back to hub-markets and distributed by street vendors throughout informal urban settlements. For instance, almost 80% of the rice and almost 60% of the fish consumed is transported less than 100 kilometres to any street vendor (Rahman et al., 2013). Maintaining such food system infrastructure contributes to livelihood security for a great proportion of the poor in the urban–exurban interface (Keck and Etzold, 2013; Mukherjee, 2015). Social-ecological memory – defined as the means by which knowledge, experiences, and practices about how to manage a local agro-ecosystem is retained in a society and revived and communicated through time (Barthel et al., 2010a) – is utilised in boat navigation and transport, in food production and commerce. Self-organised and often informal trade-links between waste handlers, boaters, food producers, food traders, street vendors, and poor urban consumers enable such sustainable resource recycling and food system infrastructure, and – owing to mutual investments and social capital – local food prices were kept stable during the global food crisis in 2007–2008, providing the urban poor with affordable food products (Keck and Etzold, 2013; Mukherjee, 2015).

Are there timeless solutions?

There is an on-going debate on the potential of food production in cities to contribute to urban sustainability (see for example Knapp et al., 2016; Maddox et al., 2016). Researchers of developing countries witness that urban farming is a widespread social practice, while the pedagogical benefits of urban agriculture at grass-roots levels may be the most obvious sustainability rationale highlighted in developed countries (Chapter 12). The latter tends to dominate the mainstream understanding of the role of urban agriculture in the contemporary discourse; rarely are the insights generated from archaeological and historical research or from contemporary developing countries integrated in an overall comparative frame of reference assessing the role localised food systems play to provide options for resilient food and livelihood security in cities.

Initially this chapter suggested that food security in cities is a timeless and permanent challenge; a challenge that was addressed in the past, remains an urgent and complex issue in the present, and is highly likely to be a challenge for future generations (Chapter 16). The underlying basic ideas driving the argument are, first, that urban food security is a challenge that motivates learning from past experiences and, second, that timeless challenges might have timeless solutions (thus being independent from the idiosyncrasies of, for instance, early 21st-century information technology argued by many to provide the solutions to our time's problems). Technological fetishism (and perhaps particularly digital, as the concept of 'Smart City' is entering the scene) (cf., Hornborg, 2001) seems to have put a veil over how people met challenges with solutions in the past. Such perspectives obscure from view that which can be evident in the records of 75,000 years (at least) of human experience in problem-solving. There are kernels to be cultivated for addressing similar challenges today and in the future. Without claiming any nomothetic validity for this proposition, the (admittedly very limited) series of cases presented here have shown that food production is, when taking a global and long-term perspective, an integrated urban practice engaging city dwellers in a wide range of different activities, often *within* the boundaries of the city. The concept of 'agro-urban landscapes' – initially coined to describe the spatial integration of farm fields, gardens, orchards, and green spaces into the settlement layout of pre-Columbian Maya cities to form a hybrid-type of city (Isendahl, 2012b) – might also be valid as a generic term to characterise several other, unrelated urban traditions and city formations, ultimately to contribute to finally putting modernist urban essentialism to rest. Hence, there are certain patterns beyond the obvious cultural and historical idiosyncrasies that offer important insights valid for understanding strategies to address sustainable food security in cities more broadly. Drawing on the archaeological and historical cases of pre-Columbian Maya cities, medieval Constantinople, and London during WWI and WWII, the insights brought up in this concluding discussion as commentaries on the contemporary case of the GBD mega region can be summarised in two keywords: diversity and flexibility.

Diversity is a key property of a resilient complex adaptive system. The food system is not any different in this regard, and local/regional food systems provide options when access to affordable food by the global food system decreases. Long distance food trade is needed especially when local or regional harvests fail due to climate anomalies, natural hazards, or political turmoil. However, global food systems introduce a number of vulnerability factors. The London and Constantinople cases elucidate vulnerability to political and military conflicts. A perhaps more latent factor is the increasing dependence on infrastructure and coordination requiring high energetic costs relative to local food production. The Maya food system relied to a significant extent on urban agriculture, which entails minimal energy expenditure for transporting produce to consumers, and is thus highly energy-efficient and insensible to fluctuations in energy availability and pricing. The dependence on fossil fuels for the operation of the currently dominating global food system relative to the low-energy requirements of the Maya system is striking.

In a sense, long-distance food trade solutions are in part a result of a social-ecological memory of how to deal with food shortages when there are disturbances to local or regional food systems. The dominance of a global food system with a few major players in the food industry, as well as the solutions and practices throughout the global-scale food system, pose particular obstacles to sustainable practice today. It is, for instance, merciless if poverty-stricken consumers cannot afford food as prices fluctuate on the global market (Chapter 25). The urban farmsteads of Maya cities were essentially functional units with the long-term flexibility to form part of and adapt to social formations of different levels of complexity. Thus, heavily taxed on surplus and labour within the Maya system of socio-economic obligations, urban farmsteads formed the economic basis of city-states. However, when most of these complex city-states experienced economic and sociopolitical collapse at the end of the first millennium CE (owing to a number of interacting factors including maximising political economy, population increase, climate instability, lowered energy gain, and so on), the farmstead institution seems in many cases to have persevered as functional, autonomous socio-economic units based on near-residential agriculture and horticulture and small-scale water management (Isendahl et al., forthcoming). Thus, autonomous and accessible food production is an important factor to build resilience against perturbations within the larger system in order to maintain livelihood security, at least for the urban poor.

As seen during the 2007–2008 economic crises, food prices may spike when there are multiple stress factors to the global food system, and often co-occur with particularly low income levels among the urban poor (Chapter 20). It is during such events that local food systems provide particularly important alternative options to provide food security in the informal urban settlements of developing countries. One strategy for the future, based not least on insights from developed countries, is to challenge the hegemony of agri-business by re-creating local and regional food systems through food hubs and value-based food supply chains that support smallholders (Jongerden et al., 2014; Berti and Mulligan, 2016; Chapters 12 and 31). Such a strategy would share many characteristics of the GBD case, where mutual investments in a local to regional food system fill a structural void and makes affordable food available to the poor. For a great proportion of dwellers in informal settlements in developing countries, however, strategies for sustainable food security cannot a priori be based on altruistic sustainability incentives of value-based food supply chains, but must focus on food and, ultimately, livelihood security. The Maya example is a case in point of the resilience of urban farmsteads to maintain functions and provide relative livelihood security through periods of fundamental disturbance to overall economic and political systems.

The self-organised food system supplying inhabitants of informal settlements in GBD with affordable food during the global crisis 2007–2008 contributed to food and livelihood security (Chapter 25). If any link in the chain of activities related to food/waste flows in such a system fails, many mutual investments fail, affecting social networks that are based on face-to-face interactions. Unfortunately, international policies and urban governance are ignorant of the resilience-building capacity of informal local food system infrastructure to level the effects of global food spikes on food security (Jongerden et al., 2014; Berti and Mulligan, 2016). Instead, these support urban expansion over canals and food producing areas and regulate street vendors in informal settlements (Chapter 20). Such policies and the lack of city-wide governance contribute to loss of urban commons, increase runoff and land subsidence, and build up vulnerability of local food infrastructure to increased flooding frequencies, storm surges, sea level rise, and other extreme events associated with global warming (Higgins et al., 2014; Chapter 18). Whereas affluent neighbourhoods often enjoy urban services and infrastructure of a high standard, in informal settlements the lack of financial capital, high corruption rates, and weak official urban governance seem often to leave governments largely impotent to develop the formal food system infrastructure necessary to keep up with population growth (Islam, 2014). Hence, in the GBD, policies – often initiated and passed with sustainability intentions in mind – aiming to regulate informal subsistence economies and close down local food systems, may undermine the resilience of local, self-organised food systems and their capacity to provide both food and livelihood security to the urban poor (Keck and Etzold, 2013; Mukherjee, 2015; Chapter 20). The dispersed settlement pattern of pre-Columbian Maya cities – integrating gardens, orchards, cultivated fields, and different systems to manage and collect surface water runoff with domestic and civic-ceremonial building complexes – is an example of urban planning adapted to manage local conditions of extreme precipitation seasonality into the urban fabric. It reminds us that there cannot be a global formula for sustainable cities, not least in areas such as the GBD where the effects of global warming are projected to have severe impacts in terms of increasing frequency and intensity of extreme events.

In many cities of Asia and Africa a significant share of foodstuffs for the poor are produced in peri-urban areas (Moustier, 2007). It is, however, possible that this share will diminish rapidly in the future (Bren d'Amour et al., 2016; Chapter 21). The spatial, economic, and social patterns that emerged in urban areas in developed countries as a result of time-space compression and industrialisation is a global and globalising process that is reproduced today in developing countries, but on a grander scale and with fewer resources available per capita. City, peri-urban, and exurban landscapes are impacted by cross-scale drivers of change related to rapid and informal demographic transitions, intense global financial speculation on land, and weak governance (Andersson and Barthel, 2016), leading to uncontrolled spatial urban expansion over food-producing landscapes (Bren d'Amour et al., 2016). In the mega region of the GBD, for instance, the annual loss rate of arable land is currently 809 square kilometres; a rate expected to increase in the future (Chaudhuri and Mishra, 2016; Dewan et al., 2012). Urban development that permanently eliminates space that can potentially be transformed for cultivation if required, may undermine, as the Constantinople and London cases suggest, the resilience capacity to respond to cuts in the global food system. A large proportion of livelihoods in the interface between informal settlements and exurban zones in developing countries will be affected if local food systems are lost. In developing countries, many people are engaged in the local to regional chains of activities connecting food production, processing, distribution, consumption, and waste management, as well as all the associated regulatory institutions and activities similar to those described in the GBD. Furthermore, many households in exurban zones produce at least part of their own food in household gardens, options likely to become displaced during such processes (Bren d'Amour et al., 2016; Chaudhuri and Mishra, 2016). The process of urban and exurban full- and part-time farmers leaving their trade for wage labour increases risk and decreases long-term urban resilience capacity and food security (Chapters 7 and 8). In Constantinople, for instance, the social-ecological memory of how to produce food in the city was maintained through the centuries, necessitated by a history of building food security during reoccurring sieges. Thus, not only is there a need to maintain land and infrastructure that can be converted to food production, but also to maintain or foster the skills necessary to produce food in cities.

At the systemic level, the loss of local food systems and their infrastructure can only be compensated by increasing imports, thus bringing additional land under cultivation elsewhere and demanding increasing yields; solutions that will snowball local dependence on the global food system (Bren d'Amour et al., 2016). Putting all eggs in the global food system basket requires efficiency and stability in one complex system. This system, under increased population growth, is and will be pushed to operate at its upper production limits in order to meet global food security needs. Such a singular approach is at best ill-advised and at worst imperils food security for billions of city dwellers. A food system at full capacity is dependent on all variables remaining 'just right', and is potentially vulnerable even to small perturbations (which, according to the non-linearity of complex system dynamics, may have unpredictable consequences on production levels, and ultimately on the coherence and resilience of the food system as a whole). If such disturbance factors are identified and recognised, they can also be addressed. However, the collapse of Maya city-states illustrates how several slowly changing social-ecological variables, difficult to discern and understand the full effects of 'in the moment', interacted to erode sustainability (Isendahl et al., 2014; Isendahl and Heckbert, 2017; Heckbert et al., 2016). Modern evidence-based science is the most powerful epistemology to have emerged over our species' 75,000 years-long history of addressing challenges. But it would be presumptuous to suggest that we have all the solutions to address what the future holds in terms of challenges to food security. Indeed, Isendahl et al. (forthcoming) argue that:

> The archaeological perspective is particularly important in focusing on the very long time perspective when thinking about the sustainability of social-ecological systems or institutions [such as food systems]. [. . .] In the mainstream discourse sustainability is hardly ever defined and commonly vaguely associated with *indefinitely*. From the archaeological record we know for certain that no system or institution works indefinitely and without social or environmental tradeoffs.

Assessing the insights from the past and comparing those to the needs of the present, these seem repeatedly to coalesce around various aspects of diversity and flexibility. Maintaining diversity of options, including local food systems that co-exist with long distance food trade solutions between and throughout pluralistic food systems, and a capacity for institutional flexibility to activate and engage in these alternatives at multiple scales, form the basis of long-term food security resilience. This is a strategy that can accommodate food system disturbances, crises, and collapses. It requires a new, unassuming paradigm – a 'safe to fail strategy' – governing food security philosophy, governance, and practice.

Acknowledgements

This chapter was prepared as part of the research network 'Pre-Columbian Tropical Urban Life: Placing the Past in Designs for Sustainable Urban Futures' (TruLife) funded by a grant from the Art and Humanities Research Council (AHRC) of the UK. We thank the editors for inviting us to contribute to this volume and for their helpful comments on our text. We are also grateful to two anonymous peer reviewers for constructive suggestions.

References

Andersson, E., & Barthel, S. (2016) Memory carriers and stewardship of Metropolitan landscapes. *Ecological Indicators*, 70, 606–614.

Barthel, S. (2014) Resilience. In: Nonini, D.M. (ed.), *A companion to urban anthropology*. Malden, MA: Wiley-Blackwell, pp. 428–446.

Barthel, S., Folke, C., & Colding, J. (2010a) Social-ecological memory in gardening: Retaining the capacity for management of ecosystem services. *Global Environmental Change*, 20, 255–265.

Barthel, S., & Isendahl, C. (2013) Urban gardens, agriculture, and water management: Sources of resilience for long-term food security in cities. *Ecological Economics*, 86, 224–234.

Barthel, S., Parker, J., & Ernstson, H. (2015) Food and green space in cities: A resilience lens on gardens and urban environmental movements. *Urban Studies*, 52(7), 1321–1338.

Barthel, S., Parker, J., Folke, C., & Colding, J. (2013) Urban gardens: Pockets of social-ecological memory. In: Tidball, K.G., & Krasny, M.E. (eds.), *Greening in the red zone: Disaster, resilience, and urgent biophilia*. Dordrecht, The Netherlands: Springer, pp. 145–158.

Barthel, S., Sörlin, S., & Ljungkvist, J. (2010b) Innovative memory and resilient cities: Echoes from ancient constantinople. In: Sinclair, P., Herschend, F., Nordquist, G., & Isendahl, C. (eds.), *The urban mind: Cultural and environmental dynamics*. Uppsala: Uppsala University, pp. 391–405.

Berti, G., & Mulligan, C. (2016) Competitiveness of small farms and innovative food supply chains: The role of food hubs in creating sustainable regional and local food systems. *Sustainability*, 8(7), 616. doi:10.3390/su8070616.

Bren d'Amour, C., Reitsma, F., Baiocchi, G., Barthel, S., Güneralp, B., Erb, K.-H., Haberl, H., Creutzig, F., & Seto, K. C. (2016) Future urban land expansion and implications for global croplands. *PNAS*, 114(34), 8939–8944.

Chase, A.F., & Chase, D.Z. (2016) Urbanism and anthropogenic landscapes. *Annual Review of Anthropology*, 45, 22.1–16. doi:101146/annurev-anthro-102215-095852.

Chase, A.F., & Scarborough, V.L. (eds.) (2014) *The resilience and vulnerability of ancient landscapes: Transforming Maya archaeology through IHOPE*. Hoboken, NJ: Wiley-Blackwell.

Chaudhuri, G., & Mishra, N.B. (2016) Spatio-temporal dynamics of land cover and land surface temperature in Ganges-Brahmaputra Delta: A comparative analysis between India and Bangladesh. *Applied Geography*, 68, 68–83.

Crouch, D., & Ward, C. (1997) *The allotment: Its landscape and culture*, 2nd ed. London: Faber and Faber.

Curtis, F., & Ehrenfield, D. (2012) The new geography of trade: Globalization's decline may stimulate local recovery. *Solutions*, 3(1), 35–40.

Deutsch, L. (2004) *Global trade, food production and ecosystem support: Making the interactions visible*. Stockholm: Department of Systems Ecology, Stockholm University.

Dewan, A.M., Yamaguchi, Y., & Rahman, Z. (2012) Dynamics of land use/cover changes and the analysis of landscape fragmentation in Dhaka Metropolitan, Bangladesh. *Geojournal*, 77, 315–330.

Dunning, N.P., & Beach, T. (2010) Farms and forests: Spatial and temporal perspectives on ancient Maya landscapes. In: Martini, I.P., & Chesworth, W. (eds.), *Landscapes and societies*. Berlin: Springer, pp. 369–389.

Eisenstadt, S.N. (1966) The basic characteristics of modernization. In: Eisenstadt, S.N. (ed.), *Modernization, protest and change*. Englewood Cliffs, NJ: Prentice-Hall, pp. 1–19.

Ernstson, H., van der Leeuw, S.E., Redman, C.L., Meffert, D.J., Davis, G., Alfsen, C., & Elmqvist, T. (2010) Urban transitions: On urban resilience and human-dominated ecosystems. *Ambio*, 39(8), 531–545.

Food and Agricultural Organization of the United Nations (FAO) (1996) *Rome declaration on World Food Security and world food summit plan of action*. Available from: www.fao.org/WFS/ [Accessed 1 June 2016].

Fletcher, R. (2009) Low-density, agrarian-based urbanism: A comparative view. *Institute of Advanced Studies Insights*, 2(4), 2–19.

Folke, C., Jansson, A., Larsson, J., & Costanza, R. (1997) Ecosystem appropriation by cities. *Ambio*, 26, 167–172.

Folke, C., Carpenter, S., Walker, B., Scheffer, M., Elmqvist, T., Gunderson, L., & Holling, C.S. (2004) Regime shifts, resilience, and biodiversity in ecosystem management. *Annual Review of Ecological Environment*, 35, 557–581.

Ford, A., & Nigh, R. (2015) *The Maya forest garden: Eight millennia of sustainable cultivation of the tropical woodlands*. Walnut Creek, CA: Left Coast Press.

Fraser, E.D.G., & Rimas, A. (2010) *Empires of food: Feast, famine and the rise and fall of civilizations*. New York: Random House.

Graham, E. (1999) Stone cities, green cities. In: Bacus, E.A. and Lucero, L.J. (eds.), *Complex polities in the ancient tropical world*, Archaeological Papers 9. Washington, DC: American Anthropological Association, pp. 185–194.

Haldon, J.F. (1990) *Byzantium in the seventh century*. Cambridge: Cambridge University Press.

Harvey, D. (1990) *The condition of postmodernity*. Malden, MA: Blackwell.

Heckbert, S., Isendahl, C., Gunn, J.D., Brewer, S., Scarborough, V.L., Chase, A.F., Chase, D.Z., Costanza, R., Dunning, N.P., Beach, T., Luzzadder-Beach, S., Lentz, D., & Sinclair, P. (2016) Growing the Precolumbian Maya social-ecological system from the bottom up. In: Isendahl, C., & Stump, D. (eds.), *The Oxford handbook of historical ecology and applied archaeology*. Oxford: Oxford University Press. Available from: DOI: 10.1093/oxfordhb/9780199672691.013.30.

Higgins, S.A., Overeem, I., Stecklaer, M.S., Syvitski, J.P.M., Seeber, L., & Akhter, S.H. (2014) InSAR measurements of compaction and subsidence in the Ganges-Brahmaputra Delta, Bangladesh. *Journal of Geophysical Research Earth Surface*, 119, 1768–1781.

Hornborg, A. (2001) *The power of the machine: Global inequalities of economy, technology, and environment*. Walnut Creek, CA: Altamira.

International Assessment of Agricultural Knowledge, Science, and Technology for Development (IAASTD) (2009) *Agriculture at a crossroads: Global report*. Washington, DC: Island Press.

Isendahl, C. (2012a) Investigating urban experiences, deconstructing urban essentialism. *UGEC (Urbanization & Global Environmental Change) Viewpoints*, 8, 25–28.

Isendahl, C. (2012b) Agro-urban landscapes: The example of Maya Lowland cities. *Antiquity*, 86(334), 1112–1125.

Isendahl, C., Dunning, N.P., & Sabloff, J.A. (2014) Growth and decline in classic Maya Puuc political economies. In: Chase, A.F., & Scarborough, V.L. (eds.), *The resilience and vulnerability of ancient landscapes: Transforming Maya archaeology through IHOPE*. Hoboken, NJ: Wiley-Blackwell, pp. 43–55.

Isendahl, C., & Heckbert, S. (2017) Pathways to sustainable development or poverty: Water security and wealth in the pre-Columbian Puuc-Nohkakab Maya lowlands. In: Cunningham, T., & Driessen, J. (eds.), *Crisis to collapse: The archaeology of social breakdown*. Louvain-la-Neuve: Presses Universitaires de Louvain, pp. 251–262.

Isendahl, C., Lucero, L.J., & Heckbert, S. (forthcoming) Sustaining freshwater security and community wealth: Diversity and change in the pre-Columbian Maya lowlands. In: Sulas, F., & Pikirayi, I. (eds.), *Water and Society from Ancient Times to the Present: Resilience, Decline, and Revival*. London: Routledge (in press).

Isendahl, C., & Smith, M.E. (2013) Sustainable agrarian urbanism: The low-density cities of the Mayas and Aztecs. *Cities*, 31, 132–143.

Islam, S. (2014) Metropolitan government: an option for sustainable development of Dhaka megacity. *Environment and Urbanization ASIA*, 5(1), 35–48.

Jongerden, J., Swagemakers, P., & Barthel, S. (2014) Connective storylines: A relational approach to initiatives in food provisioning and green infrastructures. *Spanish Journal of Rural Development*, 5, 7–18.

Keck, M., & Etzold, B. (2013) Resilience refused: Wasted potentials for improving food security in Dhaka. *Erdkunde*, 67(1), 75–91.

Knapp, L., Veen, E., Renting, H., Wiskerke, J.S.C., & Groot, J.C.J. (2016) Vulnerability analysis of urban agriculture projects: A case study of community and entrepreneurial gardens in the Netherlands and Switzerland. *Urban Agriculture & Regional Food Systems*, 1(1), 1–13.

Ljungkvist, J., Barthel, S., Finnveden, G., & Sörlin, S. (2010) The urban anthropocene: Lessons for sustainability from the environmental history of Constantinople. In: Sinclair, P.J.J., Herschend, F., Nordquist, G., & Isendahl, C. (eds.), *The urban mind: Cultural and environmental dynamics*, Studies in Global Archaeology 15. Uppsala: Uppsala University Press, pp. 367–390.

Maddox, D., et al. (2016) *Urban agriculture has many benefits: Is one of them a contribution to urban sustainability?* www.thenatureofcities.com/2016/06/30/urban-agriculture-has-many-benefits-is-one-of-them-a-contribution-to-urban-sustainability/ (accessed July 4 2016).

McKillop, H. (2002) *Salt: White gold of the ancient Maya*. Gainesville, FL: University Press of Florida.

McMichael, P. (2011) Food systems sustainability: Questions of environmental governance in the new world (dis)order. *Global Environmental Change*, 21(3), 804–812.

Moustier, P. (2007) Urban horticulture in Africa and Asia: An efficient corner food supplier. *ISHS Acta Horticulturae*, 762, 145–158.

Mukherjee, J. (2015) Beyond the urban: Rethinking urban ecology using Kolkata as a case study. *International Journal of Urban Sustainable Development*, 7(2), 131–146.

Newman, P., Beatley, T., & Boyer, H. (2009) *Resilient cities: Responding to peak oil and climate change*. Washington, DC: Island Press.

Orser, C.E. (2014) *A primer on modern world archaeology*. Clinton Corners, NY: Eliot Werner Publications.

Porter, J.R., Dyball, R., Dumaresq, D., Deutsch, L., & Matsuda, H. (2014) Feeding capitals: Urban food security and self-provisioning in Canberra, Copenhagen and Tokyo. *Global Food Security*, 3(1), 1–7.

Pothukuchi, K., & Kaufman, J.L. (2000) The food system: A stranger to the planning field. *JAPA (Journal of the American Planning Association)* 66(2), 113–124.

Reader, J. (2004) *Cities*. New York: Grove Press.

Rhaman, M., Khatun, M.B., Hossain, M.B., Hassan, M.N., & Nowsad, A.A.K.M. (2013) Present scenario of landing and distribution of fish in Bangladesh. *Pakistan Journal of Biological Sciences*, 16(22), 1488–1495.

Sassen, S. (1991) *The global city*. New York: Princeton University Press.

Select Committee on Environmental, Transport and Regional Affairs (Select Committee) (1998) *The future for allotments: Fifth report to the house of commons*. Available from: www.publications.parliament.uk/pa/cm199798/cmselect/cmenvtra/560/56002.htm [Accessed 1 June 2016].

Staller, J., & Carrasco, M. (eds.) (2010) *Pre-Columbian foodways: Interdisciplinary approaches to food, culture, and markets in ancient Mesoamerica*. New York: Springer.

Steffen, W., Persson, Å., Deutsch, L., Zalasiewicz, J., Williemas, M., Richardson, K., Crumley, C., Crutzen, P., Folke, C., Gordon, L., Molina, M., Ramanathan, V., Rockström, J., Scheffer, M., Schellenhuber, H.-J., & Svedin, U. (2011) The anthropocene: From global change to planetary stewardship. *Ambio*, 40, 739–761.

Tainter, J.A., & Allen, T.F.H (2015) Energy gain and the evolution of organization. In: Isendahl, C., & Stump, D. (eds.), *The Oxford handbook of historical ecology and applied archaeology*. Oxford: Oxford University Press. Available from: DOI: 10.1093/oxfordhb/9780199672691.013.31.

United Nations, Department of Economic and Social Affairs, Population Division (UN) (2014) *World urbanization prospects: The 2014 revision, highlights (ST/ESA/SER.A/352)*. New York: United Nations.

Warf, B. (2008) *Time-space compression*. New York: Routledge.

Wirth, L. (1938) Urbanism as a way of life. *The American Journal of Sociology*, 44(1), 1–24.

Wolman, A. (1965) The metabolism of cities. *Scientific American*, 213(3), 179–190.

Woolston, H.B. (1912) The urban habit of mind. *The American Journal of Sociology*, 17(5), 602–614.

Further reading

Barthel, S., & Isendahl, C. (2013) Urban gardens, agriculture, and water management: Sources of resilience for long-term food security in cities. *Ecological Economics*, 86, 224–234 (A comparative historical study arguing that planners need to re-imagine the city as a place where food can be grown).

Isendahl, C., & Stump, D. (forthcoming) *The Oxford handbook of historical ecology and applied archaeology*. Oxford: Oxford University Press (An eclectic collection of essays critically outlining how knowledge generated from archaeological research offers valuable insights for the planning sector).

Keck, M., & Etzold, B. (2013) Resilience refused: Wasted potentials for improving food security in Dhaka. *Erdkunde*, 67(1), 75–91 (A contemporary case elucidating the potentials and challenges to maintain urban food security).

Koc, M., MacRae, R., Mougeot, L.J.A., & Welsh, J. (1999) *For hunger-proof cities: Sustainable urban food systems*. Ottawa: International Development Research Centre (An early examination of urban food security including a wide variety of vantage points).

Scott, J.C. (1998) *Seeing like a state: How certain schemes to improve the human condition have failed*. New Haven, CT: Yale University Press (A classic critique of top-down governance, emphasising the need for local participation in planning).

Steel, C. (2008) *Hungry city: How food shapes our lives*. London: Chatto & Windus (A digestible critical outline of the food webs of 21st century cities).

5
Foraging

Jeremy Strong

Definitions

A useful starting point for any discussion of the subject is the distinction between foraging, agriculture and hunting. At a common-sense level such a separation may seem straightforward. Foragers gather wild food, while agriculture – by contrast – involves its cultivation and breeding. Hunting may be specifically understood as the killing of wild animals. Yet a sustained consideration of the fault lines between these practices reveals not sharp differentiation but zones of evident overlap. In his substantial contribution to the literature on the topic, *Food for Free* (orig. 1972), Richard Mabey describes how, through their selection of choice specimens of wild plants and subsequent disposal of their seeds around their settlements, our Neolithic ancestors may be said to have commenced the process of agriculture: "a sort of automatic selection would have taken place, with crops of the more fruitful plants growing naturally near habitation" (Mabey, 2004: 9). For Felipe Fernández-Armesto this type of landscape practice constitutes a significant phase in human civilisation, and one common to multiple early civilisations, with the "transition from gathering happen[ing] frequently and independently, in a variety of different environments" (Fernández-Armesto, 2001: 175). Relatedly, agricultural practices have clearly contributed to the forager's larder in contemporary settings where the offspring or relics of once-cultivated plants provide a harvest in places either adjacent to zones of cultivation or which have been abandoned as sites of production. This distinction between plants that are authentically wild and those arising – unplanned – but nonetheless from human practices of cultivation has long been understood. Writing in 1862, Henry David Thoreau celebrated those apple trees in the USA that, not seeded by man, had established themselves in wild settings, "in the midst of pines, birches, maples, and oaks". Nonetheless, he is at pains to stress that these trees, ancestors of European stock brought by settlers, are not native; "*our* wild apple is wild only like myself, perchance, who belong not to the aboriginal race here, but have strayed into the woods from the cultivated stock" (Thoreau, 1862). In this context it may be fruitful to think of the 'wild' aspect of 'wild food' in terms of a double meaning – encompassing both the inherently wild and that which has 'run wild'. I will also pick up later on that desire, so beautifully expressed by Thoreau, to stray into the woods.

The endeavour to strike firm and separate definitions of foraging and hunting has equivalent complications to effecting a clear divide between foraging and agriculture, and not merely in the nomenclatural sense that a gatherer of wild fungi might be styled a 'mushroom hunter'. Rather, the distinction between the forager as gatherer of wild plants and the hunter as one who pursues wild animals is itself fraught. (The contemporary consumption of road-kill would be another example of how animal-based foodstuffs might be encompassed in a roster of foraging practices.) Mabey, again, provides an apposite example of how these boundaries prove porous; his rationale for including shellfish in *Food for Free* being that "from a picker's eye view, they are more like plants than animals. They stay more or less in one place, and are gathered, not caught" (Mabey, 2004: 14). That his categorisation is entirely risible – if considered in terms of biological classification/taxonomy – is beside the point. Neither is the similarity of Mabey's wobbly collocation to other, historical and contemporary, classificatory ruses especially problematic for the would-be forager. Think, for example, of the gambits to circumvent medieval sumptuary laws (that whales, seals, and porpoises could all be considered 'fish' and therefore could be eaten on non-meat days (see Adamson, 2004)) or some of the more convoluted manoeuvres for keeping kosher (see Blech, 2008). Whilst both the latter reflect pragmatic solutions to accessing certain foods whilst simultaneously adhering to the letter (if not the spirit) of articles of faith, the parameters of foraging are not established in such a canonical, expressly prohibitive, fashion, though they may well be considered in terms of ethical considerations. Foragers are not, for example, expected to be vegetarians. Nor indeed does any entity exist to establish and police expectations, though the literature on foraging certainly offers 'good practice' guidelines focused on sustainability and safety. Instead, Mabey's inclusive definition of the forager's domain reflects an understanding of foraging not existing as *a priori* principles, but first and foremost as a human *practice* which may be charted.

A brief history of foraging

Considered from an anthropological perspective, foraging may be seen as an ancient activity that has shaped our evolution. It may reasonably be argued that we are all the descendants of competent foragers, in that "individuals with more efficient resource procurement or foraging strategies will have a reproductive advantage over individuals with less" (Ingold, 2000: 28). Reay Tannahill points to changes in the landscape itself, the northwards retreat of ice from around 11,000 BCE, and the expansion of areas of wild grain that followed, which began slowly to tip the balance away from foraging and towards new human practices of organised farming.

> Formerly, man had been a hunter, not a herdsman, a plant-gatherer, not a grower. But by just over two thousand years after the ice retreat, the deliberate cultivation of plants and domestication of livestock had begun, and the first villages had been established.
>
> *(Tannahill, 1984: 16–17)*

Equally, her account emphasises the ontological proximity of foraging and agricultural uses of the landscape in the earlier stages of human civilisation, pointing to the "almost imperceptible phases [by which] gathering developed into cultivating" (Tannahill, 1984: 34). Fernández-Armesto makes the same argument when he observes that many human societies did not have an easily identifiable juncture at which hunter-gathering was displaced by agriculture. Rather, he notes, the antecedent practices frequently contained the seeds (both literally and figuratively) of the patterns of landscape usage that would ultimately supplant them. "Hunters often manage the grazing of their prey by burning, fencing and driving. Gatherers normally replace seeds" (Fernández-Armesto, 2001: 183).

Tannahill identifies another important dimension of early (i.e. prehistoric) human foraging practices, and one that marks a distinction with contemporary foraging. Namely, that it was gendered work. While hunting was the domain of men – probably accompanied by boys acquiring the necessary skills – "the women and girls set out on their own special food-gathering tasks, looking for edible roots, greenstuffs, nuts and berries, and any small fauna which might present themselves" (Tannahill, 1984: 23). Richard Wrangham also identifies foraging/gathering as women's work in both ancient and (the few) remaining hunter-gatherer societies. "Among hunter-gatherers, gathering is mostly done by women and is often responsible for half the calories brought to camp" (Wrangham, 2009: 7). It is noteworthy how Tannahill's list of the foraging purview of the "prehistoric housewives" (Tannahill, 1984: 24) maps so neatly onto Mabey's sense of the forager's domain. Transcending modern culinary fault lines, for example, vegetarian or vegan, this is a grouping of food types defined entirely in terms of the relative ease with which the foods may be obtained. Ease, in this instance, not signifying any type of lazy idyll but rather a clear difference from the risks of tackling dangerous megafauna with flint hunting weapons. (For more on megafauna see Haynes, 2009, and for prehistoric hunting see Frison, 2004.) While such foodstuffs as snails, small crustaceans and shellfish may have fallen far short of the total food value offered by a large mammal – when a hunt was successful – they doubtless provided invaluable sustenance when hunts went badly, when preferred food species were absent, or indeed as growing human populations reduced and ultimately eradicated certain species.

Modern-day foraging

Foraging is now almost entirely supplanted (in terms of providing nourishment) by organised agricultural – and indeed industrial – practices for producing food. It may persist in a few of the remoter regions of the world as a significant contributor to the daily diet, frequently alongside hunting and primitive agriculture (see Schweitzer et al., 2000), but has otherwise been eclipsed. Given that the examination of those few pre-modern societies whose lifestyles have largely resisted external forces and 'development' lies mostly outside the scope of the present study, what remains is to consider the place of foraging in a world that may be said to have moved on. Broadly speaking, this may be divided into two categories: firstly, times and places where foraging has *re*-appeared of necessity; and secondly, the practice of foraging as a pastime and cultural choice in societies where purchasable food is abundant, easily accessible, and frequently cheap. Necessity, as seems proper, will be addressed first.

In the United Kingdom volumes such as *They Can't Ration These* (1940) and *Hedgerow Harvest* (1943) appeared in the context of the Second World War and food rationing. After many decades of the national diet moving away from home-produced goods (both in the sense of the individually domestic and the national) and towards an increasing reliance on foodstuffs sourced from around the Empire, a U-boat blockade and the interruption of global trade meant that Britain would have to return to something closer to self-sufficiency as a nation. In contemporary terms, it was a crisis of food security. Addressed in large part by increasing the area of land under cultivation and by centrally controlled measures to ensure the production of only the most necessary foodstuffs for human nourishment and their equitable distribution, an additional means by which the government sought to increase access to food involved encouraging a return to foraging. Recipe suggestions included dishes that substituted then-unavailable ingredients for ones that could be found locally, for example rose hip marmalade. *Hedgerow Harvest* asserts that "the ruby-red seed of the rose makes an excellent marmalade" (Ministry of Food leaflet, 1943). It may or may not have approximated to orange marmalade (fresh oranges only occasionally came through from America and were restricted to children) but could scarcely have been less popular than some of the manufactured 'substitute' foods, notably the ominous-sounding National Margarine. Rose hip marmalade, and equivalent relishes and side dishes, also represent a significant strand of continuity through to contemporary foraging practices in that even *in extremis* they rarely constituted the primary elements of sustenance (i.e. of calories consumed) but furnished instead a marginal, though nonetheless welcome, contribution to the diet. Whilst most present-day foragers looking to the hedgerows – for blackberries, or sloes for sloe gin – no longer do so because there is no alternative source (though, tragically, at the time of writing thousands of besieged and starving Syrians have been reduced to the consumption of grass and pets), they still engage in a practice that seasons, varies, and enlivens their overall gustatory experience.

Several writers have addressed the intersection of landscape knowledge and taste; the way in which a deep engagement that links food and place – as in the concept of *terroir* – may give rise to heightened appreciation, to "the types of pleasure that are only deepened by knowing" (Pollan, 2006: 11). Although such writing generally addresses cultivated foods, for example Tim Waterman's 2011 essay on the flavours of Payottenland (in Strong [ed.], 2011), the notion of profoundly grounded tastes may apply to foraged foods too. Whilst the Second World War did not lead to food shortages in the USA in the same fashion as in the UK, American writer M.F.K. Fisher nonetheless wrote with great verve in *How to Cook a Wolf* (orig. 1942) of the food practices occasioned by austerity. Describing the meals cooked by her friend Sue, who foraged for "little shy weeds" near her ramshackle California dwelling, the "soup dish full of sliced cactus leaves and lemon-berries and dried kelp" (Fisher, 2008: 255) seems finely poised between a culinary revelation or an ordeal.

Elio Vittorini's semi-autobiographical novel *Conversations in Sicily* (2003) includes a poignant sequence in which a mother and her adult son recall their circumstances – and, relatedly, how they ate – in his childhood, including the experience of hunger. "But we were well-off, in our house!" protests the son. "Yes," replies the mother. "Your father got paid at the end of every month, and so for ten days we were well-off [. . .] But when the first ten days were up [. . .] We ate snails" (Vittorini, 2003: 56). For some poor Sicilians in the first half of the twentieth century, foraging existed as a vital adjunct to buying and growing one's own food. Foodstuffs such as snails and wild chicory were gathered, not as part of a quaint ritual or desire to connect with the foods of the vernacular landscape, but because the alternative was going hungry. A related account of wild food occupying a more-than-marginal role in the family diet is offered by Joe Bageant in *Rainbow Pie* (2010), his angry lament for America's white rural underclass. Bageant describes how his "mountain tribe, meat hunters for two-and-a-half centuries in these hills, are truly people of the deer" (Bageant, 2010: 297). For Europeans, of course, generally occupying far more densely populated spaces, and with privately owned hunting grounds jealously guarded and invariably the preserve of a wealthy elite (Ortega y Gasset, 1995: 66–81) the notion of populist deer-hunting is nearly inconceivable. It remains a potent cultural trope in the USA, where it can be part of a blue-collar patrimony, as in Michael Cimino's 1978 film *The Deer Hunter*. However, it is probably more usually practiced as leisure and as masculine ritual, rather than as a significant and necessary contributor to the diet (see Gurven and Hill, 2009, for more on men and hunting).

Dorothy Hartley's *Food In England*, originally published in 1954, offers a magisterial survey of the nation's culinary practices, bringing together an extraordinary range of sources and references. Her account of times of food shortages in medieval England, and of the desperate foraging practices they occasioned, remain compelling:

> So comes the terrible time when berries are gathered for food and wild roots of the pig-nut and acorn are ground to make a miserable ash cake. Fish from the rivers, wild birds killed with sling stones, snakes, shellfish, braxy mutton – anything is food that will fill the belly.
>
> *(Hartley, 2006: 231)*

Her list is instructive. The fundamental difficulty for the forager of finding and preparing an acceptable form of carbohydrates is especially pointed. Poor ingredients – "wild roots of the pig nut and acorn" – are laboriously transformed into an explicitly disappointing product, "the miserable ash cake". By contrast, the berries gathered for food and the fish from the rivers sound positively Arcadian, though the fact of their being only infrequently present or available, owing to seasonality or the difficulty of catching them, reminds us that they cannot always be relied upon. Ponds of large, reared, carp would not have been accessible to the rural poor but were generally the preserve of monasteries – until their dissolution – and of aristocratic dwellings (Currie, 1991). The wild birds killed with slingshots are less likely to be plump game than blackbirds and thrushes; rather slender pickings. Thrushes and similarly sized birds are the principal quarry of one of literature's most arrestingly rendered foragers, the beautiful half-wild Manon of Marcel Pagnol's *Manon of the Springs* (1989). Shellfish should more accurately conjure a picture of chewy limpets knocked from the rocks than, say, a choice lobster, and the reference to snakes and braxy (i.e. diseased) mutton reminds us that foraging could be less about the discovery of pristine natural foodstuffs and rather more an affair of descending to foods that are despised, discarded, and sometimes outright dangerous. Ladurie's equivalent account of French peasants in the Vivarais reduced to eating "acorns, wild roots, bracken" (Ladurie, 1966: 198) in 1585–86 indicates how a return to foraging in periods of famine/shortage has been a universal human response.

Equally, Hartley's awareness of how tough rural reality could be in no way occludes her capacity to enjoy the beauty of the natural world and its edible wild products. Her often-quoted description of discovering the Chanterelle mushroom is a miniature master class in sensory evocation and poetic comparison at the same time as providing the reader with a solid guide to accurate identification, a reading of the landscape:

> You find them, suddenly, in the autumn woods, sometimes clustered so close they look like a torn golden shawl dropped down amongst the dead leaves and sticks. They are all the same clear, egg-yolk yellow, the stem coming up straight, and springing and spreading stiff as a tiny fountain spurting gold. The top surface is damp and glossy yellow; the underside crinkly matt yellow; and they smell faintly of apricots.
>
> *(Hartley, 2006: 301)*

The choice to forage

Hartley's paean to the Chanterelle mushroom makes a fine transition to our analysis of foraging conducted, not out of necessity, but for other reasons. A host of motivations may underlie the choice to forage, including, but not limited to: the desire to obtain some foods from outwith the industrial food complex (as charted in Michael Pollan's 2006 *The Omnivore's Dilemma*); a wish to connect with nature and seasonality, along the lines of Thoreau's straying into the woods; participation in historic, local (sometimes resurrected) food-gathering rituals that may have a communal dimension; the enthusiasm to find rare or otherwise unobtainable ingredients; and even the career choice, as, for example, in the case of the professional mushroom or truffle hunter supplying upscale restaurants. The latter case reminds us that not all foraged food is necessarily consumed by foragers and that a single type of foraged food may be the target of both amateur and professional foragers. Mary Hufford's essay on morel-gathering in Appalachia, for example, centres on the practice as a community-based affair where the foodstuff is perceived as a local patrimony, part of the "resources of the commons" (Hufford, 2006: 49) whereas Gary Alan Fine's *Morel Tales* (1998) also addresses the business of those who look to sell what they find. When the financial value of foraged foods is high, as with truffles for example, the potential arises for various complications. Over-exploitation may result in serious depletion, while large numbers of foragers may comprise a threat to ecosystems and species through disturbance and trampling (see Butler, n.d.). Conflict between landowners and visitors is not unknown, while there may also be competition and feuding between foragers accessing publicly accessible landscapes, for example national parks.

Perhaps one of the most resonant accounts, certainly one of the most mouth-watering, is Alix Kates Shulman's memoir of living, and foraging, out of a coastal shack in Maine over successive summers, *Drinking the Rain* (1995). A self-taught forager – guided by Euell Gibbons' *Stalking the Wild Asparagus* (1962) – her food choices combine the ethical thinker's desire to *know* about what she's eating with the gourmet's delight in savouring it. Such evocative names and ingredients as "juniper, bayberry, Irish moss", "goosefoot strawberries [. . .] and a fistful of yellow charlock blossoms" (Shulman, 1995: 133) tantalise us with their unfamiliarity – inevitably prompting the thought that the well-informed forager has access to a wider range of taste experiences. Unfamiliarity also entails risk, however, as many of the field guides expressly state. Jon Krakauer's non-fiction book *Into the Wild* (1996) describes – and to a certain extent surmises – the experiences of Christopher McCandless who, after dropping out of college, began to travel the USA, ending up in Alaska, where his body was eventually found. It is conjectured, on the basis of a final diary entry, that he died of poisoning from the plants he had foraged and stored. Krakauer postulates that McCandless was influenced by some of America's classic writing on wilderness and the human experience, notably Thoreau and Jack London. It is also known that he had purchased a guide to the edible plants of Alaska. In the new preface to *Food for Free* Mabey specifically decries "playing at survival", calling it "the unacceptable face of foraging" (2004: 7). Whilst 'playing' would be a harsh indictment of McCandless' motivations and tragic end, there may be observed a contemporary dimension of foraging as entertainment; gruesome foods forced down as a test of mettle and a thrill for squeamish onlookers. For many, popular television programmes in which presenters endure challenging landscapes and obtain nourishment from off-putting sources are likely to constitute their main – albeit mediated – exposure to foraging practices. TV adventurer Bear Grylls' book *Extreme Foods: What to Eat When Your Life Depends on It* (2015), with its Witchetty grubs and scorpions, represents the literary manifestation of this on-screen phenomenon, the author's very name fabulously suggestive of culinary machismo.

The risks of foraging

A recurring property of much of the literature on foraging is its intended function as a practical guide. Indeed some volumes on foraging are expressly sized as portable texts (e.g. Mabey) to accompany and assist foraging expeditions. Drawings and photographs, frequently in colour, accompany descriptions of plants and mushrooms (Squire, 2011, is especially useful in this regard), alongside accounts of the best times and places to find them. Accurate *identification* is the overarching aim, as with equivalent literature on, say, bird watching. Equally, since identification is, for the forager, a precursor to ingestion, the stakes are somewhat higher. This is especially the case where visual similarities exist between two organisms, one edible and the other not, as, for example, between the Horse Mushroom (*Agaricus arvensis*) and the poisonous Yellow-Staining Mushroom (*Agaricus xanthodermus*). The latter displays a yellowish hue when bruised or cut. Safety advice in such texts also extends beyond identification, for example the oft-repeated advice to pickers of watercress to avoid picking from stagnant or slow-moving water, to cut near the top, and (generally) to cook the resulting harvest. All these measures work to protect against the liver fluke parasite (*Fasciola hepatica*). Where foraging takes place around zones of agriculture and human habitation particular issues may arise in respect of contamination via faeces. All surface waters, for example, should be regarded as potentially harbouring the intestinal protozoa of humans and cattle. Cats have contaminated soils worldwide with *Toxoplasma gondii* and humans have, ourselves, proved a devastatingly effective vector of parasites. For more on human parasites see Mehlorn (2016). Foraging in urban areas, not addressed here, would pose some of the same risks as well as those that attend such environments.

Just as there is an identifiable literature of foraging, it may also be observed that foraging and foraged foods may be spotted, plucked, and relished amidst the wider field of literature. A frequently foraged food with a notable literary mention is wild garlic. With a heady, garlicky smell the plant was traditionally reviled by dairymen for its capacity, even in small amounts, to infuse the flavour of milk when consumed by cows. Chapter 22 of Hardy's *Tess of the D'Urbervilles* begins with dismay at Tess's workplace when a letter of complaint states that "the butter had a twang" (Hardy, 1985: 195). Tess and her colleagues scour the meadow, armed with small knives, and eventually find and excise no more than half-a-dozen shoots. Not a great deal, but enough to do harm, and – of course – a symbol within the wider narrative of a lurking taint, a concealed danger. For the forager, however, wild garlic is a welcome arrival. Cut in substantial quantities, carefully rinsed, it can be cooked like spinach in a large pot with butter, salt, and pepper. One of Britain's more unusual examples of a foraged food was eringo, a candied sweetmeat made from the root of Sea Holly (*Eryngium maritimum*). Celebrated in medieval times for its supposed aphrodisiac properties, it was preserved in sugar and orange-flower water, selling in the seventeenth and eighteenth centuries for more than four shillings a pound (Gordon, 2012: 90). With a welter of supposed culinary-medicinal uses, spanning the already-mentioned aphrodisiac through aromatic, diaphoretic, diuretic, expectorant, stimulant, and tonic, it is mentioned in a decidedly lusty context in Shakespeare's *The Merry Wives of Windsor* in which a pre-tryst Falstaff proclaims "let it . . . snow eringoes; let there come a tempest of provocation" (Act 5, Scene 5: 506). Charles Nicholl argues that Shakespeare's familiarity with plants and herbs was considerable, "a countryman's knowledge", and notes the likelihood that he lived very near a physic garden in the years he lodged on London's Silver Street (Nicholl, 2008. 62). Elsewhere, Shakespeare references another foraged coastal delicacy the gathering of which he describes in *King Lear* as a perilous cliff-face occupation: "Half-way down hangs one that gathers samphire; dreadful trade!" (Act 4, Scene 4: 966).

The dangers of foraging have rarely been exemplified more profoundly than in the 2004 Morecambe bay cockling disaster in North West England. At least 21 people, illegal immigrants from China who were criminally ill-prepared by their gang-master, were cut off by the incoming tide and drowned. Their tragedy speaks both to global forces – in that the victims had been transported on containers via Liverpool, and that the cockles they picked were likely destined for Spanish *paellas* – and to the necessary connection between foraging and local knowledge; thousands of miles from their native landscape, unfamiliar with local customs and risks, they headed out as it darkened and as the tide came in, unable to understand the warnings of local cocklers who tried to tell them to turn back. Foraging can still be a matter of life and death.

Conclusion

Notwithstanding the constraints that inevitably attend any attempt to summarise the key aspects of a complex human practice and its associated literature, it is hoped that that these pages provide a useful jumping off point for anyone seeking to study foraging and – perhaps – for those minded to forage themselves. Mindful that Anglo-European, as well as US, examples and texts have featured heavily it should be acknowledged that another author might have laid a greater stress upon the southern hemisphere, or indeed upon the global 'east' as well as the 'west'. Likewise, an alternative account might have dwelled at greater length upon those earlier phases in human history in which foraging was more central to our existence. As it stands, this entry should direct the reader to the enormous variety of fashions in which foraging has been and continues to be practiced, to its multiform implications, to the ways in which it has been represented, to the foodstuffs that it may encompass, and to the dishes and experiences that arise.

Key texts

Fearnley-Whittingstall, H. (1997) *A cook on the wild side: The indispensable guide to collecting and cooking wild food*. London: Boxtree.

Fernández-Armesto, F. (2001) *Civilizations: Culture, ambition, and the transformation of nature*. New York: The Free Press.

Fine, G. A. (1998) *Morel tales: The culture of mushrooming*. Cambridge: Harvard University Press

Gibbons, E. (1962) *Stalking the wild asparagus*. Chambersburg: Alan C Hood & Co.

Grylls, B. (2015) *Extreme food: What to eat when your life depends on it*. London: Transworld.

Hufford, M. (2006) Molly mooching on Bradley Mountain. The aesthetic ecology of appalacian morels. *Gastronomica: The Journal of Food and Culture*, 6(2), Spring. Berkeley: University of California Press, 49–56.

Mabey, R. (2004) *Food for free*. London: Harper Collins.

Pollan, M. (2006) *The Omnivore's dilemma: A natural history of four meals*. London: Penguin.

Further reading

Abram, D. (1996) *The spell of the sensuous: Perception and language in a more-than-human world*. New York: Pantheon.

Adamson, M. (2004) *Food in medieval times*. Westport, CT: Greenwood Press.

Bageant, J. (2010) *Rainbow pie: A memoir of redneck America*. London: Portobello Books.

Blech, Z. (2008) *Kosher food production*. London: Wiley-Blackwell.

Butler, D. (nd) Has foraging gone too far? *The Land*. Issue 20. www.thelandmagazine.org.uk/articles/has-foraging-gone-too-far.
Currie, C. (1991) The early history of the carp and its economic significance in England. *The Agricultural History Review*, 39(2), 97–107.
De Mauduit, V. (1940) *They can't ration these*. London: Michael Joseph (2008 Persephone Books).
Evelyn, J. (1699) *Acetaria: A discourse of Sallets* (1996 – Prospect Books, UK).
Fisher, M.F.K. (2008) *The art of eating*. New Jersey: Wiley.
Frison, G. (2004) *Survival by hunting: Prehistoric human predators and animal prey*. Berkeley, CA: University of California Press.
Gordon, D. (2012) *The little book of Essex*. Stroud: The History Press.
Gurven, M. & Hill, K. (2009) Why do men hunt? A reevaluation of 'Man the Hunter' and the sexual division of labor. *Current Anthropology*, 50(1), 51–74.
Hardy, T. (1985) *Tess of the D'Urbervilles*. London: Penguin.
Hartley, D. (2006) *Food in england*. London: Little, Brown.
Haynes, G. (ed.) (2009) *American megafaunal extinctions at the end of the Pleistocene*. New York: Springer.
Ingold, T. (2000) *The perception of the environment: Essays on livelihood, dwelling and skill*. Oxford: Routledge.
Krakauer, J. (1996) *Into the wild*. New York: Anchor Books.
Ladurie, E.L.R. (1966) *The peasants of languedoc*. Urbana, IL: University of Illinois Press.
Mehlorn, H. (2016) *Human parasites: Diagnosis, treatment, prevention*. New York: Springer.
Mennell, S. (1996) *All manners of food: Eating and taste in England and France from the middle ages to the present*. Urbana, IL: University of Illinois Press.
Ministry of Food [no author stated] (1943) *Hedgerow harvest*. Available from: www.recipespastandpresent.org.uk/wartime/?cat=92 (orig. Slough: Ministry of Food).
Nicholl, C. (2008) *The lodger: Shakespeare on silver street*. London: Penguin.
Ortega y Gasset, J. (1995) *Meditations on hunting*. Belgrade: Wilderness Adventures Press.
Pagnol, M. (1989) *The water of the hills: Jean de Florette & Manon of the springs*. London: Picador.
Schweitzer, P., Biesele, M., & Hitchcock, R. (eds.) (2000) *Hunters and gatherers in the modern world: Conflict, resistance and self-determination*. New York: Berghahn.
Shakespeare, W. (1990) *The complete works: Compact edition*. Oxford: Clarendon Press.
Shulman, A.K. (1995) *Drinking the rain*. New York: Farrar, Strauss and Giroux.
Squire, D. (2011) *Self sufficiency: Foraging*. London: New Holland.
Strong, J. (2011) *Educated tastes: Food, drink, and connoisseur culture*. Lincoln: University of Nebraska Press.
Tannahill, R. (1984) *Food in history*. New York: Stein and Day.
Thoreau, H. (1862) *Wild apples* www.theatlantic.com/past/docs/issues/1862nov/186211thoreau2.htm.
Vittorini, E. (2003) *Conversations in sicily: A novel*. Edinburgh: Canongate.
White, F. (1934) *Flowers as food, receipts and lore from many sources*. London: Jonathan Cape.
Wrangham, R. (2009) *Catching fire: How cooking made us human*. London: Profile Books.

6

Venison from the Bavarian forests

Linking hunters, forest diversity, and consumers through regional marketing

Gerd Lupp, Simon Tangerding, and Valerie Kantelberg

Introduction

Hunting game is an important part of landscape management practices in Bavaria. Hunting links and interacts with forest diversity, regional food, and living culture, and is reflected in traditional clothing. Game meat is considered to be of higher quality than that from farm animals. With only a very small amount of the meat market and a large share of venison provided by leisure-time hunters preferring to sell to traders of game, there are potentials for better marketing, especially on a local and regional level. New products including value-adding and creative offers such as the landscape itself can better link hunters, forest owners, farmers, and gastronomy, and thereby create new sources of income, especially by targeting quality-oriented consumers demanding (more) transparency and direct contact with venison producers.

Game and hunting

Game such as roe deer (*Capreolus capreolus*), red deer (*Cervus elaphus*), wild boar (*Sus scrofa*) and chamois (*Rupicapra rupicapra*) inhabit many Central European and Bavarian woodlands (Figure 6.1) and, not surprisingly, play an important role in food and culture in Bavaria, Germany.

During the main hunting season in autumn, various restaurants and butchers offer venison dishes and delicacies. Moreover, there are several festivals connected with forests, game, venison, and hunting such as the annual 'venison weeks' in the Neustadt–Bad Windsheim area which take place in November (www.wildbretwochen.de). The traditional 'lederhosen', previously the outdoor wear for hunters until functional synthetic garments became popular in the 1980s, are preferably made of skins from red deer. Their embroidery (Figure 6.2) often shows stags or chamois and oak leaves (Köcheler, 1991). Venison is considered of higher quality compared to red meats from farm animals by people who are conscious of healthy eating (Köttl et al., 2014). The most important characteristics are the typical taste and a lower fat content compared to beef, pork, or lamb. The nutritional value is considered high due to its high protein content, high amounts of minerals, vitamins, trace elements, and unsaturated fatty acids (Deutz, 2012). Furthermore, venison does not contain traces of antibiotics and growth hormones, which are commonly found in most other red meats (Deutz and Deutz, 2008). Moreover, animal welfare is considered a physical and mental state not only related to the absence of negative emotions but also associated with positive feelings for sentient beings (Fraser, 1989; Duncan, 1996). There is an increasing awareness for the well-being of animals in developed

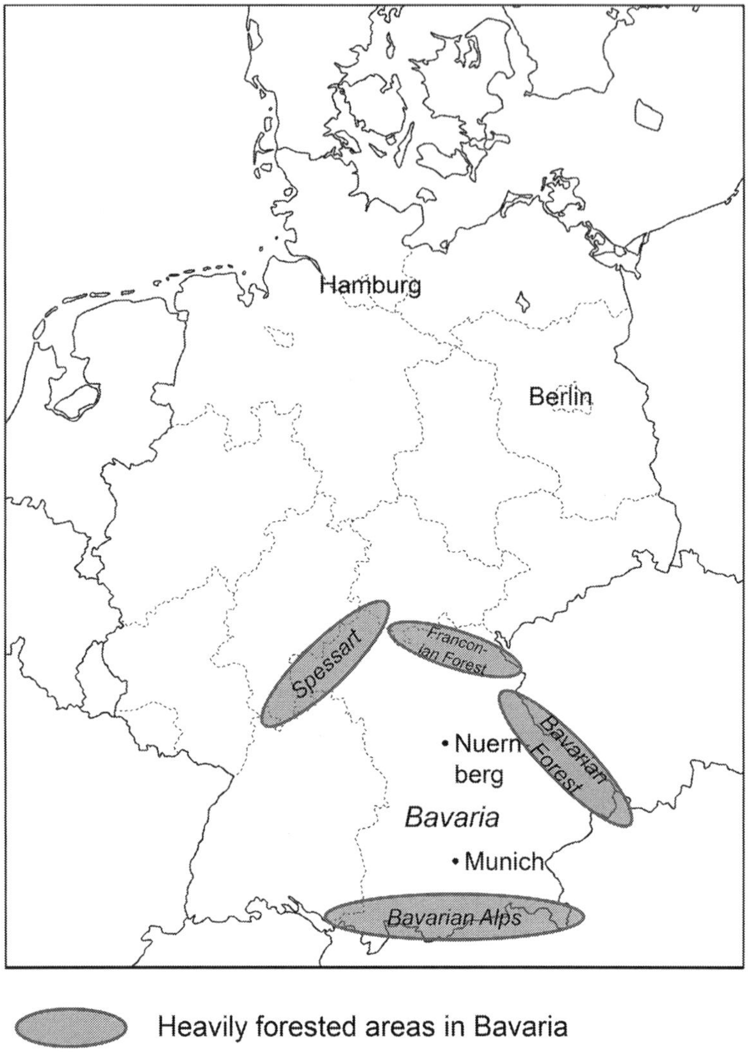

Figure 6.1 Map of Germany and Bavaria showing heavily forested areas; smaller woodlands can be found throughout Bavaria. (2,600,000 hectares or 36.9% of Bavaria is covered by woodlands.)

countries, foremost involving the avoidance of animal suffering (Bradshaw and Bateson, 2000; Boogaaard et al., 2006). Compared to farm animals, fewer concerns and issues regarding these topics arise. Game animals live in their species-appropriate natural environment. Although there is a discussion about the most appropriate hunting practices and about the best way to ensure immediate lethality (e.g. Bradshaw and Bateson, 2000; Knott et al., 2009), using good hunting practices such as a 'heart and lung shot' (as recommended by the British Association for Shooting and Conservation), the animal is killed immediately without suffering. Also intensively discussed concerns related to transport and slaughter procedures in meats (Terlouw et al., 2008) are thereby avoided. For hunters it is important to observe hygienic rules to maintain meat quality. In the forest, it is essential to quickly eviscerate the animal after shooting and to ensure quick transportation to a cooling facility. Thereafter, strict hygienic rules have to be observed in an indoor environment when the meat is being prepared for human consumption (Paulsen et al., 2011; Laaksonen and Paulsen, 2015).

Figure 6.2 Embroidery of chamois and oak leaves.
Source: Lupp

Hunting and forest ecology

Human intervention in modified forests and novel ecosystems is a crucial management practice and can provide measurable improvements in terms of ecological outcomes (Martin and Baltzinger, 2002; Hothorn and Müller, 2010). After the extinction of almost all large predators such as the wolf (*Canis lupus*) in Central Europe, the larger herbivores such as red deer have no natural enemies. Their populations would, therefore, increase to a maximum biophysical carrying capacity (Daily and Ehrlich, 1992). Even in lower numbers, these ungulates can have detrimental impacts on forest ecosystems and their ability to provide goods and services (Mayer and Ott, 1991; Stewart, 2001; Palmer and Truscott, 2003).

The effect of mechanisms connected with hunting intensity, and moreover, the effects on biodiversity and tree species composition in mature forest stands have been intensively discussed in the literature (see, for example, Akashi and Nakashizuka, 1999; Kienast et al., 1999; Pellerin et al., 2010; White, 2012). Tree diversity, growth, and quality loss of timber in forest stands can be directly correlated with the number of game (Akashi and Nakashizuka, 1999; Bradshaw et al., 2003; Didion et al., 2009). In forests, roe and red deer as well as chamois prefer to browse on young broadleaves such as beech (*Fagus sylvatica*), sycamore (*Acer pseudoplatanus*), ash (*Fraxinus excelsior*), other rare hardwoods, and white fir (*Abies alba*) (Figure 6.3). In the past thirty years, forest management in Central Europe has strived to develop more resilient forests of richly structured natural mixed stands with a broad variety of broadleaves and white fir, shifting away from Norway spruce (*Picea abies*) and Scots pine (*Pinus sylvestris*) monocultures (Burschel and Huss, 2003), which will also be the trees most affected by climate change (Lindner et al., 2010). Although there is a lack of long-term monitoring data, managing the game population by hunting can play a role to support the development of these mixed structured forests, including that of rare tree species (Mayer and Ott, 1991; Hothorn and Müller, 2010). Particularly in alpine regions, the protective function allocated to mountain

Figure 6.3 White fir tree with low vitality damaged by browsing. This tree would normally be more than 5 metres tall.

Source: Dimke

forests concerning infrastructure or even inhabited areas cannot be maintained if game populations are too high (Mayer and Ott, 1991). Trees are also damaged by stags that use tree trunks to remove the velvet from their antlers before the mating season starts (Amann, 2006).

During the past thirty years, the wild boar population has grown rapidly due to improved living conditions provided by the intensification of agriculture, mainly by the increased cultivation of maize (Petrak, 2009). Wild boars are, in principle, considered to be, above all, forest inhabitants, but especially during spring and autumn with lush plant growth on agricultural land, these omnivores can cause severe damage foremost to maize and grains but also to rapeseed (canola), or even to meadows and pastures. However, the concept of 'damage' depends on the targets and objectives set by interest groups, and there is not always a conflict between wildlife and resources (Reimoser, 2003). In order to reduce damages and economic losses for farmers, young and middle-aged boars may be hunted throughout the year, and not only in autumn (Hahn, 2014).

Hunters in Germany and Bavaria

Predominantly (e.g. 90% of wild boar), game in Germany is hunted by private and leisure-time hunters (The Bavarian Hunting Association, pers. comm.). Apart from a small number of professional hunters, for example, those managing game populations in state-owned forests or large private forests, the vast majority of them are informally called known as 'hobby hunters'. In 2015, in Germany around 369,000 persons held a hunting license (0.4% of the total population), a total of 49,000 in the Federal Free State of Bavaria alone (DJV, 2016a). The hunting bag in Bavaria during the 2015–16 season was around 313,000 roe deer, 85,000 wild boars, 12,000 red deer, and 4,000 chamois (BJV, 2017). In the following section the two most popular leasing models are described.

Hunters can lease from large forest owners and gain the right to hunt in a specific area. These leasing contracts are valid for nine or twelve years and apply to a single land holding or a forest tract outside built-up areas, usually larger than 81 hectares in the lowlands and 300 hectares in high mountains. Venison is part of the contract and belongs to the lessee. The cost of these leasing contracts varies between a few euros and more than 100 euros per hectare. In the Federal Free State of Bavaria around 15% of the public forest is leased to hunters (BaySF, 2016a). The other way to get a lease is to acquire a so-called deerstalking hunting license. Such a license is usually valid for one year only, and the hunters have the right to shoot a certain number of each game species. Venison is not part of the leasing contract, but the shooter has the general right of pre-emption. The cost of these one-year contracts varies, ranging from between 250 to 700 euros. For example, the Bavarian State forest sells that type of license to around 4,400 persons each year (BaySF, 2016a). Moreover, some hunters just buy a license for shooting a single animal. Amount, gender, and age group of the game to be shot is independent of the contract, predetermined by the public hunting authority and subject to the strict protection of the German Protection of the Animal Acts (TierSchG) and the Bavarian hunting law (BayJG), respectively.

Challenges for regional marketing of venison to consumers

Altogether venison accounts for around 1% of the total meat consumption in Germany (StMELF, 2012). Around 40% of the venison sold in Germany is imported from abroad, mainly to provide convenience products in the supermarkets (StMELF, 2012). Local game products, therefore, can be considered a niche market for an acquired taste (see Chapter 34). Its consumption, however, remains stable compared to high, but steadily decreasing rates for other kinds of meat in Germany (Statista, 2016). Fresh game meat is mainly available during the main hunting season from the beginning of May until the end of February in Bavaria. Nonetheless, venison offers various possibilities in the value chain for manufacturing, and further processing such as sausage.

As a result of various food scandals at the turn of the 21st century, an increasing number of customers demand transparency, freshness, and good production standards for meat (Dehn, 2006; Hoffman and Wiklund, 2006). In Germany, customers often perceive products with a regional denomination to be of higher quality compared with those products without such labels and from abroad. According to several surveys, the main arguments for consumers favouring regional products are the following: 'short transportation distances', support of 'local producers', and 'freshness' (DLG, 2011; Roosen, 2013). A regional product (see Chapters 35 and 36) is perceived as originating from the countryside surrounding a town or city, and, to a lesser extent, from the federal state where interviewees from these surveys live (DLG, 2011). Many customer groups also state the importance of animal welfare (Schröder et al., 2005; Banik et al., 2007; BMEL, 2014). Venison can fulfil these societal demands in several ways and offers various possibilities in the value chain for manufacturing and further processing. Some customers, however, still fear radioactive radiation doses due to the Chernobyl disaster, although even today, meat has to be tested and withdrawn from the market when radiation exceeds legal limits, although the doses attained from normal venison consumption are considered rather low compared to radiation from natural and medical sources (e.g. Henrikson and Salsebøl, 1988), but there should be a critical dialogue on maximum allowable radiation levels in provisioning ecosystem services such as venison and the impacts of such safe appropriation thresholds on human well-being (Gralla et al., 2014). Nonetheless, venison is increasingly perceived in a more positive way and is becoming more and more popular among Germans (Köttl et al., 2014).

Although the awareness for food quality and animal welfare is high, and therefore, the potential for the regional marketing of venison is increasing steadily, consumer assessments such as the Nestlé studies (see e.g. Nestlé, 2011), which include the willingness to pay more money for regional foods and the analysis of shopping behaviour, present more differentiated conclusions. Still, the majority of consumer choice is based on low prices and the convenience of 'one-stop' shopping opportunities (ibid.). Only a smaller consumer group called 'quality eaters' in the Nestlé studies spends much time and money on conscious nutrition, and visit farmers markets or specialty shops to buy high quality regional and/or organic food (ibid.). The motivating factors for this are premium demand, interest in health and nutrition, an ecologically conscious lifestyle, and also seeking the warmth of a happy family life (ibid.).

Challenges for hunters interested in marketing venison

For consumers, there is often no information available about when forest offices or hunters nearby sell venison. Venison is often marketed as a whole animal or in a few large chunks rather than in smaller pieces such as units prepared ready for cooking. Also, consumer knowledge about the wide range of options for preparing and cooking venison is limited and decreasing (Köttl et al., 2014). Therefore, the availability of regional venison meeting the needs and demands of consumers is often very limited.

In the hunting season 2014–15, hunters in Germany marketed more than 20,000 tons of venison (1% of the German meat consumption), mainly roe and red deer as well as wild boar. The amount of venison changes significantly each year due to weather conditions, but the overall hunting bag has increased. The primary value of venison paid to the hunters was around 150 million euros (DJV, 2016b). According to the TEEB Germany study for forests (TEEB DE, 2016), this value could increase significantly thanks to resale and processing.

Most amateur hunters do not have an economic need to market the meat more profitably, while for many forest owners, hunting is primarily perceived as a necessity to support silviculture and nature conservation objectives. However, the sale of venison contributes only about 1–8% in comparison to revenues gained from timber production (Borchers et al., 2008). As a result of many strict regulations and laws concerning hygiene, safety, and handling of food, selling the surplus from private consumption to traders of game is considered to be the easiest solution, although this strategy often provides the least profit compared with other options.

Marketing initiatives for venison in Bavaria

Despite the unfavourable starting conditions, a number of initiatives have begun to tackle this challenge intending to bridge the gap between consumer demands and the forest owners and hunters. Encouraged by the efforts of the public forest offices and private forest owners to improve the marketing of venison and to react to consumer demands, several projects have been launched. Some examples are described here. Also, from an economic point of view, providing products for private consumers increase the value of the products, which thus leads to better revenues (Borchers et al., 2008). According to Borchers et al. (2008), selling convenience products may increase the revenue from an average of 1.28 euros per kilo (e.g. selling it to a game trader) to 3.74 euros by offering large pieces to regional restaurants, and to 4.25 euros for convenience products such as exactly portioned vacuum-packed fillets or goulash. Cold cuts or sausages provide revenues of only 1.40 euros due to the larger investments required (such as meat grinders or a smokehouse). By marketing and selling these processed products, more parts of the animal can be used and marketed (Deutz and Deutz, 2008). With the help of value-added products such as grilling sausage or portioned marinated meat for barbeques, hunters have new opportunities for marketing venison in summer, a time of year when there is low demand for traditional game dishes such as roasts.

Direct contact with customers is considered advantageous for raising awareness, providing opportunities such as product tastings to convince critical customers, for proving authenticity and for enhancing customer loyalty. An example for a deeper commitment to providing venison for end-customers is the "Feines Wildbret aus dem bayerischen Staatswald" (Delicious Venison from the Bavarian State Forest), an initiative set up by six Bavarian State Forest Offices; Ebrach, Fichtelberg, Forchheim, Kelheim, Pegnitz, Wasserburg, and Weissenhorn (BaySF, 2016b). Investments have been made in recent years to provide a game collection centre with facilities for processing the meat. These offices now provide a salesroom and options to buy vacuum-packed convenience products. Thanks to large forest areas and reliable supply chains, a variety of products now exist for potential customers. The Fichtelberg Forest Office has also published two books on cooking and grilling with venison (Figure 6.4). In these books, foresters and hunters present recipes of their favourite venison dishes (Figure 6.5). Some of these recipes are also available online (BaySF, 2016b). Considered a success story, this concept will be put into practice by almost all forest offices of the Bavarian State Forest over the next few years.

In the Federal Free State of Bavaria, there is a political goal aiming to strengthen rural areas, including support for farmers and forest owners. The idea is to initiate employment in rural areas by encouraging the diversification of products and by adding value chains such as agricultural tourism, farm shops, and the gastronomy sector. One part of the program is to enhance education and learning for both schoolchildren and adults, raising awareness for farming, forestry, and food production as well as providing a better understanding of ecological issues (StMELF, 2010). Indirectly, this program also intends to encourage customers to buy regional products from Bavaria. An example of an extended value-adding chain is demonstrated by the Hubertushof farm, named after the patron saint of hunters. It provides a broad range of educational offers related to venison in combination with either practical hands-on experience such as cookery courses or adventure (such as guided hikes to the local forest providing information on the ecology of various game species and hunting. See www.erleben-auf-dem-hubertushof.de). With European funding schemes such as 'LEADER' (a French acronym meaning Links between Actions for the Development of the Rural Economy), rural development projects are supported at the local level in order to revitalise rural areas and

Figure 6.4 Venison cookbooks written by foresters and hunters of the Fichtelberg Forest Office.

Source: Hertel

Figure 6.5 Recipe for a wild boar sausage burger.

Source: Hertel

to create jobs. Promoting new ideas, cooperation, and networking are the core issues of such programs. Local groups manage these projects. In Bavaria, between 2014 and 2020, around 86% of the Federal Free State is eligible for funding by LEADER. An idea for LEADER in the Hubertushof region is to establish a network of hunters, authorities, and associations and to found a non-profit private limited company for a venison market centre, not only with all necessary facilities but also with teaching rooms. The centre is also a reaction to the changing demography of hunters and younger persons who nowadays live more and more in towns and cities without appropriate facilities at their homes to prepare and process the meat.

Concluding remarks

Although numerous aspects have to be taken into account and may be time-consuming, regional marketing of venison may be considered worthwhile. It helps to connect forest owners, hunters, and customers and can create a better understanding of forest ecosystems and hunting as a means to enhance forest health and biodiversity as well as provisioning valuable and healthy meat. However, regional venison will always remain a niche product due to its limited quantity and seasonal availability. Nonetheless, regional products and an awareness for healthy and well-produced foods are part of trendsetting and an increasingly holistic-conscious lifestyle. In the case of venison, due to its 'Slow Food' characteristic, a steady, growing market potential is foreseeable. Strategies should, therefore, strive to better address the expectations of these quality-oriented consumers (Kleinhückelkotten et al., 2006; Murin et al., 2015). Modern marketing approaches such as better information on availability, the setting up of salesrooms or market stalls at organic or farmers markets provide valuable support for the successful marketing of venison. Recent trends indicate (Nestlé, 2016) that dining out for conviviality is of growing importance. With creative ideas embedding game, forest, and venison into experience and adventure-oriented offers, the customer base can be broadened and expanded, and at the same time this can lead to increasing added value and moreover take into account the educational aspects for demanding consumers.

Acknowledgements

We would like to thank Martin Hertel, Bavarian State Forest office Fichtelberg, and Paul Dimke, Bavarian State Forest Institute, for providing the pictures and information; Barbara Kuhn, Hubertushof for providing her example; and Elizabeth Hamzi-Schmidt for polishing the language.

Further reading

Akashi, N., & Nakashizuka, T. (1999) Effects of bark-stripping by sika deer (Cervus nippon) on population dynamics of a mixed forest in Japan. *Forest Ecology and Management*, 113, 75–85.

Amann, G. (2006) *Säugetiere und Kaltblütler des Waldes*. Neumann-Neudamm, Melsungen; 336 p.

Banik, I., Simon, J., & Hartmann, M. (2007) *Regionale Herkunft als Erfolgsfaktor für die Qualitätswahrnehmung von Verbrauchern in Nordrhein-Westfalen*. Bonn: Landwirtschaftliche Fakultät der Universität Bonn. Schriftreihe des Lehr- und Forschungsschwerpunktes USL, 152 p.

BaySF – Bayerische Staatsforsten (2016a) *Jagdmöglichkeiten*. www.baysf.de/de/wald-bewirtschaften/jagd/jagdangebote.html (accessed 10 October 2016).

BaySF – Bayerische Staatsforsten (2016b) *Wildes Kochen*. www.baysf.de/de/wald-bewirtschaften/jagd/wildbret.html (accessed 10 October 2016).

BJV – Bavarian Hunting Association (2017) *Jagdstrecken*. www.jagd-bayern.de/index.php?id=544&type=123 (accessed 10 October 2016).

BMEL – Bundesministerium für Ernährung, Landwirtschaft und Verbraucherschutz (2014) *Einkaufs- und Ernährungsverhalten in Deutschland*. TNS-Emnid-Umfrage des BMEL.

Boogaard, B.K., Oosting, S.J., & Bock, B.B. (2006): Elements of societal perception of farm animal welfare: a quantitative study in The Netherlands. *Livestock Science*, 104(1–2), 13–22.

Borchers, J., Gommlich, M., & Wegmann, C. (2008) Wildbret – Wertschöpfungsfaktor im Erwerbsforstbetrieb? *AFZ-Der Wald*, 21, 1159–1161.

Bradshaw, E.L., & Bateson, P. (2000) Animal welfare and wildlife conservation. In: Gosling, L.M., & Sutherland, W.J. (eds.), *Behaviour and conservation*. Cambridge: Cambridge University Press, pp. 330–348.

Bradshaw, R.H.W., Hannon, G.E., & Lister, A.M. (2003) A long-term perspective on ungulate-vegetation interactions. *Forest Ecology and Management*, 181(1–2), 267–280.

Burschel, P., & Huss, J. (2003) *Grundriss des Waldbaus. Ein Leitfaden für Studium und Praxis*. Ulmer: Stuttgart, p. 487.

Daily, G.C., & Ehrlich, P.R. (1992) Population, sustainability, and Earth's carrying capacity. *BioScience*, 42(10), November, 761–771.

Dehn, S. (2006) *Wildbret erfolgreich vermarkten – Ein Ratgeber für Jäger*. Stuttgart: Kosmos, p. 114.

Deutz, A. (2012) *Wildbrethygiene heute – Beurteilung, Versorgung, Rechtslage*. München: BLV, p. 159.

Deutz, A., & Deutz, U. (2008) *Das Wildbret: Vom Aufbrechen bis zur Zubereitung*. Graz, Stuttgart: Leopold Stocker Verlag, p. 144.

Didion, M., Kupferschmid, A.D., & Bugmann, H. (2009) Long-term effects of ungulate browsing on forest composition and structure. *Forest Ecology and Management*, 258(S1), 44–55.

DJV – Deutscher Jagdverband (2016a) *Jagdscheininhaber in der Bundesrepublik Deutschland*. www.jagdverband.de/node/707jagdverband.de/sites/default/files/4811_A5_stat_05_jagdscheininhaber.pdf (accessed 10 October 2016).

DJV – Deutscher Jagdverband (2016b) *Wildbretaufkommen*. www.jagdverband.de/daten-und-fakten/jagdstatistik/wildbretstatistik (accessed 10 October 2016).

DLG – Deutsche Landwirtschafts-Gesellschaft (2011) *Regionalität aus Verbrauchersicht*. Frankfurt/Main: DLG-Studie.

Duncan, I.J.H. (1996) Animal welfare in terms of feelings. *Acta Agriculture Scandinavia*, 27, 29–35.

Fraser, A.F. (1989) Animal welfare practice: Primary factors and objectives. *Applied Animal Behaviour Science*, 22, 159–176.

Gralla, F., Abson, D., Møller, A.P., Lang, D.J., & von Wehrden, H. (2014) The impact of nuclear accidents on provisioning ecosystem services. *Ecological Indicators*, 41, 1–14.

Hahn, N. (2014) *Brennpunkt Schwarzwild – Projekt zur Entwicklung innovativer regionaler Konzepte*. Abschlussbericht erstellt im Auftrag der LWF, p. 240.

Henrikson, T., & Salsebøl, G. (1988) Fallout and radiation doses in Norway after the Chernobyl accident. *Environment International*, 14, 157–163.

Hoffman, L.C., & Wiklund, E. (2006) Game and venison – meat for the modern consumer. *Meat Science*, 74, 197–208.

Hothorn, T., & Müller, J. (2010) Large-scale reduction of ungulate browsing by managed sport hunting. *Forest Ecology and Management*, 260(9), 1416–1423.

Kienast, F., Fritschi, J., Bissegger, M., & Abderhalden, W. (1999) Modeling successional patterns of high-elevation forests under changing herbivore pressure responses at the landscape level. *Forest Ecology and Management*, 120, 35–46.

Kleinhückelkotten, S., Wippermann, C., Behrendt, D., Fiedrich, G., Schürzer de Magalhaes, I., Klär, K., & Wippermann, K. (2006) *Kommunikation zur Agro-Biodiversität. Voraussetzungen für und Anforderungen an eine integrierte Kommunikationsstrategie zu biologischer Vielfalt und genetischen Ressourcen in der Land-, Forst-, Fischerei- und Ernährungswirtschaft (einschließlich Gartenbau)*. Hannover/Heidelberg: ECOLOG-Institut/Sinus Sociovision, 189 p.

Knott, J., Gilbert, J., Green, R.E., & Hoccom, D.G. (2009) Comparison of the lethality of lead and copper bullets in deer control operations to reduce incidental lead poisoning; field trials in England and Scotland. *Conservation Evidence*, 6, 71–78.

Köcheler, A. (1991) *Die Gebirgstrachten im oberen Allgäu – Ein Streifzug durch die Arbeiten früherer Geschichtsschreiber, Maler, Graphiker und Photographen*. Sankt Ottilien: EOS Verlag, 152 p.

Köttl, B., Kantelberg, V., Seifert, H., & Friedmann, T. (2014) Wildbret regional vermarkten – Was Verbraucher wollen und worauf Erleger von Wild achten sollten. *LWF aktuell*, 103, 48–50.

Laaksonen, S., & Paulsen, P. (2015) *Hunting hygiene*. Wageningen: Wageningen Academic Publishers, 304 p.

Lindner, M., Maroschek, M., Netherer, S., Kremer, A., d, Barbati, A., Garcia-Gonzalo, J., Seidl, R., Delzon, S., Corona, P., Kolström, M., Lexer, M.J., & Marchetti, M. (2010) Climate change impacts, adaptive capacity, and vulnerability of European forest ecosystems. *Forest Ecology and Management*, 259, 698–709.

Martin, J.L., & Baltzinger, C. (2002) Interaction among deer browsing, hunting, and tree regeneration. *Canadian Journal of Forest Research*, 32(7), 1254–1264.

Mayer, H., & Ott, E. (1991) *Gebirgswaldbau. Schutzwaldpflege. Ein waldbaulicher Beitrag zur Landschaftsökologie und zum Umweltschutz*. München: Elsevier, 587 p.

Murin, I., Marková, I., Zelený, J., & Jaďuďová, J. (2015) Green marketing as a tool influencing consumers´ behavior: Slovak case study of regional mark preference. *Procedia Economics and Finance*, 34, 260–267.

Nestlé (2011) *Die Nestlé Studie 2011 – So is(s)t Deutschland*. www.nestle.de/verantwortung/nestle-studie/2011 (accessed 10 October 2016).

Nestlé (2016) *Die Nestlé Studie 2016 – So is(s)t Deutschland*. www.nestle.de/verantwortung/nestle-studie/2016 (accessed 10 October 2016).

Palmer, S., & Truscott, A.M. (2003) Browsing by deer on naturally regenerating Scots Pine (Pinus sylvestris L.) and its effect on sapling growth. *Forest Ecology and Management*, 182(1–3), 31–47.

Paulsen, P., Bauer, A., Vodnansky, M., Winkelmayer, R., & Smulders, F.J.M. (2011) *Game meat hygiene in focus: Microbiology, epidemiology, risk analysis and quality assurance*. Wageningen: Wageningen Academic Publishers, 352p.

Pellerin, M., Saïda, S., Richard, E., Jean-Luc Hamann, J.-L., Dubois-Coli, C., & Hum, P. (2010) Impact of deer on temperate forest vegetation and woody debris as protection of forest regeneration against browsing. *Forest Ecology and Management*, 260, 429–437.

Petrak, M. (2009) Schwarzwild – Gewinner in der Kulturlandschaft: Wildbiologische Grundlagen und Empfehlungen für die Praxis von Jagd und Naturschutz, Forstwirtschaft und Landwirtschaft. *Säugetierkundliche Informationen*, 7, 255–267.

Reimoser, F. (2003) Steering the impacts of ungulates on temperate forests. *Journal for Nature Conservation*, 10, 243–252.

Roosen, J., Köttl, B., & Hasselbach, J. (2013) *Regional und Bio: Kaufmotive und Kaufverhalten bei bayerischen Ökoprodukten*. https:// www.mcr.wi.tum.de/fileadmin/w00bet/www/Pictures/pdf/Projekte/Bericht_Regional_und_Bio_2016.pdf (accessed 10 October 2016).

Schröder, C., Burchardi, H., & Thiele, H. (2005) Zahlungsbereitschaft für Frischmilch aus der Region: Ergebnisse einer Kontingenten Bewertung und einer experimentellen Untersuchung. *Agrarwirtschaft*, 54(5), 244–257.

Statista (2016) *Pro-Kopf-Konsum von Fleisch in Deutschland nach Art in den Jahren 2014 und 2015 (in Kilogramm)*. https:// de.statista.com/statistik/daten/studie/311479/umfrage/pro-kopf-konsum-von-fleisch-in-deutschland-nach-arten/ (accessed 10 October 2016).

Stewart, A.J.A. (2001) The impact of deer on lowland woodland invertebrates: a review of the evidence and priorities for future research. *Forestry*, 74(3), 259–270.

StMELF – Bayerisches Staatsministerium für Ernährung, Landwirtschaft und Forsten (2010) *Professionelle Leistungen vom Bauernhof – Perspektiven für den ländlichen Raum*. www.stmelf.bayern.de/mam/cms01/allgemein/publikationen/a3_perspektiven_laendlicher_raum.pdf (accessed 10 October 2016).

StMELF – Bayerisches Staatsministeriumfür Ernährung, Landwirtschaft und Forsten (2012) *Schafe, Ziegen, Gehegewild, Geflügel, Bienen – Fakten und Daten aus Bayern* 2012, Nr. 47, München.

TEEB DE – Naturkapital Deutschland (2016) *Ökosystemleistungen in ländlichen Räumen – Grundlage für menschliches Wohlergehen und nachhaltige wirtschaftliche Entwicklung*. Hrsg. von Christina von Haaren und Christian Albert. Leibniz Universität Hannover, Helmholtz-Zentrum für Umweltforschung – UFZ. Hannover, Leipzig.

Terlouw, E.M.C., Arnould, C., Auperin, B., Berri, C., Le Bihan-Duval, E., Deiss, V., Lefevre, F., Lensink, B.J., & Mounier, L. (2008) Pre-slaughter conditions, animal stress and welfare: current status and possible future research. *Animal*, 2(10), 1501–1517.

White, M.A. (2012) Long-term effects of deer browsing: Composition, structure and productivity in a north eastern Minnesota old-growth forest. *Forest Ecology and Management*, 269, 222–228.

Cited laws and regulations

BayJG – Bavarian Hunting Right. www.gesetze-bayern.de/Content/Document/BayJG (accessed 10 October 2016).

TierSchG – German Federal Protection of the Animal Acts. www.gesetze-im-internet.de/tierschg/BJNR012770972.html#BJNR012770972BJNG000103377 (accessed 10 October 2016).

7

Sustaining Russian Old Believers

Landscapes of fish and onions in Estonia

Helen Soováli-Sepping, Anu Printsmann, and Hannes Palang

Russian Old Believers in Estonia

Russian Old Believers, also called Starovery [староверы, старообрядцы], translated literally as old believers or old ritualists or raskolniks [раскольники] as Schismatics, are a cultural minority in Estonia, with a current population of 2,600 according to the 2011 census, who, in the end of the 17th and beginning of the 18th century, migrated to the western coast of Lake Peipus from Russia (see Figure 7.1). The first written sources referring to Old Believers moving to Estonian territory go back to the end of the 17th century, when Estonia was under Swedish rule. Many Russian Orthodox Christians did not respond well to the changes in ecclesiastical life that took place under the rule of the Orthodox patriarch Nikon and was supported by the Russian Czar Alexei. Those opposing the reforms, the community of Old Believers, evaded persecution by fleeing to the edges of the Russian Empire. Later, this cultural group moved to different parts of the world (Plaat, 2016). The Old Believers who escaped to Estonian territory founded street-villages densely packed with houses on both sides of the street, and inhabited a strip of then-unused lakeshore along Lake Peipus (Kõivupuu, 2012). This uninhabited lake strip was an excellent geographical location for fishing communities to settle.

Over the centuries and through a range of different reigns and occupations, Old Believers have faced repression in Estonian territory as well, due to their position as an isolated cultural group with a number of distinct traditions, their certain conservatism towards changes in the society, and their disregard for regulations from state authorities. The 20th century left a significant mark on Old Believers' community life when, during the German Occupation (1941–1944), Old Believers were sent to concentration camps in Germany and mobilised in the German army, and subsequently, found themselves under Soviet occupation (1944–1991), when *all* religious groups were persecuted (Plaat, 2016). Despite young people leaving traditional settlements for towns and the secularising processes that have taken place over last 60 years, this religious group has nonetheless remained in existence until today largely due to their customary laws and daily activities of fishing and growing vegetables. In addition to their excellent knowledge of fishing and gardening, Old Believers are also experts in the forest. Mushrooms, for example, have an important place in Old Believers' seasonal diet because they are used to replacing meat in certain periods of the religious cycle when, according to the church calendar, meat is not allowed. This, together with foraging for nuts and berries, provides an understanding of how religious faith incorporates good ecological management into the community's teachings. Anderson (2013) proposes that religion gives communities the means to

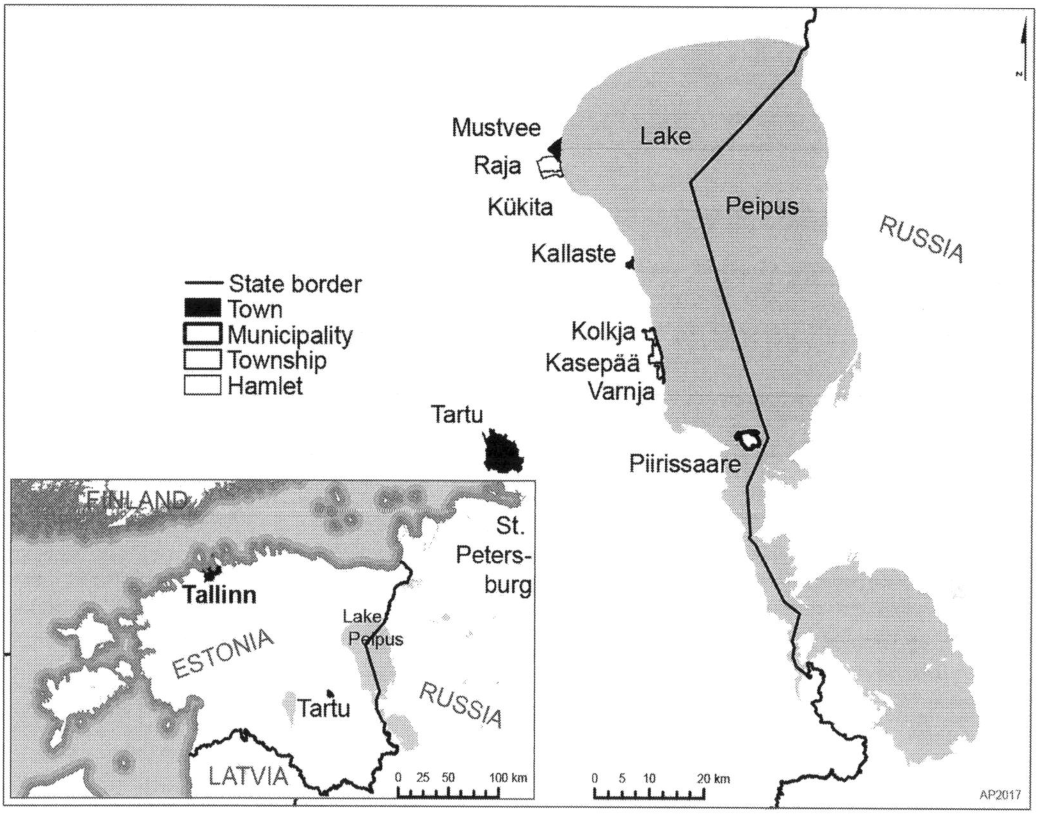

Figure 7.1 Map showing the regional context. The closely situated villages at the western shore of Lake Peipus are inhabited by Old Believers.

Source: Estonian Land Board, 2017

enforce long-term, wide-reaching strategies and regulations. It appears that such customary rights regimes help to ensure sustainable management of both resources and landscapes over the long-term, as well as distribution of resources that permits all members of a society to meet subsistence needs (see Howard, 2013).

As the land surrounding Lake Peipus was of poor quality and not desirable to any other group, the Old Believers cultivated the land for their own purposes and in accordance with their own logic. The Old Believers did not legally own their land as property until 1991, when Estonia became an independent country once again. Today, the landscapes of Lake Peipus are considered some of the most idyllic places in all of Estonia, and this area continues to be one of the most-visited places among Estonian tourists. Intriguingly, the area has not gained popularity as a second-home area among Estonia's inhabitants. The vernacular architecture of small, wooden houses that are renovated with bricolage materials, as well as the landscape itself – dotted with prayer houses, graveyards, and vegetable beds – is so distinct that it would be challenging to adapt to this area as an outsider (Figure 7.2).

Through the last century, male Old Believers have been employed primarily through fishing, while female Old Believers traditionally tended to vegetable gardens growing chicory, cucumber, and onions, both for eating and for selling. In the following sections, we present the historic shaping and functioning of these gendered landscapes as foodways binding together cultural, social, economic, religious, and landscape practices relating to the production and consumption of food in the unique heritage landscape (Harvey and Waterton, 2015; Roe, 2016).

Figure 7.2 Aerial image of Kasepää township, Peipsiääre municipality, April 19, 2014. Despite the turmoil of political and socio-economic changes the landscape has maintained its distinct character and more than 100-year-old logic of narrow and very long plots. The names on the map denote the plot names: the underlined names refer to fish and fishing (e.g. Rääbis – Vendance), others to gardening (e.g. Tsiguri – Chicory), white backdrop to prayer house as a central location binding the landscape together.

Source: Estonian Land Board, 2017

Landscape of fish

Lake Peipus – called 'the sea' by locals – is regarded as the territory of Old Believers. As defined by the late Zosima Samsonovich Jotkin, the leader of Old Believers' community in the 2000s:

> Our landscape reaches maximum five kilometres from the lake. Then forests and bogs begin. No, we are not afraid of those people living behind the forest, we do communicate with them, but there is not much to do there. Our life has face [*sic*] towards the lake and back towards the mainland.
>
> *(Printsmann and Palang, 2016: 299)*

Lake Peipus is the fourth largest lake in Europe and thereby increasingly important on the global scale as a 3,555-square kilometre freshwater reservoir, of which 44% belongs today to the Republic of Estonia and 56% to the Russian Federation. While the lake is rather shallow at an average of 7.1 metres in depth, it supports 34 species of fish (Haberman et al., 2008). Over 300,000 migratory birds use Lake Peipus as a stopover site (Luigujõe et al., 2008).

In medieval times, Lake Peipus was in common use by neighbouring Estonian and Votian village communities (Kurs, 2006). Customary law allowed everyone from the neighbourhood to fish in the lake. During the 15th century, the first fishing villages appeared on the lakeshore. The lands surrounding the lake were not well-suited for agriculture, so the lake and the activities connected with the lake provided the main source of income for village residents. Historically, the large Russian market has placed a high demand on fish, and transporting fish to nearby centres was easy due to geographic proximity. Russia, with its frequent religious bans on the consumption of all meat, was a great consumer of fish. By way of high and sustained demand, monasteries were the most important agents in the development of fisheries (Ränk, 1934).

Notes from the beginning of the 17th century suggest that Lake Peipus was so rich in fish that it, at the time, supplied great parts of Russia and half of the territory of the modern Baltic countries. The most valuable fish at that time were Peipus bream (*Abramis brama*), vendace (*Coregonus albula*), and smelt (*Osmerus eperlamus*). The fish-rich lake encouraged new settlers in search of better living conditions to inhabit the area. This, in turn, had a negative impact on the lake's ecosystem. Lake Peipus showed serious signs of overfishing at the beginning of the 19th century, and fishing quotas had to be established. Fishermen were disgruntled with these regulations and restrictions, but after some six years of quotas, fish populations began to recover (see Tammiksaar, 2014).

The lakeshore dwelling houses and backyards were designed for fish storage and outfitted with processing facilities. Fish would be cleaned and stored here before being sold in the market or transported to fishing salesmen. Among Russian Old Believers who grew up on the lake, it is said that each different sound the lake made in the morning forecast whether it would be a good fishing day or not. Young boys started to go out on the lake with their grandfathers on shorter distances, and later they would go on longer ventures with their fathers. All the navigational landmarks were learned through experience. These landmarks were protected by law: for example cutting down specific trees on the shoreline was strictly prohibited and warranted steep fines.

The fishing landscape has had a seasonal character. Seasons were defined by the spawning times of different fish. Fishermen had a distinct overview of when and where fish would come to spawn (Ränk, 1934). The most common fish were Peipus bream, vendace, pike (*Esox lucius*), smelt, and whitefish (*Coregonus maraena*). According to ethnographic notes (ibid.) from the first half of the 20th century the long-term fishing was practised in winter as ice fishing, as the ice cover of the lake lasted more than 100 days of the year. Men built light huts for winter fishing. These huts were placed on sleighs or runners and, with the help of horses, they were pulled onto the ice some 15 to 25 kilometres from the shore. A company of fishermen, 16 to 25 men together, would live in the huts for six days and come home by Saturday evening (Ränk, 1934; see also Moora, 1964). The horses were rented from Estonian farmers, as they were only needed seasonally for pulling the fish huts onto the lake. With the appearance of cars in the 1970s, horses and huts were replaced by snowmobiles and karakats [каракатица]. Karakats are machines with huge smooth wheels, self-assembled and invented by Old Believers, which enable driving onto the ice (Laviolette and Sirotina, 2015).

When Estonia declared independence in 1918, then, as a political act, this brought along a decrease in demand for fish from Russia (ibid.). As the vegetable plots were too small to allow for larger agricultural farming activities, men did summer work in nearby Estonian farms or worked as stonemasons across the Estonian territory, northern Finland and even in bigger cities like St. Petersburg or Moscow in Russia.

Fishing was also the main income for local men later during Soviet times, when local fishing *sovkhozes* (state farms) had massive fishing quotas. Fish, which at that time were cheap and plentiful, were sold to the vast Soviet market. As a result, fish stocks started to diminish in the lake again in the late 1980s. Today, fish are still caught from the lake by the male Old Believers along with other fishermen. Fish remain the primary source of income for male Old Believers inhabiting the western shore of Lake Peipus to this day.

Fish stocks in Lake Peipus have declined significantly within the last two decades. The primary reasons for this decrease are human induced and environmental challenges facing the lake. The lake's ecological condition is largely dependent on the water level which, in turn, depends on climate variability. In addition, eutrophication, which is caused by agricultural pollutants, makes it hard for fish to survive (Kangur, 2014). As of today, the two countries that own Lake Peipus have not been able to reach a mutual understanding about how to protect the lake environment (ibid.). Zoja, our interviewee in the fieldwork to Peipus villages, was born in 1934 in Varnja village. She retired to her forefathers' homestead. She is now the key person of the prayer house. She expressed her opinion about fish and property issues: "And before the lake was state-owned [Soviet period], fish was ours. Now it is the other way around. Our fish is state-owned." Lake Peipus has faced overfishing and there are more fishermen, both professional and amateur, in the area than fishing possibilities. There are strict fishing quotas in place, and when they are met, the local populations have no work. Moreover, as there is great demand for fish, fishing licenses are accordingly expensive and not always affordable for local fishermen.

In addition to consuming fresh fish, Old Believers have traditionally smoked and dried it for preservation. Old Believers have a rich collection of fish dishes starting from pies, soups, and pierogies which are cooked, baked, and fried in diverse ways (Ladõnskaja, 2011). Traditionally, every part of the fish is used, from entrails and fish heads to caviar and minnows, and everything in between. Classically, women were in charge of cooking fish, and when a man was looking for a bride, a woman's skills were appraised by way of her diversity of fish recipes. Apart from cooking, women had an equally important role in providing for the family: they tended the garden for family and economic purposes. Chicory, onions, cucumbers, and other vegetables have all been primarily economically viable commodities in these areas for 150 years and more.

Landscape of onions

As land was scarce along the lakeshore, then Old Believers' villages are compact, with plots as small as 1,000–4,000 square metres in size (compared to the smallest Estonian farms in the beginning of the 20th century which were up to 20 hectares in area). The land surrounding a house was used as a vegetable garden which yields crops for both personal consumption and commercial sale. The poor sandy soils have led to the use of locally adapted agricultural techniques that improve the fertility of the soils. All vegetable gardens behind Old Believers' houses feature raised beds built up off the ground up to 50 centimetres, as the ground water level is high (Figure 7.3). High beds dry and warm faster, allowing the gardener to sow seeds earlier during springtime. During the summertime, it is easier to weed this type of beds. Weeds are dumped between the beds one year, then a bed is made on top of the weeds the next year. This bed-piling technique, entirely made by hand, is said to be at least 200 years old. No chemicals are used for fertilising the soils. Up until the mid-20th century, horse manure was an exclusive, expensive commodity, as Old Believers did not own horses and it had to be transported by boat. Therefore, natural fertilisers were made from fish, fish offal, and mud-rich organic matter from lake sediments. The small plots of land around the houses and the lack of available grazing land limited the Old Believers to raise only one cow per family. Cows were fed with lake reeds during the summer. This, in turn, kept the lake shoreline open and thus the village's connection to the lake was so immediate that the border between the village and the lake was minimal. Wintertime hay was gathered from the distant marshes, floated in on boats.

Figure 7.3 Onion beds with a height of up to 50 centimetres behind Old Believers' houses.
Source: Anu Printsmann

Figure 7.4 Onions are sold in booths and pop-up stands.

Source: Helen Sooväli-Sepping

The Old Believers specialised in growing chicory in the late 19th and early 20th centuries, but this practice stopped when a border was drawn between independent Estonia and Soviet Russia in early 1920s. In addition, cucumbers were grown in the 1980s when the Soviet agricultural system could not produce enough food for the whole of the Soviet Union. Alternative small-scale farming was encouraged regionally in order to feed the millions of inhabitants in the fast growing second-largest town of the Soviet Union, Leningrad (now St. Petersburg), less than 200 kilometres away. It is the onion, though, which is the most renowned vegetable of Old Believers. Onion growing was customary on the western shore of Lake Peipus from the end of the 19th century, peaking in economic importance sometime between the 1930s and the 1990s. Today, the sale of Old Believers' onions is facing tough competition from cheaper Polish and Chinese onions; furthermore, the local vegetable has no state support in the grocery store. Onion growing is thus losing its popularity for this reason, and also because the younger generations do not want to do the hard manual work of making the beds. In the last decade, the quantity of onions grown has halved. There is still a great potential vegetable market available in Russia – however, the political situation between the two neighbouring countries prevents all sorts of trade.

Onions are best cultivated in fertile soils that are well-drained. Sandy loams are good as they are low in sulphur. The seed of the onion from the Lake Peipus region stems from the 1930s and has been handed down from generation to generation since then. Today, this onion species is protected by state regulations (Kalle et al., 2015). It is slightly smaller than the onions commonly sold in grocery stores, and it has a particularly strong flavour. Onion growing is a lengthy, multi-year process: the first year, seedlings are sown from the seed; the second year, the planted seedlings grow into onion bulbs; and the third year, the onion bulbs produce umbels. Seed growing is practiced in parallel with onion-bulb growing in order to ensure sufficient onion seed stock. Onions are harvested in late August and are dried in special sheds with large doors so that both wind and sun can dry them. Thereafter, the onions are taken to a sauna where they are dried for another two to three weeks. For that, the sauna is heated to 40–45°C to remove all moisture from the bulbs (Jüriado, 2014). Finally, the onions are woven into braids and sold in small booths along the roads at the western shore of Lake Peipus (Figure 7.4), as well as at numerous onion and fish festivals in the area. Old Believers' culinary traditions surrounding onions are rich, and during harvesting time Old Believers' villages are crowded with visitors buying onions and products made of onions. The Peipus onion is valued as an organic food and appreciated as an important Estonian element of culinary heritage (Bardone et al., 2016).

Future prospects

Today, fishing and onion growing are still practiced among Old Believers; nevertheless, these means of income and ways of life have transformed to some extent from livelihood to tourist attraction. In order to encourage traditional economic activities, to help preserve cultural traditions, and to assist in maintaining the landscapes of Old Believers, the Estonian government, often with the help of European structural funds, has implemented several policy programs, tourism-development schemes, regional programs, and heritage restrictions in spatial planning processes. Commodification of landscape heritage has resulted in the establishment of small food-processing companies with a state-initiated quality label that can be used to promote the museums and restaurants that support Old Believers' customs as well as the food itself. However, it remains questionable how much the support gains for the local population to sustain and improve its life quality and to maintain the traditional livelihood.

People nowadays are more mobile, and traditional beliefs have to some extent lost their regulative role in the communities. These socio-economic factors have resulted in sweeping landscape changes. Vast reed areas are now spread out on the shores of the lake due to eutrophication and lack of cattle breeding (and feeding), such that the view to the lake is now largely obscured. In many gardens, lawns have replaced onion beds. Add to this the declining fish stocks due to both environmental problems and an oversaturated fishing industry, with many factories and employees stationed on the western shore of the lake, and it is clear that the fate of traditional foodways and sustainable landscapes is uncertain.

Both the landscape and the Old Believers communities have, for centuries, been depicted as poor and obsolete. Although the landscapes and practices of Old Believers have been traditionally stable, historically, they have always been in flux. Fish stocks have declined in Lake Peipus before, as studies have shown. Likewise, gardening activities have shifted from chicory, to onion and cucumber, and now again to onion – everything depending upon the demand. As local and organic food becomes increasingly popular among Estonians, it is difficult to foresee how this will affect the price and demand of vegetables; perhaps sooner than expected, it will be profitable to grow vegetables once again. As an example, in recent years there have been some social actions whereby more affluent municipalities near the capital city of Tallinn have bought tons of onions to help Old Believers capitalise on their harvest. This trend is growing. The Old Believers community has earned respect among Estonians due to its good quality food products and this could mean there is further potential for promoting their goods as alternative, fair food. Despite the decrease in numbers of the Old Believers, the community is still strong and can still sustain themselves as a community with faith expressed in specific traditions and strong cultural identity. The continuous traditional culinary culture is just one example of that. The Old Believers foodways and landscapes will sustain, although in decreasing prominence, its character in years to come.

Landscapes reflect the ways humans have cared for, built up, and exploited their surroundings (Reinert et al., 2015). One way of talking about landscapes is by way of cultural sustainability (Dessein et al., 2015; Soini and Dessein, 2016). Food, including hunting/fishing, crops, and cultivation, is an essential part of cultural sustainability. Cultural meanings are constantly re-conceptualised, thus practices are always in transformation. A culture can be called sustainable when it is capable of absorbing elements from other cultures, while at the same time protecting and maintaining its core values.

Acknowledgements

This chapter has been supported by Estonian Ministry of Education target-financed project IUT3–2 Culturescapes in Transformation: Towards an Integrated Theory of Meaning Making.

Further reading

Anderson, E.N. (2013) Food cultures: Linking people to landscapes. In: Pilgrim, S., & Pretty, J. (eds.), *Nature and culture: Rebuilding lost connections*. London: Earthscan, pp. 185–196.

Bardone, E., Kannike, A., Põltsam-Jürjo, I., & Plath, U. (2016) *101 Eesti toitu ja toiduainet*. Tallinn: Varrak.

Dessein, J., Soini, K., Fairclough, G., & Horlings, L. (eds.) (2015) *Culture in, for and as sustainable development: Conclusions from the COST action IS1007 Investigating cultural sustainability*. Finland: University of Jyväskylä.

Estonian Land Board (2017) online map server at http://xgis.maaamet.ee/xGIS/XGis.

Haberman, J., Timm, T., & Raukas, A. (eds.) (2008) *Peipsi*. Tartu: Eesti Loodusfoto.

Harvey, D.C., & Waterton, E. (2015) Editorial: Landscapes of heritage and heritage landscapes. *Landscape Research*, 40(8), 905–910. Available from: DOI: 10.1080/01426397.2015.1086563.

Howard, P. (2013) Culture and agrobiodiversity: Understanding the links. In: Pilgrim, S., & Pretty, J. (eds.), *Nature and culture: Rebuilding lost connections*. London: Earthscan, pp. 163–185.

Jüriado, T. (2014) Retked vanausuliste juurde: kaheksaotsristid, tihekülad ja kõrgpeenrad. *Eesti loodus*, 5, 17–24.

Kalle, R., Bardone, E., & Sõukand, R. (2015) Sibul, kadakas ja küüslauk – toidutaimed ja regionaalne toidukultuur Eestis. *Mäetagused*, 59, 97–124. doi:10.7592/MT2015.59.toidutaim.

Kangur, K. (2014) Eesti ja Vene koostööpüüdlused Peipsil. *Eesti Loodus*, 5, 24–26.

Kõivupuu, M. (2012) The food culture of Estonian Starovery (Old Believers): Traditions and the present. In: Laganovska, K., & Svetlana, I. (eds.), *Art and music in cultural discourse*. Rezekne: Rezekne Augstskola, pp. 28–34.

Kurs, O. (2006) Peipsimaa majandusliku ja kultuurilise siirdevööndina. In: Kurs, O. (ed.), *Rajamaade rahvaid.* Tartu: Ilmamaa, pp. 95–109.

Ladõnskaja, V. (2011) *Peipsi veerel. Vanausulised paluvad lauda.* Tallinn: AS Ajakirjade kirjastus.

Laviolette, P., & Sirotina, A. (2015) Karakats: The bricolage of hybrid vehicles that skate and swim. *Journal of Ethnology and Folkloristics*, 9(1), 21–40.

Luigujõe, L., Kuresoo, A., van Eerden, M., & Borissov, V. (2008) Linnustik. In: Haberman, S., Timm, T., & Raukas, A. (eds.), *Peipsi.* Tartu: Eesti Loodusfoto, pp. 341–364.

Moora, A. (1964) *Peipsimaa etnilisest ajaloost.* Tallinn: Eesti Riiklik Kirjastus.

Plaat, J. (2016) *Russian Old Believers and their sacral buildings in Estonia.* Tallinn: MTÜ Variato.

Printsmann, A., & Palang, H. (2016) Vegetable garden as a source of identity. In: Collins, T., Kindermann, G., Newman, C., & Cronin, N. (eds.), *Landscape values place and praxis: Galway, 29th June–2nd July 2016.* Galway: National University of Ireland, pp. 298–303.

Ränk, G. (1934) *The Peipsi lake fisheries.* Õpetatud Eesti Seltsi kirjad II. Tartu: Õpetatud Eesti Selts.

Reinert, H., Miles-Watson, J., & Sooväli-Sepping, H. (2015) Landscapes of rupture: Introduction. In: Sooväli-Sepping, H., Reinert, H., & Miles-Watson, J. (eds.), *Ruptured landscapes: Landscape, identity and social change.* Landscape Series 19. Dordrecht: Springer, pp. 1–7.

Roe, M. (2016) Editorial: Food and landscape. *Landscape Research*, 41(7), 709–713. Available from: DOI: 10.1080/01426397.2016.1226016.

Soini, K., & Dessein, J. (2016) Culture-sustainability relation: Towards a conceptual Network. *Sustainability*, 8(2), 167. Available from: DOI: 10.3390/su8020167.

Tammiksaar, E. (2014) Uusi fakte kalapüügi korralduse ajaloost Peipsil. *Eesti loodus*, 5, 30–36.

8

Food production and the village under state socialism

The Balkan case

Mirjana Lozanovska and Alexandra Florea

Food production and the village under state socialism

Agriculture is integral to rural landscape which is understood to have resulted from the interaction between humans and nature, also termed cultural landscape (Rapoport, 1992). The village is a rural morphology that is historically tied to food production and a rural landscape. However, two phases of migration impacting on the village – firstly, under state socialism and secondly, in the recent post-communist era – have radically altered its agricultural function and the associated rural landscape as setting of the village.

The scope of this chapter is on the rural landscape of Eastern Europe and the Balkans in particular (Figure 8.1). The latter region contains Albania, Bulgaria, former Yugoslavian states, and Romania (Turnock, 1997: 6). This chapter is structured in two parts. The first examines the process of collectivisation and the formation of co-operatives as the socialist model to bring the villages and food production under the framework of centralised government. The discussion focuses on policies and programmes in that period of communist and socialist history that altered organisational structures and methods of food production, and changed the types of crops produced. The second part of the chapter elaborates on the detail and specificity of one village, Zavoj, in the Republic of Macedonia, to examine how ties between the village and agriculture were altered by migration from the village. By detailing the human story of the mountainous landscape and village of Zavoj, and the effect on human lives related to the rural landscape and food production, this second part complements the overview background of the chapter.

The impact of collectivisation on the village

Before the Second World War, "three-quarters of the Balkans' population lived in the countryside" (Berend, 1996: 210). The village was the prevalent model of settlement in the countryside and the matrix through which food production and agriculture was organised. With the changes to borderlines at the end of the First World War, governments instituted an agrarian reform aimed at redistributing land from the nobility, the Church, and Crown to the landless peasants (Thompson, 1993; Salzmann and Scheufler, 1974). In 1945, a second agrarian reform aimed to distribute more land to peasants (maximum five hectares per family) who had very little or no land in their possession, "preparing the ground for collectivisation as the five hectares was seldom enough to feed the peasant's family" (Meurs, 1999: 113). Both reforms were driven by political objectives for the state control and management of agriculture (Macours and Swinnen, 2000: 177),

Figure 8.1 Map of the Balkans showing the Republic of Macedonia and the location of Zavoj village.
Source: Map co-ordinated by M. Lozanovska, drawn by A. Florea

in the context of a rural population of over 12 million in 1930 in Romania (Trebici and Ghinoiu, 1986), and almost 10 million in 1921 in Yugoslavia (Horvat, 1973: 4).

The political scope of the communist regime also aimed to create a new egalitarian class with the peasant as an integral part (Scott, 1998). A new programme and strict control of land surfaces, crops, and quantities were enforced, transforming the village and the landscape. With the government's agenda to join the individual strips of land for administrative purposes, trees and rural demarcations of hedgerows and stones were removed, ditches and pathways were altered (Kligman and Verdery, 2011: 146). Cadastral maps and rational geometric planning of land plots, land surveying, and documenting were employed by states to make legible the countryside (Scott, 1998). In propaganda posters of the time, wide yellow fields of wheat or corn and tractors became symbols of successful co-operative work (Nowakowska and Ignatowich, 2014). Such landscape alterations already occurred in England in the seventeenth century (Chapters 28 and 30), where open field land and common land were unified into large farms (enforced by the law of enclosures) towards increased agricultural productivity (Williams, 1973).

The collectivisation process was constituted by gathering much of the agricultural land under state ownership and control in state and collective farms. Despite variations of how this reform was implemented and variations of the extent of the implementation and the practice after it, the Collective Agricultural Households or co-operatives radically changed the world of the village (Kligman and Verdery, 2011; Mihalache, 2011; Sisestean, 2012, Trebici and Ghinoiu, 1986). Food products traditionally cultivated by the peasants to sustain their families were replaced by castor, cotton, coriander, flax, hemp, sugar beet, and sunflowers, crops necessitated by the growing state industries (Socol, 2003: 213; Turnock, 1997: 35). Peasants did not always comply with the imposed crops and continued to cultivate cereals (corn and wheat), green manure crops, and vegetables. The locals perceived as leaders by the village community were imprisoned for a few months in order to set an example, which caused stigmatisation for the entire family and compromised their future (Socol, 2003: 213–215; Gruev, 2014: 345–346).

However, the impact of the new crops on the landscape was dramatic – larger tracts of land producing one crop and the distribution of new crops across the countryside – and altered the memory, ecology, and the future of the agricultural landscape. There was less impact in the mountainous area because stock rearing (cattle, sheep, pigs) and traditional stock food such as rye and maize continued (Turnock, 1997), although the ecosystem of the Carpathian Mountains was significantly transformed in the eighteenth and nineteenth centuries (Turnock, 2006).

The collectivisation process in Eastern Europe commenced in 1949 and followed the Soviet model, but due to the population's resistance to surrendering ownership of their land, collectivisation had intermediary phases known as associations. In associations, land and buildings remained private property, while mobile assets like draft animals, tools, and machinery were owned collectively (Salzmann and Scheufler, 1974: 11). In contrast to associations, in the collective system people could only keep their "homes, outbuildings and small gardens", the remaining properties, land, forest, animals, and tools being expropriated to the state (Kideckel, 1993: 81).

The traditional hierarchy of the village was based on the ownership of land and its size (Kligman and Verdery, 2011: 327): landowner and large landowner (more than 40 acres); large farmer (approximately 40 acres); middle farmer (approximately 20 acres); small farmer (approximately 10 acres); farming cottager (below three acres), cottager (modest cottage with little or no adjacent land); the farmhand (dependent on the farmer for work and board and the lowest category with no land ownership) (Salzmann and Scheufler, 1974: 15). Under communism, the first three categories were depicted as 'enemies of the people' and labelled as 'kulaks', who were imprisoned, relocated, and given menial work, while the state appropriated their assets. The next three categories were transformed into the proletariat – the working class (Adamson, 2007; Gruev, 2014). In Komarov village for example, the farming community agreed reluctantly to participate in the collective system (Salzmann and Scheufler, 1974: 51).

The co-operatives implemented a new hierarchy of food production that was dependent on the affinity to the communist party or at least on the cooperation in the collectivisation process (Kideckel, 1982). The co-operatives had a managing body (experience in agriculture was not required) and workers specialised in certain activities: accountants, mechanics, drivers, agriculture, animal care, and team leaders (see for example Kaneff, 1996). The number of staff employed and their specialisation was dependent on the size and profile of the co-operative. The co-operative paid salaries for its staff, and the profit was divided between its members according to "their original shares" (Kaneff, 1996: 94), while a fund was put aside for social events and community institutions. The salaries were paid according to a system of points value related to the task and workdays. Each task (for example picking fruit or spraying fruit trees) had a nominated point mark (0.75 point value and 1.5 point value respectively), according to its "difficulty and the special skills, training, or danger it involved" (Kideckel, 1993: 106).

The description of collectivisation in Komarov village, Czechoslovakia, shows that the law stipulated not only the size allowed for a private garden, but also what could be planted. Crops like grapes, strawberries, and vegetables could not exceed an area of 100 square metres. Also, a household was not permitted to have more than one cow and two hogs, ten beehives, and the number of small animals and poultry was described as 'reasonable', supposedly because such animals were for consumption and not agricultural work. A private garden provided "the last remaining link" to traditional forms of agriculture (Salzmann and Scheufler, 1974: 58–59). In Romania, private gardens comprised less than 13% of arable land, but produced more than half of the potatoes (Ronnas, 1989). Gardens were allowable only as complementary participation in co-operative's works (Sisestean, 2012: 243). The landscape within the footprint of the village was intensely cultivated and the crops were diverse, whereas open fields beyond the village were homogenised by monocrops.

Mechanisation, irrigation, and fertilisers were introduced to increase the productivity in agriculture. The curators of the exhibition *Crops for the State* (held in Poland in 2010), Nowakowska and Ignatowich, observed that "Mechanization of agriculture and electrification of the rural areas were key issues in the countryside during post-war era" (Nowakowska and Ignatowich, 2014: 71). The material exhibited consisted of postwar posters from Central and Eastern Europe that had been used to promote the communist agricultural reforms. Many featured the tractor and combine harvesters which had become symbols of the new economic system of co-operatives, while tractor operators were considered "responsible for introducing and fostering socialism in the countryside" (Nowakowska and Ignatowich, 2014: 71). Turnock (1997) argues that optimum watering rates corresponding to different crops were recommended to rationalise water consumption, but it was difficult to economically justify the irrigation system when other significant infrastructure was needed. He also points to the importance of fertilisers and pest control during socialism, explaining that in some cases, large quantities of fertiliser were applied and proved to be counterproductive.

Science was also called upon to contribute to the development of agriculture and stock rearing. Plant and animal breeding became important areas of interest for the socialist states (Turnock, 1997), likely influenced by the experiments of Ivan Michurin in the 1920s to create fruits and vegetables resilient to the harsh climate of central Russia (Nowakowska and Ignatowicz, 2014).

Modernisation as understood in the communist ideology meant a more efficient economic system of food production implemented by co-operatives and advanced technology and science into the agricultural field. From this point of view, it was not different from what Scott (1998: 89) labels as "high-modernism" in North America and Western Europe, although historically it occurred in the nineteenth century until the First World War. Industry and agriculture informed each other: the developments in the chemical industry made fertilisers available for agriculture, and the machinery industry contributed to the mechanisation of agriculture through tractors and combines and reduced the number of working animals (horses). The new jobs drew people away from the agriculture sector, increasing the urban population and putting pressure on food supplies. The effects of urbanisation and industrialisation on agriculture emerge from comparisons between ante-communist and communist data of people active in agriculture and industry (Mihalache, 2011: 22 for Romania) and gross national product data (Berend, 1996: 190–191 for the Balkans).

Soil degradation, a consequence of intensive industrialisation in rural areas, led to deforestation. Chemicals in the soil caused windfalls and tree collapses, which in turn facilitated pest infestations, affecting 3800 kilometres in Sudetes Mountains, Poland (Mazurski, 1986). Forests under the administration of co-operatives were not managed correctly. The number of trees cut exceeded the number of trees set out in the policy and plan. Moreover, the ecological cycles of forests were negatively affected because these activities were often implemented during summer. In many cases the forests disappeared. To counter the negative effects, campaigns for reforestation were initiated during the communist era (Sisestean, 2012; Nowakowska and Ignatowich, 2014).

In the case of mineral extraction, the technology of open pits was widely used, resulting in radical and extensive modification of the physical landscape and displacement of thousands of people, as was the case in Bohemia when 116 villages were demolished for coal surface mines (Glassheim, 2006).

Electric cables and pylons, telephone wires, and the construction of new architectural building typologies – industrial facilities, socialist housing, schools, and cultural centres – industrialised and modernised the village and aligned it with the communist model, bringing dramatic changes to the built environment, food production, and the landscape (Carter and Turnock, 1996). Other phenomena affecting the countryside were rural-to-urban migration in great numbers (Turnock, 1991; Micu, 2012: 236) and the feminisation of the working rural population (Cernea, 1978). An extreme example is Bulgaria, where the urban population increased from 25% in 1946 to 69% in 1993 (Rey and Bachvarov, 1998: 347). These processes and their impact on the rural landscape and food production will be further detailed in the following case study.

Food production and landscape in Zavoj, Republic of Macedonia

The traditional tie between the village and food production was eroded by rural-to-urban migration during state socialism. Yet histories and records of the continuity of food production, practices of agriculture, and return to the mountainous village, Zavoj in the Republic of Macedonia, and the village countryside, present a more complex and alternative perspective. This second part of the chapter looks at two phases of food production and landscape in Zavoj affected by migration: retrospectively to the collectivisation programmes during state socialism in Yugoslavia (1949–1991), and at the more recent post-communist phase (after 1991).

The village of Zavoj in the Republic of Macedonia is located at an elevation of about 1100 metres above sea level, a concentrated typology of urbanism along one main street, it is surrounded by a mountainous landscape between Pamni Mountain to the west, Bajchinov Mountain to the south, and Miloshev Mountain to the east. Zavoj is approached by a steep road, the width of one car, that winds its way through a landscape of corn plantations by the lower creek, oak and forest berries, and, depending on the season, wildflowers, green canopy, snow covered mountains, or the many shades of the deciduous forest. At the end of the road and the top of the mountain is the recently reconstructed chapel, its grounds landscaped with oak trees marking a circular plateau of manicured lawn. The view from the site of the chapel overlooks mountain peaks, layered and vanishing into the distance towards the historical town of Ohrid and its deep mountain lake. Mircea Eliade may have described it as 'sacred' – "some parts of space are qualitatively different from others" (Eliade, 1959: 20).

Noting that this region was under Ottoman rule from the mid-1400s to the 1900s, Zavoj is first cited in archival records in 1582 with a note of 69 Christian households. Archaeological findings, the foundations of two churches, a large cemetery, and ruins of about 100 houses, have identified a prosperous village, traces of Old Zavoj, a village three kilometres east of the present location. The older village ceased to exist after 1720 and it is deduced that 23 large households relocated and built up Zavoj in its present location (Mitrev, 1974). By 2015, only three people lived in the village. Many of the fields had been abandoned and there were very few livestock. The mixture of tiny new weekend houses, the rubble and ruins of traditional dwellings, the abandoned school, and the *zadruga* (co-operative store) present the hybrid multiple histories of the village from the early twentieth century to the present (Golob, 2009).

The first phase is associated with the last generation directly linked to agriculture, even though during state socialism able-bodied men left the village for arduous labour in nation-building programmes that included infrastructure, roads, and heavy industries. This will be called the 'agricultural phase' of the village. Traditional songs refer to the terms *na teren* and *pečalba* pointing to the difficulties and tragic results of this temporary migration of itinerant work. The second phase of the village, the 'migrant phase', is associated with the children of the 'agricultural generation' who were relocated for educational purposes, and thus did not live in the village. They have inherited the ancestral home and land after the passing away of their parents. The story of Zavoj illustrates a continued connection to the village, but the relationship between the village and food production in each of these periods is contrasted. Research in the village, including documentation of the village landscape, built environment, and its countryside was conducted during a series of fieldwork trips in the period 1988–2015, with 23 interviews conducted in 1988 of the inhabitants belonging to the 'agricultural generation' and ten interviews in 2010 of the members of the 'migrant generation'.

The role of rural committees, collectivisation, compulsory purchase of stock, and increased state ownership illustrate that migration was a major impact on Zavoj village during the optimistic period 1945–91 of Socialist Yugoslavia (Achkoska, 2007). Promotional literature at the height of Yugoslavia's socialist model in the 1970s illustrates both women and men at work in industries, and the industrialisation of the agricultural landscape (Kalezić, 1984; Balevski, 1981). While not as severe as the Eastern bloc due to Yugoslavia's political separation and leadership in the Non-Aligned Movement, new mechanised processes of food production transformed the landscape, with the rural reform programmes in Yugoslavia, meaning the end of the village as site for food production.

Increased permanent migration from Zavoj after Second World War reached a climax in the late 1960s and the 1970s, when severe poverty overcame the village. As shown in the first part of this chapter, modernisation of agricultural production was integral to the communist postwar programme. This was not, however, consistent. Some villages underwent dramatic change while for others the modernisation occurred at the level of organisation of labour. In Zavoj, by the early 1970s, an irrigation system that was not updated or maintained proved detrimental when a local drought hit the region. The village of Zavoj became a place of departure, rendering both the village and its agricultural landscape a story of the past. In the mid-1990s there were 41 inhabitants in Zavoj, more than half of whom were over 65, and none were less than 35 years of age (Census, Republic of Macedonia, 2004). Many of the elderly had experienced life under state socialism. The operation of the collective store – the *zadruga* – was inconsistent, and the village school was instead used as communal space. By 2010 both had become memorial ruins of the modernist socialist collectivisation history, and the industrialised landscape surrounding the school – with its intact brutalist monument – presenting an image of a bygone era. In 2002 there were 12 inhabitants.

A discussion of food production and landscape will follow firstly on the agricultural phase and then on the migrant phase of the village.

In the period of state socialism, 1945–91, the men of the 'agricultural phase', the generation born between 1920 and late 1930, engaged in itinerant work and were frequently absent from the village. As a teenager in the 1970s, later to join other migrants, Jovan recalls that the frequency of his father's return to Zavoj depended on the distance of his work location and his capacity to return by bicycle, ranging from once per week, once per month, or once every three months. Children departed each day to attend their upper primary school located in the larger village of Svinishta nearby (Jovan, pers. comm., Leskoec, 2010).

The finances gained from the itinerant work enabled the 'agricultural generation' to set up new housing for their children in or on the fringe of nearby towns where they would attend technical or secondary schools. In this early period of migration and resettlement, agricultural work and food production continued in the village as a vital necessity. The migrant children returned to the village to assist with agricultural work, and the frequency, duration, and regularity was defined by the seasonal schedule of food production. It was most frequent in the summer when the harvesting needed to be carried out. Jovan, at the technical high school in Ohrid in the 1970s, detailed the necessity of agricultural work – harvesting of wheat, rye, barley, corn, and other grains was necessary for making livestock feed. On return visits to the village, livestock was slaughtered as required, and the families cured meats in wooden barrels, smoked ribs and sausages, and made cheeses (*sirejne*) and yoghurt (Jovan, pers. comm., Leskoec, 2010).

In addition to intensive work in the summer, the migrant children of many families returned to the village regularly on weekends, transporting sugar, oil, flour, and juice to the village and returning with fresh vegetables, fruits, meats, and traditional food prepared by their mothers. Mile states, "there was tons of work" and the produce and prepared foods such as *ǵomleze*, *petuliçi*, and *komat* (pastry specialties), were brought back to their own house in town to contribute to the weekly groceries. These are pastry specialties that are time consuming to prepare. *Ġhese* is made in a special facility in which the smoke from the embers is integral to the special flavour. It is prepared by mixing a thin 'pancake' type mix; the crepe mix is poured on a *sač*, a pan that looks like a large lid (the embers are underneath). The crepes are prepared one at a time, and then layered in a baking dish to be again cooked on the embers. Despite their departure from the village, a strong narrative of the dependency on food production in the village evolves (Mile, pers. comm., Zavoj, 2010).

Young married women were left in the village with the elderly and the smaller children and they managed the necessary agricultural work in the village. Agricultural work, including tilling and preparing the

Figure 8.2 Member of the 'agricultural generation', Çveta, drying the fruits that she has collected, including apricots, plums, and peaches, in the sun. The dried fruits are used for *compote,* an infused tea drink consumed in the winter.

Source: Photograph M. Lozanovska, fieldwork, Zavoj, Republic of Macedonia, 1988

soil, sowing the crops, planting, harvesting, and herding the animals, became women's work or work managed by women (see Figure 8.2). The communal spaces of the village were characterised by women's use and a dominant female presence (Lozanovska, 1989). In the period of the men's absences, the village itself was symbolised as maternal and feminine and this reference gained social and psychoanalytic dimensions. The village as maternal space – as nurturing, nourishing, comforting – had become inscribed in the collective memory of the village such that during the 'migrant phase' it symbolised an anchoring sense of belonging for all that had migrated. The *zadruga* introduced the foods of modernised co-operatives and industries such as oil, sugar, and flour. In contrast, a masculine culture evolved at the sites of itinerant work where the men were often accommodated in dormitories, sometime temporary buildings, and in the company of other male labourers (Berger and Mohr, 1975).

Traditional wood fire ovens, located along pathways between private property boundaries, had once served as communal preparation facilities. In 1988 the wood fire ovens were used on ceremonial occasions by small groups of women who baked loaves of bread and casseroles. Another cluster of structures including the 'smoking/*sač*' room, the summer kitchen, and the drying/storage shelter were located just beyond the main house. Onion, garlic, corn, and peppers were strung up to dry in the sun and then prepared for winter foods. Food preparation work during the warmer summer months meant that the villagers only entered the house to sleep or bathe. Rural structures dot the landscape tracing its cultivated history. Subject to disuse and the forces of nature on their biodegradable materials, they have become altogether dissolved into memory landscapes.

Women told stories about walking in the mountains either while herding the animals or to collect fruits from planted orchards, wild berries, and mushrooms. Wandering in the mountains also presented a freedom and sense of liberation contrasting the burden of work. Not only the village but the landscape had become accessible and a threshold to a new agency.

By 1988 the village was inhabited by 80 households. The men of the agricultural generation had eventually returned to the village in their retirement. Once again they embarked on food production activities including herding sheep in the mountains, cattle in the village vicinity, tending to chickens in the barns and pigs in the yard, with a few households tending their fields and orchards. Traditional structures such as the '*rakija*-making' hut/distillery were coded as masculine but had been abandoned during the period of itinerant work, and except for the barn, they were in disrepair and not in use in 1988. The peasant history of long mountainous journeys searching for good animal feed manifested in shepherd huts where the shepherd would rest for the day or overnight. The mountainous landscape surrounding Zavoj is dotted with these simple structures. Their stories relay the elevated crossings over mountains, and the descent into deep valleys and streams, and reveal how agricultural activity extended beyond the traditional village and inscribed a narrative of human cultivation and history onto the wider landscape.

The migrant generation had left the village in the late 1960s and early 1970s, and many had relocated to two nearby settlements: Leskoeç, a village located 20 kilometres from Zavoj, and Železnička Naselba, a district on the fringes of the town of Ohrid. In the 1970s both lacked urban infrastructure – electricity, water, roads, and sewerage – in Leskoeç a septic tank was used for sewage waste. Peasants were attracted to the looser regional codes for planning and construction which allowed for unregulated house construction.

Members of the migrant generation went abroad to work as young adults, and like their fathers, some participated in the 'guestworker programmes' in Germany and also temporary work in the Netherlands. The funds raised financed the re-construction and completion of the house for their new families, often building literally on the footings the 'agricultural generation' had established.

Looking at the resettlement pattern of the migrant generation through the theories of spatial practice and spatial narrative reveals an entangled mixture of urban and rural trends (de Certeau, 1984). The new houses included new luxury sitting rooms and new technologies servicing the indoor kitchen, bathroom,

Figure 8.3 Communal food preparation structures. Section and plan of *rakija* (traditional distilled drink) making structure (on left). Woodfire oven for bread baking and baking of other food. Members of the 'agricultural generation', Vesa, Dola, and Sredbra, prepare foods for the holy day, Day of the Holy Mother, August 28, 1988.

Source: Architectural documentation and photograph: Lozanovska, fieldwork, Zavoj, Republic of Macedonia, 1988

and laundry facilities. Large vegetable gardens, afforded by the larger plots, were dispersed in between the dwellings. Structures and facilities for outdoor cooking and preparation of food included cool rooms for storage, wood-fire systems for cooking, with the iconic outdoor deep trough/sink located around the edges of the dwelling. Areas shaded by grapevine or kiwi fruit vine (*Actinidia* sp.), transported from Australia by visiting migrants, were used for casual social gatherings at the front of the house, encouraging social interaction with neighbours (Armstrong, 1999; Hage, 1997; Lozanovska, 2014). Food cultivation and production continued during the emigrant phase at the site of the new house as a subsidiary activity to purchasing food.

In the period 1988 to 2010 the inhabitants that had remained in Zavoj were elderly and many passed away, marking the end of the agricultural generation of the village. Their children inherited the fields, vegetable gardens, grazing land, and livestock, in addition to the houses. The village was neglected. Agriculture was not the main interest to the emigrant generation. The agricultural character of the landscape entered a phase of de-cultivation reflected in the vernacular stone and *pide* architecture turning into rubble, and over time into dirt. This process of land de-cultivation encountered an opposite process of new construction and new modes of using the landscape. Migrant members began to build new weekend houses in their ancestral village, Zavoj, such that in the period 2007–15 the once picturesque village had become a construction site characterised by building material – mounds of sand, stacks of bricks, and reinforcement rods. Incomplete dwelling structures were symptomatic of a new model of village, one not related to food production but to recreation and tourism.

Figure 8.4 Taseski house ('migrant generation' members Naum and Olga who have returned to live in the village), Zavoj. Plan (left) showing facilities at the front of the house – store, summer kitchen, tap, porch, and yard – rear vegetable garden, and modest but modern house with bathroom. Photograph of Olga and family friend preparing preserves from the summer harvest.

Source: Architectural sketch and photograph: Lozanovska, fieldwork, Zavoj, Republic of Macedonia, September 2010

Naum, a member of the 'migrant generation' is one of two participants that have returned to live in the village (see Figure 8.4). Like many other migrants he talks about his love of nature and mountain air, but his return is unusual, if not exceptional (Naum, pers. comm., Zavoj, 2010). He retired in 2009 but had returned to the village a decade earlier. While he has the same rural peasant childhood as his migrant community, his approach to rural life has combined the experience and local knowledge from his childhood with global and researched information on plants, the value of organic produce, rural economies, and environmental issues. His tertiary education (and a degree from the Faculty of Philosophy, Ss. Cyril and Methodius, Skopje, Republic of Macedonia in 1984) makes him different from migrants from Zavoj who are educated but have relocated permanently to Skopje, the capital city of the nation, and from local migrants who have not applied their education and have also not returned to Zavoj to live (Lozanovska, 2011).

Naum's story outlines a distinct 'migrant' relationship to food production and the mountainous landscape of Zavoj. He has said it is an 'ideal life', living 1100 metres above sea level with clean air and sunshine, combined with only 20 kilometres travel to Ohrid, and suggests there is potential to make a living (Naum, pers. comm., Zavoj, 2010). He was able to put a new roof on the house from the profits of his honey. More impressive is his knowledge about the forest fruits and plants specific to the region – at least four species of berries, plums, wild pear, wild cherry, and a variety of mushrooms. With one other retired migrant he has started to graft the pears in the forest, and has been discovering long-abandoned orchards, some including apples local to the area. The terrain is rugged but with his donkey, Čirko, he can transport food and firewood. His education has mediated his relationship to the embodied memory in the village and mountainous landscape from the emigrant community's predominant interest in social gathering. Rather it has re-activated an interest in agriculture and food production and a new interpretation and value of the landscape associated with this history.

Figure 8.5 Taseski house, Zavoj. 'Migrant generation' member Naum, who has returned to live in the village, talks of the clean mountain air. Photograph of rear vegetable plot – with Naum tending the bee hives – illustrates the distant views.

Source: Photograph: Lozanovska, fieldwork, Zavoj, Republic of Macedonia, September 2010

Describing the conditions of their childhood in Zavoj in the 1960s and 1970s before mechanisation and modernisation – overcrowding, little hygiene, no running or centralised water, electricity, or sewage system – the now middle-aged migrant generation have illustrated the scale of change in agriculture and in food production. Their memory of, for example, the village tap that was ornately constructed is mixed with fondness for their ancestors. As children, the 'migrant generation' assisted with agricultural work. The boys herded animals into the mountains, learning the mountain tracks, frequently staying in the mountains overnight. Mile adds that children wandered in the mountains and made the environment their playground, and in winter they played outside on sledges on the snow. Girls tended to crops and fields, and to the domestic preparation of foods for the winter. Participants have said families did not go hungry and their diets were based on their own produce, milk, cheese, fat, meat – no biscuits or sweets – though at times bread and sugar (Mile, pers. comm., Zavoj, 2010).

Since 1999, Naum has lived with his wife, Olga, in the house that was purchased by his father but has since been renovated, with the installation of a modern bathroom in 2001, a wood stove, and several combined electric/wood stoves (Naum, pers. comm., Zavoj, 2010). It is a modest but modern home. Many new houses have appeared in the village but remain closed during the weekdays of the summer months. In contrast everyday food production activities enliven the space around their house. In one corner of the front yard is a shed where all the harvested forest fruits are sorted and processed. The outdoor kitchen adjacent to this shed serves for the preparation of winter preserves, including *ajvar*, a paprika relish specialty of the region, fruit compotes, jams, and pickles. A decorative outdoor trough with tap is located near a covered porch used for sitting and viewing the main village street. This type of outdoor trough was also noted in the houses in Leskoeç and Železnička Naselba, and in studies of migrant houses in Melbourne, Australia (Lozanovska, 2014). Different from garden taps, they indicate uses related to food production, and exemplify mixed rural-urban home practices.

A panoramic photograph of Zavoj illustrates the breathtaking view from the site of the village high on the mountain (see Figure 8.5). During the summer, Naum spends the day in the mountains rediscovering old rootstock and remapping the agricultural history or in the vegetable garden at the rear. A small piece of land behind the house is packed with edible foods including new introductions such as broccoli, and traditional vegetables – tomatoes, beans, and many varieties of peppers (*piperki*), an orchard (cherry, raspberry, Fuji apple), and bees (Naum, pers. comm., Zavoj, 2010). The sloping site has not been an obstacle to the cultivation of this vegetable garden. The view of the mountainous terrain provides an indication of what 1100 metres above sea level means. Olga is mostly in the outdoors at the front productive zone of the house.

Conclusion

Societies, capitalist or communist both, employed science and technology to increase the efficiency, rationalise, and standardise food production. In addition, new management and organisation structures, such as Fordism and Taylorism, were employed internationally in order to gain the most out of the labourers. Co-operatives were implemented to create food surplus that will sustain the urban migrants engaged in industries, as well as produce crops employed by these new industries. They had an immense impact on the countryside associated with the village as the traditional core model for food production. Peasants did not have the knowledge to produce these new crops, neither did the crops have any value for the domestic economy of their household. Large tracts of land managed by co-operatives homogenised the rural landscape. Monocropping was facilitated by the use of machinery, irrigation systems, and fertilisers, and endangered the ecology of the landscape (see Chapters 9, 10, 16 and 17). In contrast, the small private gardens were intensively cultivated and used for livestock rearing and complemented the household economy.

Collectivisation was implemented differently across the Balkans. Yugoslavia was a case in point of less stringent control on land, food production, and distribution of the food products. Nevertheless, the effects on the village were similar with an erasure of its role in food production due to rural-to-urban migration.

In the post-communist era very few villages in the Balkans or Eastern Europe have returned to their former self-sufficiency in food production or to a renewed role promoting a new ecology of food production. Departure from the villages and the industrialisation of food production under state socialism caused an irreversible impact, and their edible landscapes are disappearing.

In the twenty-first century, the village is navigating its role in a global economy. Niche food products such as honey or cheese, and the activities of leisure and tourism, led to the commodification of the rural areas. However, in Zavoj, we have argued that the activities of the migrants' generation are not the same as touristic recreational private leisure activities of urban people in the countryside (Lofvers and Musch, 2004). The Zavoj migrants return to this particular village landscape because it is directly linked to their individual and collective memory. Their activities are sometimes accompanied by tending to small fields, and a cultivation of the mountains near the village, not only as owned land but as an inherited landscape.

Despite the ongoing connections of the 'migrant generation', an increasing number of villages lie abandoned in the Balkans and in Eastern Europe. The deterioration of their formal, geometric, and material order parallels how they are symbolically forgotten as places where people once cultivated their livelihood and lives. In that process, the villages transform themselves into the landscape, a landscape rewritten through rural practices of food production and rural settlement, embedding the foundations and seeds of human history.

Note: fieldwork, Zavoj, Republic of Macedonia

Research in the village, including documentation of the village landscape and built environment and its countryside, was conducted during a series of fieldwork trips in the period 1988–2015, with 23 interviews conducted in 1988 of the inhabitants belonging to the 'agricultural generation' and ten interviews in 2010 of the members of the 'migrant generation'. In the 1970s many emigrants from Zavoj relocated to Leskoeç, a village 16 kilometres from Zavoj, heading towards the town of Ohrid, and Železnička Naselba, a fringe district of the town of Ohrid, 20 kilometres from Zavoj. The fieldwork carried out in 2010 included ten interviews with Zavoj emigrants in both Železnička Naselba and Leskoeç, a tour of their homes, and architectural documentation of the houses in their new settlements.

Key texts

Kideckel, D. (1993) *The solitude of collectivism: Romanian villagers to the revolution and beyond*. Ithaca: Cornell University Press.

Kligman, G., & Verdery, K. (2011) *Peasants under siege: The collectivization of Romanian agriculture*, 1949–1962. Princeton, NJ: Princeton University Press. www.jstor.org/stable/j.ctt7rz08.10 (accessed 27 June 2016).

Salzmann, Z., & Scheufler, V. (1974) *Komatov: A Czech farming village*. USA: Holt, Rinehart and Winston.

Turnock, D. (1997) *The East European economy in context: Communism and transition*. London: Routledge.

Further reading

Achkoska, V. (2007) *Egzodusot Na Makedonskoto Selanstvo Vo Socijalizmot*. Makavej, Skopje, Republic of Macedonia.

Adamson, K. (2007) Discourses of violence and the ideological strategies of the Romanian communist party, 1944–1953. *East European Politics and Societies*, 21(4), 559–587.

Armstrong, H. (1999) Migrants' domestic gardens: A people-plant expression of the experience of migration. In: Burchett, M.D., Tarran, J., & Wood, R.A. (eds.), *Towards a New Millennium in People-Plant Relationships*, 28(35), 237–260.

Balevski, M. (1981) *Facts about the socialist Republic of Macedonia*. The secretariat for information of the socialist republic of Macedonia, Skopje.

Berend, I. (1996) *Central and eastern Europe, 1944–1993: Detour from the periphery to the periphery*. Cambridge: Cambridge University Press.

Berger, J., & Mohr, J. (1975) *A seventh man: The story of migrant worker in Europe*. Cambridge: Granta Books.

Carter, F.W., & Turnock, D. (1996) *Environmental problems in Eastern Europe*. First published in 1993, London: Routledge.

Cernea, M. (1978) Macrosocial change, feminization of agriculture and Peasant women's threefold economic role. *Sociologia Ruralis*, 18(2–3), 107–124.

de Certeau, M. (1984) *The practice of everyday life*. Berkeley, CA: University of California Press.

Eliade, M. (1959) *The sacred and the profane: The nature of religion*. San Diego: Harcourt Brace Jovanovich.

Glassheim, E. (2006) Ethnic cleansing, communism, and environmental devastation in Czechoslovakia's Borderlands, 1945–1989. *The Journal of Modern History*, 78(1), 65–92.

Golob, T. (2009) Slovenian migrants in transnational social spaces: Exploring multylayered identifications and ambivalent belongings. *Anthropological Notebooks*, 15(3), 65–77.

Gruev, M. (2014) Collectivization and social change in Bulgaria, 1940s–1950s. In: Iordachi, C., & Bauerkamper, A. (eds.), *The collectivization of agriculture in communist eastern Europe: Comparison and entanglements*. Budapest and New York: Central European University Press, pp. 329–368.

Hage, G. (1997) At home in the entrails of the West. In: Grace, H., Hage, G., & Johnson, L. (eds.), *Home/world; space, community and marginality in Sydney's west*. Annundale, NSW: Pluto Press.

Horvat, B. (1973) Agricultural population, landownership pattern, and population. *Eastern European Economics*, 2(2), 3–17.

Kalezić, D (1984), *The socialist Republic of Macedonia today*. Skopje: Samoupravna Praksa.

Kaneff, D. (1996) Responses to 'democratic' land reforms in a Bulgarian village. In: Abrahams, R. (ed.), *After socialism: Land reform and social change in eastern Europe*. Oxford: Berghahn Books, pp. 85–114.

Kideckel, D. (1982) The socialist transformation of agriculture in a Romanian Commune, 1945–62. *American Ethnologist*, 9(2), 320–340.

Lofvers, W., & Musch, M. (2004) The countryside. *OASE: Architectural Journal*, 63, 2–13.

Lozanovska, M. (1989) Gender and architecture in a Macedonian village. *Exedra, Deakin University Journal on Architecture*, 1(2), 24–31.

Lozanovska, M. (2011) Sacred time after emigration: A study of the Holy Mother Festival in Zavoj. In: Küchler, S., Kürti, L., & Elkadi, H. (eds.), *Every day's a festival!- diversity on show*. Wantage, UK: Sean Kingston Publishing, pp. 113–139.

Lozanovska, M. (2014) Edible suburbs. In: Keeffe, G., & Roggema, R. (eds.), *Why we need small cows: Ways to design for urban agriculture*. Wageningen: VHL University Press, pp. 273–292.

Macours, K., & Swinnen, J. (2000) Causes of output decline in economic transition: The case of Central and Eastern European Agriculture. *Journal of Comparative Economics*, 28(1).

Mazurski, K. (1986) The destruction of forests in the Polish Sudetes Mountains by industrial emissions. *Forest Ecology and Management*, 17 (4), 245–255.

Meurs, W. (1999) Land reform in Romania – a never ending story. *South-East Europe Review for Labour and Social Affairs*, 2(2), 109–122.

Micu, C. (2012) A unique policy of urbanization? The rural systematization in Romania. *New Europe College Stefan Odobleja Program Yearbook*, 2011(12), 229–252.

Mihalache, F. (2011) Etape recente in transformarea mediului rural romanesc (Recent stages in the transformation of the Romanian rural environment). In: Ioan-Franc, V. (ed.), *Mediul rural romanesc: evolutii si involutii. Schimbare sociala si antreprenoriat (Romanian rural environment: Evolutions and involutions, Social change and entrepreneurship)*. Bucharest: Expert, pp. 12–30.

Mitrev, A. (1974) *Memoirs, essays, articles* Митрев, А. (1974) Мемоари, Огледи, Статии Скопје/Skopje.

Nowakowska, U., & Ignatowicz, H. (2014) Crops for the state!' Agriculture and the countryside in 1950s propaganda posters of Central and Eastern Europe. *Folk Life: Journal of Ethnological Studies*, 52(1), 62–81.

Rapoport, A. (1992) On cultural landscapes. *TDSR*, 3(2), 33–47.

Rey, V., & Bachvarov, M. (1998) Rural settlements in transition- agricultural and countryside crisis in Central-Eastern Europe. *GeoJournal*, 44 (4), 345–353.

Ronnas, P. (1989) Turning the Romanian peasant into a new socialist man: An assessment of rural development policy in Romania. *Soviet Studies*, 41(4), 543–559.

Scott, J. (1998) *Seeing like a state: How certain schemes to improve the human condition have failed*. New Haven, CT and London: Yale University Press.

Sisestean, G. (2012) Un fragment de viata cotidiana in satul colectivizat din Romania anilor 1971–1985 (A fragment of everyday life in the collectivized village in Romania between 1971–1985). *Revista Romana de Sociologie*, 3–4, 237–263.

Socol, G. (2003) Cadrul de viata si calitatea vietii taranimii in perioada comunista (The way of life and the quality of life for small farmers in Romania, during communism). *Calitatea vietii*. Bucharest, Edit. Academiei Romane, 14(2), 205–218.

Statistical Office, Households and Dwellings in the Republic of Macedonia (2004) *Census of population, Republic of Macedonia, 2002*. Skopje.

Thompson, S. (1993) Agrarian reform in Eastern Europe following World War I: Motives and outcomes. *American Journal of Agricultural Economics*, 75(3), 840–844.

Trebici, V., & Ghinoiu, I. (1986) *Demografie si etnografie* (Demography and ethnography). Bucharest: Editura Stiintifica si Enciclopedica.

Turnock, D. (1991) Romanian villages: Rural planning under communism. *Rural History*, 2(1), 81–112.

Turnock, D. (2006) Settlement history and sustainability in the Carpathians in the eighteenth and nineteenth centuries. *Review of Historical Geography and Toponomastics*, 1(1), 31–60.

Williams, R. (1973) *The country and the city*. London: Chatto and Windus.

Part II
Agricultures

9

Shifts in agriculture praxis

Farm modernisation and global integration

Anders Wästfelt

Introduction

The last 100 years have seen a tremendous change in how agriculture is practised in the Western global and economically integrated world, while at the same time traditional practices are used in other parts of the world (Federico, 2009;Von Braun, 2007).These discrepancies relate to the extent of urbanisation and industrialisation. Most significant is the major shift in how farming has become dependent on the world market and how this shift has dramatically altered agriculture. It has changed not only how farming is conducted, but also the world's entire relationship with agriculture, landscape, and food production (Clapp, 2015).

This chapter focuses on the characteristics of the most important shifts in agriculture praxis during the 20th century and the beginning of the 21st century. Shifts in agriculture praxis can be interpreted as forced by both endogenous and exogenous factors in relation to farms. Mark Tauger describes this as a "dual subordination"; farmers are both dependent on local environmental conditions and subject to rules of agencies outside the farm (Tauger, 2013). Innovations and technical development have driven long-term changes on farms, but agriculture praxis is also modified by socioeconomic pressure from multiple surrounding and interacting socio-technological domains, changed food requirements, transnational policies, regional regulation, and national laws. Energy prices and labour costs in other sectors are factors that have strongly impacted farmers as they have become globally integrated and forced to adapt to the socioeconomic situation in which they are situated. Farmers at all times have adapted to new circumstances. Some become early adopters of new techniques while others maintain the traditional praxis. Sustained praxis or innovative use are often related to the transition of farms between generations and are also strongly connected to limitations set by the property structure and landscape configuration.

How geography matters and how agriculture is different

Geography matters in permanent agriculture because it is dependent on local conditions, as cultivation is primarily a bioecological process (Bowler, 2014; Federico, 2009). It differs from industrial production because it is fixed in space, and of necessity is spatially dispersed. Nomadic pastoralists, on the contrary, depend on seasonal migration of animals to good grazing areas, and extensive land use. In many developed societies, the immovability of the permanent settlement and raising of animals and growing cereals have gone hand in hand historically (Myrdal and Morell, 2011). Historically, mixed farming has been practised in many places on Earth, in different forms, and it is important to recognise the advantages of mixed farming in order to understand long lasting rural societies. In the mixed farming system, managing animals and growing a variety of crops to a large extent represent local optimisation of the use of accessible labour over the yearly crop calendar in relation to climatic conditions (Brassley and Soffe, 2016; Martiin, 2013). Besides increased output compared to pastoralism, one core benefit of mixed agriculture is that as long as farmers are themselves consumers of what they produce, then diversity in production is a benefit. Farming both animals and crops also makes it possible to transfer nutrients from where animals are confined in winter to arable parcels to improve soils with manure.

Mixed farming activities also benefit from markets that can facilitate exchange of produce types between producers. As long as transportation costs were relatively high in relation to the value of products, most farmers were restricted by food production for family or local consumption (Myrdal and Morell, 2011). This also meant that to increase production and specialise in single crops was of limited value. The emergence of high value cash crops, such as tea, sugar, and tobacco, and the introduction of transportation through deep-water vessels and railroads introduced international markets. These developments impacted farming decisions and local land use models. In industrialised and highly urbanised societies, and in regional, national, and international markets, mechanisation and industrialisation of farming and transportation superseded the spatial restrictions of the mixed farming model. The integration of agriculture into international markets has increased labour division and specialisation.

The focus in this chapter is on changes from European mixed farming praxis where consumption and production historically were geographically close to each other, to a highly globally integrated sector. Industrialisation and market integration made the aforementioned shifts possible and were aided by liberalisation of markets, intensified transportation, and increased market exchange. An example from Sweden will demonstrate the practical implications of what the shift has meant in practice. Finally, changes in European policy and post-rural farming praxis in the 21st century are described.

Shifts in agriculture practices – a Swedish example

Swedish farmers on small to midsized farms (2–30 hectares/5–75 acres) 100 years ago farmed in traditional ways to support the family with food, and exchanged their surpluses. All farms had horses or oxen for heavy field work. Other common animals on most farms were pigs, sheep, chickens, and milking cows for the provision of meat, egg, and milk products. Many different crops were sown and every family had its own vegetable garden (Knudsen, 2009). Manure stored in the winter was essential for replenishing fields with nutrients.

At the end of the 19th century, increasing urbanisation, industrialisation, and distribution infrastructure – such as railroads and roads – allowed many farmers to produce for market sale. Farmers' cooperatives were often organised to manage the processing and distribution of dairy products, meat, or milled flour. In remote rural areas, however, production remained directed towards self-sufficiency, often in combination with seasonal labour migration. In these rural societies, an important aim for families was transferring the farm and the farming skills to the next generation. This is still key in large parts of the world. Farming in the 19th century generated a diverse landscape with arable fields, meadows, and pastures (Emanuelsson et al., 2009). This pattern is still evident in large parts of the world and even in parts of Eastern Europe.

Over the last 150–200 years, four major shifts in how farming is conducted can be identified, which have impacted both farm work and the farm structure: (1) intensification by crop rotation, plant breeding, and agro chemistry have greatly increased yields; (2) mechanisation by fossil fuel-driven machinery has replaced draught animals (and makes it possible to use their grazing lands for cultivation); (3) structural adjustment of fields for adaptation to new techniques has led to many small farms being replaced by larger enterprises; and (4) to increase labour productivity, specialisation has replaced mixed farming (Bowler, 2014; Primdahl and Swaffield, 2010).

By the end of the 20th century, full-time farmers worked in a completely different context than in the early 20th century. They often bought or leased adjacent farms, transforming them into specialised businesses producing commodities for the world market. They had become managers reacting to and depending on the market (Knudsen, 2009). By using new machinery, a single farmer could manage areas that 100 years earlier might have needed 100 or more persons and an equal number of horses. Today a single farmer in Sweden, using the largest machine available, can farm about 500–700 hectares (approximately 1200–1700 acres) (Juhola et al., 2017). Some have invested in privately owned silos to enable storage of products until prices peak on the world market. Contemporary farmers normally produce meat, milk, or cereals, growing only cash crops, and they often use cropping systems involving very few crops in rotation. Today, farmers rarely consume what they produce. They act as intentional owners of small companies striving towards improving labour productivity and profitability. Adaptation to utilise European Union subsidies is essential to improve the profitability of the contemporary farm. This includes landscape preservation, such as grassland for grazing. In this context, the primary goal is not the increase in production but the acquisition of the subsidy. Many smaller farms have become amalgamated into larger farms, or have become hobby or horse farms producing recreational activities. Both large scale agriculture operations and small hobby farms are today part of the agriculture landscape, although at different scales.

In addition to the transformation towards large scale commodity producers consisting mainly of large scale full-time farmers, there is also a long-term trend where family farms are kept in the family but the former production diversity has instead changed to income generating diversity, whereby agricultural production becomes one small income stream. The differences between the family farms and the commodity farm previously described did not occur quickly or without intention. Farms were initially restructured and industrialised to increase food provisioning after World War II and to increase the wealth of European countries (Hoyle et al., 2010). As an example, before World War II, farming in Sweden continued the model of pre-industrial agriculture and very few farmers had access to tractors. Accordingly, the Swedish government, during the postwar period up until the late 1980s, encouraged agricultural restructuring through county administrative boards and their agriculture departments by facilitating the amalgamation of smaller farms (2–20 hectares) into larger farms (Jansson and Wästfelt, 2010).

The Swedish subsidy system after World War II was designed to help the farmers' income keep pace with that in other sectors of society. In the 1980s the restructuring and improvements had become so efficient in Sweden and also in Europe that they produced much more than the market demanded. Production, however, continued to increase. Since the subsidies were designed to guarantee fixed prices for farmers, the Swedish government became owners of large quantities of agricultural commodities that it could not sell on the local market. Instead, the products were sold on world markets for low prices. This also meant that, over time, costs became higher and higher for the state. To counteract this problem, in 1990, Sweden was the second country in the world (after New Zealand) to completely de-regulate national subsidies to agriculture and accept full competition on the world market (Lindberg, 2008). However, with the accession of Sweden to the European Union in 1995 and the consequent adoption of the Common Agricultural Policy (Daugbjerg, 1997), Swedish agriculture sought a short re-regulation process. Since 2005 this has included the adoption of a direct payments system and alignment with the common market organisation (Common Agriculture Policy pillar 1), and inclusion in the commonly regulated rural development policy package (Common Agriculture Policy pillar 2) (Sorrentino and Henke, 2016).

Although farming praxis changed dramatically and fundamentally after World War II, no single factor caused the changes; rather, there were a number of simultaneous processes.

Since mechanisation and industrialisation commenced, agricultural economies have been driven by the possibility of increasing labour efficiency. New technologies, new crops, and changing practices have increased the productivity from each unit of land (Federico, 2009). This increase has quadrupled the outcome in some regions due to new crops, artificial pesticides and fertilisers and often the addition of improved drainage and/or irrigation systems. The increase in labour efficiency has led to a decrease in the demand for labour in agriculture. At the same time, labour demand has increased in non-agricultural industries. This process is a strong driver behind urbanisation.

Farm economics are crucial to and inherent in agricultural management; however, farmers need to respond in new ways as they become increasingly involved in shifting commodity markets. Farm economics and technical shifts, however, cannot be perceived as independent of policy shifts and changes in the overall economies of society. The spatial integration of economies over larger and larger geographical areas, along with lower relative transportation costs, are important factors specifically affecting agriculture. Yet local agricultural practices not only need to adapt to limitations of fixed locations, but must also keep up with shifting regional and global socioeconomics.

Accordingly, contemporary farmers have had to shift from being self-sufficient owners to business managers. In this new role they need to keep up with technological development and try to increase revenue. The benefits of this process have become questioned as it has both social and environmental impacts. From a different point of view, the modernisation and industrialisation of farming are viewed as reducing the food sovereignty of farmers and regions and as being elements of the expanding modern world capitalism (Bernstein, 2016; Douwe van der Ploeg, 2008; Weis, 2010).

European policy shifts from growth to global integration

After World War II, policy in Europe, as in Sweden, was formulated to support a strong productivity increase to feed the population, but also so that farmers would have good incomes. The introduction of larger but fewer farms was rationalised by focusing on specialisation and efficiency in resource use. Until 1985, the main effect of the Common Agriculture Policy was to stimulate production, largely by maintaining domestic European prices above world market prices as income support to farmers (Hoyle et al., 2010). From a production efficiency perspective, the 1960s and 1970s can be seen as the golden age of agriculture. The prices were relatively high and farmers harvested the fruits of two decades of investment in restructuring and new machinery, but at the cost of environmental problems. An early reaction and counter-movement was the 'green movement' in the 1960s and 1970s. Organic agriculture was re-invented in opposition to the use of agriculture chemicals (see Chapter 10). This model had been in use a hundred years previously as conventional praxis. The new organic farms established were initially relatively small in scale (Padel, 2001).

As previously noted in the Swedish example, the success of European agricultural postwar policies in the 1980s in generating such large surpluses meant that it was not possible to sell it at reasonable prices. One solution was to sell it at low prices to the developing world. Eventually, however, this practice was greatly criticised as limiting the development of domestic agriculture in developing countries (McMichael, 2005).

Agriculture policy generally shifted from the support of active modernisation through the 1980s, to de-regulation in the 1990s, and then to the decoupling of subsidies. In 2003, with the Fischler reform, the European Union decided to introduce the single payment scheme and thereby drew subsidies away from the promotion of production and towards support based on the size of farmed areas (pillar 1) (Sorrentino and Henke, 2016). This policy change was, to a large extent, forced by negotiations with the World Trade Organisation (WTO) to decrease customs duties and allow farmers to compete on world markets. It is noteworthy that when the Rural Development Programme was introduced as a second pillar of the Common Agriculture Policy, it meant that the European model of agriculture passed into a post-productive phase (Wilson, 2001). A scientific interpretation of these changes is the introduction of the concept of multifunctional agriculture. Wilson (2008) argued that multifunctional agriculture could either be strong or weak, where the weakly multifunctional farms were highly globally integrated while the strong were based on mixed farming with greater independence of global commodity markets. The decoupling of European Union subsidies by the Fischler reform further changed the agricultural sector, leading to fewer and larger farm units (Giannakis and Bruggeman, 2015). European farmers also became price-takers without any possibility to influence or control market prices. They now have to compete as part of or against the globalised food system.

The Fischler reform was intentionally designed to reduce agriculture intensification as a response to overproduction. At the same time, it introduced environmental awareness into the subsidy schemes. These changes, however, have had an effect on farming structures. Farmers now have incentives to reorganise and find new ways to enlarge farms as the subsidies are delivered in relation to area, but without any obligation to increase production. So from a business point of view they have incentives to increase the size of farms.

This recent shifts in the Common Agriculture Policy, together with world market competition, have forced farmers to continue to improve labour efficiency (Baráth and Fertő, 2017). This has mainly been done through horizontal intensification (enlargement), meaning that larger farms are more economically preferable, and are promoted by the subsidy system. Strong promotion of international trade and accepting imports of products even if they do not have the quality required inside the European Union encourage intensification. Different forms of vertical integration (market integration), including highly labour efficient commodity production, niche producers, and organic producers, have been ways for farmers to try to keep pace with increasing competition.

In the early 21st century, specialisation continues since productivity of a single commodity can be technically optimised and low production costs ensured. On a global level it is possible to notice that economically richer countries, where agriculture is a small part of the gross domestic product, are expanding production of a small number of crops. Yet in economically poorer countries, labour-intensive and diverse agriculture continues to grow.

Subsequently, the farming praxis is sensitive to subsidies and market conditions at other places throughout the world. This is made possible by the steady market integration and continuous decrease in relative transportation costs.

The subsidy schemes in the European Union are today designed to steer agriculture in environmentally favourable ways (Baylis et al., 2008). Recently, researchers have been focusing on the landscape scale as a more appropriate context for the application of subsidy schemes (Lefebvre et al., 2015). In Europe, the Common Agriculture Policy is designed for and directed towards individual farms and farmers. However, this system lacks the capacity to hinder the continuing structural adjustment towards large farms. Instead the direct payment by area of land has had the indirect effect of encouraging the aggregation of farms. Therefore, profitability can increase even if the yields diminish, since the subsidy increases in line with increasing land area. Increasing farm size is supported by the new high-tech methods for managing larger areas of land, but this does not necessarily mean that land is properly managed or used in an environmentally friendly way (Wästfelt, 2017).

Farmers respond to increased competition arising from enlargement of the market by reducing product diversity to improve business and labour productivity. This process can be replicated at different levels, both at the farm level but also at the global level. Fodder, for example, can be cheaply produced on one continent and transported to feed animals on another. Such increasingly complex intersections on continental and regional scales, however, affect new local relationships, leading to an overall decrease in biodiversity, local subsistence insecurity, and increased use of fossil fuels (Müller and Munroe, 2014).

The benefit from trade comes partly from the fact that adaptation to local ecological conditions at one place can improve yields, but it also comes from the fact that labour costs can be much lower in other places. As long as the transportation cost is no larger than the difference in labour cost, trade is favourable. Paradoxically, it might be possible to produce the same product or fodder locally but in relative economic terms it is not feasible, even if it would benefit the local community and the overall environment.

Continuing globalisation and industrialisation produce new labour market opportunities and result in continuing urbanisation. In this process, societies effectively distance themselves from environment stewardship and traditional agriculture (Weis, 2010). As a counter reaction to industrialised agriculture, the discourse around food and organic production has significantly increased in the last 20 years (Reganold and Wachter, 2016; Rotz and Fraser, 2015). This discussion still largely excludes commodity agriculture, which, to a large extent, is seen as having its own development course (Klintman and Boström, 2004) (even if all individual farms have their own characteristics). The commodity agriculture sector is often compared to other industries, and like them, is seen as not embracing other social and environmental values (see Chapter 16).

Eco-modernisation

Reflecting on the history of agricultural production and its transformation in the 20th century identifies shifts in agriculture praxis towards specialisation and increasing labour efficiency. Subsequent contemporary changes, apart from converting agriculture land into areas for recreational activities such as horse farms or golf courses (see Chapters 14 and 15), also include the development of organic and niche producers. This can be seen as a reaction to industrialised agriculture and commodity production (Padel, 2001). This growing awareness of the environmental qualities of both food products and production methods has also resulted in a growing interest in community-supported agriculture as well as urban and peri-urban agriculture (Nost, 2014). Even if production organisation changes have been progressing simultaneously, these are not synchronised as they have different goals.

Organic agriculture re-conceived between World War I and World War II was inspired by organic methods (especially the use of compost) in Asia and developed in parallel in England and Germany (see Chapter 10). In the postwar era, organic agriculture was seen as an alternative to labour-efficient but environmentally degrading conventional agriculture.

Since the 1990s, organic agriculture has increased its market share in Europe (Willer and Schaack, 2015) and in some instances, farms have increased in scale to levels similar to conventional production. Today, organic farms often equal the largest conventional farms in size (Wästfelt et al., 2012). However, these farms operate with a different praxis and often strive for a high degree of self-sufficiency in manure production and on-farm fertilisation. Many organic farms raise animals and grow cereals for the market. Organic farming can be seen as a reintroduction of mixed farming in a new way. Among its comparative advantages is the elimination of the need to purchase artificial fertiliser and pesticide, and the fact that the demand is slightly higher than the produce, meaning that the prices are also higher. European Common Agriculture Policy subsidies can also make it slightly more profitable to be an organic farmer; on the other hand such farming is often characterised by lower labour efficiency (Reganold and Wachter, 2016). In organic agriculture it has been possible to see that farmers often need to organise themselves in cooperatives to make it possible to deliver larger quantities of produce when the market evolves.

Global integration

A global integration in agriculture occurred in four different waves; the first ended with the globalisation of plants which happened in the late 18th century when all known crop varieties were tested for their ability to grow in every region of the planet (Federico, 2009). The second wave of global integration was the development of long distance transportation from the 18th and 19th centuries. This transformed local agriculture by integrating it into wider markets. In Britain, for example, this process had a huge impact of partly restructuring agriculture which substituted domestic production of cereal with imports from Australia, and changed the direction of domestic production towards wool production. The third wave was the industrialisation of agriculture with artificial fertilisers, pesticides, and heavy machinery driven by fossil fuels. After the World War II, agriculture was mostly of national interest and strictly regulated. While the

US played a significant role in mechanisation, industrialisation and trade, the first steps in the liberalisation were taken in the 1970s (Tauger, 2013). The fourth wave started after the collapse of the Soviet empire in the early 1990s as a wave of intensified global market integration, when many countries signed agreements with the WTO and the world market expanded. This last wave also included the establishment of super large scale farms, often driven by financial interests (Kuns et al., 2016).

As in other sectors, globalisation has been perceived as good for the economy. In the last 20 years in the field of agriculture, a number of enacted agreements have impacted the flow of agriculture products and forced local restructuring and global specialisation. Since 1960, the transportation of agriculture goods has increased 500% (Lambin, 2012). Historically, lightweight products with high value were transported; today, bulk transportation, for example of fodder, occurs from one continent to another. This has resulted in the promotion of industrial agriculture, which in turn has been claimed to reduce local food sovereignty (Jansen, 2015; see Chapter 31).

The increased market integration is coupled with unprecedented mechanisation of agriculture and the industrialisation of society, which is synergistic with ongoing urbanisation, and was especially so during the 20th century. The relationship between urban and rural areas has been transformed, with most people no longer having contact with agriculture, even in rural areas (Clapp, 2015; Woods, 2009). The globalisation of food production is ongoing and land use is becoming more generically interrelated throughout the world (Bergmann and Holmberg, 2016).

Contemporary praxis directions

Farming strategies have never been uniform. In contemporary Europe, it is possible to recognise four characteristic farming strategies that show how different forms of farming praxis relate to inputs of labour, capital, and land: (1) In eastern and southern Europe, the economic crisis has meant that many people have continued farming small family farms that are self-sufficient in production. These farms are often in remote areas or hillsides unable to convert to a larger scale. (2) On Europe's agricultural plains, farming has mostly become highly specialised and the size of individual farms has greatly increased. (3) An alternative to conventional farming developed in the 1970s as a reaction to the environmental problems resulting from rapid modernisation after World War II. Since the 1990s, many organic farms have transformed into large enterprises equal in size to conventional farms, and it is interesting to note that they often reintroduced animals to provide cheap access to manure. (4) To counteract the global commodity market, many niche producers try to make a profit from cutting the distribution and value chain through direct selling (such as farm-gate sales and farmers' markets). A new interest in peri-urban and urban agriculture is also part of this trend (see Chapters 12, 13, 14, 15 and 22).

In the 1990s it was argued in Europe that future farming should be multifunctional, with farmers producing a number of environmental services and values in addition to food (Wilson and Rigg, 2003). Contemporary Common Agriculture Policy subsidies advance these services as a public good (Lefebvre et al., 2015). Even if this is still the case, farmers have not stopped changing their farming strategies. Instead of diversification in production, most farmers in Sweden have additional income from non-farm business or employment. However, to operate within this strategy, it is necessary to have a highly specialised farming activity or methods which do not require daily control.

Niche producers in particular endeavour to gain benefits from the comparative advantages provided by a specific place. Many farmers try to benefit by loading other values, such as environmental qualities or the recreational value of coming to the farm for self-harvesting, into the agriculture product and enterprise. Despite this, the dramatic scale increase in commodity producing farms in Sweden continues, which can be interpreted as resulting from incentives introduced by the world market to achieve increased labour efficiency, although the area is not necessarily used efficiently, as it is subsidised.

Farm modernisation and global integration

Over the last few centuries agriculture development has been driving societal development while being forced to continuously improve competitiveness from other society sectors and to adapt to rapidly integrated global market agents. Today, it is equally important to be able to store and sell the products right in time as a professional grower to make it possible to make good returns from farming. What is typical is that the farming is always dependent on unique environmental conditions which are fixed. This limitation has to a large extent been overruled by the modernised high technology agriculture that developed during the 20th century. It has resulted in an increased labour division and specialisation worldwide.

In the integrated Western world: (1) Mixed farming has lost its high importance. (2) Family farms are no longer in need of much manual work. (3) Farmers are striving to achieve labour efficiency either by horizontal or vertical integration. Existing family farmers have become transformed into business managers with increased dependency on inputs from outside the farm, as well as increased dependency on world market prices. (4) There has been a change from a situation of diverse activities on farms in the mixed farming system to income diversity, where the farm is just one source of income. (5) The increased importance of world trade agreements in recent decades in Europe has resulted in changed policies designed to promote in parallel the environmental use of land and the creation of incentives to increase farm size.

The changes in praxis follow policy shifts, which in Europe have gone from a promotion of production improvement after the World War II to promotion of post-production and subsidies designed for multifunctional agriculture.

Paradoxically, parallel to the described shifts, organic and niche production has grown steadily during the last 20 years and the products are in demand from the urban population. Organic production has often re-established mixed farming in combining cereal production and raising animals with a focus on synergies in production.

Considering the overall picture it can be noted that today there are global tendencies towards overproduction which force continuous restructuring. It has also been recognised that the ongoing processes drive apart small and large scale farms worldwide; small farms in developing countries become smaller and more dependent on self-subsistence while large farms in the industrialised world are turned into super large farms to compete with each other (Von Braun, 2007).

What is observable is that all these local to global processes have strong impacts on both farm spatiality and how farming is conducted. But all farms are still affected by dual subordination, which nowadays can be re-interpreted as triple subordination, as adaptation to both global markets and regional subsidy systems are needed along with the dependency of spatially fixed local conditions.

Further reading

Baráth, L., & Fertő, I. (2017) Productivity and convergence in European agriculture. *Journal of Agricultural Economics*, 68, 228–248.

Baylis, K., Peplow, S., Rausser, G., & Simon, L. (2008) Agri-environmental policies in the EU and United States: A comparison. *Ecological Economics*, 65, 753–764.

Bergmann, L., & Holmberg, M. (2016) Land in motion. *Annals of the American Association of Geographers*, 1–25.

Bernstein, H. (2016) Agrarian political economy and modern world capitalism: The contributions of food regime analysis. *The Journal of Peasant Studies*, 43, 611–647.

Bowler, I., (2014). *The geography of agriculture in developed market economies*. London: Routledge

Brassley, P., & Soffe, R. (2016) *Agriculture: A very short introduction*. Oxford: Oxford University Press.

Clapp, J. (2015) Distant agricultural landscapes. *Sustainability Science*, 10, 305–316.

Daugbjerg, C. (1997) Policy networks and agricultural policy reforms: Explaining deregulation in Sweden and re-regulation in the European community. *Governance*, 10, 123–141.

Douwe van der Ploeg, J. (2008) *The new peasantries: Struggles for autonomy and sustainability in an era of empire and globalization*. London and Sterling, VA: Earthscan.

Emanuelsson, U., Arding, M., Petersson, M. (2009) *The rural landscapes of Europe: How man has shaped European nature*. Stockholm: Formas

Federico, G. (2009) *Feeding the world: An economic history of agriculture, 1800–2000*. Princeton, NJ: Princeton University Press.

Giannakis, E., & Bruggeman, A. (2015) The highly variable economic performance of European agriculture. *Land Use Policy*, 45, 26–35.

Hoyle, R.W., van Bavel, B.J.P., & Brakensiek, S. (2010) *Social relations: Property and power*. Turnhout: Brepols Publishers.

Jansen, K. (2015) The debate on food sovereignty theory: Agrarian capitalism, dispossession and agroecology. *The Journal of Peasant Studies*, 24, 213–232.

Jansson, U., & Wästfelt, A. Rural landscape changes from a long-term perspective: Farming, policy, economy and society from 1750 to today. In: *Placing human geography: Sweden through time and space*. (Placing human geography: Sweden through time and space 2010, 2010)

Juhola, S., Klein, N., Käyhkö, J., & Schmid Neset, T. (2017) Climate change transformations in Nordic agriculture? *Journal of Rural Studies*, 51, 28–36.

Klintman, M., & Boström, M. (2004) Framings of science and ideology: Organic food labelling in the US and Sweden. *Environmental Politics*, 13, 612–634.

Knudsen, A.L. (2009) *Farmers on welfare*. Ithaca, NY: Cornell University Press.

Kuns, B., Visser, O., & Wästfelt, A. (2016) The stock market and the steppe: The challenges faced by stock-market financed, Nordic farming ventures in Russia and Ukraine. *Journal of Rural Studies*, 45, 199–217.

Lambin, E.F. (2012) Global land availability: Malthus versus Ricardo. *Global Food Security*, 1, 83–87.

Lefebvre, M., Espinosa, M., Gomez y Paloma, S., Paracchini, M., Piorr, A., & Zasada, I. (2015) Agricultural landscapes as multi-scale public good and the role of the Common Agricultural Policy. *Journal of Environmental Planning and Management*, 58, 2088–2112.

Lindberg, H. (2008) Politikbyte och idéernas betydelse Reformeringen av den svenska jordbrukspolitiken. *Historisk Tidskrift*, 128, 2–27.

Martiin, C. (2013) *The world of agricultural economics: An introduction*. London: Routledge.

McMichael, P (2005) Global development and the corporate food regime. In: Buttel, F.H., & McMichael, P. (eds.), *New directions in the sociology of global development* (research in rural sociology and development, Vol 11). Bingley: Emerald Group Publishing Limited, pp. 265–299.

Müller, D., & Munroe, D.K. (2014) Current and future challenges in land-use science. *Journal of Land Use Science*, 9, 133–142.

Myrdal, J. & Morell, M. (2011). *The agrarian history of Sweden: From 4000 BC to AD 2000*. Lund: Nordic Academic Press.

Nost, E. (2014) Scaling-up local foods: Commodity practice in community supported agriculture (CSA). *Journal of Rural Studies*, 34, 152–160.

Padel, S. (2001) Conversion to organic farming: A typical example of the diffusion of an innovation? *Sociologia ruralis*, 41, 40–61.

Primdahl, J., & Swaffield, S. (2010) *Globalisation and agricultural landscapes: Change patterns and policy trends in developed countries*. Cambridge: Cambridge University Press.

Reganold, J.P., & Wachter, J.M. (2016) Organic agriculture in the twenty-first century. *Nature Plants*, 2, 15221.

Rotz, S. & Fraser, E.D.G. (2015) Resilience and the industrial food system: Analyzing the impacts of agricultural industrialization on food system vulnerability. *Journal of Environmental Studies and Sciences*, 5, 459–473.

Sorrentino, A., & Henke, R. (2016) *The common agricultural policy after the Fischler Reform: National implementations, impact assessment and the agenda for future reforms*. London: Routledge.

Tauger, M.B. (2013) *Agriculture in world history*. London: Routledge.

Von Braun, J. (2007) *The world food situation: New driving forces and required actions*. Washington, D.C.: International Food Policy Research Institute.

Wästfelt, A., Saltzman, K., Gräslund Berg, E., & Dahlberg, A. (2012) Landscape care paradoxes: Swedish landscape care arrangements in a European context. *Geoforum*, 43, 1171–1181.

Wästfelt, A. (2017) Det svenska lantbrukets omvandling 1990–2014: Excemplet Uppsala län, Eriksson, C. (ed.). Uppsala: Swedish University of Agriculture Science.

Weis, T. (2010) The accelerating biophysical contradictions of industrial capitalist agriculture. *Journal of Agrarian Change*, 10, 315–341.

Willer, H., & Schaack, D. (2015) Organic farming and market development in Europe. In: *The World of organic agriculture. Statistics and emerging trends 2015*. Research Institute of Organic Agriculture (FiBL) and International Federation of Organic Agriculture Movements (IFOAM).

Wilson, G.A., & Rigg, J. (2003) 'Post-productivist' agricultural regimes and the South: Discordant concepts? *Progress in Human Geography*, 27, 681–707.

Wilson, G.A. (2001) From productivism to post-productivism . . ., & back again? Exploring the (un)changed natural and mental landscapes of European agriculture. *Transactions of the Institute of British Geographers*, 26, 77–102.

Wilson, G.A. (2008) From 'weak' to 'strong' multifunctionality: Conceptualising farm-level multifunctional transitional pathways. *Journal of Rural Studies*, 24, 367–383.

Woods, M. (2009) Rural geography: Blurring boundaries and making connections. *Progress in Human Geography*, 33, 849–858.

10

Alternative agriculture

Innovations for growing and cultivating diverse ways of knowing

Jonathan Code

Introduction

This chapter provides an overview of approaches to land stewardship and food production commonly referred to as 'alternative' agriculture. With this designation, the question naturally arises as to which type(s) of agriculture these approaches are contrasted, such that they gain the status 'alternative'? The answer may seem, when first considered, to be self-evident – that is, these are alternatives to the dominant industrial and agribusiness approaches (see Chapters 9, 16 and 31). Whereas this comparison is, in broad terms, the basis for the overview of alternative agriculture that follows, it must be noted from the outset, that the category 'alternative' becomes problematic if it is used to imply a *cohesive* set of agricultural practices that can *collectively* be juxtaposed to the agribusiness and industrial scale methods that came to dominate agricultural landscapes in the 20th century. The reality is somewhat less clear-cut than this. Considering just a handful of examples (biodynamics, organics, and permaculture for instance), noteworthy differences amongst these approaches warrant – it could be argued – further division into sub-categories of 'alternative'. This is to say that, when they are studied in some detail, these alternatives begin to emerge as alternatives to *each other*, as well as to more conventional agricultural practices. The term 'alternative' will, therefore, initially be treated as a convenient umbrella term for the following considerations of agricultural practices that do not follow the often mono-cultural and highly industrialised models previously noted. Once a selection of alternatives has been considered, however, this chapter concludes with a brief discussion that will delve further into the proposal that a distinction between the approaches outlined next may well be justified, and will begin to outline possible directions for such distinctions to be made.

One further consideration should be stated at the outset of this chapter. Whereas agribusiness and industrial scale farming techniques have become, in some parts of the world, the 'norm' for agriculture (the 'alternatives' explored in this chapter thus being contrasted to these), in several areas of the globe, practices referred to as traditional farming, akin to 'organic' farming, and forest gardening have been the 'norm' for centuries (Barker, 2007). Many of these pre-industrial forms of agriculture remain the primary approach to food production for large numbers of people to this day (Morton, 2007) and are evident alternatives to industrial intensive agriculture, though they do not necessarily sit within the context of any one of the alternative approaches that will be addressed later.

With these initial reflections in mind, this chapter focuses on a selection of some of the most evident alternatives to industrial agriculture that emerged in the 20th century and that continue to develop in the early decades of the 21st.

Approach taken

A number of entry points can be taken to a study of different examples of alternative agriculture. Such a study could begin from the outset with a comparison of their different perspectives on technical or practical considerations (such as methods of fertilisation, pest management, animal husbandry) and these could be juxtaposed to practices more prevalent in industrial agriculture. Alternatively, an overview could be provided of the current presence in the global food market for alternative agriculture, or of acreage of land under cultivation within a given set of practices (e.g. acreage under organic cultivation, or biodynamic cultivation). Such an entry point would reveal the fact that agriculture in the early 21st century, particularly in the industrial countries of the northern hemisphere, is still dominated by monocultures sustained by large amounts of synthetic fertilisers, pesticides, and a petroleum-based energy economy (Barker, 2007; Chapter 16) – but these are not the stepping off points for this chapter.

Rather than begin with specifics of technique or current impact, the entry point to this chapter is to outline different forms of alternative agriculture as they developed *in time*. This historical or developmental perspective is taken in order to situate different 'alternatives' in relation to agricultural practices that were current at the time they were instigated, and to which they were frequently a response. New initiatives in agriculture have often been sparked by events unfolding in the wake of earlier innovations, and where these were instances of disease, environmental degradation, soil loss, crop failure, or other crises in the agricultural realm, innovative solutions were sought and developed.

It will become evident through the examples that follow that quite different responses have emerged to the challenges ascribed to industrial agriculture practices. These are, on the one hand, evident as novel methods or techniques that give alternative approaches their distinctive character. An initial focus on these practical features of different approaches to alternative agriculture will be taken. Underlying these practical interventions, however, are less evident (and often less discussed) *ways of knowing* that informed the innovation of the alternative approaches in question. This aspect of alternative agriculture will be touched upon briefly in the final section of this chapter.

Setting the scene

As has been noted in other chapters in this volume, the early years of the 20th century witnessed a number of developments in agriculture that furthered a trend where small- to medium-scale, mixed farms were being supplanted by industrial scale monocultures (Chapters 9, 12, 16, 25 and 31).

In 1910 the Haber-Bosch process was developed for producing ammonia, an artificial nitrogenous fertiliser, and in 1913, BASF's plant in Germany opened with a capacity to produce 30 tonnes per day (Modak, 2002; Smil, 2001). The development of synthetic fertilisers heralded dramatic changes for global agriculture, as they soon came to replace the long-established practice of fertilising with manures and using rotations that included cover crops to restore and/or maintain soil fertility (Conford, 2001; Kuepper, 2010). The development of the tractor, such as the Fordson tractor sold in 1917 and the International Harvester Farmall tractor (released in 1923) would continue to replace draft animals and significant proportions of farm labourers with mechanical power (Dimitri et al., 2005; Duarte and Sakar, 2009; Conford, 2001). From this time onwards the agricultural landscape was evermore shaped by technological and mechanical innovation, and by intervention.

The postwar periods (following both WWI and WWII) witnessed dramatic increases in the use of synthetic chemicals for fertilisers, pesticides, herbicides, and insecticides. Although these (and other) innovations in the agricultural realm contributed to increases in yield and moderated crop loss to pests, concerns of possible detrimental effects to these new developments in agricultural practice began to be raised by farmers in central Europe in the second decade of the 20th century. It was, in fact, just as aerial applications of pesticides were beginning (the first was in August 1921 in the US – the application of lead arsenate to kill catalpa sphinx caterpillars (Abrol and Shankar, 2012)) that the first widely recognised alternative approach to agriculture addressed in this chapter was initiated.

Biological-dynamic agriculture: 1924–present

In 1923, while innovations in industrial scale agriculture were – in many respects – still very much in their infancy, a group of farmers in middle Europe approached the philosopher and social reformer Dr. Rudolf Steiner (1861–1925) with a request that he speak on the subject of the renewal of agriculture (Hurter, 2014). They were seeking advice, amongst a variety of concerns, on how to restore and increase *vitality* in soil and crop at a time when the intensive use of synthetic chemicals in agriculture had barely begun. The farmers who brought their concerns to Dr. Steiner were already noticing marked decline in seed viability and an increase in diseases such as foot-and-mouth (Pfeiffer, 1947/2004). They had already noticed the deterioration of plant and animal health due to chemical fertilisation. In developmental terms, it is remarkable that concerns were already being raised by farmers at this time, some 40 years prior to such concerns being brought starkly to the public eye in publications such as Rachel Carson's *Silent Spring* (1962). The response to these farmers' concerns (and persistent requests for advice on these matters) was a series of eight lectures, given in Koberwitz, Silesia (now Poland) that initiated what today goes by the name of *biodynamic* agriculture. The lectures were given between the 7th and the 16th of June, 1924, and have been published under the titles *Agriculture* (Steiner, 1993) and *The Agriculture Course* (Steiner, 2004).

Following on from this course of lectures, individuals set out to put Steiner's recommendations into immediate practice and a 'biological-dynamic experimental circle' (Paull, 2011; Conford, 2001) was formed to begin to undertake formal research into the proposals given in the lectures and discussions. The experimental circle was comprised of farmers, scientists, vets, and other individuals inspired by the initial indications given in Steiner's lectures (Paull, 2011).

The 'biological-dynamic' approach, based on Steiner's indications for considering the spiritual foundations of agriculture, came in time to be known as *biodynamics* and it goes under that name to this day. As of 2011, biodynamic techniques were used on 142,482 hectares in 47 countries (Paull, 2011) and at the time of writing (spring 2017) biodynamics has a global presence as an alternative approach to agriculture, with associations and members in a growing number of countries (Hurter, 2014). Biodynamics places emphasis on the following key concepts and practices.

The farm should be conceived of as an 'organism', 'individuality', or 'self-contained entity' (Steiner, 1993) with, as much as is practically possible, the needs of the various members of the farm organism being met from within the farm itself (nutrient needs, re-incorporation of waste products into the fertility cycle, careful conservation of energy and nutrient flows). This principle is proposed in order to develop resilience and long-term sustainability for the whole farm (Masson, 2012; Thornton Smith, 2009; Hurter, 2014).

Biodynamics takes a 'qualitative-ecological' (Koepf, 1989) approach to agriculture, in contradistinction to the quantitative-reductive approach of many industrial methods. This means that attention is placed by biodynamic practitioners on supporting "earthly and cosmic forces that form life" (Koepf, 1989), a notion that deeply challenges much of the theory and practice that gave rise to industrial agriculture and that continues to inform its application to this day. Biodynamic practice, in fact, hinges on the notion of *vitality* and on supporting/enhancing *life processes* in soil, plant, animal, and human. This emphasis on engaging with, and therefore *understanding*, living processes leads to questions of *ways of knowing* that will be discussed further in the final section of this chapter.

Already in 1924 Steiner drew attention to the importance of soil fertility and proposed that a healthy soil (implying soil *life*) was the basis for healthy plants and animals and, ultimately, was the basis for healthy food for human consumption (Steiner, 1993; Pfeiffer, 1947/2004). This emphasis on cultivating soil health could perhaps be seen as an example of what was also developing as *humus farming* (Kuepper, 2010) in the early 20th century. In biodynamics, however, supporting the development of the humus fraction in soil is furthered by the use of specific *preparations* whose production methods were first described in the lectures given in Koberwitz in 1924.

It is precisely the production and use of these herbal 'preparations' (coupled with the adherence to astronomical planting calendars) that has led several critics of biodynamics to dismiss it as an approach founded on pseudoscience and/or 'magical thinking' (Holger, 1994; Chalker-Scott, 2004). Newcomers to biodynamics will certainly be struck by the field and compost preparations and their method of production, and they can be all too quickly dismissed as 'muck and magic' if taken only at face value. An alternative would be to take these preparations (and the planting calendar used by biodynamic practitioners) as a stimulus for a study of paradigmatic differences in agricultural practices (Lorand, 1996) and to research not only *what* a given approach entails as practices but *how* these arose from an epistemological and ontological perspective (Code, 2014). The preparations are, in this light, one aspect of an approach to agriculture (i.e. biological-dynamic) that arises from a very different worldview and understanding of the organisms, beings and processes in nature than is evident in either conventional agriculture or, in fact, in other 'alternatives' to conventional agriculture (Code, 2014; Thornton Smith, 2009).

Formal research has been undertaken in biodynamics since its inception and although much of this research has been undertaken on the European continent, and is therefore not published in the English language, a body of research and literature available to English speaking audiences continues to be developed (see www.considera.org/reslit.html). An overview of research findings arising from studies of biodynamic cultivation methods (including the use of preparations) which have appeared in a central European context – either at universities, in collaboration with universities, or in scientific journals – is given by Fritz in the 2014 publication *Agriculture for the Future* (ed. by Hurter). Much of the research referred to in this overview was undertaken by professors, by PhD candidates, or by other qualified researchers. The publication *Agriculture for the Future* (2014) also provides an overview of current biodynamic agricultural activities (applied biodynamics, research, social development, and worldwide initiatives) globally.

Biodynamic produce and processes can be certified under the Demeter trademark. Demeter is an international body that defines and oversees standards for biodynamic products. Information about Demeter International can be found here: www.demeter.net.

Organic agriculture: 1930s–present

What are the origins of *organic* agriculture? These are, in many respects, more difficult to localise in a specific moment in time or place than are those of biodynamics. This difficulty is due to the fact that all agriculture prior to the development of synthetic fertilisers, pesticides and the intensive use of petroleum-derived products in agriculture could be considered to be 'organic' in a broad sense of the term. 'Organic', in this less formal sense, can perhaps be understood to mean 'natural' or 'traditional'. Ever since the early 20th century, however, the term 'organic' has become a much more specific identifier or rubric for a particular alternative agricultural method and practice. Philip Conford writes in *The Origins of the Organic Movement*:

> If organic *methods* have existed for centuries, the organic *movement* could begin only once an alternative to them existed. This did not happen until the nineteenth century brought the industrial production of artificial fertilizers and the rapid spread of urbanization.
>
> *(2001: 17)*

Credit is given for popularising (and formalising) this alternative to conventional methods in the UK to Sir Albert Howard (1873–1947) and to Lady Eve Balfour (1898–1990), and in the US, to J.I. Rodale (1898–1971) (Kuepper, 2010; Gillman, 2008; Conford, 2001). The term *organic farming* features in the title of Howard's seminal 1947 book *The Soil and Health: A Study of Organic Agriculture*, but its first use to designate an identified approach to agriculture appears some seven years earlier. Its first appearance in print (where it is contrasted to 'chemical farming') is in a book titled *Look to the Land*, published in 1940 by Oxford University agriculturalist Lord Northbourne (Paull, 2006). Northbourne was deeply inspired by the agriculture lectures given by Steiner as well as by a meeting with Pfeiffer, when the latter came to the UK on Nourthbourne's invitation, and he proposes in his book a number of tenets drawn from the Agriculture lectures – including the notion of treating the farm as an organism or 'living whole' (Paull, 2006).

Sir Albert Howard and his wife Gabrielle drew inspiration from the 'traditional' agricultural methods that they witnessed being practiced in India in the early 1900s (Conford, 2001). They eventually sought to bring a rigorous research approach to natural farming practices (they were both accomplished botanists and scientists) at a time when these had largely been focused on the more chemical intensive approaches. The Howards sought to place 'organics' on a scientific footing, as a fully viable alternative to food production in the 20th century (and, presumably, beyond).

Other early contributors sought to establish organic agriculture on the basis of research and comparative trials (usually comparisons with conventional approaches to food production). Lady Eve Balfour launched the Haughley Experiment in the UK in 1939, comparing organic and conventional farming methods in side-by-side trials (Gillman, 2008). Balfour's subsequent publication of *The Living Soil* contributed to the formation of the Soil Association in the UK – as she did herself, by serving as its first president.

In the US, J.I. Rodale promoted organic agricultural methods and established the Rodale Institute and the Rodale Press to research and promote organic agriculture to the general public (Conford, 2001).

Many other individuals contributed significantly to the development of organic agriculture as a formal alternative to conventional practices, but space does not permit giving due credit to them all (see Conford's *The Origins of the Organic Movement*).

Over time, the distinguishing features of organic agriculture came to include, amongst other tenets and practices, the following (from Gillman, 2008):

- The maintenance and enhancement of soil fertility through crop rotation, the planting of cover crops and green manures, the planting of legumes to fix nitrogen in the soil, and the production and application of compost made from natural materials.
- The development of polycultures (mixed plant communities) rather than monocultures (single crop production) in order to support beneficial insects, soil microorganisms and species diversity.
- Weed management through biological, mechanical and physical means that include regular crop rotation, mowing and tilling, flame weeding, the use of weed suppressing species, mulching and grazing.
- Control of insect pests, fungus and bacteria carried out through interventions such as the introduction of predatory beneficial insects, crop rotation, companion planting, insect traps, physical barriers, biologic pesticides and herbicides.
- Livestock grazed or fed on organically produced fodder, feed (grain, tubers, hay/silage) and pasture.

Where biodynamics can be identified with its references to *vitality* and supporting *life processes* in the agricultural individuality or organism, literature on organic agriculture reveals frequent references to the term *natural*. A recurring articulation of the aims of organic agriculture is that of respecting and harnessing "*natural processes* to build positive health across the ecology of the farm" (cited in numerous publications). Farmer and author Wendell Berry articulates his definition of organic agriculture as follows:

> An organic farm, properly speaking, is not one that uses certain methods and substances and avoids others; it is a farm whose structure is formed in imitation of the structure of a natural system that has the integrity, the independence and the benign dependence of an organism.
>
> *(2009: 143–144)*

The meaning and usage of term *natural* is not as self-evident or straightforward as one might think (Soper, 1995), and just as the concept *vitality* raises questions in biodynamics (in terms of its meaning, assessment, and measure) so too does the term *natural* in the realm of environmental studies, agroecology, and agriculture (also see Chapter 17). In the latter half of the 20th century, organic agriculture increasingly adopted descriptors from the burgeoning discipline of ecology and has been officially described in these terms by its largest international body of standardisation, IFOAM (International Federation of Organic Agriculture Movements):

> Organic systems should fit the cycles and ecological balances in nature. Organic management must be adapted to local conditions, ecology, culture, and scale. It is essential to maintain and improve environmental quality and conserve resources by reuse, recycling, and the efficient management of materials and energy. Organic agriculture should attain ecological balance through the design of farming systems, establishment of habitats, and maintenance of genetic and agricultural diversity. Those who produce, process, trade, or consume organic products should protect and benefit the common environment including landscapes, climate, habitats, biodiversity, air, and water.
>
> *(IFOAM, 2009: 7)*

Research into organic agriculture is quite extensive and is varied in type and results; space does not permit a comprehensive description of these. There are a number of organisations worldwide currently active in carrying out research into organic agriculture.

Organic agriculture is not without its critics. Criticisms are brought to bear on several aspects of organics. These might be with regards to claims that organic food is healthier than food produced through conventional means (Dangour et al., 2010), and opinions for and against these claims continue to develop (Woese et al., 1997). Criticism is also brought to bear on the proposals that organic agriculture is more environmentally friendly and sustainable then conventional farming (Trewas, 2001).

Counterarguments to many of these criticisms have been published by IFOAM, and can be sourced freely on www.ifoam.org; they are made available under the title *Criticisms and Frequent Misconceptions about Organic Agriculture: The Counter-Arguments* (IFOAM, 2008).

Proponents of organic agriculture at times raise criticisms of various aspects of its development, such as those articulated by Michael Pollan in his *The Omnivore's Dilemma* (2006) wherein the mega-scale of some organic producers is likened to industrial types of production whose intensive (albeit 'organic') operations call in to question the farm organism ideals of its early founders.

Standards for organic agriculture were produced in various countries throughout the latter half of the 20th century and in 1972, the International Federation of Organic Agriculture Movements (IFOAM) was founded in Versailles, France.

Permaculture: 1978–present

Along with the questions raised by Carson in *Silent Spring* regarding the detrimental effects of synthetic chemicals used in agriculture, the 20th century witnessed a wide variety of environmental crises that could be linked either directly or indirectly to the methods inherent in highly mechanised, monocultural, and chemically intensive farming practices. The Dust Bowl in the US in the 1920s and 1930s could be cited as an early and dramatic example of these (Hyams, 1976), where loss of topsoil due to the combined factors of tillage methods, climate events, and loss of native plant communities created an ecological and cultural disaster on a national scale. Critics of industrial agriculture drew attention (and continue to do so) to a host of other issues, including the heavy reliance on non-renewable resources, the energy intensive and polluting nature of agricultural machinery and practices, the loss of biodiversity which Carson's book highlighted, and the seemingly endless need to develop new chemical concoctions to address the phenomena of resistance that emerged in insect and plant populations (see Chapter 16). These aspects of conventional agriculture led its critics to posit that the very practices developed to increase agricultural production and provide food for a growing global population would be the very interventions that would ultimately lead to catastrophic ecological collapses that would, in turn, lead to cultural crises on a global scale. One response to the perspective that modern agriculture ultimately threatened the sustainability of both the environment and human culture was the proposal that a system of agriculture be developed that was both a 'permanent agriculture' and one that supported a 'permanent culture' – *permaculture* (Mollison and Holmgren, 1978, 1979; Mollison, 1988, 1991).

Permaculture is described as a systematic method, a "philosophy of working with rather than against nature" (Mollison, 1991: 1), a movement, and a worldview (Ferguson and Lovell, 2013). The philosophy of 'working with rather than against nature' arose for Mollison and Holmgren, the co-founders of Permaculture, as a direct response to the many ways in which 20th century agriculture apparently invested significant resources in 'fighting' or 'combating' natural processes. Rather then investing huge resources in trying to model agricultural production after industrial factories, that is, as 'single product systems' (Mollison, 1991), Mollison and Holmgren advocated the development of agricultural systems modelled on natural ecosystems (Mollison and Holmgren, 1978, 1979; Mollison, 1988, 1991; Mars, 2003). Antecedents and/or inspirations for Mollison and Holmgren's initiative in developing Permaculture include the works of J.R. Smith (*Tree Crops: A Permanent Agriculture*, 1988), P.A. Yeomans (*Water for Every Farm*, 1964/2008), and M. Fukuoka (*One Straw Revolution*, 1978).

In their extensive review of Permaculture publications representing over three decades of developments in this area, authors Ferguson and Lovell distil out of the body of literature on Permaculture the following aim or objective for this approach to alternative agriculture:

> With systematic site design, emphasising diversity at multiple scales, integrated water management, and access to global germplasm, (permaculturalists) can increase the productivity demonstrated by heritage agroecosystems – especially labor productivity – while retaining their most desirable attributes of sustainability and multifunctionality.
>
> *(Ferguson and Lovell, 2013: 270)*

Central to Permaculture are the notions of *design* and *systems*. Some key aspects of Permaculture include:

- An articulation and adherence to *three core tenets*: Care for the Earth, Care for the People, and Return of Surplus (at times articulated as Fare Share) (Mollison, 1988; Holmgren, 2000).
- An articulation of *design principles* to follow in the development of Permaculture systems. Holmgren describes 12 of these in his *Permaculture: Principles and Pathways Beyond Sustainability* (2013).
- The use of *layers* in design. Modelled after mature ecosystems, which develop different layers (such as canopy, understory, ground cover) and complex inter-species relationships, Permaculture systems strive to maximise the layers in the system (Mollison, 1988).
- Designed systems are conceived of in terms of *zones* that take into account the interactions in the system between the human environment and activity and the different species that dwell in that system (Burnett, 2001). Zones range from 0 to 5, where 0 is the house or home centre and 5 is wilderness or wild space with little to no human intervention.

Permaculture practitioners may involve any number of several ecologically inspired practices in the design and development of systems. Some of these include agroforestry (Bell, 2004; Whitefield, 1993; Jacke and Toensmeier, 2005), natural building, rainwater harvesting, intensive grazing, and keyline design (Mollison, 1988; Yeomans, 1964/2008).

Critical considerations of Permaculture arise from proponents of its approach as well as from those not working with its philosophy and practices. Ferguson and Lovell (2013) point to the fact that relatively little formal research has been undertaken into Permaculture with regard to productivity and realisation of its claims for self-sustaining, productive food systems. They propose, based on their review of Permaculture literature, that:

> The potential of permaculture to contribute broadly to agroecological transition is limited by several factors. Of primary importance is the general isolation of permaculture from science, both in terms of a lack of scholarly research about permaculture and neglect within the permaculture literature of contemporary scientific perspectives. This deficit is compounded by over-reaching and oversimplifying claims made by movement adherents and the absence of any systematic multisite assessment of permaculture's impacts. Additionally, the difficulty of providing a clear and distinguishing description of permaculture can cause confusion and hinder rigorous and systematic discussion.
>
> *(2013: 2)*

Robert Scott, in his article *A Critical Review of Permaculture in the Unites States* (unpublished) asks the question as to why Permaculture in the United States seems to generate a large number of graduates from design courses and publications, but relatively few examples of established and productive Permaculture sites. He asks:

> Given that permaculture grew out of an alternative approach to agriculture and the design of human settlements, it seems reasonable to ask: why does the permaculture movement in the United States appear to have a stronger tendency toward education than agriculture?
>
> *(Scott, 2010)*

Other criticisms about Permaculture raise questions about its potential to realise the claims made by its early proponents with regard to the productive capacity of its 'forest garden' foodscapes and other 'idealised' design proposals that are put forth in the literature (see for instance Williams et al., 2001; Romanowski, 2007).

Although these criticisms are raised, and often by proponents of Permaculture themselves, this alternative to conventional industrial agriculture continues to grow worldwide and attract interest both at grassroots level and more formally in a range of sustainable agriculture circles.

There are no standards for Permaculture produce in the way that these were developed for biodynamic produce (Demeter) and organic produce (organic certified).

Discussion

The following statement appears in the UNCTAD (United Nations Conference on Trade and Development) Trade and Environment Review (2013) titled *Wake Up Before It Is Too Late*:

> The world needs a paradigm shift in agricultural development: from a 'green revolution' to an 'ecological intensification' approach. This implies a rapid and significant shift from conventional, mono-cultural based and high-external-input-dependent industrial production towards mosaics of sustainable, regenerative production systems that also considerably improve the productivity of small-scale farms. We need to see a move from a linear to a holistic approach in agricultural management, which recognizes that a farmer is not only a producer of agricultural goods, but also a manager of an agro-ecological system that provides quite a number of public goods and services (e.g. water, soil, landscape, energy, biodiversity, and recreation). The required transformation is much more profound than simply tweaking the existing industrial agricultural systems.
>
> *(2013: 2)*

The small selection of alternative approaches to the industrial model of agriculture that have been outlined in this chapter are presented as possible contributions to the paradigm shift" – the move from "linear to holistic approaches" – which the UNCTAD report advocates. The methods they employ and approaches they take offer (though not without contention or their share of critics) a "radical transformation of agriculture . . . one guided by the notion that ecological change in agriculture cannot be promoted without comparable changes in the social, political, cultural and economic arenas that also conform agriculture" (UNCTAD, 2013: 27).

The notion of a 'paradigm' shift implies transformation on quite fundamental levels, that is, not merely in practical approaches or applied techniques. These must be seen as the outer manifestation of changes in actual worldview or orientation – in ways of knowing and perceiving. It could be argued that a study and engagement with alternative approaches to agriculture that doesn't address fundamental questions regarding *ways of knowing* falls into the danger of merely presenting practical 'tweaks' to existing agricultural methods while failing to address the deeper ways of knowing that inform these. It must also be considered that for an adequate evaluation (or well-founded criticism) of different alternatives to 'chemical agriculture' this dimension must also be made explicit, otherwise these alternatives will likely only be assessed from the perspective of the current paradigm, which will then fail to comprehend the new theories and practices that they represent (Code, 2014).

The sciences that so deeply influenced the rise of industrial agriculture often applied methods of analytical reduction to the complexity of natural phenomena in order to gain clarity about their constituent parts and the relation of these each to the other. Liebig's identification of nitrogen, potassium, and phosphorus as essential elements in plant nutrition can be seen as a classic example of this approach. Liebig saw:

> his agricultural work as proof of chemistry's ability to transform an empirical activity into a science. Botanists and physiologists had neglected plant chemistry, but the processes of transformation were purely chemical and could not be understood without laboratory experiments.
>
> *(Conford, 2001: 17)*

Since the 18th century, this reductive and analytical approach to science has come to inform more and more aspects of the natural sciences (and eventually the social sciences as well), and has greatly influenced developments in agriculture – but it is not without its critics. An exhaustive list of contributors to this discussion would be beyond the bounds of this chapter and its primary focus, though readers will find an entry into this significant area of study through the works of Chalmers (1990, 1999), Barfield (1988), Bateson (1972,1979), Burt (2003), Steiner (1983,1985), Capra (2014), and Brady (2008) – to name but a few. It could be proposed, however, that one of the primary differences between "conventional, mono-cultural based and high-external-input-dependent industrial production" (UNCTAD, 2013: 27) and the alternatives outlined in this chapter is the way in which the relationship between the 'parts' and the 'whole' of the farm (or organisms within the farm) are conceived. Grasping the relationship between 'part' and 'whole' is anyway a challenge for cognition (Bortoft, 1996), let alone for attempts to intervene with living systems, and a key challenge for agriculture (and for science) is how this challenge is addressed.

In *A Guide for the Perplexed* (1995), E.F. Schumacher invites a well considered and highly lucid consideration of the differences between ways of knowing *adequate* to comprehending the mineral, plant, animal and human aspects of nature and concludes the following:

> Physics and chemistry deal with the lowest level, 'mineral'. At this level . . . life, consciousness, and self-awareness – do not exist (or, in any case, are totally inoperative and therefore cannot be noticed). Physics and chemistry can tell us nothing, absolutely nothing about them. These sciences possess no concepts relating to such powers and are incapable of describing their effects.
>
> *(1995: 29)*

Schumacher proposes that different types of science, different ways of knowing, need to be developed in order that our comprehension be *adequate* to the fullness of the phenomena in question. Rather than reject physics, chemistry or other sciences wholesale (as they are evidently suited to the study of the inorganic realm), Schumacher advocates greater attention to the boundaries of *any* way of knowing such that these remain appropriate and best suited to the phenomena under investigation. He highlights the tendency for ways of knowing suited to one realm of phenomena (for example the mineral realm) to be unquestioningly applied to the study of *all* phenomena with this quote from the psychologist Victor Frankl:

> The present danger does not really lie in the loss of universality on the part of the scientist, but rather in his pretense and claim of totality. . . . What we have to deplore therefore is not so much that *scientists are specializing*, but rather the fact that *specialists are generalizing*.
>
> *(Schumacher, 1995: 15)*

In an agricultural context, it can be inferred that Schumacher would invite caution if attempts to manage the complexity of mineral, plant, animal, and human interactions that constitute a 'farm' (or farm organism, from a biodynamic perspective) were to be undertaken only on the basis of knowledge and practice derived from attention to inorganic realms.

This perspective has been proposed by a number of scientists, and one who warrants particular attention in light of the focus on *vitality* and *life processes* in biodynamics is J.W. von Goethe. A comprehensive and contemporary articulation of Goethean science is provided by physicist and philosopher of science Henri Bortoft, whose book *The Wholeness of Nature* (1996) makes a very strong case for its cultivation, particularly where science seeks to understand the living world. Rudolf Steiner referred to Goethe as "the Copernicus and the Kepler of the organic world" (Conford, 2001: 67; Zajonc and Seamon, 1998). Steiner himself spent nearly a decade editing and commenting on Goethe's method and it can be inferred from his concerted attention to Goethe's work that elements of this way of knowing contributed to his own later work, including his insights into agriculture. Several contemporary studies, in fact, explicitly take a Goethean scientific approach as the basis for an investigation of aspects of a biodynamic approach to agriculture (Bockemuhl, 2006; Code, 2014).

It has been noted that where biodynamics places emphasis on life processes, and thereby presents the challenge that we develop sciences adequate to investigating the phenomena of life, permaculture places a good deal of attention on developing a systems approach to designing productive landscapes. Reference to systems is also evident in the UNCTAD report, wherein we find mention of 'agricultural systems' and 'regenerative production systems'. Systems theory and systems science evidently represent a further area of study and consideration in terms of paradigmatic shifts in ways of knowing that can then inform shifts in ways of doing such as are needed for a 'transformed' agriculture. Publications that engage with and articulate aspects of systems theory and its application in diverse fields continues to grow (see for instance Bateson, 1972; Capra, 1997/2014; Bertanalffy, 1968; Checkland, 1999; Luhmann, 2013). The degree to which permaculture has been informed by systems theory is discussed by several authors (Rothe, 2014; Smith et al., 2007). Systems theory seeks to address the complexity of living organisms and their interactions, and strives to overcome the one-sided emphasis on the individual elements of a system (its parts) by maintaining an awareness of the self-organising wholeness of the organisation being investigated.

Conclusion

To have given space to considerations of *ways of knowing* in an overview of alternatives to industrial agriculture is to acknowledge, after Parker Palmer, that "every way of knowing becomes a way of living" (1993). To contextualise this statement further for the themes this chapter addresses, it could be re-phrased as 'every way of knowing becomes a way of growing'. Proposals for paradigm shifts or fundamental transformations in the way we grow lead to the insight that these can only come about if we also address the ways that we 'know'. A twofold development is thus offered in any of the approaches to agriculture outlined in this chapter. On the one hand this is a question of grasping (and perhaps not initially understanding) new techniques and methods of working with the land and its co-inhabitants. On the other hand practitioners of alternative approaches to agriculture are invited to pay due attention to the cultivation of consciousness as well as to the creation of compost.

Further reading

General

Abrol, D.P., & Shankar, U. (ed.) (2012) *Integrated pest management: Principles and practice.* Jammu, India: CABI.
Barfield, O. (1988) *Saving the appearances: A study in idolatry.* Middletown, CT: Wesleyan University Press.
Barker, D. (2007) *The rise and predictable fall of globalized industrial agriculture.* San Francisco: IFG (The International Forum on Globalization).
Bateson, G. (1972) *Steps to an ecology of mind: Collected essays in anthropology, psychiatry, evolution, and epistemology.* Chicago: University of Chicago Press.
Bateson, G. (1979) *Mind and nature: A necessary unity (Advances in systems theory, complexity, and the human sciences).* New York: Hampton Press.
Berry, W. (2009) *The gift of good land.* Berkeley, CA: Counterpoint Press.
Bertalanffy, L. (1968) *General system theory: Foundations, development, applications.* New York: George Braziller.
Bockemuhl, J. (2006) *Extraordinary plant qualities for biodynamics.* Edinburgh: Floris Books.
Bortof, H. (2008) *Taking appearance seriously: The dynamic way of seeing in Goethe and European thought.* Edinburgh: Floris Books.
Bortoft, H. (1996) Authentic and counterfeit wholes. In: *The wholeness of nature: Goethe's way toward a conscious participation in nature.* New York: Lindisfarne Press.
Bortoft, H. (1996) *The wholeness of nature: Goethe's way toward a conscious participation in nature.* New York: Lindisfarne Press.
Brady, R., Maier, G., & Edelglass, S. (2008) *Being on Earth: Practice in tending the appearances.* http://natureinstitute.org/txt/gm/boe/.
Burt, E.A. (2003) *The metaphysical foundations of modern science.* New York: Dover Publications.
Capra, F. (1997/2014) *The systems view of life.* Cambridge: Cambridge University Press.
Carson, R. (1962) *Silent spring.* 1st Edition. Boston: Houghton Mifflin.
Chalmers, A. (1990) *Science and its fabrication.* Maidenhead, UK: Open University Press.
Chalmers, A. (1999) *What is this thing called science?* Maidenhead, UK: Open University Press.
Checkland, P. (1999) *Systems thinking, systems practice: Includes a 30-year retrospective.* New York: Wiley.
Dimitri, C., Effland, A., & Conklin, N. (2005) *The 20th century transformation of U.S. agriculture and farm policy.* Electronic Report from the Economic Research Service, www.ers.usda.gov.
Duarte, V., & Sarkar, S. (2009) *A Cinderella story: The early evolution of the American tractor industry.* www.cefage.uevora.pt (accessed 11 April 2016).
Houser, J.S. (1922) The airplane in catalpa sphinx control. *Ohio Agricultural Experiment Station, Monthly Bulletin* (7), 126–136.
Luhmann, N. (2013) *Introduction to systems Theory.* New York: Polity.
Mars, R. (2003) *The Basics of Permaculture Design.* East Meon, UK: Permanent Publications.
Modak, J. (2002) *Haber process for ammonia synthesis.* Berlin: Research Gate.
Morton, J. (2007) The impact of climate change on smallholder and subsistence agriculture. *PNAS* (104)50. http://www.pnas.org/content/104/50/19680.
Palmer, P. (1993) *The violence of our knowledge.* www.21learn.org/arch/articles/palmer_spirituality.html.
Rothe, K. (2014) *Permaculture design: On the practice of radical imagination.* http://scholarworks.umass.edu/cpo/vol3/iss1/4/.
Schumacher, E.F. (1995 ed.) *A guide for the perplexed.* New York: Vintage Books.

Smil, V. (2001) *Enriching the Earth: Fritz Haber, Carl Bosch, and the transformation of world food.* Cambridge, MA: MIT Press.

Smith, T., Willetts, J., & Mitchell, C. (2007) *Implications of the synergies between systems theory and permaculture for learning about and acting towards sustainability.* ANZSEE Conference "Re-inventing Sustainability: A climate for change" 3–6.

Steiner, R. (1983) *The boundaries of natural science.* New York: Anthroposophic Press.

Steiner, R. (1985) *The origins of natural science.* New York: Anthroposophic Press.

UNCTAD (2013) *Wake up before it is too late: Make agriculture truly sustainable now for food security in a changing climate.* United Nations Publication. unctad.org/en/PublicationsLibrary/ditcted2012d3_en.pdf.

Zajonc, A. & Seamon, D. (eds.) (1998) *Goethe's way of science.* New York: SUNY Press.

Biodynamics

Code, J. (2014) *Muck and mind: Encountering biodynamic agriculture.* Great Barrington, MA: Lindisfarne Press.

Chalker-Scott, L. (2004) *The myth of biodynamic agriculture: Horticultural myths.* Washington, DC: Washington State University.

Fritz, J. (2014) Results of scientific trials. *Agricutlture for the future.* Verlag am Goetheanum, 201–214.

Holger, K. (1994) Biological dynamic farming – an occult form of alternative agriculture? *Journal of Agriculture and Environmental Ethics,* 7(2), 173–187.

Hurter, U. (ed.) (2014) *Agriculture for the future: Biodynamic agriculture today.* Dornach: Verlag am Goetheanum.

Koepf, H.H. (1989) *The biodynamic farm: Developing a holistic organism.* Great Barrington, MA: Steiner Books.

Lorand, A (1996) Biodynamic agriculture: A paradigmatic analysis. http://www.triptolemos.net/wp-content/uploads/2015/07/Lorand-short-Dissertation.pdf.

Masson, P. (2012) *A biodynamic manual: Practical instructions for farmers and gardeners.* Edinburgh: Floris Books.

Paull, J. (2011) Biodynamic agriculture: The journey from Koberwitz to the world, 1924–1938. *Journal of Organic Systems,* 6(1), 27–41.

Pfeiffer, E. (1947/2004) *Soil fertility, renewal and preservation: Biodynamic farming and gardening.* 2nd edition. Sussex, UK: Lanthorn Press.

Steiner, R. (1993) *Agriculture.* Pennsylvania, PA: Biodynamic Farming and Gardening Association Inc.

Steiner, R. (2004) *Agriculture course: The Birth of the Biodynamic Method.* East Sussex, UK: Rudolf Steiner Press.

Thornton Smith, R. (2009) *Cosmos, Earth and nutrition: The biodynamic approach to agriculture.* Forest Row: Sophia Books.

Organics

Conford, P. (2001) *The origins of the organic movement.* Edinburgh, UK: Floris Books.

Dangour, A.D., Lock, K., Hayter, A., Aikenhead, A., Allen, E., & Uauy, R. (2010) Nutrition-related health effects of organic foods: A systematic review. *American Journal of Clinical Nutrition,* 92(1), 208–210.

Kuepper, G. (2010) *A brief overview of the history and philosophy of organic agriculture.* Kerr Center for Sustainable Agriculture. http://kerrcenter.com/publications/organic-philosophy-report.pdf.

Gillman, P. (2008) *The truth about organic gardening: Benefits, drawbacks and the bottom line.* Portland, OR: Timber Press.

Howard, L. (1947/2011) *The soil and health: A study of organic agriculture.* Oxford: Oxford City Press.

IFOAM (2008) *Criticisms and frequent misconceptions about organic agriculture.* http://infohub.ifoam.bio/sites/default/files/page/files/misconceptions_compiled.pdf.

IFOAM (2009) *One earth, many hands: 2008 annual report.* Bonn, Germany. https://www.ifoam.bio/sites/default/files/page/files/ifoam_annual_report_2008.pdf.

Paull, J. (2006) The farm as organism: The foundational idea of organic agriculture. *Journal of Bio-Dynamics Tasmania,* 83, 14–18.

Paull, J. (2011) The Beteshanger Summer School: Missing link between biodynamic agriculture and organic farming. *Journal of Organic Systems,* 6(2), 13–26.

Pollan, M. (2006) *The Omnivore's dilemma.* New York: Penguin Books.

Soper, K. (1995) *What is nature?: Culture, politics and the non-human.* New Jersey: Wiley.

Seufert, V., Ramankutty, N., & Foley, J.A. (2012) Comparing the yields of organic and conventional agriculture. *Nature,* 485, 229–232.

Trewavas (2001) Urban myths of organic farming. *NATURE*, 410. www.nature.com.
Willer, H., & Yussefi, M. (2006) The world of organic agriculture: Statistics and emerging trends. IFOAM, Bonn, Germany & FiBL, Frick, Switzerland, p. 28.
Woese, K., Lange, D., Boess, C., & Bogl, K.W. (1997) A comparison of organically and conventionally grown foods: Results of a review of the relevant literature. *Journal of the Science of Food and Agriculture*, 74. 281–293.

Permaculture

Bell, G. (2004) *The permaculture way*, 2nd ed. East Meon, UK: Permanent Publications.
Bell, G. (2004) *The permaculture garden*. East Meon, UK: Permanent Publications.
Burnett, G. (2001) *Permaculture: A beginner's guide*. Spiralseed. https://spiralseed.co.uk/product-category/publications/.
Ferguson, R.S., & Lovell, S.T. (2013) *Permaculture for agroecology: Design, movement, practice, and worldview. A review*. www.resilience.org/stories/2014-01-23/the-trouble-with-permaculture.
Fukuoka, M. (1978) *The one straw revolution*. Emmaus, PA: Rodale Books.
Fukuoka, M. (1987) *The natural way of farming – the theory and practice of green philosophy*, rev ed. Tokyo: Japan Publications.
Hart, R. (1996) *Forest gardening*. Cambridge, UK: Green Books.
Hemenway, T. (2009) *Gaia's garden: A guide to home-scale permaculture*. Vermont: Chelsea Green.
Holmgren, D. (2000) *Permaculture*. Holmgren Design Services. http://www.holmgren.com.au/contact/.
Holmgren, D. (2006) *Collected writings and publications*. Holmgren Design Services. http://www.holmgren.com.au/contact/.
Holmgren, D. (2003) *Permaculture: Principles and pathways beyond sustainability*. Holmgren Design Services. http://www.holmgren.com.au/contact/.
Hyams, E. (1976) *Soil and Civilization*. New York: Harper Collins.
Jacke, D. & Toensmeier, E. (2005) Edible forest gardens Volume 1 and Volume II. *Edible Forest Gardens*. www.edibleforestgardens.com.
King, F.H. (ed.) (2004) *Farmers of forty centuries*. Mineola, NY: Dover Publications.
Law, B. (2005) *The woodland house*. East Meon, UK: Permanent Publications.
Mollison, B., & Holmgren, D. (1978) *Permaculture one*. Tasmania: Tagari.
Mollison, B., & Holmgren, D. (1979) *Permaculture two*. Tasmania: Tagari.
Mollison, B. (1988) *Permaculture: A designer's manual*. Tasmania: Tagari.
Mollison, B. (1991) *Introduction to permaculture*. Tasmania: Tagari.
Morrow, R. (ed.) (2010) *Earth user's guide to permaculture*. East Meon, UK: Permanent Publications.
Paul, W. (2011) *Symbols and patterns: Interview with Owen Hablutzel, Director, Permaculture Research Institute, USA*. http://planetshifter.com/node/1954.
Romanowski, N. (2007) *Sustainable freshwater aquaculture: The complete guide from backyard to investor*. Australia: UNSW Press.
Mars, R. (2005) *The basics of permaculture design*. Vermont: Chelsea Green Publishing.
Scott, R. (2010). *A critical review of permaculture in the United States*. Self-published, http://robscott.net/2010/comments/.
Smith, J.R. (1988) *Tree crops: A permanent agriculture*. 2nd edition. Washington, D.C.: Island Press.
Whitefield, P. (1993) *Permaculture in a nutshell*. East Meon, UK: Permanent Publications.
Whitefield, P. (1996) *How to make a forest garden*. East Meon, UK: Permanent Publications.
Whitefield, P. (2004) *Earth care manual*. East Meon, UK: Permanent Publications.
Williams, G., Polk, E., & Warshall, P. (2001) Permaculture: hype or hope? *Whole Earth*. Winter, 90–92.
Yeomans, P.A. (1964/2008) *Water for every farm*. Createspace. https://www.createspace.com/.

11

Seascapes

Food from the marine landscape

Maggie Roe

"A seascape is a landscape without end in both space and time."

(Waterman, 2011: 48)

"The Humpbacks use the rotational forces of the planet,
The azimuth of the sun,
The taste and temperature of the tides,
The contours of the sea-bed:
Canyons, plains, vaults;
The mountainous summits of the mid-Atlantic ridge, twice the width of the Andes,
Which stretches for ten thousand miles, from Iceland to Patagonia;
Guided by hydrothermal vents in the earth's crust,
The topography of coral reefs,
The position of the moon, and its tidal pull;
Navigating with a lodestone disc made of polarised magnetite,
A compass in the brain,
Sensitive to the geomagnetic flow of force-fields
Under phosphorescent seas,
They find their way to the mating grounds."

(Williams, 1989: 34)

"As the sun rises over a sleepy beach, the reflections of shimmering slender plants can be seen dancing in the shallows; as the tide recedes a wealth of plant life and shellfish can be found patiently clinging to the rocks. Seaweed and shellfish – Neptune's jewels – are both highly nutritious, versatile and easily accessible . . . wild food is far superior in flavour and nutritional content compared to shop-bought, cultivated and farmed food (and it's free)."

(Christian, 2013: 5)

Introduction

The productive landscape is often one that is regarded as primarily mono-cultural and land-based. Much of the concentration on food production policy and planning is either on large-scale agriculture in rural areas (see Chapter 9), sea fisheries, or on the growing interest in urban food production (Chapters 12, 13, 14 and 15). On the consumption side, the focus is on health, global food security, and justice (Chapters 26, 31 and 32). In landscape studies there is an increasing interest in the cultural aspects of food (Roe et al., 2016; Viljoen and Wiskerke, 2012) but this rarely extends into the marine realm. Coastal and marine environments are areas that are often not regarded as 'landscape' or in terms of the way they have been manipulated over time to provide food. However, these are often multifunctional environments where the interactions between humans and natural processes in order to produce or gather food have resulted in a rich cultural heritage (Figure 11.1), and considerable aesthetic and sensory relevance based around thinking in terms of food, as Christian (2013) and a number of authors suggest.

Seafood is now traded throughout the world and is seen as a perishable and complex commodity with high nutritional value that is poorly understood by most consumers (Pieniak et al., 2013). Globally we are all consumers of fish in one way or another since 25% of global fish catch "ends up as fertilizer, food for pigs and poultry, as fish oil supplements and [. . .] as processed food for fish" (Probyn, 2016: 5). In the Earth's oceans, pressures and impacts as a result of fishing are now severe, undocumented, or hidden as a result of commercial competition and political manoeuvring (ibid.). This has significant implications for ecological processes and human communities that depend on the marine environment for their livelihoods and cultural wellbeing. While industrial fisheries policy and practice are often politically motivated and in the focus of governments around the world, there is a rich alternative history of food production, gathering, and

Figure 11.1 The commercial seascape of Staithes village, Yorkshire, UK. Tourism has become the primary function of many fishing village seascapes that were created in response to the gathering of food from the sea.

consumption from marine environments that has moulded our understanding, use, cultures, and ecology of these environments that is not often told – or is invisible (O'Connor, 2016). Historic studies show that humans often survived in coastal areas as a result of the access to food in the marine environment when land-based food was scarce (Pauly and Watson, 2013; Chapter 3). As the interest in informal foraging grows and as food security issues become more pronounced, cultural practices relating to many marine foods are also being more widely revealed (e.g. Probyn, 2016). In the UK, for example, informal foraging is now becoming formalised through the activities of small businesses creating artisan products and the provision of tourist experiences based on the land-sea interface and 'wild' marine foods. An example of this is *Mara* artisan seaweeds, harvested at low tide from Scottish coastal waters under licence from the Crown Estate and Scottish Natural Heritage (www.maraseaweed.com) (see Chapter 5). If, as Salmon (2012: 8) suggests, "eating is not only a political act but also a cultural act" and all landscapes are 'cultural' to one degree or another (Roe and Taylor, 2014), then understanding the food cultures of coastal and marine landscapes (seascapes) are important for cultural, sociopolitical, and economic reasons.

This chapter does not attempt to examine the global industrialised fishing industry as a food source. Rather it uses an examination of people's interaction with seascapes with the objective of gathering and growing food in order to understand how the marine environment can be regarded as 'landscape'. Barthes (1975) considers that an entire 'world' or social environment is present in and signified by food and there is increasing interest in 'thick' analyses of our relationship with food, environments, culture, and society (see for example Panelli and Tipa, 2009). Therefore, thinking about food provides a useful lens to consider changing human relationships with natural processes; in particular the complex socio-cultural relations and cultures that grow up as a result of interactions with the marine environment and the human need for food. This idea can be conceptualised in the term 'seascape', which is used throughout the chapter.

This concept of seascape and the opportunities or affordances of the marine environment as a cultural 'landscape' where food is grown and gathered, and cultural meanings and associations have developed, is poorly articulated in policy and research. This is in spite of a large body of literature concerning fisheries and marine sciences and systems research. Lowitt's (2013, 2014) study of rural and coastal foodscapes provides a rare and interesting example of the potential of the study of marine and coastal landscapes through a food lens. This study indicates the importance of local seafood although there are increasing constraints on access to the food resource. There is some evidence of the development of more holistic thinking as a result of Marine Protected Area (MPA) research that, for example, understands the deep interdependencies between humans and seascapes (see for example Barr, 2014), the unevenness of global fish stocks (Worm and Branch, 2012) and the need to consider the human relationship with oceans in different ways to those with the land. Part of the problem is the lack of information on the myriad different ways people interact with the marine environment. Part of the picture can be gained through sources such as the Food and Agricultural Organisation of the United Nations (FAO) (2016a) figures which suggest that 56.6 million people around the world are engaged in the fish-capture and aquaculture sectors (marine and inland, 2014 figures). This bare information tells us about commercial fishing but nothing about what Probyn (2016: 5) describes as the "human-fish entanglement".

The following sections examine the connections between food and seascapes. First, the key idea of the sea as landscape is developed. A number of important food seascape types are then identified to consider potential future areas of research which, like the oceans themselves, are potentially large, highly complex, and require multiple skills to navigate.

The sea as landscape

As is suggested in 'Underwater Seascapes' (Musard et al., 2014) the underwater can be described as 'landscape'; its physical and intangible values are as rich, if less well-understood, as those relating to the land. The European Landscape Convention (ELC) includes marine areas as well as land and inland waters in its identification of areas as landscape types that are considered under its Articles (CoE, 2000). The ELC also indicates that all landscapes are constantly changing. The flux in coastal and marine environments is highly complex and many of the ecological processes are poorly understood. As is often stated, we know less about parts of the world's oceans than we do about some of the dry landscapes of other planets within our solar system; although the Earth's ocean floor has been mapped to a 5 km resolution, the detail is not as well mapped as that of the Moon, Venus, or Mars (Copely, 2014). The detailed maps of the oceans that do exist at 100 km resolution only cover about 15%, and these provide limited information on physical forms and nothing about processes or the character of the spaces or species that live there.

The growing interest in seascapes (Álvarez et al., 2014) is partly as a result of more holistic approaches to the land-sea interface in both policy and academic research, but is also because of a growing realisation of the pressures on marine environments. Ritchie and Ellis (2010: 703) describe "the decline in fish stocks, increased levels of marine pollution, deterioration of marine ecosystems and reducing biodiversity, coupled with intensification and competition for marine space and uses". There is increasing concern about the impact of climate change which compounds the problems caused by human impacts (Graham et al., 2011; Chapter 18). Ocean acidification is having severe negative impacts on zooplankton, shellfish, fish, and coral reefs, and global warming is affecting water temperature, sea currents, and nutrient availability (Sumaila et al., 2011; Vincentnathan, 2012). Meanwhile there is growing demand for development of the seabed, a growth in commercial shipping and a loss of traditional knowledge and cultures based on livelihoods with direct interaction with the sea.

Seascape analysis is now being used as a planning tool to understand the visual impact of development such as offshore wind turbines and the conservation of underwater heritage (see for example Gee and Burkhard, 2012; Hill et al., 2001; Pungetti, 2012; Scott et al., 2005). Bremner (2014) documents and develops ideas emerging from cultural geography and Lefebvre's (1991) views of space to consider the nature of the ocean as a "culture – nature hybrid composed by specific intersections of social structure, individual behaviours, institutional arrangements and natural features" (Bremner, 2014: 20). Archaeological studies in particular have used the term 'maritime cultural landscape' to suggest how findings on the land-sea interface indicate important interactions between natural environmental processes and human use or management akin to that found on land (Westerdahl, 1992). This literature provides understandings of the relationships between past communities and the marine space, often with vestiges of traditions and practices evident in the form of memories and current practice as well as archaeological artefacts.

Examining seascapes through a food lens can bring alive community histories and can also help fill the gaps in knowledge relating to attachment to place, identities, and traditions. The study of middens around the world provide much evidence of past marine food consumption and cultures (Figure 11.2). In Japan, kitchen middens reveal that Japanese oysters (*Crassostrea gigas* and *Ostrea denselamelloa*) have been eaten since the prehistoric period (Kusuki, 1991). Similarly, in coastal areas of Australia, middens dating to well over 10,000 years ago show that oysters were an important part of everyday life (Probyn, 2016; Chapter 3). The midden at Sand, near Applecross on the west coast of Scotland, is one of many indicating transhumance in Mesolithic communities (9,000–10,000 years ago) in order to take advantage of seasonal seafood availability. Shellfish were used for many domestic uses as well as consumption. These areas of study provide extraordinary insights into the use of the shoreline and marine environments for food collection and processing and also raise numerous questions concerning the nature of the use of such places as temporary responses to environmental conditions or long-term reliance as part of wider strategies for survival. The location of sites can indicate where the sea was used as a highway "which both united and separated different areas and islands and [where] it allowed for wider vistas and a clear sight of other lands" (Hardy and Wickham-Jones, 2002: 832). The west coast of Scotland has numerous islands and would have been a rich food resource for hunter-gatherers moving around the inlets in small boats, exploiting the various ecological niches of the land-sea interface "including salt and fresh water, shallow and deep water, coastal, lowland and upland land units" (ibid.: 832).

More recently, David Attenborough's extraordinary explorations for the BBC have brought marine riches to the public view via television. David Attenborough (born 1926), naturalist and broadcaster, has now made numerous documentaries which have brought previously poorly understood wildlife to the

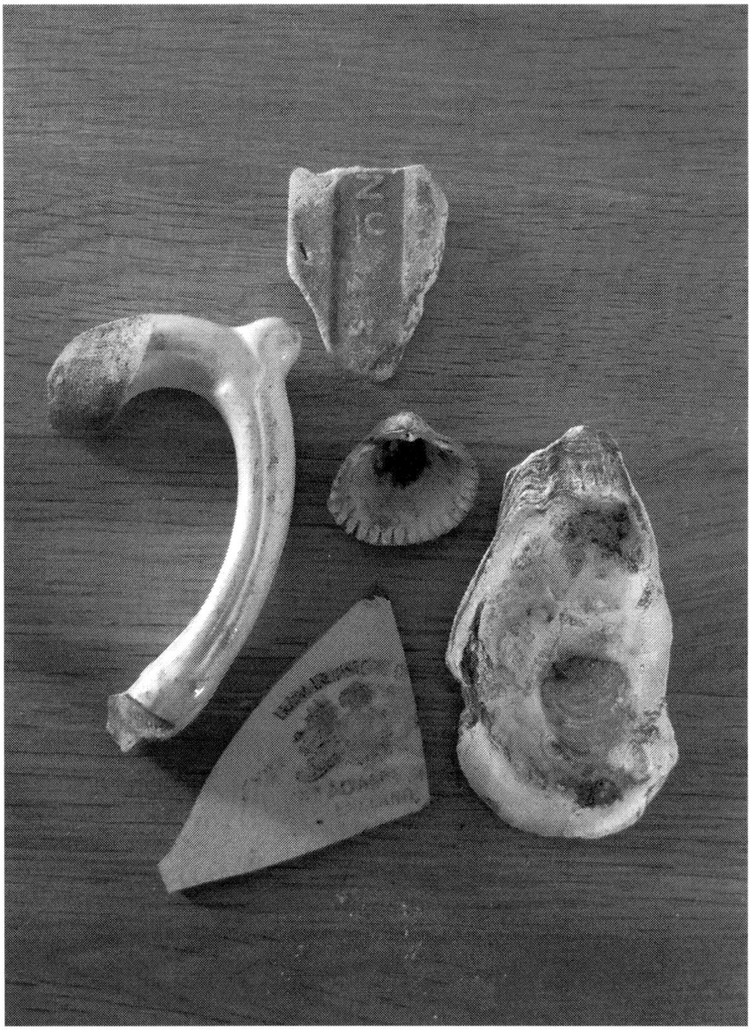

Figure 11.2 Midden finds in Barnard Castle, County Durham, UK, include oyster and cockle shells. The excavation of the 15th century kitchen drains at Barnard Castle indicate "the wide variety of food on the tables of the aristocracy, especially marine fish transported 60 km or more, [which] would be difficult to match in our most expensive modern restaurants."

Source: Rackham, 1994: 55

public view. His series *The Blue Planet* (2001) focused on the marine environment showing both the beauty and complexity of the landscape of the sea and the species living in it. He is a strong advocate for improving public awareness of the increasing danger to species and damage to marine environments of particular types of fishing. Jacques Cousteau (1910–1997) was a pioneer of this kind of work in the 1950s and 1960s. Sailing around the world on his ship *Calypso*, he was particularly famous as an undersea researcher, explorer, and filmmaker. He was instrumental in the invention and development of the aqua-lung in 1943. For those in the UK, Cousteau revealed fish as creatures far more diverse than the content of the ubiquitous fish finger might have suggested at a time when postwar food rationing had only recently ended and the culinary environment was very basic (Figure 11.3). The television stories of Cousteau followed in the footsteps of

Figure 11.3 The magic of the underwater seascape has been revealed by commentators such as David Attenborough and Jacques Cousteau and a current media focus on loss of marine species and issues such as the coral bleaching of the Great Barrier Reef as a result of climate change.

Source: Iona Jenkins

marine expeditions told in stories and myths around the world. These myths are often about gathering food as well as exploring the unknown, recounting human interactions with the sea where both perceptual and physical boundaries are fluid and the environment is inhabited by magical creatures such as selkies, talking fish, sea monsters, and strange natural phenomena. The selkies (or selchies) are mythological creatures in Scottish, Irish, and Faroese folklore traditions regarded as half woman, half seal. Such beliefs may have arisen from encounters between early Scottish settlers and Inuit, Finnish, and Sami people who, while foraging for food, commonly used skin-kayaks and wore animal skins which were taken off and dried out on the rocks (see Bond and Sufflield, 2012).

The seascape provides rich opportunities for examining people's relationships with food both over time and in contemporary practice. However, the terminology commonly used in relation to land-based food and marine-based food differs, and this can be confusing; it indicates the different attitudes to food and means of gathering in these environments. 'Harvesting' on land refers primarily to gathering cultivated food. 'Foraging' is used for gathering wild food (including hunting; see Chapters 5 and 6). In the marine environment, the term 'harvesting' is used for all types of food gathering including hunting, foraging, and aquaculture (cultivation). The following sections examine these three types of human food relationships in the marine environment.

Hunting

Hunting food at sea – primarily fishing – has always been an unpredictable way of obtaining food, as it not only depends on ocean climatic systems but also on "solar supported primary productivity in marine systems" (Pauly and Watson, 2013: 2). The size of the wild-fish catch from the sea is difficult to predict, not only because of the physical conditions but also because nations often lie about the size of their catch (Probyn, 2016). However, FAO figures indicate that the global commercial marine wild-fish capture was 81.5 million tonnes in 2014 and has remained almost static over the last six years with 31.4% of all fish stocks being fished at a 'biologically unsustainable level' (FAO, 2016a: 6). A high level (25%) of the global fish catch is not eaten directly, but used primarily for fertiliser and oil production (FAO, 2016a; Probyn, 2016).

Species such as seals and whales have been hunted historically for food and other products such as oil, pelts and bones. Hunting in the marine environment remains a difficult and dangerous occupation. One of the most famous stories in western traditions that reflect these dangers is *Moby-Dick*, by Herman Melville (1851). This is the story of the obsessive Captain Ahab who was prepared to sacrifice his ship and the lives of those on board in the search of a great white whale. Gathering food from the sea and coastal areas demands very particular skills, knowledge, and engagement in the hunt. Pálsson (1994) provides insights to Icelandic fisher skills. Immersion in the practical world of fishing demands a unity of "emotion and cognition, body and mind" (ibid.: 901) and a collective way of working and experience that takes the participants out of their 'comfort zone' into the nature of the hunt. The experienced skippers in this study spoke of knowing the details and patterns of the sea bottom "as well as their fingers" (ibid.: 910) so they could perceive the fish and the nature of the sea bed as if it were landscape. Also important was an ability to read and predict the weather conditions. The descriptions here indicate that 'fishing enskilment' is not just cognitive; "rather, it involves the whole person interacting with the social and natural environment" (ibid.: 919; Chapter 34).

There is, of course, a great difference between commercial or industrial fishing and subsistence fishing in terms of practice, impact, and cultural significance. Industrial fishing has become increasingly competitive and remote from the home base of the fishers. This requires large, robust vessels with processing and storage capabilities (Pauly and Watson, 2013) which, in effect, replicate land-based factory food-processing and production systems. Studies of traditional communities show very different patterns of seafood gathering and living. The Bajau Laut people or 'sea gypsies' of Malaysia and the SamaDilaut or SamaMandilaut of the Philippines are nomadic sea-going people who spend most of their lives on or in the sea. Many now live in stilted houses in shallow water rather than on boats, living by trading and subsistence fishing. Time is measured by the tides (Lane, 2011). The origin of these people is disputed. However, their traditional myths indicate a long history of relying on the sea for sustenance. As with nomads in many countries, these people suffer much from discrimination, legal persecution, and incursion into their fishing grounds by commercial fishers, poverty, and attempts by government at cultural assimilation and resettlement. The traditional fish-selling from boats is being replaced by settled markets, which have affected the people's lifestyles and ability to subsist. Increasing pressures on marine ecosystems globally, particularly over-fishing, are affecting the lifestyles, cultures, and seascapes of ocean-living peoples such as the Sama-Bajau (also see Chapter 7). Illegal over-exploitation of fisheries resources or 'protein mining' of the world's oceans is a particular problem (Pauly and Watson, 2013) and this is being exacerbated by climate change and other human activities which threaten the food security of small communities who rely directly on the oceans for their food (Field et al., 2009).

In other parts of the world, changes in the morphology and use of coastal landscapes can be seen by studying how the nodes and links related to food seascapes evolved into fishing villages, towns, ports, markets, and the routes between them. Heathcote Williams (1988) reminds us of the large number of coastal settlements set up as a result of whale hunting. These then became cities such as Boston, New York, Tokyo, Leningrad, Lisbon, Reykjavik, Vancouver, Sydney, Durban, Liverpool, and Bergen. Whales provided food and myriad other products. For example, the Japanese use the sperm whale (*Physeter macrocephalus*) for various oils, including high grade machine oils, shoe cream, pencil leads, crayons, candles, detergents, soaps and cosmetics, also alcohol production, glue, gelatine for drug capsules, hormone supplements, photographic film production, bonemeal for fertilizer, perfumes, parchment from skin, racket strings from sinews, brooches and buttons from teeth. Food products are meat, oils, and margarine. Baleen whale meat is highly valued raw for sashimi, as steaks, sausages, stews, pickled and sliced. The earliest records of commercial whaling are of the North Atlantic right whales (*Eubalaena glacialis*) by the Basque people in 1000 CE (Roman et al., 2014). In the thousand years following this, tens of millions of whales have been killed.

The search for seafood has encouraged people to live in remote places such as Greenland's North-eastern coast (Figure 11.4). Here, the seascape turns to an ice landscape in the winter and boats are replaced by dog-sleds, skis, and snowmobiles for transport. Some of these places are now destinations for upmarket tourist industries and one of the attractions is the food that is produced. Paley (2014) reports on the traditional meat-based diet of a remote village in Greenland. This includes polar bear head, seal, and Atlantic cod. In the Arctic, the Inuit traditionally gained up to 99% of their calories from hunting and eating seals, narwhals, and fish (Gibbons, undated). Frink and Giordano (2015) provide a fascinating account of various food pre-preparation (such as air-drying, smoking, boiling, salting, fermenting, and pounding) and storage strategies in the arctic where ground cultivation was not possible. The material and cultural food practices of Inuit communities have shaped and continue to define their identities politically and culturally, as well as moulded a sharing culture and landscape so that people can survive in harsh conditions (Searles, 2002). In a study by Blanchet and Rochette (2008) around the fourteen coastal villages of the Nunavik Inuit of Quebec in Canada, fish and seafood were found to account for 32% of the diet. Marine mammal meat accounted for 12% of traditional food, mostly beluga, but also fat, meat, offal of seal and walrus, *muktuk* (frozen whale skin and blubber), and *igunak* (fermented seal meat). Other marine food included seaweed (marine algae), mussels, and scallops. In this study, fish and seafood were eaten about three times a week and marine mammal meat about once a week. The marine food was particularly important in providing high omega-3 (or n-3) fatty acid intake giving protection against cardiovascular disease (CVD) and a lower incidence of thrombotic disease, high blood pressure, diabetes, certain types of cancer, and problems with mental health. Food is traditionally eaten raw so as not to lose important vitamins that are at a premium in

Figure 11.4 The frozen seascape: hunters of a small remote Greenland town of Ittoqqortoormiit travel through the Arctic landscape.

Source: Jacques Ducoin

such places (such as vitamin C). Raw blubber is metabolised by the human body slowly, so is important for hunters who may have to survive for long periods on small amounts of food (Searles, 2002).

Similar traditional practices are reported in the past in early 19th century St Kilda, a remote island off the west coast of Scotland. On St Kilda there was reliance on the communal hunting and gathering of whatever food was available (Maclean, 1989). The majority of the diet consisted of seabirds and their eggs which were gained from the island's cliffs. Fulmars (*Fulmarus glacialis*), puffins (*Fratercula arctica*), and young gannets (*Morus bassanus*) or 'guga' were reported as food as early as 1698. In the 17th century the catch of gannets was more than 22,000 per year. Between 1829 and 1843 the average catch of fulmars was as many as 12,000 per year. These were dried and salted and stored in barrels for the long winter. Fulmar oil was used for its medicinal properties. Although fish were abundant, this was primarily a bird culture and risking your life on the cliff ropes to gather eggs and birds was seen as part of a way of life, whereas risking it in fishing from a boat was not (ibid.). Nicolson's (2001) engaging descriptions of the Shiant islands (nearer to the Scottish mainland) describe how the inhabitants primarily ate fish, but they also had a taste for seabirds and travelled between islands to gather puffins. In such cultures, adult seabirds were caught in nets, while young birds were taken out of the nest. Different food cultures developed even within relatively small distances as a result of the nature of the seascape, particularly the remoteness and adversity of the elements. Most of these traditions have now disappeared in the UK, as many seabirds are protected. A conservation tradition (see Chapter 17) has now overtaken the seabird-as-food tradition; seabirds are caught to be ringed in such places rather than eaten. There are some exceptions, such as on the isle of Ness, where in 1954 the guga hunters gained a licence to continue a tradition from the 15th century to collect 2,000 young gannets per annum for consumption (Baldwin, 2009; Herald Scotland, 2014).

Foraging and gathering

Foraging can be a group or individual activity, can involve tools, seasonal knowledge and prediction, co-operative working, and trade-off between food gained and energy expended (Danchin et al., 2008; Chapters 3 and 5). It is the gathering of deliberately enhanced or wild living resources – such things as plants, berries, fruit, nuts, mushrooms, seaweeds, bird's eggs, and crustaceans for individual or household consumption. The environmental costs are often not calculated or able to be calculated (Hartill et al., 2005), and the benefits (whether nutritional or otherwise) are often not taken into account as the actual scale of foraging may be small. Foraging raises social as well as environmental issues and it is sometimes seen as an act of subversion outside the law and regulations (Lee and Garikipati, 2011). Today there is a common view that foraged food from the coast or marine environments is public property, freely available for individual gathering and use (ibid.). This does not, however, address concerns about managing the marine environment as a common resource where individual benefits are less important than the community benefits and a cooperative approach needs to be taken for sustainable management. This thinking does vary between countries and cultures. Common pool resource is a term that developed in economics studies referring to a resource (such as fish stocks) that is seen to benefit a group of people, but which may be diminished by individual action where self-interest is more important than conserving the resource for common benefit. The discussion of the marine environment as a CPR has generally examined the impacts of commercial fish stocks and fishing rights where the impacts are clearer (see Ostrom et al., 2006).

In studies on Maori food traditions, Panelli and Tipa (2009) suggest that gathering food is a place-specific part of a complex and multi-layered relationship with the environment relating to long-established customary practices, myths, and seasonal rhythms that are seen to sustain health and wellbeing. There is a growing interest in the UK in foraging evidenced by a range of academic and other material such as Christian's (2013) *Eat the Beach: A Guide to the Edible Seashore*, which captures his personal experience as a chef and his reflections on the aesthetic qualities of the food landscape. Such practical handbooks follow in the tradition of foraging guides from naturalists such as Mabey (1989, 1996) and information from more recent 'bushcraft' and 'wilderness' 'experts' primarily disseminated online and through TV programmes (Mears, 2001). Christian (2013) recognises the potential of foraging as an immersive sensory and cultural landscape experience. In the UK, foraging is primarily carried out as a leisure or lifestyle pursuit, while in many other countries it is still important or essential for food, nutrition, and financial security, with little recognition of its value within food policy research or discourses (Hickey et al., 2016).

Many foods, including seafoods, which were once seen as staple because they were gathered freely from the wild, and are identifiably from a particular location, are now expensive and regarded as exotic. The desire for such food has changed radically due to a variety of drivers, not just the free availability of them. The consumption of oysters in the UK, for example, has risen and fallen over time according to fashion and availability. For example, consumption reportedly rose from 5,000 to 30,000 tonnes between 1980 and 2012, mostly as a result of successful marketing (Probyn, 2016). Many shellfish and crustaceans around the world have had a history of varying abundance and consumption due to changing consumption patterns (Figure 11.5). The Maine lobster (*Homarus americanus*) was considered a low-status food for many years by European settlers in America due to its abundance. Lewis (1998) discusses how lobsters became part of a gourmet culture when lobster eating was adopted by wealthy summer visitors to the Boston area during the 1870s. These tourists associated the lobster with the taste of holiday and the Maine landscape. A similar story is told about oysters in Sydney, Australia, that were an important source of protein for poor immigrants in the mid-19th century. Oysters were gathered from the harbour and were cheaper than eggs or milk (Probyn, 2016). This situation changed as wild oysters become scarce due to pollution, disease, and over-harvesting. Today, the rarity value of wild oysters means they are regarded as a 'posh' food.

Menzel's (1991) edited volume provides a fascinating picture of the range of bivalve molluscs (such as oysters, clams, mussels, and scallops) presently used for human food and their cultivation, gathering, preparation, and consumption. Molluscs have many additional uses, for example as domestic animal food, in

Figure 11.5 Shellfish that were commonly foraged and once regarded as cheap and part of the ordinary food culture in the UK are now increasingly expensive and, other than the gourmet restaurant market, not commonly eaten.

construction, chemical manufacturing, as farmland fertiliser, ornaments, and jewellery. The rise in demand for bivalves has encouraged the development of bivalve aquaculture systems, some of which are private hatcheries and some are controlled public beds supplied with wild seed. Oysters need clean water and are therefore considered an indicator of the quality of the environmental conditions. Much is written about this and other benefits of the oyster to coastal areas, such as their colonising habit which can protect coastal zones from the erosive impacts of storms (Probyn, 2016). However, it is the way oysters connect people to seascapes through their taste that is most commented upon in the food literature: "like wines which vary with their terroir, oysters can be sweet, salty, earthy, nutty, silky, steely or even melony" (Vallely, 2010: n.p.). Oysters are often described as connecting the eater to the sea through their taste which is a "briny, nectar, and zinky combination like no other" (Shellfish Association of Great Britain, 2010: 3) and where the process and relationships that get them from the bottom of the ocean to the plate in front of you is described as an "artisanal sea ecology" (Probyn, 2012: 71; also see Chapter 36).

Shellfish collection forms a large part of the food harvested by the traditional haenyeo, also sometimes called 'jamnyeo' or 'ama' (women divers) of Jeju Island, South Korea. The haenyeo forage for sea urchins, shellfish – particularly abalone – and seaweed while men collect fin fish (SUN-AE, 2012). This is an example of close interaction between humans and the sea created primarily in response to the need for gathering food. In a fascinating study Hong and Rabn (1967) suggest that the tradition of gathering the food riches from the ocean floor on the Korean and Japanese coasts can be traced back over 1,500 years. On Jeju, the practice was probably in response to the difficulty of growing food on this volcanic island. In the mid-17th century both men and women gathered seafood and were referred to as 'haein' or 'person of the sea'. For various reasons this became women's work only (SUN-AE, 2012). The women sometimes dive to 20 metres and hold their breath for up to two minutes (Kang-hyun, 2014). The *sumbisori* is a practice of slowing down exhalation, which makes the required deep diving possible. It sounds like a whistle and is also used by the haenyeo to communicate to each other. It is a "complex cultural sound-object" (Karikis, 2015: n.p.) or an individually identifiable acoustic tool to indicate a diver's location which can be visually difficult to ascertain in the marine environment. Similar sonic communication and identification techniques are used by whales and other cetaceans in watery environments. A key issue for the haenyeo is the temperature drop experienced by the body in diving. The *bulteok* is traditionally an open air stone structure with a fireplace used by the haenyeo on the coast at Jeju as a warming hut, but also a community space where clothes are changed and other work and social activities are carried out (Byun et al., 2015). Around these gathering areas there are middens in creation: the beaches are littered with abalone shells and other evidence of foraging.

The haenyeo have become a powerful symbol of women's identity in Jeju and their way of life is now celebrated through the Haenyeo Museum on the Island and through the recognition given by the tentative inscription of the 'culture of Jeju Haenyeo (women divers) (01068)' on the UNESCO Intangible Cultural Heritage list 'in urgent need of safeguarding'. At the museum the lives, songs, traditions and relationship between women and the sea and their collection of food is celebrated and memorialised. Theirs is an extraordinary story of endurance, perseverance, and unusual gender role-reversal. The intangible heritage is particularly interesting. The haenyeo are often referred to as the 'Korean Mermaids'. The links between diving women around the world and the mermaid myth is explored by Bond and Sufflield (2012), and important gender implications, by SUN-AE (2012). There are numerous stories and legends as well as songs relating to the way of life and the relationship of the women to the sea. No diver ever works alone and there is a strong divers' union that regulates activities including the boundaries of search, the type of catch, the methods of gathering, and the seasons for harvesting. As with many of those still involved in traditional activities, the haenyeo are now mostly elderly and there are few younger women prepared to take on this kind of food gathering, particularly when food traditions and availability of food have changed.

The case of the haenyeo is fascinating because of the gendered nature of this tradition for food gathering and also that it relies on total immersion and an understanding of the underwater that is hard to imagine being passed on in any way except through direct experience, by oral tradition and over many years. This culture, and similar ama (sea-women) cultures found in the region is a "strong women-led hereditary collectivism" (McDonald, 2011: 47) – or an example of thinking of the marine environment as a common resource – that has led to long-term sustainability of the resource and of the community economy.

Growing: aquaculture systems and seascapes

Aquaculture systems range from factory-type salmon farming familiar in the Scottish sea lochs in the UK to more traditional 'homegarden' type systems. The Food and Agriculture Organisation (FAO) defines aquaculture as:

> the farming of aquatic organisms such as fish, molluscs, crustaceans, and aquatic plants. [. . .] with intervention in the rearing process to enhance production [. . .] and individual or corporate ownership of the stock being cultivated, the planning, development and operation of aquaculture systems, sites, facilities and practices, and the production and transport.
>
> *(FAO, 2016b: n.p.)*

Many aquaculture systems occur in inland freshwater situations and are small-scale (Campbell and Pauly, 2013). Ferreira et al. (2014) provide an indication of the complexity of the issues, concerns, and policy relating to marine aquaculture systems including environmental pollution and visual, commercial, and shipping impacts. It is difficult to estimate the amount of food that is produced by global aquaculture systems and fisheries catch, particularly as reports by some major producers are notoriously unreliable. About 50–70% of the global consumption of fish comes from aquaculture systems which are rapidly moving offshore into deeper waters (ibid.).

Understanding environmental impacts in marine systems is difficult, and public acceptability of impacts is related to the perceived benefits of the aquaculture system in question (Whitmarsh and Palmieri, 2009). In the UK the visual impact on scenery of fish farms in inshore waters is a particular concern (Falconer et al., 2013; Nimmo et al., 2011) and large-scale modern aquaculture systems can result in severe ecological impacts including pollution from faecal matter that affects the oxygen in the water, the development of diseases resulting from intensive practices, and the potential for interbreeding with wild species (Menzel, 1991; Probyn, 2016). Visual assessment methods for seascapes are often based on those developed for landscape character analysis methods on land and offshore wind turbine impact assessment (Hill et al., 2001; Scott et al., 2005) which incorporate consideration of environmental values, capacity, and sensitivity.

Aquaculture systems change the supply and demand of food types and also the cultural relationship between people and food. Aquaculture systems can also provide opportunities for people to eat food – such as prawns or salmon – that was formerly seen as rare or special. However, as with land-based environmental impacts, the marine-based impacts of such 'farming' methods may be hidden.

Sea vegetables – seaweeds – are often forgotten in the discussion of aquaculture systems. Many types of seaweed are foraged or gathered from the wild, while others have long been cultivated and managed such as in Japan. In 2014, approximately 28.5 million tonnes of seaweeds and other algae were harvested, mostly in Japan, the Republic of Korea, and China (FAO, 2016a). Seaweed is used for direct consumption or food processing, but it is also used in the pharmaceutical, cosmetic industry and for fertiliser. Sand (2015) traces the management of fish and seaweed for sushi in Tokyo bay and the critical relationship with the city's sewage system, originally for supplying nutrients and then toxic pollutants: "the bay and the city thus literally nourished one another, just as the city's location on the bay nurtured the culture of sushi" (ibid.: 232). Commercial cultivation of nori seaweed (*Porphyra tenera*) was developed in the 18th century and transformed the bay into a lake with extensive seaweed beds which survived until the port extensions in the 1960s. Nearer home in the UK there is a tradition of seaweed culture in Wales and O'Connor (2016) provides interesting links between the Japanese and Welsh seaweed growing cultures. Seaweed growing is under a revival in a number of countries, not only as an artisan superfood, but also as part of integrated aquaculture systems; seaweeds are particularly useful because they have the potential to remove nitrogen, phosphorous, and carbon and to reoxygenate water (Probyn, 2016).

Seascape management: liminal manipulation for food

There are many examples of the marginal or liminal manipulation and management by humans of coastal landscapes to produce food. In Lymington, on the south coast of England, the landscape shows a clear manipulation of 2,000 years of human activity in relation to the production of salt. Here, seawater was impounded in shallow clay-lined lagoons or 'salterns' and then left to evaporate. Such systems are found globally. In the UK, salt production needed a little help because the climate was not hot enough to rely on evaporation alone. At Lymington there is evidence that since the 11th century wind-pumps were used to send the salty solution into storage cisterns, then large metal boiling pans where the liquid was heated and evaporated off to leave salt crystals (James, 2016). Docks were created to allow for transportation of the salt and the historic view of a peaceful landscape (now a nature reserve) would have been very different: an industrial landscape of lagoons, canal, docks, wind pumps, piles of coal, salt barns, and boiling houses, emitting smoke and steam with carriers and sailing barges coming and going taking the sea salt to Canada, USA, Scandinavia, and Europe (Figure 11.6). This salt industry disappeared in the late 19th century, but at its peak in the early 18th century it supported Lymington as a rich coastal town (James, 2016). Since then, over the years the lagoons have been used as a fish farm and for growing oysters.

Figure 11.6 The salterns of Lymington, now a tranquil haven for wildlife, but formerly an industrial seascape.

In Japan, the salt industry has also been important as a food-based industry that linked land and sea management. McDonald (2011) reports that there is new interest in the local knowledge and methods of traditional salt making on the Noto peninsular of the west coast of Honshu. This required integrated management of the seawater and the land-based timber supply to provide fires to aid evaporation. This system at Honshu is similar to that found at Lymington and is an example of *satoumi*. In Japanese *sato* means 'the area where people live', while *umi* means 'the sea' (Yanagi, 2008). The translated definition of this term is given as "a coastal sea area in which there is a harmonious coexistence of Nature and human-beings" or "high productivity and biodiversity in the coastal sea area with human interaction" (ibid.: 351). Embedded within the literature about *satoumi* is the idea that it is an area of seascape that is actively managed, based on centuries-old customary methods (Berque and Matsuda, 2013; Yanagi, 2008). The term *satoumi* seems to have three key meanings: it is used to describe a land-sea interface area itself, which may include seashores, rocky shores, tidal flats, coral reefs, and seaweed/grass beds; it also refers to the human management approaches to the mosaic of ecosystem types; and the term also refers to the inherent interlinkages and mutual dependencies between ecosystem types and human management (Kohsaka et al., 2013; Yanagi, 2008). The focus is particularly on urban coastal zones where human influence is high. The management is based on a number of principles: an understanding of traditional knowledge; the establishment of networks of communities that are connected by water in areas stretching from rivers and river basins to the coast; and, the maintenance of biodiversity-rich habitats. The key to the success and to the principles which are being adopted into environmental management policy is long-term co-management of these important habitats and food resources with local fishermen and other stakeholders, rather than an exclusionary or top-down approach. The adoption of a *satoumi* approach into Japanese environmental planning has developed into an integrated ecosystem approach to coastal strategies, plans and policies. A good example is the management of seaweed and oyster beds in Akita in order to improve the populations of sandfish which spawn in the area (Akimichi and Sugiyama, 2011). This action is in conjunction with agreement with fishers on their catch levels, and land-based tree-planting strategies that help to control the way nutrients travel through groundwater into the coastal waters. The key is in understanding the link between land and sea processes and inshore-deep sea migration patterns and drivers, and in gaining stakeholder agreement. *Satoumi* approaches are generally seen to increase biodiversity as a result of human intervention and interaction with ecosystems. The benefits of human interaction in creating richer coastal environments is supported by other research which emphasises the role that small scale fishing operated by local fishers has in constructing the cultural and natural dimensions of specific landscapes that reflect the communities that maintain them (Álvarez et al., 2014; Berkes et al., 2001; Berkes, 2003).

Conclusions

Steel (2012) has written extensively on using food as a lens to interpret human civilisations as various forms of *sitopia*: "food-places whose diverse characteristics share a common thread" (ibid.: 3). She suggests that in order to understand fields not traditionally considered within the (broadly speaking) planning fields, we need to "make effective use of the creative capacity of spatial imagination" (ibid.), and food can provide a conceptual tool to do this. Olwig (2015) suggests that the contemporary focus on foods that can be seen as 'alternative' to those produced by agribusiness may have its roots in centuries of common cultural traditions where the meaning of food is attached to differing senses of landscape, and differing ways of sensing landscape. Studying food consumption, production, and cultures where land and sea 'scapes' are regarded as seamless helps to illustrate this idea and provides myriad opportunities for considering human-environmental relationships. There are numerous and fascinating models for growing and gathering food from the coast and sea from around the world. There are concerns about commercial interests which exploit the seascape as a source of particular produce and do not take an integrated and multidimensional approach to human-ecological processes. As Bharucha and Pretty (2010: 2922) suggest: "Edible species provide more than just food and income. In communities with a tradition of wild food use, it is part of a living link with the land, a keystone of culture." Our knowledge of human impact on the oceans and coasts is generally poor, and there has been little focus on the total ecosystem and landscape/seascape impact. For example, Roman et al. (2014) suggest that the near hunting to extinction of whales in the 1970s (66–90% estimated killed) damaged the marine ecosystem in ways that we are only just beginning to understand. Whales are an important part of the nutrient cycling system upon which the global marine ecosystem depends. A complete picture of the cultural connections and interactions between humans and the sea is too complex to cover in this chapter, but this is not surprising; terrestrial life emerged from the salt waters of the primordial seas and much of our makeup surely derives from that relationship. One intriguing idea that links food gathering and human evolution is the 'aquatic ape' hypothesis developed by Elaine Morgan (1972, 1982) who suggests that upright walking and other aspects of human evolution may have been as a result of shellfish gathering. This idea is highly contested and its detractors suggest that it is not a coherent theory (see Langdon, 1997). Today, examining the human search for food in the oceans reveals a highly complex web of stakeholder concerns encompassing fisheries, conservation, recreational, and commercial interests that extend to an even more complex exchange between land and sea and that begin to extend our very ideas of what 'landscape' is. Wylie (2007) suggests that examining landscape through a variety of frames in this way provides considerable potential for new research.

Williams' (1988) wonderful ode to the whale – 'Whale Nation' – reminds us that seas cover seven-tenths of the Earth's surface, and that from space the planet looks predominantly blue. The Apollo cameras of the 1970s gave us the first images of this earth as primarily a planet of water – of the sea; a blue planet where the interconnectedness of life on land, sea, and in the air seemed obvious (Lovelock, 1979; Roe, 2012). Bremner (2014: 18) explains that the ocean is "a fluid space where many transnational systems, practices and imaginaries intersect and overlap". Thinking in these terms and of Steinberg's (2001: 10) argument that that the ocean is "a space that, like land, shapes and is shaped by social and physical processes" helps to envisage the sea as much more than just a "resource to be managed, a transportation surface to be crossed or as an ungovernable and undifferentiated void" (Bremner, 2014: 20). It is easy to look at the ocean and just see an expanse of water that only appears to change with the non-human processes and forces of weather, tides, light and temperature, but if we consider the concept of seascapes, then the importance of the associations, meanings and traditions and the interactions between human and non-human species and processes is illumined as an extraordinary, vast, and poorly understood part of the global landscape.

Legislation in the UK such as the Marine and Coastal Access Act (2009), the Marine Scotland Act (2010), and the establishment the Marine Management Organisation (MMO) relating to the marine environment (see JNCC, 2014) suggests that the time is ripe for a focus on seascapes for landscape research. The marine 'problem' as an institutional one is represented in a number of papers (for example Gazzola et al., 2015; Ritchie and Ellis, 2010) and there is much evidence to indicate that the degradation of the marine environment by those searching for food in an unsustainable way is continuing and is still poorly controlled or ignored by successive governments (see for example McKie, 2014). However, the problem is not solely one of institutional responsibility, planning designation or regulation, but also of conceptualisation. The seascape is generally regarded as a common resource to pillage, or rather a resource that no-one owns (see Gazzola et al., 2015; Ostrom et al., 2006); it is ecologically complex and biologically rich, mysterious, poorly researched and explored, culturally significant, physically dangerous, and its health is ultimately fundamental to life on the planet.

The oceans not only provide food, but for many, their ability to do this is critical. The ama of Hegura Island, Japan have collected seafood by traditional means for more than 1,000 years (McDonald, 2011). They observe increasing change in the marine environment from external forces that are affecting the food source. The ama are working with scientists who are documenting that change. The picture that is painted is very depressing: of an environment under deep stress that may be beyond recovery (ibid.). However, there is also hope. This example brings together traditional food and place-based knowledge and practices, with technologies and scientific methods. This kind of collaboration plus global co-management agreements, raising knowledge and awareness about seascapes, as well as policy and marine food regulation, may help change our approach to the common resource of the oceans.

Acknowledgements

Part of this chapter concerning the haenyeo of South Korea was first published under the Hydrocitizenship research project, funded by The Arts & Humanities Research Council (AHRC): 'Towards Hydrocitizenship, Connecting Communities and Nature Through Interdependent Multiple Water Issues' www.hydrocitizenship. Source material for this chapter was gathered during various field visits to: Jeju Island (Korea), Osaka (Japan), Applecross (Scotland), and Lymington (England). I would like to thank all those who engaged in conversations on sea, food, and landscapes during that fieldwork.

References

Akimichi, T., & Sugiyama, H. (2011) Satoumi to integrate resource conservation and use: Sandfish fisheries in Akita Prefecture. In: United Nationals University Institute of Advanced Studies Operating Unit Ishikawa/Kanazawa (ed.), *Biological and cultural diversity in coastal communities: Exploring the potential of Satoumi for implementing the ecosystem approach in the Japanese Archipelago*, Technical Series No. 61. Montreal: Secretariat of the Convention on Biological Diversity, pp. 24–29.

Álvarez, P., Seingier, G., Bocco, G., Espejel, I., & Norieg, J. (2014) Regional landscape change in fishing communities of the Mexican North Pacific. *Landscape Research*, 40(7), 855–874. Available from: DOI: 10.1080/01426397.2015.1031095.

Baldwin, J. (2009) Harvesting seabirds and their eggs on the Irish Sea Islands (Part 1: The Welsh Islands, Lundy and Scilly. *Folk Life*, 47(1), 76–96. Available form: DOI: 10.1179/175967009X422837.

Barr, B.W. (2014) Observations on developing a maritime cultural landscape approach to managing US National Marine Sanctuaries. In: Van Tilburg et al. (eds.), *Proceedings of the 2nd Asia Pacific underwater cultural heritage conference*, The Museum of Underwater Archaeology, online resource. Available from: www.themua.org.

Barthes, R. (1975) Toward a psychosociology of contemporary food consumption. In: Forster, E., & Forster, R. (eds.), *European diet from pre-industrial to modern times*. New York: Harper and Row, pp. 47–59.

Berkes, F. (2003) Alternatives to conventional management: Lessons from small-scale fisheries. *Environments*, 31(1). http://umanitoba.ca/institutes/natural_resources/canadaresearchchair/Alternatives%20to%20Conventional%20 Management%20-%20Lessons%20from%20Small-Scale%20Fisheries.pdf.

Berkes, F., Mahon, P., MacConney, P., Pollnac, R., & Pomeroy, R. (2001) *Managing small scale fisheries: Alternative directions and methods*. Ottawa, Canada: IDRC.

Berque, J., & Matsuda, O. (2013) Coastal biodiversity management in Japanese satoumi. *Marine Policy*, 39, 191–200. Available from: http://dx.doi.org/10.1016/j.marpol.2012.10.013.

Bharucha, Z., & Pretty, J. (2010) The roles and values of wild foods in agricultural systems. *Philosophical Transactions of the Royal Society B*, 365, 2913–2926. Available from: DOI: 10.1098/rstb.2010.0123.

Blanchet, C., & Rochette, L. (2008) *Nutrition and food consumption among the inuit of Nunavik. Nunavik Inuit Health Survey 2004, Qanuippitaa? How are we?* Quebec: Institut national de santé publique du Québec (INSPQ) and Nunavik Regional Board of Health and Social Services (NRBHSS). www.inspq.qc.ca/pdf/publications/762_ESI_Nutrition_Report_MA.pdf (accessed 8 July 2016).

Bond, W., & Sufflield, P. (2012) *The origins of the Mermaid myth*. www.womanthouartgod.com/mermaids.php (accessed 4 August 2016).

Bremner, L. (2014) Folded ocean: The spatial transformation of the Indian Ocean world. *Journal of the Indian Ocean Region*, 10(1), 18–45. Availablle from: DOI: 10.1080/19480881.2013.847555.

Byun, K., Kang, E-J, Yoo, C., & Kim, K-H. (2015) Spatial transformation and functions of Bulteok as space for Haenyeo on Jeju Island. *Korea Journal of Asian Architecture and Building Engineering*, 14(3), 533–540. Available from: DOI: 10.3130/jaabe.14.533.

Campbell, B., & Pauly, D. (2013) Mariculture: A global analysis of production trends since 1950. *Marine Policy*, 39, 94–100. Available from: DOI: 10.1016/j.marpol.2012.10.009.

Christian, F. (2013) *Eat the beach: A guide to the edible seashore*, 2nd ed., Coastal Survival Handbook No. 1. Marston Gate: Practical Inspiration/Amazon.

CoE (Council of Europe) (2000) *The European landscape convention text*. www.coe.int/fr/web/conventions/full-list/-/conventions/treaty/176 (accessed 18 August 2016).

Copely, J. (2014) Mapping the deep, and the real story behind the "95% unexplored" oceans. http://moocs.southampton.ac.uk/oceans/2014/10/04/mapping-the-deep-and-the-real-story-behind-the-95-unexplored-oceans/ (accessed 19 January 2017).

Danchin, E., Giraldeau, L., & Cezilly, F. (2008) *Behavioural ecology*. New York: Oxford University Press.

Falconer, L., Hunter, D-C., Telfer, T.C., & Ross, L.G. (2013) Visual, seascape and landscape analysis to support coastal aquaculture site selection. *Land Use Policy*, 34, 1–10. Available from: DOI: 10.1016/j.landusepol.2013.02.002.

FAO (Food and Agriculture Organisation of the United Nations) (2016a) *The state of world fisheries and aquaculture 2016: Contributing to food security and nutrition for all*. Rome: FAO. www.fao.org/3/a-i5555e.pdf.

FAO (Food and Agriculture Organisation of the United Nations) (2016b) *FAO term portal, fisheries glossary, fisheries and aquaculture department* (Entry 1222). www.fao.org (accessed 3 August 2016).

Ferreira, J.G., Saurel, C., Lencart e Silva, J.D., Nunes, J.P., & Vazquez, F. (2014) Modelling of interactions between inshore and offshore aquaculture. *Aquaculture*, 426–427, 154–164. Available from: http://dx.doi.org/10.1016/j.aquaculture.2014.01.030.

Field, I.C., Meekan, M.G., Buckworth, R.C., Corey, J.A., & Bradshaw, C.J.A. (2009) Protein mining the world's oceans: Australasia as an example of illegal expansion-and-displacement fishing. *Fish and Fisheries*, 10(3), 323–328. Available from: DOI: 10.1111/j.1467-2979.2009.00325.x.

Frink, L., & Giordano, C. (2015) Women and subsistence food technology: The arctic seal poke storage system. *Food and Foodways*, 23(4), 251–272. doi: 10.1080/07409710.2015.1099906

Gazzola, P., Roe, M.H., & Cowie, P.J. (2015) Marine spatial planning and terrestrial spatial planning: reflecting on new agendas. *Environment and Planning C: Government and Policy*, 33(5), 1156–1172. doi: 10.1177/0263774X15612317.

Gee, K., & Burkhard, B. (2012) Offshore wind farming on Germany's North Sea coast: Tracing regime shifts across scales. In: Pleninger, T., & Bieling, C. (eds.), *Resilience and the cultural landscape: Understanding and managing change in human-shaped environments*. Cambridge: Cambridge University Press, pp. 185–202.

Gibbons, A. (undated) The evolution of diet. *National Geographic*. www.nationalgeographic.com/foodfeatures/evolution-of-diet/#close (accessed 27 May 2016).

Graham, N.A.J., Chabanet, P., Evans, R.D., Jennings, S., Letourneur, Y., MacNeil, M.A., McClanahan, T.A., Őhman, M.C., Polunin, N.V.C., & Wilson, S.K. (2011) Extinction vulnerability of coral reef fishes. *Ecology Letters*, 14, 341–348. Available from: DOI: 10.1111/j.1461-0248.2011.01592.x.

Hardy, K., & Wickham-Jones, C. (2002) Scotland's first settlers: The mesolithic seascape of the inner sound, skye and its contribution to the early prehistory of Scotland. *Antiquity*, 825–833. Available from: DOI: 10.1017/S0003598X00091304.

Hartill, B., Cryer, M., & Morrison, M. (2005) Estimates of biomass, sustainable yield, and harvest: Neither necessary nor sufficient for the management of non-commercial urban intertidal shellfish fisheries. *Fisheries Research*, 71(2), 209–222. Available from: DOI: 10.1016/j.fishres.2004.08.032.

HeraldScotland (2014) Island delicacy is a rare remioner of hard days, limited larder options and the bravery of the touch, Hebridean men of Ness. 8 November 2014. www.heraldscaotland.com (accessed 18 August 2016).

Hickey, G., Pouliot, M., Smith-Hall, C., Wunder, S., & Nielsen, M.R. (2016) Quantifying the economic contribution of wild food harvests to rural livelihoods: A global-comparative analysis. *Food Policy*, 62, 122–132. Available from: DOI: 10.1016/j.foodpol.2016.06.001.

Hill, M., Briggs, J., Minto, P., Bagnall, D., Foley, J., & Williams, A. (2001) *Guide to best practice in seascape assessment*. Dublin: The Marine Institute.

Hong, S.K., & Rabn, H. (1967) The diving women of Korea and Japan. *Scientific American*, 34–42.

James, J. (2016) *The salt industry of lymington and the solent coast*. Lymington: Lymington Museum Trust/St Barbe Museum and Art Gallery.

JNCC (Joint Nature Conservation Committee) (2014) http://jncc.defra.gov.uk/marine (accessed 29 January 2017).

Kang-hyun, J. (2014) A sad history behind, Jeju Women prove their fortitude through diving. *Koreana: Journal of the Korean Foundation*, Special Issue: Haenyeo: Iconic Female Divers of Jeju. Available from: http://issuu.com/the_korea_foundation/docs/koreana_summer_2014__english__215792b9ea6138 [Accessed 8 February 2016].

Karikis, M. (2015) *Artist's statement: The vanishing community of haenyeo all-women sea workers on a Pacific island*. www.culture24.org.uk/art/photography-and-film/art540967-mikhail-seawomen-pacific-earsthetic-brighton-dome (accessed 14 August 2016).

Kohsaka, R., Shih, W., Saito, O., & Sadohara, S. (2013) Local assessment of Tokyo: Satoyama and Satoumi: Traditional landscape and management practices in a contemporary urban environment. In: Elmqvist, T., Fragkias, M., Goodness, J., Güneralp, B., Marcotullio, P.J., McDonald, R.I., Parnell, S., Schewenius, M., Sendstad, M., Seto, K.C., & Wilkinson, C. (eds.), *Urbanization, biodiversity and ecosystem services: Challenges and opportunities: A global assessment*. Dordrecht, The Netherlands: Springer, pp. 93–106. Available from: DOI: 10.1007/978-94007-7088-1_8.

Kusuki, Y. (1991) Oyster culture in Japan and adjacent countries: Crassostrea gigas (Thunberg). In: Menzel, W. (ed.), *Estuarine and Marine Bivalve Mollusk culture*. Boca Raton, FL: CRC Press, pp. 227–243.

Lane, M. (2011) What freediving does to the body. *BBC News*, 12 January 2011. www.bbc.com/news/science-environment-12151830 (accessed 14 August 2016).

Langdon, J.H. (1997) Umbrella hypotheses and parsimony in human evolution: A critique of the Aquatic Ape Hypothesis. *Journal of Human Evolution*, 33(4), 479–494.

Lee, J. & Garikipati, S. (2011) Negotiating the non-negotiable: British foraging law in theory and practice. *Journal of Environmental Law*, 23(3), 415–439. Available from DOI: 10.1093/jel/eqr014.

Lefebvre, H. (1991) *The production of space*. Oxford: Blackwell.

Lewis, G.H. (1998) The Maine lobster as regional icon: Competing images over time and social class. In: Shortridge, B.G., & Shortridge, J.R. (eds.), *The taste of American place: A reader on regional and ethnic foods*. Lanham, USA: Rowman and Littlefield, pp. 65–83.

Lovelock, J.E. (1979) *Gaia: A new look at life on earth*. London and New York: Oxford University Press.

Lowitt, K. (2013) *An examination of rural and coastal foodscapes: Insights for the study of community food security and sustainable food system*. Unpublished PhD thesis, Memorial University of Newfoundland, Canada. www.curra.ca (accessed 18 August 2016).

Lowitt, K. (2014) A coastal foodscape: Examining the relationship between changing fisheries and community food security on the west coast of Newfoundland. *Ecology and Society*, 19(3), 48. Available from: DOI: 10.5751/ES-06498-190348.

Mabey, R. (1989) *Food for free*. London: Collins.

Mabey, R. (1996) *Flora Britannica*. London: Sinclair-Stevenson.

Maclean, C. (1989 [orig. 1972]) *Island on the edge of the world: The story of St Kilda*. Edinburgh: Cannongate.

McDonald, A. (2011) The Ama-san of Hegura Island: Carrying on the traditions of her ancestors – over 1,4000 years of community-based resource management. In: United Nationals University Institute of Advanced Studies Operating Unit Ishikawa/Kanazawa (ed.), *Biological and cultural diversity in coastal communities: Exploring the potential of Satoumi for Implementing the ecosystem approach in the Japanese Archipelago*. Technical Series No. 61. Montreal: Secretariat of the Convention on Biological Diversity, pp. 45–53.

McKie, R. (2014) 'An absolute disgrace': How intensive fishing is killing our oceans. *The Observer Tech Monthly*, 9 February, 27–28.

Mears, R. (2001) *Outdoor survival handbook*. London: Ebury Press.

Melville, H. (1851) *Moby-Dick: or, The Whale*. New York: Harper & Brothers/London: Richard Bentley.

Menzel, W. (ed.) (1991) *Estuarine and Marine Bivalve Mollusk culture*. Boca Raton, FL: CRC.

Morgan, E. (1972) *The descent of woman*. London: Souvenir Press.

Morgan, E. (1982) *Aquatic ape hypothesis*. New York: Stein & Day.

Musard, O., Le Dû-Blayo, L., Fancour, P., Beurier, J-P, Feunteun, E., & Talassinos, L. (eds.) (2014) *Underwater seascapes: From geographical to ecological perspectives*. Heidelberg: Springer.

Nicolson, A. (2001) *Sea room: An island life*. London: Harpercollins.

Nimmo, F., Cappell, R., Huntingdon, T., & Grant, A. (2011) Does fish farming impact on tourism in Scotland? *Aquaculture Research*, 42, 132–141.

O'Connor, K. (2016) Invisible foodscapes: Into the blue. In: Abbots, E-J., & Lavis, A. (ed.), *Why we eat, how we eat: Contemporary encounters between foods and bodies*. Abingdon: Routledge, 2nd ed., 15–34.

Olwig, K. (2015) The culinary landscape heritage of the festive carnivalesque and the religious feast. Presentation at *The role of food and food culture for a sustainable landscape heritage*, Symposium, Stora Sessionssalen, Swedish National Heritage Board Stockholm, 22 May 2015.

Ostrom, E., Gardner, R., & Walker, J. (2006) *Rules, games, and common-pool resources*. Ann Arbor, MI: University of Michigan Press.

Paley, M. (2014) We are what we eat. *National Geographic*. http://proof.nationalgeographic.com/2014/09/30/we-are-what-we-eat-in-the-arctic-its-all-about-meat/ (accessed 27 May 2016).

Pálsson, G. (1994) Enskilment at sea. *Man*, 29(4), 901–927. Availabel from: DOI: 128.240.225.59.

Panelli, R., & Tipa, G. (2009) Beyond foodscapes: Considering geographies of indigenous well-being. *Health and Place*, 15, 455–465. Available from: DOI: 10.1016/j.healthplace.2008.08.005.

Pieniak, Z., Vanhonacker, F., & Verbeke, W. (2013) Consumer knowledge and use of information about fish and aquaculture. *Food Policy*, 40, 25–30. Available from: http://dx.doi.org/10.1016/j.foodpol.2013.01.005.

Probyn, E. (2012) In the interests of taste and place: Economies of attachment. In: Pratt, G., & Rosner, V. (eds.), *The global and the intimate: Feminism in our time*. New York: Columbia University Press, pp. 57–84.

Probyn, E. (2016) *Eating the ocean*. Durham, NC and London: Duke University Press.

Pungetti, G. (2012) Islands, culture, landscape and seascape. *Journal of Marine and Island Cultures*, 1, 51–54. Available from: http://dx.doi.org/10.1016/j.imic.2012.11.007.

Rackham, J. (1994) *Animal bones*. Berkley, CA and Lost Angeles: University of California Press/British Museum.

Ritchie, H., & Ellis, G. (2010) 'A system that works for the sea'? Exploring stakeholder engagement in marine spatial planning. *Journal of Environmental Planning and Management*, 53(6), 701–723. Available from: doi.org/10.1080/09640568.2010.488100.

Roe, M.H. (2012) Making sense of place at the landscape scale. In: Convery, I., Dutson, T., Davies, P., & Corsane, G. (eds.), *Making sense of place: An edited collection*, Newcastle University Heritage Matters series. Woodbridge: Suffolk, Boydell and Brewer, pp. 191–206.

Roe, M.H., Sarlov-Herlin, I., & Speak, S., (2016) Identity, food and landscape character in the urban context. *Landscape Research*, 41(7), 757–772. doi: 10.1080/01426397.2016.1212324.

Roe, M.H., & Taylor, K. (eds.) (2014) *New cultural landscapes*. London: Routledge.

Roman, J., Estes, J.A., Morissette, L., Smith, C., Costa, D., McCarthy, J., Nation, J., Nicol, S., Pershing, A., & Smetacek, V. (2014) Whales as marine ecosystem engineers. *Frontiers in Ecology and the Environment*, 12, 377–385. Available from: DOI: 10.1890/130220.

Salmon, E. (2012) *Eating the landscape: American Indian stories of food, identity and resilience*. Tuscon, AZ: University of Arizona Press.

Sand, J. (2015) How Tokyo invented Sushi. In: Imbert, D. (ed.), *Food and the city: Histories of culture and cultivation*. Washington, DC: Dumbarton Oaks, pp. 223–248.

Schulp, C.J., Thuiller, W., & Verburg, P. (2014) Wild food in Europe: A synthesis of knowledge and data of terrestrial wild food as an ecosystem service. *Ecological Economics*, 105, 292–305. Available from: DOI: 10.1016/jecolecon.2014.06018.

Scott, K., Anderson, C., Dunsford, H., Benson, J., & MacFarlane, R. (2005) *An assessment of the sensitivity and capacity of the Scottish seascape in relation to offshore windfarm*. Inverness: Scottish Natural Heritage.

Searles, E. (2002) Food and the making of modern inuit identities. *Food and Foodways: Explorations in the History and Culture of Human Nourishment*, 10(1–2), 55–78. Available from: DOI: 10.1080/07409710212485.

Shellfish Association of Great Britain (2010) *Cooking with oysters*. www.shellfish.org.uk/files/Cooking-With/6891SAGB%20oyster%20recipe%20book%20final%202%20LowRes.pdf.

Steel, C. (2012) Sitopia – harnessing the power of food. In: Viljoen, A., & Wiskerke, J.S.C. (eds.), *Sustainable food planning: Evolving theory and practice*. Wageningen, WAP, pp. 37–45.

Steinberg, P.E. (2001) *The social construction of the ocean*. Cambridge: Cambridge University Press.

Steinberg, P.E. (2001) *The social construction of the ocean*. Cambridge: Cambridge University Press.

Sumaila, U.R., Cheung, W.W.L., Lam, V.W.Y., Pauly, D., & Herrick, S. (2011) Climate change impacts on the biophysics and economics of world fisheries. *Nature Climate Change*, 1–8[online] doi: 10.1038/NCLIMATE1301

SUN-AE II (2012) Why do Korean women dive? A discussion from the viewpoint of gender (Special Issue Gender in Aquaculture and Fisheries: Moving the Agenda Forward), *Asian Fisheries Science*, 25S, 47–58.

The Blue Planet: Seas of Life (2001) [TV nature documentary] UK:BBC Natural HistoryUnit/Discovery Channel.

Vallely, P. (2010) Shucks! Why British oysters are off the menu. *The Independent* [online]. www.independent.co.uk/life-style/food-and-drink/features/shucks-why-british-oysters-are-off-the-menu-2080162.html (accessed 26 January 2017).

Vincentnathan, L. (2012) Food rights, food frugality, and climate change. *Food and Foodways*, 20(1), 76–92. doi:10.1080/07409710.2012.652028

Viljoen, A., & Wiskerke, J.S.C. (eds.) (2012) *Sustainable food planning: Evolving theory and practice*. Wageningen, WAP.

Waterman, T. (2011) A word. . . . *Landscape: The Journal of the Landscape Institute* (Seascape architecture, why we should treat the land and sea as a continuum), 48.

Watson, R.A. & Pauly, D. (2013) The changing face of global fisheries: The 1950s vs the 2000s. Marine Policy, 42, 1–4. http://dx.doi.org/10.1016/j.marpol.2013.01.022.

Westerdahl, C. (1992) The maritime cultural landscape. *International Journal of Nautical Archaeology*, 21(1), 5–14. Available from: doi: 10.1111/j.1095-9270.1992.tb00336.x.

Whitmarsh, D., & Palmieri, M.G. (2009) Social acceptability of marine aquaculture: The use of survey-based methods for eliciting public and stakeholder preferences. *Marine Policy*, 33, 452–457. Available from: DOI: 10.1016/j.marpol.2008.10.003.

Williams, H. (1988) *Whale nation*. London: Jonathan Cape.

Worm, B., & Branch, T.A. (2012) The future of fish. *Trends in Ecology and Evolution*, 27(11), 594–599. doi:10.1016/j.tree.2012.07.005.

Wylie, J. (2007) *Landscape*. London: Routledge.

Yanagi, T. (2008) "Sato-Umi" – a new concept for sustainable fisheries. In: Tsukamoto, K., Kawamura, T., Takeuchi, T., Beard Jr, T.D., & Kaiser, M. (eds.), *Fisheries for global welfare and environment*, 5th World Fisheries Congress. Terrapub, pp. 351–358. Available from: www.terrapub.co.jp/onlineproceedings/fs/wfc2008/pdf/wfcbk_351.pdf [Accessed 19 August 2016].

Further reading

Atkinson, D., & Hale, A. (eds.) (2012) *Source to sea: ScARF Marine and Maritime Panel report*, Scottish Archaeological Research Framework: Society of Antiquaries of Scotland. www.scottishheritagehub.com/ (accessed 24 May 2016).

British Seafishing website. http://britishseafishing.co.uk/.

Carson, R.L. (1952) *Under the sea-wind: A naturalist's picture of ocean life*. London: Staples Press.

Cetinkaya, G. (2009) Challenges for the maintenance of traditional knowledge in the satoyama and satoumi ecosystems, Noto Peninsula, Japan. *Human Ecology Review*, 16(1), 27–40.

Cooney, G. (2003) Introduction: Seeing land from sea. *World Archaeology*, 35(3), 323–328.

Fleming, A. (2005) *St Kilda and the wider world: Tales of an iconic island*. Bollington: Windgather Press.

Government Office for Science (2011) *The future of food and farming: Challenges and choices for global sustainability*. London: CMO.

Indrawan, M., Yabe, M., Nomura, H., & Harris, R., (2014) Deconstructing satoyama – the socio-ecological landscape in Japan. *Ecological Engineering*, 64, 77–84. http://dx.doi.org/10.1016/j.ecoleng.2013.12.038.

Kracker, L.M. (1999) The geography of fish: The use of remote sensing and spatial analysis tools in fisheries research. *The Professional Geographer*, 51(3), 440–450.

McNiven, I.J. (2003) Saltwater people: Spiritscapes, maritime rituals and the archaeology of Australian indigenous seascapes. *World Archaeology*, 35(3), 329–349.

Milner, N., Craig, O.E., & Bailey, G.N. (eds.) (2007) *Shell middens in Atlantic Europe*. Oxford: Oxbow.

Natural England (2012) *An approach to seascape character assessment*. Bristol: Natural England.

Natural England (2012) *Seascape characterisation around the English coast (marine plan areas 3 and 4 and Part of area 6 pilot study)*. Bristol: Natural England.

Ostrom, E. (1990) *Governing the commons: The evolution of institutions for collective action*. Cambridge: Cambrige University Press.

12

Dimensions of urban agriculture

Joshua Zeunert

Introduction

Agricultural production's close proximity to settlements proved crucial to food supply security until very recent times (see Chapter 4). For the past 12,000 years since the Neolithic (Agricultural) Revolution, agriculture and civilisation have grown symbiotically, with food surpluses enabling permanent settlements and the development of urban life (Chapters 1 and 2). Over the past century, and particularly since the misleadingly named 'green revolution' (Chapter 16 outlines the negative environmental ramifications), a transformational shift in agricultural practices has increased food production through large-scale mechanisation, high-yield crop varieties and breeds, and artificial fertilisers and pesticides (Bowler, 1992; Steel, 2008; Zeunert, 2017: 170; Chapter 9). This phenomenal reconfiguration of food and landscape systems has largely been enabled by harnessing extraordinarily concentrated fossil fuel energy sources, particularly oil, natural gas, and their by-products (Chapters 10 and 16). Subsequent hyper population growth and rapid urbanisation have dramatically reconfigured thousands of years of relative consistency in the evolution and relationships between settlements, the food that sustains them, and the proximity of food production.

The past two decades has witnessed a resurgence in urban agricultural interest and discourse. This timeframe forms the focus of this chapter's discussion, which combines a number of useful definitions to position urban agriculture (UA) before examining it more closely. Smit's widely cited definition describes UA as:

> An industry that produces, processes and markets food and fuel, largely in response to the daily demand of consumers within a town, city or metropolis, on land and water dispersed throughout the urban and peri-urban area, applying intensive production methods, using and reusing natural resources and urban wastes, to yield a diversity of crops and livestock.
>
> *(Smit et al., 1996: 3)*

Using Butler and Maronek's (2002) definition, Philips (2013: 48) includes "a multiplicity of other benefits and services [. . .] recreation and leisure; economic vitality and business entrepreneurship; individual health and well-being; community health and well-being; landscape beautification; and environmental restoration and remediation". Knowd et al. (2005: 2) rightly argues that UA can be "a response to modernisation that has cultural and philosophical dimensions". This chapter inclusively defines 'urban' as encompassing peri-urban regions, consistent with leading UA (as in Smit, above) and well-articulated by Urban Agriculture Europe (COST):

> Urban agriculture spans all actors, communities, activities, places and economies that focus on biological production in a spatial context, which-according to local standards – is categorized as 'urban'. Urban agriculture takes place in intra- and periurban areas [. . .] urban agriculture is structurally embedded in the urban fabric; it is integrated into the social and cultural life, the economics, and the metabolism of the city.
>
> *(Lohrberg et al., 2016: 21)*

This chapter therefore encompasses both peri, intra and inner urban regions as constituting 'urban' agriculture (noting that Chapters 14 and 15 focus specifically on peri-urban contexts). Notwithstanding McClintock et al.'s (2013: 55) observation of the difficulty of capturing UA's multiplicity, this chapter frames its diversity through six key dimensions: *spatial*, *economic*, *social*, *environmental*, *practical*, and *intellectual* (Figure 12.1). While UA's characteristics do not neatly fit into six discrete categories without overlap, discussion of each of these dimensions follows herein.

Dimension one: spatial

This dimension examines urban agriculture's spatial characteristics and the related forces shaping its physical makeup and scale. Agriculture's transition from local, to regional, and now global was and is boosted by fossil fuel-powered farm machinery and synthetic nitrogen fertiliser, with today's 'industrial agriculture' further facilitated by petroleum, distillate, aviation fuel, and fuel oils (Chapters 16 and 25; Bowler, 1992; Norberg-Hodge et al., 2001; Weis, 2010; Stierand, 2012: 67–68; Berry, 2015). This is the "kinetic expressionism era" (German philosopher Peter Sloterdijk, cited in Fücks, 2015: 190–191) – representing a colossal acceleration of the historic time/space continuum (Harvey, 1989). As agriculture becomes further distanced

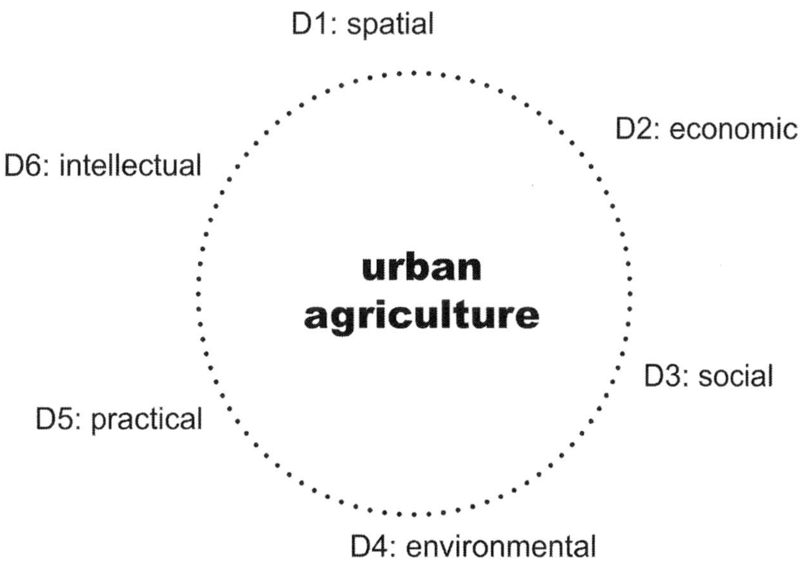

Figure 12.1 The primary dimensions of urban agriculture as defined by the author.

from expanding urban settlements the world over, operations diminish in number and labour force while increasing in landholding size.

Urban agriculture is often deemed trivial in the face of this dominant, global industrialised food system (see for example Badami and Ramankutty, 2015) – associated with diminutive raised planting beds, isolated and temporary interventions, failure-prone vertical gardens and ailing food plants in hardscape dominated inner-city environments. Notwithstanding its recognised social dimensions (discussed in *Dimension Three: Social*), when measured against the Goliath of global agriculture, such small-scale UA measures seem mere "band-aids" (Meyer, in Zeunert, 2017: 212), essentially inconsequential to global food security (Ramankutty et al., 2008, World Bank, n.d.b; Chapter 25). This view is comprehensible when considering the size of individual agricultural landholdings in large countries such as the USA, Canada, China, Brazil, and Australia, which can exceed that of some countries – the state of South Australia's 34,000 square kilometre Anna Creek cattle station, for example, is greater than Belgium's entire 30,528 square kilometre landmass.

Yet urban agriculture is far from marginal, whether as measured by standard financial metrics (see *Dimension Two: Economic*), in the immeasurably important role it plays for communities worldwide (*Dimension Three: Social*), or in its high volume contribution to health and nutrition. Historically, UA was substantially employed in public green spaces, vacant and private land during twentieth century conflicts, wars and food crises. In many cases it provided significant yields: a substantial 40% of vegetable consumption in America (Brown and Jameton, 2000) and around half of Britain's fruit and vegetables (Garnett, 1996) during the Second World War – both considerable given their respective populations of 140 and 48 million people in 1945. More recently, global cropland analysis indicates that urban areas contain 11% of all irrigated and 4.7% of rainfed croplands; when extending the capture to within 20 kilometres of urban areas, this jumps to 60% of irrigated and 35% of rain-fed croplands (Thebo et al., 2014). In the USA, Burton et al. (2013: 87) state that UA contributes 40% of total food produced on just 10% of agricultural land.

Land losses and gains

Peri-urban farmers provide the bulk of UA production, both spatially and financially (Thebo et al., 2014; Chapter 15). Peri-urban agriculture is, however, diminishing (Chapter 14). Australian census data, for example, reveals 47.2% of perishable vegetables grown in peri-urban areas in 2010–11 (Sinclair, 2015: 2) but this is waning and likely to further reduce (Chapter 15). Ongoing urban sprawl and resultant land-value increase is largely responsible for this loss, mirrored in myriad cities worldwide. Studies identify UA expansion opportunities to help offset such reductions: Ackerman (2012: 3) pinpoints around 5,000 acres of potentially suitable vacant faming land (2,023 hectares, six times Central Park) in New York City (NYC), with Meier (2013: S-1) suggesting that agricultural land in NYC's Metropolitan Statistical Area could support between 58–89% of the 18,897,109 population's fruit and vegetable needs "if that land was dedicated entirely to production for local markets". Substantial city green belts devoted to fruit and vegetables offer notable production potential (Garnett, 1996), such as those encircling many English cities (Zeunert, 2017: 52). Large public green spaces offer further spatial and financial opportunities (Zeunert, 2016).

Production suitability

Certain vegetables, herbs, fruits, nuts, and more spatially efficient, compact staple crops can meaningfully contribute to total urban food demands. Long-time UA proponents, the *RUAF Foundation*, cites significant percentages of selected produce (leafy vegetables/all vegetables/eggs/poultry/milk/pork/fruit) grown in urban and peri-urban agriculture in sixteen (mainly developing) cities (de Zeeuw and Dubbeling, 2009: 11). More recent data for the study's two most developed cities (Singapore and Hong Kong), indicates significantly lower volumes in the two decades elapsed (NZTE, 2014; Tse, 2015), possibly due to uptake of global over local agriculture, and urban population growth.

Food footprints

It is worth noting that current urban food footprints (see Rees, 1992; Wackernagel, 1994; Weller, 2009: 440–441; Roggema and Keeffe, 2014: 12) well exceed even the most low-density, vacant, and/or shrinking cities' (such as Detroit) capacity for food self-sufficiency within their area (see Grewal and Grewal, 2012; McClintock et al., 2013; Porter et al., 2014; Badami and Ramankutty, 2015). This unfeasibility is often cited by UA's critics (see for example Badami and Ramankutty, 2015). UA's possible production contribution is, however, substantially dependent on a range of variable factors. These include: dietary/purchasing choices (Chapter 16), population numbers/densities, suitable land availability and deployment, economic subsidies/paid labour arrangements, centralised/decentralised infrastructure sophistication (primarily for nutrient supplies through wastewater and soil conditioners (Chapter 23), and irrigation water/recycled water supplies (see Chapter 19).

Dietary choices hugely affect UA's spatial production significance. Spatially, livestock and many staple crops like rice, maize (corn), and wheat cover expansive tracts of arable and marginal lands well beyond an urban capacity. Extensive embodied resources are required for meat, dairy, and seafood production (see Pimentel and Pimentel, 2003; Weis, 2010; Chapter 16). Despite design visions such as 'Pig City' (MVRDV, 2009; see also Gorgolewski et al., 2011: 56–57; Roggema, 2014: 44–47), urban environments are unsuited to and avoided by industrial animal agriculture, and methods like factory farming and CAFOs (concentrated animal feeding operations, see for example Gurian-Sherman, 2008). Urban contexts typically present increased regulations, expense, odours due to intensive methods, as well as potential public outcry and reduced consumer demand following more visible animal welfare concerns. In fact, a key reason for UA's dispersal from 'modernising' urban regions is due to animal husbandry conflicts and subsequent bans (Grace et al., 2015: 255). Accordingly, UA is better suited to perishable, diversified, labour intensive, and high-income specialty crops (Zeunert, 2016); organic/traditional production methods (to reduce conflict from spraying – see Chapter 10); and spatially efficient vegan and vegetarian diets (Jeavons, 2006: 248–249; Peters et al., 2016; Chapter 16).

Spatially efficient solutions

For many regions, however, especially mega cities and those allowing continuing low-density sprawl, on-ground UA land availability can be limited. Accordingly, many observers have noted that rooftops present an under-realised spatial typology (see for example Mandel, 2013; Ackerman et al., 2014; Plakias, 2016; Novak, 2016). Prominent examples of rooftop UA include several in NYC, such as *Brooklyn Grange* (www.brooklyngrangefarm.com/) and *Eagle Street Rooftop Farm* (http://rooftopfarms.org/). Green roof implementation is considerably aided by increasing local, state, and national bylaws mandating inclusion on new (often commercial) buildings (such as in Munich, Basel, Toronto, Copenhagen, and France – see Zeunert, 2017: 129). Food production nevertheless remains a low priority in the built environment (noting that it requires 'intensive', deeper soil thus more expensive structure than 'extensive' (shallow) systems).

Intensive horticultural production methods such as rooftop greenhouses and vertical plant crops (such as 'Gothan Greens' in New York and Chicago, http://gothamgreens.com/, and 'Sky Greens' in Singapore, www.skygreens.com/) demonstrate high-tech and spatially efficient approaches to increase UA's economic viability in dense, urbanised areas where in-ground food production opportunities are limited and/or land expensive. Other compact approaches include aquaculture and mushroom production. It is currently unclear as to what extent emerging and innovative compact methods such as algae, insect, and animal protein manufacturing could impact consumption and thus spatial patterns.

Visions

UA has been a central component in spatial planning visions at urban and metropolitan scales. Examples include: new city planning such as Howard's *Town and Country* (1898–1902); Lloyd-Wright's *Broadacre City*

(1932–35); Le Corbusier's *Ferme Radieuse* (Radiant Farm, 1934–1937, see McLeod, 2015); Hilberseimer's *New Regional Pattern* (1944–49, Waldheim, 2010); Migge's *Everyman Self-sufficient!* (1918, Haney, 2010); as well as more recently Branzi's *Agronica* (1993–94, ibid.); Weller's *Food City* scenario (Weller, 2009: 274–289); and Lim and Liu's (2010) reintegration of agriculture in urban environments. Unfortunately, these and other large spatial visions (Integrated Design Commission, 2011: 12–20; SPUR, 2012; Roggema, 2014; Zeunert, 2016) are either wholly unrealised or have only had small aspects manifest. Most western UA continues to occurs at the site scale (Viljoen et al., 2005; Gorgolewski et al., 2011; Miazzo et al., 2013; Philips, 2013; Viljoen and Bohn, 2014; Roggema and Keeffe, 2014; Roggema, 2016; Zeunert, 2017: 166–195), yet planning can and indeed needs to play a more significant role in facilitating UA's spatial presence within the urban matrix (see *Dimension Six: Intellectual*; Chapter 26; Aubry et al., 2012; Popovitch, 2014; Viljoen et al., 2015).

Dimension two: economic

Dimension Two's discussion of UA's economic dimensions focuses on financial aspects (see Gudeman (2007) for a wider dialogue regarding value). As an important initial point, UA is financially significant: in 1989, over a third of the USA's agricultural production by value occurred within Metropolitan Statistical Areas (Gorgolewski et al., 2011: 232, citing Smit et al., 2001, citing Heimlich, 1989). Over two decades later, US metropolitan counties still accounted for 38.7% of total agricultural sales' value (USDA, 2012; T. Parker, USDA, pers. comm. 20 April 2017). This value is primarily generated in peri-urban regions from, for example, irrigated vegetable production, dairying, and pig and poultry raising.

Discourse

UA varies across cultural contexts, reflecting different patterns of economic development (as well as values and aesthetic legacies – for example Chapters 13, 14, 15 and 22). Until the last decade, UA discourse (without explicit acknowledgement) has largely been divided into developed or developing economies, with academic discussion prioritising the latter. The majority of developing world UA is subject to different economic dynamics than that of the developed world and is outside this chapter's scope (for further reading, explore Smit et al., 1996; Mougeot, 2005; Mougeot, 2006; Pearson et al., 2010; Dubbeling et al., 2010; Sahakian, 2016; Chapter 21).

Economic literature specifically focusing on UA has been "very limited" (see Golden's US review (2013: 13)), but discourse is expanding (Lazarus, 2000b; Kaufman and Bailkey, 2000; Moustier and Danso, 2006; Masi, 2008; Brown, 2008; van der Schans and Wiskerke, 2012; Caputo, 2012; Martin, 2013; Schouw and Roggema, 2014; Cabannes, 2015; Zeunert, 2016; Stone, 2016). UA is usually an "additive, as opposed to substitutive, form of economic development, creating value from sites, which for whatever reason are not suitable or attractive for other uses" (Meier, 2013: S-2). Much non-pre-existing UA in developed economies is catalysed by short-term government, foundation or charitable grants; and fallow, vacant, temporary, 'meanwhile' (thus cheap) leasing or no-cost land use (see for example, Gorgolewski et al., 2011; Miazzo et al., 2013; Philips, 2013; Stone, 2016). Maintaining short-term and intergenerational financial viability in developed world contexts is almost always challenging. Innovative, high-profile examples defying short-term challenges (see for example, Allen and Wilson, 2012; Plakias, 2016; Stone, 2016) are exceptions to the greater disappearance of peri-urban farms under more immediately lucrative land use developments such as housing (see Chapters 14 and 15).

Broader landscape and food discourse critiques agro-industrialisation and neoliberal economic agendas continuing to hinder small agricultural operations' economic viability in many developed and increasingly, developing economies (Bolwer, 1992; McMichael, 2009; Wolf, 2014; Holt-Giménez, 2015; Springer et al., 2016; Chapters 21 and 31). Scholars are increasingly using the term 'agribusiness' rather than 'agriculture' to reflect agriculture's deregulation, facilitation of corporate power, privatisation, creeping acquisitions, and loss of individual rights (Morgan et al., 2006; McMichael, 2015). Such trends prioritise economies of scale, favouring corporate agribusiness and food giants.

Global agribusiness's continued expansion partly explains increased UA interest (Knowd et al., 2005: 2) and food sovereignty and justice concerns in discourse (Chapters 31 and 32; Gottlieb and Joshi, 2010; Berry, 2015; Claeys, 2015; McMichael, 2015; Kennedy and Liljeblad, 2016). This is based to varying degrees on objections to aggressive food system commodification and resultant homogeneity, and loss of agrobiodiversity and artisanal agriculture (see Table 31.1). Food sovereignty also rejects ongoing enclosure, land grabs, and loss of common land and other resources (Chapter 31; Linn, 1999; FAO, 1999; Norberg-Hodge et al., 2001; Freyfogle, 2003; Freyfogle, 2007; Linn, 2007; St. Martin, 2009; Tornaghi, 2012; SPUR, 2012; Wolford et al., 2013; Gilbert, 2014; Claeys, 2015; Holt-Giménez, 2015; Gaarde, 2017).

Key themes

Three key economic themes relevant to UA include subsidies, land prices, and green gentrification. Agricultural subsidies can greatly affect UA's viability. Minimal Australian subsidies (as in New Zealand, Chile and South Africa amongst others) leave operations little choice but to 'get big or get out' (Kelly, 2002, cited in Cebon, 2010; Burton et al., 2013: 31–32, 89; Chapter 9). Subsequently, total farm numbers decrease, while remaining farms increase in size and technology, heightening challenges for small UA interventions to compete with mega-producers. In the mega-scale, green-revolution-method-led USA food system, less than 1% of the population is listed as engaged in agriculture (Bomford, 2010). The World Bank confirms similar statistics for other developed nations, with 1% in the UK (World Bank, n.d.a). Contrastingly, in poorer countries, figures as high as 75% in Guinea (ibid.) indicate widespread small-scale, subsistence, family and artisanal agriculture. Such polarised proportions correlate with agriculture's contribution to gross domestic product – only 1% in high income countries versus 27% in low income countries (UN, 2016: 146).

Global economic emphasis reduces regional production diversity and multifunctionality, in turn diminishing national/local food systems' resilience through dependence on global agriculture, uninterrupted supply chains and maintaining amicable diplomatic arrangements (Chapters 9 and 16; Burton et al., 2013). Where small subsidies are more substantial, such as in Japan, statistics paint a very different picture. Almost one-third of all Japanese agricultural output is generated by UA and urban farmers, accounting for 25% of farming households (Moreno-Peñaranda, 2011), and helping to explain (along with sociocultural reasons) the number of small farms in Tokyo (see Niwa, 2012). The scale of subsidies and concessions is fundamental to UA's viability beyond hobbyist, subsistence, volunteer, and recreational modes (de la Salle and Holland, 2010). Subsidisation processes are routinely political and beyond the scope of this section (it should be noted that subsidies can and often do support economically unviable industries and rural land uses such as sheep grazing in Britain (Monbiot, 2013)).

Land prices also hugely affect UA's viability and scale. Cities with ongoing GDP and land-cost increases struggle to implement and/or maintain viable UA (as in numerous cities in South-East Asia and China, most Australian capital cities, and London, as opposed to the majority of the UK). This contrasts economically stagnant and shrinking areas with vacant and low cost land (Stone, 2016; Oswalt and Rieniets, 2006), as in the well-documented example of Detroit, USA (Giorda, 2012). In these contexts, UA is gaining collective momentum and diversifying beyond traditional typologies like community gardens, temporary and small-scale interventions.

When people join together to improve their communities – a regular outcome of UA (see *Dimension Three: Social*) – they can unwittingly become agents of their area's gentrification, potentially then affecting UA's on-going financial viability. Social activation stimulates development, increasing land, rent and living costs and leading to displacement. While this 'green gentrification' can be seen as a measure of UA's success (Dunn, in Finkel, 2012), it is also symptomatic of urban development and governance's commercial focus and a lack of policy and planning protection measures to prevent 'pricing-out' and subsequent displacement of socially (and economically) 'regenerative' activities. It is notoriously challenging to convince decision makers of social and environmental value over financial, despite the success of preventative planning and policies in cities like Berlin (Curry et al., 2014; Popovitch, 2014). UA solutions can include built-in transportation and disassembly design measures (such as raised planters on wheels or pallets), but ultimately address symptoms rather than causes. Furthermore, the age-long success of ground/soil-based crop growth and investments in improving soil quality help to remind us of permanent spatial interventions' importance (Jackson, 2010).

Lateral viability

Where a lack of sympathetic financial mechanisms recognise its wider potential, urban agriculture must be innovative and opportunistic, engaging highly competent, entrepreneurial, dedicated practitioners and value-added forms of income generation (Zeunert, 2016; Stone, 2016). These measures can combat potentially low-income returns versus time inputs (Urban Partners, 2007). Rare but potentially replicable initiatives by city mayors and political leaders utilising *complimentary currencies* have been successfully demonstrated in Curitiba's 1970's Cambio Verde (Green Change; sorted garbage/recyclables in exchange for local food program), replicated in Mexico's 'Mercado de Trueque' (Barter Market) (Zeunert, 2017: 233, 286) and adapted in New York City's food stamp exchange at farmers markets (van der Valk, 2016). UA also maintains viability through a variety of measures such as food pacts (see the Milan Pact (2016), *Dimension Six: Intellectual*), local, health and city region food procurement policies and 'short-chain' strategies (such as requiring a percentage of supplied food to be sourced from a specified proximity), demonstrating a high multiplier effect and significantly aiding local employment (MacLeod and Scott, 2007; Stefani et al., 2015; Morgan, 2015: 1387; Baker and de Zeeuw, 2015; Moustier and Renting, 2015; Dubbeling et al., 2016).

"Not everything that can be counted counts, and not everything that counts can be counted" (Cameron, 1963: 13). Beyond financial considerations, UA's role invokes considerations of ethics (Chapter 24), values, intrinsic worth, and ultimately, survival (Kunstler, 2006; Roberts, 2008; Chapter 16). It is important to note that global land enclosure and land grabs displace numerous populations from the land and agricultural livelihoods, having a commodifying effect on humanity itself (Linn, 1999; Kugelman et al., 2009; Wolford et al., 2013). In order to strengthen its hold in urban regions, UA must increase advocacy, education, improve comprehension and continue to bypass reductive financial quantification. Positively, UA offers potential to reduce food expenditure, generate secondary income, stimulate small and medium-sized enterprises (SME's), and create employment (Smit et al., 1996; Bakker et al., 2000; de Zeeuw and Dubbeling, 2009; Cockrall-King, 2012; de Zeeuw and Drechsel, 2015; Dubbeling et al., 2016).

Dimension three: social

Dimension three encompasses UA's social influences, challenges, and possibilities, noting that multiple social topics are well covered in this book's chapters. UA is vitally important worldwide, especially for subsistence-based lifestyles and for those outside of the labour market. At the turn of the millennium, an estimated 200 million urban residents produced food for the urban market, contributing 15–20% of total food production (Armar-Klemesu, 2000: 103; Burton et al., 2013: 87). According to Smit (et al., 2001: Ch. 2: 1), based on 1993 data, approximately 800 million people were in some way involved in UA across the globe. In 2015, Grace (et al., 2015: 255) estimates that figure rose, with "at least one billion intra- and peri-urban dwellers" practicing urban agriculture.

Given its impact on so many lives, social dimensions dominate UA discourse, regularly reflecting positive and transformative stories of community cohesion and engagement, social inclusion, poverty and crime reduction, intergenerational education, and meaningful work opportunities (see Patel, 1991; Malakoff, 1995; Garnett, 1996; Pretty, 2002; Butler and Moronek, 2002; Mason, 2006; Sumner et al., 2010; Fox, 2011; Nettle, 2014). Benefits are well documented, ranging from physical health to psychological wellbeing (Kaplan,

1973; Brown and Jameton, 2000; Flores, 2006; Lee-Smith and Prain, 2006; Hassink et al., 2007; Duchemin et al., 2008; Blaine et al., 2010; Hodgson, 2010; Firth et al., 2011; DLA et al., 2011; Allen and Wilson, 2012; McClintock et al., 2013). Challenges, on the other hand, involve food security and nutrition (see Chapters 20 and 25), the right to access common land to grow food (Chapters 31 and 32), achieving aesthetic design (Creasey, 2010; Zeunert, 2011; Gyertyán, 2014), and restrictive cultural and social perceptions.

Typologies

UA practices vary worldwide, subject to cultural and geographical nuances and complexities: as part of an ingrained Mediterranean tradition (Chapter 13), or cultivated in precious arable land patchworked between modern infrastructure and revered forested mountains in densely populated South Korea and Japan. Overwhelmingly though, 'working' agricultural landscapes are increasingly rare in urban contexts. When UA *is* present, it commonly takes the form of allotments and community gardens (see Chapter 30; Bell, 2016), school gardens, temporary events and installations, and exhibitions (Viljoen et al., 2005; de la Salle and Holland, 2010; Gorgolewski et al., 2011; Miazzo et al., 2013; Philips, 2013; Viljoen and Bohn, 2014; Zeunert, 2017: 166–195). These are frequently located on less visible, discarded, degraded, 'meanwhile', potentially contaminated, flood-prone, and low-value urban leftover sites. Such socially oriented UA typologies are generally small, ranging between 0.01–3 hectares (0.025–7.4 acres) and often 'grassroots' community-led or small growing interventions rather than commercial operations (Kaufman and Bailkey, 2000; Zeunert, 2016). Yields may not be the primary focus, with individuals or groups often involved for social, recreational, or health/psychologically related reasons (Blaine et al., 2010; Firth et al., 2011; Nettle, 2014).

Positively, UA is increasingly occupying privately owned spaces – rooftops, balconies, courtyards and suburban front and back yards, street medians, and leftover areas (Viljoen et al., 2005; de la Salle and Holland, 2010; Gorgolewski et al., 2011; Mobbs, 2012; Miazzo et al., 2013; Philips, 2013; Viljoen and Bohn, 2014; Zeunert, 2017: 166–195). High-tech, intensive and entrepreneurial new forms of micro and SME's in developed regions are navigating modern urban contexts and brownfield sites (see for example, Gorgolewski et al., 2011: 158–193; Miazzo et al., 2013; Viljoen and Bohn, 2014). In the USA, UA is on the increase due to food security and nutritional concerns, enabled by greater provision for community and private gardens, urban homesteading, public domain design and utilisation of vacant land (McClintock et al., 2013).

Aesthetics

As with any socially engaged activity, UA does not exist in a vacuum and aesthetic perceptions are affected by a suite of influences. Meyer describes aesthetics as "the art and science of perception, not just a formal category of how something looks" (Zeunert, 2017: 212). City-based food production challenges traditional Western urban/rural separations. The fields, orchards, and vineyards belonging to a rural, pastoral idyll rarely carry through to the modern city, even as spatially truncated typologies (fruit trees rather than orchards, or fruiting grape vines in place of full-scale vineyards). The absence of these small-scale agricultural emblems cannot easily be explained away by arguments of land value but of decorative preference, hinting at persistent, culturally embedded aesthetic programming (see Chapter 13 for exceptions). Historical-cultural British and Western European garden-architecture traditions, favouring ornamental plants, the groomed lawn of elite estates and eschewing 'common' agricultural associations in public spaces has profoundly affected design tastes in post-colonial nations and spread throughout the world. There is, for example, an astounding twenty million acres of non-native lawn in North America alone, reflecting this largely unconscious impact of cultural entrenchment (see Stone, 2016; Pollan, 2008). Formal, ordered plantings continue to dominate urban green spaces worldwide (Zeunert, 2016). As an additional barrier, formal aesthetics are far more difficult to achieve with edible and annual species (albeit adeptly demonstrated in instances such a Château de Villandry in France (Zeunert, 2011: 13–14)). Food producing plants by comparison are routinely considered messy, unkempt, maintenance intensive or too hard to formalise, signifying a lack of recognition for potential cultural and economic gain as well as for design and social engagement possibilities (Zeunert, 2011; Gyertyán, 2014; Zeunert, 2016).

Perception

Beyond aesthetics, targeting socio-cultural perception is essential to reduce demand for environmentally intensive 'foodstuffs' and strengthen local food systems. Food marketing is a $US1.6 trillion global industry (Murray, 2007; with others valuing it at over $2 trillion in the same time-frame). Highly processed, mass-produced, packaged, refrigerated, and frozen food is distributed globally through supply chain logistics, with accompanying marketing promoting Western aspirational lifestyles as part of a trend where "uselessness, leisure, and adornment have taken over as part of a larger overwhelming urge to appear "modern" and sophisticated" (Yu, 2010: 3). For most urban regions, twentieth-century modernisation has and continues to sweep UA aside (Sumner et al., 2010; Weis, 2010; Morgan, 2013: 20). Mechanised agricultural systems continue to remove unprecedented proportions of the population from agricultural livelihoods (Weis, 2010; Langridge et al., 2014; Lowder et al., 2016), as well as having a homogenising effect on cultures, practices, landscapes, crop varieties, and food (Norberg-Hodge et al., 2001; Steel, 2008; Roy and Ong, 2011; Speak and Kumar, 2016; Chapter 31). Farmers in developed countries are typically advanced in age (such as in Japan, where in 2011 they averaged 65.9 years old) indicating failing intergenerational food production security and an alarming inability to recruit the younger generation. UA's social potential includes shifting this dynamic through engaging the highly urbanised younger generations in food systems.

Changing consumer perception and demand can re-shape agricultural and UA practices across all six dimensions identified in this chapter. Local, artisan, Slow Food and 'locavorism' movements have made small but important inroads into expanding understanding of food systems, justice and associated social impacts (Nabhan, 2002; Moustier and Danso, 2006; Ladner, 2011; Chapters 31, 31, 35 and 36). DuPuis and Goodman (2005: 359, 364), however, critique local food movements as realms of "inequality and hegemonic domination", geared to elite socio-economic social echelons (see for example Strong, 2011) as opposed to an "effective social movement of resistance to globalism". It should be noted, however, that alternative/small-scale food systems' higher prices can be byproducts of industrial-scale production and governing financial measures such as subsidisation.

Overcoming idealised perceptions of UA may be also be required, where romanticised small scale agricultural ideals fail to take into account time and labour intensive farming practices and spartan realities of subsistence-based lifestyles. An increasing number of books such as Carpenter's popular *Farm City* (2011) assist in dispelling UA myths or crystallising support. Other influences include: famous figures (Michelle Obama's expansion of the White House vegetable garden in 2009); boutique, ethical food tourism (aiding food system transparency); and cooking and food documentary television (revealing consumption/production links and increasing demand for local, high-quality produce – though may conversely reinforce rural/urban binaries or conceal agro-industrial ecological and animal rights realities (Pretty, 2002; Sumner et al., 2010; Berry, 2015; Imbert, 2015; Chapters 31 and 35)).

Social viability

UA's viability and longevity can help to be secured by creating meaningful social community webs (Allen, 2010; Allen and Wilson, 2012). These accessible, interconnected networks are able to provide multiple gains (Blaine et al., 2010), including, for example, facilitating gratis/temporary vacant land use, collective, community and volunteer labour arrangements, and sharing knowledge, skillsets and resources. These benefits are especially crucial when financial support is lacking, as social capital and community cohesion can make the difference between UA viability and failure.

Despite a proliferation of social UA activities in many developed countries, discourse citing an urban "food revolution" (Ladner, 2011; Tracey, 2011; Cockrall-King, 2012) is perhaps exaggerated. These new and emerging forms of UA landscapes are yet to command prominent spatial significance or widespread social acceptance and have not been seriously addressed and implemented by planners, landscape architects, architects, municipalities, and those responsible for planning and designing the urban public realm (see *Dimension Six: Intellectual*).

Dimension four: environmental

This section briefly summarises environmental dimensions of UA, with focus on urban site-specific factors (Chapter 16 examines and incorporates many environmental dimensions relevant to urban agriculture). In brief, several wider environmental topics relating to UA include:

- National, city or individual food self-sufficiency and self-reliance (Hopkins, 2008; Markham, 2010; Grewal and Grewal, 2012);
- Ecological and spatial footprints (Rees, 1992; Wackernagel, 1994; Maas, 1999: 61–63, 94–115; Weller, 2009; 440–441; MVRDV, 2009, MVRDV, 2010);
- Energy and seasonal intensity, greenhouse gas emissions, fossil fuel use across the food and agricultural spectrum, and life-cycle analysis (growing, processing, distribution, see for example Heinberg, 2005; Steel, 2008; Pimentel, 2009; Denny, 2012; Moreau et al., 2012, Jansma et al., 2012);
- Agrobiodiversity (FAO, 1999; Kremen and Miles, 2012);
- Climate change, resilience, and adaptation (Chapters 18 and 19; Holmgren, 2002; Burton et al., 2013; Lwasa and Dubbeling, 2015);
- Biodiversity and agroecology (Chapter 17; Kremen and Miles, 2012; Méndez et al., 2015); and
- Food miles, local and short food supply chains (Garnett, 2003; Halweil, 2004; Doron, 2005; Gaballa and Abraham, 2008; Steel, 2008: 72–78; Roberts, 2008; Bomford, 2010; Ladner, 2011; Mobbs, 2012: 227; Jansma et al., 2012; Canfora, 2016).

Misguided myths powerfully influence UA's rejection or uptake. Site concerns range from contaminated urban-grown produce (ignoring rural agriculture's synthetic chemicals, toxic pesticides, fertilisers, and other processes (such as irradiation) (Norberg-Hodge, 2001; Nordahl, 2009; Zeunert, 2011)) to animal and human defecation (despite manure and excrement being age-old and fundamental agricultural applications (see Chapter 23)). Industrial contamination of soil, groundwater, water used for irrigation, materials and structures, and air pollution (see Griffiths, 2003) are, however, prevalent in many urban (and rural) areas. Solutions to soil contamination can be as simple as planting in raised beds/planters with membrane barriers preventing roots from entering soil (although this suits smaller areas), and/or planting edible species that do not transmit pollutants to their edible parts. Phytoremediation (using plants to purify soils and surface water) can be a relatively low-cost option for cleanup, though it may be slow or unfeasible depending on the severity of contamination (Kennen and Kirkwood, 2015). Responses to community demand for UA on a particular site may prompt soil tests, measures to work around constraints or site clean-up, or choosing alternative site locations. Ideally, a strategic approach from local municipalities is needed, for example, identifying and keeping an inventory of suitable land areas, prioritising these for deployment.

There are multiple contextual and environmental synergies for UA, existing to varying degrees in different regions. These include increasing the use of abundant, continually generated urban 'wastes', many of which are actually potential resources, such as:

- Material synergies: excess and 'waste' materials:
    - Non-mains water sources (rain, storm, grey, black and recycled). These may have beneficial nutrients and/or harmful pollutants present (Chapters 15 and 19);

    - Compostable materials returning nutrients to soils (food and garden waste, cardboard and paper, crop and food processing wastes);
    - Urine and sewage sludge (Chapter 23);
    - Excess heat from industry and buildings to heat greenhouses;

- Improved microclimates:

    - Beneficial urban heat island conditions can reduce frosts, extend growing seasons, and accommodate a broader range of crops (although this reduces chill hours required for some crops and extreme heat conditions may also result in damage);
    - Wind conditions may be improved (or worsened) depending on the specific site conditions and adjoining building heights; and

- Increased biodiversity and agrobiodiversity in small areas. Most UA involves a diverse range of crops when compared to industrial agriculture's monocultures and this also increases pollination (Kremen and Miles, 2012).

To increase environmental effectiveness across a range of city systems, UA requires strategic integration through systems planning, design, and execution in order for it to be a connected component of green and blue infrastructures rather than ad hoc and unsystematic.

Dimension five: practical

Dimension five encompasses UA's practical rather than theoretical aspects. There is a substantial amount of published work outlining food production methods in technical, horticultural and gardening-related discourse, as well as in community-based digital resources such as websites, instructional videos, online networks, and blogs. Practical discourse can blur the boundaries between strictly urban agriculture and smaller rural enterprises and vary focus from individual, family, small-scale, collective, community, and shared growing forms. Several key practice areas and resources include:

- Horticultural publications (such as the Royal Horticultural Society, see for example Baker, 1999; Blazey and Varkulevicius, 2006; Klein, 2007; Pollock and Anderson, 2012; Fewster and Shackleton, 2014);
- Market gardening, peri-urban farming, and multi-functional urban farming, which are showing a resurgence in some regions (Coleman et al., 1995; Dawling, 2013; Fortier, 2014; Stone, 2016; Volk, 2017);
- Non-industrial production methods such as organic and biodynamic growing and permaculture (Chapter 10);
- Supplementing and providing for family and local food needs through urban homesteading and urban self-sufficiency (see for example Jeavons, 2006; Madigan, 2009; Markham, 2010; Mobbs, 2012);
- Aesthetic and design considerations using edible plants (Creasey, 2010; Zeunert, 2011; Gyertyán, 2014);
- Novel approaches, such as planting edible species on road verges and medians, in front yards, and through guerrilla gardening (depending on focus, this category can blur into social dimensions and discourse) (see for example Flores, 2006; Houbein, 2008; Haeg and Balmori, 2008; Reynolds, 2009; Hardman and Larkham, 2014a); and
- Rooftop farming (open air and greenhouse), indoor farming and growing under lights, aquaponics, hydroponics, and aeroponics (see for example Bernstein, 2011; Resh, 2012; Sloper, 2013; Novak, 2016).

Competency, sound growing skills and dedicated time are essential for UA's success, especially at key stages during seasonal growing calendars. Aforementioned social synergies and networks, such as free labour provided by volunteers and corporate sustainability programs can assist UA (such as at CERES Organic Market Garden in Melbourne and Camperdown Commons in Sydney, Australia). All forms of practical engagement in UA increase understanding of natural and urban processes, reinforcing understanding and strengthening its intellectual and theoretical dimensions.

Dimension six: intellectual

UA's intellectual significance is tied to that of its perceived worth – a quantitative and qualitative value spectrum traversing governance and policy, discourse, professional decision-making (by planners, urban designers, architects, and landscape architects) and community interests. Intellectual dimensions thus incorporate many of the previous five dimensions, enabled or hindered by governance and knowledge systems, including policy, bylaws, guidelines, charters, urban planning, civic design disciplines, community engagement, and activism (Mougeot, 2011; Aubry et al., 2012; Curry et al., 2014; Hardman and Larkham, 2014b; Hunt, 2015; Kennedy and Liljeblad, 2016).

Governance

The essentially irreversible loss of easily built on, cleared agricultural land consumed by development routinely occurs despite supposed protection within municipal and state planning frameworks (Chapters 14, 15 and 26). Protecting agriculturally valuable metropolitan land, regardless of whether it is active or fallow, is of critical importance to sustainability and food security (Chapters 16 and 25) and hangs on convincing decision makers (governments, planners, and designers) of its significance – often in the face of powerful development lobbies (APA, 2007; Hodgson, 2010). The UN Habitat World Cities Report (2016) makes numerous references to the need to protect agricultural and common land from urban development. It is worth noting that such requests are not new, with landscape architects Leberecht Migge (1918) (see Haney, 2010) and Jens Jensen (1920) calling for cities to devote tracts of land for growing fruits and vegetables a century ago.

Leading cities are increasingly attempting to protect peri-UA, curtail urban sprawl, and make municipal land available for UA (La Rosa et al., 2014). Further action can be spurred on by urban food pacts like the Milan Urban Food Policy Pact (2016), signed by over 130 major cities, and earlier declarations in Quito, Medellin, and Bonn, amongst others. Other relevant initiatives include Mexico City's 'Community Dining Rooms Program', the 'Vancouver Food Strategy', the 'Baltimore Food Policy Initiative', Birmingham's 'Sustainable Diets and Nutrition', Lusaka's 'Women Groups Economic Empowerment', Quito's 'AGRUPAR' UA programme, Toronto's 'Grab Some Good', and Riga's Food Waste strategies. UA is more pronounced in regions where commercial forms of intra urban agriculture have been enabled as a legal land use, as in several cities in the USA and Europe. In New York City, for example, UA is facilitated by a range of policy, planning, and social initiatives, such as bylaws allowing urban beekeeping and small-scale animal husbandry (Popovitch, 2014). Lobbying and community feedback loops shape these processes from 'top-down' influencers (such as decision makers, elected officials, public figures and key professionals) to 'bottom-up' stakeholders (the urban public at large, communities, groups, and individuals).

The usually slow and cumbersome urban planning process must increase its efficiency, flexibility, and nimbleness in order to strengthen urban food systems' resilience (Tornaghi, 2012). Currently, municipalities may facilitate UA through sympathetic ordinances (Popovitch, 2014), covenants, planning/zoning overlays, land leases, and trusts – all to varying degrees of success. While many regions provide for allotments, community gardens, and other forms of socially oriented UA, very few offer employment scope or actively nurture a local food industry (the grant-funded *Garden Enfield* program in north London (2013) seeks to do so, but outcomes at the time of writing are unclear). Urban planners must recognise that properly facilitated and regulated city-based agriculture forms a crucial part of the urban system, linking food security, social, environmental and economic urban policy objectives, and contributing not only to food resilience but wider urban sustainability (de Zeeuw, Pers. Comm. 16 Dec 2016).

Discourse

UA discourse varies significantly in scale and scope. Within design discourse, project visions are often at an individual scale, focusing on elements such as vertical farms and installations like green roofs and vertical walls (see for example Viljoen et al., 2005; Gorgolewski et al., 2011; Fox, 2011; Fell, 2011; Miazzo et al., 2013; Philips, 2013; Viljoen and Bohn, 2014). Focus is therefore small, object-based and fragmented, rather than big-picture and interconnected (Roggema, 2014: 50–59). This reflects UA's current 'bottom-up' practices and site-scale approach and a need for more systematic initiatives at local, municipal and city levels for leverage and interconnection (Dubbeling et al., 2010).

Literature overviews provide insights into UA's multidimensional makeup, such as several annotated bibliographies and literature reviews, including the 32-page *URS/IDRC Global Urban Agriculture: an annotated bibliography* (Canada, 1984); Bruinsma and Hertog's substantial 804 page annotated bibliography (Sweden, 2003); Golden's 19-page literature review (USA focused, 2013); and regional reviews such as Guzman's (Philippines/Manilla, 2005) amongst others. Much UA discourse, however, focuses on social concerns (see *Dimension Three: Social*) and small-scale initiatives. Even, for example, in Viljoen and Wiskerke's (2012) comprehensive and useful volume, *Sustainable Food Planning: Evolving Theory and Practice*, a majority of its planning section (Chapters 21–30) examines the well-covered social aspects of small scale and community UA in developed nations.

For several decades, UA discourse has been generated by international institutions and funding agencies that incorporate UA programmes, not limited to the International Development Research Centre (IDRC), United Nations Habitat (UN Habitat), United Nations Development Programme (UNDP), The Food and Agriculture Organization of the United Nations (FAO), the United Nations University and Local Governments for Sustainability (ICLEI) (Lee-Smith, 2010; Mougeot, 2011; FAO, n.d.; Badami and Ramankutty, 2015: 8).

In a scholarly context, UA discourse has been routinely excluded from academic curriculums and typically exists as a sub-branch of landscape architecture, planning, architecture, horticulture, gardening, environmental studies, social sciences, or history. Some institutions are recognising this gap and increasing numbers of programs, and qualifications are slowly becoming available at both further (vocational) and higher (university) education levels.

Interest in UA within the built environment and design disciplines and their discourse has increased in the past ten years. UA planning literature has been on the rise over the last two decades and includes important contributions from Cabannes and Dubbeling (2000); APA (2007); Dubbeling et al. (2010); Hodgson (2010); LA and GLA (2010); Viljoen and Wiskerke (2012); Morgan (2013); Philips, (2013); de Zeeuw and Drechsel (2015); Roggema (2016); and Ilieva (2016). In design disciplines, relevant texts to note include: Nordahl (2009 [2014]); Alon-Mozes (2009); Haney (2010); Zeunert (2011); Solomon (2012); Philips (2013); Parham (2015); Zeunert (2016); Napawan (2016) in landscape architecture, urban design and urbanism; in architecture, Viljoen et al. (2005); Doron (2005); Lim and Liu (2010); Duany (2011); and Creasy (2010) in garden design.

It is important to recognise that UA discourse incorporates a range of differing terminology. Related terms include CPULs (continuous productive urban landscapes) (Viljoen et al., 2005; Viljoen and Bohn, 2014); agricultural urbanism and agrarian urbanism (de la Salle and Holland, 2010; Waldheim, 2010; Duany, 2011); metropolitan agriculture (Heimlich and Brooks, 1989); city region food systems (FAO, n.d.; Dubbeling et al., 2016); food urbanism (Parham, 2015); peri-urban agriculture (Opitz et al., 2016); community agriculture; foodscapes and aesthetic foodscape design (Zeunert, 2011; Solomon, 2012); food sensitive planning and design (FSPUD) (DLA et al., 2011; Chapter 26); transition towns and networks (Hopkins, 2008; https://transitionnetwork.org/); and incredible edibles (http://incredibleediblenetwork.org.uk/). All contain

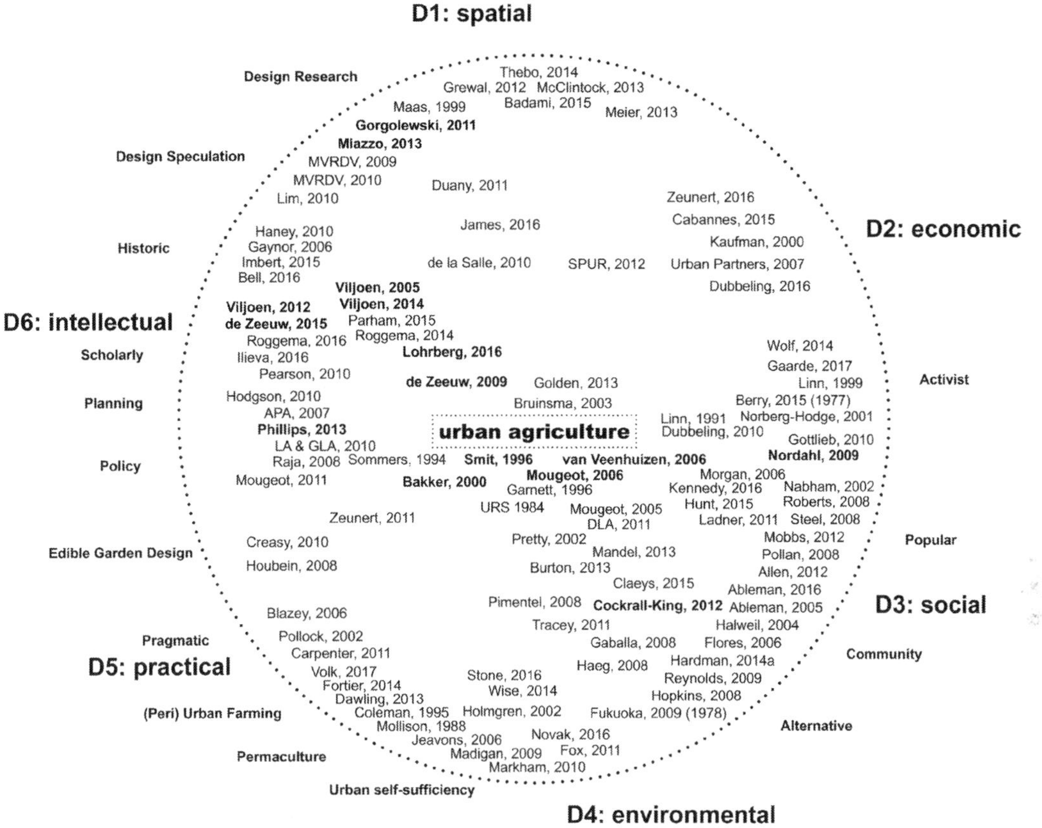

Figure 12.2 Urban agriculture's discourse spectrum as estimated by the author. Literature (with an emphasis on books) relevant to the theory and practice of urban agriculture is categorised according to the dominant theme of focus. Key texts are in bold. Please note that only the first author is listed.

UA as an integral element, and there are multiple others of which UA forms a minor part. While new visions, terminology, and modes of practice are important for progress and uptake, it is also crucial that UA does not splinter into disparate factions leading to horizontal hostility (Kennedy, 1970), thereby losing impact.

As a final note, critics of UA (perhaps rightly so), have noted that much UA discourse is motivated by an advocacy agenda, rather than being subject to analytical rigour (see for example Ellis and Sumberg, 1998; Zezza and Tasciotti, 2010; Badami and Ramankutty, 2015). This is evident in UA's varying statistics as noted throughout this chapter. Increased scholarly attention, such as within academic journals, may help to address this shortcoming. While this chapter attempts to overview UA in developed world contexts, a more substantial global and current review would be highly beneficial for further research. Key themes and works in UA literature (as based on the proposed six dimensions (Figure 12.1)) are illustrated in the discourse spectrum diagram, Figure 12.2.

Conclusion

An inclusive conception of urban agriculture is key to capturing its breadth of scope and improving understanding of its diversity and values. Even when viewed within this chapter's more limited focus on recent decades and developed world contexts, UA evades clear cut categorisation. Notwithstanding its multidimensional make-up, UA may be viewed as comprising six primary dimensions: *spatial, economic, social, environmental, practical* and *intellectual* (Figure 12.1), and significant discourse framed within this structure (Figure 12.2).

A greater understanding of UA's worth within governance, academia and society is imperative to overcoming routine perceptions of UA as incidental with negligible value. UA is far from trivial, unsuited to urban contexts, and inconsequential to market-agricultural systems and food security (van der Schans and Wiskerke, 2012; Wolf, 2014; Springer et al., 2016) – particularly given that over a third of US agricultural production by financial value takes place within metropolitan statistical areas (Gorgolewski et al., 2011: 232; T. Parker, USDA, pers. comm. 20 April 2017). UA offers significant health, social, economic and environmental benefits to the lives of millions, if not over a billion, urban inhabitants worldwide (Smit et al., 2001: Ch. 2: 1; Grace et al., 2015: 255). Increasing awareness encompasses dissolving prevalent bias towards large-scale industrial agriculture, unhelpful urban/rural dichotomies, and 'either-or' locations.

UA is regularly displaced and lost due to urban development, peri-urban city expansion, agricultural land rezoning, attempts to appear 'modern', green gentrification, and food system industrialisation. This not only results in loss of precious agricultural land, but also situated, embedded, and embodied knowledge (Chapter 34). At the same time, various UA typologies are increasing: in parks; on vacant land; parking lots; private gardens and lawns; green belts; green spaces; infrastructure corridors; street verges and medians; and structurally suited rooftops. These offer 'low-hanging fruit' for UA's expansion. If UA is to endure and have meaningful impact, it must address immediate concerns of placement, scale, makeup, aesthetics, risk levels, and maintenance requirements. Economic and knowledge-based resources are needed to reconcile and achieve these ongoing challenges.

There is great potential to pursue wider, more ambitious, professional and systematic UA ambits, establishing and expanding the field beyond mere legitimate urban land use and practice to become an interconnected and serviced element in hard and soft urban infrastructures. UA visions and practices must become increasingly systematic in order to address numerous environmental, regulatory, economic, organisational, and technical challenges. Urban conditions offer multiple synergies and opportunities for UA to be coupled with elements such as water and nutrient recycling, social networks and labour bases, site and precinct activation, remediation, nature engagement, and increasing biodiversity. This increased environmental embedding calls for wider appreciation of UA as an integral sustainability and resilience contributor (Chapter 16). Understanding personal and collective footprints, as well as having a better comprehension of what rural agriculture systems and their practices entail, will continue to assist UA's uptake.

UA offers powerful potential to pursue meaningful economic arrangements and employment if skilfully executed across expanded spatial scales. A recent proliferation of practices in certain developed regions is already increasing commercial activities through SMEs, incubator programmes, commercially functioning urban farms, food hubs, retail operations, and websites. It is vital that these and other new forms are facilitated by sympathetic planning, regulatory and financial frameworks that recognise their ability to catalyse positive social, environmental, and economic change.

Acknowledgements

The author would like to thank Alys Daroy for her critical feedback and rigorous editing. Grateful appreciation is given to the two anonymous peer reviewers, as well as to Henk de Zeeuw and Joe Nasr for their very helpful review, suggestions, and time. Thanks also to Tim Waterman for his review and editing.

Key texts

Bakker, N., Dubbeling, M., Guendel, S., Sabel Koschella, U., & de Zeeuw, H. (eds.) (2000) *Growing cities, growing food, urban agriculture on the policy agenda*. Feldafing: DSE.

Cockrall-King, J. (2012) *Food and the city: Urban agriculture and the new food revolution.* Amherst, NY: Prometheus Books.

de Zeeuw, H., & Dubbeling, M. (2009) *Cities food and agriculture: Challenges and the way forward.* RUAF: The Netherlands.

de Zeeuw, H., & Drechsel, P. (2015) *Cities and agriculture: Developing resilient urban food systems.* New York: Routledge.

Gorgolewski, M., Komisar, J., & Nasr, J. (2011) *Carrot city: Creating places for urban agriculture.* New York: Monacelli Press.

Lohrberg, F., LičKa, L., Scazzosi, L., & Timpe, A. (2016) *Urban agriculture Europe.* Berlin: Jovis.

Miazzo, F., Minkjan, M., & CITIES (2013) *Farming the city: Food as a tool for today's urbanization.* Haarlem, the Netherlands: Valiz/Trancity.

Mougeot, J. (2006) *Growing better cities: Urban agriculture for sustainable development.* Ottawa, Canada: International Development Research Centre.

Nordahl, D. (2009 [2014]) *Public produce: The new urban agriculture.* Washington, DC: Island Press.

Philips, A. (2013) *Designing urban agriculture: A complete guide to the planning, design, construction, maintenance and management of edible landscapes.* Hoboken: Wiley.

Smit, J., Ratta, A., & Nasr, J. (1996/2001) *Urban agriculture: Food, jobs and sustainable cities,* New York: UNDP. Second edition (2001) is accessible at: www.jacsmit.com/book.html

van Veenhuizen (ed.) (2006) *Cities farming for the future: Urban agriculture for green and productive cities.* Philippines: International Institute of Rural Reconstruction and ETC Urban Agriculture.

Viljoen, A., Bohn, K., & Howe, J. (2005) *Continuous productive urban landscapes: Designing urban agriculture for sustainable cities.* Amsterdam: Elsevier.

Viljoen, A., & Wiskerke, J. (eds.) (2012) *Sustainable food planning: Evolving theory and practice.* Wageningen, the Netherlands: Wageningen Academic Publishers.

Viljoen, A., & Bohn, K. (2014) *Second nature urban agriculture: Designing productive cities.* New York: Routledge.

Key websites

Carrot City <www.carrotcity.org>
City Farmer <www.cityfarmer.info/>
City Region Food Systems <http://cityregionfoodsystems.org/r>
International Network for Urban Agriculture <www.inuag.org/>
International Urban Food Network <www.iufn.org/en/>
Jac Smit <www.jacsmit.com/>
RUAF Foundation <www.ruaf.org>
RUAF Urban Agriculture Magazine <www.ruaf.org/publications/magazines>

Further reading

Ableman, M. (2005) *Fields of plenty: A farmer's journey in search of real food and the people who grow it.* San Francisco: Chronicle Books.

Ableman, M. (2016) *Street farm: Growing food, jobs, and hope on the urban frontier.* Vermont: Chelsea Green Publishing.

Ackerman, K. (2012). *The potential for urban agriculture in New York City: Growing capacity, food security, & green infrastructure.* Edition II. New York: Urban Design Lab at the Earth Institute Columbia University.

Ackerman, K., Conard, M., Culligan, P., Plunz, R., Sutto, M.-P., & Whittinghill, L. (2014) Sustainable food systems for future cities: The potential of urban agriculture. *Economic and Social Review*, 45, 189–206.

Allen, E. (2010) *Growing community food systems.* In: Heinberg, R., & Lerch, D. (eds.), *The post carbon reader: Managing the 21st century's sustainability crises.* Healdsburg, CA: Watershed Media, pp. 140–150.

Allen, W., & Wilson, C. (2012) *The good food revolution: Growing healthy food, people, and communities.* New York: Gotham Books.

Alon-Mozes, T. (2009) Landscape architecture and agriculture: Common seeds and diverging sprigs in Israeli practice. *Landscape Journal*, 28(2), 166–180.

American Planning Association (APA) (2007) *Policy guide on community and regional food planning.* www.planning.org/policy/guides/adopted/food.htm (accessed 23 May 2014).

Armar-Klemesu, M. (2000) *Urban agriculture and food security, nutrition and health,* Thematic Paper 4. In: Bakker, N., et al. (eds.), *Growing cities, growing food: Urban agriculture on the policy agenda.* Fefdafing: DSE, pp. 99–117.

Aubry, C., Ramamonjisoa, J., Dabat, M.-H., Rakotoarisoa, J., Rakotondraibe, J., & Rabeharisoa, L. (2012) Urban agriculture and land use in cities: An approach with the multi-functionality and sustainability concepts in the case of Antananarivo (Madagascar), *Land Use Policy*, 29, 429–439.
Badami, M., & Ramankutty, N. (2015) Urban agriculture and food security: A critique based on an assessment of urban land constraints. *Global Food Security*, 4, 8–15.
Baker, H. (1999) *Growing fruit*. London: Octopus Pub. Group.
Baker, L., & de Zeeuw, H. (2015) *Urban food policies and programmes: An overview*. In: de Zeeuw, H., & Drechsel, P. (eds.), *Cities and agriculture: Developing resilient urban food systems*. New York: Routledge.
Bell, S., Fox-Kämper, R., Benson, M., Keshavarz, N., Noori, S., Caputo, S., & Voigt, A. (2016) *Urban allotment gardens in Europe*. New York: Routledge.
Bernstein, S. (2011) *Aquaponic gardening: A step-by-step guide to raising vegetables and fish together*. Gabriola: New Society Publishers.
Berry, W. (2015) *Unsettling of America: Culture and agriculture*. ZULU: Counterpoint.
Blaine, T., Grewal, P., Dawes, A., & Snider, D. (2010) Profiling community gardeners. *Journal of Extension* (online), 48(6), 6FEA6. Available from: www.joe.org/joe/2010december/a6.php.
Blazey, C., & Varkulevicius, J. (2006) *The Australian fruit and vegetable garden: Grow the best fruit and vegetables for good health and flavour*. Dromana: Digger's Club.
Bomford, M. (2010) *Getting fossil fuels off the plate*. In: Heinberg, R., & Lerch, D. (eds.), *The post carbon reader: Managing the 21st century's sustainability crises*. Healdsburg, CA: Watershed Media, pp. 119–127.
Bowler, I. (1992) *The geography of agriculture in developed market economies*. Harlow: Longman Scientific and Technical.
Brown, K., & Jameton, A. (2000) Public health implications of urban agriculture. *Journal of Public Health Policy*, 21(1), 20–39.
Brown, S. (2008) *Urban agriculture: Is there now an opportunity for a viable small-scale sustainable agriculture to emerge in brisbane, Australia?* Master of Sustainable Agriculture Dissertation, Orange, Australia: Charles Sturt University. http://community-wealth.org/sites/clone.community-wealth.org/files/downloads/paper-brown_0.pdf (accessed 27 April 2013).
Bruinsma, W., & Hertog, W. (eds.) (2003) *Annotated bibliography on urban agriculture*, Leusden, the Netherlands: ETC. www.ruaf.org/sites/default/files/annotated_bibliography.pdf (accessed 27 April 2013).
Burton, P., Lyons, K., Richards, C., Amati, M., Rose, N., Des Fours, L., Pires, V., & Barclay, R. (2013) *Urban food security, urban resilience and climate change*. Gold Coast: National Climate Change Adaptation Research Facility.
Butler, L., & Moronek, D. (eds.) (2002) *Urban and agriculture communities: Opportunities for common ground*. Ames, IA: Council for Agricultural Science and Technology.
Cabannes, Y. (2015) Financing urban agriculture: What do we know and what should we know. In: de Zeeuw, H., & Drechsel, P. (eds.), *Cities and agriculture: Developing resilient urban food systems*. New York: Routledge.
Cabannes, Y., & Dubbeling, M. (2000) *Urban agriculture and urban planning: What should be taken into consideration to plan the city of the 21st century*. Presented at Urban Agriculture and Horticulture, the Linkage with Urban Planning, July 7–9, Berlin, Germany.
Cameron, W. (1963) *Informal sociology, a casual introduction to sociological thinking*. New York: Random House.
Canfora, I. (2016) Is the short food supply chain an efficient solution for sustainability in food market? *Agriculture and Agricultural Science Procedia*, 8, 402–407.
Caputo, S. (2012) The purpose of urban food production in developed countries. In: Viljoen, A., & Wiskerke, J. (eds.), *Sustainable food planning: Evolving theory and practice*. Wageningen, The Netherlands: Wageningen Academic Publishers, pp. 259–270.
Carpenter, N., & Rosenthal, W. (2011) *The essential urban farmer*. New York: Penguin Books.
Cebon, M. (2010) *The impacts of Australian trade and agriculture policies for farmers (and everyone else)*, online proceedings of the From Plains to Plate Conference, Adelaide, February 12, 2010. http://futureoffoodsa.ning.com/page/presentations-1 (accessed 4 August 2010).
Claeys, P. (2015) *Human rights and the food sovereignty movement: Reclaiming control*. Oxon: Routledge.
Coleman, E., Field, M., Amsel, S., & Hawken, P. (1995) *The new organic grower: A master's manual of tools and techniques for the home and market gardener*. White River Junction, VT: Chelsea Green.
Creasy, R. (2010) *Edible landscaping*. San Francisco: Sierra Club Books.
Curry, N., Reed, M., Keech, D., Maye, D., & Kirwan, J. (2014) Urban agriculture and the policies of the European Union: The need for renewal. *Spanish Journal of Rural Development*, 91–106.

David Lock Associates, University of Melbourne and National Heart Foundation of Australia (2011) *Food-sensitive planning and urban design: A conceptual framework for achieving a sustainable and healthy food system*. Melbourne. www.vichealth.vic.gov.au/~/media/ResourceCentre/PublicationsandResources/healthy%20eating/Food%20systems%20-%20food%20supply/FoodSensitivePlanning_UrbanDesign_FullReport.de la (accessed 21 October 2011).

Dawling, P. (2013) *Sustainable market farming: Intensive vegetable production on a few acres*. Canada, BC: New Society Publishers.

de la Salle, J., & Holland, M. (2010) *Agricultural urbanism: Handbook for building sustainable food and agriculture systems in 21st century cities*. Manitoba: Green Frigate Books.

Denny, G. (2012) Urban agriculture and season food footprints: An LCA study of tomato production and consumption in the UK. In: Viljoen, A., & Wiskerke, J. (eds.), *Sustainable food planning: Evolving theory and practice*. Wageningen, The Netherlands: Wageningen Academic Publishers.

Doron, G. (2005) Urban agriculture: Small, medium, large. *Architectural Design*, 75(3), 52–59.

Duany, A. (2011) *Garden cities: Theory and practice of agrarian urbanism*. London: The Prince's Foundation for the Built Environment.

Dubbeling, M., Carey, J., & Hochberg, K. (2016) *The role of private sector in city region food systems*. Leusden, The Netherlands: RUAF Foundation.

Dubbeling, M., Veenhuizen, R., & de Zeeuw, H. (2010) *Cities, poverty and food: Multi-stakeholder policy and planning in urban agriculture*. Rugby, Warwickshire: RUAF Foundation, Practical Action Publishing.

Duchemin, E., Wegmuller, F., & Legault, A-M. (2008) Urban agriculture: Multi-dimensional tools for social development in poor neighbourhoods, FACTS reports: Field ACTions science reports. *The Journal of Field Actions*, 1. Available from: http://factsreports.revues.org/113 (accessed 11 May 2011).

DuPuis, E., & Goodman, D. (2005) Should we go "home" to eat?: Toward a reflexive politics of localism. *Journal of Rural Studies*, 21, 359–371.

Ellis, F., & Sumberg, J. (1998) Food production, urban areas and policy responses. *World Development*, 26, 213–225.

FAO (1995) *Dimensions of need – an atlas of food and agriculture* www.fao.org/docrep/u8480e/u8480e07.htm (accessed 16 May 2016).

FAO (1999) *What is happening to agrobiodiversity? Women: Users, preservers and managers of agro-biodiversity*. www.fao.org/docrep/007/y5609e/y5609e02.htm (accessed 11 April 2011).

FAO (nd) *Food for the cities program*. www.fao.org/in-action/food-for-cities-programme/approach/crfs/en/ (accessed 24 March 2017).

Fell, D. (2011) *Vertical gardening: Grow up, not out, for more vegetables and flowers in much less space*. Emmaus, PA: Rodale.

Fewster, H., & Shackleton, B. (2014) *RHS grow fruit and veg guide*. London: DK.

Finkel, E. (2012) *A growthful enterprise on Perry Street*. LISC Chicago's New Communities Program. www.newcommunities.org/news/articleDetail.asp?objectID=2434 (accessed 18 May 2014).

Firth, C., Maye, D., & Pearson, D. (2011) Developing "community" in community gardens. *Local Environment: The International Journal of Justice and Sustainability*, 16(6), 555–568.

Flores, H. (2006) *Food not lawns: How to turn your yard into a garden and your neighborhood into a community*. White River Junction, VT: Chelsea Green Publishing.

Fortier, J.-M. (2014) *The market gardener: A successful grower's handbook for small-scale organic farming*. Gabriola Island: New Society Publishers.

Fox, T. (2011) *Urban farming: Sustainable city living in your backyard, in your community, and in the world*. Irvine, CA: BowTie Press.

Freyfogle, E. (2003) *The land we share: Private property and the common good*. Washington, DC: Island Press.

Freyfogle, E. (2007) *On private property: Finding common ground on the ownership of land*. Boston, MA: Beacon Press.

Fücks, R. (2015) *Green growth, smart growth: A new approach to economics, innovation and the environment*. London: Anthem Press.

Gaarde, I. (2017) *Peasants negotiating a global policy space*. Oxon: Routledge.

Gaballa, S., & Abraham, A. (2008) *Food miles in Australia: A preliminary study of Melbourne, Victoria*. Melbourne: Centre for Education and Research in Environmental Strategies.

Garnett, T. (1996) Farming the city: The potential of urban agriculture. *Ecologist*, 26(6), 299–307.

Garnett, T. (2003) *Wise moves: Exploring the relationship between food, transport and CO_2*. London: Transport 2000 Trust.

Gaynor, A. (2006) *Harvest of the suburbs*. Crawley, WA: UWA Press.

Gilbert, J. (2014) *Common ground: Democracy and collectivity in an age of individualism*. London: Pluto Press.

Giorda, E. (2012) Farming in Mowtown: Competing narratives for urban development and urban agriculture in Detroit. In: Viljoen, A., & Wiskerke, J. (eds.), *Sustainable food planning: Evolving theory and practice*. Wageningen, The Netherlands: Wageningen Academic Publishers, pp. 271–281.

Gobster, P., Nassauer, J., Daniel, T., & Fry, G. (2007) The shared landscape: What does aesthetics have to do with ecology? *Landscape Ecol*, 22, 959–972.

Golden, S. (2013) *Urban agriculture impacts: Social, health, and economic: A literature review*. UC Sustainable Agriculture Research and Education Program. http://asi.ucdavis.edu/resources/publications/UA%20Lit%20Review-%20Golden%20Reduced%2011-15.pdf (accessed 18 June 2014).

Gottlieb, R., & Joshi, A. (2010) *Food justice*. Cambridge, MA: MIT Press.

Grace, D., Lindahl, J., Corream, M., & Kakkar, M. (2015) Urban livestock keeping. In: de Zeeuw, H., & Drechsel, P. (eds.), *Cities and agriculture: Developing resilient urban food systems*. New York: Routledge.

Grewal, S., & Grewal, P. (2012) Can cities become self reliant in food? *Cities*, 29, 1–11.

Griffiths, H. (2003) *Effects of air pollution on agricultural crops*. Ontario Ministry of Agriculture, Food and Rural Affairs. www.omafra.gov.on.ca/english/crops/facts/01-015.htm (accessed 2 April 2011).

Gudeman, S. (2007) *The anthropology of economy: Community, market, and culture*. Malden, MA: Blackwell.

Gurian-Sherman, D. (2008) *CAFOs uncovered: The untold costs of confined animal feeding operations*. Cambridge, MA: Union of Concerned Scientists.

Gyertyán, F. (2014) *Growing delight – the aesthetic dimension of urban gardening in Amsterdam*. Masters thesis, Wageningen University https://issuu.com/frrrr/docs/report_fruzsina_final_small_format (accessed 5 January 2015).

Haeg, F., & Balmori, D. (2008) *Edible estates: Attack on the front lawn*. New York: Metropolis Books.

Halweil, B. (2004) *Eat here: Reclaiming homegrown pleasures in a global supermarket*. New York: W.W. Norton and Company.

Haney, D. (2010) *When modern was green: Life and work of landscape architect Leberecht Migge*. New York: Routledge.

Hardman, M., & Larkham, P. (2014a) *Informal urban agriculture: The secret lives of guerrilla gardeners*. Cham: Springer.

Hardman, M., & Larkham, P. (2014b) The rise of the 'food charter': A mechanism to increase urban agriculture. *Land Use Policy*, 39, 400–402.

Harvey, D. (1989) *The Condition of postmodernity*. Oxford, UK: Blackwell.

Harvey, D. (2013) *Rebel cities: From the right to the city to the urban revolution*. New York: Verso.

Hassink, J., Zwartbol, C., Agricola, H.J., Elings, M., & Thissen, J.T.N.M. (2007) Current status and potential of care farms in the Netherlands. *NJAS—Wageningen Journal of Life Sciences*, 55(1), 21–36.

Heimlich, R. (ed.) (1989) *Land use transition in urbanizing areas: Research and information needs*. Washington, DC: The Farm Foundation in Cooperation with the U.S. Department of Agriculture Economic Research Service.

Heimlich, R., & Brooks, D. (1989) *Metropolitan growth and agriculture: Farming in the city's shadow*. Washington, DC: U.S. Dept. of Agriculture Economic Research Service.

Heinberg, R. (2005) *What will we eat as the oil runs out?*, Conference Paper, FEASTA Conference, June 23–25, Dublin, Ireland www.feasta.org/events/foodconf/foodconfmedia.htm#heinberg (accessed 27 August 2008).

Heinberg, R., & Lerch, D. (eds.) (2010) *The post carbon reader: Managing the 21st century's sustainability crises*. Healdsburg, CA: Watershed Media.

Hodgson, K., Campbell, M., & Bailkey, M. (2010) *Urban agriculture: Growing healthy, sustainable communities*. Chicago: APA Planners Press.

Holmgren, D. (2002) *Permaculture: Principles and pathways beyond sustainability*. Hepburn: Holmgren Design Services.

Holmgren, D. (2009) *Future scenarios: How communities can adapt to peak oil and climate change*. Canada: Chelsea Green.

Holt-Giménez, E. (2015) Racism and capitalism: Dual challenges for the food movement. *Journal of Agriculture, Food Systems, and Community Development*, 23–25.

Hopkins, R. (2008) *The transition handbook*. White River Junction, VT: Chelsea Green.

Houbein, L. (2008) *One magic square: Grow your own food on one square metre*. Kent Town: Wakefield Press.

Hunt, A. (2015) *Civic engagement in food system governance: A comparative perspective of American and British local food movements*. Oxon: Routledge.

Integrated Design Commission (2011) *Life on the edge: An exercise in solicited optimism*, Adelaide: 5000+ Integrated Design Strategy for inner Adelaide.

Ilieva, R. (2016) *Urban food planning: Seeds of transition in the global north*. Oxon: Routledge.

Imbert, D. (2015) *Food and the city: Histories of culture and cultivation*. Washington, DC: Dumbarton Oaks Research Library and Collection.

Jackson, W. (2010) Tackling the oldest environmental problem: Agriculture and its impact on soil. In: Heinberg, R., & Lerch, D. (eds.), *The post carbon reader: Managing the 21st century's sustainability crises*. Healdsburg, CA: Watershed Media, pp. 128–139.

James, S. (2016) *Farming on the fringe: Peri-urban agriculture, cultural diversity and sustainability in Sydney*. Cham: Springer International Publishing.

Jansma, J., Wijnand, S., Stilma, E., van Oost, A., & Visser, A. (2012) The impact of local food production on food miles, fossil energy use and greenhouse gas emission: The case of the Dutch city of Almere. In: Viljoen, A., & Wiskerke, J. (eds.), *Sustainable food planning: Evolving theory and practice*. Wageningen, The Netherlands: Wageningen Academic Publishers, pp. 307–321.

Jarosz, L. (2009) Energy, climate change, meat, and markets: Mapping the coordinates of the current world food crisis. *Geography Compass*, 3/6, 2065–2083.

Jeavons, J. (2006) *How to grow more vegetables: And fruits, nuts, berries, grains, and other crops than you ever thought possible on less land than you can imagine*. Berkeley CA: Ten Speed Press.

Jensen, J. (1920) *A greater west side park system*. Chicago: West Park Commission.

Kaplan, R. (1973) Some psychological benefits of gardening. *Environment and Behavior*, 5(2), 145–162.

Kaufman, J., & Bailkey, M. (2000) *Farming inside cities: Entrepreneurial urban agriculture in the United States*. Cambridge, MA: Lincoln Institute of Land Policy.

Kennedy, A., & Liljeblad, J. (2016) *Food systems governance: Challenges for justice, equality and human rights*. Oxon: Routledge.

Kennedy, F. (1970) Institutionalized oppression vs. the female. In: Morgan, R. (ed.), *Sisterhood is powerful: An anthology of writings from the women's liberation movement*. New York: Random House.

Kennen, K., & Kirkwood, N. (2015) *Phyto: Principles and resources for site remediation and landscape design*. New York: Routledge.

Klein, C. (2007) *Grow your own veg*. London: Mitchell Beazley.

Knowd, I., Mason, D., & Docking, A. (2005) *Urban agriculture: The new frontier*. State of Australian Cities Research Network http://apo.org.au/node/60366 (accessed 20 October 2016).

Kremen, C., & Miles, A. (2012) Ecosystem services in biologically diversified versus conventional farming systems: Benefits, externalities, and trade-offs. *Ecology and Society*, 17.

Kugelman, M., Levenstein, S.L., & Atkin, C. (2009) *Land grab?: the race for the world's farmland*. Washington, DC: Woodrow Wilson International Center for Scholars.

Kunstler, J. (2006) *The long emergency: Surviving the converging catastrophes of the 21st century*. London: Atlantic Books.

La Rosa, D., Barbarossa, L., Privitera, R., & Martinico, F. (2014) Agriculture and the city: A method for sustainable planning of new forms of agriculture in urban contexts. *Land Use Policy*, 41, 290–303.

Ladner, P. (2011) *The urban food revolution: Changing the way we feed cities*. Gabriola: New Society Publishers.

Langridge, P., Cordell, D., & D'Occhio, M. (2014) Agriculture in Australia: Growing more than our farming future. *The Conversation*, July 15. https://theconversation.com/agriculture-in-australia-growing-more-than-our-farming-future-22843 (accessed 19 September 2016).

Lazarus, C. (2000a) Urban agriculture: Join the revolution. *New Village Journal* (2, Community Scale Economics), 64–73. Available from: www.newvillage.net/Journal/Issue2/2urbanagriculture.html (accessed 11 May 2013).

Lazarus, C. (2000b) Profiles in urban agriculture: The urban agriculture network, Washington, DC. *New Village Journal* (2, Community Scale Economics), 76. Available from: www.newvillage.net/Journal/Issue2/2urbanagnet.html (accessed 11 May 2013).

Lee-Smith, D., & Prain, G. (2006) Urban agriculture and health. In: Hawkes, C., & Ruel, M. (eds.), *Understanding the links between agriculture and health*, 2020 Vision Focus 13. Washington, DC: International Food Policy Research Institute (Brief 13 of 16), pp. 27–28. Available from: www.ifpri.org/publication/understanding-links-between-agriculture-and-health (accessed 19 May 2013).

Lee-Smith, D. (2010) Cities feeding people: An update on urban agriculture in equatorial Africa, *Environment and Urbanization*, 22, 483–499.

Lim, C.J., & Liu, E. (2010) *Smartcities + eco-warriors*. Abingdon: Routledge.

Linn, K. (1991) *From rubble to restoration: Sustainable habitats through urban agriculture*. San Francisco, CA: Urban Habitat Program of Earth Island Institute.

Linn, K. (1999) Reclaiming the sacred commons. *New Village Journal* (1, Community Revitalization), 42–49. Available from: www.newvillage.net/assets/docs/linn.pdf (accessed 11 May 2013).

Linn, K. (2007) *Building commons and community*. Oakland, CA: New Village Press.

Locke, A. Wiggins, S., Henley, G., & Keats, S. (2013) *Diverting grain from animal feed and biofuels*. London: Overseas Development Institute.

London Assembly and Greater London Authority (LA and GLA) (2010) *Cultivating the capital: Food growing and the planning system in London*. London: Greater London Authority.

Louv, R. (2005) *Last child in the woods: Saving our children from nature-deficit disorder*. Chapel Hill, NC: Algonquin Books of Chapel Hill.

Lowder, S., Skoet, J., & Raney, T. (2016) The number, size, and distribution of farms, smallholder farms, and family farms worldwide. *World Development*, 87, 16–29.

Lozanovska, M. (2014) Edible suburbs. In: Roggema, R., & Keeffe, G. (eds.), *Why we need small cows: Ways to design for urban agriculture*. Velp, The Netherlands: VHL University of Applied Sciences, pp. 273–292.

Lwasa, S., & Dubbeling, M. (2015) Urban agriculture and climate change. In: de Zeeuw, H., & Drechsel, P. (eds.), *Cities and agriculture: Developing resilient urban food systems*. New York: Routledge.

Maas, W. (1999) *Metacity datatown*. Rotterdam: MVRDV/010 Publishers.

MacLeod, M., & Scott, J. (2007) *Local food procurement policies: A literature review*. Halifax: Department of Energy.

Madigan, C. (2009) *The backyard homestead*. North Adams, MA: Storey Pub.

Malakoff, D. (1995) What good is community greening? *Community Greening Review: Journal of the American Community Gardening Association*, 5, 4–11.

Mandel, L. (2013) *Eat up: The inside scoop on rooftop agriculture*. Gabriola: New Society Publishers.

Markham, B. (2010) *Mini farming: Self sufficiency on a 1/4 acre*. New York: Skyhorse Pub.

Martin, J. (2013) Growing better, not bigger: How one Quebec farmer is making $100,000 on just 1.5 acres. *Small Farm Canada*, May/June, 34–6.

Masi, B. (2008) *Defining the urban – agrarian space*. In: Rugare, S., Schwarz, T., and Cleveland Urban Design Collaborative (eds.), *Cities growing smaller*, Urban Infill, Vol. 1. Cleveland, OH: Kent State University's Cleveland Urban Design Collaborative, pp. 85–102, Chapter 7. Available from: www.cudc.kent.edu/publications/urban_infill/cities_growing_smaller/

Mason, D. (2006) *Urban agriculture: A report of an overseas study trip by David Mason, 2006 Churchill fellow, to identify how sustainable urban agriculture can benefit the quality of life of Australian communities*. Canberra: Winston Churchill Memorial Trust of Australia. Available from: www.parliament.vic.gov.au/images/stories/committees/osisdv/sustainable_development_of_agricultural_business/submissions/T._Budge_Attach_4.pdf (accessed 18 April 2012).

Mason, D., & Knowd, I. (2010) The emergence of urban agriculture: Sydney, Australia. *International Journal of Agricultural Sustainability*, 8(1–2), 62–71.

McClintock, N., Cooper, J., & Khandeshi, S. (2013) Assessing the potential contribution of vacant land to urban vegetable production and consumption in Oakland, California. *Landscape and Urban Planning*, 111(March), 46–58.

McLeod, M. (2015) "The country is the other city of tomorrow": Le Corbusier's Ferme Radieuse and Village Radieux: In: Imbert, D. (ed.), *Food and the city: Histories of culture and cultivation*. Washington, DC: Dumbarton Oaks Research Library and Collection.

McMichael, P. (2009) A food regime genealogy. *The Journal of Peasant Studies*, 36(1), 139–169.

McMichael, P. (2015) The land question in the food sovereignty project. *Globalizations*, 12(4), 434–451.

Meier, S. (2013) *Sustainable urban agriculture: Confirming viable scenarios for production*. New York: New York State Energy Research and Development Authority and Columbia University. Available from: http://purl.org/net/nysl/nysdocs/892964623 (accessed 8 March 2016).

Méndez, V., Bacon, C., Cohen, R., & Gliessman, S. (2015) *Agroecology: A transdisciplinary, participatory and action-oriented approach*. Boca Raton: CRC Press.

Merson, J., Attwater, R., Arnpt, P., Wildman, H., & Chapple, R. (2010) The challenges to urban agriculture in the Sydney basin and lower Blue Mountains region of Australia. *International Journal of Agricultural Sustainability*, 8(1–2), 72–85.

Meyer, E. (2008) Sustaining beauty: The performance of appearance. *Journal of Landscape Architecture*, (3), 6–23.

Meyer, E. (2015) Beyond sustaining beauty: Musings on a manifesto. In: Deming, E. (ed.), *Values in landscape architecture and environmental design*. Baton Rouge: LSU Press.

Milan Urban Food Policy Pact (2016) http://www.milanurbanfoodpolicypact.org/ (accessed 3 March 2017).

Mobbs, M. (2012) *Sustainable food*. Sydney: Choice Books.

Monbiot, G. (2013) Why Britain's barren uplands have farming subsidies to blame. *The Guardian*. www.theguardian.com/environment/georgemonbiot/2013/may/22/britain-uplands-farming-subsidies (accessed 30 June 2014).

Moreau, T., Adams, T., Mullinix, K., Fallick, A., & Condon, P. (2012) Recommended practices for climate-smart urban and peri-urban agriculture. In: Viljoen, A., & Wiskerke, J. (eds.), *Sustainable food planning: Evolving theory and practice*. Wageningen, The Netherlands: Wageningen Academic Publishers, pp. 295–306.

Moreno-Peñaranda, R. (2011) *Japan's urban agriculture: Cultivating sustainability and well-being*. United Nations University. Available from: https://unu.edu/publications/articles/japan-s-urban-agriculture-what-does-the-future-hold.html (accessed 5 November 2016).

Morgan, K. (2013) Feeding the city: The challenge of urban food planning. In: Miazzo, F., & Minkjan, M. (eds.), *Farming the City: Food as a tool for today's urbanisation*. Haarlem, The Netherlands: Valiz/Trancity, pp. 18–21.

Morgan, K. (2015) Nourishing the city: The rise of the urban food question in the Global North. *Urban Studies*, 52, 1379–1394.

Morgan, K., Marsden, T., & Murdoch, J. (2006) *Worlds of food: Place, power and provenance in the food chain*. Oxford: Oxford University Press.

Mougeot, L. (2005) *Agropolis: The social, political and environmental dimensions of urban agriculture*. London: International Development Research Centre.

Mougeot, L. (2011) *International support to research and policy on urban agriculture (1996–2010): Achievements and challenges*. Ottawa: International Development Research Centre.

Moustier, P., & Danso, G. (2006) Local economic development and marketing of urban produced food. In: van Veenhuizen, R., & ETC – Urban Agriculture (eds.), *Cities farming for the future: Urban agriculture for green and productive cities*. Silang, Cavite: Philippines: International Institute of Rural Reconstruction, and ETC Urban Agriculture, pp. 173–195. Available from: www.ruaf.org/sites/default/files/Chapter%207_1.pdf (accessed 10 September 2016).

Moustier, P., & Renting, H. (2015) Urban agriculture and short chain food marketing in developing countries. In: de Zeeuw, H., & Drechsel, P. (eds.), *Cities and agriculture: Developing resilient urban food systems*. New York: Routledge.

Murray, S. (2007) The world's biggest industry. *Forbes*. www.forbes.com/2007/11/11/growth-agriculture-business-forbeslife-food07-cx_sm_1113bigfood.html (accessed 10 September 2016).

MVRDV (2009) *City pig*. www.youtube.com/watch?v=x1r5GutrXX4andindex=5andlist=PL7E60F6F32AFB9DB3 (accessed 10 August 2016).

MVRDV (2010) *Food city*. www.youtube.com/watch?v=5JRDHKAJqtAandindex=8andlist=PL7E60F6F32AFB9DB3 (accessed 10 August 2016).

Nabhan, G. (2002) *Coming home to eat: The pleasures and politics of local foods*. New York: Norton.

Napawan, N. (2016) Complexity in urban agriculture: The role of landscape typologies in promoting urban agriculture's growth. *Journal of Urbanism: International Research on Placemaking and Urban Sustainability*, (9), 19–38.

Nasr, J., Komisar, J., & Gorgolewski, M. (2009) Designing for food and agriculture: Recent explorations at Ryerson University. *Open House International*, 34(2).

Nettle, C. (2014) *Community gardening as social action*. London: Ashgate.

Niwa, N. (2012) Why is there agriculture in Tokyo? In: Viljoen, A., & Wiskerke, J. (eds.), *Sustainable food planning: Evolving theory and practice*, Wageningen, The Netherlands: Wageningen Academic Publishers, pp. 283–294.

Norberg-Hodge, H., Goering, P., Page, J., & Goering, P. (2001) *From the ground up: Rethinking industrial agriculture*. New York: Zed Books.

Novak, A. (2016) *The rooftop growing guide: How to transform your roof into a vegetable garden or farm*. Berkeley, CA: Ten Speed Press.

NZTE (2014) *Sustainability trends for food and beverage in Asia*, May 2014. www.nzte.govt.nz/en/export/market-research/food-and-beverage/sustainability-trends-for-food-and-beverage-in-asia-may-2014/#toc-hong-kong-and-macao (accessed 10 August 2016).

Oles, T., & Lickwar, P. (2015) Why so serious, landscape architect? *LA+*, Pleasure/Fall, 82–83.

Opitz, I., Berges, R., Piorr, A., & Krikser, T. (2016) Contributing to food security in urban areas: Differences between urban agriculture and peri-urban agriculture in the Global North. *Agriculture and Human Values: Journal of the Agriculture, Food, and Human Values Society* (33), 341–358.

Oswalt, P., & Rieniets, T. (2006) *Atlas of shrinking cities*. Ostfildern: Hatje Cantz.

Parham, S. (2015) *Food and urbanism: The convivial city and a sustainable future*. London: Bloomsbury Academic.

Patel, I.C. (1991) Gardening's socioeconomic impacts. *Journal of Extension*, 29(4), 7–8.
Pearson, C., Pilgrim, S., & Pretty, J. (2010) *Urban agriculture: Diverse activities and benefits for city society*. Abingdon, UK: Earthscan.
Peters, C., Picardy, J., Darrouzet-Nardi, A., Wilkins, J., Griffin, T., & Fick, G. (2016) Carrying capacity of U.S. agricultural land: Ten diet scenarios. *Elementa: Science of the Anthropocene*, (4), 000116.
Pimentel, D. (2009) Energy inputs in food crop production in developing and developed nations. *Energies*, 2, 1–24.
Pimentel, D., & Pimentel, M. (2003) Sustainability of meat-based and plant-based diets and the environment. *The American Journal of Clinical Nutrition* (78).
Pimentel, D., & Pimentel, M. (2008) *Food, energy, and society*. Boca Raton, FL: CRC Press.
Plakias, A. (2016) *The farm on the roof: What Brooklyn Grange taught us about entrepreneurship, community, and growing a sustainable business*. New York: Avery.
Pollan, M. (2008) *Second nature: A gardener's education*. New York: Grove Press.
Pollock, M., & Anderson, P. (2012) *RHS vegetable and fruit gardening*. London: Dorling Kindersley.
Popovitch, T. (2014) *10 American cities lead the way with urban agriculture ordinances*. http://seedstock.com/2014/05/27/10-american-cities-lead-the-way-with-urban-agriculture-ordinances/ (accessed 23 May 2016).
Porter, J., Dyball, R., Dumaresq, D., Deutsch, L., & Matsuda, H. (2014) Feeding capitals: Urban food security and self-provisioning in Canberra, Copenhagen and Tokyo. *Global Food Security*, 3, 1–7.
Potteiger, M., & Purinton, J. (1998) *Landscape narratives: Design practices for telling stories*. New York: Wiley.
Pretty, J. (2002) *Agri-culture: Reconnecting people, land and nature*. London: Earthscan.
Pudup, M. (2008) It takes a garden: Cultivating citizen-subjects in organized garden projects. *Geoforum*, 39(3), 1228–1240.
Raja, S., Born, B., & Russell, J. (2008) *A planners guide to community and regional food planning: Transforming food environments, facilitating healthy eating*. Chicago, IL: American Planning Association.
Ramankutty, N., Evan, A., Monfreda, C., & Foley, J. (2008) Farming the planet: 1. Geographic distribution of global agricultural lands in the year 2000. *Global Biogeochemical Cycles*, 22.
Rees, W. (1992) Ecological footprints and appropriated carrying capacity: What urban economics leaves out, *Environment and Urbanization*, 4, 121–130.
Resh, H. (2012) *Hydroponic food production: A definitive guidebook for the advanced home gardener and the commercial hydroponic grower*. Boca Raton, FL: CRC Press.
Reynolds, R. (2009) *On guerrilla gardening: A handbook for gardening without boundaries*. London: Bloomsbury.
Roberts, P. (2008) *The end of food*. Boston, MA: Houghton Mifflin.
Roggema, R. (2014) Finding space for productive cities. In: Roggema, R., & Keeffe, G. (eds.), *Why we need small cows: Ways to design for urban agriculture*. Velp, The Netherlands: VHL University of Applied Sciences, pp. 37–61.
Roggema, R. (ed.) (2016) *Sustainable urban agriculture and food planning*. Oxon: Routledge.
Roggema, R., & Keeffe, G. (2014) *Why we need small cows: Ways to design for urban agriculture*. Velp, The Netherlands: VHL University of Applied Sciences.
Roy, A., & Ong, A. (2011) *Worlding cities: Asian experiments and the art of being global*. Chichester, West Sussex: Wiley-Blackwell.
Sahakian, M., Saloma, C., & Erkman, S. (2016) *Food consumption in the city: Practices and patterns in urban Asia and the Pacific*. Oxon: Routledge.
Schouw, J., & Roggema, R. (2014) Financing urban agriculture. In: Roggema, R., & Keeffe, G. (eds.), *Why we need small cows: Ways to design for urban agriculture*. Velp, The Netherlands: VHL University of Applied Sciences, pp. 311–321.
Sinclair, I. (2015) *Growing food in a residential landscape*, State of Australian Cities Conference, The Gold Coast, Queensland.
Sloper, C. (2013) *The LED grow book: Better, easier, less watts*. [S.l.]: CreateSpace Independent Publ. Platform.
Solomon, D. (2012) Urbaniahoeve: Expanded urban agriculture. In: Viljoen, A., & Wiskerke, J. (eds.), *Sustainable food planning: Evolving theory and practice*. Wageningen, The Netherlands: Wageningen Academic Publishers, pp. 365–382.
Sommers, P., & Smit, J. (1994) *Promoting urban agriculture: A strategy framework for planners in North America, Europe, and Asia*, Cities Feeding People Series, Report 9. Ottawa: IDRC. Available from: https://idl-bnc.idrc.ca/dspace/bitstream/10625/13680/1/103014.pdf [Accessed 18 November 2016].
Speak, S., & Kumar, A. (2016) Fit and misfit – ordinary urban injustice in a global context. In: Davoudi, S., & Bell, D. (eds.), *Justice and fairness in the city*. Bristol: University of Bristol, Policy Press.
Spiaggi, E. (2005) Urban agriculture and local sustainable development in Rosario, Argentina: Integration of economic, social, technical and environmental variables. In: Mougeot, L. (ed.), *Agropolis: The social, political and environmental*

dimensions of urban agriculture. London Earthscan; IDRC, pp. 187–202, Chapter 7. Available from: http://idl-bnc.idrc.ca/dspace/bitstream/10625/28341/47/IDL-28341.pdf [Accessed 18 January 2017].

Springer, S., Birch, K., & MacLeavy, J. (2016) *The handbook of neoliberalism*. New York: Routledge.

San Francisco Bay Area Planning and Urban Research Association (SPUR) (2012) *Public harvest: Expanding the use of public land for urban agriculture in San Francisco*, SPUR Report 04/2012. San Francisco, CA: SPUR. Available from: www.spur.org/sites/default/files/publications_pdfs/SPUR_Public_Harvest.pdf [Accessed 18 February 2013].

St. Martin, K. (2009) Toward a cartography of the commons: Constituting the political and economic possibilities of place. *Professional Geographer*, 61(4), 493–507.

Steel, C. (2008) *Hungry city: How food shapes our lives*. London: Random House.

Stefani, G., Tiberti, M., & Lombardi, G. (2015) Public food procurement: A systematic literature review, Working Paper N. 2/2015 – Management, Marketing and Organization Sciences, Università degli Studi di Firenze. www.academia.edu/18183807/Public_Food_Procurement_A_Systematic_Literature_Review (accessed 19 March 2017).

Stewart, B., & Duane, T. (2009) Easement exchanges for agricultural conservation: A case study under the Williamson Act in California. *Landscape Journal* (28), 181–197.

Stierand, P. (2012) Food policy councils: Recovering the local level in food policy. In: Viljoen, A., & Wiskerke, J. (eds.), *Sustainable food planning: Evolving theory and practice*. Wageningen, The Netherlands: Wageningen Academic Publishers.

Stone, C. (2016) *The urban farmer: Growing food for profit on leased and borrowed land*. Gabriola Island: New Society Publishers.

Strong, J. (2011) *Educated tastes: Food, drink, and connoisseur culture*. Lincoln: Bison.

Sumner, J., Mair, H., & Nelson, E. (2010) Putting the culture back into agriculture: Civic engagement, community and the celebration of local food. *International Journal of Agricultural Sustainability*, 8(1–2), 54–61.

Thebo, A., Drechsel, P., & Lambin, E. (2014) Global assessment of urban and peri-urban agriculture: Irrigated and rain-fed croplands. *Environmental Research Letters*, 9.

Thompson, I. (2012) Ten tenets and six questions for landscape urbanism. *Landscape Research*, 37, 7–26.

Tornaghi, C. (2012) Public space, urban agriculture and the grassroots creation of new commons: Lessons and challenges for policy makers. In: Viljoen, A., & Wiskerke, J. (eds.), *Sustainable food planning: Evolving theory and practice*. Wageningen, The Netherlands, Wageningen Academic Publishers, pp. 349–363.

Tracey, D. (2011) *Urban agriculture: Ideas and designs for the new food revolution*. Gabriola: New Society Publishers.

Tse, B. (2015) *Why Hong Kong is so far behind Singapore in agriculture*. www.ejinsight.com/20150213-hk-far-behind-sing-in-farming/ (accessed 6 July 2016).

UN Habitat (2016) *World cities report 2016: Urbanization and development: Emerging futures*, Nairobi: United Nations Human Settlements Programme.

Urban Partners (2007) *Farming in Philadelphia: Feasibility analysis and next steps*. Philadelphia, PA: Institute for Innovations in Local Farming. Available from: www.spinfarming.com/common/pdfs/STF_inst_for_innovations_dec07.pdf [Accessed 15 April 2012].

Urban Resource Systems (URS) and International Development Research Centre (1984) *Global urban agriculture: An annotated bibliography*. San Francisco, CA: Urban Resource Systems.

USDA (2012) *Census of Agriculture*, United States Department of Agriculture, https://www.agcensus.usda.gov/Publications/2012/ (accessed 15 May 2015).

van der Schans, J., & Wiskerke, J. (2012) Urban agriculture in developed economies. In: Viljoen, A., & Wiskerke, J. (eds.), *Sustainable food planning: Evolving theory and practice*. Wageningen, The Netherlands: Wageningen Academic Publishers, pp. 245–258.

van der Valk, A. (2016) Foodscape gastropolis New York city. In: Roggema, R. (ed.), *Sustainable urban agriculture and food planning*. London: Routledge, pp. 53–72.

van Leeuwen, E., Nijkamp, P., & de Noronha Vaz, T. (2010) The multifunctional use of urban greenspace. *International Journal of Agricultural Sustainability*, 8(1–2), 20–25.

van Veenhuizen, R., & Danso, G. (2007) *Profitability and sustainability of urban and peri-urban agriculture*. Rome: Food and Agriculture Organization of the United Nations.

Viljoen, A., Schlesinger, J., Bohn, K., & Drescher, A. (2015) Agriculture in urban design and spatial planning. In: de Zeeuw, H., & Drechsel, P. (eds.), *Cities and agriculture: Developing resilient urban food systems*. New York: Routledge.

Volk, J. (2017) *Compact farms: 15 proven plans for market farms on 5 acres or less*. North Adams, MA: Storey Publishing.

Wackernagel, M. (1994) *Ecological footprint and appropriated carrying capacity: A tool for planning toward sustainability*. PhD thesis, School of Community and Regional Planning, The University of British Columbia.

Waldheim, C. (2010) Notes toward a history of Agrarian urbanism. *Places Journal*, November. Available from: https://placesjournal.org/article/notes-toward-a-history-of-agrarian-urbanism/ (accessed 3 April 2012).

Weis, T. (2010) The accelerating biophysical contradictions of industrial capitalist agriculture. *Journal of Agrarian Change* (10), 315–341.

Weller, R. (2009) *Boomtown 2050*. Perth: UWA.

Weller, R. (2014) Stewardship now? Reflections on landscape architecture's Raison d'etre in the 21st century. *Landscape Journal* (33), 85–108.

Wilson, E. (1984) *Biophilia*. Cambridge, MA: Harvard University Press.

Wiltshire, R., & Geoghegan, L. (2012) Growing alone, growing together, growing apart? Reflections n the social organisation of voluntary urban food production in Britain. In: Viljoen, A., & Wiskerke, J. (eds.), *Sustainable food planning: Evolving theory and practice*. Wageningen, The Netherlands: Wageningen Academic Publishers, pp. 337–348.

Wise, P. (2014) *Grow your own: The potential value and impacts of residential and community food gardening*. A.C.T.: Australia Institute.

Wolf, S. (2014) *The neoliberal regime in the agri-food sector: Crisis, resilience and restructuring*. New York: Routledge.

Wolford, W., Borras, S.M., Hall, R., Scoones, I., & White, B. (2013) *Governing global land deals: The role of the state in the rush for land*. Chichester, UK: Wiley Blackwell.

World Bank (nda) *Employment in agriculture (% of total employment)*. http://data.worldbank.org/indicator/SL.AGR.EMPL.ZS (accessed 9 May 2016).

World Bank (ndb) *Agricultural land (% of land area)*. http://data.worldbank.org/indicator/AG.LND.AGRI.ZS (accessed 9 May 2016).

Yu, K. (2010) Beautiful big feet: Toward a new landscape aesthetic. *Harvard Design Magazine*, Fall/Winter 2009/10, 48–59.

Yu, K., & Padua, M. (2006) *The art of survival: Recovering landscape architecture*. Mulgrave: Images Publishing Group.

Zeunert, J. (2011) *Eating the landscape: Aesthetic foodscape design and its role in Australian landscape architecture*. Proceedings of AILA National Conference Transform, Brisbane, 13 August 2011. www.researchgate.net/publication/297730109_Eating_the_Landscape_Aesthetic_Foodscape_Design_and_its_role_in_Australian_Landscape_Architecture?ev=prf_pub (accessed 15 November 2016).

Zeunert, J. (2016) Urban agriculture upscaled: Economically and socially productive public green space. In: Roggema, R. (ed.), *Sustainable urban agriculture and food planning*. Oxon: Routledge, pp. 107–125.

Zeunert, J. (2017) *Landscape architecture and environmental sustainability: Creating positive change through design*. London and New York: Bloomsbury.

Zezza, A., & Tasciotti, L. (2010) Urban agriculture, poverty, and food security: Empirical evidence from a sample of developing countries. *Food Policy*, 35, 265–273.

13

Mediterranean urban agriculture

Tal Alon-Mozes and Efrat Eizenberg

"... *every man under his vine and under his fig-tree* ..."

(Kings 1, 5:5)

Prologue

The Mediterranean urban landscape is dotted with small to medium sized agricultural plots: from several square metres to hundreds of hectares. They are part of the traditional urban fabric of the old inner cities of Casablanca and Istanbul, as well as the peri-urban landscapes of sprawling Beirut and Barcelona (see Figure 13.1). Small vegetable gardens decorate the roofs of millennials in Tel Aviv-Jaffa, and new greenhouses around Tunis serve the city with fresh produce year-round. Most of these gardens produce fresh vegetables: tomatoes, cucumbers, lettuce, beans, eggplants, and, with availability of space, also grape vines, and olive and fruit trees, amongst others. Rarely, the gardens may include livestock: chickens, or goats and more. Shaping the region's cultural rituals and based on its unique climatic conditions and long history of agricultural development, makes Mediterranean urban agriculture (MUA) a common, identifiable, practice. In all other aspects, however, the Mediterranean is an uneven region for exploration. Mediterranean urban gardens and farms vary in size, cultivation methods, social, cultural, and economic significance, and in modes of interaction with municipal, state and international authorities. Generally put, according to some scholars, UA in the northern Mediterranean (that is, the EU countries), or in what can be termed 'developed countries' differs significantly from UA in the 'developing countries' on the southern shores of the Mediterranean (Mougeot, 2010; Orsini, 2013; Zezza and Tasciotti, 2010). However, the constant movement of people, artefacts, and knowledge between Europe, Asia, and Africa makes MUA a unique topic for examination, blurring this common dichotomised notion.

Due to the diverse examples of MUA, our survey adopts an inclusive definition of urban agriculture, following the 2016 Urban Agriculture Europe (COST) definition:

> Urban agriculture spans all actors, communities, activities, places and economies that focus on biological production in a spatial context, which – according to local standards – is categorised as 'urban'. Urban agriculture takes place in intra- and peri-urban areas and one of its key characteristics is that it is more deeply integrated in the urban system compared to other agriculture. Urban agriculture is structurally embedded in the urban fabric; it is integrated into the social and cultural life, the economics, and the metabolism of the city.
>
> *(Lohrberg et al., 2016: 21)*

Figure 13.1 Jardí Comestible de les Glòries community garden in Barcelona.
Source: Laura Calvet Mir

Surveying urban agriculture initiatives in the Mediterranean basin reveals their commonalities as well as differences and provides a unique entry point to view the ongoing exchanges between developed and developing countries. The examples surveyed herein are drawn from a rich archive of academic and non-academic literature, including documented case studies of Mediterranean urban gardens and farms. This chapter first presents the history of MUA, discussing the characteristics of the region that promote UA and the relation between rural and urban agriculture in the Mediterranean basin. The second part presents several case studies. The discussion summarises our findings and explores future possibilities for MUA.

A short history and the unique characteristics of the Mediterranean basin

The Mediterranean region is characterised by arid and semi-arid climates, seasonal precipitation, indigenous crops, fragility of agriculture lands, and a long history of urbanisation (Nasr and Kaldjian, 1997). The first agricultural crops were domesticated not far from the Mediterranean basin, in South Eastern Turkey and North Syria (Gopher and Abbo, 2016). From there, they moved westward to Cyprus, Turkey, and the Balkans, as well as southward toward Palestine. Annual cereals were first domesticated, followed by fruit trees such as vines, figs, olives, pomegranates, and dates, all species that are still dominant in contemporary Mediterranean agricultural landscapes. The agricultural revolution of the late Pleistocene period saw new relations with nature, as well as new forms of social and cultural organisation; novel agricultural practices, and permanent settlements, which later developed into concentrated townships (ibid.).

Agricultural production and urban centers around the Mediterranean are unique to the region and they maintain special relationships. According to *Medi-terra: The Future of Agriculture and Food in Mediterranean Countries* (Hervieu, 2008), Mediterranean agriculture had escaped, to a certain degree, the modernisation processes that characterise agricultural practices in central Europe and the UK. It is identified by a plurality of family farms, including mixed farms, and it is connected to more sustainable agricultural practices, quality food production, and the integration of other activities such as agro-tourism alongside 'pure' agricultural activities.

The Mediterranean city is also unique, escaping a clear typology. According to Cavallo et al. (2014), it is distinguished by informal expansion of human settlements; planning that trails after, rather than directs expansion; compact urban form (as compared to the more dispersed structure of the continental European model); and a distinctive relationship between city and countryside. In addition, due to the importance of the Mediterranean diet in structuring the local culture, it has a significant role in configuring the Mediterranean public space of food exchange, consumption, and enjoyment.

Since the late 18th century, scholars perceived the Mediterranean basin as a distinct geo-cultural entity, different from other parts of the world (Braudel, 1996; Mendelson, 2002; Niola, 2006). This concept, which persists to the present day, perceives the Mediterranean Sea as an internal one surrounded by 'washed by sun', 'joyful' countries. It is worth mentioning that even in the glorious days of this perception, not all countries were part of the comprehensive concept; North Africa was part of the Mediterranean idea only as part of colonial Europe, and the Islamic states were considered by Braudel as the blocking force, preventing the Mediterranean from reaching its full cultural potential (Hochberg, 2011: 44).

Currently, the gap between North Mediterranean and its southern and eastern shores is becoming wider, leaving the concept of 'Mediterraneanism' as an empty shell. The north and south of the Mediterranean basin differ in topography, soil, rainfall, available lands for cultivation due to population size, as well as cultural and religious characteristics. While Christianity dominates the western and northern parts of the Mediterranean, Islam dominates its eastern and southern parts. However, ongoing movement between its different parts characterise the region. Europeans moved southward during the colonial era, disseminating modern practices including modern agriculture in the colonised countries, while currently, millions of refugees make the opposite journey. These opposing movements of people, artefacts, and knowledge are tightly connected to the expansion of UA and will be discussed later.

Mediterranean urban agriculture (MUA)

Urban agriculture in the Mediterranean has a long history. It was practiced in Pompeii during the Roman times (Jashemski, 1974), and according to Danner (1997), during the middle ages, each household in the Mediterranean basin had an allotment of 100–120 square metres. Charlemagne (742–814 CE) made urban agriculture a common European practice, issuing an edict that spelled out which crops were to be grown within towns (Smit et al., 2001).

During the modern era, when people left their villages and moved to towns, cultivating allotments on the outskirts of the crowded industrial towns became common practice (Keshavarz and Bell, 2016). In less-developed regions, historic urban gardens escaped hectic urbanisation and urban fruit and vegetable gardens were an integral part of the city since ancient times. In Istanbul in 1900, for example, there were more than 1,200 'Bostans' (market gardens) covering 12 square kilometres on both sides of the Bosporus, serving most of the town with fresh produce (Quinn, 2013). During times of economic or political crisis, urban agriculture became prominent all over the Mediterranean. War gardens and small family farms were intended to serve local populations with substantial produce.

MUA today: the survey

Table 13.1 summarises information about MUA as documented by various scholars and from internet sources. Except for the Israeli case studies, which are based on primary sources, all other case studies are secondary. This makes the survey limited in its scope and missing data, especially on traditional urban gardening and farming in the developing countries that escape documentation. Despite this, the existing data enables general demarcation of the phenomenon, while emphasising the need for a more comprehensive survey in the future.

Table 13.1 presents a summary of UA in each Mediterranean country, starting with Turkey and going clockwise. The table is followed by a map (Figure 13.2) that presents the character and types of UA in

Table 13.1 A summary of UA in each Mediterranean country.

Istanbul-Turkey, traditional urban agriculture exists within the city and in peri-urban areas. A unique example is the urban farm next to the Theodosian Land walls of 413 AD, which is in danger of destruction due to recent and accelerating urbanisation. The site is protected as a world heritage site by UNESCO, however, the gardens are not safe (see Golonu, 2013; The Observer, 2016; Quinn, 2013).	**Beirut-Lebanon**, traditional urban agriculture exists within the city and in peri-urban areas (see Harrigan, 2014; Hamadeh et al., 2014; ANERA, 2013). The Environment and Sustainable Development Unit in the American University of Beirut which serves as a regional interdisciplinary research center, provides training, services, and consultancies and hosts several large community-development projects, including roof gardens in refugee camps (www.aub.edu.lb/fafs/esdu/projects).	**Multiple locals*-Israel**, UA is an emerging practice. It multiplied its extent from 30 sites in 2004 to more than 300 in 2014. It is practiced mainly within cities as a social practice and supported by municipalities and environmental NGOs. While it is practiced as a leisure activity for the affluent, for the poor (new immigrants), it enhances food security and attachment to the local landscape (see Eizenberg and Fenster, 2015). *Including Jerusalem, Tel Aviv-Jaffa, Rishon le Zion, Haifa, Kfar Saba, and others.	**Gaza-Palestine**, Very dense urban area and agricultural hinterland which used to be a prospering agricultural area. However, success is limited due to lack of water, land, equipment, fertilisers, and political instability. Since 1998 several projects were initiated, integrating aquaponics with roof gardening. It also includes raising chickens, rabbits, and pigeons on rooftops (see Bartlett, 2010; Laeremans and Sourani, 2005; Longari; Smit et al., 2001:12–13; Zurayk et al., 2012). Various projects currently supported by PARC-Palestinian Agricultural Relief Committees, FAO – Food and agriculture organization of the United Nations, and local NGO, GUAC-Gaza Urban Agriculture Committee (www.parc.ps).
Alexandria-Egypt, no documentation, except of anecdotal information on people keeping sheep and goats on rooftops for the holiday feasts.	**Tunisia-Tunisia**, various initiatives such as commercial farms within cities, heritage food production and commercial trade in peri-urban areas, non-irrigated olive orchards and irrigated vegetables plots, greenhouses and rain water harvesting. In addition, agriculture as a leisure activity. The initiatives are supported by a 1996 amendment recognising agriculture as component of urban planning. However, and due to rapid urbanisation, it is impossible to monitor and implement the regulations; 80% of urban growers cannot make a living from agriculture only. The projects are supported by the Sousse University in collaboration with Euro-Med universities to promote an "urbanisme vert" (see Rejeb, 2011; Rejeb et al., 2012).	**Algiers-Algeria**, Low-priority for urban and peri-urban agriculture and no supporting policy. Currently, buds of change: national awareness of the need to conserve the country's agricultural heritage for future generations. In 2014, initiative to create 500 ha of greenhouses to grow three successive crops of vegetables each year. In parallel, accelerated urbanisation brought a loss of 250,000 ha of agricultural land since the 1960s (see Lakrouf, 2012; Naïli, 2011).	**Casablanca-Morocco,*** From 2005 to 2014 an international project: Urban Agriculture as an Integrative Factor of Climate-Optimized Urban Development, Casablanca/Morocco, which has four Pilot Project components: 1: Urban Agriculture + Industry 2: Urban Agriculture + Informal Settlements 3: Urban Agriculture + Peri-Urban Tourism 4: Urban Agriculture + Healthy Food Production The project is a collaboration of UN institutions, the German Government, and Munich University (Tu Munchen) with local NGOs. While some projects have succeeded, others find it hard to take off (see Giseke, 2015). * Although Casablanca is not situated along the Mediterranean shores, we include this case study as it represents many of the characteristics of Mediterranean practice.

Barcelona-Spain, hundreds of local initiatives flourish despite intensive urbanisation. Various stakeholders are involved including the Catalan Government, Municipalities such as 14 within Barcelona Provincial Council, NGOs such as the Farmers' Union in Catalonia, and individual family farms (see www.urbanagricultureeurope.la.rwth-aachen.de/online-atlas.html).	**Montpellier-France**, A long tradition of urban agriculture supported by the municipality. A great variety of initiatives within the city and the peri-urban area. Integration of urban agriculture with a recreation park-agripark. Despite the great variety of activities, professional farms are diminishing while new forms of specifically urban agriculture proliferate. The projects are supported mainly by the municipality (see Scheromm et al., 2012; http://prodinra.inra.fr/?locale=fr#!ConsultNotice:176976).	**Rome-Italy**, A strong traditional link between urban populations and local agriculture which currently is manifesting in new modes of urban-rural relations that are not just about food consumption. City-region food systems create a complex system of social, economic and political behaviour. UA as a device of resilience for the city. The cultivated area within the City of Rome has increased between 2000 and 2010, as well as the number of farmers. The Increasing importance of alternative and local food networks that promote short chain initiatives. The projects are supported by the Municipality and NGOs (see Antonelli and Lamberti, 2011; Cavallo et al., 2014).	**Athens-Greece**, No tradition of allotment gardens. Recently, various projects emerged mainly as a response to the current economic crisis (see Anthopoulou, 2016; Tsiambaos, 2016).

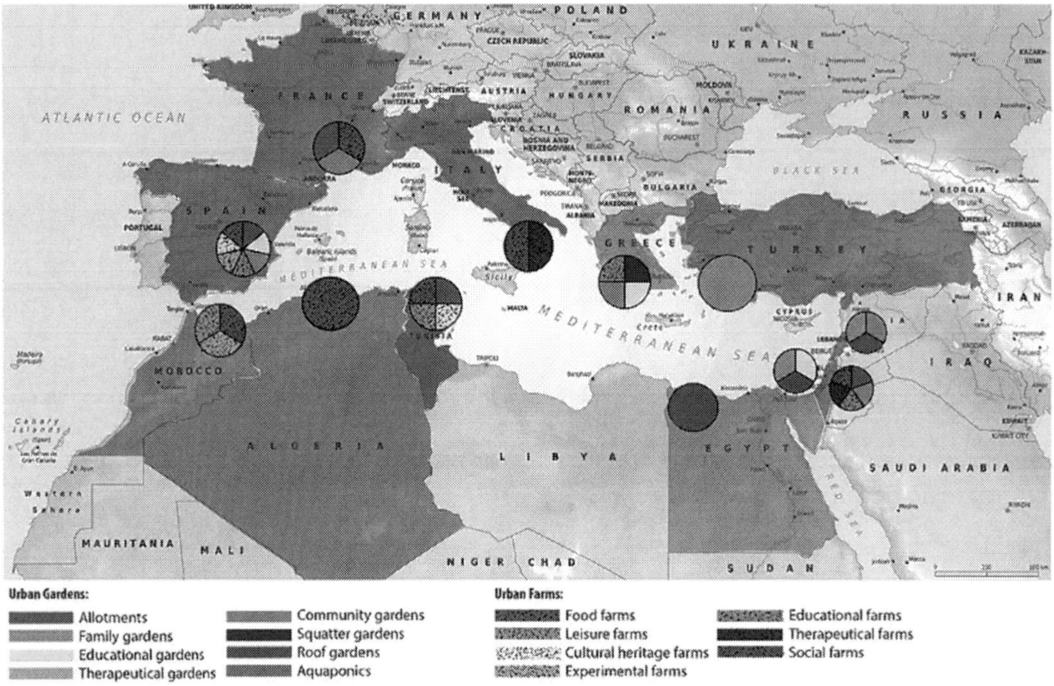

Figure 13.2 Types of Mediterranean urban agriculture.
Source: Graphic design: Hilla Michowiz Setton

each country based on our survey (for additional information, see UA COST Atlas www.urbanagricultureeurope.la.rwth-aachen.de/online-atlas.html).

Discussion

Our chapter explores MUA as a unique example of UA in developed and developing countries which share a large and distinct geographical unit. The literature emphasises the importance of UA for increasing food security in both regions, it also notes differences between them. The contribution of UA practice to social well-being, to physical health, its contribution to development and progress, as well as the risk of chemical pollutants it poses, seem to be different in the developed countries and the developing ones (see Mok et al., 2014 for USA and Canada, United Kingdom, Australia, and Japan; Hamilton et al., 2014 for Latin America and the Caribbean, Africa, Asia, Eastern Europe, and the Pacific Island countries; Orsini et al., 2013). Based on the review of the MUA case studies presented above, this chapter seeks to overcome the common split between developing and developed countries and looks at similarities, connectivity, and mutual influences along the Mediterranean shore.

MUA is diverse and rich, ranging between various opposing poles, including between areas of moderate weather and those suffering increasing drought due to climate change, between practices of crisis and practices of leisure, and between bottom-up initiatives and top-down projects. However, the geographical proximity and historical connections between developed and developing countries around the Mediterranean Sea enables a fertile exchange of people, artefacts, and knowledge between them. Examining MUA via these lenses of movements and differences sheds light on the uniqueness of MUA, as well as on modes of cultural, economic, and political exchange between the developed and the developing worlds.

MUA in the developed countries is currently primarily prospering as a leisure activity, linked to initiatives of sustainability, resilience, and preservation of local cultural heritage, via education and practice (see Chapters 12 and 16). It represents the local population's desire to return to a slower, more immediate way of life, and forms a platform for pro-environmental behaviours. The various models of urban agriculture prospering around Barcelona, Montpellier, or Rome demonstrate this trend (see Figure 13.3).

In parallel, some UA in the developed countries is an outcome of movement of people from the developing countries to the developed ones, as demonstrated by the farm and community gardens in Bat-Yam, and Rishon-LeZion, Israel which serve the Ethiopian immigrants in Israel (Eizenberg, 2010) (Figure 13.4). In such gardens and farms, UA is tightly connected to food security and to the preservation of a traditional diet. Historically, such projects helped South American immigrants (Eizenberg, 2016) and South Asian refugees (Corlett et al., 2003; Airriess and Clawson, 1994) in the USA, and they are currently implemented in various American towns (Kripke, 2015). Their benefits for the waves of refugees from North Africa, and since 2011 from the Mediterranean Middle East, who move westward and northward, is beyond question. However, despite the initiative to implement UA in Turkey, which currently hosts more than two million Syrian refugees, it seems that without municipal, state and international support, these projects cannot materialise (Buzaurd, 2016).

UA practices face various challenges on both the North (Eizenberg et al., 2016) and South shores of the Mediterranean (as the examples from Istanbul, Algiers, and Tunis demonstrate). In countries on the south shore of the Mediterranean, with the utmost need for food security, work, and access to a healthy diet, UA is shrinking and urban residents are increasingly denied the right to this urban practice. However, top-down projects originating in the developed countries and implemented in the developing ones often have better chances of success as is evident in the Casablanca example. This project is a collaboration of UN institutions, the German government and universities and the local community of gardeners. It brings advanced Western technologies of wastewater purification and Western trends such as agro-tourism and organic farming to the outskirts of town. Numerous reports describe its successful implementation but also point to the need for new regulations and long-term social practices to better support it (see www.

Figure 13.3 Orto Errante, community/guerrilla garden in Rome.

Source: Chiara Certomà

Figure 13.4 Allotment gardens in Rishon-LeZion, Israel.
Source: Tal Alon-Mozes

uac-m.org). Whether such flagship projects can serve as role models in other locales without the intensive involvement of international experts and supporting bodies is a central issue for the future of MUA in developing countries.

At the same time, in countries on the north shore of the Mediterranean, UA is perceived and offered mostly for sustainable recreation. As Morgan (2015) stresses, however, nutrition insecurity or malnutrition (a name replacing food (in)security since the latter is attached to the Global South) is a major problem of the Global North. Thus, he continues, the challenge of integrated UA for cities of the Global North (or the North shore of the Mediterranean in our case) through policy and planning is still pending.

Our examples emphasise the increasing role of policy in the success of MUA. They demonstrate the importance of international as well as local policies in making urban agriculture part of the Mediterranean landscape. Without favourable policies, such projects are doomed to disappear due to the market's failure to assimilate urban agriculture's various benefits. Such policies, at times responding to grassroots and community initiatives, are increasingly adopted in developed countries, but they are only emerging in the developing countries, and are not always implemented. International initiatives, such as the greenhouses in Tunis, demonstrate the potential to bridge these gaps.

Conclusion

MUA demonstrates more complex networks of movement, beyond the well-documented and funded export of European knowledge. Urban agriculture contests modern industrial agricultural practices and advances a model of 'civic agriculture' and a close relationship between the gardener/farmer/labour, and

Figure 13.5 Ethiopian immigrants' community garden in Rishon-LeZion, Israel.
Source: Tal Alon-Mozes

produce (Delind, 2002). To a certain extent, urban agriculture aims towards a sustainable future by the use of traditional ecological practices, levels of intensity, and cultural diversity.

Thus, despite the dichotomy between the northern and the southern Mediterranean, MUA emphasises the numerous points of exchange, where a cross-pollination of ideas, practices, and people is taking place, blurring the binary of developing and developed worlds. This spatial practice that migrates in both directions and is constantly being reshaped and transformed in the movement is another challenge to the dominant idea of the developed world as leading progress and the developing world as following it.

Key texts

Bell, S., Fox-Kämper, R., Keshavarz, N., Benson, M., Caputo, S., Noori, S., & Voigt, A. (eds.) (2016) *Urban allotment gardens in Europe*. New York, NY: Routledge.

Lohrberg, F., Licka L., & Scazzosi, L., & Timpe, A. (2016) *Urban agriculture Europe*. Berlin: JOVIS Publishers.

Nasr, J., & Kaldjian P. (1997) Agriculture in Middle Eastern cities: Commonalities and contrasts. *Arid Lands Newsletter*, 42. https://ag.arizona.edu/OALS/ALN/aln42/nasr.html (accessed 16 June 2017).

Smit, J., Nasr, J., & Ratta, A. (2001). Urban agriculture yesterday and today. *Urban Agriculture Food, Jobs and Sustainable Cities*, 1–37. www.jacsmit.com/book/Chap02.pdf (accessed 16 June 2017).

Further reading

Airriess, C.A., & Clawson, D.L. (1994) Vietnamese market gardens in New Orleans. *Geographical Review*, 84(1), 16–31.

Alon-Mozes, T. (2014) Urban agriculture in contemporary Israel: Place, community, and values: Two case studies. In: Roggema, R., & Keeffe, G. (eds.), *AESOP 6: Finding Space for PRoductive cities: Proceedings of AESOP 6, 5–7 July 2014, Leeuwarden, Netherlands*, pp. 961–978.

ANERA (2013) Rooftop gardening in Nahr El Bared Refugee Camp, Lebanon. ANERA: Improving Life in the Middle East. www.anera.org/stories/rooftop-gardening-in-nahr-el-bared-refugee-camp-lebanon/ (accessed 16 June 2017).

Anthopoulou, T. (2016) Agros in Hellinikon, greater Athens area, Greece. In: Bell, S., Fox-Kämper, R., Keshavarz, N., Benson, M., Caputo, S., Noori, S., & Voigt, A. (eds.), *Urban allotment gardens in Europe*. New York: Routledge, pp. 67–68.

Anthopoulou, T. (2016) Claiming the right to public space and neighborhood stabilization in Maroussi, Greece. In: Bell, S., Fox-Kämper, R., Keshavarz, N., Benson, M., Caputo, S., Noori, S., & Voigt, A. (eds.), *Urban allotment gardens in Europe*. New York: Routledge, pp. 96–97.

Antonelli, A., & Lamberti, L. (2011) Multifunctionality of periurban agriculture in Italy. *Watch Letter of the International Centre for Advanced Agronomic Studies (CIHEAM)*, 18, 7–10.

Bartlett, E. (2010) Gaza: The roof is now the field. Uruknet.info. www.uruknet.info/?p=72459 (accessed 16 June 2017).

Braudel, F. (1996) *The mediterranean and the mediterranean world in the age of Philip II Volume I*. Berkeley, CA: University of California Press.

Buzaurd, J. (2016) Development of urban agriculture projects: A tool to foster livelihoods and social inclusion of Syrian refugees in Turkey. *M4D COMMUNITY* website. www.migration4development.org/en/M4D/knowledge/blogs/development-urban-agriculture-projects-tool-foster-livelihoods-and-social (accessed 16 June 2017).

Cavallo, A., Di Donato, B., Guadagno, R., & Marino, D. (2014) The agriculture in Mediterranean urban phenomenon: Rome foodscapes as an infrastructure. In Roggema, R., & Keeffe, G. (eds.), *AESOP 6: Finding space for productive cities: Proceedings of AESOP 6, 5–7 July 2014, Leeuwarden, Netherlands*, pp. 213–230.

Corlett, J.L., Dean, E.A., & Grivetti, L.E. (2003) Hmong gardens: Botanical diversity in an urban setting. *Economic Botany*, 57(3), 365–379.

Danner, P. (1997) Megara, Megara Hyblaea and Selinus: The relationship between the town planning of a mother city, a colony and a sub colony in archaic period. In: Andersen, H.D. (ed.), *Urbanization in the Mediterranean in the 9th to 6th Centuries BC*. Copenhagen: Museum Tusculanum Press, pp. 143–166.

DeLind, L. (2002) Place, work, and civic agriculture: Common fields for cultivation. *Agriculture and Human Values*, 19, 217–224.

Eizenberg, E. (2010) Report on the Ethiopian immigrants farm in Bat-Yam. Scientific report written for the JDC, Jerusalem, Israel.

Eizenberg, E. (2016) *From the ground up: Community gardens in New York City and the politics of spatial transformation*. New York and London: Routledge.

Eizenberg, E., & Fenster, T. (2015) Reframing urban controlled spaces: Community gardens in Jerusalem and Tel Aviv-Jaffa. *ACME*, 14(4), 1132–1160.

Eizenberg, E., Tappert, S., Thomas, N., & Zilans, A. (2016) Political-economic urban restructuring. In: Bell, S., Fox-Kämper, R., Keshavarz, N., Benson, M., Caputo, S., Noori, S., & Voigt, A. (eds.), *Urban allotment gardens in Europe*. New York and London: Routledge, pp. 94–116.

Giseke, U. (ed.) (2015) *Urban agriculture for growing city regions: Connecting urban-rural spheres in Casablanca.* New York: Routledge.

Golonu, B. (2013) Istanbul's urban farms in Peril. *Mother Farmer.* http://modernfarmer.com/2013/08/istanbuls-disappearing-urban-farms (accessed 16 June 2017).

Gopher, A., & Abbo, S. (2016) *Plant domestication and the origins of agriculture in the Near East.* Tel Aviv: Resling Publishing (Hebrew).

Hamadeh, S., Tawk, S., & Abi Said, M. (2014) Urban agriculture and food security in the Middle Eastern context: A case study from Lebanon and Jordan. In: Babar, Z., & Mirgani, S. (eds.), *Food security in the Middle East.* Oxford: Oxford University Press.

Hamilton, A.J., Burry, K., Mok, H.F., Barker, F., Grove, J.R., & Williamson, V.G. (2014) Give peas a chance? Urban agriculture in developing countries. A review. *Agronomy for Sustainable Development*, 34(1), 45–73.

Harrigan, J. (2014) An economic analysis of national food sovereignty policies in the Middle East: The case of Lebanon and Jordan. In: Babar, Z., & Mirgani, S. (eds.), *Food security in the Middle East.* Oxford: Oxford University Press.

Hervieu, B. (ed.) (2008) *Medi-terra: The future of agriculture and food in Mediterranean countries.* Paris: Presses de la Fondation Nationale des Sciences Politiques (OFCE).

Hochberg, G.Z. (2011) The Mediterranean option: On the politics of regional affiliation in current Israeli cultural imagination. *Journal of Levantine Studies*, 1, 41–65.

Jashemski, W.F. (1974) The discovery of a market-garden orchard at Pompeii: The garden of the "House of the Ship Europa". *American Journal of Archaeology*, 78(4), 391–404. Available from: www.jstor.org/stable/502753?seq=1#page_scan_tab_contents [Accessed 7 June 2017].

Keshavarz, N., & Bell, S. (2016) A history of urban gardens in Europe. In: Bell, S., Fox-Kämper, R., Keshavarz, N., Benson, M., Caputo, S., Noori, S., & Voigt, A. (eds.), *Urban allotment gardens in Europe.* New York: Routledge, pp. 8–32. Available from: www.google.com/books?hl=en&lr=&id=_RO4CwAAQBAJ&oi=fnd&pg=PA8&dq=%22a+history+of+urban+gardens+in%22&ots=9ciCCWEA9t&sig=zbIsH3vXw3Pi2nL827uDOLSuwEU [Accessed 16 June 2017].

Kripke, G. (2015) Can an urban farm help solve the refugee crisis? *Oxfarm America.* http://politicsofpoverty.oxfamamerica.org/2015/11/can-an-urban-farm-help-solve-the-refugee-crisis (accessed 16 June 2017).

Laeremans, L., & Sourani, A. (2005) Urban agriculture in the Gaza Strip, Palestine. *Urban Agriculture Magazine: Multiple Functions of Urban Agriculture*, 15.

Lakrouf, R. (2012) Algeria. In Thomas, G. (ed.), *Growing greener cities in Africa: First status report on urban and peri-urban horticulture in Africa.* Food and Agriculture Organization of the United Nations, pp. 35–38. www.fao.org/docrep/016/i3002e/i3002e.pdf (accessed 16 June 2017).

Lohrberg, F., Licka, L., & Scazzosi, L., & Timpe, A. (2016) *Urban agriculture Europe.* Berlin: JOVIS Publishers, pp. 16–17.

Longari, M. *Urban agriculture in the Gaza Strip through vertical gardens and aquaponics.* FAO report: Food and Agriculture Organization of the United Nations. www.fao.org/fileadmin/templates/FCIT/PDF/Fact_sheet_on_aquaponics_Final.pdf (accessed 16 June 2017).

Mendelson, D. (2002) The idea of the Mediterranean in early nineteenth-century French literature. *Mediterranean Historical Review*, 17(1), 25–48.

Mok, H.F., Williamson, V.G., Grove, J.R., Burry, K.S., Barker, F., & Hamilton, A.J. (2014) Strawberry fields forever? Urban agriculture in developed countries: a review. *Agronomy for Sustainable Development*, 34(1), 21–43.

Morgan, K. (2015) Nourishing the city: The rise of the urban food question in the global North. *Urban Studies*, 52(8), 1379–1394.

Mougeot, L.J. (ed.) (2010) *Agropolis: "The social, political and environmental dimensions of urban agriculture".* London and New York: Routledge.

Naïli, M. (2011) Feeding cities in Algeria: A multidimensional challenge. *Watch Letter of the International Centre for Advanced Agronomic Studies (CIHEAM)*, 18, 11–13.

Nasr, J., & Kaldjian, P. (1997) Agriculture in Middle Eastern cities: Commonalities and contrasts. *Arid Lands Newsletter, 42.* https://ag.arizona.edu/OALS/ALN/aln42/nasr.html (accessed 16 June 2017).

Niola, M. (2006) The invention of the Mediterranean. In: Ascari, M., & Corrado, A. (eds.), *Sites of exchange: European crossroads and faultiness.* Amsterdam and New York: Editions Rodopi BV, pp. 76–96.

The Observer (2016) Istanbul's farmers fight to keep historic urban agriculture. *The Observer.* http://observers.france24.com/en/20160205-istanbul-farmers-historic-urban-agriculture (accessed 16 June 2017).

Orsini, F., Kahane, R., Nono-Womdim, R., & Gianquinto, G. (2013) Urban agriculture in the developing world: A review. *Agronomy for Sustainable Development*, 33(4), 695–720.

Quinn, A. (2013) *Bostans: Istanbul's urban gardens. Growing culture*. www.agrowingculture.org/bostans-istanbuls-urban-gardens-draft/ (accessed 16 June 2017).

Rejeb, H. (2011) Urban and periurban agricultural issues in Tunisia. *Watch Letter of the International Centre for Advanced Agronomic Studies (CIHEAM)*, 18, 4–7.

Rejeb, H., Hammami, S., & Sayari, N. (2012) Tunisia. In Thomas, G. (ed.), *Growing greener cities in Africa: First status report on urban and peri-urban horticulture in Africa*. Food and Agriculture Organization of the United Nations, pp. 83–85. www.fao.org/docrep/016/i3002e/i3002e.pdf [accessed 16 June 2017).

Scheromm, P., Perrin, C., & Soulard, C.T. (2012) A dual trend for urban agriculture: The case of Montpellier. In AgUrb 2012: Agriculture in an urbanizing society, 1–4 April 2012, Wageningen, The Netherland. http://prodinra.inra.fr/record/176976 (accessed 16 June 2017).

Smit, J., Nasr, J., & Ratta, A. (2001) Urban agriculture yesterday and today. *Urban Agriculture Food, Jobs and Sustainable Cities*, 1–37. www.jacsmit.com/book/Chap02.pdf (accessed 16 June 2017).

Tóth, A., & Feriancova, L. (2013) *Urban agriculture Barcelona: Space, use, perception. Conference: Zborník príspevkov zo študentskej vedeckej konferencie FZKI 2013, at Nitra: SPU*, pp. 253–259, 267 p., ISBN: 978-80-552-1039. www.academia.edu/6731238/Urban_Agriculture_Barcelona_Space_Use_Perception (accessed 16 June 2017).

Tsiambaos, K. (2016) Condominium in central Athens, Greece. In: Bell, S., Fox-Kämper, R., Keshavarz, N., Benson, M., Caputo, S., Noori, S., & Voigt, A. (eds.), *Urban allotment gardens in Europe*. New York: Routledge, p. 240.

Zezza, A., & Tasciotti, L. (2010) Urban agriculture, poverty, and food security: Empirical evidence from a sample of developing countries. *Food Policy*, 35(4), 265–273.

Zurayk, R., Gough, A., Sourani, A., & Al Jaajaa, M. (2012) *Food security challenges and innovation: The case of Gaza*. High-level Expert Forum, FAO. www.fao.org/fileadmin/templates/cfs_high_level_forum/documents/FS_Challenges_Innovation_-Gaza_Arab_Group.pdf (accessed 16 June 2017).

14

Peri-urban food production as means towards urban food security and increased urban resilience

E. Gunilla A. Olsson

Introduction to urban and peri-urban food challenges

Urban food security and its relation to urban vulnerability and resilience

Cities all over the world share a number of challenges in their aspiration to build sustainable and viable livelihoods for their inhabitants. There is no official, generally accepted, definition of 'sustainable city' but very often, transportation systems, energy and water supply, waste management, buildings and architecture, green spaces, health and recreation, and adaptation to climate change are included (see for example, The London Sustainable Development Commission (n.d.) and Jepson Jr. and Edwards (2010)). Food supply for the urban population is almost always taken for granted in the examination of sustainable cities, and the issue of *urban food security* is strangely absent in the discussion of urban risks and vulnerability (see Chapters 4, 12, 15 and 16). Food security means that citizens should "have physical, social and economic access to sufficient, safe and nutritious food which meets their dietary needs and food preferences for an active and healthy life" (FAO, 2008; see Chapters 20 and 25). Food security is directly related to planning for food provisioning (Chapter 26), which, in turn, is linked to the different steps in food systems: production, processing, distribution, marketing, selling (Ericksen, 2008; Misselhorn et al., 2012). Today, the food consumed in urban as well as rural regions is to a large extent imported from other, often distant, parts of the world (Porter et al., 2014). European food production has become part of the global agro-food market as well as part of a global food system. The functioning and efficiency of the global food system requires fulfilment of a number of preconditions such as access to land, resources, and input products for agriculture and food production; effective and reliable transportation and accessibility along communication routes in air, water, land, and cyberspace; access to energy for transport, production, processing and distribution; absence of disturbances from political and military conflicts or terror attacks; absence of environmental disturbances like climate change effects (Sage, 2013; Porter et al., 2014; Dyball, 2015; Morgan, 2015; also see Chapters 16 and 18). A lack of local control and substantial uncertainty in all the previously mentioned variables not only threaten food security in general but also urban food security (LA and GLA, 2010; Kirwan and Maye, 2013; Dyball, 2015).

Many countries, such as Sweden, have emergency preparedness programmes in place for disturbances in energy supply and for flooding, with plans for evacuation routes (LSU and SDMI, n.d.). However, no emergency preparedness or strategies exist for interruptions in delivery and distribution of food to consumers in the city and for disturbances of inputs to Swedish agriculture. Additionally, the national storage of food for emergency situations for citizens has been abandoned, in Sweden in 1995 at the entry into the European Union, with the argument of its not being cost-effective.

The issue of food security is neglected or under-prioritised in community resilience planning, and this becomes particularly explicit in urban regions where the majority of the human population lives today (Chapters 16, 20, 25 and 31). This situation contributes to increased vulnerability and can be seen as an important national security issue (Morgan, 2015). The vulnerability of a city is counterbalanced by the resilience conditions of the same city. Resilience generally refers to the capacity of a system to react to disturbances by restructuring in such a way that its capacity for maintaining key functional processes is upheld (Folke, 2006; Gallopin, 2006; Scott, 2013; Chapter 16). Resilience implies transformative capacity and active capacity building (Walker et al., 2004; Darnhofer et al., 2014). *Urban resilience* was recently elaborated in an extensive literature review (Meerow et al., 2016), where the following definition was proposed:

> Urban resilience refers to the ability of an urban system and all its constituent socio-ecological and socio-technical networks across temporal and spatial scales – to maintain or rapidly return to desired functions in the face of a disturbance, to adapt to change, and to quickly transform systems that limit current or future adaptive capacity.
>
> *(ibid.: 45)*

In terms of increasing urban resilience in the food security dimension, it is thus desirable to obtain a larger local influence in food provisioning to increase current low levels of local food provision (Babu and Blom, 2014). Local food provision systems with short chains between producers and consumers provide a higher level of food security and preparedness for disturbances (Blay-Palmer et al., 2013). Such disturbances can be catastrophic weather events such as floods and drought periods, or discontinuities in supply of inputs to agriculture from distant sources (for example fossil fuels/petroleum/distillate, fertilisers, animal feed). By having access to a diversity of crops and food products adapted to local conditions and seasons – instead of being dependent on the global food chains – and by direct consumer influence on food production, options are increased for adaptation and resilient responses to possible changes (ibid.).

Accomplishing urban food security will not mean achieving complete local self-sufficiency, however it nonetheless requires large areas of suitable land that can partly be found in peri-urban areas, as in Gothenburg (Olsson et al., 2015; Olsson et al., 2016). Peri-urban food production connected to an urban market can potentially link the urban and rural regions and can counteract ongoing polarisation between the regions (Marsden, 2013; Chapter 29). Current land-use developments in peri-urban regions, however, demonstrate different trends (also see Chapter 15).

Dependence on the global food system is so great in western Europe that previous peri-urban food production and agricultural landscapes for local populations have been substantially substituted with leisure and recreational demands such as horseback riding activities, golf, and other sports (Piorr et al., 2011). An 'equinised' peri-urban landscape: dominated by horse-related activities such as horse grazing, cultivation of horse feed on arable fields, training and competition arenas, and tracks for horseback riding, can be seen in many urban regions of western cities, for example, Malmö (Elgåker et al., 2010), Berlin (Zasada et al., 2013), Copenhagen (Zasada et al., 2011), and Gothenburg (Olsson et al., 2015; Olsson and Olsson, 2016).

Urban sustainability and food justice

The challenge of obtaining food security for an increasing human population, especially in urban regions, is sometimes met by arguments stating the need for more efficient and more intensive as well as high-tech food production (Godfray et al., 2010; Tester and Langridge, 2011). This requires well-functioning systems for operation and security, which can be expected to develop only in limited number of global regions

(Kuyper and Struik, 2014). Such development would increase dependence on disturbance-free global transportation and logistic systems and on global food systems (Morgan and Sonnino, 2010; Chapter 16). *Food justice* (Chapter 32) is defined as "the benefits and risks of where, what and how food is grown and produced, transported and distributed, and accessed and eaten, are shared fairly" (Gottlieb and Joshi, 2010: 141). Food justice relates directly to social inequalities and their link to structural inequalities (Kolb, 2015) and various forms of social, cultural, economic, and spatial exclusion (Hope and Agyeman, 2011; Hochedez and Le Gall, 2016; Chapter 31).

Food security as well as food justice would be enhanced by access to locally produced food, with short supply chains along the preferences of citizens, thereby easing direct dependence on global food systems (Dyball, 2015). This would simultaneously increase green spaces and access to other ecosystem services (beyond food) in urban as well as in peri-urban regions, which is also beneficial for health and well-being (Kabisch and Haase, 2014; Carrus et al., 2015).

The urban *segregation* challenge relates to increasing gaps among groups of citizens along social, economic, and environmental variables, and access to resources (Chapter 32). Such segregation phenomena are related in particular to habitation (degree of gentrification), degree of cultural and economic homogeneity, access to social and cultural services, green spaces (gardens and other green spaces), and to recreation opportunities (Durand and Jaglin, 2012; Kolb, 2015). These unequal distribution phenomena converge with issues of unequal capacity for participation in communication with public authorities and in community organised activities for the transformation of living conditions. Such inequalities may lead to increasing mistrust between groups and increased social unrest as observed in segregated areas, as evident in Gothenburg (Abrahamsson, 2015). High unemployment among underprivileged groups and gender inequalities (OECD/EU, 2015; Brunello and De Paola, 2017) are worsening the situation by accelerating segregation and contributing to a vicious circle of declining social sustainability of the city. Engaging immigrants in local food production and preparation is viewed as a potential mechanism for cultural integration, identity, sense of place, and belonging (Harris et al., 2014; Chapter 39). Many immigrants have rural backgrounds and expertise in farming/gardening/cooking but at present are not in the position to use that expertise (see Chapter 22). There is potential for knowledge exchange between different groups and joint knowledge production and transdisciplinary learning in cultivation and other food production activities (Minkoff-Zern, 2014; Sonnino and Hanmer, 2016). Other benefits would be cultural integration and stimulation to participate in various community organised activities that would promote democratic activity and social cohesion (Krasny and Tidball, 2009). Thus, local food production not only contributes to enhanced urban food security and urban food justice but also has potential to positively influence social interactions and influence urban resilience (Sonnino and Hanmer, 2016; Chapters 22 and 39).

The global context of urban and peri-urban food production

The vulnerability of urban food security and need for local food production in urban regions are hardly new insights. There are numerous examples from European cities during periods of trade blockades related to the world wars in the 20th century on extensive food production activities, for example, in London (Zweiniger-Bargielowska, 2000; Cook's Info, n.d.). Such initiatives are not only related to food security but also good examples of *food sovereignty* (Chapter 31). Food sovereignty is "the right of peoples and sovereign states to democratically determine their own agricultural and food policies" (IAASTD, 2008: 5) which resulted from bottom-up initiatives by urban citizens to control and influence their own food security. A sign of the growing awareness of the urgency of food security and food sovereignty in the Global North is a recent survey on urban food security in developed countries (Opitz et al., 2016). Following are examples from cities in the industrialised world and their activities on urban food security.

The city of Stockholm Sweden (approximately one million inhabitants) is working on a new regional development plan for the period until 2050. Here, the importance of local and regional food production is emphasised – from the angle of strategic security (Stockholms läns landsting, 2015). The connection between current food provision, vulnerability, and links to the global food systems is also highlighted (Carlsson-Kanyama, 2015). The Stockholm region does not possess large agricultural land areas, which makes planning for local food production especially urgent.

The city of Cleveland, Ohio, USA, has approximately 400,000 inhabitants. It was found that if all food, at 100% self sufficiency, was to be produced by intensive urban agriculture in the city (such as roof gardens etc.), inhabitants would only need half of the land required by conventional agriculture to produce a similar amount of food (Grewal and Grewal, 2012). The significantly smaller area needed is explained by the fact that production methods in urban and peri-urban agriculture are generally more intensive, have higher crop diversity, have different harvest periods, employ co-production schemes (Chappell and LaValle, 2011; Björklund et al., 2012), and can include agroforesty systems in peri-urban lands. However, the figures provided on areas required in Grewal and Grewal's study (2012) should be read as estimations to be related to cultivation systems, involvement of animal breeding, and other production factors. In spite of its approximate character, the paper gives a thought provoking insight on possibilities for alternatives to the global food production systems.

Studies on urban food security in Canberra, Australia, (550,000 inhabitants) (Pearson and Dyball, 2014) are based on food production in urban and peri-urban Canberra. The results show that self-sufficiency exists for some food products such as meat (mutton and beef), fruit, and dairy products, but there is also a desire to import a number of food items to satisfy consumer demand for diversity, annual supply, and variation. A prerequisite for urban food security is that cities are directly linked to their peri-urban regions and there is a mutual collaboration in terms of food production, distribution, marketing, and sales (Pearson and Dyball, 2014).

Several cities and urban regions in Canada are hosting vibrant local food system activities run by local communities, as very intentional measures to counteract the global food system and to increase their own food sovereignty and food security (Blay-Palmer et al., 2013). This happens, for example, in Toronto (Campbell and MacRae, 2013) and in Guelph, Ontario (Hayhurst et al., 2013). Local food hubs have also been developed in rural settings in Ontario involving indigenous communities and including a diversity of wild foods from local landscapes in the food system (Stroink and Nelson, 2013). Barriers to development of local food initiatives were studied over a broad range of Ontario's community food projects (Mount et al., 2013). It was found that in spite of factors like location (urban, peri-urban, and rural), regional politics and economics could be barriers. Moreover, there was a large degree of convergence across organisational forms on 'intra-regional collaboration' and 'awareness and education' among consumers. Interpretation of factors, however, varied by rational desire (motivation for the project such as 'community action', 'health food', and 'organic food') rather than by organisational form (Mount et al., 2013).

Development of local food policies and charters is an ongoing process for urban food security, and they are seen in a number of urban environments in developed and developing countries (Hardman and Larkham, 2014; Chapter 12). Food charters are mutual agreements (letters of intent, contracts, etc.) to guide the work among different actors such as enterprises, practitioners, organisations, and authorities on food provision activities. If the charter contains binding signatures, the food charter becomes a symbol for cooperation and augmentation of links between key actors and urban and peri-urban regions in the work for food security (Hardman and Larkham, 2014). Cities like New York, Detroit, London, and particularly Toronto have enhanced urban and peri-urban agriculture through their food policies and charters, with links to the city planning (Hardman and Larkham, 2014).

It is well known that urban and peri-urban agriculture has an important role for the urban population in many developing countries (Chapters 12, 13 and 15). Authors de Zeeuw and Dubbeling (2009) show that local production of urban and peri-urban vegetables has attained 80–100% self-sufficiency in several cities, both in Africa and Asia. Local food production and shorter food chains help to handle risks in urban environments, including food security and hunger as well as respond to cultural needs, identities, and traditions (Lerner and Eakin, 2011).

A case study of peri-urban food production in Gothenburg, western Sweden

The development and current situation in the peri-urban region of Gothenburg, Sweden, will serve as an example, which, to some degree, can be generalised to other west-European cities during the 21st century.

The case of Hisingen in Gothenburg, Sweden

The city of Gothenburg on the Swedish west coast is the second largest in Sweden with 550,000 inhabitants. The urban region is constituted by 13 municipalities, with a population of approximately one million citizens (Figure 14.1). Until the 1950s (when the population was approximately 350,000), much of the consumed food in the city was produced within the region, both from the landscape in situ and from large numbers of greenhouses. The peri-urban northern part of Gothenburg, the Hisingen region, was considered the kitchen garden of the city, and the region was famous for its good quality of vegetables, for example, 'Hisingen potatoes' were highly sought after (Olsson et al., 2015; Olsson and Olsson, 2016). The region was characterised by mixed forms of agriculture with livestock, dairy, and grain production that resulted in shaping a cultural landscape with a large number of habitats and high biological diversity. The origin of this cultural landscape traces back to prehistory, with the present existence of numerous archaeological remains, particularly from the Bronze Age (Nilsson et al., 2014).

The main part of the agricultural land in Hisingen was bought by the city in 1967 with the intention of securing land for urban expansion. Such expansion occurred, but not at the pace expected, and still much

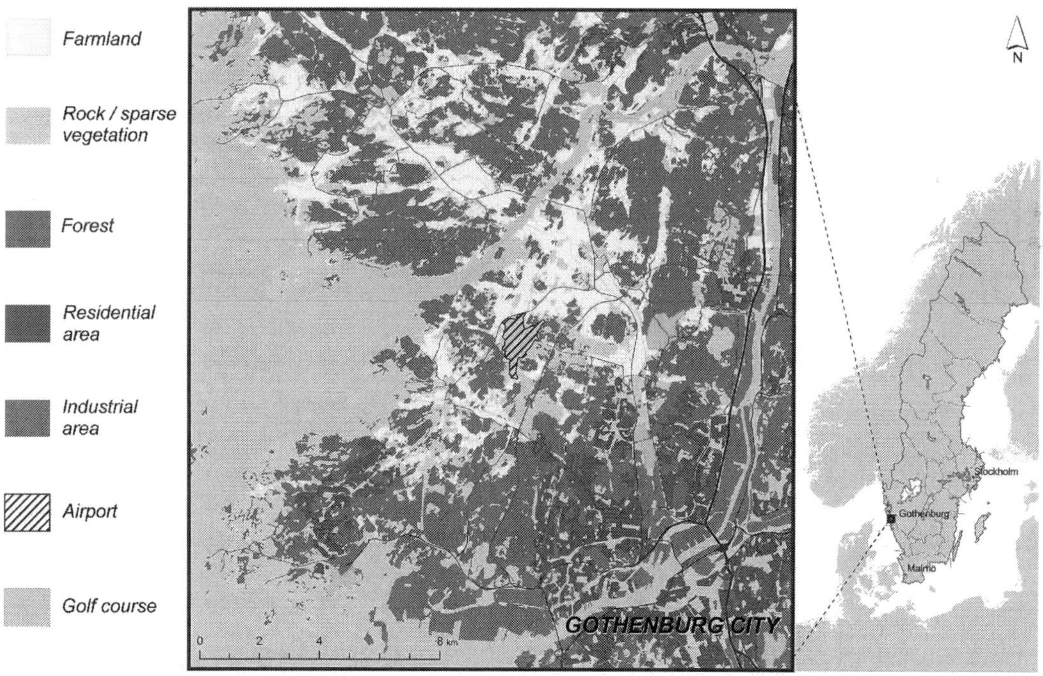

Figure 14.1 Location of the case study area, Hisingen, part of Gothenburg municipality. Outside this map but within 60 km from Gothenburg City are the remaining 12 municipalities within the Gothenburg region.

Source: Adapted from Wästfelt and Zhang, 2016

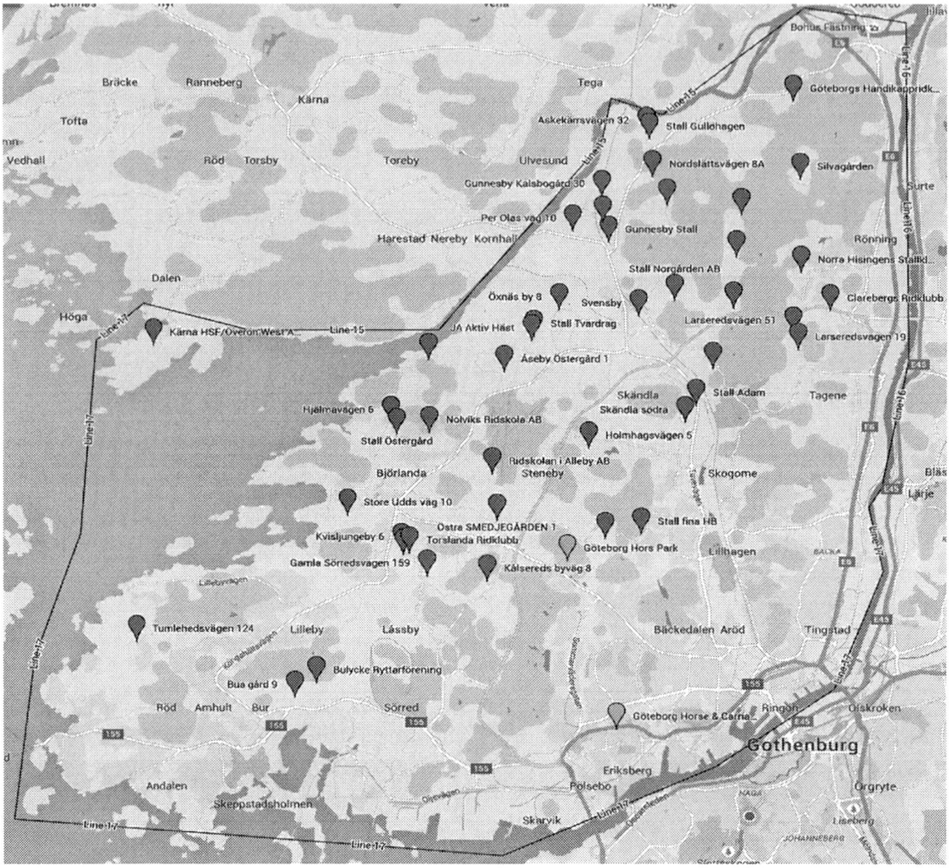

Figure 14.2 Hisingen, part of the Gothenburg municipality, is an equinised peri-urban landscape. Horse farms and stables (black dots) with own homepage, Hisingen 2015.

Source: Olsson et al., 2016

of this land is not exploited by building activities. Most food production, however, has been substituted by leisure related activities for the urban population such as golf courses and other sports (Olsson et al., 2016). The major land use on former agricultural land is now horseback riding (Figure 14.2), which is relevant to the aforementioned term, 'equinised peri-urban landscape' (Olsson et al., 2015). The short-term leasing contracts of the land favours horse farms, since they are often combined with other economic activities, in contrast to food production farms that demand full time engagement. The arable land is either used for feed production for the horses or for horse grazing only. The implications of this land use can be summarised as:

- Food production for humans is substituted by feed and hay production for horses.
- Horses graze on arable fields only, not on the outlands and natural grasslands that are overgrowing and turning into shrub lands (Olsson et al., 2016). This results in a polarised landscape with negative consequences for biological diversity and impaired access for visitors.
- Increased vehicle traffic by horse owners.

The total land use change of the peri-urban agricultural land in Hisingen is shown in Figure 14.3. This development reflects the equinisation trend and the abandonment of food production for the city. It

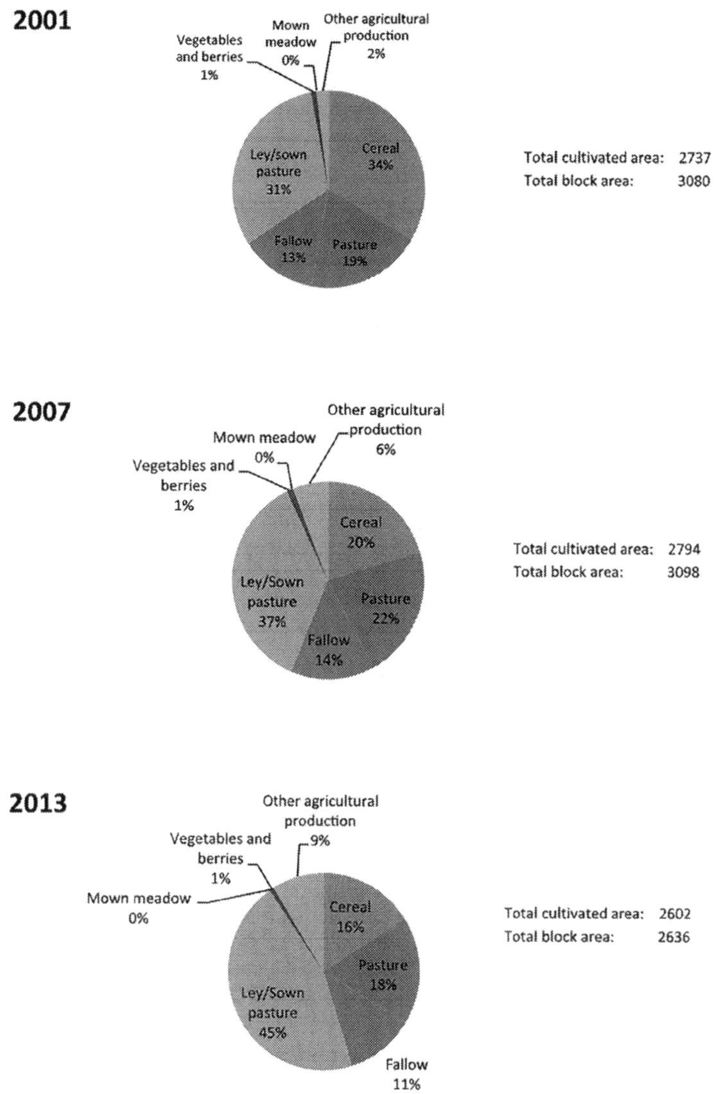

Figure 14.3 Agricultural land use in Hisingen, 2001–2013.
Source: Olsson et al., 2015

includes a decrease in agricultural land with grain crop production where grain is either sold outside the region or used as animal feed in the region (Olsson et al., 2015). There is an increase in grass production on arable fields due to the demand for horse feed and also increases in permanent pasture area on former arable land (Olsson et al., 2015). Currently in Hisingen, there are two cattle farms for meat production, of which one provides direct local sale. Both farms utilise free-range grass-fed meat production, where livestock graze both on arable land and on semi-natural grasslands in the former outlands (see Chapter 2) and this practice maintains the traditional coastal cultural landscape. Interestingly, much of the land areas used for this meat production are protected areas, nature reserves, and Natura2000 areas. The livestock perform important labour in the management of protected areas, which here are composed of habitats in the cultural landscape shaped by earlier agricultural activities such as livestock grazing or mowing. This combination of management of protected areas and livestock raising provides a number of ecosystem services such as meat for human food consumption, biological and cultural diversity from the habitats and landscape, and recreational and leisure services for visitors to those landscapes (Ahlm, 2015; Olsson et al., 2015).

There is only one local enterprise in Hisingen producing vegetables for direct sale. Of the vegetables consumed in the city today, only 10% are produced within the region of the 13 municipalities (Olsson and Olsson, 2016); for other food products, the share of regional production is currently unknown. An interesting comparison of the attractiveness to consumers of local production and sale is shown from the municipality of Kungälv, a peri-urban area close to Gothenburg. Here are two large food reseller chains that offer locally produced and organic certified milk and yoghurt to consumers who bring their own glass containers for their purchases – and the activity is economically beneficial. This successful activity is an indicator of largely untapped potential in locally produced food products and opportunities for other similar initiatives.

Urban agriculture in Gothenburg

Today, there are numerous urban agriculture activities in local urban food production in Gothenburg. Such activities and initiatives include, for example, community gardens, cultivation on temporarily free spaces like demolition sites and even production forms that combine vegetable and berry cultivation with pig farming (see Stadsjord, n.d.). Some activities are driven by the city and some are self-organised; several also include food production in peri-urban areas (see Box 14.1). Other activities are intentionally organised to engage and mix different groups of the urban population with the intention of countering segregation, to increase social cohesion and move towards sustainability (Melander, 2013). These activities are important in both stimulating and maintaining urban citizens' interest in influencing their own food availability and nutrition, and in using food issues as tools for aspirations of social integration (Abrahamsson, 2015). This is another untapped potential for local food production- increasing social sustainability.

. . . and visions for the future?

The plans for future land use in Hisingen, Gothenburg can be seen in Figure 14.4. Some land will be used for further urban expansion but particularly interesting is the agricultural land, now used for agriculture and horse farms with the weakest protection. Those areas are reserved for agriculture for the next 5–20 years, at maximum up to 2034 (Municipality of Gothenburg, 2015) and might be used for building activities long before then. Here exists a visible environmental and sustainability conflict since the city has decided to support local food production, and it would be logical to reserve the agricultural land for food production. At the time of writing (2017), there is ongoing work on a strategy for local sustainable food production for the city within a framework of urban food security and sustainable development, and the municipally owned land will be considered for that purpose. As a scoping study for the local food strategy, the author recently estimated the potential for local food production of vegetables on the land owned by the city in the peri-urban Hisingen (3,100 hectares). It was found that the current consumption of vegetables in Gothenburg could be produced on a fourth (23%) of the available land, applying conventional cultivation forms (Olsson and Olsson, 2016). Furthermore, only a fifth (19%) of the city's land is needed if bio-intensive cultivation forms are applied (Colasanti et al., 2010; McClintock et al., 2013; Olsson and Olsson, 2016). Thus, this illustrates a substantial potential of relatively limited peri-urban land areas for contribution to urban food security. The existence of of agricultural land owned by the municipality and the vast amount of agricultural land within the 13 municipalities – peri-urban areas of the City of Gothenburg – contribute to a unique situation for the Gothenburg urban region in terms of potential for local food production. This is very different from many other European cities like Berlin (Doernberg et al., 2016) and Gent (Kerselaers, 2012).

Box 14.1 Important urban and peri-urban food producing activities in Gothenburg in 2016. Partly adapted from Olsson, M. 2015.

Stadsnära odling (City-Near Agriculture) www.stadsnaraodling.se
Political assignment for Land and Property Office in the City (2011) to support cultivation initiatives for private consumption from groups and individuals on urban and peri-urban cultivation and particularly organic production. During the period 2011–2015, 20 different projects have been performed.

Stadsbruk Göteborg (Urban Agriculture, Gothenburg)
Political assignment for Land and Property Office in the City (2014) to support urban commercial organic food production activities in collaboration with universities in Malmö, Växjö, Kristianstad, SLU Alnarp, and the private agricultural facilitator Xenophilia into a Vinnova research project with the name Stadsbruk Göteborg.

Stadslandet inom Utveckling Nordost (Urban-Rural Land Development North-East) www.utveckling-nordost.se:
A collaborative effort from City of Gothenburg and European Development Fund, run by Utveckling Nordost AB, owned by City of Gothenburg.

The work is performed by local city councils, different city offices (property, nature, traffic offices) and Business Region Göteborg. Included activities are food cultivation and food production and production with integration dimensions.

Andelsjordbruk (Community Supported Agriculture):
Collaboration between City of G-g, the G-g Region and the County of Western Sweden. Aiming to scale up urban and peri-urban food cultivation to include the G-g Region and to create new business models linking producers and consumers and to stimulate Community Supported Agriculture (CSA) (Berg and Andersson, 2015).

Stadsjord (Urban Agriculture-Gothenburg, NGO) www.stadsjord.se
NGO on urban and peri-urban agriculture. The fundamental concept is Cultivation – Food – Culture – Sustainability, and it is built on bottom-up initiatives and includes a strong element of integration of different cultural groups. Included in the work is an obvious aspiration to link urban food production to surrounding rural regions.

Föreningen Göteborgs Koloniträdgårdar (Association of Allotment Gardens, Gothenburg) fgk.dinstudio.se
There is a long historical tradition in Gothenburg of local food production through the garden allotment movement, which has had the utmost importance, not solely for food provisioning, but for collaboration, social cohesion, and social resilience (Ambjörnsson, 2015). The city has one of the oldest garden allotment organisations, which began in 1905 and is still in existence today with 19 local branches, of which some also have animal breeding (chicken, doves, pigs) (FGK, 2016).

What would sustainable and resilient food production for the city of Gothenburg look like? How could the specific preconditions in Gothenburg be used for such goals? A possible vision for sustainable food production based on local prerequisites is briefly sketched out in Box 14.2 (Olsson and Olsson, 2016). It is outlined with some generality to be relevant for other cities and urban regions.

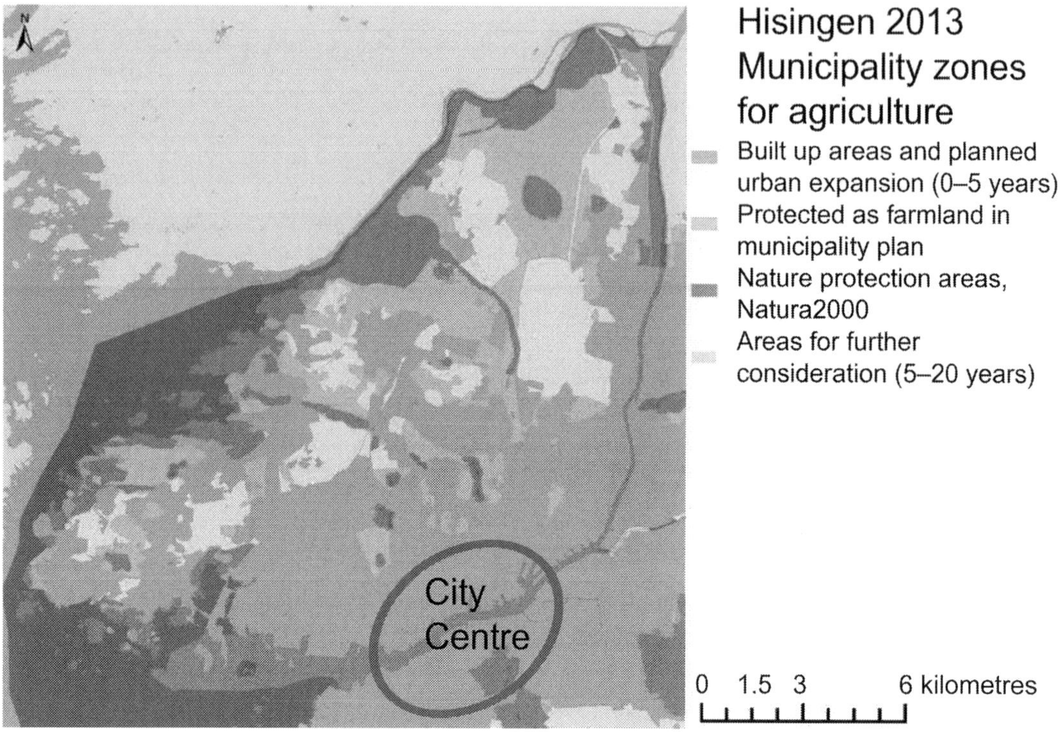

Figure 14.4 Zoning of land use and plan 2013–2034 for Hisingen, peri-urban Gothenburg.
Source: Adapted from municipality of Gothenburg, 2014

Local food production as a vehicle towards increased urban resilience

Which views and values govern societal goals and strategies for food security and sustainable food provisioning policies? In some countries, for example the United Kingdom, the current food security discourse is driven by ideas of 'sustainable intensification' of agriculture, increasing biotechnology and market liberalisation, leaving little room for local food systems influenced by local producers and consumers (Kirwan and Maye, 2013). It has been argued that local food supply chains can provide balance against the global agribusiness paradigm and that local food systems founded on local participation from 'below' also have better possibilities to address the challenges of social and food justice (ibid.; Chapter 32).

Segregated urban environments show connections between socio-economic and settlement variables such as housing quality, space, and access to green spaces (Kabisch and Haase, 2014; Kolb, 2015). This is evidence of yet another positive potential in local food systems of the possibility of active participation by all groups of consumers in different food production activities, which offers prerequisites for both food justice (Gottlieb and Joshi, 2010; Chapter 32) and food sovereignty (Chapter 31). Simultaneously, such joint work and engagement can be sources of empowerment and can function in the process of breaking segregation and social inequality (Levkoe, 2006; Hochedez and Le Gall, 2016).

Participation in food production activities has significant pedagogical function and provides possibilities of distribution of knowledge across cultural borders (Krasny and Tidball, 2009). Further, it contributes to the understanding of connections and limitations between food production, consumption, and dietary patterns and also contributes directly to sustainability efforts and indirectly to urban resilience (Colding and Barthel, 2013). An increase of green spaces in urban and peri-urban areas, resulting from food production activities

Box 14.2 A possible vision of sustainable food production for Gothenburg (from Olsson and Olsson, 2016).

Vision for sustainable and resilient food production in Gothenburg

- The dependence on the global food system is broken by access and availability to locally and regionally produced food products for all inhabitants in. It is available either through local sale or through food stores in the city. Locally produced food makes significant contributions to the food demand but does not cover full needs. Food is also imported from national and European producers. Some niche products are imported from non-European producers.
- Many citizens in Gothenburg are involved in food production activities, directly or indirectly. This is as natural and self-evident as the present habit (2016) to buy food in the food stores on a weekly basis.
- The population in Gothenburg is culturally and ethnically diverse and has different food habits and preferences. This situation is reflected in large diversity of food products and ingredients for different dishes. Most of these food items are produced locally, and there is specialised niche cultivation of spices and vegetables in situ as well as in fossil-free greenhouses.
- Food production and food processing are arenas for knowledge exchange, integration, and social activities linking cultures, groups, and generations.
- Food production and food processing provide new jobs and opportunities for commercial business enterprises.
- Current land use for sport activities (such as horseback riding) on peri-urban land is combined with food production in several different ways, for example, by letting horses graze in the outlands and thus freeing the arable lands for food production for humans. Horse manure is used for arable cultivation, resulting in improved nutrient circulation, soil quality, soil organic matter, etc.
- The protected areas in peri-urban Gothenburg are all part of the agricultural landscape and as such are shaped by food production. Organic food production on arable lands and livestock grazing in outlands will again provide food for the city. This gives synergies by maintaining bio-cultural values in the protected areas, producing food for human consumption and reducing management costs for the reserves and simultaneously making the landscape attractive and accessible to visitors.
- The local food production links the urban and regional regions. Food production is maintained by contracts between peri-urban producers and consumers in the city. A significant share of the provisioning is performed via Community Supported Agriculture (CSA) enterprises. This creates mutual dependence between producers and consumers, which counteracts with the polarisation between urban and rural regions.

also provides biological diversity (Barthel et al., 2013). Other benefits include additional ecosystem services from green spaces such as carbon sequestration, protection against flooding by improved water infiltration through reduction of impermeable areas, and recreation and other cultural services. The sum effect of all those functions and services is a direct contribution to the sustainability of a region (Andersson et al., 2015).

Urban and peri-urban food system activities can be the start of a transition to an agricultural paradigm shift so that food production – again – becomes directly linked to social and environmental sustainability and public health work (McMichael, 2011). Such direction of food production and the local system may have far reaching effects:

> This in a future new world order, revaluing agriculture as the foundation of civilization (in the material, rather than the chronological, sense) is arguably the basis for overcoming ultimately short-sighted neoliberal principles of environmental governance in the current world order.
>
> *(ibid.: 811)*

A suitable development model for peri-urban food production is presented by Marsden (2013) as the Eco-Economic Paradigm, EEP. It is put forward as an alternative to the bio-economic model of agricultural production (ibid.), where the eco-economy encourages small enterprises ('micro-economic behaviour and practices') and involves complex networks of actors, having potential to realign urban and rural regions (Marsden, 2013). Food production activities utilise resources in ecologically efficient ways and aim to be sustainable, for example, by using renewable energy, social enterprises, agro-tourism and recovery of nutrient wastes and wastewater (ibid.). This can provide benefits and additional value in urban and rural regions from ecological, economic, and social perspectives (ibid.). This type of EEP-system is already in function in 12 European countries as well as in several regions in China (ibid.). It is emphasised that EEP systems can be seen as a counterforce against ongoing globalisation of food systems and loss of power and influence of food production (ibid.), as well as erosion of food security. An important dimension of such food systems is that they function as a direct link between urban and rural regions and thus, in contrast to the polarised land use resulting from globalised agro-industrial systems (Marsden and Sonnino, 2012).

Concluding thoughts

In planning for sustainable cities there are many good arguments – as discussed in this chapter – to include spaces for urban and peri-urban food cultivation. Additionally, there is demand for and engagement in urban agriculture from increasing numbers of urban citizens.

The varying activities promoting urban and peri-urban food production and local food systems emerging in different places around the world have originated from people's interest and engagement in influencing access to food based on their own priorities as well as the possibility to influence livelihoods, fuelled by the encouragement of collective actions (Blay-Palmer et al., 2016). In this way, local food system activities can be ways to question and challenge global food systems and global market mechanisms that govern food production and to seek change in institutional contexts, founded on active collaboration between producers and consumers (Horlings and Marsden, 2011; Dyball, 2015). There is an obvious connection between food system activities and political ecology since the power dimension is evident. From a perspective of political ecology, it is also imperative to identify the links between resource use, unsustainable dependence on fossil fuels in agri-food systems, and climate change, as well as the social, environmental, and economic injustices within the existing global food system (Galt, 2013). In such a context, the link between the work for sustainable food systems and a sustainable society becomes apparent.

Further reading

Abrahamsson, H. (2015) Göteborg i världen. Stadens historia förklarar nuet och hållbarheten. *Mistra Urban Futures Report* 2015:6.

Ahlm, E. (2015) *Natura 2000- områden som hållbarhetsresurser: En fallstudie om hur bevarande av kulturlandskap kan integreras med producerande och kulturella ekosystemtjänster.* Globala Studier, Humanekologi, Göteborgs universitet.

Ambjörnsson, R. (2015) *Den hemliga trädgården. Om trädgårdar i litteratur och verklighet.* Albert Bonniers förlag.

Andersson, E., Tengö, M., McPhearson, T., & Kremer, P. (2015) Cultural ecosystem services as a gateway for improving urban sustainability. *Ecosystem Services*, 12, 165–168.

Babu, S.C., & Blom, S. (2014) Capacity development for resilient food systems. *Building Resilience for Food and Nutrition Security*. 2020 Conference Paper 6, May 2014, IFPRI, Washington.

Barthel, S., Crumley, C., & Svedin, U. (2013) Bio-cultural refugia – safeguarding diversity of practices for food security and biodiversity. *Global Environmental Change*, 23, 1142–1152.

Berg, M., & Andersson, J. (2015) Stadsnära matproduktion – hur skapar vi förutsättningar för lokal mat? *Mistra Urban Futures Report* 2015:4.

Björklund, J., Araya, H., Edwards, S., Goncalves, A., Höök, K., Lundberg, J., & Medina, C. (2012) Ecosystem-based agriculture combining production and conservation – a viable way to feed the world in the long term? *Journal of Sustainable Agriculture*, 36, 824–855.

Blay-Palmer, A., Landman, K., Knezevoc, I., Hayhurst, R. (2013) Constructing resilient, transformative communities through sustainable "food hubs". *Local Environment. The International Journal of Justice and Sustainability*, 18(5), 521–528.

Blay-Palmer, A., Sonnino, R. & Custot, J. (2016). A food politics of the possible? Growing sustainable food systems through networks of knowledge. *Agriculture and Human Values* 33:27-43.

Brunello, G. and De Paola, M. (2017). School segregation of immigrants and its effects on educational outcomes in Europe. EENEE Analytical Report No 30. Prepared for the European Commission. Luxembourg: Publications Office of the European Union, 2017.

Campbell, A.M., & MacRae, R. (2013) Local food plus: The connective tissue in local/sustainable supply chain development. *Local Environment*, 18(5), 557–566.

Carlsson-Kanyama, A. (2015) Säkerhet och sårbarhet i framtida Stockholmsregioner. Perspektiv som kan bidra till stärkt hållbarhet. I: Stockholms läns landsting 2015. In: *Sju perspektiv på hållbarhet. Om hur hållbarhetsperspektiv kan stärkas i en ny regional utvecklingsplan för Stockholmsregionen*, Arbetsmaterial, Tillväxt- och regionplaneförvaltningen. Stockholm: Stockholms läns landsting.

Carrus, G., Scopelliti, M., Lafortezza, R., Colangelo, G., Ferrini, F., Salbatano, F., Agrimi, M., Portoghesi, L., Semenzato, P., & Sanesi, G. (2015) Go greener, feel better? The positive effects of biodiversity on the well-being of individuals visiting urban and peri-urban green areas. *Landscape and Urban Planning*, 134, 221–228.

Chappell, M.J., & LaValle, L.A. (2011) Food security and biodiversity: Can we have both? An agro-ecological analysis. *Agriculture and Human Values*, 28, 3–26.

Colasanti, K.J.A., & Hamm, M.W. (2010) Assessing the local food supply capacity of Detroit, Michigan. *Journal of Agriculture, Food Systems and Community Development*, 1(2), 41–58.

Colding, J., & Barthel, S. (2013) The potential of 'Urban Green Commons' in the resilience building of cities. *Ecological Economics*, 86, 156–166.

Cook's Info (n.d.) *British wartime food* website. www.cooksinfo.com/british-wartime-food (accessed 15 December 2016).

Darnhofer, I., de Los Rios, I., Knickel, K., Koopmans, M., Lamine, C., Olsson, G.A., de Roest, K., Rogge, E., Sumane, S., Tisenkops, T. (2014) *Analytical framework for RETHINK project*; RURAGRI, ERA-NET, European Commission: Brussels, Belgium.

de Zeeuw, H., & Dubbeling, M. (2009) *Cities food and agriculture: Challenges and the way forward*. Discussion paper for the Technical Consultation "Agriculture, Food and Cities", 24–25 September 2009, Rome, Food and Agriculture of the United Nations – Food for the Cities. Multi-disciplinary Action (FAO-FCIT) and RUAF Foundation (International Network of Resource centres on Urban Agriculture and Food security).

Doernberg, A., Zasada, I., Bruszewska, K., Skoczowski, B., & Piorr, A. (2016) Potentials and limitations of regional organic food supply: A qualitative analysis of two food chain types in berlin Metropolitan region. *Sustainability*, 8, 1125–1145.

Durand, M., & Jaglin, S. (2012) Inégalités environnementales et écologiques: quelles applications dans les territoires et les services urbains? *Flux*, 89–90, 4–4.

Dyball, R. (2015) From industrial production to biosensitivity: The need for a food system paradigm shift. *Journal of Environmental Studies and Science*, 5, 560–572.

Elgåker, H., Pinzke, S., Lindholm, G., & Nilsson, C. (2010) Horse keeping in urban and Periurban areas: New conditions for physical planning in Sweden. *Danish Journal of Geography*, 110, 81–98.

Ericksen, P. (2008) Conceptualizing food systems for global environmental change research. *Global Environmental Change*, 18, 34–245.

FAO (2008) *An introduction to the basic concepts of food security*. Rome: FAO. www.fao.org/docrep/013/al936e/al936e00.pdf (accessed 1 November 2015).

FGK. (2016) *Föreningen Göteborgs Koloniträdgårdar*. www.fgk.dinstudio.se/news_1.html.

Folke, C. (2006) Resilience: The emergence of a perspective for social – ecological systems analyses. *Global Environmental Change*, 16, 253–267.

Gallopin, G.C. (2006) Linkages between vulnerability, resilience, and adaptive capacity. *Global Environmental Change*, 16, 293–303.

Galt, R.E. (2013) Placing food systems in first world political ecology: A review and research agenda. *Geography Compass*, 7/9, 637–658.
Godfray, H.C.J., Beddington, J.R., Crute, I.R., Haddad, L., Lawrence, D., Muir, J.F.,. . . Toulmin, C. (2010) Food security: The challenge of feeding 9 billion people. *Science*, 327, 812–817.
Göteborgs Stad. (2013) *Göteborgs stads miljöproram* (Environmental plan and program for Gothenburg City). Gothenburg.
Gottlieb, R., & Joshi, A. (2010) *Food justice*. Cambridge, MA: MIT Press.
Grewal, S.S., & Grewal, P.S. (2012) Can cities become self-reliant in food? *Cities*, 29, 1–11.
Hardman, M., & Larkham, P.J. (2014) The rise of a 'food charter': A mechanism to increase urban agriculture. *Land Use Policy*, 39, 400–402.
Harris, N., Rowe Minniss, F., & Somerset, S. (2014) Refugees connecting with a new country through community food gardening. *International Journal of Environmental Research and Public Health*, 11, 9202–9216.
Hayhurst, R.D., Dietrich-O'Connor, F., Hazen, S., & Landman, K. (2013) Community based research for food system policy development in the City of Guelph, Ontario. *Local Environment*, 18(5), 606–619.
Hochedez, C., & Le Gall, J. (2016) Food justice and agriculture. *Spatial Justice*, 9, 1–31.
Hope, A.A., & Agyeman, J. (2011) *Cultivating food justice*. Cambridge, MA: The MIT Press.
Horlings, L.G., & Marsden, T.K. (2011) Towards the real green revolution? Exploring the conceptual dimensions of a new ecological modernisation of agriculture that could 'feed the world'. *Global Environmental Change*, 21, 441–452.
IAASTE (2008) *Agriculture at a crossroads,* Executive Summary of the Synthesis Report. Washington, DC: Island Press.
Jepson Jr, E.J., & Edwards, M.M. (2010) How possible is sustainable urban development? An analysis of planners' perceptions about new urbanism, smart growth and the ecological city. *Planning Practice and Research*, 25, 417–437.
Kabisch, N., & Haase, D. (2014) Green justice or just green? Provision of urban green spaces in Berlin, Germany. *Landscape and Urban Planning*, 122, 129–139.
Kerselears, E. (2012) *Participatory development of a land value assessment tool for agriculture to support rural planning in Flanders*. PhD thesis, Gent University, Gent, Belgium.
Kirwan, J., & Maye, D. (2013) Food security framings within the UK and the integration of local food systems. *Journal of Rural Studies*, 29, 91–100.
Kolb, V. (2015) *Geographical analysis of environmental and ecological inequalities in coastal urban territories*. PhD thesis, University of La Rochelle and CNRS, France.
Krasny, M., & Tidball, K. (2009) Community gardens as context for science, stewardship and advocacy learning. *Cities and the Environment*, 2, 8–18.
Kuyper, T.W., & Struik, P.C. (2014) Epilogue: Global food security, rhetoric, and the sustainable intensification debate. *Current Opinion in Environmental Sustainability*, 8, 71–79.
LA and GLA. London Assembly and Greater London Authority (2010) *Cultivating the capital: Food growing and the planning system in London*. London: Greater London Authority.
Lerner, A.M., & Eakin, H. (2011) An obsolete dichotomy? Rethinking the rural – urban interface in terms of food security and production in the global south. *The Geographical Journal*, 177(4), 311–320.
Levkoe, C.Z. (2006) Learning democracy through food justice movements. *Agriculture and Human Values*, 23, 89–98.
LSU and SDMI (n.d.) *Making cities resilient* website. http://sdmi-resilient-cities.com/ (accessed 1 March 2017).
Marsden, T. (2013) Sustainable place-making for sustainability science: The contested case of agri-food and urban-rural relations. *Sustainability Science*, 8, 213–226.
Marsden, T., & Sonnino, R. (2012) Human health and wellbeing and the sustainability of urban-regional food systems. *Current Opinion in Environmental Sustainability*, 4, 427–430.
Mayor of London and London Assembly (nd) *London Sustainable Development Commission* (LSDC) website. www.london.gov.uk/what-we-do/environment/leadership-and-policy/london-sustainable-development-commission (accessed 23 June 2017).
McClintock, N., Cooper, J., & Khandeshi, S. (2013) Assessing the potential contribution of vacant land to urban vegetable production and consumption in Oakland, California. *Landscape and Urban Planning*, 111, 46–58.
McMichael, M. (2011) Food system sustainability: Questions of environmental governance in the new world (dis)order. *Global Environmental Change*, 21, 804–812.
Meerow, S., Newell, J.P., & Stults, M. (2016) Defining urban resilience: A review. *Landscape and Urban Planning*, 147, 38–49.

Melander, D. (2013) Future Göteborg grows in Northeast. Stadslandet Göteborg. Utveckling Nordosts rapport 2013. www.utvecklingnordost.se (accessed 23 June 2017).

Minkoff-Zern, L.A. (2014) Knowing "Good Food": Immigrant knowledge and the racial politics of farmworker food insecurity. *Antipode*, 46, 1190–1204.

Misselhorn, A., Aggarwal, P., Ericksen, P., Gregory, P., Horn-Phathanothai, L., Ingram, J., & Wiebe, K. (2012) A vision for attaining food security. *Current Opinion in Environmental Sustainability*, 4, 7–17.

Morgan, K. (2015) Nourishing the city: The rise of the urban food question in the global North. *Urban Studies*, 52, 1379–1394.

Morgan, K., & Sonnino, R. (2010) The urban foodscape: World cities and the new food equation. *Cambridge Journal of Regions, Economy and Society*, 3, 209–224.

Mount, P. et al., (2013) Barriers to local food movement: Ontario's community food projects and the capacity for convergence. *Local Environment*, 18(5), 592–605.

Nilsson, H., Nordström, K., & Ragnesten, U. (2014) *Värdefulla odlingslandskap. Natur- och kulturvärden i Göteborgs jordbruksområden*. Göteborgs Stadsmuseum. Kulturmiljörapport 2014:1. Göteborg.

OECD/EU (2015), Indicators of Immigrant Integration 2015: Settling In, OECD Publishing, Paris. http://dx.doi.org/10.1787/9789264234024-en.

Olsson, E.G.A., Bruckmeier, K., & Wästfelt, A. (2015) *Peri-urban agricultural transformations in Gothenburg, Sweden*, RETHINK Case Study Report, February. Gothenburg: School of Global Studies, University of Gothenburg.

Olsson, E.G.A., Kerselaers, E., Kristensen, L.S., Primdahl, J., Rogge, E., & Wästfelt, A. (2016) Peri-urban agriculture and its relation to urban resilience. *Sustainability*, 8, 1340–1361.

Olsson, E.G.A., & Olsson, M. (2016) Matproduktion och urban hållbarhet – fallstudie från Hisingen och Göteborgs framtida möjligheter. *Mistra Urban Futures Papers 2016:1*.

Olsson, M. (2015). Markanvändningen på Hisingen (in Swedish). School of Global Studies. University of Gothenburg, Gothenburg.

Opitz, I., Berges, R., Piorr, A., & Krikser, T. (2016) Contributing to food security in urban areas: Differences between urban agriculture and peri-urban agriculture in the Global North. *Agriculture and Human Values*, 33(2), 341–358.

Pearson, C., & Dyball, R. (2014) City food security. In: Pearson, L., Newton, P., & Roberts, J.P. (eds.), *Resilient sustainable cities: A future*. Florence: Taylor and Francis, pp. 113–122.

Piorr, A., Ravetz, J., & Tosics, I. (eds.) (2011) *Peri-urbanisation in Europe: Towards European policies to sustain urban-rural futures*. Copenhagen: University of Copenhagen; Academic Books Life Sciences.

Porter, J.R., Dyball, R., Dumaresq, D., Deutsch, L., & Matsuda, H. (2014) Feeding capitals: Urban food security and self-provisioning in Canberra, Copenhagen and Tokyo. *Global Food Security*, 3, 1–7.

Sage, C. (2013) The interconnected challenges for food security from a food regimes perspective: Energy, climate and malconsumption. *Journal of Rural Studies*, 29, 71–80.

Satterthwaite, D., McGranahan, G., & Tacoli, C. (2010) Urbanization and its implications for food and farming. *Philosophical Transactions of the Royal Society B*, 365, 2809–2820.

Scott, M. (2013) Resilience: A conceptual lens or rural studies? *Geography Compass*, 7, 597–610.

Sonnino, R., & Hanmer, O. (2016) Beyond food provision: Understanding community growing in the context of food poverty. *Geoforum*, 74, 213–221.

Stadsjord (n.d.) *Stadsjord* website. www.stadsjord.se (accessed 15 December 2016).

Stockholms läns landsting. (2015) *Stärkt hållbarhet I Europas mest attraktiva storstadsregion*. Program för regional utvecklingsplan för Stockholmsregionen, RUFS 2050. Stockholms läns landsting. Stockholm.

Stroink, M.L., & Nelson, C.H. (2013) Complexity and food hubs: Five case studies from Northern Ontario. *Local Environment*, 18(5), 620–635.

Tester, M., & Langridge, P. (2011) Breeding technologies to increase crop production in a changing world. *Science*, 327, 817–819.

Walker, B., Holling, C.S., Carpenter, S.R., & Kinzig, A. (2004) Resilience, adaptability and transformability in social-ecological systems. *Ecology and Society*, 9, 5.

Wästfelt, A. & Zhang, Q. (2016) Reclaiming localisation for revitalising agriculture: A case study of peri-urban agricultural change in Gothenburg, Sweden. *Journal of Rural Studies,* 47, 172–185.

Zasada, I., Berges, R., Hilgendorf, J., & Piorr, A. (2013) Horsekeeping and the peri-urban development in the Berlin metropolitan region. *Journal of Land Use Science*, 8, 199–214.

Zasada, I., Fertner, C., Piorr, A., & Sick Nielsen, T. (2011) Peri-urbanisation and multifunctional adaption of agriculture around Copenhagen. *Geografisk Tidskrift*, 111, 59–72.

Zweiniger-Bargielowska, I. (2000) *Austerity in Britain: Rationing, controls, and consumption 1939–1955*. Oxford and New York: Oxford University Press.

Unpublished source

Municipality of Gothenburg 2015 – planning maps for agricultural land within the municipality.

15

Peri-urban agriculture in Australia

Pressure on the urban fringe

Rachel Carey and Sarah James

Introduction

Australia is highly urbanised, with a large proportion of its population residing in five major state capital cities (Department of Infrastructure and Regional Development, 2015). These cities have some of the highest rates of population growth in the Organisation for Economic Co-operation and Development (OECD) (Department of Infrastructure and Regional Development, 2015) and low rates of urban density with much of their growth occurring on the city fringe (Spencer et al., 2015).

Australia's capital cities have historically had significant areas of peri-urban food production (Bilson, 2007; Houston, 2005), which are important to the nation's vegetable supply, particularly to the supply of highly perishable vegetables (Sinclair et al., 2004; Sinclair, 2015a; Sheridan et al., 2015). As Australia's cities have sprawled to accommodate rapid population growth, this has led to the loss of fertile agricultural land on the urban fringe (Budge, 2013). While the food needs of Australia's cities are increasing in line with population growth, the capacity of their peri-urban fringes to meet food demand is falling (Carey et al., 2016; Institute for Sustainable Futures, 2015).

The peri-urban regions of Australia's cities could play an important role in increasing the resilience of their food supply in the face of growing pressures on food production from climate change, particularly water scarcity (Carey et al., 2015; Buxton and Carey, 2014; Chapters 18 and 19). However, Australia's metropolitan planning strategies have historically been relatively ineffective in realising protection of agricultural land on the urban fringe (Budge, 2013), and sound planning decisions are routinely hampered by development pressure (Freestone, 1992) and a lack of reliable data (James and O'Neill, 2016).

Small-scale farmers on the peri-urban fringe also face threats to their economic viability from a tight 'cost-price' squeeze (James, 2014), exacerbated by high levels of corporate concentration in Australia's food retail sector (Andrée, 2014; ACCC, 2008). Some peri-urban farmers have left farming or plan to (Taylor -et al., 2015), and solutions to address the long-term viability of peri-urban food growing areas around Australia's major cities require new, more integrated approaches.

This chapter describes the significance of Australia's peri-urban areas for food production, and the factors that affect food production around Sydney and Melbourne. It then explores policy approaches for protection of peri-urban food production (also see Chapters 14 and 26) and its role in increasing the resilience and sustainability of the food supply (Chapter 16).

Peri-urban agriculture in Australia

Peri-urban agriculture is production that occurs in the fringe areas of towns and cities, which represent 'transition zones' between urban and rural regions (Opitz et al., 2016). The peri-urban market gardens of Australia's major cities have long been an important source of food, particularly for fresh vegetables. Market gardens spread out around Australia's cities in the 1800s, and have played an important role in feeding city populations (Bilson, 2007; Gaynor, 2006). Fruit and vegetables have typically been grown close to cities worldwide in order to minimise transportation (Steel, 2009). In Australia, the transportation of perishable foods is complicated by significant distances between major cities, and until the mid-1900s, by the availability of refrigerated transport (Cahn, 1977).

The peri-urban regions around Australia's cities are still important for fresh food production, and are among the most highly productive agricultural areas in the country. In the mid-1990s, the peri-urban regions of Australia's five largest mainland states produced almost 25% of the nation's gross agricultural value from less than 3% of the agricultural land (Houston, 2005). Peri-urban areas grew almost half of the perishable vegetables in Australia in 2010–11 (such as broccoli, capsicum, cauliflower, coriander, herbs, lettuce, mushrooms and fresh tomatoes), and also produced a significant amount of poultry meat and eggs (Sinclair, 2015a, 2015b). Emerging research indicates that peri-urban areas produce a wide variety of foods, particularly in the outer hinterland areas, and have the capacity to meet 20–40% of the current food needs of the population in Sydney and Melbourne (Sheridan et al., 2015; Institute for Sustainable Futures, 2015).

Australia is relatively self-sufficient in fresh foods (Edwards et al., 2011; Spencer and Kneebone, 2012), as confirmed by estimates that over 90% of the fresh food consumed is grown in Australia (PMSEIC, 2010). The nation is also a net food exporter (DAFF, 2014), with agricultural exports representing around 20% of the total export value (Hamblin, 2009). Food imports are mostly processed foods, although a small quantity of fresh food is imported (DAFF, 2014). Australia's food supply is, furthermore, generally perceived to be quite secure (see, for example, PMSEIC, 2010). However, the nation's position as a significant food exporter masks underlying vulnerabilities in the domestic food supply, particularly for fruit and vegetables. There are also long-term risks to food supply from climate change and declining supplies of the natural resources that underpin food production (PMSEIC, 2010). Several studies have shown that there are likely to be future deficits in fruit and vegetable supply, due to population growth, urban sprawl and other pressures on the peri-urban fringe (Larsen et al., 2011; Soebels et al., 2011; Sheridan et al., 2015). The following sections explore these pressures.

Population growth, urban sprawl and loss of peri-urban farmland

Almost 80% of Australia's 24 million people live in cities (Department of Infrastructure and Regional Development, 2015), and around two-thirds live in the capital cities of its states and territories (Infrastructure Australia, 2015). Australia also has one of the highest rates of population growth in the OECD – around 1.4% per annum, compared to an OCED average of 0.5% (Department of Infrastructure and Regional Development, 2015). Most of this growth is occurring in Australia's cities (Infrastructure Australia, 2015), increasing the rate of urbanisation (Buxton et al., 2015).

Historically, much of the population growth in Australia has been accommodated in the outer suburbs of cities on the peri-urban fringe. Australia's cities are characterised by a high degree of urban sprawl and relatively low rates of urban density (Spencer et al., 2015). Rapid population growth occurred in the capital cities as a result of migration following the two World Wars (ABS, 1996). Market gardens in state capitals like Sydney and Melbourne were pushed further out as the cities expanded (James, 2016a; Parker and Jarecki, 2003). Rapid growth of the capital cities continued in the decades following the Second World War, putting further pressure on the peri-urban fringe (James, 2016a).

It is unclear how much farmland has been lost on the urban fringe of Australia's state capitals as a result of urban sprawl, or the impact of this loss on peri-urban food production, due to a lack of data (James and

O'Neill, 2016). However, Ramsey and Gallegos (2011) estimate that between 2000 and 2006, the cities of Sydney, Melbourne, Adelaide and Perth experienced decreases in the area of fruit production of around 4–11%, while land available for vegetable production fell by about 28% in Brisbane and 14% in Perth. Agricultural land in six Local Government Areas to the south-east and north-east of Melbourne is of the highest quality in areas close to its Urban Growth Boundary and is particularly at risk from urban development (VEIL, 2013). At the rate of current urban expansion, significant areas of vegetable production are likely to be lost around Sydney and Melbourne (Institute for Sustainable Futures, 2015; Sheridan et al., 2015).

The significance of the loss of fertile land on the peri-urban fringe is increased because only 6% of Australia's land is arable (World Bank, 2016). Moreover, only a small amount of this arable land is high quality and suitable for fruit and vegetable production. Australian soils are ancient, fragile, relatively infertile and lacking in phosphorous and nitrogen (ABS, 2012). Much of the best quality agricultural land and most reliable rainfall is on the coastal fringe, where the capital cities are situated (Budge, 2013; McKeon, 2008). However, these expanding cities are displacing some of the nation's best agricultural soils in areas with reliable precipitation.

The rapid expansion of Australia's capital cities leads to conflict between farmers and new residents (Henderson, 2003; Houston, 2005), and the peri-urban fringe has become an area of intense land-use disputes (Holmes, 2008). New residents may object to the unfamiliar noises, smells and aesthetics of modern agriculture (Steven, 2005), lodging complaints with local or state governments (Henderson, 2003). Farmers in a number of Australian states have called for 'Right to Farm' laws to protect them from litigation arising from agricultural operations, although only the state of Tasmania has enacted such laws (Griffith, 2015).

Economic factors affecting farm viability

In addition to urban encroachment, small-scale farmers on the peri-urban fringe of Australian cities face threats to their economic viability (James, 2016b). Farmers report low returns on products and little annual increase (James, 2016b). Australia's retail sector is highly concentrated, with two corporate conglomerates' supermarket chains (Coles and Woolworths) dominating the market. Therefore farmers are often compelled to accept unfavourable prices and conditions set by retailers (Higgins, 2001; James, 2016a), with claims farmers receive as little as 10% of the final price of produce (Neales, 2014). The market concentration of these giant retailers means that they are increasingly able to control the food supply chain from 'seed to supermarket', imposing strict, cosmetic product specification standards (Burch and Lawrence, 2005). Produce may be left to spoil on the tree or in the field due to the unwillingness of retailers to pay farmers the arranged prices for their goods and the rejection of goods that fail to meet strict cosmetic standards (James, 2016a).

Farmers face increasing costs for farm inputs – such as fertilisers, water, energy and fuel (Millar and Roots, 2012; Lawrence et al., 2013) – without equivalent increases in product prices (Cebon, 2010). The price of peri-urban land also rises with continued urban expansion, making it difficult for new farmers to purchase farmland (Buxton and Taylor, 2011). Furthermore, city fringe farms are often relatively small-scale operations. Sydney's farms under production are estimated to have around 1–4 hectares (Sinclair et al., 2004). There is also little opportunity to increase in size as a means to improve farm viability, due to the cost of peri-urban land and surrounding residential development. In this regard, and in contrast to governments in countries such as Canada (Eagle et al., 2015), there has been little concrete action by Australian state governments to intervene.

Small-scale farmers are in a weak negotiating position in relation to product prices, competing with other growers across Australia as well as cheaper international imports (Sharp, 2009; Hattersley et al., 2013). Increasingly there is pressure to 'get big or get out', with smaller farms being pushed out of the market (Lawrence et al., 2013). If peri-urban farmers don't enter into direct contracts with supermarkets, they may have to rely on more precarious distribution channels, such as selling directly through central city markets or, to a lesser degree, through alternative markets, such as farmers markets, farm-gate sales or community supported agriculture (James, 2016a).

On the peri-urban fringes of Sydney, many farmers who once grew vegetables have turned to more profitable and reliable 'crops', such as turf (Malcolm and Fahd, 2009). Some growers also regard the relatively low returns of peri-urban farming as less viable than selling land for residential development. The lack of planning measures to protect peri-urban farmland (Buxton and Carey, 2014; Chapter 14) and subsequent high land prices may be particularly attractive for farmers who may be struggling financially (James, 2016a). The sub-division of agricultural land is also seen by farmers as a way of achieving secure retirement funds (Llausàs et al., 2016). These circumstances point to the need for 'dual' sustainability criteria (Aubry et al., 2012). Peri-urban farmland can sustain the city only if it is also economically sustainable for the farmers who care for the land withheld from urban development.

Climate change and water scarcity

Australia is forecast to be particularly affected by climate change and is already experiencing its impacts (Reisinger et al., 2014), with adverse effects on agriculture (Hughes et al., 2015). The average temperature and number of extreme temperature days has increased since 1930, and rainfall has decreased over Southern Australia in recent decades (BoM, 2014). These trends are expected to continue to produce higher temperatures, more days of extreme heat, and decreased rainfall across Southern Australia by up to 20% in a low emissions scenario and by up to 30% in a high emissions scenario. The frequency and severity of droughts and storms is also expected to increase (BoM, 2014).

Climate change is expected to reduce agricultural productivity across Australia (Hughes et al., 2015), and to particularly affect production in the nation's most important agricultural region, the Murray-Darling Basin, in the south-east of Australia (Resinger et al., 2014). One of the main ways that climate change is likely to affect food production in Southern Australia is through a reduction in the availability of water (Lawrence et al., 2013), exacerbating pressure on the existing resource. Southern Australia is a region already facing water scarcity (Molden, 2007), where there is significant competition for water from agriculture, urban uses and the restoration of environmental flows in the Murray-Darling Basin (Wahlquist, 2008). Water availability is likely to be the main resource constraint on agricultural production in Australia in future (Hughes et al., 2015).

Water scarcity will also affect food production in peri-urban regions of Southern Australia, which have benefited less from investments in improving the efficiency of water delivery than the Murray-Darling Basin (Sheridan et al., 2016). However, the peri-urban regions around Australia's major cities have the advantage of access to stormwater and recycled wastewater from city water treatment plants, which represent some of the most secure sources of water for food production during prolonged drought (Carey et al., 2015). Southern Australia has some of the most extensive projects delivering recycled water for agriculture in the world, although only a fraction of the available recycled water is currently used, due primarily to a lack of infrastructure to deliver the water to farmers (Sheridan et al., 2016). The pressures on peri-urban production previously described are explored next in case studies of Australia's two largest cities, Sydney and Melbourne.

Sydney

Sydney is Australia's largest city with five million residents and is forecast to reach 8–8.9 million by 2061 (ABS, 2016a). It is located in the south-east of Australia in the state of New South Wales. Sitting in a natural basin created by the Blue Mountains to the west, the Illawarra escarpment to the south and the Hornsby Plateau to the north, Sydney arguably faces the greatest pressure for space of any Australian city (Bunker and Houston, 2003).

Sydney has been fed by peri-urban market gardens from the earliest days of colonisation (Nixon, 1976; Humphreys et al., 2001). Satellite towns were created to support and service farming in areas that now form the peri-urban fringe of Sydney (State Library of NSW, 2012). These areas are still key agricultural areas but are under the greatest threat from urban development, especially the denoted growth areas in Figure 15.1.

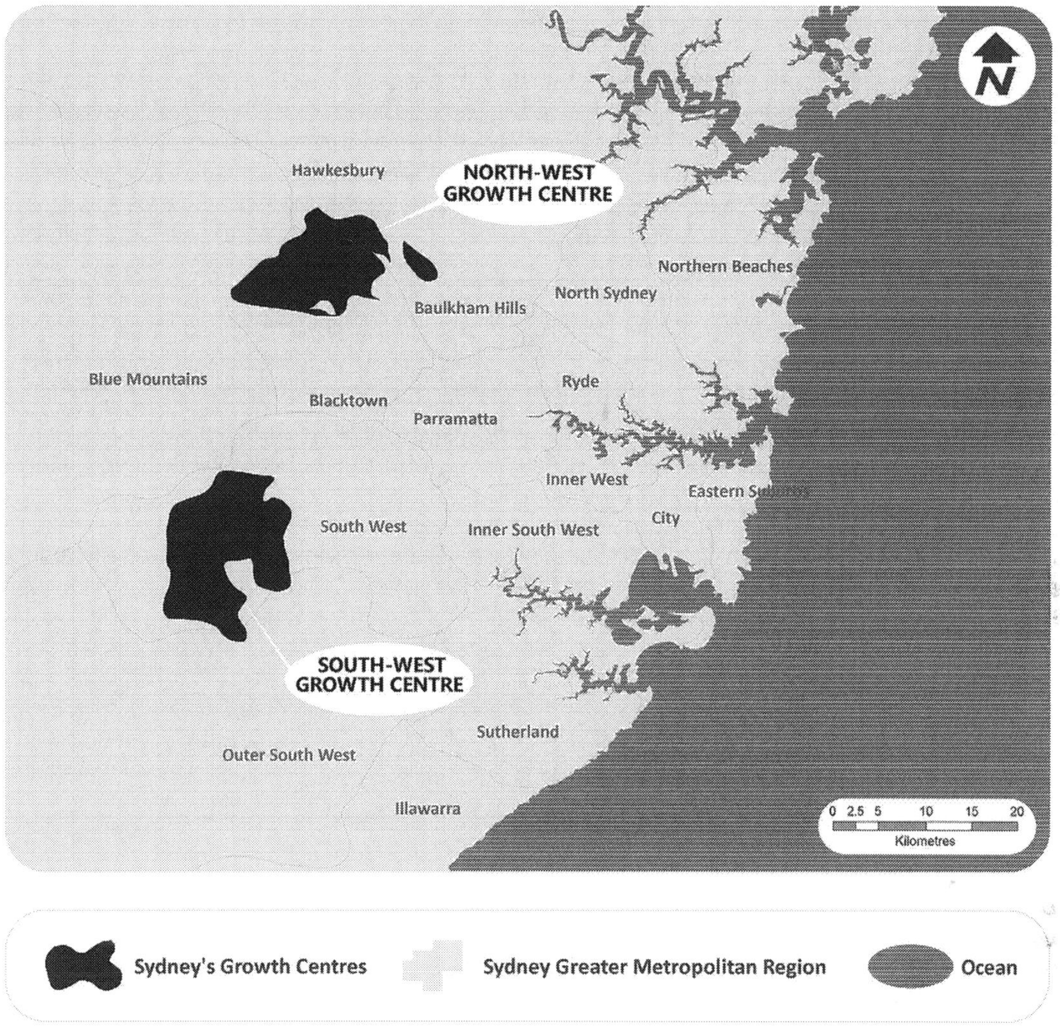

Figure 15.1 Sydney's peri-urban fringe and growth areas.

Sydney's peri-urban farms have the capacity to meet around 20% of the city's total food needs, although it is unclear how much of the food they produce is consumed in the city (Institute for Sustainable Futures, 2015). They primarily supply foods that benefit most from being close to market, namely perishable vegetables and fruits, as well as poultry and eggs. Sydney's production represents a significant amount of the overall state production of these foods, including up to 45% by value of vegetables, as well as 45% of chicken meat and 50% of eggs, depending on the year (ABS, 2010). For some varieties of vegetables, Sydney often produces the majority of the state's production in any given year. In 2008–2009, Sydney produced 99% of the Asian vegetables, 83% of the mushrooms, 58% of the herbs other than parsley, 84% of the cabbages, 62% of the fresh tomatoes and 72% of cucumbers in New South Wales (ABS, 2010).

As a result of on-going development in planned 'Growth Centres', it is probable that Sydney will lose over 50% of all vegetable farms on its peri-urban fringe (Malcolm and Fahd, 2009). Some estimates suggest that planned urban growth on Sydney's fringe could threaten up to 90% of the city's fresh vegetable production, and could reduce the total capacity of Sydney's peri-urban farms to feed the city from 20% to 6% (Institute for Sustainable Futures, 2015).

Drought is a significant issue for farmers in New South Wales (Power, 2015), although it hasn't affected farmers in Sydney to the same extent as those in Melbourne and Adelaide. Farmers in peri-urban Sydney are relatively water secure due to the regular and constant supply of water that the city provides, but are among those most under threat from urban encroachment (James, 2016a). They are also interested in access to recycled urban water for food production to secure their water supply, as is already in place in several urban regions across Australia.

In addition to environmental pressures, socio-cultural factors also play a significant role in the nature of land-use contestation on the urban fringe. Sydney's peri-urban farmers are very culturally diverse, with an estimated 80% of vegetable farmers coming from culturally and linguistically diverse backgrounds (James, 2016a). This has created a further challenge in terms of protection of fringe farmland through planning policies, as these growers are often reluctant to engage with government or oppose changes (Parker and Jarecki, 2003; James, 2016a). Indeed, they have been described as 'Sydney's invisible farmers' (Galvin, 2006), as they have been largely neglected by planners and policy makers in land use planning processes. Despite these challenges, however, these culturally diverse farmers feel that they are making an important contribution to a sustainable future for the city (James, 2016a).

To date, Sydney's land use planning has paid little attention to peri-urban farming and food security. An exception to this was the 1951 Sydney Greenbelt, which sought to protect peri-urban land for agricultural, amenity and environmental purposes. This visionary scheme, however, was slowly destroyed as development consumed proposed greenbelt land, until the greenbelt plan was impossible to implement (Freestone, 1992). Some sections of the greenbelt have been retained (or repurchased) to form what is now known as the 'Western Sydney Parklands' (Gilbert, 2007). In 2012 160 hectares of the parklands were designated for agricultural production, with around 36 hectares under production in 2017 (Western Sydney Parklands Trust, 2017). However, while subsequent metropolitan plans following the creation of the 1951 Greenbelt for Sydney contain statements in support of protecting 'agriculture and resource lands', there has been little evidence of plans or actions to substantively stop development on the urban fringe (James, 2016a).

In the 2005 City of Cities Metropolitan Strategy, however, there was an apparent change in this regard. 'Green Zones' were planned to protect pockets of agricultural and bushland within 'growth areas' on the urban fringe (Parliament of New South Wales, 2006). However, a lack of consultation and attention to concerns about loss of perceived 'development rights' allowed these zones to disappear under a wave of community protest (James, 2014). Among those rejecting the green zones were small-scale farmers, one of the key groups this planning initiative was (ostensibly) intended to protect. The repeal of the green zones indicated a lack of vision on behalf of the Department of Planning and a failure to involve farmers in how to best address the complex challenges of protecting peri-urban farms (James, 2014). Interviews with Sydney farmers revealed that alternative approaches might have garnered more support. Transfer of development rights (TDR), purchase of development rights (PDR) or 'agricultural parks' (where farmland could be leased) were approaches of more interest to farmers (James, 2014). In areas of intense competition for land, these approaches offer a way of moving beyond the dominant farming/housing dichotomy in which privileging housing development leaves no place for farms (James, 2014).

Melbourne

Melbourne, in the state of Victoria, is Australia's second largest city, with a population of around 4.7 million. It is also the nation's fastest growing city (ABS, 2016b), predicted to reach a population of 7–8 million by 2050 (Buxton et al., 2015) and expected to overtake Sydney to become Australia's largest city by 2061 (at between 7.6 and 9.8 million) (ABS, 2016b; Department of Infrastructure and Regional Development, 2015).

Market gardens were established in peri-urban areas of Melbourne in the mid-1800s. The area to the south of the city was established as the main vegetable growing area, while orchards and dairy farming spread out to the east (Cole, 1980). Peri-urban Melbourne (see Figure 15.2) is still a significant region of food production, and currently has the capacity to meet around 41% of the food needs of Greater Melbourne's population, although it is unclear exactly how much of this food is consumed in the city (Sheridan et al., 2015). The peri-urban region is particularly important for fresh vegetable production. It produces almost half of the vegetables grown in the state of Victoria, and grows a high proportion of some types of vegetables, including 92% of cauliflower, 66% of broccoli and 66% of lettuce. The peri-urban region also produces 96% of Victoria's berries, 81% of chicken meat and 67% of eggs (Sheridan et al., 2015). Production in the peri-urban region is scattered over many relatively small areas around the city, and some areas are important for the production of particular types of fruit and vegetables, such as Koo Wee Rup, which produces over 90% of Australia's asparagus (Buxton and Carey, 2014).

Melbourne was largely self-sufficient in vegetable production until the Second World War (Harcourt, 1939), but by the 1950s, Melbourne's market gardens were being pushed further out of the city by urban development (Greenlees, 1951). Displacement of market gardens and fringe farmland accelerated in subsequent decades, and between 1986 and 2001, the area of agricultural land in peri-urban Melbourne decreased by around 18% (Parberry et al., 2008). Around 53,000 hectares has been lost since through multiple expansions of the city's Urban Growth Boundary (Buxton et al., 2016). If Melbourne continues to expand as it has in the past, following the long-term trend of urban density, the capacity of peri-urban Melbourne to meet the city's food needs could fall from 41% in 2015 to around 18% by 2050 (see Figure 15.3). Vegetable production is likely to be particularly affected. The capacity of the

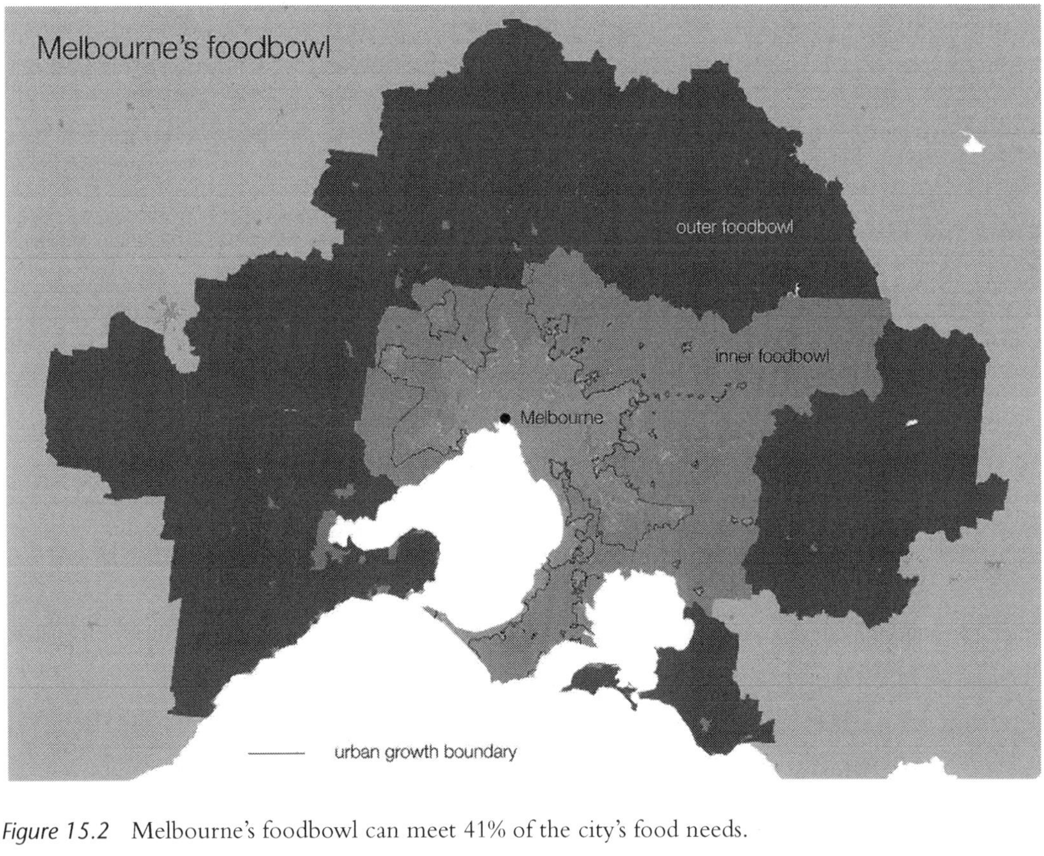

Figure 15.2 Melbourne's foodbowl can meet 41% of the city's food needs.

Figure 15.3 Estimated impact of urban expansion on Melbourne's peri-urban food production.

peri-urban region to meet the city's vegetable needs is estimated to fall from around 82% to 21% by 2050 (Sheridan et al., 2015).

South-east Australia is a water-scarce region (Molden, 2007), and water availability is a major constraint on agricultural production (DELWP, 2016). Peri-urban regions of Melbourne have benefited less from irrigation upgrades and water trading than rural areas of food production. However, Melbourne's peri-urban region has access to recycled water from the city's water treatment plants. Werribee South, one of Melbourne's main vegetable growing areas (responsible for around 10% of the state's vegetable production), is adjacent to the city's Western Treatment Plant, and the district became entirely dependent on recycled water towards the end of Australia's Millennium Drought, which lasted from 1996 to 2010 (Sheridan et al., 2016). Recycled water initiatives have the potential to increase the resilience of Melbourne's food supply in the face of increasing water scarcity (Carey et al., 2015). However, only a small proportion of the available recycled water is currently used to support peri-urban production, and areas of fertile farmland around the city's water treatment plants are among the areas of peri-urban Melbourne most at risk from urban development (Sheridan et al., 2016).

There have been a number of attempts to protect agricultural land around Melbourne through planning legislation since the rapid postwar expansion of the city in the 1950s (McFarland, 2015; Chapter 26). These have included the introduction of legislation in 1971 to create a 'green belt' that protects non-urban areas between the city's growth corridors (known as Melbourne's 'Green Wedges'), the establishment of regional planning authorities for peri-urban regions of particular significance, the introduction of rural planning zones and the creation of a legislated Urban Growth Boundary (Buxton et al., 2016). However, the loss of significant areas of peri-urban agricultural land has continued despite these measures (McFarland, 2015), with several expansions of the Urban Growth Boundary since its introduction in 2002. The 2014 metropolitan planning strategy *Plan Melbourne* proposed the introduction of a new 'high value agricultural food overlay' to protect significant areas of agricultural land (Buxton and Carey, 2014). Yet no commitment has been made to implement this proposal, and in 2017, the Victorian state government announced the development of seventeen new suburbs in designated growth areas on the city fringe (Victorian State Government, 2017). The city continues to lack an effective strategy for addressing protection of peri-urban agricultural land, and there are calls for a broader, more integrated approach (McFarland, 2015; Buxton and Carey, 2014; Buxton et al., 2016).

The future of peri-urban agriculture in Australia's cities: policy and planning

State governments in Australia have primary responsibility for land use planning, and they have introduced a variety of measures over recent decades to protect peri-urban agricultural land, as the Sydney and Melbourne case studies illustrate. These measures have likely slowed the loss of peri-urban agricultural land, but have largely failed to stop urban encroachment (Buxton, 2014). Sydney's Greenbelt and Melbourne's Green Wedges illustrate the need for strong political will to uphold legislation against ongoing pressure for urban development of fringe areas (Freestone, 1992; Buxton and Goodman, 2008). There is also a need to develop a broader social and political consensus on the importance of protecting peri-urban agricultural land (Buxton and Carey, 2014), as peri-urban agriculture has been consistently undervalued in Australia, both in economic terms (Houston, 2005) and for its social and environmental contributions (McFarland, 2015).

Arguments for preserving farmland on the city fringe typically ascribe a range of values to peri-urban farming landscapes, and often emphasise the amenity value of agricultural landscapes for urban dwellers (Brinkley, 2012). The concept of 'greenspace' preservation tends to conflate agricultural and conservation land use values in planning and policy. The 'heritage' value of the rural landscapes of small-scale farmers is also often invoked as a reason for their protection (Lennon, 2015). The values of 'greenspace' conservation, amenity value, and rural heritage have featured strongly in arguments for the protection of peri-urban farming in Australia. Indeed Freestone (1992) argues that in the development of the Sydney Greenbelt, aesthetic (rather than production) values prevailed in the case for protection. In contemporary discourse, the development of agri-tourism is regarded as a means to leverage the aesthetic and heritage values ascribed to these spaces by urban populations in seeking farmland protection and improving farm incomes (Mason and Knowd, 2010; Stapleton, 2008; Chapter 14).

However, the reality of modern agriculture is often at odds with notions of the rural idyll (Bunce, 1994; Bunce, 1998). Many small-scale lots in Sydney and Melbourne use greenhouses or poly-tunnels, and intensive poultry and egg production occurs in large sheds that may clash with the aesthetics (and scents) demanded for amenity and heritage values. The competing values of production versus aesthetics lead to conflict between farmers and residential neighbours in peri-urban Australia (Taylor et al., 2017). Conflict also arises between production and conservation values in the conservation of 'greenspace', and similar conflicts have been noted internationally in attempts to protect peri-urban agriculture using the notion of a rural landscape (Cadieux et al., 2013; Chapter 14).

This conflict between notions of green space and productive landscapes points to the need for a revaluation of peri-urban agricultural land. New challenges to food production from climate change and concerns about the sustainability of city food supplies (Brunori and Di Iacovo, 2014; Larsen et al., 2011) highlight the need to re-value productive peri-urban landscapes beyond amenity, heritage or nature conservation (Cadieux, 2008). A more integrated approach to peri-urban landscapes is needed that brings together environmental and productive values, as has been explored elsewhere in Australia (van Oosterzee et al., 2014; Wynne et al., 2016).

Effective regulation is critical to prevent land fragmentation (such as the subdivision of larger properties into smaller lots), as this fuels land speculation, drives up land prices and increases development pressure (Buxton et al., 2016). Similarly, metropolitan planning strategies must address food security as a key issue (Budge, 2013). There is also a need to address the economic pressures on farmers (Sinclair, 1999; Harman et al., 2015; James, 2014; Harman and Choy, 2011). Farmers on Sydney's fringe have been shifting from food to turf (grass) production, due to greater reliability of farming income (Malcolm and Fahd, 2009). This highlights the need for greater exploration of market-based approaches, such as tradeable development rights, to compliment measures such as land use zoning (Elton Consulting, 2009).

Tradeable development rights enable landowners to sell the development rights on their land to developers who then use those rights at another location. These programs have been widely used in the United States, but are uncommon in Australia (Harman et al., 2015). Peri-urban farmers have indicated that tradeable development rights have the potential to address concerns about long-term economic viability and any potential opposition to planning measures (James, 2014, 2016a). However, it is unclear how tradeable development rights might work in an Australian context, since there are political, institutional and legislative barriers that would need to be overcome (Harman and Choy, 2011).

There is growing recognition of the need for a more integrated approach to protection of peri-urban farmland in Australia (McFarland, 2015; Fielke and Bardsley, 2014; Buxton and Carey, 2014; Buxton et al., 2016). State planning policies in New South Wales and Victoria have become narrowly focused on economic growth since the 1980s, as a conduit for a neoliberal policy agenda (McFarland, 2015). This has de-emphasised other values of peri-urban agricultural land, including its contribution to social well-being, environmental sustainability and food security. However, these broader values of peri-urban agriculture are receiving increasing attention internationally in the face of emerging challenges to the resilience and sustainability of food systems, particularly from the impacts of climate change and population growth (De Schutter, 2014; Dubbeling and de Zeeuw, 2011; Chapter 16).

Australia's cities are geographically isolated and particularly vulnerable to the impacts of climate change, including sudden shocks, such as storms and floods, and chronic stresses, such as drought. The potential impacts of such events were highlighted by the 2010–2011 Brisbane floods (DAFF, 2012) and the Millennium Drought, which saw interruptions to food supply and spikes in fresh food prices (Quiggin, 2007). Peri-urban and urban food growing areas (see Chapters 12, 13, 14 and 22) could contribute to a more resilient food supply that is better able to withstand and recover from sudden shocks and chronic stresses in the context of climate pressures and declining availability of the natural resources that underpin food production (Burton et al., 2013; FAO/RUAF, 2015; Carey et al., 2016). The peri-urban regions of cities have the potential to be some of the most water-secure areas for food production in Southern Australia because of their access to urban wastewater streams, particularly recycled water from city water treatment plants (Carey et al., 2015).

These emerging challenges to the resilience of food systems highlight the need for a comprehensive approach to land use planning, water policy and food security (see Chapter 26). Critical to this is the implementation of planning policies and processes that enable the protection of agricultural areas close to cities (Buxton and Carey, 2014). The newly designated Environment and Food Production Areas in Adelaide is one such example (Government of South Australia, 2015). Stronger statutory planning measures will need to be accompanied by a range of other strategies to strengthen the resilience of city food systems, including measures to address farm viability (Buxton et al., 2016). The development of effective policies will also require access to reliable data, including maps of high value agricultural land around Australia's cities and monitoring loss (or gain) of peri-urban farmland, with existing data gaps needing to be addressed (James and O'Neill, 2016). Adoption of the holistic and integrated approach outlined here is urgently required to ensure the protection and preservation of Australia's productive peri-urban regions in an uncertain future.

Further reading

ABS (1996) *4102.0 Australian social trends, 1996*. Canberra: Australian Bureau of Statistics.

ABS (2010) *Value of agricultural commodities produced, Australia, 2008–09*. Canberra: Australian Bureau of Statistics.

ABS (2012) *1301.0 Year book Australia 2012*. Canberra: Australian Bureau of Statistics.

ABS (2016a) *Population projections, Australia 2012 to 2101: New South Wales*. Canberra: Australian Bureau of Statistics. Available from: www.abs.gov.au/ausstats/abs@.nsf/Lookup/3222.0main+features72012%20(base)%20to%202101 (accessed 18 May 2017).

ABS (2016b) *Population projections, Australia 2012 to 2101: Victoria*. Canberra: Australian Bureau of Statistics. Available from: www.abs.gov.au/ausstats/abs@.nsf/Lookup/3222.0main+features72012%20(base)%20to%202101 (accessed 18 May 2017).

ACCC (2008) *Report of the ACCC inquiry into the competitiveness of retail prices for standard groceries, July 2008*. Canberra: Commonwealth of Australia.

Andrée, P. (2014) Citizen-farmers: The possibilities and the limits of Australia's emerging alternative food networks. In: P. Andrée, J. Ayres, M. Bosia, & M. Massicotte (eds.), *Globalization and food sovereignty: Global and local change in the new politics of food*. Toronto: University of Toronto Press.

Aubry, C., Ramamonjisoa, J., Dabat, M., Rakotoarisoa, J., Rakotondraibe, J., & Rabeharisoa, L. (2012) Urban agriculture and land use in cities: An approach ith the multi-functionality and sustainability concepts in the case of Antananarivo (Madagascar). *Land Use Policy*, 29, 429–439.

Bilson, G. (2007) *One continuous picnic: A gastronomic history of Australia*. Melbourne: Melbourne University Press.

BoM (2014) *State of the climate 2014*. Melbourne: Bureau of Meteorology.

Brinkley, C. (2012) Evaluating the benefits of Peri-urban agriculture. *Journal of Planning Literature*, 27(3), 259–269.

Brunori, G., & Di Iacovo, F. (2014) Urban food security and landscape change: A demand-side approach. *Landscape Research*, 39, 141–157.

Budge, T. (2013) Is food a missing ingredient in Australia's metropolitan planning strategies? In: Farmar-Bowers, Q., Higgins, V., & Millar, J. (eds.), *Food security in Australia: Challenges and prospects for the future*. New York: Springer, pp. 367–380.

Bunce, M. (1994) *The countryside ideal: Anglo-American images of landscape*. London: Routledge.

Bunce, M. (1998) Thirty years of farmland preservation in North America: Discourses and ideologies of a movement. *Journal of Rural Studies*, 14(2), 233–247.

Bunker, R., & Houston, P. (2003) Prospects for the rural-urban fringe in Australia: Observations from a brief history of the landscapes around Sydney and Adelaide. *Australian Geographical Studies*, 41, 303–323.

Burch, D., & Lawrence, G. (2005) Supermarket own brands, supply chains and the transformation of the agri-food system. *International Journal of Sociology of Agriculture and Food*, 13, 1–28.

Burton, P., Lyons, K., Richards, C., Amati, M., Rose, N., Des Fours, L., Pires, V., & Barclay, R. (2013) *Urban food security, urban resilience and climate change*. Gold Coast: National Climate Change Adaptation Research Facility.

Buxton, M. (2014) The expanding urban fringe: Impacts on Peri-urban areas, Melbourne, Australia. In: Maheshwari, B., Purohit, R., Malano, H., et al. (eds.), *The security of water, food, energy and liveability of cities: Challenges and opportunities for Peri-urban futures*. Dordrecht, The Netherlands: Springer Netherlands, pp. 55–70.

Buxton, M., & Carey, R. (2014) The use of planning provisions and legislation to protect Peri-urban agricultural land. *Australian Environment Review*, September, 191–195.

Buxton, M., Carey, R., & Phelan, K. (2016) The role of Peri-urban land use planning in resilient urban food systems: A case study of Melbourne, Australia. In: B. Maheshwari, V. Singh, and B. Thoradeniya (eds.), *Balanced urban development: Options and strategies for liveable cities*. New York: Springer.

Buxton, M., & Goodman, R. (2008) Protecting Melbourne's green wedges- the fate of a public policy. In: Amati, M. (ed.), *Urban green belts in the twenty-first century*. Hampshire: Ashgate, pp. 61–82.

Buxton, M., Phelan, K., & Hurley, J. (2015) *Melbourne at 8 million: Matching land supply to dwelling demand*, October. Melbourne: RMIT University.

Buxton, M., & Taylor, E. (2011) Urban land supply, governance and the pricing of land. *Urban Policy and Research*, 29, 5–22.

Cadieux, K. (2008) Political ecology of exurban "Lifestyle" landscapes at Christchurch's contested urban fence. *Urban Forestry & Urban Greening*, 7, 183–194.

Cadieux, K., Taylor, L., & Bunce, M. (2013) Landscape ideology in the greater golden horseshoe greenbelt plan: Negotiating material landscapes and abstract ideals in the city's countryside. *Journal of Rural Studies*, 32, 307–319.

Cahn, A. (1977) Australians in the early twentieth century. In: Wood, B. (ed.), *Tucker in Australia*. Melbourne: Hill of Content Publishing.

Carey, R., Larsen, K., & Sheridan, J. (2015) *The role of cities in climate-resilient food systems: A foodprint Melbourne briefing*. Melbourne: Victorian Eco-Innovation Lab.

Carey, R., Larsen, K., Sheridan, J., & Candy, S. (2016) *Melbourne's food future: Planning a resilient city foodbowl*. Melbourne: Victorian Eco-Innovation Lab.

Cebon, M. (2010) *The impacts of Australian trade and agriculture policies for farmers (and everyone else)*, online proceedings of the From Plains to Plate Conference Adelaide, February 12, 2010. http://futureoffoodsa.ning.com/page/presentations-1 (accessed 4 August 2010).

Cole, C. (1980) *Melbourne markets 1841–1979: The story of the fruit and vegetable markets in the city of Melbourne*. Melbourne: Melbourne Wholesale Fruit and Vegetable Market Trust.

DAFF (2012) *Resilience in the Australian food supply chain*. Canberra: Department of Agriculture, Fisheries and Forestry.

DAFF (2014) *Australian food statistics 2012–13*. Canberra: Department of Agriculture, Fisheries and Forestry.

De Schutter, O. (2014) *Report of the special rapporteur on the right to food, Olivier de Schutter. Final report: The transformative potential of the right to food*. New York: United Nations General Assembly.

DELWP (2016) *Managing extreme water shortage in Victoria: Lessons from the millennium drought*. Melbourne: Department of Environment, Land, Water and Planning.

Department of Infrastructure and Regional Development (2015) *State of Australian cities 2014–2015: Progress in Australian regions*. Canberra: Department of Infrastructure and Regional Development.

Dubbeling, M., & de Zeeuw, H. (2011) Urban agriculture and climate change adaptation: Ensuring food security through adaptation resilient cities. In: Otto-Zimmermann, K. (ed.), *Resilient cities: Cities and adaptation to climate change. Proceedings of the Global Forum 2010*. Netherlands: Springer, pp. 441–449.

Eagle, A., Eagle, D., Stobbe, T., & van Kooten, G. (2015) Farmland protection and agricultural land values at the urban-rural fringe: British Columbia's agricultural land reserve. *American Journal of Agricultural Economics*, 97, 282–298.

Edwards, F., Dixon, J., Friel, S., Hall, G., Larsen, K., Lockie, S., Wood, B., Lawrence, M., Hanigan, I., Hogan, A., & Hattersley, L. (2011) Climate change adaptation at the intersection of food and health. *Asia-Pacific Journal of Public Health*, 23(2), 91S–104S.

Elton Consulting. (2009) *Sydney's agriculture – planning for the future (Forums outcomes report)*. Sydney: New South Wales Department of Primary Industries.

FAO/RUAF (2015) *City region food systems: Building sustainable and resilient city regions*. Rome: Food and Agricultural Organization of the United Nations.

Fielke, S., & Bardsley, D. (2014) Regional agricultural governance in Peri-urban and rural South Australia: Strategies to improve multi-functionality. *Sustainability Science*, 10, 231–243.

Freestone, R. (1992) Sydney's Green Belt 1945–1960. *Australian Planner*, 30, 70–77.

Galvin, N. (2006) Sydney's invisible farmers. *Sydney Morning Herald Science Blog*. http://blogs.smh.com.au/science/archives/2006/11/where_does_our.html (accessed 17 November 2007).

Gaynor, A. (2006) *Harvest of the suburbs: An environmental history of growing food in Australian cities*. Crawley: UWA Press.

Gilbert, H. (2007) Private property rights and the public interest in land use conflicts: The case of Sydney's lost Green-belt. *State of Australian Cities Conference*, 28–30 November 2007, Adelaide, SA.

Government of South Australia (2015) *Proposed environment and food production area*. Adelaide: Department of Planning, Transport and Infrastructure.

Greenlees, D. (1951) Wanted – a vegetable plan. *The Argus*, 24 September 1951.

Griffith, G. (2015) Right to farm laws. *NSW Parliament E-Brief*. June 2015. NSW Parliament Research Service, Sydney: NSW Parliament.

Hamblin, A. (2009) Policy directions for agricultural land use in Australia and other post-industrial economies. *Land Use Policy*, 26, 1195–1204.

Harcourt, J. (1939) How Melbourne gets its vegetables. *The Argus*, 6 May 1939.

Harman, B., & Choy, D. (2011) Perspectives on tradeable development rights for eco-system service protection: Lessons from an Australian Peri-urban region. *Journal of Environmental Planning and Management*, 54(5), 617–635.

Harman, B., Pruetz, R., & Houston, P. (2015) Tradeable development rights to protect Peri-urban areas: Lessons from the United States and observations on Australian practice. *Journal of Environmental Planning and Management*, 58, 357–381.

Hattersley, L., Isaacs, B., and Burch, D. (2013) Supermarket power, own-labels, and manufacturer counterstrategies: International relations of cooperation and competition in the fruit canning industry. *Agriculture and Human Values*, 30, 225–233.

Henderson, L. (2003) Regulating land use conflict on the urban fringe: Two contrasting case studies from the Australian Poultry Industry. *Australian Geographer*, 34(1), 3–17.

Higgins, V. (2001) Governing the boundaries of viability: Economic expertise and the production of the 'Low-Income Farm Problem' in Australia. *Sociologia Ruralis*, 41, 358–375.

Holmes, J. (2008) Impulses towards a multifunctional transition in rural Australia: Interpreting regional dynamics in landscapes, lifestyles and livelihoods. *Landscape Research*, 33, 211–223.

Houston, P. (2005) Re-valuing the fringe: Some findings on the value of agricultural production in Australia's Peri-urban regions. *Geographical Research*, 43(2), 209–223.

Hughes, L., Steffen, W., Rice, M., & Pearce, A. (2015) *Feeding a hungry nation: Climate change, food and farming in Australia*. Sydney: Climate Council of Australia.

Humphreys, A., Wright, M., Fowler, A., Tinslay, P., Gojak, D., & Tooker, M. (2001) *Rockdale market gardens conservation management plan*, NSW department of urban affairs and planning. Sydney: Architectural History Services.

Infrastructure Australia. (2015) *Population estimates and projections*. Australian Infrastructure Audit Background Paper. April 2015. Canberra: Infrastructure Australia.

Institute of Sustainable Futures (2015) *Sydney's food futures*. Sydney: University of Technology Sydney, Institute of Sustainable Futures. Available from: www.sydneyfoodfutures.net/ [Accessed 15 June 2016].

James, S. (2016a) *Farming on the fringe: Peri-urban agriculture, cultural diversity and sustainability in Sydney*. Switzerland: Springer.

James, S., & O'Neill, P. (2016) Planning for Peri-urban agriculture – a geographically-specific, evidence based approach from Sydney. *Australian Geographer*, 47, 173–194.

James, S.W. (2014) Protecting Sydney's Peri-urban agriculture: Moving beyond a housing/farming dichotomy. *Geographical Research*, 52, 377–386.

James, S.W. (2016b) Beyond 'Local' food: How supermarkets and consumer choice affect the economic viability of small-scale family farms in Sydney, Australia. *Area*, 48, 103–110.

Larsen, K., Turner, G., Ryan, C., & Lawrence, M. (2011) *Victorian food supply scenarios: Impacts on availability of a nutritious diet*. Melbourne: Victorian Eco-Innovation Lab.

Lawrence, G., Richards, C., & Lyons, K. (2013) Food security in Australia in an era of neoliberalism, productivism and climate change. *Journal of Rural Studies*, 29, 30–39.

Lennon, J. (2015) Changes to continuing landscapes: Industrialisation of Australia's productive rural lands. *Landscape Research*, 40(6), 684–700.

Llausàs, A., Buxton, M., & Beilin, R. (2016) Spatial planning and changing landscapes: A failure of policy in Peri-urban Victoria, Australia. *Journal of Environmental Planning and Management*, 59(7), 1304–1322.

Malcolm, P., & Fahd, R. (2009) *Ground truthing of Sydney vegetable industry in 2008*. Sydney: Horticulture Australia Limited, NSW Department of Primary Industries.

Mason, D., & Knowd, I. (2010) The emergence of urban agriculture: Sydney, Australia. *International Journal of Agricultural Sustainability*, 8(1–2), 62–71.

McFarland, P. (2015) The Peri-urban land-use planning tangle: An Australian perspective. *International Planning Studies*, 20(3), 161–179.

McKeon, G. (2008) *Australia's variable climate*. Sydney: Australian State of the Environment Committee, Department of the Environment and Energy. Available from: www.environment.gov.au/node/22581 [Accessed 15 August 2016].

Millar, J., & Roots, J. (2012) Changes in Australian agriculture and land use: Implications for future food security. *International Journal of Agricultural Sustainability*, 10, 25–39.

Molden, D. (ed.) (2007) *Water for food, water for life: A comprehensive assessment of water management in agriculture*. London: Earthscan, and Colombo: International Water Management Institute.

Neales, S. (2014) Fair go for farmers or else, supermarket giants told. *The Australian*, 21 October 2014. www.theaustralian.com.au/national-affairs/fair-go-for-farmers-or-else-supermarket-giants-told/news-story/5616ec423af87da740d3bcd2faed967c (accessed 20 February 2017).

Nixon, R.E. (1976) *The Chinese community of Camden: 1910 Onwards*. Library, C. (ed.). Camden: Camden Historical Society.

Opitz, I., Berges, R., Piorr, A., & Krikser, T. (2016) Contributing to food security in urban areas: Differences between urban agriculture and Peri-urban agriculture in the global North. *Agriculture and Human Values*, 33, 341–358.

Parberry, P., Wilkinson, R., & Karunaratne, K. (2008) *Square pegs in green wedges? Landholders and natural resource management in Melbourne's rural Hinterland*, Summary Report. Melbourne: Department of Primary Industries.

Parker, F., & Jarecki, S. (2003) Peri-urban agriculture and public policy: Sydney's forgotten farmers. Paper presented at *Social Inequality and the Knowledge Society Conference*, 12 November 2003, Macquarie University, Sydney.

Parliament of New South Wales. (2006) "South-Western Sydney land rezoning." *Legislative Council (Hansard)*. Sydney: Government of New South Wales (accessed 2 February 2007).

PMSEIC (2010) *Australia and food security in a changing world*. Canberra: The Prime Minister's Science, Engineering and Innovation Council.

Power, J. (2015) As drought enters fourth year NSW farmers call for it to be declared natural disaster. *The Sydney Morning Herald*, 7 July 2015. Sydney, Australia: Fairfax.

Quiggin, J. (2007) *Drought, climate change and food prices in Australia*. Brisbane: University of Queensland.

Ramsey, R., & Gallegos, D. (2011) What are the implications of Peri-urban agriculture on food security in Australian cities? Paper presented at: *2nd National Food Futures Conference*, 22–23 November 2011, Hobart, TAS. (Unpublished).

Reisinger, A., Kitching, R.L., Cheiw, F., Hughes, L., Newton, P., Shuster, S.S, Tait, A., Whetton, P., et al. (2014) Australasia. In: Barros, V.R., Field, C.B., Dokken, D.J., Mastrandrea, M.D., Mach, K.J., Bilir, T.E., Chatterjee, M., Ebi, K.L., Estrada, Y.O., Genova, R.C., Girma, B., Kissel, E.S., Levy, A.N., MacCracken, S., Mastrandrea, P.R., & White, L.L. (eds.), *Climate change 2014: Impacts, adaptation, and vulnerability. Part B: Regional aspects. Contribution of working group II to the fifth assessment report of the intergovernmental panel on climate change*. Cambridge and New York: Cambridge University Press, pp. 1371–1438.

Sharp, A. (2009) Pay up or lose shelf space: Coles turns $500m screws on suppliers. *The Age*, 3 March 2009. Melbourne, Australia: Fairfax.

Sheridan, J., Carey, R., & Candy, S. (2016) *Melbourne's foodprint: What does it take to feed a city?* Melbourne: Victorian Eco-Innovation Lab.

Sheridan, J., Larsen, K., & Carey, R. (2015) *Melbourne's foodbowl: Now and at 7 million*. Melbourne: Victorian Eco-Innovation Lab.

Sinclair, I. (2015a) Growing food in a residential landscape. Paper presented at the *State of Australian Cities Conference 2015*, 9–11 December 2015, The Gold Coast, Queensland.

Sinclair, I. (2015b) *Sydney Peri-urban network issues paper*. Sydney: Edge Land Planning.

Sinclair, I., Docking, A., Jarecki, S., Parker, F., & Saville, L. (2004) *From the outside looking in: The future of Sydney's rural land: Background issues and workshop outcomes*. Sydney: University of Western Sydney.

Sinclair, I. (1999) *Is there a future for Australia's agricultural land?* Paper presented to RAPI National Congress, Darwin, 19–23 September 1999. http://www.ruralplanning.com.au/library/papers/rapinat99.pdf.

Soebels, J., Richardson, S., Turner, G., Maude, A., Tan, Y., Beer, A., & Wei, Z. (2011) *Summary of research into the long-term physical implications of net overseas migration Australia in 2050.* Adelaide: Flinders University, National Institute of Labour Studies.

Spencer, A., Gill, J., & Schmahmann, L. (2015) Urban or suburban? Examining the density of Australian cities in a global context. Paper presented at *The State of Australian Cities Conference 2015*, 9–11 December 2015, The Gold Coast, Queensland.

Spencer, S., & Kneebone, M. (2012) *FOODmap: An analysis of the Australian food supply chain.* Canberra: Department of Agriculture, Fisheries and Forestry.

Stapleton, J. (2008) Eat-local movement provides shot in the arm for struggling farmers, *The Australian*, August 4, p. 31. Sydney, Australia: News Corp.

State Library of NSW (2012) *The macquarie towns.* Sydney: Government of New South Wales. Available from: http://www2.sl.nsw.gov.au/archive/discover_collections/history_nation/macquarie/governor/towns.html [Accessed 7 June 2016].

Steel, C. (2009) *Hungry city: How food shapes our lives.* London: Vintage Books.

Steven, R.H. (2005) Managing land-use conflict around urban centres: Australian poultry farmer attitudes towards relocation. *Applied Geography*, 25, 97–119.

Taylor, E., Butt, A., & Amati, M. (2017) Making the blood broil: Conflicts over imagined rurality in peri-urban Australia. *Planning Practice and Research*, 32(1), 85–102.

van Oosterzee, P., Dale, A., & Preece, N. (2014) Integrating agriculture and climate change mitigation at landscape scale: Implications from an Australian case study. *Global Environmental Change*, 29, 306–317.

VEIL. (2013) Mapping Melbourne for land capability: Assessing the agricultural capability of Melbourne's Peri-urban area. A preliminary evaluation. Report prepared for VicHealth by the Victorian Eco-Innovation Lab. September 2013. Melbourne: Victorian Eco-Innovation Lab.

Victorian State Government (2017) *Unlocking new communities and affordable housing.* Media release, 28 February 2017. www.premier.vic.gov.au/unlocking-new-communities-and-affordable-housing/ (accessed 2 March 2017).

Wahlquist, A. (2008) *Thirsty country: Options for Australia.* Sydney: Allen and Unwin.

Western Sydney Parklands Trust (2017) *Western Sydney parklands trust.* Sydney: Western Sydney Parklands Trust. Available from: www.westernsydneyparklands.com.au/ [Accessed 2 February 2017].

World Bank. (2016) *Arable land – % of land area.* Washington, DC: World Bank. http://data.worldbank.org/indicator/AG.LND.ARBL.ZS (accessed 12 June 2016).

Wynne, L., Cordell, D., Chong, J., & Jacobs, B. (2016) *Planning tools for strategic management of Peri-urban food production.* Sydney: RICS Research Trust.

Part III

Ecology, resources, sustainability, and climate change

16

Challenges in agricultural sustainability and resilience

Towards regenerative practice

Joshua Zeunert

Introduction

The ongoing global expansion of industrial agriculture and agribusiness continues to rapidly transform global food systems from localised to globalised, small to expansive in scale, low to high tech, and from natural and organic to industrial and synthetic (Berry, 1977; Bowler, 1992; Roberts, 2008; Weis, 2010; Stierand, 2012: 67–68). In 1960, agriculture's 'green revolution' was in the early stages of rapid deployment; at this time, the world's population stood at three billion with 34% urbanised. This inappropriately named approach to agriculture (due to its use of the term 'green' as discussed throughout this chapter) has led to large increases in food production. These are achieved through use of synthetic fertilisers, chemical pesticides, large irrigation systems, plant breeding, increased mechanisation, and utilisation of efficiency-oriented technologies. Consequently, less than half a century later in 2008, the global population exceeded seven billion, with over 50% living in urban areas. Feeding four billion extra people in this timeframe is an extraordinary agricultural feat, but some observers have questioned whether this system of agriculture can be sustained (Lowe, 2005: 60; LA and GLA, 2010: 9; Cribb, 2010).

Serious questions of sustainability and resilience are notwithstanding the push for continued deployment and intensification of industrial agriculture to supply ongoing population (and economic) growth projections (FAO, 2011a: vii, ix; Chapters 19 and 25). Neoliberalism is a central force behind continual growth (Harvey, 2005; Rosin et al., 2012; Wolf, 2014; Springer et al., 2016), with agribusiness a key component of this market-led approach. Yet finite resource-dependent systems – industrial agriculture being the focus in this chapter – are ultimately temporary when viewed across longer time-scales (Heinberg, 2005; Kunstler, 2006; Erisman et al., 2008; Pimentel, 2009; Heingerg and Lerch, 2010; Zeunert, 2011; Moreau et al., 2012). Accordingly, food systems must shift from consumptive, energy intensive, ecologically destructive, and unsustainable to restorative, wherein widespread implementation of regenerative practices rebuilds lost biocapacity, increases resilience, and establishes equilibrium between natural and cultural systems.

This chapter examines the concepts of *sustainability*, *resilience*, and *regenerative practice* as applied to agricultural landscape and food systems. It is consequently structured in three sections that define and discuss these topic areas. A brief 'flyover' identifying key literature is necessary due to the sheer amount of discourse within these broad topics. Accordingly, reference throughout is made to key chapters in this volume and other relevant sources. Primary focus is given to developed countries and prominence placed on terrestrial food production (see Chapter 11 for Seascapes). This chapter addresses medium and long-term challenges through a holistic, systems-thinking approach oriented at collective and common interests for natural and cultural wellbeing rather than isolated, individual, or corporate interests (Chapter 24; Bowler, 1992; Linn, 1999; Freyfogle, 2007; Tornaghi, 2012; Claeys, 2015; Lowder et al., 2016; Cresswell Riol, 2016; Gaarde, 2017).

Sustainable agricultural landscapes

It is useful to examine the macro characteristics of food systems' sustainability rather than the minutiae. Firstly, *sustainability* must be defined; the most widely used is the Brundtland Commission's Report 'Our Common Future' (WCED, 1987: 8), being "development that meets the needs of the present without compromising the ability of future generations to meet their own needs". Despite the report's thinly veiled economic bias (van der Ryn and Cowan, 1996: 6–7) and it being over three decades since it was coined, this definition provides a useful intergenerational perspective (albeit generational perceptions of what constitutes a 'need' continue to shift). It is also important to consider both prior timescales and possible future scenarios beyond short-term observations and the current cheap fossil fuel era for a wider view of sustainability (Holmgren, 2002; Heinberg, 2005; Kunstler, 2006; Holmgren, 2009; Heinberg and Lerch, 2010; Weis, 2010; Herbert, 2011; Nikiforuk, 2012; Leggett, 2014; Fücks, 2015; Morgan, 2015).

Decades of discourse have blurred the term *sustainability* (Cronon, 1995; van der Ryn and Cowan, 1996; Nordhaus and Shellenberger, 2004; Freyfogle, 2006; Jensen and McBay, 2009), with economic and social sustainability factors skewing its central meaning from the ability to balance human consumption with planetary biocapacity. The usefulness of *sustainability* is increased when combined with the concepts of *resilience* and *regenerative* practices, both examined later in this chapter.

Agricultural, food, and landscape systems concern multiple stakeholders with, at times, conflicting perspectives. In order to understand agricultural and food sustainability across molecular to global scales and timeframes ranging from seconds to centuries (Lichtfouse, 2012), we must cultivate a 'systems perspective'. As defined by the author, sustainability is particularly relevant when food systems are seen as:

> *Our collective and intergenerational ability to feed ourselves and manage landscape, natural and social systems for food in perpetuity.*

'*Feeding ourselves*' can refer to the individual, family, community, state, national, or global scale, depending on perspective. Ultimately, however, sustainability necessitates global measurement in view of our collective reliance on Earth's biocapacity and resilience. The following subsections *environmental*, *economic*, and *social* examine agricultural and food sustainability factors, adapting the 'triple bottom line' themes (Elkington, 2004). While the merits of these categories are debated in the discourse, they nonetheless provide a useful framework for discussing sustainability challenges of landscape and food.

Environmental sustainability

There are several key concepts and measures – albeit contested – regarding environmental sustainability and resource management relevant to food and landscape systems:

- Carrying capacity (also known as ecological or biological limits), posits that the maximum population size that can be sustained is based on the food, habitat, water, and other resources available in the environment and Earth (Vogt, 1948; Odum, 1971; Catton, 1980; Brown, 1996; Barrett and Odum, 2000; Marten, 2001; Lane, 2010; Graymore et al., 2010);
- Overshoot (Ehrlich, 1968; Meadows, 1972; Catton, 1980; Marten, 2001; Meadows et al., 2004; MacLean, 2008) occurs when a population exceeds the long-term carrying capacity of its environment, leading to collapse, crash, or die-off (and is known as being 'Malthusian');

- Ecological and food footprints (Rees, 1992; Wackernagel, 1994; Khan and Hanjra, 2009; Khan et al., 2009; Oppenlander, 2012; Oppenlander, 2013; WWF, 2016; Peters et al., 2016) are based on appropriating carrying capacity for an individual or population to determine the area of productive land (in acres/hectares) and water required for resource provision, or the number of 'Earths' required to support and assimilate wastes for all activities (or for specific factors such as food itself). See for example, GFN, www.footprintnetwork.org/en/. In the year 2000, for example, London was calculated as consuming 293 times its spatial footprint and 42 times its biocapacity (Mostafavi and Doherty, 2010: 216–217); with the city accommodating over 1.5 million new inhabitants since then. Weller (2009) documents the escalation of 'McMansion' housing in *Boomtown*, demonstrating that an average household in Perth (Australia) requires 14.5 hectares of productive land, with food needing over three times as much area than any other factor (Weller, 2009: 440–441). Roggema and Keeffe (2014: 12) calculate that a single McDonalds restaurant would require a 30 kilometre high urban farm on its roof to accommodate its food footprint.

Seven decades ago (and with a global population of around 2.5 billion), Vogt (1948) noted that:

> We must realise that not only does every area have a limited carrying capacity, but also that this carrying capacity is shrinking and the demand growing. Until this understanding becomes an intrinsic part of our thinking and wields a powerful influence on our formation of national and international policies we are scarcely likely to see in what direction our destiny lies.
>
> *(Vogt, 1948: 80)*

Human ingenuity has temporarily and very successfully circumvented carrying capacity through utilising finite fossil fuel energy sources in agriculture (Lane, 2010), as explained by Odum (1971),

> A whole generation of citizens thought that the carrying capacity of the earth was proportional to the amount of land under cultivation and that higher efficiencies in using the energy of the sun had arrived. This is a sad hoax, for industrial man no longer eats potatoes made from solar energy, now he eats potatoes partly made of oil.
>
> *(Odum, 1971: 115)*

Fukuoka (1978: 94), a farmer himself, reinforces Odum's argument, asserting that industrial agriculturalists are profit-oriented "manufacturers" producing "watery chemical concoctions", mixtures of "nitrogen, phosphorus, and potash" using "synthetic feed, chemicals, and hormones". Kunstler (2006) presents a wide perspective on civilisation's fossil fuel dependence for sustenance – what he terms a 'long emergency' – and here, specifically highlights the acute predicament presented by industrial food systems:

> To put it simply, [we] have been eating oil and natural gas for the past century, at an ever-accelerating pace. Without the massive 'inputs' of cheap gasoline and diesel fuel for machines, irrigation, and trucking, or petroleum-based herbicides and pesticides, or fertilizers made out of natural gas, [we] will be compelled to radically reorganize the way food is produced, or starve.
>
> *(Kunstler, 2006: 239)*

Erisman et al. (2008: 637) quantify Kunstler's confronting assertion by estimating that almost half (48%) of the people on the Earth (in 2008) were fed as a result of food dependent upon synthetic nitrogen fertiliser use (made from natural gas – see Figure 16.1). Adding to this predicament, synthetic nitrogen fertiliser accumulates in soils and results in high acidity; a United Nations Food and Agricultural Organization (UN FAO) estimate states that around one-third of global arable land is now so acidic that it cannot support high-yielding crops (cited in Roberts, 2008: 214).

Figure 16.1 Modern fertiliser plants use the Haber–Bosch process to create synthetic nitrogen for extensive global use in agriculture (see Chapter 10). Thus, seeking to expand industrial agriculture, for example, into regions in Africa (see Chapter 21), requires significant investment in infrastructure, machinery, knowledge, and their associated costs (see, for example, Chapter 19). Traditional fertiliser utilises either human (Chapter 23), animal, or compostable wastes, or their combination.

Source: Joshua Zeunert

With respect to measures of human labour 'inputs', green revolution-based agriculture appears extraordinarily productive, for example, only 1% or less of the population in the UK and USA are engaged in agriculture (World Bank, n.d.a). Yet in terms of energy input (fossil fuel 'calories') for energy returned (nutritional calories), industrial agriculture is staggeringly inefficient (Weis, 2010). In some countries (primarily those incorporating green revolution agricultural methods), Pimentel (2009: 19) notes that the use of fossil energy has increased more than a hundredfold as compared to the early 1950s, also noting that the developed world consumes 70% of the world's fossil energy. Such energy use is both inequitable and intergenerationally unsustainable.

Heinberg (2005) provides a tacitly simple indicator of food system sustainability:

> Food is energy. And it takes energy to get food. These two facts, taken together, have always established the biological limits to the human population and always will.
>
> *(Heinberg, 2005: 1)*

The Ehrlich's controversial book *The Population Bomb* (1968) (influenced by Vogt's (1948) *Road to Survival* (Desrochers and Hoffbauer, 2009)) posited human population overshoot. Marten clearly articulates this concept:

> Sometimes a population grows so rapidly that it overshoots carrying capacity before negative feedback can stop the increase. If a population overshoots, it usually depletes its food so severely that negative feedback in the form of more deaths and fewer births quickly reduces it below carrying capacity.
>
> *(Marten, 2001: 24)*

The green revolution's success has resulted in widespread dismissal of *overshoot* and *carrying capacity* concepts from 'Cornucopian' perspectives, although much counter debate results from 'Malthusian' viewpoints (see for example a special issue on the topic in *The Electronic Journal of Sustainable Development* (2009) 1(3)). Cornucopian critics of overpopulation believe that Malthusian notions misunderstand "the inherent capacity of market economies to tap into creative human brains and to continually deliver innovative solutions to pressing problems" (Desrochers, 2009: n.p.). Yet cornucopian solutions still usually require industrial and technological processes that rely on finite energy sources. Apt, perhaps is Lovelock's assertion that "If there were a billion people living on the planet, we could do whatever we please. But there are nearly seven billion. At this scale, life as we know it today is not sustainable" (Lovelock in Goodell, 2010: 91). When using a broader timescale that accounts for limited and diminishing remaining supplies of oil, natural gas, and phosphorus (that form the basis of industrial agriculture), a limited timeframe exists for food systems (and indeed market economies) to transition to sustainable resource based modes of operation (Lane, 2010). Yet continuing population growth (Gerland et al., 2014; UN, 2015) and increasing consumption patterns heighten the likelihood of widespread future food crises (Brown, 1996).

Dismissal of overshoot for the human population is usually based on short-term and econocentric perspectives, anthropocentrism, self and/or corporate interests, persistent dualisms (such as the separation of nature and culture), and even human arrogance. As noted by Klein (2014: 21), unlike economic systems, nature's laws are not malleable. Die-off and overshoot are notoriously difficult to predict (Ehrlich and Ehrlich, 2009), yet this does not obviate their likelihood and therefore the need to plan and act for their avoidance.

The sustainability of food systems would result when the resources invested in agriculture and food (their ecological footprint or embodied energy) are balanced with the amount of biocapacity that is being naturally or intentionally replenished and regenerated. This 'net balance' needs to be inclusive of all associated processes in agricultural and food system inputs and outputs, for example: not drawing more freshwater from the Earth than what is naturally replenished; returning more carbon and nutrients into the soil than taken out; maintaining or increasing (top) soil cover/depth and health; not drawing more nutrients and fertilisers than are being replenished; and not discarding more wastes than can be assimilated through healthy terrestrial, aquatic, and seascape ecologies. Net balance is far from achieved in current food and agricultural systems, which create substantial impacts and demonstrate various unsustainable outcomes. Conventional agriculture and food systems are:

- The largest global consumer of freshwater resources at 70% (Pimentel et al., 2004: 909; FAO, 2011a: 26);
- Responsible for over 20% of global greenhouse gas emissions from livestock and crop production and conversion of forests to farmland (FAO, 2016: vi);
- Occupying nearly 40% of the world's ice-free land surface in the last several decades (Ramankutty et al., 2008, World Bank, n.d.b);
- Using and releasing enormous amounts of pollutants and chemicals into terrestrial, aquatic and atmospheric environments and being the leading source of pollution in many countries (Carson, 1962; WWF, n.d.);
- Sending regular food items enormous distances to reach consumers ('food miles', see for example Garnett, 2003; Halweil, 2004; Gaballa and Abraham, 2008; Pimentel, 2009; Bomford, 2010; Jansma et al., 2012; Denny, 2012);
- Wasting 20–50% of food in developed countries and one-third globally throughout the production and supply chain (FAO, 2011b: v); and
- Failing to capture and return food and compostable wastes in production cycles, for example, in Australia food waste makes up around 40% of the domestic waste stream (Boland, 2010: 2), with 'waste miles' also problematic (also see Chapter 23).

Vast journeys are undertaken by industrial foodstuffs trafficked through global freight logistics and accordingly many producers and retailers use durable, horticulturally bred fruit and vegetable varieties (often at the expense of flavour, see Blazey and Varkulevicius, 2006: 7; Steel, 2008). Increased interest in food system sustainability has been assisted by illuminating studies describing and quantifying 'food miles' that can readily demonstrate global foodstuffs containing more embodied fossil fuel calories than nutrient calories (Garnett, 2003; Halweil, 2004; Heinberg, 2005; Gaballa and Abraham, 2008; Roberts, 2008; Pimentel, 2009; Bomford, 2010; Ladner, 2011; Jansma et al., 2012; Denny, 2012). Gaballa and Abraham (2008) calculated that the 'food miles' of Australia's 29 most common food items cover a total distance of 70,803 kilometres, which is equivalent to travelling nearly twice around the circumference of the Earth (40,072 kilometres), or travelling three times around Australia's coastline. Such bizarre and yet commonplace scenarios exploit cheap energy sources, inexpensive labour markets, and trade agreements, meaning that global food can routinely be sold more cheaply than local food. Food miles are further enabled through post-harvest preservation and produce treatments (such as irradiation, picking unripe produce and later gassing with ethylene; optimising packaging) and a complex array of refrigerated and frozen food transportation logistics in trucks, trains, boats, containers, ships and planes. Supermarkets and produce retailers have been noted for a lack of seasonality – a 'permanent global summer time' (PGST) (see Blythman, 2004) – whereby global shipping supersedes local availability and facilitates year-round supply. All of these techniques depend on (cheap) fossil fuels for economic and logistical viability.

Economic sustainability

The dominant global economic systems since the late 1970s embrace free-market, capitalist, and neoliberal models whose key characteristics include extensive economic liberalisation; free trade; privatisation; open markets; deregulation; a reduction in government and public sector spending; activities and intervention. These enhance the role of the private and corporate sector in the economy (Bowler, 1992; Harvey, 2005; McMichael, 2009; Rosin et al., 2012; Wolf, 2014; Holt-Giménez, 2015; Springer et al., 2016). Such economic models are underpinned by natural resources and ecological systems in order to generate financial returns, gross domestic product (GDP) and necessary perpetual growth – yet these environmental factors are subjugated or altogether disregarded 'externalities' based on the fallacy of natural resources' infinite supply (Saul, 2005; Weis, 2010; Heinberg, 2011; Bauwens cited in Zeunert, 2017: 230). This is demonstrated in *State of the Environment* reporting (see for example Seager et al., 1995; Millennium Ecosystem Assessment, 2001) and *The State of Food and Agriculture* (see for example FAO, 2016), which have for decades documented diminishing biocapacity, renewable and non-renewable resources.

There have been numerous unsuccessful efforts to improve capitalist and neoliberal economic systems' sustainability outcomes through treaties, arrangements, policies, reworking and balancing parameters (see for example Hawken et al., 2000; Elkington, 2004; Daly and Farley, 2011). Economic factors, however, continue to dominate and subjugate social and environmental concerns. Accordingly, our economic system and our environment are in fundamental opposition, which is well articulated by Klein (2014):

> Our economy is at war with many forms of life on earth, including human life. What the climate needs to avoid collapse is a contraction in humanity's use of resources; what our economic model demands to avoid collapse is unfettered expansion. Only one of these sets of rules can be changed, and it's not the laws of nature.
>
> *(Klein, 2014: 21)*

The resulting perpetual and ever-increasing output and efficiency 'treadmill' substantially challenges existing and future agricultural systems operating within global neoliberalism (Harvey, 2005; Rosin et al., 2012; Wolf, 2014; Springer et al., 2016) and is articulated by Abbey (1994):

> We are slaves in the sense that we depend for our daily survival upon an expand-or-expire agro-industrial empire – a crackpot machine – that the specialists cannot comprehend and the managers cannot manage. Which is, furthermore, devouring world resources at an exponential rate.
>
> *(Abbey in Cooley, 1994: 21)*

To offset agriculture's perpetual growth challenge some countries provide economic subsidisation (Blandford et al., 2015: 418–419) (for example Indonesia, Finland, Switzerland, Japan, South Korea, Iceland – at the time of writing) and these can help to support the viability of smaller land holdings and more diverse operations. In contrast, countries with minimal or lower subsidies (New Zealand, Australia, Chile, South Africa) typically require larger, industrial scales of operation and precise technologies to facilitate financial survival (for example image sensing and drone technology). Lack of subsidisation can create heightened pressure for agriculturalists – combined with the intrinsic exclusion of environmental factors – resulting in routine exploitation of the environment to maintain economic output and undermining long-term sustainability (FAO, 2016: 98). Yet the FAO also notes that, for example, input subsidies can create inefficient use of agrochemicals and actually increase the emissions from production (FAO, 2016: xvi). The predicament of agriculture and food systems' increasing assimilation into neoliberal economic structures is even retrospectively lamented by a key sponsor:

> We need the World Bank, we need the IMF, we need all the big foundations, we need all the governments to admit that for 30 years we all blew it, including me, when I was president. We blew it. We were wrong to believe that food is like some other product in international trade. And we all have to go back to a more environmentally responsible, sustainable form of agriculture. We should go back to a policy of maximum food self-sufficiency.
>
> *(Clinton, 2008, cited in Cebon, 2010: 30)*

As noted by Newell, however, transition to sustainable economic and social systems is substantially challenged by:

> conflicts, trade-offs and compromises that are implied by a fundamental restructuring of the economy and the relations of power which will determine which pathway is chosen. The 'incumbent' regime of existing actors and interests that benefit from on-going reliance on a fossil fuel economy and that have played such a decisive role in the development of capitalism over the last century and beyond will not give up their position easily. Nor in likelihood will states that depend indirectly on the revenues generated by these actors and that have, so far, shown little appetite for initiating structural change.
>
> *(Newell, 2015: 71)*

Social sustainability

Sustainability aspirations for food systems require more than sound natural resource management and economic systems. Are agricultural and food systems able to be socially sustainable? This important question is primarily examined in Parts 5–7 of this book (see for example Chapters 31 and 32) with only a brief overview provided here.

Agribusiness is a more befitting term than *agriculture* to describe how food is grown, processed, transported, and consumed in developed (and increasingly, developing) countries (see Chapters 20 and 21). The 'culture' of agriculture has been placed under corporatised pressure; commodified, processed, marketed, and traded for maximum profit for shareholders (Norberg-Hodge et al., 2001; Chapter 31). Roberts (2008) presents a significant range of social and economic challenges in *The End of Food*, outlining, for example, the 'grotesque symmetry' of the overnourished one billion individuals versus one billion undernourished (Roberts, 2008: 84; statistics confirmed by the UN FAO (FAO, 2011a: ix)). Thus while an adequate volume of food is produced under the current food system, it is not justly or fairly distributed and consumed, nor necessarily of optimum quality and nutritional value. Accordingly, Pimentel (2009: 1–2) uses World Health Organisation data stating that 60% of the world population is malnourished, that is, not having the right balance of nutrients.

Failure to equitably distribute food has been and continues to be an ongoing issue, despite continued population growth potentially heightening this quandary. Most agricultural (and indeed much wider) literature assumes ongoing population growth as a predetermined outcome (see for example Godfray et al., 2010; Gerland et al., 2014, UN, 2015; Chapter 19). That is, this discourse presumes that current government policies and attitudes remain pro-population growth and that no mechanisms are introduced to disincentivise or discourage it. Ongoing population growth effectively protracts 'business as usual'; continued and increased consumption, new markets in emerging economies, expansion of western dietary patterns, and thus growth in industrial agriculture. Assuming and indeed ensuring manifestation of these scenarios are hallmarks of neoliberalism and its insatiable appetite for deregulated growth (Harvey, 2005; McMichael, 2009; Rosin et al., 2012; Wolf, 2014; Holt-Giménez, 2015; Springer et al., 2016). Yet as noted by Lowe (2005: 108), the future is something we are creating, not somewhere we are going. Similarly, O'Hear (1999) deconstructs the notion of 'progress' as being desirable, optimal or inevitable. Our daily choices and decisions collectively determine the course of future outcomes and their sustainability; whether to endorse and contribute to these potential outcomes, or whether to lobby and make sure that personal practices do not contribute to their continuation.

Accordingly, it is important to recognise that consumption patterns and agriculture are interconnected; supply and demand are symbiotic. Individuals and communities can yield collective influence in shaping food systems to either support or challenge powerful global agribusiness and governance structures (see Chapter 33). Citizens' purchasing habits underpin food system producers, their practices and ethics (see Chapter 24), whether a local producer or global agribusiness corporation. Subsequent collective demand determines the viability of food and agricultural operations and their associated landscape practices.

Steel (2008) presents a wide spectrum of social sustainability challenges in *Hungry City*, such as obesity stemming from sedentary Western lifestyles, over-consumption, and increased processed food (with recent obesity spikes evident in non-Western regions such as the Pacific Islands and Middle East). There are a multitude of consumption trends and social drivers creating a spectrum of health issues and sustainability impacts (Kearney, 2010).

Dietary consumption significantly determines sustainability, animal welfare, ecological footprints and accordingly shapes and degrades vast landscape and seascape areas (Maas, 1999; Pimentel and Pimentel, 2003; FAO, 2006; Jeavons, 2006; Jarosz, 2009; Goodland, 2010; Oppenlander, 2012; Oppenlander, 2013). Maas (1999: 61–63, 94–115) and MVRDV (2010) visually communicate the additional spatial area of productive land required for meat eating societies compared with vegetarian. Animal products such as beef, seafood, and dairy, for example, require immense productive areas and embody resources and practices such as freshwater consumption, grain crops (fed to animals – 40% of total cereals (Locke et al., 2013)) and ocean dredging (to feed farmed fish and for fertiliser). Substantial adoption of plant-based diets would greatly reduce agriculture's spatial and ecological footprint and deliver significant sustainability benefits (Pimentel and Pimentel, 2003; FAO, 2006; Oppenlander, 2012; Oppenlander, 2013; Peters et al., 2016; FAO, 2016), while freeing land areas and resources for other uses and concerns such as conservation (see Chapter 17; Wilson, 2016; Gordon et al., 2017).

Pimentel and Pimentel (2003: 661S) state that the volume of grain fed to US livestock is sufficient to feed about 840 million people who follow a plant-based diet (the US population was 290 million in 2003). Jeavons (2006: 248–249) documents the spatial areas required by dietary choices (in fossil fuel and post-fossil fuel eras), stating that a high animal product diet requires 31,000–63,000 square feet per person (2,880–5,853 square metres) versus 7,000 square feet (650 square metres) for a vegan under a 'biointensive' production method (both account for fossil fuels availability); this demonstrates spatial variations by factors of four to over eight. Another measure to articulate animal products' intensive spatial impacts is New Zealand's (the world's largest dairy exporter) average accommodation of 2.85 cows per hectare (LIC, 2016: 7); one hectare can accommodate 15 vegan diets under Jeavons' production methods (2006; 248–249). Peters' comparison of ten US diet types found that "carrying capacity was generally higher for scenarios with less meat and highest for the lacto-vegetarian diet" (Peters et al., 2016: 1).

In recent decades, increased availability of dietary environmental data, as well as scrutiny and exposure of food system practices – animal welfare in documentaries for example – help to explain increases in vegetarianism, veganism, and local and short-chain diets. These trends, however, are largely situated in developed countries. Accordingly, they are greatly overshadowed by increased uptake of meat and dairy-heavy western diets (and thus expansion of these agricultural systems – see Chapters 9 and 25) in highly populated developing countries such as India, China, and Brazil (Pimentel and Pimentel, 2003; Pingali, 2007; Jarosz, 2009; MVRDV, 2009; MVRDV, 2010; Popkin et al., 2012; Chapter 20).

Resilient agricultural landscapes

The concept of *resilience* is applied in this section to agricultural resources and practices, associated infrastructures and pertinently, to the social systems that provide the operational basis and functioning of food systems. In this context, the author defines this as:

> *The ability and capacity for agricultural landscape and food systems, through their synergistic social, governance, economic structures and processes, to withstand challenges, survive, adapt, and quickly recover from disturbance, chronic stresses, and acute shocks.*

Numerous chapters in this book relate to food systems' resilience, most pertinently Food Sovereignty (Chapter 31), Food Justice and Politics (Chapter 32), Alternative Agriculture (Chapter 10), Food Security (Chapters 20 and 25), Peri-Urban Agriculture (Chapters 14 and 15), Water and Irrigation (Chapter 19), Nutrients and Waste (Chapter 23), and Climate Change (Chapter 18; Fedoroff et al., 2010: 833). As outlined in Table 16.1, resilience relates to and can be framed by short-term and medium- to long-term challenges, disturbances, stresses, and shocks.

Oft-cited projections and models of continued population (for example UN, 2015) and economic growth posit substantially increasing food outputs, such as a 50% increase by 2030 and 100% by 2050 (see Chapters 19 and 25; Conway, 1998; Weis, 2010; FAO, 2016). These proponents suggest that food system growth and resilience can be achieved through continued technological pursuits, advancements and intensification, citing that increased yields and 'efficiencies' will continue an upward trend (see for example Simon, 1996; DeGregori, 2002; Desrochers, 2009; Godfray et al., 2010; Fedoroff et al., 2010; Chapters 18, 19, 23 and 25). Such perspectives usually use recent historical data (several decades up to the past half-century) and do not interrogate the longevity and quality of landscapes and the economic costs of required resources to underpin their models for increased yields, efficiency, and technological advancements. The last five to seven decades of remarkably increased food production is no guarantee of continued production or growth (Brown, 1996; Norberg-Hodge et al., 2001; Lowe, 2005: 60; LA and GLA, 2010: 9; Cribb, 2010). In 2009, for example, Pimentel (2009: 1) calculated that per capita availability of world cereal grains had been declining for 24 years.

It is difficult to conceive how a global food system and population so fundamentally reliant on plentiful cheap oil and natural gas (Erisman et al., 2008; Weis, 2010; Herbert, 2011) will obtain the necessary substitute energy sources to facilitate unhindered upward growth. It seems likely, therefore, that at some point in

Table 16.1 Key short-term and medium- to long-term factors that relate to the resilience of food systems.

Short-Term Resilience Challenges	*References*
Natural disasters: extreme weather events and their increased volatility under climate change (drought/ heat waves/inundation/storms/extreme heat/ earthquakes)	Burton et al., 2013; Leal Filho, 2015; FAO, 2016; Chapter 18
Shocks and geopolitical conflicts: relating to finite natural resources required in food systems (fresh water, natural gas (e.g. as synthetic nitrogen), phosphorus and fertilisers, oil (petrol, diesel))	Wright, 2009; FAO, 2016
Economic downturns and recession: social disturbance, shocks, and potential collapse (which can also be an ongoing occurrence)	Kunstler, 2006; Berlin, 2016
Trade and diplomatic arrangements: affecting the flow of resources, commodities and food	Harvey, 2005; Cebon, 2010; Howard, 2016
Medium- to Long Term Resilience Challenges	*References*
Climate change impacts: such as climatic variation and shifting of weather patterns in relation to landholdings, soils, fixed settlements and infrastructures that are built around stable and predictable climate patterns for yields, efficiency, and returns on investment	Flannery, 2005, Lovelock, 2006; Holmgren, 2009; FAO, 2011a; Nabhan, 2013; Burton et al., 2013; Leal Filho, 2015; FAO, 2016; Chapter 18
Availability, affordability, and longevity of finite natural resources and fossil fuels required in food systems: fresh water reserves, topsoil, natural gas, phosphorus, fertilizers, pesticides, crude oil, diesel and petrol	Barrett and Odum, 2000; Holmgren, 2002; Meadows et al., 2004; Heinberg, 2005; Campbell, 2005; Goodstein, 2005; Kunstler, 2006; Heinberg, 2007; Erisman et al., 2008; Hopkins, 2008; Holmgren, 2009; Pimentel, 2009; Jarosz, 2009; Cribb, 2010; Weis, 2010; Deffeyes, 2010; Heinberg and Lerch, 2010; Herbert, 2011; FAO, 2011a; Nikiforuk, 2012; Leggett, 2014; FAO, 2016
Availability and exhaustion of arable land suited to agricultural production: exhaustion of soil nutrients; loss of topsoil; erosion; soil health and soil pollution (and resultant issues of soil salinity, acidification, nutrient depletion, compaction, inability to support edible and/or high yielding crops)	Jackson, 1980; FAO, in Roberts, 2008: 214; Niles, 2008; Cribb, 2010; Jackson, 2010; Lichtfouse, 2012; FAO, 2016
Ongoing population growth and overpopulation	Catton, 1980; Brown, 1996; Meadows et al., 2004; Pimentel, 2009; Ehrlich and Ehrlich, 2009; Lovelock in Goodell, 2010: 91; Cribb, 2010; Gerland et al., 2014, UN, 2015
Food sovereignty, control of food systems and parasitic corporate modes: across all stages of food systems (profitability; bioengineering; market domination; seed patents; terminator genes; lawsuits and litigation; creeping acquisitions; concentrated power structures; monopolisation)	Steel, 2008; Roberts, 2008; Jarosz, 2009; Cribb, 2010; Cebon, 2010; Ladner, 2011; ABC, 2011; Rosin et al., 2012; Wolf, 2014; Claeys, 2015; McMichael, 2015; Howard, 2016; Cresswell Riol, 2016; Gaarde, 2017
Overconsumption and dietary choice: such as increasing meat and dairy consumption	Maas, 1999; Pimentel and Pimentel, 2003; FAO, 2006; Jeavons, 2006; Pingali, 2007; Jarosz, 2009; MVRDV, 2009; Goodland, 2010; Cribb, 2010; MVRDV, 2010; Popkin et al., 2012; Oppenlander, 2012; Oppenlander, 2013; Peters et al., 2016; FAO, 2016

Medium- to Long Term Resilience Challenges	*References*
Diversion of food (grains) for meat production	Pimentel and Pimentel, 2003; Pimentel, 2009; Locke et al., 2013
Diversion of food (grains) for biofuels/gas/petrol/ ethanol production	Pimentel, 2009; Kullander, 2010
Urban sprawl, real estate development and land speculation: consuming urban and peri-urban agricultural land and subsequent economically unviable agricultural land	Stewart and Duane, 2009; Chapters 14 & 15
Significant distances: presented between agricultural production areas and urban consumption, heightened through fossil fuel price rises and subsequent increased food prices	Gaballa and Abraham, 2008
Decreased population proportions: engaged and skilled in food production, loss of intergenerational knowledge and skill transfer (average age of farmers, lack of youth in developed agriculture, dwindling labour bases)	Berry, 1977; Allen, 2010; Allen and Wilson, 2012
Loss of cultural traditions: practices, varieties, homogenisation of food landscapes, food crops and loss of agrobiodiversity	FAO, 1999; Norberg-Hodge, 2001; Steel, 2008
Lack of visibility of food production: in daily life by urban inhabitants and resultant disconnection from perceived importance	Gobster, 2007; Zeunert, 2011
Perceptions, prevalent binary dualisms: town/country; urban/rural; culture/nature and production/ consumption, reinforcing marginalisation rather than the integration of food production	Chapter 12
Resistance to visual character: of urban and peri-urban agriculture in many nations seeking to express 'modernity' and wealth	Zeunert, 2011; Gyertyán, 2014; Morgan, 2015; Chapters 12 & 30
Lack of land protection: facilitation mechanisms and planning zoning for peri-urban and urban agriculture	LA and GLA, 2010; Chapters 12, 13, 14, 15 & 26

the foreseeable future there will be a return to agriculture systems less reliant on fossil fuel inputs (Kunstler, 2006; Odum and Odum, 2008; Chapter 10).

Many developed cities only possess three or four days of food stocks, revealing production and distribution vulnerability and small margins for interruptions and errors. A London Assembly and Greater London Authority report described this as "sleepwalking into a major problem" (LA and GLA, 2010: 9). Continual food supplies are enabled by transportation and logistics networks linking producers, processors, distributors, and retailers. Accordingly, a resilient food supply is not only vulnerable to finite natural resource dependence (as examined in the *Sustainability* section) but also to agreeable global relations and the resilience of multiple co-dependent infrastructural, economic, trade, and diplomacy networks. Increasingly volatile climatic conditions, which are widely forecast to increase into the 21st century, present substantial additional challenges across this production and supply chain (FAO, 2016).

The importance of *biological diversity* in providing increased ecological resilience is well documented and understood (see for example, the UN International Convention of Biological Diversity). For agricultural systems, however, the concept of *agricultural biodiversity* (agrobiodiversity) is poorly or little understood outside of some 'alternative' approaches (Chapters 10 and 24). Industrial agriculture's reductive and controlling system obviates diversity (Carson, 1962; Jackson, 1980; Norberg-Hodge et al., 2001; Steel, 2008). The resultant homogenisation amalgamates cultures, practices, landscapes, crop varieties, and foodstuffs (Norberg-Hodge et al., 2001; Steel, 2008; Roy and Ong, 2011; Speak and Kumar, 2016; Chapters 21 and 31). This is well articulated by a FAO report:

> Since the 1900s, some 75 percent of plant genetic diversity has been lost as farmers worldwide have left their multiple local varieties and landraces for genetically uniform, high-yielding varieties [. . .] Today, 75 percent of the world's food is generated from only 12 plants and five animal species.
>
> *(FAO, 1999)*

Signatory nations to the United Nations Convention on Biological Diversity (signed by 150 government leaders at the 1992 Rio Earth Summit) possess a legal obligation to preserve agrobiodiversity. 'Natural' and diversified farming methods can encourage agrobiodiversity and biodiversity, such as through intercropping, polycultures, companion planting, and use of natural predators, while also providing multiple ecosystem services, reduced environmental externalities and need for off-farm inputs (Sandhu et al., 2010; Kremen and Miles, 2012; Chapters 10 and 17). It is also crucial to ensure that seed collection, seed use, and seed swaps maintain independence from corporate control and patents (Chapter 31). Such traditional farming practices, however, make up an increasingly small proportion of industrial agriculture (Chapters 9 and 10). Subsequently, many nations are failing to address agrobiodiversity issues, and certainly beyond storing seeds in seed banks. These factors indicate a broader need to tackle sociocultural, power and economic structures within the global, industrial food system to increase sustainability, resilience, and diversity (Howard, 2016; Chapter 31).

Humankind has become substantially urbanised since the enclosure of common land, industrialisation, and, in particular, during the rapid economic development of emergent economies (such as China and India) (Zeunert, 2017: 8–37). This rural-urban shift is evident in the aforementioned (less than) 1% of the population engaged in agriculture in countries early to urbanise such as the USA and UK. While agricultural production creates flow-on employment through processing, value-adding, consumption, and so forth, proportions engaged in the production of food in developed economies are minor. In contrast, lower GDP per-capita countries such as Guinea in Africa are in fact more resilient in respect to the ratio of food producers to food consumers, with 75% of their workforce engaged in agriculture (World Bank, n.d. a). This ratio is declining globally (Satterthwaite et al., 2010), demonstrating intergenerational vulnerability in the loss of pre-industrial agricultural knowledge and skills (see Chapter 21). Accordingly, large percentages of populations in highly urbanised countries possess minimal to no access to land for food production and/or lack necessary skills, with consumers wholly reliant on the resilience of their food systems. There are a small number of communities attempting to prepare for post carbon, climate change era agriculture, and future food scenarios (Odum and Odum, 2008; Holmgren, 2009; Heinberg and Lerch, 2010; Berlin, 2016), largely without government support or funding (such as Transition Towns and Networks (Hopkins, 2008; Audet, 2016)).

Regenerative agricultural landscapes

> Edible, *adj.*: Good to eat, and wholesome to digest, as a worm to a toad, a toad to a snake, a snake to a pig, a pig to a man, and a man to a worm.
>
> (Bierce, 1911)

Despite decades of theory and discourse, sustainability practice has failed to deliver viable and resilient long-term food systems. Can *regenerative* agriculture provide a blueprint for enduring food systems that, as defined by the author:

Work with, rather than against, natural processes to promote ecosystem vitality, rebuild biocapacity and achieve ongoing food system outputs exceeding their environmental food footprint?

Regenerative practices differ from *sustainability* pursuits (which essentially seeks net balance, continuance, zero harm, or simply to increase efficiency) in that they actively strive to give back, renew, restore, and achieve net benefit; to do more good than harm. Sustainability can arguably be anthropocentrically focussed, whereas a regenerative philosophy is more ecocentric, viewing humankind as guests or stewards rather than masters, or, as is evident in Bierce's quote above, part of the cycle of life (also see Chapter 24). Regenerative agriculture may be defined as aiming to grow as much food using as few resources as possible in a way that revitalises soil and sequesters carbon (Toensmeier and Herren, 2016). Through biologically attuned production methods, it aims to achieve increasingly complex ecological structures over time, so that yields increase while external inputs decrease (Falk, 2013).

Regenerative practices can be traced over thousands of years of agriculture, land design, and practice, yet are relatively rare within the current global food system (and indeed the history of agriculture) (Jackson, 1980). Nevertheless, certain practitioners have historically sought 'beyond sustainability' outcomes without necessarily identifying as 'regenerative' (with 20th century proponents including George Washington Carver, Albert Howard, William Albrecht, Evelyn Balfour, J.I. and Robert Rodale, Masanobu Fukuoka, John Hamaker, Bhaskar Save, Bill Mollison and David Holmgren, Sepp Holzer, Takao Furuno, and John Jeavons). Permaculture (Mollison and Holmgren, 1978; see Chapters 10 and 24) is one such holistic expression, providing a theoretical and ethical framework for 'permanent agriculture' and living. Co-originator Holmgren (2002) recognises the need to be regenerative in a capacity beyond sustainability, and Mollison's (1988) substantial text is one of many outlining its practical application. Other aspects of regenerative agriculture incorporate restorative agriculture/farming (Shepard, 2013), natural farming (Fukuoka, 1978), agroecology (Tull et al., 1987; Méndez et al., 2015; Chapters 17 and 24), forest gardening (Hart, 1996), water harvesting (Yeomans, 1965), perennial agriculture (Jackson, 1980; Shepard, 2013; Toensmeier and Herren, 2016), grazing landscape and animal management (Salatin, 2011), and some organic and biodynamic practices (Chapter 10). Consistent with these practices, regenerative agriculture encourages, develops, and utilises rather than suppresses biodiversity, returning multiple ecosystem services and agricultural benefits (Kremen and Miles, 2012).

Terminology's constant flux, reinvention, and reuse has seen various incarnations of regenerative agriculture, indicating its enduring principles and that it is, in essence, not a new construct. Early texts explicitly utilising the term *regenerative* include Madden's (1984) book (also using 'organic' and 'sustainable') and several of the Rodale Institute's (USA) mid 1980s texts (see for example Francis and Harwood, 1985; Tull et al., 1987). The term has recently resurfaced following a period of adaptation, demonstrated by the 2015 Organic Consumers Association's 'Regeneration International' and their online annotated bibliography (http://regenerationinternational.org/annotated-bibliography/). This resurgence is in part due to the ability to sequester greenhouse gas emissions through building soil carbon and carbon farming (Niles, 2008; White, 2013; Ohlson, 2014; White, 2014; Eisenstein, 2015; Toensmeier and Herren, 2016).

The notion of being regenerative in built environment and landscape design has been attributed to landscape architect John Lyle's *Regenerative Design for Sustainable Development* (1994) and developed through a number of subsequent works (such as Van der Ryn and Cowan, 1996; McDonough and Braungart, 2002; Capra, 2002; Reed, 2007; Benson and Roe, 2007; France, 2008; Papanek, 2009; McDonough and Braungart, 2013). Capra expresses design's need to be regenerative through "bridg[ing] the wide gap between human design and the ecologically sustainable systems of nature" (2002: 233). McDonough and Braungart have achieved success through articulating regenerative principles in relation to design of the built environment, industrial processes and products through their books *Cradle to Cradle* (2002), which advocated design and construction of the built environment to emulate the abundance found in 'nature', and *The Upcycle: Beyond Sustainability – Designing for Abundance* (2013), where they advocate for circulating energy flows that maintain or increase. Truly regenerative design can be difficult to achieve given the built environment's commercial constraints within the current market-led model.

Landscape architecture arguably offers greater potential for regenerative processes than architecture and construction due to its direct engagement with living systems, however, it comprises a small proportion of built environment activity and focus (Bélanger in Zeunert, 2017: 123). The Sustainable SITES landscape rating tool (launched in 2009) demonstrates an awareness of regenerative landscape processes within the landscape discipline, however, quantification to date has proven challenging (Zeunert, 2017: 274–291). Landscape's regenerative capacity is evident in an historic and ongoing legacy of design projects (Zeunert, 2017). As in regenerative agriculture, the study of whether or not net energy and regenerative benefits have been achieved is yet to be adequately quantified in quantitative science. The Landscape Architecture Foundation's 'Performance Series' (since 2010; see https://landscapeperformance.org/) contributes to this process (Zeunert, 2017: 288), however, is not specifically focused on regenerative aspects. A range of new urban agriculture discourse and practice is exploring regenerative systems through planning, landscape architecture, architecture, urban design, and other disciplines (see Chapters 12, 26 and 27).

Both regenerative agriculture and design require further quantifiable research in order to determine their measurable outcomes. Nevertheless, "there is overwhelming evidence both that the methods work and they may offer the means to address a number of prevailing environmental challenges" (Rhodes, 2012: 345). Soil is a more quantifiable element of regenerative agriculture (relevant to soil building, health, carbon, and minerals – see for example Jeavons, 2006; Jackson, 2010; Rhodes, 2012). Regenerative outcomes may be achieved through minimising soil disturbance and tillage (conservation and zero tillage), composting and mulching, perennial agriculture, crop rotation, companion planting, utilising cover crops and green manures. The foundation of soil for life embodies this regenerative systems perspective:

> The soil is the great connector of lives, the source and destination of all. It is the healer and restorer and resurrector, by which disease passes into health, age into youth, death into life. Without proper care for it we can have no community, because without proper care for it we can have no life.
>
> *(Berry, 1977: 86)*

The notion of being *regenerative* as framed in this section provides an antidote to unsustainable industrial food systems and the rapacious agribusiness model (see Chapters 9 and 31). For regenerative design and agriculture to substantially expand beyond fringe and marginal activities, however, limiting modes of perception and quantification require alignment with the notion of (re)building more natural capital than what is extracted and ultimately, increasing biocapacity. While there is abundant ongoing discourse related to personal well-being and social concerns (such as whether organic food is more nutritious than conventional farming), maintaining a purely anthropocentric focus ultimately fails to encompass the holistic systems perspective required to achieve enduring, resilient, and regenerative food and landscape systems.

Conclusion

> To say the world is running out of time then, [is] to say the world is running out of usable energy. In the words of Sir Arthur Eddington, 'Entropy is time's arrow'.
>
> (Rifkin and Howard, 1980: 48)

Dwindling finite fossil fuel energy reserves have led to increasingly desperate attempts for their extraction from dangerous and low quality sources (such as deep water and arctic oil drilling, tar sands, and fracking).

Our heavy reliance on these resources – particularly natural gas, crude oil, and phosphorus – for daily food intake indicates that our industrial food systems are indeed running out of time and energy (Brown, 1996; Holmgren, 2002; Heinberg, 2005; Kunstler, 2006; Erisman et al., 2008; Hopkins, 2008; Roberts, 2008; Steel, 2008; Holmgren, 2009; Pimentel, 2009; Jarosz, 2009; Cribb, 2010; Weis, 2010; Deffeyes, 2010; Heinberg and Lerch, 2010; Herbert, 2011; FAO, 2011a; Rosin et al., 2012; Nikiforuk, 2012; FAO, 2016). Much agriculture globally is heavily and increasingly dependent upon diminishing and finite energy sources, and is thus fundamentally unsustainable. Industrial agriculture devours copious volumes of freshwater (Brown, 1996) and discards polluted wastewater, mines and degrades agricultural soil and land (Jackson, 1980; Fedoroff et al., 2010: 833), while unethically distributing output calories for consumption, and exhibiting high leakage from food wastage (Roberts, 2008: 84; FAO, 2011a: ix; FAO, 2011b). Cornucopian and neoliberal advocates for perpetual economic and population growth and scientific advancement continue to push business-as-usual practices through increasing industrialisation and intensification of global food production and further infiltration of agribusiness into 'undeveloped' regions (Fedoroff et al., 2010: 833; Chapter 21). Such processes will likely be assisted through ongoing molecular techniques, plant breeding (Tester and Langridge, 2010), technologies (such as drones), and robotics that may help to stall population die-off in the short-term, however, ultimately these approaches (with the exception of plant breeding) have demonstrated that they deplete planetary carrying capacity, heighten climate change, and undermine intergenerational sustainability. The western agricultural and agribusiness model is far from sustainable – a key focus discussed throughout chapter – so how can its continued advancement possibly present a viable long-term solution?

To avoid population collapse it is essential to rapidly and extensively deploy various key measures in agricultural systems to increase resilience and regeneration. These include: incentivising population de-growth; departing from perpetual growth economic systems (Klein, 2014; Newell, 2015), or at least decoupling food systems for maximum self-sufficiency (Clinton, in Cebon, 2010); switching to plant-based diets (FAO, 2016; Peters et al., 2016); increasing food sovereignty (McMichael, 2015; Chapter 31), food security (Morgan, 2015; Chapters 20 and 25), food justice (Chapter 32), and access to common land (Chapters 28 and 30); adapting practices and crops to unstable climatic conditions (Chapters 18 and 19); increasing implementation of perennial agriculture and extensive means of topsoil preservation; increasing the instances of alternative and ethical agricultures (Chapters 10 and 24); carbon farming; ceasing food waste; increasing cultural flexibility in perceiving foodstuffs and food sources; improving water and nutrient cycling (Chapters 19 and 23); expanding short-chain food networks through producing, eating and shopping locally; and actively implementing regenerative agriculture. These primary measures would make substantial inroads in the transition and reconfiguration of ensuring that agricultural, food and natural systems are increasingly regenerative and this needs to be accompanied by social change and resilient social systems.

The innate regenerative capacity of natural systems, plants and soil, when executed through systems-based environmental design processes, and implemented and managed by ethical and fair social structures and stewardship, present significant capacity and hope to provide sustainable food systems and resilient landscapes. Practices that harness and direct 'natural' processes and use regenerative techniques can both rebuild biocapacity and provide a potentially enduring blueprint for ongoing production methods that do not undermine the foundation for ongoing habitation of the Earth.

Acknowledgements

The author would like to thank Alys Daroy for her critical review and editing. Grateful appreciation is given to the two anonymous peer reviewers, as well as to Beau Beza, David Jones, and Tim Waterman for their review and feedback.

Further reading

ABC (2011) *Hungry Beast: Coles & Woolies*, http://www.youtube.com/watch?v=lnQMOyGeP_I (accessed 24 May 2011).

Allen, E. (2010) Growing community food systems. In: Heinberg, R., & Lerch, D. (eds.), *The post carbon reader: Managing the 21st century's sustainability crises*. Healdsburg, CA: Watershed Media, pp. 140–150.

Allen, W., & Wilson, C. (2012) *The good food revolution: Growing healthy food, people, and communities*. New York: Gotham Books.

Audet, R. (2016) Transition as discourse. *International Journal of Sustainable Development* (19), 365.

Barrett, G., & Odum, E. (2000) The twenty-first century: The world at carrying capacity. *BioScience*, 50, 363.

Benson, J., & Roe, M. (2007) *Landscape and sustainability*. London: Taylor and Francis.

Berlin, K. (2016) Alternative economies for the anthropocene: Change, happiness and future scenarios. *Ecozon@* 7(2), 149–164.

Berry, W. (1977) *The unsettling of America: Culture and agriculture*. San Francisco: Sierra Club books.

Bierce, A. (1911) *The devil's dictionary*. New York: Neale.

Blandford, D., Gaasland, I., & Vårdal, E. (2015) Trade liberalization versus climate change policy for reducing greenhouse gas emissions in agriculture: Some insights from Norway. *Applied Economic Perspectives and Policy*, 37, 418–436.

Blazey, C., & Varkulevicius, J. (2006) *The Australian fruit and vegetable garden: grow the best fruit and vegetables for good health and flavor*. Dromana, Vic: Digger's Club.

Blythman, J. (2004) Permanent global summertime. *The Ecologist* www.theecologist.org/investigations/food_and_farming/268630/permanent_global_summertime.html (accessed 15 May 2017).

Boland, J. (2010) *Food waste, what is it, how do we count it and how do we reduce it?*, online proceedings of the From Plains to Plate Conference Adelaide, February 12, 2010, http://futureoffoodsa.ning.com/page/presentations-1 (accessed 4 August 2010).

Bomford, M. (2010) Getting fossil fuels off the plate. In: Heinberg, R., & Lerch, D. (eds.), *The post carbon reader: Managing the 21st century's sustainability crises*. Healdsburg, CA: Watershed Media, pp. 119–127.

Bowler, I. (1992) *The geography of agriculture in developed market economies*. Harlow: Longman Scientific and Technical.

Brown, L. (1996) *Tough choices: Facing the challenge of food scarcity*. New York: Norton.

Burton, P., Lyons, K., Richards, C., Amati, M., Rose, N., Des Fours, L., Pires, V., & Barclay, R. (2013) *Urban food security, urban resilience and climate change*. Gold Coast: National Climate Change Adaptation Research Facility.

Campbell, C. (2005) *Oil crisis*. Brentwood, UK: Multi-Science Publishing.

Capra, F. (2002) *The hidden connections: Integrating the biological, cognitive, and social dimensions of life into a science of sustainability*. New York: Doubleday.

Carson, R. (1962) *Silent spring*. Boston: Houghton Mifflin.

Catton, W. (1980) *Overshoot: The ecological basis of revolutionary change*. Urbana, IL: University of Illinois Press.

Cebon, M. (2010) *The impacts of Australian trade and agriculture policies for farmers (and everyone else)*, online proceedings of the From Plains to Plate Conference Adelaide, February 12, 2010. http://futureoffoodsa.ning.com/page/presentations-1 (accessed 4 August 2010).

Claeys, P. (2015) *Human rights and the food sovereignty movement: Reclaiming control*. Oxon: Routledge.

Conway, G. (1998) *The doubly green revolution: Food for all in the 21st century*. Ithaca, NY: Cornell University Press.

Cooley, J. (1994) *Earthly words: Essays on contemporary American nature and environmental writers*. Ann Arbor, MI: University of Michigan Press, pp. 19–28.

Cresswell Riol, K. (2016) *The right to food guidelines, democracy and citizen participation*. Oxon: Routledge.

Cribb, J. (2010) *The coming famine: The global food crisis and what we can do to avoid it*. Berkeley, CA: University of California Press.

Cribb, J. (2017) *Surviving the 21st century: Humanity's ten great challenges and how we can overcome them*. Switzerland: Springer.

Cronon, W. (ed.) (1995) *Uncommon ground: Toward reinventing nature*. New York: W.W. Norton and Co.

Daly, H., & Farley, J. (2011) *Ecological economics: Principles and applications*. Washington, DC: Island Press.

de la Salle, J., & Holland, M. (2010) *Agricultural urbanism: Handbook for building sustainable food and agriculture systems in 21st century cities*. Manitoba: Green Frigate Books.

de Zeeuw, H., & Drechsel, P. (2015) Cities and agriculture: Developing resilient urban food systems. New York: Routledge.
Deffeyes, K. (2010) *When oil peaked*. New York: Hill and Wang.
DeGregori, T. (2002) *Bountiful harvest: Technology, food safety, and the environment*. Washington DC: Cato Institute.
Denny, G. (2012) Urban agriculture and season food footprints: An LCA study of tomato production and consumption in the UK. In: Viljoen, A., & Wiskerke, J. (eds.), *Sustainable food planning: Evolving theory and practice*. Wageningen, The Netherlands: Wageningen Academic Publishers.
Desrochers, P. (2009) Deconstructing the population bomb. *PERC Report*, 27(3). www.quebecoislibre.org/09/090915-11.htm (accessed 3 December 2016).
Desrochers, P., & Hoffbauer, C. (2009) The post war intellectual roots of the population bomb. Fairfield Osborn's 'Our Plundered Planet' and William Vogt's 'Road to Survival' in retrospect. *The Electronic Journal of Sustainable Development*, 1(3), 37–61.
Diamond, J. (2005) *Collapse: How societies choose to fail or survive*. London: Penguin.
Edwards, A. (2005) *The sustainability revolution: Portrait of a paradigm shift*. Gabriola: New Society Publishers.
Ehrlich, P. (1968) *The population bomb*. New York: Ballantine Books.
Ehrlich, P., & Ehrlich, A. (2009) The population bomb revisited. *Electronic Journal of Sustainable Development*, 1(3), 63–71.
Eisenstein, C. (2015) We need regenerative farming, not geoengineering. *The Guardian*. www.theguardian.com/sustainable-business/2015/mar/09/we-need-regenerative-farming-not-geoengineering (accessed 24 November 2016).
Elkington, J. (2004) Enter the triple bottom line. In: Henriques, A., & Richardson, J. (eds.), *The triple bottom line, does it all add up? Assessing the sustainability of business and CSR*. London: Earthscan, pp. 1–16.
Erisman, J., Sutton, M., Galloway, J., Klimont, Z., & Winiwarter, W. (2008) How a century of ammonia synthesis changed the world. *Nature Geoscience*, October, 1, 637.
Falk, B. (2013) *The resilient farm and homestead: An innovative permaculture and whole systems design approach*. Vermont: Chelsea Green.
FAO (1995) *Dimensions of need – an atlas of food and agriculture*. www.fao.org/docrep/u8480e/u8480e07.htm (accessed 16 May 2016).
FAO (1999) *What is happening to agrobiodiversity? Women: Users, preservers and managers of agro-biodiversity*. www.fao.org/docrep/007/y5609e/y5609e02.htm (accessed 11 April 2011).
FAO (2006) *Livestock's long shadow: Environmental issues and options*. Food and Agriculture Organization of the United Nations. www.fao.org/docrep/010/a0701e/a0701e00.HTM (accessed 14 January 2011).
FAO (2011a) *The state of the world's land and water resources for food and agriculture*. Oxon: Routledge.
FAO. (2011b) *Global food losses and food waste – extent, causes and prevention*. Rome: Food and Agriculture Organization of the United Nations.
FAO (2016) *The state of food and agriculture*. Rome: Food and Agriculture Organization of the United Nations.
Fedoroff, N.V., Battisti, D.S., Beachy, R.N., Cooper, P.J.M., Fischhoff, D.A., Hodges, C.N., Knauf, V.C., Lobell, D., Mazur, B.J., Molden, D., Reynolds, M.P., Ronald, P.C., Rosegrant, M.W., Sanchez, P.A., Vonshak, A., & Zhu, J.-K. (2010) Radically rethinking agriculture for the 21st century. *Science*, 327, 833–834.
Flannery, T. (2005) *The weather makers: How man is changing the climate and what it means for life on Earth*. New York: Atlantic Monthly Press.
France, R. (2008) *Handbook of regenerative landscape design*. Boca Raton: CRC Press.
Francis, C., & Harwood, R. (1985) *Enough food: Achieving food security through regenerative agriculture*. Emmaus, PA: Rodale Institute.
Freyfogle, E. (2006) *Why conservation is failing and how it can regain ground*. New Haven, CT: Yale University Press.
Freyfogle, E. (2007) *On private property: Finding common ground on the ownership of land*. Boston, MA: Beacon Press.
Fücks, R. (2015) *Green growth, smart growth: A new approach to economics, innovation and the environment*. New York: Anthem Press.
Fukuoka, M. (1978) *The one-straw revolution: An introduction to natural farming*. Emmaus: Rodale Press.
Gaarde, I. (2017) *Peasants negotiating a global policy space*. Oxon: Routledge.
Gaballa, S., & Abraham, A. (2008) *Food miles in Australia: A preliminary study of Melbourne, Victoria*. Melbourne: Centre for Education and Research in Environmental Strategies.
Garnett, T. (2003) *Wise moves: Exploring the relationship between food, transport and* CO_2. London: Transport 2000 Trust.
Gerland, P., et al. (2014) World population stabilization unlikely this century. *Science*, 346, 234–237.

Gobster, P., Nassauer, J., Daniel, T., & Fry, G. (2007) The shared landscape: What does aesthetics have to do with ecology? *Landscape Ecol*, 22, 959–972.

Godfray, C., Beddington, J., Crute, I., Haddad, L., Lawrence, D., Muir, J., Pretty, J., Robinson, S., Thomas, S., & Toulmin, C. (2010) Food security: The challenge of feeding 9 billion people. *Science*, 327, 812–818.

Godfray, H., Crute, I., Haddad, L., Lawrence, D., Muir, J., Nisbett, N., Pretty, J., Robinson, S., Toulmin, C., & Whiteley, R. (2010) The future of the global food system. *Philosophical Transactions of the Royal Society of London. Series B, Biological Sciences*, (365), 2769–2777.

Goodell, J. (2010) *How to cool the planet: Geoengineering and the audacious quest to fix earth's climate*. Boston, MA: Houghton Mifflin Harcourt.

Goodland, R. (2010) The overlooked climate solution: Joint action by governments, industry, and consumers. *Journal of Human Security* (6), 50–60.

Goodstein, D. (2005) *Out of gas: The end of the age of oil*. New York: WW Norton.

Gordon, I.J., Prins, H.H.T., & Squire, G.R. (2017) *Food production and nature conservation: Conflicts and solutions*. New York: Routledge.

Graymore, M., Sipe, N., & Rickson, R. (2010) Sustaining human carrying capacity: A tool for regional sustainability assessment. *Ecological Economics*, 69 459–468.

Gyertyán, F. (2014) *Growing delight – the aesthetic dimension of urban gardening in Amsterdam*. Master's thesis, Wageningen University. https://issuu.com/frrrr/docs/report_fruzsina_final_small_format (accessed 5 January 2015).

Halweil, B. (2004) *Eat here: Reclaiming homegrown pleasures in a global supermarket*. New York: W.W. Norton and Company.

Handmer, J., & Dovers, S. (2013) *Handbook of disaster policies and institutions: Improving emergency management and climate change adaptation*. Abingdon: Routledge/Earthscan.

Hart, R. (1996) *Forest gardening: Rediscovering nature and community in a post-industrial age*. Totnes: Green Earth.

Harvey, D. (2005) *A brief history of neoliberalism*. Oxford: Oxford University Press.

Hawken, P., Lovins, A., & Lovins, L. (2000) *Natural capitalism: Creating the next industrial revolution*. New York: Little Brown and Company.

Heinberg, R. (2005) *What will we eat as the oil runs out?*, Conference Paper, FEASTA Conference, June 23–25, Dublin Ireland. www.feasta.org/events/foodconf/foodconfmedia.htm#heinberg (accessed 27 August 2008).

Heinberg, R. (2007) *Peak everything: Waking up to the century of declines*. BC Canada: New Society Publishers.

Heinberg, R. (2011) *The end of growth: Adapting to our new economic reality*. Gabriola Island: New Society Publishers.

Heinberg, R., & Lerch, D. (eds.) (2010) *The post carbon reader: Managing the 21st century's sustainability crises*. Healdsburg, CA: Watershed Media.

Herbert, B. (2011) *Age of cheap fuel is over: IEA*. www.abc.net.au/news/stories/2011/04/28/3202209.htm (accessed 4 June 2011).

Holmgren, D. (2002) *Permaculture: Principles and pathways beyond sustainability*. Hepburn: Holmgren Design Services.

Holmgren, D. (2009) *Future scenarios: How communities can adapt to peak oil and climate change*. Canada: Chelsea Green.

Holt-Giménez, E. (2015) Racism and capitalism: Dual challenges for the food movement. *Journal of Agriculture, Food Systems, and Community Development*, 23–25.

Hopkins, R. (2008) *The transition handbook*. White River Junction, VT: Chelsea Green.

Howard, P. (2016) *Concentration and power in the food system: Who controls what we eat?* London: Bloomsbury.

Jackson, W. (1980) *New roots for agriculture*. San Francisco: Friends of the Earth.

Jackson, W. (2010) Tackling the oldest environmental problem: Agriculture and its impact on soil. In: Heinberg, R., & Lerch, D. (eds.), *The post carbon reader: Managing the 21st century's sustainability crises*. Healdsburg, CA: Watershed Media, pp. 128–139.

Jansma, J.E., Wijnand, S., Stilma, E. van Oost, A., & Visser, A. (2012) The impact of local food production on food miles, fossil energy use and greenhouse gas emission: The case of the Dutch city of Almere. In: Viljoen, A., & Wiskerke, J. (eds.), *Sustainable food planning: Evolving theory and practice*. Wageningen, The Netherlands: Wageningen Academic Publishers, pp. 307–321.

Jarosz, L. (2009) Energy, climate change, meat, and markets: Mapping the coordinates of the current world food crisis. *Geography Compass*, 3/6, 2065–2083.

Jeavons, J. (2006) *How to grow more vegetables: And fruits, nuts, berries, grains, and other crops than you ever thought possible on less land than you can imagine*. Berkeley CA: Ten Speed Press.

Jensen, D., & McBay, A. (2009) *What we leave behind*. New York: Seven Stories Press.

Kearney, J. (2010) Food consumption trends and drivers. *Philosophical Transactions of the Royal Society of London. Series B, Biological Sciences*, (365–2), 793–807.

Khan, S., & Hanjra, M.A. (2009) Footprints of water and energy inputs in food production – global perspectives. *Food Policy*, 34, 130–140.

Khan, S., Khan, M.A., Hanjra, M.A., & Mu, J. (2009) Pathways to reduce the environmental footprints of water and energy inputs in food production. *Food Policy*, 34, 141–149.

Klein, N. (2014) *This changes everything: Capitalism vs. the climate*. London: Allen Lane.

Kremen, C., & Miles, A. (2012) Ecosystem services in biologically diversified versus conventional farming systems: Benefits, externalities, and trade-offs. *Ecology and Society*, 17.

Kullander, S. (2010) Food security: Crops for people not for cars, *Ambio*, 39, 249–256.

Kunstler, J. (2006) *The long emergency: S urviving the converging catastrophes of the 21st century*. London: Atlantic Books.

LA and GLA (2010) *Cultivating the capital: Food growing and the planning system in London*. London: London Assembly and Greater London Authority.

Ladner, P. (2011) *The urban food revolution: Changing the way we feed cities*. Gabriola: New Society Publishers.

Lane, M. (2010) The carrying capacity imperative: Assessing regional carrying capacity methodologies for sustainable land-use planning. *Land Use Policy*, 27, 1038–1045.

Leal Filho, W. (2015) *Handbook of climate change adaptation*. Heidelberg: Springer Reference.

Leggett, J. (2014) *The energy of nations: Risk blindness and the road to renaissance*. New York: Routledge.

LIC (2016) *New Zealand dairy statistics 2015–16*. Hamilton, NZ: Livestock Improvement Corporation Limited and DairyNZ Limited.

Lichtfouse, E. (2012) *Farming for food and water security*. Dordrecht, The Netherlands: Springer.

Linn, K. (1999) Reclaiming the sacred commons. *New Village Journal* (1, Community Revitalization), 42–49. Available from: www.newvillage.net/assets/docs/linn.pdf>

Locke, A., Wiggins, S., Henley, G., & Keats, S. (2013) *Diverting grain from animal feed and biofuels*. London: Overseas Development Institute.

Lovelock, J. (2006) *The revenge of Gaia: Earth's climate in crisis and the fate of humanity*. New York: Basic Books.

Lowder, S., Skoet, J., & Raney, T. (2016) The number, size, and distribution of farms, smallholder farms, and family farms worldwide. *World Development*, 87, 16–29.

Lowe, I. (2005) *A big fix*. Melbourne: Black Inc.

Lyle, J. (1994) *Regenerative design for sustainable development*. New York: John Wiley.

Maas, W. (1999) *Metacity datatown*. Rotterdam: MVRDV/010 Publishers.

MacLean, A. (2008) *Over: The American landscape at the tipping point*. New York: Abrams.

Madden, J. (1984) *Regenerative agriculture: Beyond organic and sustainable food production*. East Lansing, MI: Cooperative Extension Service, Michigan State University.

Marten, G. (2001) *Human ecology: Basic concepts for sustainable development*. Sterling, VA: Earthscan.

McDonough, W., & Braungart, M. (2002) *Cradle to cradle*. New York: North Point Press.

McDonough, W., & Braungart, M. (2013) *The upcycle: Beyond sustainability – designing for abundance*. New York: North Point Press.

McHarg, I. (1969) *Design with nature*. Garden City, NY: Natural History Press.

McMichael, P. (2009) A food regime genealogy. *The Journal of Peasant Studies*, 36(1), 139–169.

McMichael, P. (2015) The land question in the food sovereignty project. *Globalizations*, 12(4), 434–451.

Meadows, D. (1972) *The limits to growth; a report for the Club of Rome's project on the predicament of mankind*. New York: Universe Books.

Meadows, D., Randers, J., & Meadows, D.L. (2004) *The limits to growth: The 30-year update*. White River Junction VT: Chelsea Green.

Méndez, V., Bacon, C., Cohen, R., & Gliessman, S. (2015) *Agroecology: A transdisciplinary, participatory and action-oriented approach*. Boca Raton: CRC Press.

Millennium Ecosystem Assessment (Program) (2001) *Millennium ecosystem assessment*. www.millenniumassessment.org/en/index.html (accessed 3 April 2006).

Mollison, B. (1979) *Permaculture 2: Practical design for town and country in permanent agriculture*. Stanley TAS: Tagari.

Mollison, B. (1988) *Permaculture: A designers' manual*. Tyalgum: Tagari.

Mollison, B., & Holmgren, D. (1978) *Permaculture 1: A perennial agricultural system for human settlements*. Melbourne: Corgi.
Monbiot, G. (2014) *The pricing of everything*. SPERI Annual Lecture, Sheffield, 29 April 2014. www.youtube.com/watch?v=ni1tX0bpTR8 (accessed 25 May 2014).
Moreau, T., Adams, T., Mullinix, K., Fallick, A., & Condon, P. (2012) Recommended practices for climate-smart urban and peri-urban agriculture. In: Viljoen, A., & Wiskerke, J. (eds.), *Sustainable food planning: Evolving theory and practice*. Wageningen, The Netherlands: Wageningen Academic Publishers, pp. 295–306.
Morgan, K. (2013) Feeding the city: The challenge of urban food planning. In: Miazzo, F., M. Minkjan, & CITIES (eds.), *Farming the city: Food as a tool for today's urbanization*. Haarlem, The Netherlands: Valiz/Trancity.
Morgan, K. (2015) Nourishing the city: The rise of the urban food question in the global North. *Urban Studies*, 52: 1379–1394.
Mostafavi, M., & Doherty, G. (2010) *Ecological urbanism*. Baden: Lars Müller, pp. 216–217.
MVRDV (2009) *City pig*. www.youtube.com/watch?v=x1r5GutrXX4andindex=5andlist=PL7E60F6F32AFB9DB3 (accessed 10 August 2016).
MVRDV (2010) *Food city*. www.youtube.com/watch?v=5JRDHKAJqtAandindex=8andlist=PL7E60F6F32AFB9DB3 (accessed 10 August 2016).
Nabhan, G.P. (2013) *Growing food in a hotter, drier land: Lessons from desert farmers on adapting to climate uncertainty*. White River Junction, Vermont: Chelsea Green Publishing.
National Research Council (2010) *Toward sustainable agricultural systems in the 21st century*. Washington, DC: National Academies Press.
Newell, P. (2015) Green transformations in capitalism. In: Scoones, I., Leach, M., & Newell, P. (eds.), *The politics of green transformations*. New York: Routledge.
Nikiforuk, A. (2012) *The energy of slaves: Oil and the new servitude*. Vancouver: Greystone Books.
Niles, M. (2008) Sustainable soils: Reducing, mitigating, and adapting to climate change with organic agriculture. *Sustainable Development Law and Policy*, 9(1), 19–69.
Norberg-Hodge, H., Goering, P., Page, J., & Goering, P. (2001) *From the ground up: Rethinking industrial agriculture*. New York: Zed Books.
Nordhaus, T., & Shellenberger, M. (2004) *The death of environmentalism*. www.thebreakthrough.org/images/Death_of_Environmentalism.pdf (accessed 13 December 2013).
Odum, H. (1971) *Environment, power, and society*. New York: Wiley.
Odum, H. (1996) *Environmental accounting: Emergy and environmental decision making*. New York: Wiley.
Odum, H., & Odum, E. (2008) *A prosperous way down: Principles and policies*. Boulder, CO: University Press of Colorado.
O'Hear, A. (1999) *After progress: Finding the old way forward*. London: Bloomsbury.
Ohlson, K. (2014) *The soil will save us: How scientists, farmers, and foodies are healing the soil to save the planet*. New York: Rodale.
Oppenlander, R. (2012) *Comfortably unaware: What we choose to eat is killing us and our planet*. New York: Beaufort Books.
Oppenlander, R. (2013) *Food choice and sustainability: Why buying local, eating less meat, and taking baby steps won't work*. Minneapolis, MN: Langdon Street Press.
Papanek, V. (2009) *Design for the real world: Human ecology and social change*. London: Thames and Hudson.
Peters, C., Picardy, J., Darrouzet-Nardi, A., Wilkins, J., Griffin, T., & Fick, G. (2016) Carrying capacity of U.S. agricultural land: Ten diet scenarios. *Elementa: Science of the Anthropocene*, 4, 000116.
Philips, A. (2013) *Designing urban agriculture: A complete guide to the planning, design, construction, maintenance and management of edible landscapes*. Hoboken, NJ: Wiley.
Pimentel, D., Berger, B., Filiberto, D., Newton, M., Wolfe, B., Karabinakis, E., Clark, S., Poon, E., Abbett, E., & Nandagopal, E. (2004) Water resources: Agricultural and environmental issues. *BioScience*, 54(10), 909–918.
Pimentel, D. (2009) Energy inputs in food crop production in developing and developed nations. *Energies*, 2, 1–24.
Pimentel, D., & Pimentel, M. (2003) Sustainability of meat-based and plant-based diets and the environment. *American Journal of Clinical Nutrition* (78), 660S–663S.
Pimentel, D., & Pimentel, M. (2008) *Food, energy, and society*. Boca Raton, FL: CRC Press.
Pingali, P. (2007) Westernization of Asian diets and the transformation of food systems: Implications for research and policy. *Food Policy*, 32, 281–298.

Popkin, B., Adair, L., & Ng, S. (2012) Now and then: The global nutrition transition: The pandemic of obesity in developing countries. *Nutrition Reviews*, 70(1), 3–21.

Ramankutty, N., Evan, A., Monfreda, C., & Foley, J. (2008) Farming the planet: 1. Geographic distribution of global agricultural lands in the year 2000. *Global Biogeochemical Cycles*, 22.

Reed, B. (2007) Shifting from 'sustainability' to regeneration. *Building Research and Information*, 35, 674–680.

Rees, W. (1992) Ecological footprints and appropriated carrying capacity: What urban economics leaves out. *Environment and Urbanization*, 4, 121–130.

Rhodes, C. (2012) Feeding and healing the world: Through regenerative agriculture and permaculture. *Science Progress*, 95, 345–446.

Rifkin, J., & Hopward, T. (1980) *Entropy: A new world view*. New York: Viking Press.

Roberts, P. (2008) *The end of food*. Boston, MA: Houghton Mifflin.

Roggema, R., & Keeffe, G. (2014) *Why we need small cows: Ways to design for urban agriculture*. Velp, The Netherlands: VHL University of Applied Sciences.

Rosin, C., Campbell, H., & Stock, P. (2012) *Food systems failure: The global food crisis and the future of agriculture*. Abingdon, Oxon: Earthscan.

Roy, A., & Ong, A. (2011) *Worlding cities: Asian experiments and the art of being global*. Chichester, West Sussex: Wiley-Blackwell.

Salatin, J. (2011) *Folks, this ain't normal: A farmer's advice for happier hens, healthier people, and a better world*. New York: Center Street.

Sandhu, H., Wratten, S., & Cullen, R. (2010) Organic agriculture and ecosystem services. *Environmental Science and Policy*, 13, 1–7.

Satterthwaite, D., Mcgranahan, G., & Tacoli, C. (2010) Urbanization and its implications for food and farming. *Philosophical Transactions of the Royal Society B: Biological Sciences*, 365, 2809–2820.

Saul, J.R. (2005) *The collapse of globalism: And the reinvention of the world*. Toronto: Viking Canada.

Scienceman, D. (1987) Energy and emergy. In: Pillet, G., & Murota, T. (eds.), *Environmental economics: The analysis of a major interface*, Geneva: R. Leimgruber, pp. 257–276.

Seager, J., Read, C., & Stott, P. (1995) *The state of the environment atlas*. London: Penguin Books.

Shepard, M. (2013) *Restoration agriculture: Real-world permaculture for farmers*. Austin, TX: Acres U.S.A.

Simon, J. (1996) *The ultimate resource 2*. New York: Princeton University Press.

Smit, J., Ratta, A., & Nasr, J. (1996) *Urban agriculture: Food, jobs and sustainable cities*. New York: UNDP.

Speak, S., & Kumar, A. (2016) Fit and misfit – ordinary urban injustice in a global context. In: Davoudi, S., & Bell, D. (eds.), *Justice and fairness in the city*. Bristol: University of Bristol: Policy Press.

Springer, S., Birch, K., & MacLeavy, J. (2016) *The handbook of neoliberalism*. New York: Routledge.

Steel, C. (2008) *Hungry city: How food shapes our lives*. London: Random House.

Stewart, B., & Duane, T. (2009) Easement exchanges for agricultural conservation: A case study under the Williamson Act in California. *Landscape Journal*, 28, 181–197.

Stierand, P. (2012) Food policy councils: Recovering the local level in food policy. In: Viljoen, A., & Wiskerke, J. (eds.), *Sustainable food planning: Evolving theory and practice*. Wageningen, The Netherlands: Wageningen Academic Publishers.

Tester, M., & Langridge, P. (2010) Breeding technologies to increase crop production in a changing world. *Science*, 327, 818–822.

The Electronic Journal of Sustainable Development (2009) 1(3). geog.utm.utoronto.ca/desrochers/The_Population_Bomb.pdf (accessed 1 February 2017).

Toensmeier, E., & Herren, H. (2016) *The carbon farming solution: A global toolkit of perennial crops and regenerative agriculture practices for climate change mitigation and food security*. Vermont: Chelsea Green.

Tornaghi, C. (2012) Public space, urban agriculture and the grassroots creation of new commons: Lessons and challenges for policy makers, In: Viljoen, A., & Wiskerke, J. (eds.), *Sustainable food planning: Evolving theory and practice*. Wageningen, The Netherlands: Wageningen Academic Publishers, pp. 349–363.

Tull, K., Sands, M., & Altieri, M. (1987) *Experiences in success: Case studies in growing enough food through regenerative agriculture*. Emmaus, PA: Rodale International.

UN International Convention of Biological Diversity. www.cbd.int (accessed 23 March 2013).

UN (2015) *World population prospects: The 2015 revision, key findings and advance tables*, ESA/P/WP.242, United Nations, Department of Economic and Social Affairs, Population Division. https://esa.un.org/unpd/wpp/Publications/Files/Key_Findings_WPP_2015.pdf (accessed 2 December 2016).

van der Ryn, S., & Cowan, S. (1996) *Ecological design*. Washington, DC: Island Press.

Viljoen, A., & Wiskerke, J. (eds.) (2012) *Sustainable food planning: Evolving theory and practice*. Wageningen, The Netherlands: Wageningen Academic Publishers.

Vogt, W. (1948) *Road to survival*. New York: W. Sloane Associates.

Wackernagel, M. (1994) *Ecological footprint and appropriated carrying capacity: A tool for planning toward sustainability*. PhD thesis, Vancouver, Canada: School of Community and Regional Planning, The University of British Columbia.

Weis, T. (2010) The accelerating biophysical contradictions of industrial capitalist agriculture, *Journal of Agrarian Change*, (10), 315–341.

Weller, R. (2009) *Boom town*. Western Australia: UWA Press.

White, C. (2014) *Grass, soil, hope: A journey through carbon country*. Vermont: Chelsea Green.

White, W. (2013) *Biosequestration and ecological diversity: Mitigating and adapting to climate change and environmental degradation*. Boca Raton, FL: Taylor and Francis.

Wilson, E. (2016) *Half Earth: Our planet's fight for life*. New York: W.W. Norton.

Wolf, S. (2014) *The neoliberal regime in the agri-food sector: Crisis, resilience and restructuring*. New York: Routledge.

World Bank (nd a) *Employment in agriculture (% of total employment)*. http://data.worldbank.org/indicator/SL.AGR.EMPL.ZS (accessed 9 May 2016).

World Bank (nd b) *Agricultural land (% of land area)*. http://data.worldbank.org/indicator/AG.LND.AGRI.ZS (accessed 9 May 2016).

World Commission on Environment and Development (1987) *Our common future*. Oxford: Oxford University Press.

Wright, J. (2009) *Sustainable agriculture and food security in an era of oil scarcity: Lessons from Cuba*. London: Earthscan.

WWF (2016) *Living planet report 2016: Risk and resilience in a new era*. Gland: WWF International.

WWF (nd) *Farming: Pollution*. wwf.panda.org/what_we_do/footprint/agriculture/impacts/pollution/ (accessed 27 January).

Yeomans, P.A. (1965) *Water for every farm*. Sydney: Murray.

Zeunert, J. (2011) *Eating the landscape: Aesthetic foodscape design and its role in Australian landscape architecture*. Proceedings of AILA National Conference Transform, Brisbane, 13 August 2011.

Zeunert, J. (2016) *New natures*, Conference Presentation, AILA Festival of Landscape Architecture, Canberra 28 October, National Gallery of Australia.

Zeunert, J. (2017) *Landscape architecture and environmental sustainability: Creating positive change through design*. London and New York: Bloomsbury.

17

Conservation and ecology

Gary Austin

Agricultural context

This chapter adopts an ecological-centric perspective and discusses ecosystem health including the preservation of native species of plants, animals, and insects. When discussing the production of food and its impact on ecological systems either within or adjacent to farms, it is useful to separate agriculture into two scales: (1) commodity or industrial scale agriculture, and (2) agriculture practiced within urban and peri-urban zones. Commodity agriculture is the mass production of a limited number of products (for wheat, rice, corn, soybeans, potatoes, and other staples) using entire regions, industrial scale machinery, and artificial inputs to maximise yield and labour productivity (Robertson and Swinton, 2005).

These intensive commodity systems, growing food for humans and livestock, are especially responsible for the destruction of natural ecosystems through conversion of wetlands, grasslands, and forests at the regional and sub-continental scale (Foley et al., 2011). This chapter presents effective planning, restoration and management measures within and around farms necessary for stable ecological functions and reduced rates of species extinction. Preservation or restoration of biodiversity within and adjacent to farmland has intrinsic value but also supports anthropocentric self-interest directly and indirectly (Robertson and Swinton, 2005).

This book links landscape and agriculture, suggesting that food is an ecosystem service of the landscape. We absolutely derive human benefit from soil, water, sunlight, and space as they interact to produce food. However, about 10,000 years ago we embarked on an increasingly artificial food production system. That system is incredibly important since virtually every human being relies on it for at least a portion of their food. Today, agricultural problems of soil with insufficient nutrients, water scarcity, insect damage, or plant diseases are overcome with the inputs of artificial fertiliser, artificial irrigation, artificial pesticides, fossil fuels, technology, and other approaches (see Chapter 16). Generally, all of these deficiencies and others occur together in large-scale industrial agriculture. Despite the artificial inputs, the benefit of this model is astounding productivity where the landscape is defined only as a substrate for food production and where conservation of resources is not relevant in the short-term. Such a starkly characterised food production system is not sustainable when enlarged to dominate regions. In fact, about one-third of the land area of the Earth is dedicated to the production of crops while another quarter is intensively grazed for livestock production (Scherr and McNeely, 2007). We must acknowledge that production of food commodities requires the irrevocable subtraction of billions of acres from natural ecosystems. Furthermore, there is a limited capacity to sustain additional losses of ecosystem area and functions as discussed in following sections in this chapter. However, there are planning and management measures that nations can take (and are already implementing) to mitigate ecosystem losses without significantly impacting food production, but these policies will change where and how agriculture is practiced (Fischer et al., 2008).

In contrast to commodity agriculture, within peri-urban and urban zones agriculture is practiced at a different scale and with different methods including organic farming, permaculture, hydroponic, and other production systems (see Chapters 10, 12, 13, 14 and 15). In addition, the farmers in these zones must respond to air and soil contaminants and water issues often ignored in commodity agriculture (Wortman and Lovell, 2013; Chapters 19 and 23). Peri-urban and urban farms are smaller but a far greater number of products are cultivated. In contrast to the global economy of commodity sales, these products are often sold locally or within the region. Because of the proximity to residential areas and as a matter of economic viability, organic productions methods are often employed (Martinez et al., 2010). At this scale and proximity to human settlements, the biodiversity focus is on species that are relatively tolerant of human activity and the movement of species through farms to large habitat areas in the region (Melles et al., 2003). The particular habitat and corridor requirements of these species are discussed later in the chapter as are the supportive on-farm measures that benefit biodiversity.

Conservation

Conservation and ecology in relationship to food systems are the topics of this chapter. In a simple understanding, conservation is the safeguarding of resources. Therefore, conservation is an ethic (see for example, Leopold, 1949; Chapter 24), a social construct, but conservation has been operationalised in conservation science and the legal system where policies have landscape impacts. We can speak of successes and failures in the conservation of soils or biological diversity (biodiversity) in the agricultural environment. Indeed, there has been a great deal of research on the agricultural benefits of utilising ecosystem services, such as the employment of beneficial insects as a substitute for artificial inputs that have adverse impacts on biodiversity (Scherr and McNeely, 2007). Seeking this win-win relationship between biodiversity and agricultural productivity reduces the devastating impact wrought by centuries of agricultural development (Chapters 9, 10 and 16). Farmers are generally receptive to the practical conservation of resources understood to improve production (and incidentally benefit biodiversity), but often adopt an adversarial position with regard to wildlife that can diminish production volumes or kill livestock. The terms 'eco-agriculture' and 'agroecology' (Chapters 24 and 33), and 'land sharing' are often used to mean efforts to sustain agricultural productivity while using fewer artificial inputs (Gliessman, 2007) or by enhancing non-production areas of farms for biological diversity enhancement.

A conservation perspective allows us to posit that the land, waters, and other resources dedicated to food production are human resources that we must conserve (safeguard) from degradation, land use change, and from environmental protections that might diminish the area or inputs required for food production. According to this perspective the conservation of farmland can be antithetical to ecosystem health.

On-farm measures to reduce soil erosion, water pollution, pesticide use, and other efforts to employ natural processes and organisms to foster agricultural productivity are essential for restoring greater biodiversity to agricultural lands. By themselves, however, they are inadequate for the restoration of ecosystem health or native biological diversity at the landscape scale when compared to a pre-agricultural condition. Furthermore, non-lethal farm and ranch management practices to reduce production losses are adopted more reluctantly as are financial compensation schemes (Robertson et al., 2014).

Ecology

In contrast to conservation, ecology is the study of organisms interacting with their environment, and as a system, its health can be measured, or at least characterised according to species abundance, diversity, energy transfer between tropic levels, biochemical complexity, nutrient cycling, and other measures (Farina, 2006). Ecosystems can be modelled at any scale but the landscape scale is necessary to include all species and ecosystems. At the landscape scale the conservation of healthy ecosystems has a different meaning than on-farm conservation of resources necessary for production. At the scale of the farm parcel, generally there will be insufficient habitat size and connectivity to satisfy the needs of most species. Nevertheless, some species with small territory requirements and with tolerance of human activity can be supported by on-farm measures (Robertson et al., 2014). We can speak of the 'conservation' of ecosystems but this likely requires a regional approach.

Food security

There are real human crises in food security, nutrition, and health in many countries (see Chapters 20, 21 and 22). Today about one billion people are hungry and malnourished due to cost and poor access to food, while demand for food is expected to double by 2050 (Foley et al., 2011). Increased production of commodity crops will be important, but local food production is critical for areas where hunger is endemic and as a rural economic development strategy (Foley et al., 2011). The global supply chain of commodity crops poorly serves parts of the world due to a poorly developed trade infrastructure or the lack of funds to purchase the products (Chapters 19, 21, 25 and 31). In this case, participation in local agriculture or purchasing local products with lower prices are critical to eliminating hunger. In fact, in some poor nations as much as two thirds of the population participates in the agricultural system while in developed countries a much smaller fraction (less than 2% in the USA) is involved (Hassan et al., 2005; The World Bank, 2016).

Local access to wild food including fish, game, and plants is important for the nutrition of poor people in many parts of the world (see Chapters 7 and 21), especially during times of drought and civil disruptions. This access is diminished as agriculture intensifies and on-farm biological diversity is eliminated. The production of cattle, chicken, pigs, other livestock, and fish in concentrated animal feeding operations (CAFOs) has increased the production of these foods but with substantial negative ecosystem impacts. The increased demand for the use of land to produce animal feed, increased discharge of excess nitrogen and phosphorus, and the outbreak of mad cow disease (bovine spongiform encephalopathy, BSE), severe acute respiratory syndrome (SARS), and avian flu are negative consequences of this intensification (Hassan et al., 2005).

Addressing the chronic hunger and malnutrition tragedies around the world must be coupled with restoring ecosystem functions. We are past the point where food production expansion results in ecosystem losses and species extinctions (Foley et al., 2011; IUCN, 2016). This is not an issue of whether humans are more important than any single endangered species. It is about sustaining enough ecosystem health to protect ourselves and all other organisms.

Defining the problem

Land dedicated to agriculture

Agriculture is the largest category of land use on the planet. Of the land surface not covered by ice, about 12% (crops) and 26% (pasture) is dedicated to agriculture (Foley et al., 2011). In the last 300 years, forests, and natural grasslands and savannas have decreased 20% and 40%, respectively, while croplands and pastures have increased 390% and 460%. Burning forest tracts, tilling natural grasslands, draining wetlands, and irrigating semi-arid and arid land accomplished the conversion. While about 18% of cultivated agriculture utilises irrigation, it is responsible for 40% of the global food production (Klein Goldewijk, 2013). Intensification is a current and future agricultural trend that results in highly productive monocultures, the diversion of surface water, pumping of groundwater, and the application of technology (artificial fertilisers, insecticides, herbicides, fungicides, plant hybridisation, and genetic modification), but also significant ecological damage. Overgrazing has similarly led to land use change and biodiversity loss. About 6% of the land has been converted to desert by agricultural practices (Klein Goldewijk, 2013).

In Europe, there has been nearly 400% increase in farm yield, but a 65% decrease in the number of farms, and a 77% decrease in farm jobs since 1945 (Robinson and Sutherland, 2002). Crop specialisation and use of machinery resulted in a 50% reduction in ecologically beneficial hedgerows. This intensification, including increased pesticide use and removal of autumn field stubble resulted in the loss of habitat area and quality as well as diminished soil quality (Robinson and Sutherland, 2002). In Canada, a similar pattern is evident. In Ontario land dedicated to row crops increased from 46–53%, supported by increased artificial fertiliser, pesticides and fast-growing crops, whose early harvest destroyed ground nesting birds. Therefore, farmland habitat declined 94% (Eilers, 2011).

Globally, there was a 56% increase in productivity between 1965 and 1985 but between 1985 and 2005 productivity increased 20%. The increase in productivity was the result of external inputs, fewer crop failures, and less fallow land rather than a large conversion of natural land to agricultural use (Foley et al., 2011). This intensification was the result of both agricultural technology and shorter rotation cycles. Land most suitable for agriculture is already in production but so is about 38% that is marginal. So expansion of agriculture is likely to be on marginal land (where yields are 40% lower than on suitable land) including the clearing of forested land (Alexandratos and Bruinsma, 2012).

Direct consumption of crops by humans represents 62% of the agricultural production, while 35% is to feed livestock and 3% is for biofuels and other purposes (Foley et al., 2011). These percentages vary regionally, with North America and Europe consuming about 40% directly while the percentage in Africa is double this amount. When arable farmland is converted to production of animal feed, this represents a reduction in the global food supply (Foley et al., 2011). Some livestock rangeland will be converted to manmade grasslands and some rangeland will be converted for row crops in the future. Both of these intensification processes will result in biodiversity losses. Despite negative ecosystem and food supply impacts, land dedicated to grazing is expected to increase in the coming decades (Alexandratos and Bruinsma, 2012).

Impact of modern agriculture on biological diversity

The incremental but extensive conversion of natural habitat to agriculture, commercial forestry, and urban development damages or eliminates ecosystem extent and quality until viable local populations become extinct. Between 1986 and 2005 there was a 3% increase in the amount of land dedicated to agriculture. However, most of this increase was within the tropical zone where 27% of the tropical forest has already been converted to agriculture (Foley et al., 2011). However, the negative impact on biological diversity and greenhouse gas emissions is substantial while the conversion of tropical forest to agriculture contributes little to the global food supply (Foley et al., 2011).

As suggested earlier, intensification of agriculture is the second great cause of biodiversity loss including losses in aquatic ecosystems and marine fisheries due to water pollution caused by intensive agriculture (fertiliser and pesticides) (Foley et al., 2011). Eventually, loss of habitat causes local extinction of plant and animal species and when local extinctions of populations pass a threshold, then the species becomes globally extinct (Chopra, 2005). Although climate change or predation (such as hunting) can cause species extinction, most extinctions since 1900 are the result of habitat loss or degradation. According to the International Union for Conservation of Nature and Natural Resources (IUCN), 23,250 of the nearly 80,000 assessed species are threatened with extinction. Habitat loss or degradation is responsible for the threatened status of more than 80% of the threatened species (IUCN, 2016).

Since commodity agriculture (including tree farming) is responsible for most of this habitat loss, it is correct to focus on this sector to arrest or mitigate current trends. Furthermore, the means of planting, harvesting, processing, and transporting food products is increasingly dependent on the combustion of fossil fuels that are changing the global climate. Some of the food production technologies that we have adopted are benign, while others are highly destructive, but ultimately we have failed to conserve soils, safeguard the climate, protect habitats or populations of native plants and animals, and clean agricultural tail water that creates ever-growing ecological dead zones in our coastal waters (La Point, 2007).

Species extinction

Like climate change, extinction of some species is a natural process that occurs at a measurable rate through various periods of earth history, but its exacerbation due to human acts is not natural. Recent research suggests that the extinction rate increase caused by humans is hundreds to thousands of times greater than the fossil record suggests it should be (He, 2011). Species extinctions are an unfolding catastrophe on the same scale as unchecked climate change. This catastrophe touches all taxa, threatens entire ecosystems and will ultimately adversely impact humans in ways we have yet to imagine. There is evidence that we are within the sixth mass extinction era in Earth history but this is the first time that one species threatened the existence of almost all other organisms (including itself, perhaps) (Rockström et al., 2009).

Focusing conservation efforts on extinction threats to individual species is a necessary, reactive, and crisis-driven approach. However, responses to extinction threats have been successful. For example, of the 235 species of ungulate (hoofed mammals) assessed in 1996 at least 65% would have become more highly subject to extinction if conservation programs were not implemented (IUCN, 2016).

Assessing the threat to species *groups* (rather than focusing on single species) provides a more comprehensive sense of where to dedicate conservation efforts. Preservation of ecosystems threatened by agriculture expansion and intensification and other land use change is an even more effective biological diversity protection strategy. Agricultural impacts on high visibility threatened species are well known, including the giant panda, orangutan, tiger, elephant, rhinoceros, and mountain gorilla. A host of less charismatic species are similarly threatened by land use change and intensification such as the creation of palm oil plantations and felling forests for grazing or crops (Associated Press, 2017; IUCN, 2016; Hassan et al., 2005).

The exacerbated extinction rate and the assessment of extinction threats provided below shows the urgent need to respond to this crisis. Figure 17.1 is an assessment of the historical threat to four groups of species (the initial position of the left end of the line) and the estimated increase in the threat of extinction for species in the group. The figure shows that coral species are trending toward extinction most rapidly while amphibians are currently the most highly endangered group with a steadily increasing extinction threat.

The IUCN assess the populations of 79,837 species around the world and reviews each species periodically for changes in the status of the species and its group. Figure 17.2 shows the status of species groups as of 2015. The amphibians group is not included, but an estimated 41% of amphibian species are threatened with extinction. A species is defined as 'critically endangered' or 'endangered' if its population diminished by 90% and 70%, respectively, over the last ten years.

The extinction of an entire species group is a devastating ecosystem loss. For example, 63% of all of the Cycad (*Cycas* sp.) species are threatened with extinction (Table 17.1).

As illustrated earlier, species extinctions are linked to human activities including food production, commercial forestry and fossil fuel consumption leading to climate change, and other activities. The exploding human population is expected to exceed nine billion by 2050 compared to a 2.5 billion population in 1950. Governments and industry have been eager to address the 'food production' problem but the 'population problem' has been avoided by almost every national government. We urgently need to gauge the capacity of our planet to sustain a population of humans that does not result in the human-mediated extinction of any other species (Vačkář et al., 2012). We are capable of adopting short-term planning and policy measures to manage the current food crises and establish a long-term balance between human population, food production, and ecosystem preservation and health (Lovell et al., 2010).

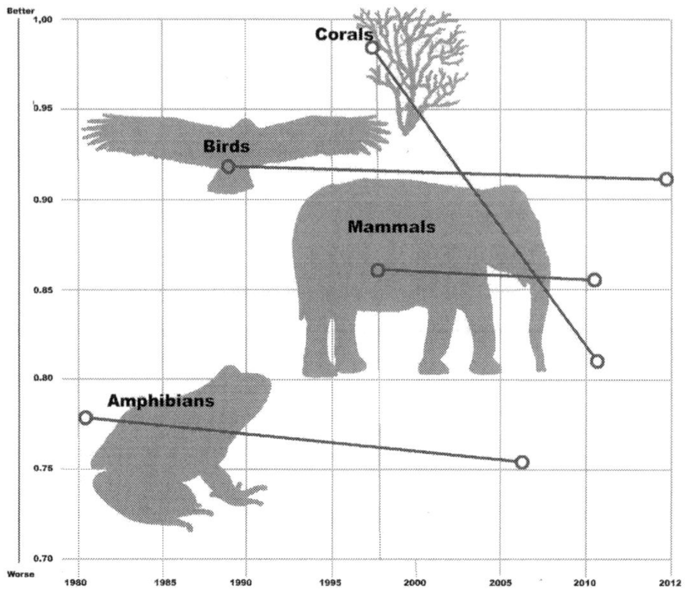

Figure 17.1 Extinction threat and historical trend. If there were no extinction pressure on all species in the group then the initial position of their line would be at 1 on the vertical axis. Downward trending lines indicate increasing extinction threat over time.

Source: IUCN, 2016

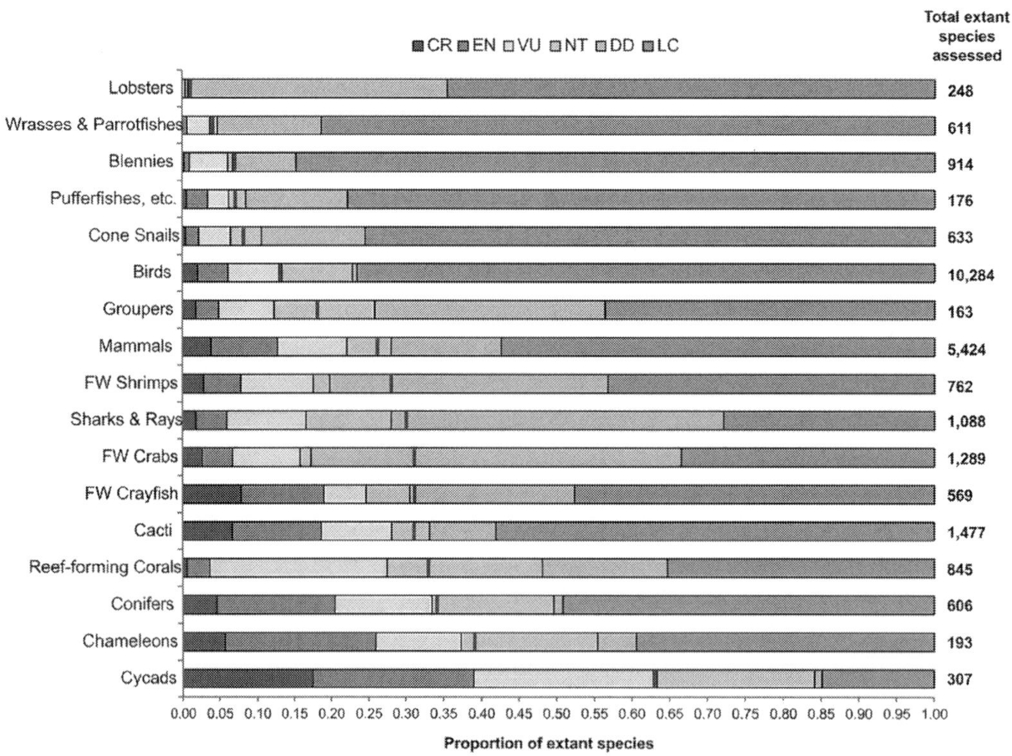

Figure 17.2 Status of species. CR – Critically Endangered, EN – Endangered, VU – Vulnerable, NT – Near Threatened, DD – Data Deficient, LC – Least Concern.

Source: IUCN, 2016

Table 17.1 Species within groups that are threatened with extinction.

Group	*% Threatened*
Cycads	63
Chameleons	39
Conifers	34
Reef-forming corals	33
Cacti	31
Sharks and rays	31
Freshwater crayfishes	31
Freshwater crabs	31
Freshwater shrimps	28
Mammals	26
Groupers	18
Birds	13
Cone snails	8
Pufferfishes, etc.	7
Blennies	7
Wrasses	4
Lobsters	<1

Source: IUCN, 2016

What can be done?

Implementing four strategies can satisfy the demand for food of an enlarging world population while reducing the current environmental and social impacts of the current food production and distribution system. The first strategy is to halt the conversion of land to agricultural use, especially in the tropics. It is ecologically critical that the demands for increased agricultural productivity occur only on land already under agricultural management (Fahrig, 2001). However, within the areas currently managed for agriculture, farmers might reconfigure the cultivated area to enhance biological diversity through ecological corridors and habitat creation (Harvey, 2007; Lovell et al., 2010). This can be accomplished without significant reduction in agricultural land use or productivity, particularly if this is coordinated by regional ecologists and associated with land swapping programs.

The second strategy includes the management of intensive cropping systems, improved agricultural species, and improved management of soil and water. Another promising approach is agriculture/forestry systems that produce a wider range of products while enhancing biological diversity (Scherr and McNeely, 2007). The widespread adoption of proven integrated pest management systems by subsistence and commercial operators reduces external inputs and increases diversity of plants, insects, and birds (Foley et al., 2011).

About 40% of available food is wasted in developed countries. The third strategy to increase the food supply is the improvement of methods for handling and storage. Another resource allocation strategy is managing 16 major crops (barley, cassava, groundnut, maize, millet, potato, oil palm, rapeseed, rice, rye, sorghum, soybean, sugar beet, sugarcane, sunflower, and wheat) for the production of human food, rather than for animal feed and biofuels. This strategy yields a 28% increase in food production (and 49% increase in food calories). Dietary reductions of animal protein to healthful levels, or better, substitution of plant based protein for most animal protein would serve the dual purposes of reducing obesity and chronic disease, and more efficient use of agricultural resources (Foley et al., 2011). A shift from the western, meat intensive diet to a more vegetarian diet holds great potential for meeting the future needs of the growing human population without converting more land to agricultural production (Peters, 2009; Chapter 16).

The last strategy is improving the yield potential of the 16 major food and feed crops. In some parts of the world the cultivation of these crops results in yields far below the possible maximum. However, use of improved seed varieties, and better nutrient and water management (see Chapters 19 and 23) can improve production without resorting to industrial production methods based on artificial fertilisers, pesticides, and energy. Achieving a yield of 95% maximum for these crops would increase the quantity of product by 58% (2.3 billion metric tons) (Foley et al., 2011).

The four strategies discussed above must be implemented simultaneously to increase the food supply to feed the anticipated growth in human population without increasing the cultivated area to include marginal farmland and diminish habitat or ecological corridors. Environmental goals for climate change mitigation, biodiversity, air and water pollution, and water use have direct or indirect implications for agriculture. These goals can be realised using the methods described in this and other chapters of this book while achieving food security (Foley et al., 2011).

Habitat size

Since habitat destruction, fragmentation, and decreased habitat quality are the primary reasons for loss of biological diversity (Mace et al., 2005), understanding the minimum habitat area required to support all organisms is an important policy and planning question. The territory requirement for species varies widely and some species tolerate some human presence or activity while others are intolerant of even low densities of human population. For example, about 25,000 acres is the home territory requirement for a single black bear, while much less is required for a bobcat (410 acres), fox (150 acres), deer (43 acres), bobwhite (20 acres), or cardinal (12 acres). However, the number of acres of suitable habitat is generally decreasing (Cerulean, 2008). For instance, in Florida an estimated 2.3 million acres of currently existing black bear habitat will be lost by 2060 (Cerulean, 2008). The extraordinary size of the territory needed by large predators and other large animals presents a serious challenge in densely populated regions of our increasingly urbanised planet. Even in areas of the western United States, where a large percentage of the land is publicly owned, the creation of a viable network of habitat preserves and ecological corridors is a challenge.

In the group of 11 western states, the US government owns 46.9% of the land (Figure 17.3) but only 4% of the land in the remaining states is in federal management. This accounting suggests that an international network of habitat and connecting corridors is feasible in the western US but is significantly more difficult elsewhere in the country (The Wildlands Network, 2017). As a whole, 28% (640 million acres) of the United States is owned and managed by the federal government (Vincent et al., 2014).

However, little of the federal land is managed exclusively as habitat. The only federal agencies with mandates to manage resources primarily for biological diversity are the Fish and Wildlife Service and the National Park Service. Nevertheless, portions of the federal land owned by the Forest Service, Bureau of Land Management, and the Department of Defense could be managed for optimum biological diversity to create a national or international network. Figure 17.3 illustrates that creating a continuous route from Canada to Mexico is feasible.

In 1967, *The Theory of Island Biogeography* and subsequent research documenting the extinction of species within national parks suggested that nature preserves are insufficient to sustain native biological diversity (MacArthur and Wilson, 2001). Conservation biologists in the 1980s and 1990s advocated for dedicating 50% of the US land area to ecosystem preservation, resulting in a biodiversity plan for the United States that was referenced by the United Nations Convention on Biological Diversity (Soulé and Noss, 1998). Environmental activists took up this cause, promoting the 'rewilding' of North and South America and other areas of the planet (Foreman, 2004). The proposal that 50% of the land and ocean area must be dedicated to ecosystem restoration and preservation in order to avoid the loss of one half to three quarters of planetary

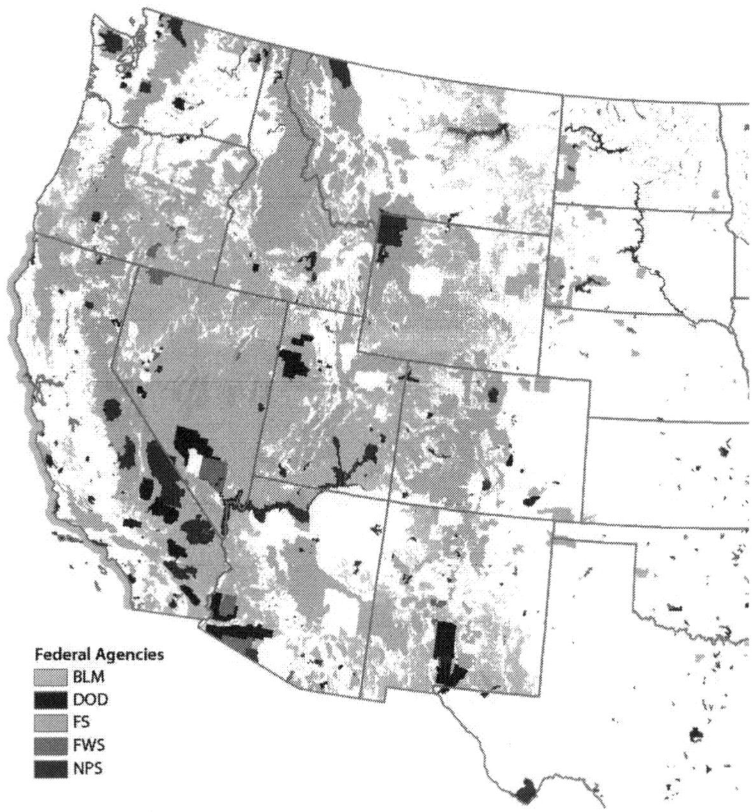

Figure 17.3 Federal ownership of land in the western United States. BLM – Bureau of Land Management, DOD – Department of Defense, FS – Forest Service, FWS – Fish and Wildlife Service, NPS – National Park Service.

Source: Map boundaries and information generated by CRS using federal lands GIS data from the National Atlas, 2005, and an ESRI USA Base Map (Vincent et al., 2014)

species has been recently reasserted by E.O. Wilson, Harvard scientist and Pulitzer Prize winning author (Wilson, 2016). The prospect of dedicating 50% of the land to habitat and corridors led to a significant political backlash from agriculture and industry groups and those that depend on the use of public land for private profit from grazing, timber harvest, and other pursuits. In North America 16.12% of the land area is within a protected preserve, while in Europe this amounts to 12.37%. Globally, about 11.5% of the land surface was dedicated to wildlife preserves in 2003. This is far less than the estimated need and also represents only nine of the 14 major earth biomes (IUCN, 2016).

International corridors

Political and economic difficulties make meeting the 50% habitat preservation benchmark in the North America and Europe unlikely. The European Union established Natura 2000, a wildlife preserve and corridor plan with more than 27,000 sites representing 18% of the land area. Limited habitat area, therefore, requires that the preservation of biological diversity depends on connecting habitat preserves at the national and international scale. These proposed corridors might greatly impact grazing areas but cropland would be less involved. There are four international corridors proposed for the North America (Figure 17.4) and others being implemented in Europe, Australia, and elsewhere (Figure 17.5) (Dowling, 2008).

Farm corridors

In agricultural areas, the creation of corridors is necessary to maximise the effectiveness of the remaining and fragmented habitat areas to reduce the extinction pressure on animals, birds, insects, and to a lesser extent, plants. Corridors often exist on farms as the edges of irrigation canals, drainage channels, fence-rows, windbreaks, creeks, slopes too steep to farm, and roads. These can be enhanced by connecting them to habitat areas and improving their quality through revegetation with native species and the removal of invasive species (Harvey, 2007).

Most sources scrupulously avoid recommendations for corridor width. This is because the width is influenced by the characteristics of the target species, interactions among species, edge effects, species dispersal characteristics, the habitat quality of the corridor, corridor length, topographic position, gaps in the corridor, small habitat anchors, the number of alternate corridors, and the contrast between the matrix and the corridor vegetation (Lindenmayer and Franklin, 2002). In addition, the corridor width is limited within existing agricultural areas to avoid significant loss of growing ground.

The reality of the context limitation has led to corridor width recommendations for regional (500 metres, 1,640 feet), intermediate (300 metres, 1,000 feet), and local (50 metres, 164 feet) contexts (Department of Environment and Conservation, NSW, 2004; Kubei, 1996). For existing agriculture land within the suburban or urban context, achieving even the 50 metre width is often impossible, but can be easily incorporated into the planning for new peri-urban agriculture. The corridor widths suggested here are for planning purposes and should be modified according to the extensive set of qualifications listed earlier (MacFadyen, 2013).

Regional corridors should connect wildlife preserves and correspond to major ecological routes, such as migratory pathways through agricultural land. The minimum regional corridor width is best estimated

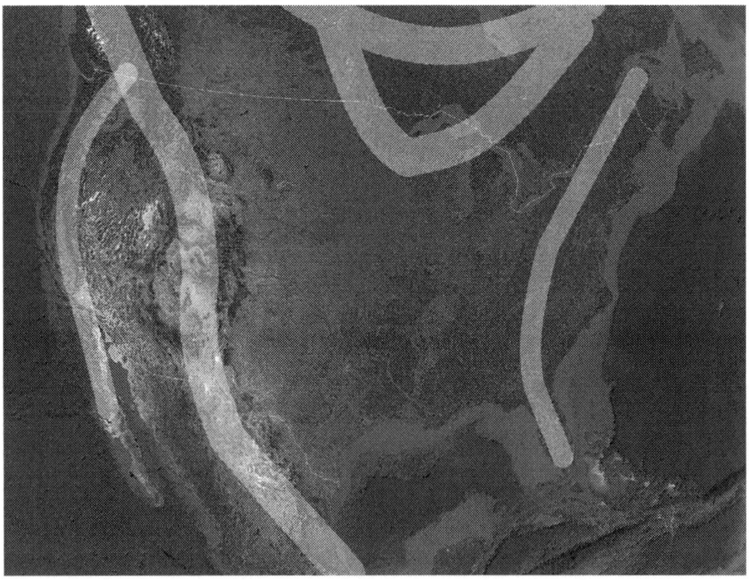

Figure 17.4 Four corridors proposed for connecting habitat preserves in Canada, USA, and Mexico.
Source: Image by author, adapted from https://wildlandsnetwork.org

Figure 17.5 The plan for the European ecological corridor. An east-west corridor through the Alps is also necessary to sustain species such as the European lynx.

Source: Smaack, License: CC-BY-SA-3.0

by identifying a guild (group of species) that is most sensitive to human disturbance and measuring the distance from the edge of a habitat patch to where the first individuals are found (Anderson and Jenkins, 2006). Twice this distance is adopted as the minimum corridor width. The intermediate corridor width would be planned to connect larger landscape features, such as vegetated ridgelines and valley floors. These cross-topography ecological corridors that connect streams or rivers to ridgelines are sometimes of great value to certain species (Lindenmayer and Franklin, 2002). The local corridors in the agricultural landscape would connect remnant habitat along gullies, drainage swales, streams, and ridgelines, steep slopes, hedgerows, wetlands, or marginal agricultural land.

Despite the recommended corridor widths, particularly when attempting to recover habitat connectivity on existing farms, research shows (Lyle, 1991) that any corridor width is better than no corridor and is beneficial to some species. For this reason, presumably, the Indiana (USA) Division of Fish and Wildlife recommends corridor widths between 15 metres (50 feet) and 61 metres (200 feet) through farmland between on-farm habitat patches for species such as deer, fox, raccoon, turtles, and reptiles. Combining corridors with shelterbelts is recommended where the planting is two rows of berry producing shrubs (dogwood

(*Cornus* sp.), crab apple (*Malus sylvestris*), chokecherry (*Prunus virginiana*), etc.) planted 1.8–2.4 metres (6–8 feet) apart between rows of coniferous trees, such as pines or spruce, spaced at 3–3.7 metres (10–12 feet) apart. If the ground is bare the agency suggests planting warm season grasses, legumes such as ladino clover (*Trifolium repens*), and native vegetation along with providing piles of brush or rock to support ground birds, songbirds, reptiles and small mammals (Indiana Division of Fish and Wildlife, 2004).

Streams and rivers are habitat areas but also natural corridors that are important since they can link habitat fragments. In the agricultural landscape the riparian areas are often deliberately narrowed to expand the arable or grazing property. Ecologically, this is highly destructive since it reduces or eliminates native habitat, isolates other riparian and upland habitat fragments, and encourages discharge of agricultural pollutants including pesticides, sediment, and excess nitrogen and phosphorus that can devastate aquatic ecosystems.

A study in Europe recognised the many ecological benefits of riparian corridors and proposed a stream and river network using GIS mapping of riparian vegetation that defined edge and core habitat along the corridors (Clerici and Vogt, 2013). The definition of edge habitat as 50 metres, 100 metres, or 200 metres resulted in three alternatives that depicted differing proportions of edge and core habitat. A wider edge habitat definition relates to species especially sensitive to human activity or predation by edge species. Of great value was the definition of 'bridging' riparian corridor segments needed to connect larger core habitat patches. The bridging sections often indicate areas where the riparian zone has been displaced by agriculture. These GIS maps (Figure 17.6) established conservation priorities for critical habitat and connectivity (Clerici and Vogt, 2013). Coupled with on-farm practices the implementation of the riparian conservation measures would preserve and restore corridors linked to regional preserves and mitigate the historic loss of habitat and species due to conversion of habitat to farmland.

Corridor length

The length of ecological corridors is related to their width. Long corridors must be punctuated by habitat patches of 50 hectares (124 acres) or more, spaced not more than 2,000 metres (6,335 feet) apart (Kubei, 1996).

Corridor gaps

Continuous corridors with vegetation consistent with the habitat areas are most effective. However, roads, fences, and fields often interrupt farm corridors and create conditions hostile to wildlife. Nevertheless, discontinuous corridors are also of value depending on the species and the size and character of the corridor gap and the mortality associated with it. Small species are thwarted by even narrow gaps, but larger and more mobile animals are generally more capable of negotiating wider corridor gaps (Hilty et al., 2006). If there are threatened or endangered species on the farm, then habitat, corridor, and gap studies are part of recovery plans in order to provide functional success and species distribution (Vanderree et al., 2004). These studies are also necessary to reduce human/wildlife conflicts if large animals are involved. Ecological corridors through farms can direct wildlife to the safest road crossing points (Hilty et al., 2006).

Farm corridors and native plants

Where there is great contrast between landscape types, as is characteristic of intensively managed agriculture, corridors of native vegetation are likely to foster distribution of associations of native plants and discourage or slow the establishment of invasive plant species (Damschen, 2006). A corridor study in northern France (49.81°N, 4.14°E) addressed the effect of stream corridors connecting a large forest ecosystem with fragments embedded in an agricultural landscape. Plants with limited dispersal potential colonised habitat fragments with plants 15 times more rapidly in connected patches than in unconnected ones. The

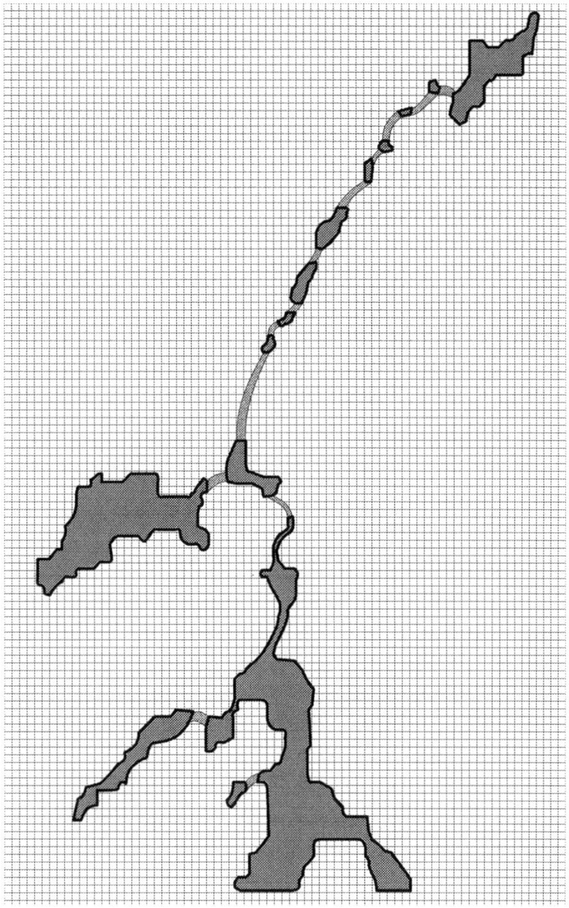

Figure 17.6 Plan diagram of a structural riparian corridor in Europe. Black = edge habitat; Dark grey = Core habitat; Light grey = bridging sections between core habitat areas; Checkerboard = agricultural matrix. Isolated islands, looping segments, perforated corridors, and branch segments connecting two core areas are four additional planning categories not shown in this illustration.

Source: Graphic by author, adapted from Clerici and Vogt, 2013

connecting streams were small (about 5 metres wide) and featured narrow strips of riparian vegetation (Araujo Calçada et al., 2013).

Ecology and farming practices

The amount of biological diversity on farms is dependent on the same characteristics as in natural ecosystems: landscape composition, landscape structure, landscape management, and also regional context (Harvey, 2007). The spatial characteristics leading to improved ecological performance of habitat patches and corridors are well understood (Dramstad, 1996) and can be applied to farmland. Measures taken on the farm should improve the variety of vegetation associations, their spatial arrangement, and management practices. These measures can be categorised according to one of four on-farm contexts: (1) within the field, (2) whole field, (3) field perimeter, and (4) landscape contexts (Kremen and Miles, 2012).

Within the field measures include the use of compost and manure (rather than artificial fertiliser), intercropping, agroforestry, and insectary strips. The application of natural fertiliser and other composted material to the field has the benefit of increasing diversity both above ground and below ground, improvement of soil quality and nutrient cycling, water retention, resilience to drought and heavy rain, carbon sequestration, increased productivity, and efficient energy use (Kremen and Miles, 2012).

Intercropping is a measure within the field that has similar benefits, in addition to weed, disease, and pest control. Since intercropping adds variety to the landscape mosaic, it tends to increase biological diversity (Harvey, 2007). Yields may also increase since different crop species, such as grain and nitrogen fixing legumes planted together, have differing rooting depths and differing access to water and nutrients. However, yield may be lower compared to industrial agriculture which is dependent on substantial external inputs. Where farmers can't afford the cost of pesticides there is greater knowledge and reliance on mixing crops to reduce the impact of insects and disease. Similarly, planting a variety of modern and traditional seed stocks for the same crop reduces risk of crop failure. On-farm diversity test blocks, seed fairs, and seed banks are methods that encourage farmers to expand the diversity of crops. The addition of underutilised crops, such as quinoa and millet, can allow the farmer to match the plant to the microclimate or soil in a way not available when using common commodity monocultures. The greater labour and complexity of growing a diverse set of crops is offset by enhanced resistance to unexpected weather and biological risks. Greater diversity in crops can also yield greater diversity in pollinators and soil microorganisms (Thompson et al., 2007).

Agroforestry is the third within-the-field measure to improve biological diversity and has the greatest potential to increase ecosystem services. Developing from orchard systems experiments (Smith, 2016) and permaculture practices (Mollison, 1992; Chapters 10, 24 and 26), there are many variants of this combination of crop or livestock production coupled with concentrated or dispersed tree plantations. An open canopy providing broken shade to coffee trees is a well-known example of agroforestry. Elsewhere, groves of trees in association with goats, cattle, chickens, or sheep are effective in increasing production of livestock, pasture diversity, and enhancing of farm biological diversity. A similar system with widely dispersed trees and crops, such as sorghum, creates a park-like agricultural setting that enriches bird populations that reduce insect numbers. The great potential of agroforestry is illustrated by *Gliricidia sepium* (quickstick) which is a leguminous tree (fixes nitrogen) that can be planted in rows between strips of crops. The tree is pruned to crop height and its leaves and green stems are buried to increase the nitrogen available to maize or other crops (Akinnifesi et al., 2006). Rows of tree crops, such as nut and fruit trees, protecting strips of crops is effective for reducing wind and water erosion and increasing organic material to the cropland as well as providing a secondary crop. Hedgerows and windrows of forested land can dramatically improve biodiversity (Dramstad, 1996), as does the use of insectary vegetation strips within the field. These are intended to increase the number of beneficial insects and native pollinators (Kremen and Miles, 2012).

The four whole field measures leading to greater diversity are no-till or low-till systems, crop rotation, cover or green manure crop, and fallowing. Of these rotation systems and cover crops and green manure crops offer the largest set of ecosystem services to the farmer (Kremen and Miles, 2012). The development of perennial grains has not yet been fully realized (Glover et al., 2010), but decades of plant breeding research is bringing this prospect closer. The erosion and pollution reduction potential of perennial grain crops would be a great ecosystem benefit both on the farm and off-site. An intermediate wheatgrass (the trade name is Kernza) is a perennial grain now in commercial production, and it may be the first crop with sufficient yields and suitable for commodity agricultural techniques (The Land Institute, 2017) (Figure 17.7).

Like hedgerows, field perimeter measures include border planting, which greatly increases the variety of species on the farm and provide habitat or corridor opportunities (Kremen and Miles, 2012). The greatest biodiversity benefit accrues from the preservation of existing native vegetation and habitats. If these have been lost or degraded, then, from an ecosystem perspective, their restoration and connection should be the

Figure 17.7 The perennial grain, *Thinopyrum intermedium* (Kernza) may become a major food source while greatly reducing erosion, runoff, and pollution due to pesticides and runoff of excess nitrate and phosphorus. Many other grains as well as crops such as chick peas and hazelnuts are under development as perennials that can substitute for commodity annuals.

Source: Dehaan, License: CC-BY-3.0

highest on-farm priority (Scherr and McNeely, 2007). Assessments of changes in the natural ecosystem by farmers increase in importance as the on-farm diversity of crops increase and the timing of irrigation, insect and disease threats, planting, and harvesting become more variable (Thompson et al., 2007). This would be a striking contrast to the system of industrial agriculture where farmers are particularly ignorant of natural ecosystems at the edges of their fields. When coordinated by a regional plan, riparian buffers, woodlots, meadows, and forests form a landscape scale habitat and especially corridors for the full range of indigenous species as discussed above (Kremen and Miles, 2012). For example, the state of Florida, USA, has undertaken a planning effort to integrate habitat and corridors that are effective statewide, on public and selected private lands (Florida Fish and Wildlife Conservation Commission, 2012).

In general, organic production of crops increases biological diversity, although some taxa benefit more than others. Compared to conventional farming methods, organic farming increases species abundance by 50% and species richness by 30% (Bengtsson et al., 2005). Organic farming supports higher biodiversity but this is especially true for small-scale organic farming where there is a variety of habitat types, such as wood lots and hedgerows. Under large-scale intensive management organic framing has more modest biodiversity levels but they can be greatly increased by incorporating within the field, whole field, field perimeter, and landscape measures as discussed earlier (Bengtsson et al., 2005).

Conclusion

Food production and natural landscapes interact to create ecosystem services for humans, but ecosystem *dis*services, such as pests and crop disease, and the desire for maximum crop yields inspire an adversarial commodity production system that destroys ecosystems. Fortunately, ecologically beneficial planning and management approaches in both agriculture and conservation science, such as ecological corridors and on-farm biological diversity measures, can reverse much of the historic ecological loss while increasing on-farm benefits, such as natural pollination, pest control, and reduced artificial inputs.

Loss of habitat and subsequent local extinction of species has a present and increasingly negative impact on the natural world and on humans. Most ecosystems that include commodity agriculture can sustain no additional loss of habitat without exacerbating the species extinction rate. Similarly, the rapidly expanding human population and demand for food, especially meat, requires a steady increase in the human food supply unless behavioural changes are widely implemented. The tension between these realities requires continued multidisciplinary research and the application of measures that benefit both agriculture and the environment. What this reality cannot abide is a wilful ignorance of either the food system or ecosystem health. Instead, a clear understanding of present and future threats and the application of effective planning and management and emerging research is a collaborative way forward involving reimagined agricultural landscapes and ecosystem planning.

Key texts

Dramstad, W.E. (1996). *Landscape ecology principles in landscape architecture and land-use planning*. Cambridge, MA: Harvard University Graduate School of Design; Washington, DC: American Society of Landscape Architects. Island Press.

Fahrig, L. (2001). How much habitat is enough? *Biological Conservation*, 100(1), 65–74.

Farina, A. (2006). *Principles and methods in landscape ecology: Toward a Science of Landscape*. New York: Springer.

Foley, J.A., Ramankutty, N., Brauman, K.A., Cassidy, E.S., Gerber, J.S., Johnston, M., & Zaks, D.P.M. (2011). Solutions for a cultivated planet. *Nature*, 478(7369), 337–342.

Harvey, C. (2007). Designing agricultural landscapes for biodiversity conservation. In: Scherr, S.J., & McNeely, J.A. (eds.), *Farming with nature: The science and practice of ecoagriculture*. Washington, DC: Island Press.

Hilty, J.A., Lidicker, W.Z., & Merenlender, A.M. (2006). *Corridor ecology: The science and practice of linking landscapes for biodiversity conservation*. Washington, DC: Island Press.

IUCN (International Union for Conservation of Nature and Natural Resources). (2016). *Summary statistics*. www.iucnredlist.org/about/summary-statistics (accessed 13 May 2016).

Kremen, C., & Miles, A. (2012). Ecosystem services in biologically diversified versus conventional farming systems: Benefits, externalities, and trade-offs. *Ecology and Society*, 17(4).

Millennium Ecosystem Assessment. (2005). *Ecosystems and human well-being: Biodiversity synthesis*. Washington, DC: Island Press.

Further reading

Akinnifesi, F.K., Makumba, W., & Kwesiga, F.R. (2006) Sustainable Maize production using Gliricidia/Maize intercropping in Southern Malawi. *Experimental Agriculture*, 42(04), 441.

Alexandratos, N., & Bruinsma, J. (2012) *World agriculture towards 2030/2050: The 2012 revision* (No. ESA Working paper No. 12–03). Rome, Italy: Food and Agriculture of the United Nations. www.fao.org/docrep/016/ap106e/ap106e.pdf.

Anderson, A.B., & Jenkins, C.N. (2006) *Applying nature's design corridors as a strategy for biodiversity conservation*. New York: Columbia University Press.

Araujo Calçada, E., Closset-Kopp, D., Gallet-Moron, E., Lenoir, J., Rêve, M., Hermy, M., & Decocq, G. (2013) Streams are efficient corridors for plant species in forest metacommunities. *Journal of Applied Ecology*, n/a–n/a.

Associated Press. (2017, March 31). China plans panda preserve 3 times size of Yellowstone Park. *Associated Press*. http://staging.hosted.ap.org/dynamic/stories/A/AS_CHINA_PANDA_PRESERVE?SITE=APandSECTION=HOMEandTEMPLATE=DEFAULTandCTIME=2017-03-31-06-26-51.

Bengtsson, J., AhnströM, J., & Weibull, A.-C. (2005) The effects of organic agriculture on biodiversity and abundance: A meta-analysis: Organic agriculture, biodiversity and abundance. *Journal of Applied* Ecology, 42(2), 261–269.

Cerulean, S. (2008) *Wildlife 2060: What's at stake for Florida?* Florida Fish and Wildlife Conservation Commission. http://myfwc.com/media/129053/FWC2060.pdf.

Clerici, N., & Vogt, P. (2013) Ranking European regions as providers of structural riparian corridors for conservation and management purposes. *International Journal of Applied Earth Observation and Geoinformation*, 21, 477–483.

Damschen, E.I. (2006) Corridors increase plant species richness at large scales. *Science*, 313(5791), 1284–1286.

Department of Environment and Conservation, NSW. (2004) *Wildlife corridors*. Department of Environment and Conservation. www.environment.nsw.gov.au/resources/nature/landholderNotes15WildlifeCorridors.pdf.

Dowling, S. (2008, May 13). *From death strip to life line: Former "Inner German Border" provides haven for wildlife*. www.spiegel.de/international/germany/from-death-strip-to-life-line-former-inner-german-border-provides-haven-for-wildlife-a-552953.html (accessed 28 July 2016).

Dramstad, W.E. (1996) *Landscape ecology principles in landscape architecture and land-use planning*. Cambridge, MA and Washington, DC: Island Press, Harvard University Graduate School of Design, American Society of Landscape Architects.

Eilers, W. (2011) *Environmental sustainability of Canadian agriculture*. Agriculture and Agri-Food Canada, Ottawa, Ontario. www.deslibris.ca/ID/228511.

Fahrig, L. (2001) How much habitat is enough? *Biological Conservation*, 100(1), 65–74.

Farina, A. (2006) *Principles and methods in landscape ecology: Toward a science of landscape*. New York: Springer.

Fischer, J., Brosi, B., Daily, G.C., Ehrlich, P.R., Goldman, R., Goldstein, J., . . . Tallis, H. (2008) Should agricultural policies encourage land sparing or wildlife-friendly farming? *Frontiers in Ecology and the Environment*, 6(7), 380–385.

Florida Fish and Wildlife Conservation Commission. (2012) *Florida's wildlife legacy initiative: Florida's state wildlife action plan*. Tallahassee, FL. http://myfwc.com/media/2663010/StateWildlifeActionPlan.pdf.

Foley, J.A., Ramankutty, N., Brauman, K.A., Cassidy, E.S., Gerber, J.S., Johnston, M., & Zaks, D.P.M. (2011) Solutions for a cultivated planet. *Nature*, 478(7369), 337–342.

Foreman, D. (2004) *Rewilding North America: A vision for conservation in the 21st century*. Washington, DC: Island Press.

Gliessman, S.R. (2007) *Agroecology: The ecology of sustainable food systems*, 2nd ed. Boca Raton: CRC Press.

Glover, J.D., Reganold, J.P., Bell, L.W., Borevitz, J., Brummer, E.C., Buckler, E.S., & Xu, Y. (2010) Increased food and ecosystem security via perennial grains. *Science*, 328(5986), 1638–1639.

Harvey, C. (2007) Designing agricultural landscapes for biodiversity conservation. In: Scherr, S.J., & McNeely, J.A. (eds.), *Farming with nature: The science and practice of ecoagriculture*. Washington, DC: Island Press.

Hassan, R.M., Scholes, R.J., Ash, N., & Millennium Ecosystem Assessment (Program) (eds.) (2005) *Ecosystems and human well-being: Current state and trends: Findings of the condition and trends working group of the millennium ecosystem assessment* (Millennium Ecosystem Assessment, Vol 1). Washington, DC: Island Press.

He, F. (2011) Species – area relationships always overestimate extinction rates from habitat loss. *Nature*, 473, 368–371.

Hilty, J.A., Lidicker, W.Z., & Merenlender, A.M. (2006) *Corridor ecology: The science and practice of linking landscapes for biodiversity conservation*. Washington, DC: Island Press.

Indiana Division of Fish and Wildlife. (2004) *Wildlife corridors. State of Indiana*. www.in.gov/dnr/fishwild/files/HMFSCorridors.pdf.

IUCN (International Union for Conservation of Nature and Natural Resources). (2016) *Summary statistics*. www.iucnredlist.org/about/summary-statistics (accessed 13 May 2016).

Klein Goldewijk, K. (2013) Land-use issues. In: *Encyclopedia of biodiversity* (pp. 555–568). Elsevier. www.researchgate.net/publication/256484320_Land-Use_Issues.

Kremen, C., & Miles, A. (2012) Ecosystem services in biologically diversified versus conventional farming systems: Benefits, externalities, and trade-offs. *Ecology and Society*, 17(4).

Kubei, J. (1996) Biocentres and corridors in a cultural landscape. A critical assessment of the territorial system of ecological stability. *Landscape and Urban Planning*, 35, 231–240.

La Point, T.W. (2007) Understanding one's place in the watershed: How earth science can inform perceptions about the future of the New Orleans region. *Technology in Society*, 29(2), 197–203.

The Land Institute. (2017) *Perennial crops: New hardware for agriculture*. https://landinstitute.org/our-work/perennial-crops/ (accessed 10 April 2017).

Leopold, A. (1949) *A sand county almanac*. New York: Oxford University Press.

Lindenmayer, D., & Franklin, J.F. (2002) *Conserving forest biodiversity: A comprehensive multiscaled approach*. Washington, DC: Island Press.

Lovell, S.T., DeSantis, S., Nathan, C.A., Olson, M.B., Ernesto Méndez, V., Kominami, H.C., . . . Morris, W.B. (2010) Integrating agroecology and landscape multifunctionality in Vermont: An evolving framework to evaluate the design of agroecosystems. *Agricultural Systems*, 103(5), 327–341.

Lyle, J. (1991) Ecological corridors in urban Southern California. In: *Wildlife conservation in metropolitan environments*. Columbia, MD: National Institute for Urban Wildlife.

MacArthur, R.H., & Wilson, E.O. (2001) *The theory of island biogeography*. Princeton, NJ: Princeton University Press.

Mace, G., Masundire, H., Baillie, J.E.M., & Millenium Ecosystem Assessment. (2005) Chapter 4 – Biodiversity. In: *Millenium ecosystem assessment: Current state and trends assessment*. www.millenniumassessment.org/documents/document.273.aspx.pdf.

MacFadyen, H. (2013, January 24). Citizen's guide to three sisters wildlife corridor. www.rmoutlook.com/Citizen%E2%80%99s-guide-to-Three-Sisters-wildlife-corridor-20130124 (accessed 14 March 2016).

Martinez, S., Hand, M., Da Pra, M., Pollack, S., Ralston, K., Smith, T., . . . Newman, C. (2010) *Local food systems: Concepts, impacts, and issues*. Washington, DC: U.S. Department of Agriculture.

Melles, S., Glenn, S., & Martin, K. (2003) Urban bird diversity and landscape complexity: Species – environment associations along a multiscale habitat gradient. *Conservation Ecology*, 7(1). www.consecol.org/vol7/iss1/art5.

Chopra, K., & Millennium Ecosystem Assessment Responses Working Group (2005) *Ecosystems and human well-being: Policy responses: Findings of the responses working group of the millennium ecosystem assessment* (Millennium Ecosystem Assessment series, Vol 3). Washington, DC: Island Press.

Mollison, B. (1992) *Permaculture: A designers' manual* (Repr). Tyalgum: Tagari Publ.

Peters, N.E. (2009) Effects of urbanization on stream water quality in the city of Atlanta, Georgia, USA. *Hydrological Processes*, 23(20), 2860–2878.

Robertson, G.P., Gross, K.L., Hamilton, S.K., Landis, D.A., Schmidt, T.M., Snapp, S.S., & Swinton, S.M. (2014) Farming for ecosystem services: An ecological approach to production agriculture. *BioScience*, 64(5), 404–415.

Robertson, G.P., & Swinton, S.M. (2005) Reconciling agricultural productivity and environmental integrity: Aa grand challenge for agriculture. *Frontiers in Ecology and the Environment*, 3(1), 38–46.

Robinson, R.A., & Sutherland, W.J. (2002) Post-war changes in arable farming and biodiversity in Great Britain. *Journal of Applied Ecology*, 39(1), 157–176.

Rockström, J., Steffen, W., Noone, K., Persson, Å., Chapin, F.S., Lambin, E.F., . . . Foley, J.A. (2009) A safe operating space for humanity. *Nature*, 461(7263), 472–475.

Scherr, S.J., & McNeely, J.A. (eds.) (2007) *Farming with nature: The science and practice of ecoagriculture*. Washington, DC: Island Press.

Smith, J.R. (2016) *Tree crops: A permanent agriculture*. Pathfinder Book reprint edition.

Soulé, M., & Noss, R.F. (1998) Rewilding and biodiversity: Complementary goals for continental conservation. *Wild Earth*, 8, 19–28.

Thompson, J., Hodgkin, T., Atta-Krah, K., Jarvis, D., Hoogendoorn, C., & Padulosi, S. (2007) Biodiversity in agroecosystems. In: Scherr, S.J., & McNeely, J.A. (eds.), *Farming with nature: The science and practice of ecoagriculture*. Washington, DC: Island Press.

Vačkář, D., ten Brink, B., Loh, J., Baillie, J.E.M., & Reyers, B. (2012) Review of multispecies indices for monitoring human impacts on biodiversity. *Ecological Indicators*, 17, 58–67.

Vanderree, R., Bennett, A., & Gilmore, D. (2004) Gap-crossing by gliding marsupials: Thresholds for use of isolated woodland patches in an agricultural landscape. *Biological Conservation*, 115(2), 241–249.

Vincent, C., Bjelopera, J., & Hanson, L. (2014) *Federal land ownership: Overview and data* (No. 7–5700 R42346). Congressional Research Service. www.fas.org/sgp/crs/misc/R42346.pdf.

The Wildlands Network. (2017) https://wildlandsnetwork.org/ (accessed 18 April 2017).

Wilson, E.O. (2016) *Half-earth: Our planet's fight for life*, 1st ed. New York: Liveright Publishing Corporation, a division of W.W. Norton and Company.

The World Bank. (2016) *The World Bank*. http://data.worldbank.org/indicator/SL.AGR.EMPL.ZS?locations=US (accessed 3 March 2017).

Wortman, S.E., & Lovell, S.T. (2013) Environmental challenges threatening the growth of urban agriculture in the United States. *Journal of Environment Quality*, 42(5), 1283.

18

Food systems and climate change

Impact and adaptation in cropping and livestock

Afshin Ghahramani and Saman Seneweera

Impact of climate change on wheat production

Wheat is one of the main grains produced across the globe and it accounts for 21% of human food worldwide (Oritz et al., 2008). Averaged across the world, a 5% decrease is predicted for global crop production of all crops by 2030 due to the effects of climate change (Challinor et al., 2014). This decline is projected to be even greater for wheat, with its global production estimated to decline by 6% for each one-degree increase in average temperature (Asseng et al., 2015). In addition, there are significant variabilities predicted for these impacts between regions and farm systems around the globe. In Australia, despite the likely resilience of the wheat production at the national scale projected for 2030 (Ghahramani et al., 2015), there would be a decline in production and profit at current farms in the majority of low rainfall regions (Ghahramani et al., 2015) which will require local adaptations for the resilience of current rural communities. To design effective climate change adaptation options, it is necessary to identify likely impacts on plant production by changes in different climate factors. Atmospheric CO_2 is input into the plant photosynthesis process and elevated CO_2 (eCO_2) is expected to stimulate photosynthesis to result in greater productivity and enhancement in water and nutrient use efficiencies (see for example Nowak et al., 2004; Seneweera and Norton, 2011). In the last decades, many field experiments around the world, such as free-air carbon dioxide enrichment (FACE) indicated a significant increase in plant biomass and yield under eCO_2 (Tubiello et al., 2007; Fitzgerald et al., 2016). But this effect on wheat production is uncertain for eCO_2 concentrations and large temperature increases (Asseng et al., 2013).

Amthor (2001) in an open-top field chamber experiment reported wheat yield increases in the range of 7% to 30% (average of 15%) for 440 ppm of atmospheric CO_2 (projected for 2030). A similar increase in yield was reported from an experiment in Western Australia with a median of about 15% (Asseng et al., 2013), and by Ainsworth and Long (2005), but both results were achieved without an average increase in temperature. Increase in temperature may often limit the direct and indirect fertilisation effect of eCO_2 effects on wheat yield, for example, a negative effect has been found during the critical flowering period, reducing grain quality and increasing water demand (see Tubiello et al., 2007, Fernando et al., 2014). Recently, FACE experiments were conducted by Fitzgerald et al. (2016) to assess the effects of temperature and eCO_2 on Mediterranean cropping systems (AGFACE). In their experiments, temperature was controlled by growing wheat in two contrasting temperature regimes (having different grain filling temperature), and a prolonged sowing treatment was introduced to capture CO_2 and temperature interaction on plant growth and development. This is the first report that has found that mean grain yield stimulation due to the effect of eCO_2 was 24% at a low-temperature site (Horsham, Australia) and 53% at a high-temperature site (Walpeup, Australia). Fitzgerald et al. (2016) observed that under some treatments yield responses exceeded by up to 70%.

The interaction of eCO_2 and temperature on crop yield varies depending on the genetic plasticity of the genotype (Seneweera et al., 2002). This is largely determined by the sink capacity of the plant (Thilakarathne et al., 2013). These findings suggest that temperature also plays a major role in determining the eCO_2 response in Mediterranean cropping systems. The low temperature reduces the sink development while high temperature accelerates the development (Ziska et al., 1996). It has been suggested (e.g. Ziska et al., 1996) that at low temperature, plant response to eCO_2 is determined by the sink development while at high temperatures, biomass accumulation is determined by the rate of the carbon assimilatory capacity. Thus plant biomass response to eCO_2 is determined by the magnitude of the interaction of these environmental variables. The optimum temperature for plant physiological processes such as photosynthesis and respiration is related to the genotypes and also the capacity of the plant to acclimate to the climate variables such as CO_2 and temperature (Fernando et al., 2014). These interactions are highly complex and further compounded by water supply (Fitzgerald et al., 2016). The details of these interactions are required to predict crop response to climate stress.

In a modelling study by Ghahramani et al. (2015) for Australian wheat production in 2030, eCO_2 shifted the system to a more productive one, and more so with moderate temperature increases and decreases in rainfall. The research also reported that the simulated fertilisation effects of eCO_2 on yield would be large enough not only to offset the negative impacts of changes in rainfall and temperature but also to increase yields for current farms. But this pattern is likely to be negatively affected by a greater increase in temperature beyond 2030 as projected yield decreases at local scales when temperature increase exceeds 1.1°C, even given the implementation of adaptation options.

Effect of eCO_2 on grain yield and quality

In some of the studies conducted in growth chambers, glasshouses, open top chambers and FACE studies, eCO_2 increased the grain yield of many cereals and pulses (Seneweera and Norton, 2001). Using different experimental methodologies, a range of experiments were carried out under various predicted future scenarios. For example, with wheat, the greatest yield stimulation was observed at 20°C, and the yield stimulation was much less when plants were under lower temperatures (Mitchell et al., 1993). Increase in grain number, single grain weight and a number of tillers partly contributed to the increase in yield (Fitzgerald et al., 2016). Amongst all the attributes, grain number per plant has an important role in determining yield. Grain quality was substantially decreased in most cereals due to eCO_2 (Myers et al., 2014, Fernando et al., 2012). The reduction in grain protein concentration, Zn and Fe concentrations were more prominent when influenced by eCO_2. Consequently, bread quality was diminished under a projected future with high atmospheric CO_2 concentration (Myers et al., 2014, Fernando et al., 2012). In terms of starch quality, the amylose to amylopectin ratio increased in rice at eCO_2 (Seneweera et al., 1996). Simultaneously, the size of the starch granules also increased with eCO_2. An increase in the granule size of starch is considered a positive outcome (Blumenthal et al., 1996). For wheat, reduction in protein concentration has a major impact on the nutritional quality and rheological characteristics of bread and, by extension, the sustainability of the industry from grain production to milling and baking, and ultimately to the consumer. Therefore, immediate adaptation to climate stress, in particular to maintain grain quality, is essential.

Impact of climate change on pasture and livestock

The response of grasslands to climate change is more complex than other agricultural systems such as cropping or forestry. In addition to the commonly studied responses of plant monocultures to eCO_2 and to changes in temperature and precipitation, many managed grasslands are composed of so many multiple plants of different functional types that their responses can differ widely (Ghahramani et al., submitted).

Legumes are observed to perform relatively better than grasses under eCO_2 (Schenk et al., 1995). The C4 plants (photosynthesis pathways of tropical and warm season plants) are also reported to be favoured by eCO_2 and warmer temperature (DaMatta et al., 2010). However, generalisations about the responses of grassland plant functional types to eCO_2 should be treated with caution (Ghahramani et al., submitted) as there are observations showing a variation of growth response to eCO_2 (Bolger et al., 1997). Campbell et al. (2000) stated that doubling CO_2 stimulated grassland production on average by 17% in field experiments at several sites in Australia, Europe, and the USA, while Tubiello et al. (2007) concluded that observed increases in above-ground production of C3 (photosynthesis pathways of cool season, temperate plants) pasture grasses and legumes are of the order of +10% and +20%, respectively. Plant-animal interactions and 'progressive nitrogen limitation' (PNL) can add more complexity to climate change impact in pastures and livestock. The N in grassland soil declines progressively under elevated CO_2 (Luo et al., 2004; Newton et al., 2006; Newton et al., 2010). A meta-analysis by Dieleman et al. (2012), reported that warming is likely to accelerate organic matter mineralisation and offsets PNL by the effect of eCO_2. However, their result does not include assessment of grasslands managed by cutting or plant-animal interactions. Overall, moderate eCO_2 is expected to increase plant photosynthetic rates but this may be limited by soil water and nitrogen deficit which are, as yet, uncertainly associated with plant photosynthetic rates.

Climate change impacts on livestock will occur through the direct and indirect effects on grassland production in grassland based livestock, decline in grain production in Intensive livestock system, and the direct and indirect effects on animals. The most important direct effects of climate change on animals will be thermal stress (cold and heat) and water availability, while key indirect effects are expected to result from livestock diseases (IPCC, 2014; Özkan et al., 2016). Climate changes in climate factors are predicted to decline forage quality, for example, protein concentrations in grasslands across the globe (Craine et al., 2010; Crimp et al., 2010; Izaurralde et al., 2011; Graux et al., 2011). This will affect the health and productivity of grassland-based livestock.

Modelling studies by Moore and Ghahramani (2013) in southern Australia (including New South Wales, Victoria, South Australia, and south of Western Australia) suggest that climate change impacts on pasture growth result in amplified impacts on livestock production. Moore and Ghahramani (2013) modelled increases in livestock conversion efficiency (due mainly to increased pasture legume content), but predicted that increased feed conversion rate is unlikely to be large enough to maintain incomes resulting from reduced pasture consumption by smaller stocking rate. When, upscaled for all southern Australia, projected declines in meat production from Australian temperate pastures by 20%, 26%, and 38% at 2030, 2050, and 2070, respectively, in the absence of adaptation measures (Ghahramani and Moore, 2015).

Climate change adaptation

The Intergovernmental Panel on Climate Change (IPCC, 2014: 5) defines climate adaptation as "The process of adjustment to actual or expected climate and its effects." Thus, the effectiveness of adaptation can be evaluated by comparing adapted future climate with adapted historical climate, and this will avoid considering the closure of the existing production gap (gap between current and potential production) with adaptations under climate change (Ghahramani et al., 2015). However, if adaptation is considered as capacity building which pursues all opportunities in addition to those gained from changes in climate, we may evaluate effectiveness in comparison with the current climate. In many cases, though, particularly under severe climate change, there may not be many opportunities to gain from changes in climate. There is a large range of possible adaptation options in both cropping and grazing systems. In cropping systems (including wheat farms), options can include the reconfiguration of systems using current technologies, for example, altering the sowing window and cultivar choice to those which cope with new thermal conditions, altering fertiliser rates, and altering water use management (Howden et al., 2007; Ghahramani et al., 2015). New technologies will also be important, including the introduction of new cultivars and climate forecasting techniques.

On average, globally adapted crop yields are projected to be around 7% greater than non-adapted crops in coming decades (Challinor et al., 2014), but there are great regional variabilities in different parts of the world. For example in Western Australia, the range of adaptation effects on crop production varies substantially between sites, crops, projected future climate, and management (Crimp et al., 2016).

Climate change adaptations in grazing systems can be grouped into feed base adaptations and/or livestock adaptations (Ghahramani and Moore, 2015). The grassland adaptations required are to increase grassland production or minimise periods and frequencies of days with low ground cover to reduce the potential risk of soil erosion (Ghahramani and Moore, 2013). Examples of grassland management adaptations evaluated in southern Australia are the increase of soil fertility by using phosphorus fertiliser (a finite resource, the implications of which are discussed in the reference), confinement feeding of animals, the replacement of a portion of grass with lucerne, and the removal of annual legumes from pastures (to slow the rate of loss of ground cover over dry summers) (Ghahramani and Moore, 2013). The livestock adaptations are mostly improvements in livestock genetics, based on historical trends, to increase forage conversion efficiencies (see Jeyaruban et al., 2009). Those examined by Moore and Ghahramani (2014) among sheep in southern Australia are: increase in animal body size, conception rate, potential fleece weight, and ram size.

In 2030 and compared to the current systems, changes in Australian national scale wheat yield are projected to decrease by 1% when using existing technology and practices but increase by 18% through application of optimal adaptation strategies with current technologies (Ghahramani et al., 2015). The potential economic value of climate change adaptation in the livestock industries of southern Australia are projected to be +69% by 2030, +69% by 2050, and +51% by 2070, compared the current systems (Ghahramani and Moore, 2015). These values are at the full adaptability of options or at the efficiency frontier of the system, all of which simply can't be achievable.

Uncertainty

There are a range of uncertainties in the evaluation of climate change impacts, the most common concern relates to technical uncertainty in a model's capability (Vermeulen et al., 2013). The impact of climate change and effectiveness of adaptations differ widely among projected future climate scenarios, a further demonstration of overall uncertainty (Patt et al., 2005; Ghahramani and Moore, 2013). As yet, none of the climate projection methods address the full range of future change in variability. Trends for changes in future price and cost levels of farm inputs and commodities are also uncertain (Moore and Ghahramani, 2013). There is a range of options to decrease some of the aforementioned uncertainties, for example, the use of an ensemble of biophysical models (Asseng et al., 2013), the use of a capacity approach with stakeholders (Regmi and Karki, 2010) to validate models for economic baselines, and the analysis of an ensemble of possible future models/projections for climate change. Overall, a greater concern generated by uncertainty could be the economic and social complexities of the agricultural systems which can outweigh aforementioned uncertainties in designing suitable adaptations (Vincent, 2007).

Upscaling

Effects of eCO_2 and other climate factors on plant physiology have been measured at a plant canopy scale. But farm-level responses may be different as many other local factors such as pests, weeds, nutrients, and competition between species can potentially alter the results. To understand the effects of climate change factors at a regional and continental scale, Ghahramani et al. (2015) employed a multipurpose model-based survey estimation methodology (Bardsley and Chambers, 1984). Their results have shown that climate change impacts and the effectiveness of the adaptations at industry and national level can give different perspectives than those encountered at local scales. This is important for making climate change policies to form effective plans for mitigation of climate change or to gain benefits where they are likely or possible.

Conclusion

The negative effects of climate change on crop and animal production are evident across the world, slowing growth rates in agricultural productivity. The decline in quality of the grain, for example, decreased concentrations of iron, zinc, protein and forage are observed in conditions of climate change. It is fortunate that there are potential adaptation options to reconfigure existing agricultural systems (grassland based, intensive feedlot, small farms) and offset or decrease negative impacts of climate change. Further, there are opportunities to be found in changes in climate. However, to prevent negative effects and provide food for a global population, in the coming decades there will be a requirement for radical transformations in agricultural systems.

Key texts

Amthor, J.S. (2001). Effects of atmospheric CO_2 concentration on wheat yield: review of results from experiments using various approaches to control CO_2 concentration. *Field Crops Research*, 73(1), 1–34.

Ainsworth, E.A., & Long, S.P. (2005). What have we learned from 15 years of free-air CO_2 enrichment (FACE)? A meta-analytic review of the responses of photosynthesis, canopy properties and plant production to rising CO_2. *New Phytologist*, 165(2), 351–372.

Asseng, S., Ewert, F., Rosenzweig, C., Jones, J.W., Hatfield, J.L., Ruane, A.C., Boote, K.J., Thorburn, P.J., Rötter, R.P., Cammarano, D., & Brisson, N. (2013). Uncertainty in simulating wheat yields under climate change. *Nature Climate Change*, 3(9), 827–832.

Asseng, S., Ewert, F., Martre, P., Rötter, R.P., Lobell, D.B., Cammarano, D., Kimball, B.A., Ottman, M.J., Wall, G.W., White, J.W., & Reynolds, M.P. (2015). Rising temperatures reduce global wheat production. *Nature Climate Change*, 5(2), 143–147.

Bardsley, P., & Chambers, R. (1984). Multipurpose estimation from unbalanced samples. *Applied Statistics*, 290–299.

Bolger, T.P., Lilley, J.M., Gifford, R.M., & Donnelly, J.R. (1997), June. Growth response of Australian temperate pasture species to CO_2 enrichment. In *Proceedings of the XVIII International Grasslands Congress'*. Winnipeg and Saskatoon, Canada, *Section* (Vol. 9, pp. 19–20).

Borlaug, N.E., & Dowswell, C.R. (2003), May. Feeding a world of ten billion people: a 21st century challenge. In *Proceedings of the international congress in the wake of the double helix: From the green revolution to the gene revolution* (pp. 27–31).

Campbell, B.D., Smith, D.S., & Pastures, G.C.T.E. (2000). A synthesis of recent global change research on pasture and rangeland production: Reduced uncertainties and their management implications. *Agriculture, Ecosystems and Environment*, 82(1), 39–55.

Challinor, A.J., Watson, J., Lobell, D.B., Howden, S.M., Smith, D.R., & Chhetri, N. (2014). A meta-analysis of crop yield under climate change and adaptation. *Nature Climate Change*, 4(4), 287–291.

DaMatta, F.M., Grandis, A., Arenque, B.C., & Buckeridge, M.S. (2010). Impacts of climate changes on crop physiology and food quality. *Food Research International*, 43(7), 1814–1823.

Dieleman, W.I., Vicca, S., Dijkstra, F.A., Hagedorn, F., Hovenden, M.J., Larsen, K.S., Morgan, J.A., Volder, A., Beier, C., Dukes, J.S., & King, J. (2012). Simple additive effects are rare: A quantitative review of plant biomass and soil process responses to combined manipulations of CO_2 and temperature. *Global Change Biology*, 18(9), 2681–2693.

Fitzgerald, G.J., Tausz, M., O'Leary, G., Mollah, M.R., Tausz-Posch, S., Seneweera, S., Mock, I., Löw, M., Partington, D.L., McNeil, D., & Norton, R.M. (2016). Elevated atmospheric [CO_2] can dramatically increase wheat yields in semi-arid environments and buffer against heat waves. *Global Change Biology*.

Ghahramani, A., & Moore, A.D. (2013). Climate change and broadacre livestock production across southern Australia. 2. Adaptation options via grassland management. *Crop and Pasture Science*, 64(6), 615–630.

Ghahramani, A., & Moore, A.D. (2015). Systemic adaptations to climate change in southern Australian grasslands and livestock: Production, profitability, methane emission and ecosystem function. *Agricultural Systems*, 133, 158–166.

Ghahramani, A., Kokic, P.N., Moore, A.D., Zheng, B., Chapman, S.C., Howden, M.S., & Crimp, S.J. (2015). The value of adapting to climate change in Australian wheat farm systems: farm to cross-regional scale. *Agriculture, Ecosystems and Environment*, 211, 112–125.

Porter, J.R., L. Xie, A.J. Challinor, K. Cochrane, S.M. Howden, M.M. Iqbal, D.B. Lobell, and M.I. Travasso (2014): Food security and food production systems. In: *Climate change 2014: Impacts, adaptation, and vulnerability*. Part A: Global and Sectoral Aspects. Contribution of Working Group II to the Fifth Assessment Report of the Intergovernmental Panel on Climate Change [Field, C.B., V.R. Barros, D.J. Dokken, K.J. Mach, M.D. Mastrandrea, T.E. Bilir, M. Chatterjee, K.L. Ebi, Y.O. Estrada, R.C. Genova, B. Girma, E.S. Kissel, A.N. Levy, S. MacCracken, P.R. Mastrandrea, and L.L. White (eds.)]. Cambridge University Press, Cambridge, United Kingdom and New York, NY, USA, pp. 485–533.

Seneweera, S. and Norton, R.M. (2011). Plant responses to increased carbon dioxide. In: *Crop adaptation to climate change*. Wiley-Blackwell Publishing Ltd., Chichester, West Sussex. United Kingdom, 198–217. ISBN 978-081382016-3.

Thilakarathne, C.L., Tausz-Posch, S., Cane, K., Norton, R.M., Tausz, M., & Seneweera, S. (2013). Intraspecific variation in growth and yield response to elevated CO 2 in wheat depends on the differences of leaf mass per unit area. *Functional Plant Biology*, 40(2), 185–194.

Further reading

Blumenthal, C., Rawson, H.M., & McKenzie, E. (1996) Changes in wheat grain quality due to doubling the level of atmospheric CO (2). *Cereal Chemistry*, 73(6), 762–766.

Craine, J.M., Elmore, A.J., Olson, K.C., & Tolleson, D. (2010) Climate change and cattle nutritional stress. *Global Change Biology*, 16(10), 2901–2911.

Crimp, S.J., Stokes, C.J., Howden, S.M., Moore, A.D., Jacobs, B., Brown, P.R., Ash, A.J., Kokic, P., & Leith, P. (2010) Managing Murray – Darling basin livestock systems in a variable and changing climate: Challenges and opportunities. *The Rangeland Journal*, 32(3), 293–304.

Crimp, S.J., et al. (2016) *Evaluating transformative adaptation options for Australian extensive farming*. A Final Technical Report prepared for DAWR as part of the Carbon Farming Initiative, Filling the Research Gaps Program.

Fernando, N., Panozzo, J., Tausz, M., Norton, R.M., Neumann, N., Fitzgerald, G.J., & Seneweera, S., (2014) Elevated CO 2 alters grain quality of two bread wheat cultivars grown under different environmental conditions. *Agriculture, Ecosystems and Environment*, 185, 24–33.

Fitzgerald, G.J., et al. (2016) Elevated atmospheric [CO_2] can dramatically increase wheat yields in semi-arid environments and buffer against heat waves. *Global Change Biology*, 22, 2269–2284.

Ghahramani, A., Howden, M.S., del Prado, A., Thomas, D.T., & Moore, A.D. (Submitted) *Climate change impact, adaptation, and mitigation in managed temperate grasslands and livestock*. Review.

Godfray, H.C.J., Beddington, J.R., Crute, I.R., Haddad, L., Lawrence, D., Muir, J.F., Pretty, J., Robinson, S., Thomas, S.M., & Toulmin, C. (2010) Food security: The challenge of feeding 9 billion people. *Science*, 327(5967), 812–818.

Graux, A.I., Gaurut, M., Agabriel, J., Baumont, R., Delagarde, R., Delaby, L., & Soussana, J.F., (2011) Development of the Pasture Simulation Model for assessing livestock production under climate change. *Agriculture, Ecosystems and Environment*, 144(1), 69–91.

Howden, S.M., Soussana, J.F., Tubiello, F.N., Chhetri, N., Dunlop, M., & Meinke, H., (2007) Adapting agriculture to climate change. *Proceedings of the National Academy of Sciences*, 104(50), 19691–19696.

IPCC (2014) *Climate change 2014: Impacts, adaptation, and vulnerability. Contribution of working group II to the fifth assessment report of the intergovernmental panel on climate change*, Barros, V.R., Field, C.B., Dokken, D.J., Mastrandrea, M.D., Mach, K.J., Bilir, T.E., Chatterjee, M., Ebi, K.L., Estrada, Y.O., Genova, R.C., Girma, B., Kissel, E.S., Levy, A.N., MacCracken, S., Mastrandrea, P.R., & White, L.L. (eds.). Cambridge and New York: Cambridge University Pres, 688 pp.

Izaurralde, R.C., Thomson, A.M., Morgan, J.A., Fay, P.A., Polley, H.W., & Hatfield, J.L. (2011) Climate impacts on agriculture: Implications for forage and rangeland production. *Agronomy Journal*, 103(2), 371–381.

Jeyaruban, M.G., Johnston, D.J., & Graser, H.U. (2009) Estimation of genotype× environment interactions for growth, fatness and reproductive traits in Australian Angus cattle. *Animal Production Science*, 49(1), 1–8.

Luo, Y., et al. (2004) Progressive nitrogen limitation of ecosystem responses to rising atmospheric carbon dioxide. *Bioscience*, 54(8), 731–739.

Mitchell, R.A.C., Mitchell, V.J., Driscoll, S.P., Franklin, J., & Lawlor, D.W., (1993) Effects of increased CO_2 concentration and temperature on growth and yield of winter wheat at two levels of nitrogen application. *Plant, Cell and Environment*, 16(5), 521–529.

Moore, A.D., & Ghahramani, A., (2013) Climate change and broadacre livestock production across southern Australia. 1. Impacts of climate change on pasture and livestock productivity, and on sustainable levels of profitability. *Global Change Biology*, 19(5), 1440–1455.

Moore, A.D., & Ghahramani, A., (2014) Climate change and broadacre livestock production across southern Australia. 3. Adaptation options via livestock genetic improvement. *Animal Production Science*, 54(2), 111–124.

Myers, S.S., et al. (2014) Increasing CO_2 threatens human nutrition. *Nature*, 510(7503), 139–142.

Newton, P.C., Lieffering, M., Bowatte, W.S.D., Brock, S.C., Hunt, C.L., Theobald, P.W., & Ross, D.J., (2010) The rate of progression and stability of progressive nitrogen limitation at elevated atmospheric CO_2 in a grazed grassland over 11 years of Free Air CO_2 enrichment. *Plant and Soil*, 336(1–2), 433–441.

Newton, P.C.D., Allard, V., Carran, R.A., & Lieffering, M., (2006) Impacts of elevated CO_2 on a grassland grazed by sheep: The New Zealand FACE experiment. In: Nösberger J., Long, S.P., Norby R.J., Stitt M., Hendrey, G.R., Blum, H. (eds.), *Managed ecosystems and CO_2. Ecological Studies (Analysis and Synthesis)*, Vol. 187. Berlin, Heidelberg: Springer, pp. 157–171.

Nowak, R.S., Ellsworth, D.S., & Smith, S.D., (2004) Functional responses of plants to elevated atmospheric CO_2 – do photosynthetic and productivity data from FACE experiments support early predictions?. *New Phytologist*, 162(2), 253–280.

Ortiz, R., et al. (2008) Climate change: Can wheat beat the heat? *Agriculture, Ecosystems and Environment*, 126(1), 46–58.

Özkan, Ş., et al. (2016) Challenges and priorities for modelling livestock health and pathogens in the context of climate change. *Environmental Research*, 151, 130–144.

Parry, M.L. (ed.) (2007) *Climate change 2007-impacts, adaptation and vulnerability: Working group II contribution to the fourth assessment report of the IPCC*, Vol. 4. Cambridge: Cambridge University Press.

Patt, A., Klein, R.J., & de la Vega-Leinert, A., (2005) Taking the uncertainty in climate-change vulnerability assessment seriously. *Comptes Rendus Geoscience*, 337(4), 411–424.

Peters, G.P., Marland, G., Le Quéré, C., Boden, T., Canadell, J.G., & Raupach, M.R. (2012) Rapid growth in CO_2 emissions after the 2008–2009 global financial crisis. *Nature Climate Change*, 2(1), 2–4.

Regmi, B.R., & Karki, G. (2010) Local adaptation plans in Nepal. *Tiempo*, 76, 21–25.

Rickards, L., & Howden, S.M. (2012) Transformational adaptation: Agriculture and climate change. *Crop and Pasture Science*, 63(3), 240–250.

Schenk, U., Manderscheid, R., Hugen, J., & Weigel, H.J. (1995) Effects of CO_2 enrichment and intraspecific competition on biomass partitioning, nitrogen content and microbial biomass carbon in soil of perennial ryegrass and white clover. *Journal of Experimental Botany*, 46(8), 987–993.

Seneweera, S.P., Blakeney, A.B., & Milham, P. (1996) Influence of rising atmospheric CO_2 and phosphorus nutrition on the grain yield and quality of rice (Oryza sativa cv. Jarrah). *Cereal Chemistry*, 73(2), 239–243.

Seneweera, S.P., Ghannoum, O., Conroy, J.P., Ishimaru, K., Okada, M., Lieffering, M., Kim, H.Y., & Kobayashi, K. (2002) Changes in source – sink relations during development influence photosynthetic acclimation of rice to free air CO_2 enrichment (FACE). *Functional Plant Biology*, 29(8), 947–955.

Smit, B., & Wandel, J., (2006) Adaptation, adaptive capacity and vulnerability. *Global Environmental Change*, 16(3), 282–292.

Tubiello, F.N., Soussana, J.F., & Howden, S.M. (2007) Crop and pasture response to climate change. *Proceedings of the National Academy of Sciences*, 104(50), 19686–19690.

Vermeulen, S.J., et al. (2013) Addressing uncertainty in adaptation planning for agriculture. *Proceedings of the National Academy of Sciences*, 110(21), 8357–8362.

Vincent, K., (2007) Uncertainty in adaptive capacity and the importance of scale. *Global Environmental Change*, 17(1), 12–24.

Ziska, L.H., Manalo, P.A., & Ordonez, R.A. (1996) Intraspecific variation in the response of rice (Oryza sativa L.) to increased CO_2 and temperature: Growth and yield response of 17 cultivars. *Journal of Experimental Botany*, 47(9), 1353–1359.

19

Investing in water management in rural and urban landscapes to achieve and sustain global food security

Munir A. Hanjra, Dennis Wichelns, and Pay Drechsel

Introduction

This century will be characterised by notable challenges to food and water security, due partly to climate change, and also to the increasing global population and to higher incomes in developing countries (Hanjra and Qureshi, 2010; Sadoff et al., 2015; Rosenzweig et al., 2017). Climate change can complicate current challenges regarding water resources and food systems, thus increasing the importance of intersectoral impacts (see Chapter 18). Food production and water use are highly influenced by changes in economic sectors (energy, transport, tourism, insurance, financial markets) and climate policies (Hanjra et al., 2017a; Rosenzweig et al., 2017). Climate change also will complicate efforts to achieve household, national, and global food security.

The global population is projected to reach 9 or 10 billion by 2050, with most of the net increase occurring in urban areas of developing countries (Gerland et al., 2014). Feeding the global population in 2050 will require an estimated 60% increase in food production, from current production levels (Unver et al., 2016). Most of that increase, perhaps 80%, will be achieved through increases in crop and livestock yields per hectare, rather than increases in production areas (Hanjra and Qureshi, 2010). Reductions in calorie intake (primarily in high-income countries), dietary shifts to less meat, and reductions in food wastage have the potential to improve global food security and mitigate the required increases in land and water use (Ridoutt et al., 2012). Yet, satisfying future global demand for food through changes in diets and lifestyles, and by intensifying farming technologies will require policy initiatives that motivate such changes (de Fraiture et al., 2007; Peters et al., 2016).

Investments in both irrigated and rainfed agriculture are needed to achieve the required increase in food production. Investment is defined as the commitment of money, resources, labour, policy, societal and cultural development, and political will to support the transformational change toward a better future. Currently, irrigation accounts for about 20% of crop production area, worldwide, while accounting for 40% of crop production (Portmann et al., 2016). Investments in irrigation development in developing countries can greatly increase crop yields per hectare, yet most farmers in the world operate in rainfed conditions where yields are relatively low and variable due to climate variability (de Fraiture et al., 2007; Williams, 2015). Investments in both activities are needed to increase aggregate output, while also enhancing household food and nutritional security and social wellbeing, particularly in developing countries.

Much of the needed increase in food production will be achieved through some form of intensification of agriculture with the goal of this being sustainable (Chartres and Noble, 2015; Rockström et al., 2016), including yield increases and intercropping in low productivity areas and sustainable yield, or even some land retirement in overexploited landscapes (Molden et al., 2010). Higher yields and incomes can be achieved with investments in irrigation and other inputs, and with training programs that enhance skills and social and cultural wellbeing in agriculture (Ward, 2014; Unver et al., 2016). Investments in urban agriculture also are essential to improve household incomes and achieve food security in urban communities (Drechsel and Hanjra, 2018; also see Chapters 12, 13, 20 and 22). Urban and peri-urban agriculture (within 20 km of urban centres) accounts for an estimated 6% of global cropland, and 35% of the urban cropland is irrigated (Thebo et al., 2014; also see Chapters 14 and 15). Much of the cropland in urban and peri-urban areas contributes to the production of vegetables for sale in local markets, for example, in Africa and Asia, thus supporting smallholder livelihoods and providing affordable nutrition for urban consumers (Drechsel et al., 2014; Hamilton et al., 2014). Just as in many rural areas, farmers operating in urban and peri-urban areas require affordable access to irrigation water and other inputs. Investments in public policies and infrastructure are needed also to optimise the use of wastewater (Hanjra et al., 2012), which is used by many farmers to irrigate vegetables and other crops (Drechsel and Keraita, 2014; Minhas et al., 2015).

Population growth and urbanisation to 2050 will be most notable in South Asia and sub-Saharan Africa (United Nations, 2015), where extensive poverty and food insecurity already pose substantial challenges to policy makers and donors investing in agriculture, water supply, and sanitation (Drechsel and Hanjra, 2016; Duchin, 2016). Policies and investments are needed in South Asia to slow the pace of groundwater pumping in some areas to achieve resource sustainability, while extending surface and ground water supplies in others to address the need for enhanced food production. There is notable potential to develop groundwater irrigation in sub-Saharan Africa (Villholth, 2013) but cost sharing models are needed (Gebregziabher et al., 2013). Investments in rainfed agriculture, such as improvements in water harvesting capacity and surface water storage systems will also enhance productivity (Barron et al., 2015; Cofie and Amede, 2015). On both of these continents, policies and investments are needed to better manage the use of untreated and partially treated wastewater in urban and peri-urban settings (Fuhrimann et al., 2016; Hanjra et al., 2017b; Chapter 23).

The need for investments that spur productivity in smallholder agriculture is particularly acute for greater wellbeing. Most farmers in Asia and Africa control fewer than 2 hectares of land, have limited access to extension services and financial credit, are unable to afford the optimal array of productive inputs each season, and largely rely on rainfall for the success or failure of their crops (see Chapter 21). Investments in water management have reduced poverty and enhanced food security in such settings (Namara et al., 2010; Pingali, 2012; Burney et al., 2013; Gohar et al., 2015). Additional investments are needed in rehabilitating and constructing new irrigation schemes, improving rainwater management, and developing groundwater resources when opportunities exist. Irrigation capacity development can be helpful in adapting to climate change (Kirby et al., 2015; Ward and Crawford, 2016). Yet, deployment of pressurised irrigation and sprinklers can increase energy inputs and carbon emissions from agricultural water use, while also increasing the farm-level cost of irrigation (Wichelns, 2007; Sanchis-Ibor et al., 2016). Therefore, climate policy solutions must combine *water efficiency improvement* and *carbon intensity reduction* targets for greater sustainability, and support farmer innovations such as solar-powered irrigation, where feasible and sustainable.

This chapter presents an overview of sustainable irrigation water management and wastewater use for food production in achieving and sustaining global food security, with emphasis on pertinent investments across landscapes in Africa and Asia.

Taking a landscape perspective

The greatest challenge for planning investments in agriculture is to match the technology, institutions, and policy packages in ways that meet the needs of the poor and smallholder farmers and enhance landscape sustainability and resilience (Molden, 2007a; also see Chapter 16). Irrigated and rainfed mosaics across settings are diverse. Water is not just a sector, but also a connector across sectors and landscapes that include surface water and groundwater resources in watersheds and catchments (Figure 19.1; Vertessy, 2007, modified). Pertinent issues in such landscapes include land use change, upstream-downstream resource flows, rural and urban linkages, safe use of wastewater, and urban water management. Climate change is an overarching issue that influences water and food security globally (Chapter 18). These links have implications for inter-sectoral water competition, sustainable intensification in agriculture, and investments in agricultural water management for *people*, *planet* and *prosperity* (Hanjra et al., 2017a). In this section, we review recent literature regarding the landscape perspective and we describe some of the promising ways to improve landscape resilience in rural and urban areas.

Water resource issues that require attention within a landscape perspective across catchments or river basins include upstream-downstream supply and demand interactions, rainwater harvesting, waterlogging and salinisation, and the movement of nutrients from rural to urban areas in the course of food production and consumption. Substantial research and policy outreach pertaining to these issues have been conducted (Fox and Rockström, 2000; Barbier, 2003; Wichelns and Oster, 2006; Barron et al., 2015). Investments and interventions are needed that will enhance water security and encourage productive use of resources, while minimising negative impacts across landscapes (Grey and Sadoff, 2007; Ward, 2014; Davis et al., 2015; Sadoff et al., 2015). Much of the discussion of integrated water resources management (IWRM) in the literature is pertinent to the landscape perspective, although the IWRM framework is not sufficiently comprehensive (Merrey et al., 2005). Political considerations, institutional constraints, and the lack of explicit consideration of livelihoods and commerce limit the usefulness of IWRM in many settings (Allan, 2003; Forster et al., 2017). In some cases, efforts to promote IWRM can prevent public officials from determining and implementing practical solutions to policy challenges in a timely fashion (Giordano and Shah, 2014; Mitchell et al., 2015).

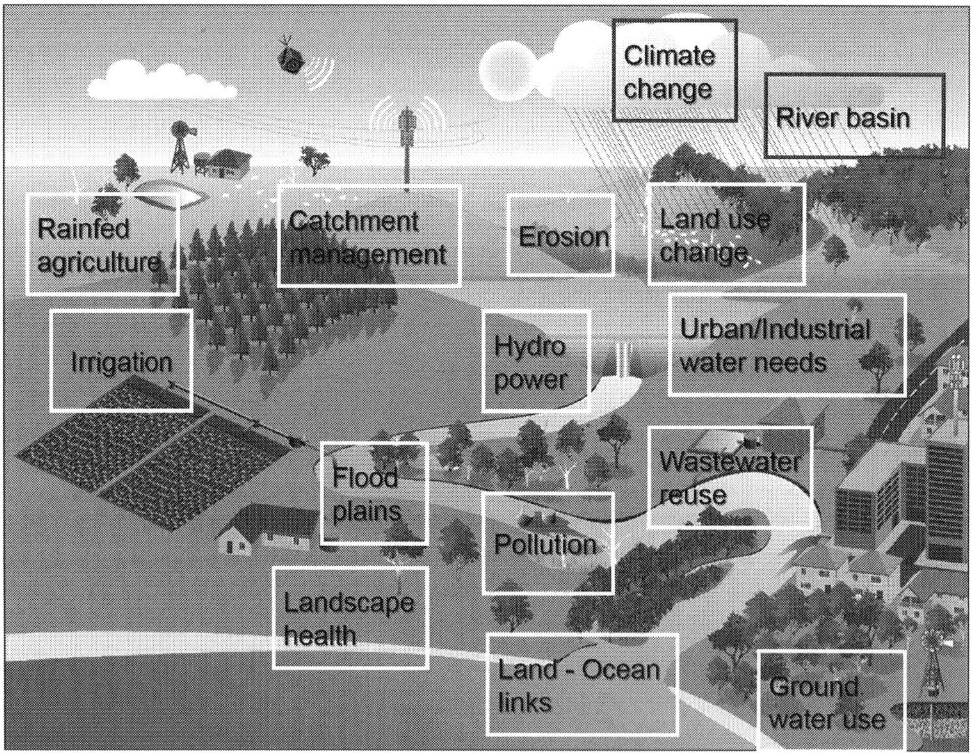

Figure 19.1 Linkages involving water in rural and urban landscapes.

Source: Vertessy, 2007, modified

Upstream-downstream interactions

In irrigated basins or catchments, competition between upstream and downstream irrigators is well known and widely studied (Barbier, 2003). Social issues and instances of anarchy are well documented (Wade, 1984). Farmers in upstream reaches of river basins or irrigation schemes have better access to water supply than do farmers in downstream reaches, particularly in arid regions and during the dry season in monsoonal regions. In such areas, economic incentives, water rights assignments, or regulations often are required to ensure that downstream farmers receive their fair share of irrigation supplies (Gaur et al., 2008). Upstream developments also can impact flora and fauna downstream of large dams and irrigation diversions (de Voogt et al., 2000; Schlüter et al., 2009). Policies that require upstream diverters to account for potential impacts on downstream irrigators, instream water users, and natural resources are needed to enhance the net benefits derived from water resources across agricultural landscapes. Programs involving payments by downstream water users to upstream diverters, in exchange for improvements in water management practices also have been proposed and implemented in some settings (Van Hecken et al., 2012; Zheng et al., 2013).

Rainwater harvesting

Rainwater harvesting for *in-situ* use for irrigation is often promoted in rainfed production areas, to increase crop yields and improve livelihoods (Fox and Rockström, 2000; Karpouzoglou and Barron, 2014; Botha et al., 2015). Ridge tilling, contour plowing, and the construction of small ponds can enhance farm production opportunities, particularly in upstream portions of watersheds (Biazin et al., 2012). However, such activities can reduce the amount of water available to farmers in lower portions of watersheds and to farmers reliant on shallow groundwater (Masih et al., 2011). In a hydrologic sense, rainwater harvesting in semi-arid regions can be viewed as a zero-sum operation. Rainwater captured and transpired in irrigation in upstream or upslope areas will not be available for groundwater recharge or for use at lower elevations (Van Meter et al., 2016). Incentives or restrictions regarding rainwater capture in upstream areas might be needed in some settings to achieve an equitable distribution of rainwater among competing users and enhance groundwater recharge (Pavelic et al., 2015).

Waterlogging and salinisation

Large areas of agricultural landscapes have become waterlogged or salinised, over time, due largely to inappropriate land and water management practices and vegetation/tree clearance (Hillel et al., 2008; Khan and Hanjra, 2008). When excessive irrigation water is applied on lands with limited natural drainage, shallow groundwater can accumulate, over time, and result in high water tables that limit plant growth and crop yield (Karimov et al., 2014; Singh, 2016). In arid areas, the high water tables can become saline, as water evaporates and salts are left behind. Tree/vegetation plantation schemes can help address the issue or artificial drainage systems must be installed to collect and remove the saline water, but such systems can be expensive and require some degree of regional coordination for success (Ritzema and Schultz, 2011). Only an estimated 22% of the world's irrigated land is served by a drainage system, and an estimated 10% to 16% of the world's irrigated land is impaired by waterlogging and salinisation (Ritzema, 2016), Alleviating this impairment will require additional investments in artificial drainage systems, improvements in land and water management practices, and the retirement from agriculture of lands for which the incremental cost of providing adequate drainage service exceeds the incremental gain (Wichelns and Qadir, 2015; Unver et al., 2016).

Climate change and saline intrusion in coastal areas can contribute to groundwater salinisation and degrade the quality of water for agriculture and for household use. In some areas of Bangladesh, saline intrusion has impaired rice production and caused the loss of agricultural land (Dasgupta et al., 2014). Improvements are needed in coastal land and water management, such as replanting mangroves and building polders.

Recovering nutrients in wastewater

With increasing population and urbanisation in many countries, food consumption in cities will increase, as will the volume of wastewater generated in urban areas. The untreated wastewater contains many nutrients that can be captured and used in agricultural production (see Chapter 23). Efforts to capture and reuse the water, energy, and nutrients, while costly, might represent an efficient use of resources in many settings in Asia and Africa (Grant et al., 2012; Drechsel et al., 2015b) and provide many other health benefits to urban residents in these regions (Dickin et al., 2016). Phosphorus is of particular interest, given that the global supply is limited, and much of the phosphorus lost in wastewater will eventually move via waterways to the ocean floor, where the recovery cost will be substantial (Mayer et al., 2016; Firmansyah et al., 2017). Other nutrients available in urban wastewater for capture and use in agriculture include nitrogen and potassium. Capturing the nutrients and using them in crop production will also improve water quality in streams and rivers that normally serve as sinks for untreated wastewater in some urban areas, and improve the resilience of urban food systems (Grant et al., 2012; de Zeeuw and Drechsel, 2015; also see Chapter 12). Efforts to capture, recover, and apply the nutrients in urban wastewater will become more common in the future, given the increasing demand for plant nutrients and the broad interest in achieving sustainable forms of agricultural production (Tran et al., 2016)

Considering urban and peri-urban agriculture

Benchmarking the environmental footprint of urban food consumption can be helpful in identifying the scale, direction, and magnitude of the water and energy resources required for sustainable urbanisation (see Chapter 16). Urban food consumption has received limited attention in urban policy (see Chapters 12, 14, 15 and 26). The urban food footprint is the third largest source of mass flows (0.8 tons per capita per annum) and carbon footprints (1.9 tons of CO_2 equivalent per capita per annum), and the largest driver of the urban ecological footprint (1.2 global ha per capita per annum) (Goldstein et al., 2016). Among foods, meat and dairy are the largest drivers of both the ecological footprint and global warming (Peters et al., 2016).

Most of the urban food footprint is contributed by imported foods, yet urban areas can invest in wastewater use and integrated water-food-energy systems to improve recycling and reduce waste. Public officials also can support peri-urban agriculture and small-scale farmers in ways that close the water and nutrient loops, in part by promoting the use of water and nutrients embedded in wastewater for local food production. Partnerships with industries that co-invest in irrigation system modernisation can enhance food security and promote landscape resilience (Zabaniotou et al., 2014).

Researchers at the International Water Management Institute (IWMI) have shown that business models promoting the use and sale of wastewater and other waste-based products, such as biogas, energy pellets, and nutrients can encourage private entrepreneurs to provide sanitation and wastewater services in peri-urban areas across Asia and Africa (Otoo and Drechsel., 2018). This will provide opportunities for financing sanitation (Drechsel and Hanjra, 2016) and improving public confidence and safety in wastewater use, while minimising the potentially negative impacts on human health and the environment.

The hydrologic reach of cities extends outward, such that the broader landscape faces the consequences of improper urban management. Pertinent issues include acid rain, eutrophication of water bodies, contamination of farmland with heavy metals, and groundwater quality degradation. Paving of land surfaces and drainage system installation have impacted floodwater flows and groundwater systems, while also presenting positive opportunities for water reuse, through low-impact development (also known as WSUD and SuDS) and urban water management (Fletcher et al., 2015).

Rural transformation

Amid increasing urbanisation and population growth, smallholder agriculture will continue to be the major sector for employment, income, and food security for poor households in developing countries (Abro et al., 2014; Wichelns, 2015; Chapters 20 and 21). Yet, agriculture also faces significant transformational challenges. Smallholder agriculture will evolve along somewhat different trajectories in Africa and Asia. The average farm size is increasing in Asia and shrinking in Africa (Wichelns, 2015). Rural-to-urban food flows will intensify in the future, motivating new policy responses. Smallholder farmers and rural areas will see greater links with commercial traders and value chains in local and urban markets, supporting diversification and value addition, but also creating the need to acquire new skills for operating in formal markets and technological and institutional innovations for marginalised smallholders (Gatzweiler and von Braun, 2016).

Policy questions regarding out-migration, rural transformation, and sustainable water resources management are already arising across Asia (such as Nepal and Bangladesh) and Africa (Ghana, Kenya, Malawi, Nigeria, Uganda, and Zimbabwe) (Gray, 2011; Jaquet et al., 2015). Data from Nepal shows that male out-migration increases women's workload, which leads to land abandonment and poor land management, resulting in the growth of invasive species harmful to livestock, while the lack of manure reduces soil fertility in uphill terraces (Jaquet et al., 2015).

In many developed regions, declining participation of youth in agriculture and the exit of populations from regional agricultural landscapes to, for example, coastal cities in Australia, impacts intergenerational viability of some rural communities. Key drivers include increased educational participation, structural change in the farm sector, and water scarcity (Barr, 2014). While not a threat to Australia's food security or to future growth in the sector, the changes in rural demographics and water scarcity likely will impact food importing countries (Qureshi et al., 2013). Investments are needed in the quality of rural life in developed countries. Better access to education, technology, and healthcare must be provided, to encourage residents to remain on the land, in support of intensive and ecological farming.

Although irrigation enhances food security and reduces poverty in both rural and urban areas (Datt and Ravallion, 1998), agriculture will be asked to 'do more with less water' in many regions in future. Domestic and municipal water needs generally have priority over other uses, such that water transfers from agriculture to cities are likely. However, urban growth generally is not constrained by water competition with agriculture. In general, rather than using a narrow financial criterion, cities select options that reflect a "path of least resistance" (Molle and Berkoff, 2006), such that economic, social, and political costs are considered together. The authors suggest that reallocating irrigation water from farms to cities, to satisfy increasing urban demands, is not necessarily optimal. In their view, a larger concern is that small agriculturalists lack the power and capital to compete with urban areas for limited water resources, due to well established relations in local political economies and the distribution of power in society. Urbanisation will continue to intensify water completion in agriculture and linked agricultural, ecological, and urban systems (Molden and Bos, 2005; Molden, 2007b; Drechsel and Hanjra, 2018).

Policy measures that encourage adoption of water-conserving irrigation technologies are widely believed to increase the water available for cities and the environment, but energy inputs and impacts on downstream landscapes are often not considered (Jackson and Hanjra, 2014; Cremades et al., 2016). Water conservation subsidies in the United States generally have not reduced water use in the conditions that occur in many river basins (Ward and Pulido-Velazquez, 2008; Ward, 2014). Policy instruments and effective delivery platforms are needed to achieve water savings, such as in the Murray Darling Basin of Australia, to deliver those savings to cities and the environment, and to promote well designed and well managed landscapes.

Implications for sustainable agriculture

Global estimates of the additional production required to ensure food security by 2050 range from 60% to 100% above the production levels of 2000 (Springer and Duchin, 2014). Data from inter-regional models of the world economy show that satisfying the food and water requirements of 9 to 10 billion people is possible with changes in diets and technology (Springer and Duchin, 2014) (on US diets see Peters et al., 2016). Relying on more extensive use of arable land and freshwater would require clearing forestland, and would impact regional water security. Water withdrawals increase by 93% in an extensive land use scenario, and 77% in the sustainability scenario from a baseline in 2000 (Springer and Duchin, 2014).

There are notable dependencies involving water use and energy consumption in agricultural landscapes, particularly as the modernisation of irrigation infrastructure and technology requires increased pumping and pressurising. In surface water dependent systems, irrigation modernisation comes at an energy cost. Water savings are achieved, but carbon emissions are increased. In groundwater dependent systems, enhanced water use efficiency, energy cost, and emissions reduction are possible when converting from surface irrigation to pressurised systems (Jackson and Hanjra, 2014). In conjunctive use systems, the proportion of surface water to groundwater largely determines the potential energy savings from reductions in water application, while the scarcity of surface water influences farm-level decisions to invest in groundwater extraction.

Analysis of water use efficiency and energy use for four provinces in China (Cremades et al., 2016) and in the southern Murray Darling Basin of Australia (Jackson and Hanjra, 2014) shows that the expansion of micro irrigation and sprinklers increased energy inputs and carbon emissions in agriculture, yet emissions decreased in areas irrigated largely with groundwater. Often, investments in modern irrigation technology do not reduce consumptive water use, but energy savings are achieved. More energy intensive technologies produce more emissions such that water savings are traded with carbon emissions. Therefore, policy solutions must combine water efficiency improvement and carbon intensity reduction targets, to encourage technologies that will reduce water use and contribute to climate change adaptation.

Groundwater development has a social value across landscapes where groundwater use is critical in sustaining millions of rural livelihoods or when groundwater is used for basic human needs such as drinking water (IWMI/WLE, 2015). However, in countries like India, China, Mexico, and Iran, energy is highly subsidised and these subsidies often become the prime drivers of groundwater overdraft (Wada et al., 2010). About 70% of the irrigated area in northern China relies on groundwater. The estimated carbon emissions from irrigation in eleven provinces (half of China's irrigated land) are 24.13 Mt CO_2 equivalent (Wang et al., 2012). This is equivalent to one third of all emissions from motor vehicles in Australia in 2010. From 1998 to 2012, groundwater pumping for irrigation in Punjab, India increased by 23%, the average pumping depth increased by 5.47 m, and energy requirements increased by 67%, such that carbon emissions increased by 110% (Kaur et al., 2016). Carbon emissions have increased from 33 to 55g per m^3 of groundwater used, and from 43 to 78 g carbon per kg of grain.

Ensembles of ten global hydrological models and six global climate models show negative impacts of climate change on global food security. Models project losses to maize, soybean, wheat, and rice ranging from 8% to 24% of kcal produced, with CO_2 fertilisation effects, and from 24% to 43% of kcal produced, without those effects (Elliott et al., 2014). Other modelling efforts (Chaturvedi et al., 2015) show that global irrigation water demand could more than double in the absence of climate policy, and policies to mitigate climate change might further increase agricultural water demand.

Integrated crop water management in both irrigated and rainfed settings might reduce the global food gap by half (Wiener et al., 2010; Jägermeyr et al., 2016). Results from twenty climate models show that efficiency improvements can reduce irrigation demand in many river basins. Globally, if 48% of non-productive water consumption were rerouted to neighbouring rainfed systems, calorie production could be increased by 26% (Jägermeyr et al., 2016). Low-cost solutions for small scale farmers show similar potential to improve yields in rainfed agriculture. In combination, global production could increase by 41%, while closing the water-related food gap by 62% (Jägermeyr et al., 2016). Replacing surface irrigation with sprinklers or drip systems could reduce non-beneficial consumption at river basin level by 54% and 76%, respectively, while maintaining current crop yields. Crop production improves by 9% and 15%, respectively, and much more in the Indus and Ganges basins (Jägermeyr et al., 2016). Improving water availability to the tail-end farmers can increase productivity by 50% (Hussain et al., 2004), yet further study is needed regarding the sustainability of sprinklers on small farms and the feasibility of drip systems for rice-wheat production across large basin landscapes.

Investments in water management

Irrigation schemes, large and small

Irrigated agriculture has enhanced food security and reduced poverty in Asia. Government and donor investments in irrigation systems and supportive policy measures across Asia during the 1960s insulated rural communities from recurring episodes of hunger and food insecurity and improved their social well-being (Hussain and Hanjra, 2004; Pingali, 2012). Today, more investment and funding are needed to rehabilitate and modernise existing surface irrigation systems (Hussain and Hanjra, 2003; Mateos et al., 2010). Renewed investments are needed also in irrigation infrastructure and water development in high potential areas and in product processing in Asia to empower the poor and raise them out of poverty.

Financing climate resilient water infrastructure is the greatest challenge. Recently, there has been a surge in private financing of large water sector infrastructure, and small-scale irrigation systems. Smallholder investments in tubewells and microirrigation systems also are increasing in Asia, as appropriate small-scale systems have become more affordable in recent years.

Similar trends are emerging in sub-Saharan Africa, due partly to the generally abundant, but largely underdeveloped land and water resources (Villholth, 2013). Investments in smallholder irrigation pose several challenges regarding equity, efficiency, and sustainability (Giordano and de Fraiture, 2014; Wichelns, 2014). In Africa, the potential for irrigated agriculture is estimated at 43 million ha, but only 13 million ha are presently irrigated. Expansion of irrigated agriculture and water development are possible in some countries, especially in sub-Saharan Africa. Expansion has been one focus of the Comprehensive Africa Agriculture Development Programme (CAADP), which sought to double the area under irrigated agriculture in the Southern African Development Community (SADC) by 2015. The program had limited success, as political commitments did not materialise into annual financial allocations or actual investments, due partly to the high cost of irrigation development (up to $15,000 per ha) and capacity constraints in some member states. SADC members need to increase investments in irrigation projects, to achieve the potential of irrigation for enhancing food security (Williams, 2015; Hanjra et al., 2016).

New models offer solutions to financing challenges in such settings. Community developed centralised irrigation systems can encourage cooperative association among farmers located in isolated farmworker communities. Individual farmers might lack capacity to invest in lifting or diverting water, but can participate in community developed irrigation schemes, with specific rules for water distribution and cost sharing (Hanjra et al., 2015). However, access to energy, development of by-laws for scheme sustainability, legal ownership of assets, and siltation due to upstream agriculture and livestock activities are emerging challenges.

Distributed irrigation, in which water access (via human power, gravity, or diesel/ solar pump), distribution (via furrows, watering cans, sprinkler, drip lines) and use (for food or high valued crop production) occurs onsite or near the same location, also can enhance farm-level access to water for crop production. Investments in distributed irrigation systems come from the farmers themselves. At US$1,000 per pump, the capital cost is competitive due to longer lifetimes and desirable locally in Sudano-Sahel (Burney et al., 2013). Large-scale uptake of such distributed irrigation systems requires water resource identification, groundwater mapping, improved local supply chains for pumps, and perception of risks among smallholders regarding diversified horticultural cropping systems, improved storage capacity, and market development (Burney et al., 2010).

In centralised irrigation schemes, smallholders pool their land and water resources to produce high-value crops (tomatoes, chillies, table grapes) for direct market sale. Contracts with local supermarkets remove the middleman and can reduce the power of supermarkets over producers. Some producers undertake their own processing and marketing to sell their raw or branded produce, such as tomato paste and bottled grape juice in Tanzania (groundwater based facility), chillies and beans in Zimbabwe (surface water dependent), and branded cherry tomatoes in Botswana (wastewater reuse facility and training incubator). Commercial banks have supported investments in pressurised drip and sprinkler irrigation systems. Local political leadership provides strong support, and also legal guarantees to banks on behalf of smallholder farmers, who still work on their commercial farms as labourers or service providers. Net annual benefits are distributed equally among all the farmers (Hanjra et al., 2015).

A major challenge in national investment strategies is designing policies to promote pro-poor and equitable development. Examples include policies favouring cheap pumps, while restricting over-pumping, such as limiting or withholding energy subsidies that encourage excessive groundwater extraction. Support for innovations, such as small solar panels and portable pumps, and plastic greenhouses suited to smallholder agriculture, also are important for the food and nutrition security of urban consumers.

Groundwater investments and policies

Groundwater is the primary source for irrigation on about 112 million ha of the 275 million ha (40%) of irrigated land in the world (Villholth, 2013). In South Asia, groundwater accounts for more than 50% of irrigated area, while groundwater resources in sub-Saharan Africa are least developed, due largely to limited potential for high yielding wells (more than 5 litres per second) (MacDonald et al., 2012), the higher cost of groundwater development ($6,000 to $23,000 per well), the age of drilling equipment, and the challenging hydrogeological formations that impede the lowering of drilling costs in Africa. By comparison, well drillers and artisans are available in many villages in Asia, where they provide low cost and affordable services to farmers ($1,000 to $1,500 per well).

Policies and investments are needed to monitor groundwater use and changes in aquifers to ensure sustainability (Box 19.1). Sustainable diversion limits on surface water resources and sustainable yield assessments for aquifers can guide investments and policy adjustments to achieve better environmental and social outcomes. Such research is needed in many countries across Central Asia, South Asia, the Middle East, and North Africa.

Box 19.1 Subsoil Water Act helps to conserve groundwater in Punjab, India.

The contribution of groundwater to total irrigation in India increased to 61% in 2010, from 28% in 1950, due largely to the rapid growth in tubewells. Many Indian states including Gujarat, Haryana, and Punjab face severe groundwater shortages and falling water tables. The Punjab Preservation of Subsoil Water Act was enacted in 2009 to slow groundwater depletion. Panel data for 1985 to 2011 reveal (Tripathi et al., 2016):

- the robust effect of the Act on reducing groundwater depletion, and
- a significant improvement in water table depth, across seasons and over time, since implementation of the Preservation of Subsoil Water Act.

Recent mapping (Altchenko and Villholth, 2013) has identified 80 shared aquifers and aquifer systems in Africa, that represent 42% of the area and 30% of the continental population. Such mapping helps to identify options for jointly managing the 35 transboundary aquifers shared by the SADC countries. Investments in water development in shared river basins and aquifers are particularly challenging. Water accounting and cost-benefit analysis must acknowledge not only the surface water flows, but also groundwater fluxes and linkages involving surface water and groundwater, to determine rules for sharing benefits among the countries.

Interventions must target issues at various scales, including local, national, and river basins. For example, capacity development is needed for farmers and for actors across irrigated landscapes and value chains, including policy makers and irrigation authorities. Investment in building the capacity of professionals and service providers engaged in irrigation is essential in Africa. Examples include initiatives for raising capacity through promotion of reference centres, certification for management, business training for irrigators, legal aid for water user associations, better links with funding agencies, and engagement in studies to guide decision making on irrigation rehabilitation projects (Hanjra et al., 2015).

Wastewater use for food security

Globally, about 356 cubic km of wastewater are generated annually, but only half of the wastewater is treated (Sato et al., 2013). The proportions are notably small in the Middle East and North Africa (8%), Latin America (18%), and Asia (32%) (Sato et al., 2013). Global annual tertiary treatment and reuse is estimated at 24 cubic km (GWI, 2014). Amid progress toward the Sustainable Development Goals (Duchin, 2016), most wastewater will remain untreated, as the global economic costs of treatment amount to several trillion dollars annually (Cazcarro et al., 2016).

An estimated 24 million ha of irrigated cropland are located within urban agglomerations, worldwide, and 130 million ha of irrigated cropland are within 20 km of urban areas (Thebo et al., 2014). Wastewater is frequently used for irrigation in urban and peri-urban agriculture, which comprise about 11% of all irrigated cropland (Thebo et al., 2014).

Urban agriculture thrives on wastewater use and enhances food security, particularly in developing countries, while allowing adaptation to increasing urbanisation, and linking cities with rural landscapes. This reduces stress on farmland, providing a buffer for pollution and improving socio-ecological resilience and sustainability (Thornbush, 2015). The direct benefits of wastewater use in food production include nutrients in the water, and the water itself, with a reliable and low cost supply (Chapter 23). It is estimated that 1,000 m^3 of municipal wastewater enough for irrigating one ha can contribute 16 to 62 kg total nitrogen, 4 to 24 kg phosphorus, 2 to 69 kg potassium, 18 to 208 kg calcium, 9 to 110 kg magnesium, and 27 to 182 kg sodium (Qadir et al., 2010). In Guanajuato, Mexico, the estimated cost to farmers for replacing the nitrogen and phosphorus lost during wastewater treatment is $900 per ha (Scott et al., 2000).

Studies in Hubli–Dharwad, India, show that wastewater allows farming in the dry season when farmers can sell their produce at three to five times the prices received during the wet season (Bradford et al., 2002). In Haroonabad, Pakistan, reliable untreated wastewater supply allows farmers to cultivate even-priced, high-value, and short-duration crops (Hussain et al., 2002). In Ghana, the reliability of free wastewater allows urban farmers to intensify vegetable production to multiple cycles year-round, while in Dakar, Senegal, untreated wastewater allows 8 to 12 harvests per year, compared to five to six harvests per year without access to wastewater (Drechsel et al., 2015a). Policy frameworks promoting safe wastewater use must consider not only the financial gains, but also the economic benefits for society, and other value propositions pertaining to the safe use of wastewater in other sectors (Figure 19.2).

The largest concerns regarding wastewater recovery and use are the potential risks to human health and the environment, especially when wastewater is used in food production (Drechsel et al., 2010; Elgallal et al., 2016). These risks pertain especially to crop production for urban markets, where along the food

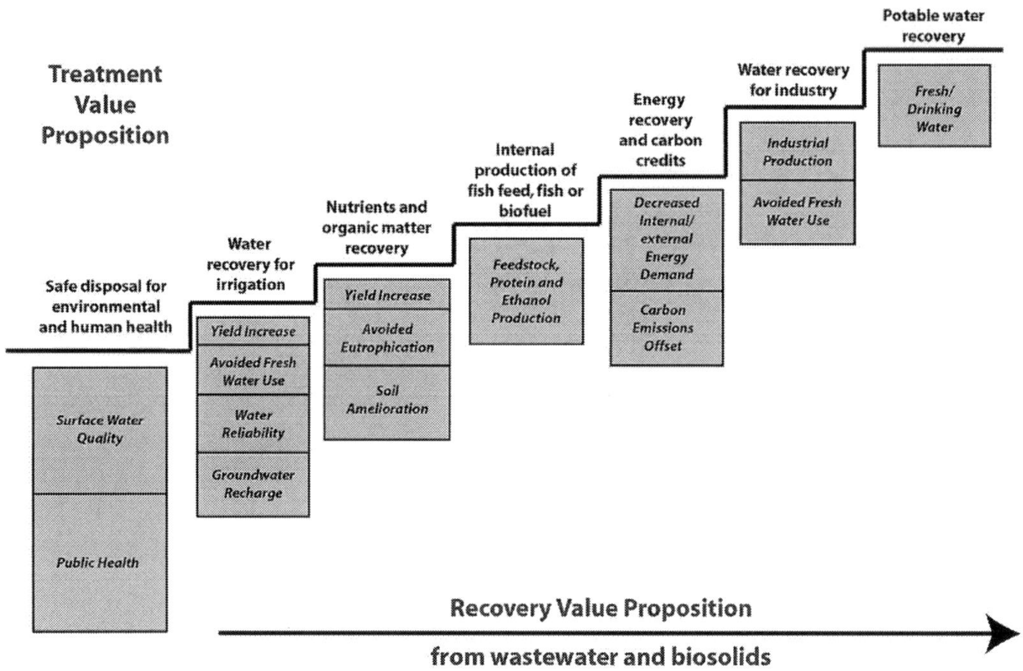

Figure 19.2 Value propositions related to wastewater use in agriculture.
Source: Drechsel et al., 2015b, modified

Table 19.1 Estimated number of urban farmers, street food kitchens, and urban consumers along the lettuce and cabbage value chain in Ghana, based on survey and sector data.

Urban farmers producing lettuce and cabbage	*Street restaurants offering salad side dishes*	*Daily consumers of salad side dishes in Ghana cities*
1,700 to 2,000	3,600 to 5,300	500,000 to 800,000

Source: Drechsel et al., 2014

chain the number of people at risk is increasing. In urban Ghana, for example, up to 2,000 urban vegetable farmers produce salad greens consumed eventually by up to 800,000 urban dwellers daily (Table 19.1).

Wastewater use policy must also consider public perceptions and address behaviour change, when promoting greater uptake. Where public perception is positive, such as in Australia, the United States, and the MENA region, the right business plan can combine several revenue streams for a higher cost-recovery rate, as in the case of the Drarga plant near Agadir in Morocco. The municipality there generates revenue from the sale of (1) treated wastewater to crop farmers, (2) reed grass from the constructed wetlands, (3) sludge compost, and (4) methane gas from energy recovery (Rao et al., 2015). Some of these components have not yet been implemented, but this special arrangement demonstrates how one municipality has responded to the challenge of obtaining public financing for wastewater treatment (Choukr-Allah et al., 2005).

Conclusions

We discussed the investments in water management needed to achieve and sustain global food security, while taking a landscape perspective regarding food production and consumption in rural and urban areas.

We have described many ideas put forth in recent literature and we have presented results from modelling efforts and case studies from Africa, Asia, and Latin America. Thus, we arrive at three conclusions regarding rural and urban landscapes: (1) Population growth and other socio-economic drivers and climate change will continue to impact food security and landscape resilience. (2) These and other socioeconomic factors will intensify the competition for land, water, energy, and labour in agriculture in both rural and urban areas. (3) Wastewater offers economic benefits and business opportunities, particularly in agriculture in urban and peri-urban settings. Utilising wastewater in agriculture also can reduce stress on freshwater resources and the environment, while returning plant nutrients back to farms, rather than allowing them to pollute receiving waters. Wastewater use in agriculture, including sewage and stormwater, can enhance both rural and urban landscapes, while contributing also to achieving sustainable forms of agricultural production.

Investments in landscape resilience will be helpful also in achieving and sustaining global food security. In particular, we find the following investments to be pertinent:

- investments in surface water and groundwater irrigation capacity, and the modernisation of existing irrigated systems, complemented by improved water management in rainfed systems;
- investments in climate adaptation and mitigation actions to maintain current productivity on irrigated and rainfed landscapes;
- investments and policies promoting the safe use of wastewater in agriculture; and
- investments in social institutions and policy frameworks for managing vulnerability to the water scarcity and climate change induced risks to food security.

We conclude that global food security can indeed be achieved and sustained beyond 2050, provided we make the right investments in landscape resilience and we address climate change with sufficient zeal for the millions of livelihoods that depend on our success.

Acknowledgements

This chapter was financially supported by the International Water Management Institute (IWMI), funded through the CGIAR Research Program on Water, Land and Ecosystems (WLE). We appreciate the helpful comments provided by several reviewers.

Key texts

Drechsel, P., & Hanjra, M.A, (2018) Section IV: Wastewater for agriculture, forestry and aquaculture: An overview of presented business cases and models & Chapter 17: Business models on rural–urban water trading. In: Otoo, M., Drechsel., P. (eds.), *Resource recovery from waste: business models for energy, nutrients and water reuse in low- and middle-income countries*. New York: Routledge.

Grey, D., & Sadoff, C.W. (2007) Sink or swim? Water security for growth and development. *Water Policy*, 9, 545–571.

Hanjra, M.A., Noble, A., Langan, S., & Lautze, J. (2017a) Feeding the 10 billion within the sustainable development goals framework (chapter 2). In: Gordon, I.J., Prins, H.H.T., & Squire, G.R. (eds.), *Food production and nature conservation: Conflicts and solutions*. London UK and New York USA: Earthscan.

Molden, D. (ed.) (2007a) *Comprehensive assessment of water management in agriculture. Water for food, water for life: A comprehensive assessment of water management in agriculture*. Colombo: International Water Management Institute: London: Earthscan. Individual chapters can be downloaded from the IWMI website at www.iwmi.cgiar.org/assessment/Publications/books.htm.

Molden, D., Oweis, T., Steduto, P., Bindraban, P., Hanjra, M.A., & Kijne, J. (2010) Improving agricultural water productivity: Between optimism and caution. *Agricultural Water Management*, 97, 528–535.

Molle, F., & Berkoff, J. (2006) *Cities versus agriculture: Revisiting intersectoral water transfers, potential gains and conflicts*. Colombo, Sri Lanka: International Water Management Institute (IWMI), Comprehensive Assessment Secretariat. vi, 70p. (Comprehensive Assessment of Water Management in Agriculture Research Report 010).

Rockström, J., Williams, J., Daily, G., Noble, A., Matthews, N., Gordon, L., Wetterstrand, H., DeClerck, F., Shah, M., Steduto, P., de Fraiture, C., Hatibu, N., Unver, O., Bird, J., Sibanda, L., & Smith, J. (2016) Sustainable intensification of agriculture for human prosperity and global sustainability. *Ambio*, 1–14.

Ward, F.A., & Pulido-Velazquez, M. (2008) Water conservation in irrigation can increase water use. *Proceedings of the National Academy of Sciences of the United States of America*, 105, 18215–18220.

Further reading

Abro, Z.A., Alemu, B.A., & Hanjra, M.A. (2014) Policies for agricultural productivity growth and poverty reduction in rural Ethiopia. *World Development*, 59, 461–474.

Allan, T.A. (2003) *IWRM/WRAM: A new sanctioned discourse?* SOAS Water Issues Study Group Occasional Paper 50. London: School of Oriental and African Studies, University of London, 27 pp.

Altchenko, Y., & Villholth, K.G. (2013) Transboundary aquifer mapping and management in Africa: A harmonised approach. *Hydrogeology Journal*, 21, 1497–1517.

Barbier, E.B. (2003) Upstream dams and downstream water allocation: The case of the Hadejia-Jama'are floodplain, northern Nigeria. *Water Resources Research*, 39, 1311.

Barr, N. (2014) *New entrants to Australian agricultural industries – where are the young farmers?* Rural Industries Research and Development Corporation, Australian Government, Canberra, Australia.

Barron, J., Kemp-Benedict, E., Morris, J., de Bruin, A., Wang, G., & Fencl, A. (2015) Mapping the potential success of agricultural water management interventions for smallholders: Where are the best opportunities? *Water Resources and Rural Development*, 6, 24–49.

Biazin, B., Sterk, G., Temesgen, M., Abdulkedir, A., & Stroosnijder, L. (2012) Rainwater harvesting and management in rainfed agricultural systems in sub-Saharan Africa – a review. *Physics and Chemistry of the Earth, Parts A/B/C*, 47–48, 139–151.

Botha, J.J., Anderson, J.J., & Van Staden, P.P. (2015) Rainwater harvesting and conservation tillage increase maize yields in South Africa. *Water Resources and Rural Development*, 6, 66–77.

Bradford, A., Brook, R., & Hunshal, C. (2002) Crop selection and wastewater irrigation, Hubli – Dharwad, India. *Urban Agriculture Magazine*, 8, 31–32.

Burney, J.A., Naylor, R.L., & Postel, S.L. (2013) The case for distributed irrigation as a development priority in sub-Saharan Africa. *Proceedings of the National Academy of Sciences of the United States of America*, 110, 12513–12517.

Burney, J., Woltering, L., Burke, M., Naylor, R., & Pasternak, D. (2010) Solar-powered drip irrigation enhances food security in the Sudano-Sahel. *Proceedings of the National Academy of Sciences of the United States of America*, 107, 1848–1853.

Cazcarro, I., López-Morales, C.A., & Duchin, F. (2016) The global economic costs of the need to treat polluted water. *Economic Systems Research*, 28, 295–314.

Chartres, C., & Noble, A. (2015) Sustainable intensification: Overcoming land and water constraints on food production. *Food Security*, 7, 235–245.

Chaturvedi, V., Hejazi, M., Edmonds, J., Clarke, L., Kyle, P., Davies, E., & Wise, M. (2015) Climate mitigation policy implications for global irrigation water demand. *Mitigation and Adaptation Strategies for Global Change*, 20, 389–407.

Choukr-Allah, R., Thor, A., & Young, P. (2005) Domestic wastewater treatment and agricultural reuse in Drarga, Morocco. In: Hamdy, A.B. (ed.), *The use of non-conventional water resources*, CIHEAM/EU DG Research, Options Méditerranéennes: Série A. Séminaires Méditerra éens, 66, 147–155.

Cofie, O., & Amede, T. (2015) Water management for sustainable agricultural intensification and smallholder resilience in sub-Saharan Africa. *Water Resources and Rural Development*, 6, 3–11.

Cremades, R., Rothausen, S., Conway, D., Zou, X., Wang, J., & Li, Y. e. (2016) Co-benefits and trade-offs in the water – energy nexus of irrigation modernization in China. *Environmental Research Letters*, 11, 054007.

Dasgupta, S., Hossain, M.M., Huq, M., & Wheeler, D. 2014. *Facing the hungry tide: Climate change, livelihood threats, and household responses in coastal Bangladesh.* Washington, DC: World Bank.

Datt, G., & Ravallion, M. (1998) Farm productivity and rural poverty in India. *The Journal of Development Studies*, 34, 62–85.

Davis, J., O'Grady, A.P., Dale, A., Arthington, A.H., Gell, P.A., Driver, P.D., Bond, N., Casanova, M., Finlayson, M., Watts, R.J., Capon, S.J., Nagelkerken, I., Tingley, R., Fry, B., Page, T.J., & Specht, A. (2015) When trends intersect: The challenge of protecting freshwater ecosystems under multiple land use and hydrological intensification scenarios. *Science of The Total Environment*, 534, 65–78.

de Voogt, K., Kite, G., Droogers, P., & Murray-Rust, H. (2000) Modelling water allocation between a wetland and irrigated agriculture in the Gediz basin, Turkey. *International Journal of Water Resources Development*, 16, 639–650.

de Fraiture, C., Wichelns, D., Rockström, J., Kemp-Benedict, E., Eriyagama, N., Gordon, L.J., Hanjra, M.A., Hoogeveen, J., Huber-Lee, A., & Karlberg, L. (2007) Looking ahead to 2050: Scenarios of alternative investment approaches. In: Molden, D. (ed.), *Comprehensive assessment of water management in agriculture. Water for food, water for life: A comprehensive assessment of water management in agriculture*, chapter 3, pp. 91–145. Colombo: International Water Management Institute; London: Earthscan.

de Zeeuw, H., & Drechsel, P. (eds.) (2015) *Cities and agriculture: Developing resilient urban food systems*. Oxon: Routledge – Earthscan, pp. 162–191.

Dickin, S., Schuster-Wallace, C., Qadir, M., & Pizzacalla, K. (2016) A review of health risks and pathways for exposure to wastewater use in agriculture. *Environmental Health Perspectives*, 124, 900–909.

Drechsel, P., Adam-Bradford, A., & Raschid-Sally, L. (2014) Irrigated vegetable farming in urban Ghana: A farming system between challenges and resilience. In: Drechsel, P., & Keraita, B. (eds.), *Irrigated urban vegetable production in Ghana: Characteristics, benefits and risk mitigation*, 2nd ed. Colombo: International Water Management Institute (IWMI), pp. 1–6.

Drechsel, P., & Hanjra, M.A. (2016) Green opportunities for urban sanitation challenges through energy, water and nutrient recovery. In: Dodds, F., & Bartram, J. (eds.), *The water, food, energy and climate nexus: Challenges and an agenda for action*. London and New York: Earthscan from Routledge.

Drechsel, P., & Keraita, B. (eds.) (2014) *Irrigated urban vegetable production in Ghana: Characteristics, benefits and risk mitigation*, 2nd ed. Colombo, Sri Lanka: International Water Management Institute (IWMI), 247p. Available from: www.iwmi.cgiar.org/Publications/Books/PDF/irrigated_urban_vegetable_production_in_ghana.pdf (accessed 17 September 2016).

Drechsel, P., Keraita, B., Cofie, O.O., & Nikiema, J. (2015a) Productive and safe use of urban organic wastes and wastewater in urban food production systems in low-income counties. In: de Zeeuw, H., & Drechsel, P. (eds.), *Cities and agriculture: Developing resilient urban food systems*. Oxon: Routledge – Earthscan, pp. 162–191.

Drechsel, P., Qadir, M., & Wichelns, D. (eds.) (2015b) *Wastewater: Economic asset in an urbanizing world*. Dordrecht, Heidelberg New York and London: Springer, on behalf of IWMI/ CGIAR/ UNU INWEH.

Drechsel, P., Scott, C.A., Raschid-Sally, L., Redwood, M., & Bahri, A. (2010) *Wastewater irrigation and health: Assessing and mitigating risk in low-income countries*. London: Earthscan.

Duchin, F. (2016) A global case-study framework applied to water supply and sanitation. *Journal of Industrial Ecology*, 20, 387–395.

Elgallal, M., Fletcher, L., & Evans, B. (2016) Assessment of potential risks associated with chemicals in wastewater used for irrigation in arid and semiarid zones: A review. *Agricultural Water Management*, 177, 419–431.

Elliott, J., Deryng, D., Müller, C., Frieler, K., Konzmann, M., Gerten, D., Glotter, M., Flörke, M., Wada, Y., Best, N., Eisner, S., Fekete, B.M., Folberth, C., Foster, I., Gosling, S.N., Haddeland, I., et al. (2014) Constraints and potentials of future irrigation water availability on agricultural production under climate change. *Proceedings of the National Academy of Sciences of the United States of America*, 111, 3239–3244.

Firmansyah, I., Spiller, M., de Ruijter, F.J., Carsjens, G.J., & Zeeman, G. (2017) Assessment of nitrogen and phosphorus flows in agricultural and urban systems in a small island under limited data availability. *Science of The Total Environment*, 574, 1521–1532.

Fletcher, T.D., Shuster, W., Hunt, W.F., Ashley, R., Butler, D., Arthur, S., Trowsdale, S., Barraud, S., Semadeni-Davies, A., Bertrand-Krajewski, J.-L., Mikkelsen, P.S., Rivard, G., Uhl, M., Dagenais, D., & Viklander, M. (2015) SUDS, LID, BMPs, WSUD and more – the evolution and application of terminology surrounding urban drainage. *Urban Water Journal*, 12, 525–542.

Forster, J.J., Downsborough, L., & Chomba, M.J. (2017) When policy hits practice: Structure, agency, and power in South African water governance. *Society and Natural Resources*, 1–16.

Fox, P., & Rockström, J. (2000) Water-harvesting for supplementary irrigation of cereal crops to overcome intra-seasonal dry-spells in the Sahel. *Physics and Chemistry of the Earth, Part B: Hydrology, Oceans and Atmosphere*, 25, 289–296.

Fuhrimann, S., Pham-Duc, P., Cissé, G., Tram, N.T., Thu Ha, H., Dung, D.T., Ngoc, P., Nguyen-Viet, H., Anh Vuong, T., Utzinger, J., Schindler, C., & Winkler, M.S. (2016) Microbial contamination along the main open wastewater and storm water channel of Hanoi, Vietnam, and potential health risks for urban farmers. *Science of the Total Environment*, 566–567, 1014–1022.

Gatzweiler, W.F., & von Braun, J. (2016) *Technological and institutional innovations for marginalized smallholders in agricultural development*, Gatzweiler, W.F., & Von Braun, J. (eds.). Cham: Springer International Publishing.

Gaur, A., Biggs, T.W., Gumma, M.K., Parthasaradhi, G., & Turral, H. (2008) Water scarcity effects on equitable water distribution and land use in a major irrigation project – case study in India. *Journal of Irrigation and Drainage Engineering*, 134.

Gebregziabher, G., Villholth, K.G., Hanjra, M.A., Yirga, M., & Namara, R.E. (2013) Cost-benefit analysis and ideas for cost sharing of groundwater irrigation: Evidence from north-eastern Ethiopia. *Water International*, 38, 852–863.

Gerland, P., Raftery, A.E., Ševčíková, H., Li, N., Gu, D., Spoorenberg, T., Alkema, L., Fosdick, B.K., Chunn, J., Lalic, N., Bay, G., Buettner, T., Heilig, G.K., & Wilmoth, J. (2014) World population stabilization unlikely this century. *Science*, 346, 234–237.

Giordano, M., & de Fraiture, C. (2014) Small private irrigation: Enhancing benefits and managing trade-offs. *Agricultural Water Management*, 131, 175–182.

Giordano, M., & Shah, T. (2014) From IWRM back to integrated water resources management. *International Journal of Water Resources Development*, 30, 364–376.

Gohar, A.A., Amer, S.A., & Ward, F.A. (2015) Irrigation infrastructure and water appropriation rules for food security. *Journal of Hydrology*, 520, 85–100.

Goldstein, B., Birkved, M., Fernández, J., & Hauschild, M. (2016) Surveying the environmental footprint of urban food consumption. *Journal of Industrial Ecology*, n/a–n/a.

Grant, S.B., Saphores, J.-D., Feldman, D.L., Hamilton, A.J., Fletcher, T.D., Cook, P.L.M., Stewardson, M., Sanders, B.F., Levin, L.A., Ambrose, R.F., Deletic, A., Brown, R., Jiang, S.C., Rosso, D., Cooper, W.J., & Marusic, I. (2012) Taking the "Waste" out of "Wastewater" for human water security and ecosystem sustainability. *Science*, 337, 681–686.

Gray, C.L. (2011) Soil quality and human migration in Kenya and Uganda. *Global Environmental Change*, 21, 421–430.

Grey, D., & Sadoff, C.W. (2007) Sink or swim? Water security for growth and development. *Water Policy*, 9, 545–571.

GWI (2014) *Global water market 2014*, Vol. 1. Oxford, UK: Global Water Intelligence.

Hamilton, A.J., Burry, K., Mok, H.F., Barker, S.F., Grove, J.R., & Williamson, V.G. (2014) Give peas a chance? Urban agriculture in developing countries. A review. *Agronomy for Sustainable Development*, 34, 45–73.

Hanjra, M.A., Barron, J., Johnston, R., & Williams, T. (2015) *Local level food and nutrition security: Case studies of smallholder irrigation schemes in sub-Saharan Africa*. EUROPEAN COMMISSION Joint Research Centre Workshop on local level food and nutrition security and the role of subsistence/smallholder farms, Seville, Spain, 9–10 September 2015.

Hanjra, M.A., Blackwell, J., Carr, G., Zhang, F., & Jackson, T.M. (2012) Wastewater irrigation and environmental health: Implications for water governance and public policy. *International Journal of Hygiene and Environmental Health*, 215, 255–269.

Hanjra, M.A., Drechsel, P., & Masundire, H.M. (2017b) Urbanization, water quality and water reuse. In: Lautze, J., Phiri, Z., Smakhtin, V., & Saruchera, D. (eds.), *Zambezi river basin: Pathways for sustainable development*. New York/Colombo: Earthscan/IWMI, pp. 158–174.

Hanjra, M.A., & Qureshi, M.E. (2010) Global water crisis and future food security in an era of climate change. *Food Policy*, 35, 365–377.

Hanjra, M.A., Zawe, C., Barron, J., & Williams, T. (2016) Public private partnership investments in irrigation and sustainable agricultural water management solutions, Zimbabwe. Draft Report, January 2016. IWMI/WLE project on Small Scale Irrigation Schemes in sub-Saharan Africa. International Water Management Institute, Colombo, Sri Lanka.

Hillel, D., Braimoh, A.K., & Vlek, P.L.G. (2008) Soil degradation under irrigation. In: Braimoh, A.K., & Vlek, P.L.G. (eds.), *Land use and soil resources*. Dordrecht, The Netherlands: Springer Netherlands.

Hussain, I., & Hanjra, M.A. (2003) Does irrigation water matter for rural poverty alleviation? Evidence from South and South-East Asia. *Water Policy*, 5, 429–442.

Hussain, I., & Hanjra, M.A. (2004) Irrigation and poverty alleviation: Review of the empirical evidence. *Irrigation and Drainage*, 53, 1–15.

Hussain, I., Mudasser, M., Hanjra, M.A., Amrasinghe, U., & Molden, D. (2004) Improving wheat productivity in Pakistan: Econometric analysis using panel data from Chaj in the upper Indus basin. *Water International*, 29, 189–200.

Hussain, I., Raschid, L., Hanjra, M.A., Marikar, F., & Van Der Hoek, W. 2002. *Wastewater use in agriculture: Review of impacts and methodological issues in valuing impacts. (With an extended list of bibliographical references). Working Paper 37, Colombo, Sri Lanka, International Water Management Institute* [Online].

IWMI/WLE. (2015) *Groundwater and ecosystem services: A frameowk for managing smallholder groundwater-dependent agrarian socio-ecologies – applying an ecosystem services and resilience approach*. Colombo, Sri Lanka: International Water Management Institute (IWMI). CGIAR Research Program on Water, Land and Ecosystems (WLE). 25p.

Jackson, T., & Hanjra, M.A. (2014) Energy, water and food: Exploring links in irrigated cropping systems. In: Bundschuh, J., & Chen, G. (eds.), *Sustainable energy solutions in agriculture*. Boca Raton, FL: CRC Press.

Jägermeyr, J., Gerten, D., Schaphoff, S., Heinke, J., Lucht, W., & Rockström, J. (2016) Integrated crop water management might sustainably halve the global food gap. *Environmental Research Letters*, 11, 1–14.

Jaquet, S., Schwilch, G., Hartung-Hofmann, F., Adhikari, A., Sudmeier-Rieux, K., Shrestha, G., Liniger, H.P., & Kohler, T. (2015) Does outmigration lead to land degradation? Labour shortage and land management in a western Nepal watershed. *Applied Geography*, 62, 157–170.

Karimov, A.K., Šimůnek, J., Hanjra, M.A., Avliyakulov, M., & Forkutsa, I. (2014) Effects of the shallow water table on water use of winter wheat and ecosystem health: Implications for unlocking the potential of groundwater in the Fergana Valley (Central Asia). *Agricultural Water Management*, 131, 57–69.

Karpouzoglou, T., & Barron, J. (2014) A global and regional perspective of rainwater harvesting in sub-Saharan Africa's rainfed farming systems. *Physics and Chemistry of the Earth, Parts A/B/C*, 72–75, 43–53.

Kaur, S., Aggarwal, R., & Lal, R. (2016) Assessment and mitigation of greenhouse gas emissions from groundwater irrigation. *Irrigation and Drainage*, 65, 762–770.

Khan, S., & Hanjra, M.A. (2008) Sustainable land and water management policies and practices: A pathway to environmental sustainability in large irrigation systems. *Land Degradation and Development*, 19, 469–487.

Kirby, M., Connor, J., Ahmad, M.-u. D., Gao, L., & Mainuddin, M. (2015) Irrigator and environmental water management adaptation to climate change and water reallocation in the Murray – Darling Basin. *Water Economics and Policy*, 01, 1550009.

MacDonald, A., Bonsor, H., Dochartaigh, B., & Taylor, R. (2012) Quantitative maps of groundwater resources in Africa. *Environmental Research Letters*, 7, 1–7.

Masih, I., Maskey, S., Uhlenbrook, S., & Smakhtin, V. (2011) Impact of upstream changes in rain-fed agriculture on downstream flow in a semi-arid basin. *Agricultural Water Management*, 100, 36–45.

Mateos, L., Lozano, D., Baghil, A.B.O., Diallo, O.A., Gómez-Macpherson, H., Comas, J., & Connor, D. (2010) Irrigation performance before and after rehabilitation of a representative, small irrigation scheme besides the Senegal River, Mauritania. *Agricultural Water Management*, 97, 901–909.

Mayer, B.K., Baker, L.A., Boyer, T.H., Drechsel, P., Gifford, M., Hanjra, M.A., Parameswaran, P., Stoltzfus, J., Westerhoff, P., & Rittmann, B.E. (2016) Total value of phosphorus recovery. *Environmental Science and Technology*. 10.1021/acs.est.6b01239.

Merrey, D.J., Drechsel, P., de Vries, F.W.T.P., & Sally, H. (2005) Integrating "livelihoods" into integrated water resources management: Taking the integration paradigm to its logical next step for developing countries. *Regional Environmental Change*, 5, 197–204.

Minhas, P.S., Khajanchi, L., Yadav, R.K., Dubey, S.K., & Chaturvedi, R.K. (2015) Long term impact of waste water irrigation and nutrient rates: I. Performance, sustainability and produce quality of peri urban cropping systems. *Agricultural Water Management*, 156, 100–109.

Mitchell, B., Bellette, K., & Richardson, S. (2015) 'Integrated' approaches to water and natural resources management in South Australia *International Journal of Water Resources Development*, 31, 718–731.

Molden, D. (2007a) Water responses to urbanization. *Paddy and Water Environment*, 5, 207–209.

Molden, D., & Bos, M.G. (2005) Improving basin water use in linked agricultural, ecological and urban systems: Seeking new flow paths and avoiding dead ends. *Water Science and Technology*, 51, 147–154.

Molle, F., & Berkoff, J. (2006) *Cities versus agriculture: Revisiting intersectoral water transfers, potential gains and conflicts*. Colombo, Sri Lanka: International Water Management Institute (IWMI), Comprehensive Assessment Secretariat. vi, 70p. (Comprehensive Assessment of Water Management in Agriculture Research Report 010).

Namara, R.E., Hanjra, M.A., Castillo, G.E., Ravnborg, H.M., Smith, L., & Van Koppen, B. (2010) Agricultural water management and poverty linkages. *Agricultural Water Management*, 97, 520–527.

Otoo, M., & Drechsel., P. (eds.) (2018) *Resource recovery from waste: Business models for energy, nutrients and water reuse in low- and middle-income countries*. Abingdon: Routledge.

Pavelic, P., Brindha, K., Amarnath, G., Eriyagama, N., Muthuwatta, L., Smakhtin, V., Gangopadhyay, P.K., Malik, R.P.S., Mishra, A., Sharma, B.R., Hanjra, M.A., Reddy, R.V., Mishra, V.K., Verma, C.L., & Kant, L. (2015) *Controlling floods and droughts through underground storage: From concept to pilot implementation in the Ganges River Basin*, IWMI Research Report 165. Colombo, Sri Lanka: International Water Management Institute (IWMI), 33p. Available from: DOI: 10.5337/2016.200.

Peters, C., Picardy, J., Darrouzet-Nardi, A.J.W., Griffin, T., & Fick, G. (2016) Carrying capacity of U.S. agricultural land: Ten diet scenarios *Elementa: Science of the Anthropocene*, 4, 000116.

Pingali, P.L. (2012) Green revolution: Impacts, limits, and the path ahead. *Proceedings of the National Academy of Sciences*, 109, 12302–12308.

Portmann, F.T., Siebert, S., & Döll, P. (2016) MIRCA2000 – Global monthly irrigated and rainfed crop areas around the year 2000: A new high-resolution data set for agricultural and hydrological modeling. *Global Biogeochemical Cycles*, 24, GB1011.

Qadir, M., Wichelns, D., Raschid-Sally, L., McCornick, P.G., Drechsel, P., Bahri, A., & Minhas, P.S. (2010) The challenges of wastewater irrigation in developing countries. *Agricultural Water Management*, 97, 561–568.

Qureshi, M., Hanjra, M.A., & Ward, J. (2013) Impact of water scarcity in Australia on global food security in an era of climate change. *Food Policy*, 38, 136–145.

Rao, K., Hanjra, M.A., Drechsel, P., & Danso, G. (2015) Business models and economic approaches supporting water reuse. In: Drechsel, P., Qadir, M., & Wichelns, D. (eds.), *Wastewater: Economic asset in an urbanizing world*. Dordrecht, The Netherlands, Heidelberg, New York, and London: Springer, on behalf of IWMI/ CGIAR/ UNU-INWEH.

Ridoutt, B.G., Sanguansri, P., Nolan, M., & Marks, N. (2012) Meat consumption and water scarcity: Beware of generalizations. *Journal of Cleaner Production*, 128, 127–133.

Ritzema, H.P. (2016) Drain for gain: Managing salinity in irrigated lands – a review. *Agricultural Water Management*, 176, 18–28.

Ritzema, H., & Schultz, B. (2011) Optimizing subsurface drainage practices in irrigated agriculture in the semi-arid and arid regions: Experiences from Egypt, India and Pakistan. *Irrigation and Drainage*, 60, 360–369.

Rosenzweig, C., et al. (2017) Assessing inter-sectoral climate change risks: the role of ISIMIP. *Environmental Research Letters*, 12, 010301.

Sadoff, C.W., Hall, J.W., Grey, D., Aerts, J.C.J.H., Ait-Kadi, M., Brown, C., Cox, A., Dadson, S., Garrick, D., Kelman, J., McCornick, P., Ringler, C., Rosegrant, M., Whittington, D., & Wiberg, D. (2015) *Securing water, sustaining growth: Report of the GWP/OECD task force on water security and sustainable growth*. Oxford: University of Oxford, 180pp.

Sanchis-Ibor, C., García-Mollá, M., & Avellà-Reus, L. (2016) Effects of drip irrigation promotion policies on water use and irrigation costs in Valencia, Spain. *Water Policy*, 10.2166/wp.2016.025.

Sato, T., Qadir, M., Yamamoto, S., Endo, T., & Zahoor, A. (2013) Global, regional, and country level need for data on wastewater generation, treatment, and use. *Agricultural Water Management*, 130, 1–13.

Schlüter, M., Leslie, H., & Levin, S. (2009) Managing water-use trade-offs in a semi-arid river delta to sustain multiple ecosystem services: A modeling approach. *Ecological Research*, 24, 491–503.

Scott, C.A., Zarazua, J.A., & Levine, G. (2000) *Urban wastewater reuse for crop production in the water-short Guanajuato River Basin, Mexico*, Research Report 41. Colombo, Sri Lanka: International Water Management Institute.

Singh, A. (2016) Hydrological problems of water resources in irrigated agriculture: A management perspective. *Journal of Hydrology*, 541(Part B), 1430–1440.

Springer, N.P., & Duchin, F. (2014) Feeding nine billion people sustainably: Conserving land and water through shifting diets and changes in technologies. *Environmental Science and Technology*, 48, 4444–4451.

Thebo, A.L., Drechsel, P., & Lambin, E.F. (2014) Global assessment of urban and peri-urban agriculture: Irrigated and rainfed croplands. *Environmental Research Letters*, 9, 1–15.

Thornbush, M. (2015) Urban agriculture in the transition to low carbon cities through urban greening. *AIMS Environmental Science*, 2, 852–867.

Tran, Q.K., Schwabe, K.A., & Jassby, D. (2016) Wastewater reuse for agriculture: Development of a Regional Water Reuse Decision-Support Model (RWRM) for cost-effective irrigation sources. *Environmental Science and Technology*, 50, 9390–9399.

Tripathi, A., Mishra, A.K., & Verma, G. (2016) Impact of preservation of Subsoil Water Act on groundwater depletion: The case of Punjab, India. *Environmental Management*, 58, 48–59.

United Nations. (2015) United Nations, Department of Economic and Social Affairs, Population Division (2015) World urbanization prospects: The 2014 Revision, online at http://esa.un.org/unpd/wup/ (accessed 27 April 2016).

Unver, O., et al.. 2016. *Key and smart actions to alleviate hunger and poverty through irrigation and drainage*. FAO background paper on sub-theme 3: key and smart actions to alleviate hunger and poverty through irrigation and drainage. Second World Irrigation Forum, Chiang Mai, Thailland.

Van Hecken, G., Bastiaensen, J., & Vásquez, W.F. (2012) The viability of local payments for watershed services: Empirical evidence from Matiguás, Nicaragua. *Ecological Economics*, 74, 169–176.

Van Meter, K.J., Steiff, M., McLaughlin, D.L., & Basu, N.B. (2016) The socioecohydrology of rainwater harvesting in India: Understanding water storage and release dynamics across spatial scales. *Hydrology and Earth System Science*, 20, 2629–2647.

Vertessy, R. (2007) *Australia's water resources information imperative and the role of the Water Resources Observation Network (WRON)*. Land and Water Resources Division, CSIRO, Canberra. Seminar on 'Harnessing the power of water information, Wagga Wagga/ Canberra, Australia.

Villholth, K.G. (2013) Groundwater irrigation for smallholders in Sub-Saharan Africa – a synthesis of current knowledge to guide sustainable outcomes. *Water International*, 38, 369–391.

Wada, Y., van Beek, L.P.H., van Kempen, C.M., Reckman, J.W.T.M., Vasak, S., & Bierkens, M.F.P. (2010) Global depletion of groundwater resources. *Geophysical Research Letters*, 37, L20402.

Wade, R. (1984) Irrigation reform in conditions of populist anarchy: An Indian case. *Journal of Development Economics*, 14, 285–303.

Wang, J., Sabrina, G.S.A.R., Declan, C., Lijuan, Z., Wei, X., Ian, P.H., & Yumin, L. (2012) China's water – energy nexus: Greenhouse-gas emissions from groundwater use for agriculture. *Environmental Research Letters*, 7, 014035.

Ward, F.A. (2014) Economic impacts on irrigated agriculture of water conservation programs in drought. *Journal of Hydrology*, 508, 114–127.

Ward, F.A., & Crawford, T.L. (2016) Economic performance of irrigation capacity development to adapt to climate in the American Southwest. *Journal of Hydrology*, 540, 757–773.

Wichelns, D. (2007) Economic implications of microirrigation. *Developments in Agricultural Engineering*, 13, 221–258.

Wichelns, D. (2014) Investing in small, private irrigation to increase production and enhance livelihoods. *Agricultural Water Management*, 131, 163–166.

Wichelns, D. (2015) Achieving water and food security in 2050: Outlook, policies, and investments. *Agriculture*, 5, 188–220.

Wichelns, D., & Oster, J.D. (2006) Sustainable irrigation is necessary and achievable, but direct costs and environmental impacts can be substantial. *Agricultural Water Management*, 86, 114–127.

Wichelns, D., & Qadir, M. (2015) Achieving sustainable irrigation requires effective management of salts, soil salinity, and shallow groundwater. *Agricultural Water Management*, 157, 31–38.

Wiener, T., Thenkabail, P., Montgomery, J., Pearlman, J., Lawford, R., & Hanjra, M.A. (2010) *A blueprint for water for the world: Pathway to a blue revolution*, Virginia, USA: Water for the World program, IEEE/USGS/NASA.

Williams, T.O. (2015) Reconciling food and water security objectives of MENA and sub-Saharan Africa: Is there a role for large-scale agricultural investments? *Food Security*, 7, 1199–1209.

Zabaniotou, A., Rovas, D., Libutti, A., & Monteleone, M. (2014) Boosting circular economy and closing the loop in agriculture: Case study of a small-scale pyrolysis-biochar based system integrated in an olive farm in symbiosis with an olive mill. *Environmental Development*. Available from: DOI: 10.1016/j.envdev.2014.12.002.

Zheng, H., Robinson, B.E., Liang, Y.-C., Polasky, S., Ma, D.-C., Wang, F.-C., Ruckelshaus, M., Ouyang, Z.-Y., & Daily, G.C. (2013) Benefits, costs, and livelihood implications of a regional payment for ecosystem service program. *Proceedings of the National Academy of Sciences*, 110, 16681–16686.

Part IV
Developing worlds

20

Food security, landscape, urban change, and poverty in the developing world

Suzanne Speak

Introduction

This chapter explores the relationship between landscape and food security of the urban poor in the contemporary context of urbanisation in the developing world. The main aim of the chapter is to describe the ways in which changes to rural and urban landscapes conspire to undermine food security for the poor. The chapter draws on the author's research in the 'developing world'. That research has used ethnographic methods to critically explore the lives and livelihood strategies of the urban poor, especially the homeless and those in informal settlements (see Chapter 22). The chapter particularly draws on research in India and Bangladesh.

The term 'developing world' is used here to refer to those countries experiencing very rapid social and economic change, rapid urbanisation, insecure political situations and significant economic disparities between their poorer and more affluent populations. In these countries unsustainable urban and rural development are changing both rural and urban landscapes (Pauchard et al., 2006). This change is one factor driving rural to urban migration, which also has an impact on the food security of the increasing numbers of urban poor. However, the author acknowledges that the specific and nuanced contexts of these countries vary greatly between Asia, Latin America and sub-Saharan Africa especially (see Chapter 21).

Despite the fact that levels of poverty and hunger in the developing world are higher in rural areas than in urban areas, some scholars have noted that urban food insecurity is an increasing problem for the poor in many developing cities (see for example Ahmed et al., 2008; Cohen and Garrett, 2010). However, there is very little research or data available specifically on urban food insecurity and under-nutrition, and even less on its relationship with landscape. This is problematic, because as the world's urban population increases, so does urban poverty – and so too do the stresses on urban services and systems. Under these stresses, it seems likely that urban food insecurity will increase also. Access to food in urban areas is dependent on functioning distribution chains to deliver food to all parts of the urban and peri-urban arena and accessible markets. However, urban distribution chains can be insecure, particularly in times of environmental stress, economic or political crisis, war or civil unrest (Auyero and Moran, 2007). In recent years the political power of urban food shortages to dislodge governments has been witnessed, not least with the food riots which preceded the Arab Spring of the early 21st century (Perez, 2013). Moreover, food security in urban areas is dependent on regular, adequate earned income to pay for food. However, much of the employment available to the urban poor is low paid and insecure putting the available food out of reach. Even without such stresses, urban planning policy can lead to uneven distribution of food around urban areas, and thus can also lead to food insecurity, especially for those in peripheral areas (Speak, 2014; 2012). However, as Steel (2008) notes, urban planners are slow to recognise this or to acknowledge the relationship between food and the physical and institutional urban landscape (see Chapters 12, 14, 15, 26 and 27).

The chapter next discusses the conceptual frameworks on which it draws. It then explains the context of urbanisation in developing countries and how that impacts the landscape. It moves on to unpack the implications of that change for the food security of the urban poor.

This chapter uses two specific conceptual approaches to understanding the relationship between food, landscape and poverty. First, it draws its structure from the pillars of food security as developed by the United Nations Food and Agriculture Organisation (FAO) during the World Summit of Food Security in 1996. Three 'pillars of food security' were originally identified with a fourth on 'stability' added by the FAO in 2006 to reflect increasing concerns about sudden shocks and longer-term issues of sustainability for future generations.

These pillars can be unpacked as follows:

- Availability: sufficient quantities of food available on a consistent basis.
- Access: having access to sufficient resources (income, markets, transport) to obtain appropriate foods.
- Utility: appropriate use based on knowledge of basic nutrition and care, as well as adequate water, sanitation and cooking facilities.
- Stability: access to adequate food at all times without risk of losing access to food as a consequence of sudden shocks (e.g. an economic or climatic crisis) or cyclical events (e.g. seasonal food insecurity) (Barrett, 2010).

The second conceptual approach used in this chapter is that of 'food sovereignty' (see Chapter 31). Within the structure of the four pillars of food security, the work highlights how the food-landscape-poverty relationship intersects with the concept of food sovereignty. The concept can be thought of broadly as the "right of nations and peoples to control their own food systems, including their own markets, production modes, food cultures and environments" (Wittman et al., 2010: 2). It was introduced in 1996 by La Via Campesina as an alternative to the dominant food security discourse, which focused on improving supply through intensification and industrialisation of agriculture. Based on this alternative discourse, in 2007, the Nyéléni Declaration argued that the power, domination and control of large-scale, multi-national agricultural industries and transnational corporations (TNCs) over the global food system can actively inhibit the ability of smaller-scale producers to make a living. That power also manifests in changes to the landscape through intensification and industrialisation of food production (Jonsson et al., 2012).

Contexts of change

The European Landscape Convention (2000) understands that change is an integral part of all landscapes. They are not static, inert entities but formed both to accommodate human behaviours and activities, and also as a result of them. However, difficulties arise when the management and control of that change is directed to the benefit of only a specific goal or group of people to the detriment of others, or to the landscape. Some scholars raise concerns that this occurs when governments seek to present a sanitised, contemporary urbanised face to the world in order to attract inward investment (Roy and Ong, 2011; Speak, 2014). This face cannot be seen to have the blemishes of poverty or an old agrarian past.

The historic landscapes of most cities of the world were deeply connected to food production out of necessity as, without rapid transport, food needed to be produced where people lived (Chapters 4, 9, 12, 14 and 29). Prior to the industrial age cities were mapped by the arrival, production and processing of the food which fed their inhabitants. Meat, in the form of cattle, sheep, pigs, and poultry, largely walked into the city along drove roads from the rural areas. In London, the area known as Poultry is so named because it was where poultry was sold in the 17th century, having been walked in from the surrounding farm land. The area of Cornhill refers back to its history as a location of grain markets and stores. Pudding Lane still stands testament to the site of the great fire of London which began in a bakery in 1666 (Mills, 2010). The foods produced, slaughtered and sold in these locations, and the sounds and smells which such activities produced, conveyed a sense of place and defined the urban landscape.

Even more recently, the importance of creating spaces for food production within and around cities was recognised by urban planners and designers such as Raymond Unwin, Ebenezer Howard and Frank Lloyd Wright. Le Corbusier's 'Contemporary City' of 1922 offered three zones for food production (Vitello and Brinkley, 2014). The two World Wars of 1914–18 and 1939–45 also emphasised the importance of urban food growing (Howe and Wheeler, 1999). In recent decades improved food transportation, urbanisation and economic growth, combined with industrialisation and commercialisation of food systems, has radically altered food's relationship with urban landscapes and with urban food consumers, and exacerbated the rural/urban dichotomy (Chapter 9). However, the contemporary model of rural production feeding urban consumption is failing to ensure urban food security for all, especially the poor, whose numbers are increasing in urban areas (Cohen and Garrett, 2010; Chapter 22).

The urbanisation of poverty in many parts of the developing world is being driven, in part, by rural migrants moving to cities out of aspiration for an improved livelihood or desperation at the loss of livelihood (De Haan, 1999, 2002; De Haan and Yaqub, 2010). Much of this migration has a direct link to landscape change and thus to food. Migration can be cyclical, short or longer term, or permanent. For some it is part of a planned household economic strategy to supplement their livelihood and food security from a rural landscape (De Haan, 2002; Tipple and Speak, 2009). For others, however, migration can arise from a point of crisis, when a household's productive rural landscape is lost to environmental disaster, war or intensification of farming (Speak, 2011). Many of these changes can have a direct relationship with food production, as will be shown. Indeed, the act of migration itself can lead to loss of land and changes to landscape, as a tipping point is reached at which the remaining rural labour is insufficient to make maintaining and working the land economically viable (Chapter 8).

On arriving in the city, many rural migrants move from a landscape of fragile food supply to a landscape ostensibly of abundance, where food is ever present and available – at a cost. Having lost all productive capacity, however, low income urban households must now buy all their food, probably their water, and pay for fuel to cook with. They can also find themselves in a position of greater poverty than that which they left, dependent on low paid, insecure work for their livelihoods (De Haan, 1999). Moreover, as Ruel et al. (2010) highlight, they can be disproportionately more affected by food and fiscal crises than their rural counterparts.

This urban poverty is exacerbated by planning and economic development policies which seek to expand cities into rural areas and reimagine urban areas as modernised, homogenous, globally recognisable sites for investment and business (Roy and Ong, 2011). In doing this, urban planning policy is denying the urban poor the landscapes and behaviours which support their everyday lives and underpinned their prior food security, committing them to landscapes which limit their access to food and their use of it (Speak and Kumar, 2016). This is exemplified by the case of Gurgaon in the state of Haryana, near Delhi, India.

Land use change and food security in Gurgaon

In India tensions are increasing over the changing use of rural land. The longstanding prioritisation of land for food production is giving way to the need for urban expansion to accommodate international investment and modernisation, upon which India's economic growth has been built. In support of this, embedded in the Delhi 2021 master plan is the concept of Delhi as a 'world class city' (Delhi Development Authority, 2010: 2). This identity is built upon a remodelling of the city and its hinterland to attract investment and to capture and retain India's new, more affluent middle class. Such remodelling requires land, which is in short supply in Delhi and not easily acquired by private developers who have not generally been given permission to buy and convert agricultural land. However, in the early 1990s, across the border in the adjacent state of Haryana, traditionally one of India's most important food producing areas, the State Government passed a number of laws which made rural land acquisition by private developers much easier (Narain, 2009; Rajagopalan and Tabarrok, 2014). They had also established the Haryana Urban Development Authority (HUDA). This generated a streamlined procedure with minimal conditions for private developers to acquire land and change its use from agriculture to urban development.

The change came at exactly the right time to capture the population growth and economic boom which was to hit Delhi in the coming decades. Now, just 25 kilometres south of Delhi sitting in former agricultural land rises Gurgaon satellite city, in an area that was once little more than a rural village with minimal services and infrastructure. The original village of Gurgaon, with its fields and traditional village houses of mud brick, is all but gone. In its place stand shopping malls, luxury gated estates, and manicured golf courses, as shown in Figures 20.1 and 20.2.

The development of new urban areas often produces a change in the surrounding peri-urban landscape. This is not simply because of subsequent urban sprawl into the periphery of the new urban area. Change also occurs because the new urban market increases demand for food produce and makes commercial production more viable than it previously was, displacing subsistence farming (Narain, 2009). In the case of the Gurgaon area, however, many farmers, especially those whose lands were in single large parcels rather than numerous smaller scattered plots, have lost substantial amounts of land. Amid allegations of non-payment of compensation for appropriated lands, some have been reduced to subsistence farming (Narain, 2009). This is increased further as those who were once paid farm labourers lose their jobs and are also forced into subsistence farming on any land they can lay claim to.

The landscape is changed further still by the growth of informal housing. As the market for cheap accommodation to house the influx of construction labourers gets stronger, it becomes economically more viable to build dwellings for rent rather than to farm the land. Ultimately, the food security and food sovereignty of the local farming population has been radically reduced and many have migrated to the city (Chapters 22 and 31).

Figure 20.1 Traditional village buildings of Gurgaon.

Figure 20.2 M.G. Road, Gurgaon.

Landscape and the pillars of food security

The next section of the chapter draws on empirical work by the author in India and Bangladesh to describe the relationship between landscape and food security for the urban poor. It is structured using the FAO's four pillars of food security as set out in the introduction.

The availability of food

The drive to increase the availability of food, especially for global export markets, can have perversely negative impacts on food security for the poor as it brings landscape change which is driving rural people from their land and livelihoods. In Bangladesh, for example, the recent global increase in demand for shrimp has completely transformed the rural landscape of the environmentally delicate coastal area along the Bay of Bengal.

Shrimp aquaculture began in Bangladesh in the early 1970s. It was originally undertaken in *ghers*, earthen ponds contained by dikes and situated in riversides smallholdings (Islam et al., 2005; Ahmed et al., 2008). Traditionally rice was grown in the ghers from August to December or January, and shrimp was produced from February to July or August (Ahmed, 2003; Islam et al., 2005) (see Figures 20.3 and 20.4). Water quality was maintained by the cyclical exchange of tidal water and fresh waters from river floods (Ahmed et al., 2008).

The traditional smallholdings accommodated rice, shrimp, vegetables, and small livestock. In many respects, this was a landscape in tune with the ethos of food sovereignty. Through these smallholdings and ghers, communities and households had access to the productive capacity to supply much of their own food. However, it was also a landscape of entrenched poverty, which had not benefitted greatly from the Government's investment in agricultural research and grain production from the 1970s onwards. It remained subject to the uncertainties of climate change and prone to extreme flooding. To protect the land from cyclones, the Government constructed a large number of embankments in the coastal area. These prevented the refreshing of water in the ghers, which became stagnant and waterlogged, making crop rotation problematic.

Figure 20.3 Traditional gher on right during rice season.

Figure 20.4 Landscape of traditional smallholding.

However, the growing international market for shrimp did benefit from the landscape change which helped to drive intensification and support a massive increase in production of shrimp for export (Deb, 1998; Ito, 2002). The result has been a total transformation of the landscape, as shown in Figures 20.5 and 20.6.

There has also been a significant socio-economic impact on the smallholders and landless farm labourers. The original smallholders have either sold their land to large companies or, if they refused to sell, have been forced to leave, as intensification changes the surrounding landscape so radically as to make traditional

Figure 20.5 Landscape after intensification of shrimp production.

Figure 20.6 Landscape after intensification of shrimp production.

farming impossible. There is evidence that in some cases, investors flood the land of those who will not sell or leave of their own accord (Deb, 1998; Ito, 2002). Many smallholders have migrated to the surrounding cities, or to the capital, Dhaka, where they seek shelter in informal settlements and are dependent on paid employment for their food security (Primavera, 1997: Chapter 22).

In the case of the Bay of Bengal area we see a convergence of the conceptual framings 'food security' and 'food sovereignty', on which this chapter is based. In this case it is an increase in the productive capacity of the land, through intensification to increase national and global *availability* of shrimp, which has changed the landscape. This has reduced the food sovereignty of subsistence and small scale commercial farmers by excluding them from their productive resource – the land. As will be discussed later, many who migrate to urban areas will find themselves in new locations and settlements which conspire to reduce their food security and food sovereignty still further.

Access to food, services and housing

The rapid urbanisation of the developing world radically alters urban landscapes, as vacant or unsafe land is appropriated by rural migrants for informal settlements. Many of these settlements are exceptionally poor in services and housing. Being densely built up, without services, water or sanitation, these may be portrayed as landscapes of squalor. However, even in the poorest inner urban slums, people can make a living selling fresh produce bought at wholesale markets. Informal settlers frequently also benefit from a wider urban landscape of fresh food street markets, street vendors of cheap fast food, cafés which make leftover food available to the poor, mosques and temples that feed the poor, and outreach locations of NGOs and charities offering food and support. Informal settlements are, therefore, also very often set within foodscapes of the smells, sounds, colours, cultures, richness and delight of an adequate, affordable diet (Speak, 2014). This is exemplified by the scene in Figure 20.7 of a young boy standing by his family's vegetable stall in a Delhi informal settlement in 2003.

As the availability of urban land diminishes, these informally settled landscapes, once of little development value, are required for more profitable uses. The urban poor are evicted or, in some cases, relocated into formally planned low income housing areas. However, the landscapes into which they are relocated are very often truly inhospitable. They can be many kilometres from the city centre and unsupportive of an adequate diet.

Figure 20.7 A vegetable stall in a central Delhi informal settlement.

This is exemplified in Delhi, India, where low income, informal settlements along the banks of the Yamuna River, including the one pictured above, were evicted from what had become prime development land.

In response to pressure from private owners of the land, and the Government's desire to develop an internationally recognisable contemporary urban landscape of golf courses, gated residential estates, and investment sites, Delhi Municipality moved settlers some 15–20 kilometres north of the city. It relocated them into a hard, dirty, barren landscape of waste land, known as the Bhalaswa relocation colony, as shown in Figure 20.8.

The colony is sandwiched between a polluted lake and a municipal rubbish tip which leaches toxic heavy metals into the groundwater (Jhamnani and Singh, 2009; Sidhu et al., 2015). Although a major road now passes the colony, at the time of the initial relocation it took well over an hour to reach the area by bus from the city. It offered no employment opportunities, no productive capacity nor any food retail facilities at all, and the nearest market was several kilometres away and very difficult for low income people to reach.

Residents made several attempts to address their new food shortages. At one point they suggested that the Municipality and an NGO might help them cultivate a large area of land in the centre of the settlement, seen in Figure 20.8.

The value of urban agriculture for both food security and household livelihoods in the developing world is now well recognised, as is its contribution to sustainable development and community cohesion (Hernández-Garcia, 2016; Chapter 22). Using this waste land for such a purpose would have placed the productive capacity back in the hands of rural migrants and been entirely consistent with the ethos of the food sovereignty. Permission was refused. When the author questioned the refusal with one Municipal official the response was:

> Bhalaswa is an urban extension, not a rural place, it's not appropriate [. . .] they can go to the market to buy their food [. . .] beside, the land is not good.
>
> *(personal communication between author and Municipal official, 2003)*

Figure 20.8 Land at the centre of the Bhalaswa relocation colony in 2008, now built upon.

Indeed, the land is not good and in this case refusal could have been entirely justified on the basis of the toxicity of the land and groundwater. However, this land and water is still used by the settlers for multiple household tasks involving food, as will be discussed in the following section on the use of food.

In the early days of the settlement residents established makeshift market stalls at the side of the road to sell food brought in from the city to both passing travellers and to other settlers. The Municipality had these demolished. Again, their justification for this was that the area was zoned for residential use, not for retail, and settlers should go to the local market – although there wasn't one. This highlights what Duncan (1989) argues, that urban landscapes are culturally produced models of how the environment should look and also how it should behave. However, forcing household or community behaviour to fit into a contemporary cultural model of the urban landscape runs contrary to the lived reality and food security strategies of the poor.

Even where settlements are not as isolated or toxic as Bhalaswa, their landscapes can make it difficult to establish or maintain access to food. For example, Tipple (2005a) noted the importance of small home-based enterprises, like shops and cafés, in maintaining good access to food in low income settlements. However, narrow, densely packed and deeply pot-holed streets make customer access and goods deliveries by motorised vehicles virtually impossible, making it difficult for small enterprises to flourish.

Poor terrain can also make moving around settlements difficult for all but the most able bodied. In their work on enabling environments for people with disabilities, Maart et al. (2007) noted that, in informal settlements, the elderly, frail, or people with even minor mobility difficulties can experience extreme challenges traversing poor terrain to access services and accommodate basic needs like food shopping.

Use, storage and preparation of food

Recognising the way in which the urban landscape can influence the appropriate and safe use of food requires an empirical understanding of the environments of the very low income, un-serviced settlements which abound in so many cities of the developing world. Many, if not most, lack adequate water and sanitation, have unmade roads and unusable open spaces, are densely inhabited and may be on low lying, swampy, or contaminated land. Such settlements are often landscapes of dust, dirt, stagnant water, untreated human waste, flies, pollution, and other airborne vectors of disease (Vlahov et al., 2007). These problems undermine the best efforts of householders to maintain hygiene within the house and its surroundings, with implications for the safety of food.

This was the situation of most of the slums along the banks of the Yamuna River in Delhi. However, the colony of Bhalaswa, to which people were relocated under the rhetorical agenda of 'adequate housing', is equally bad if not worse. Rubbish from the municipal tip, and from the recycling activities by which people make a living, blows across the flat landscape constantly. It is a swamp in the wet season, requiring the Municipality to spray insecticide to kill malarial mosquitoes. In the dry season it is a dust bowl with clouds of dust and sand blowing into the dwellings. Most importantly, the area is badly serviced. Groundwater is contaminated (Jhamnani and Singh, 2009) and potable water is delivered daily by tanker but is rationed, sanitation is poor, and energy is generally from diesel generators (Speak, 2012). It is within this landscape that people must store, prepare, cook and eat their food.

Housing plots allocated to the settlers of Bhalaswa were a meagre 12 square metres. In such small dwellings there is no room for dedicated or adequate food storage or preparation space. As dwellings have little ventilation, cooking – generally on polluting kerosene stoves or open fires – is often done outside amongst the dust, flies and toxins that the landscape hosts, leading to frequent outbreaks of food poisoning (Tipple, 2005b). In the absence of adequate water supply, people must either use unsafe water for cooking rice, their basic staple or spend their limited household budget on bottled water for cooking.

Outbreaks of food poisoning within a single household may be containable. However, Rane (2011) and Nema et al. (2007) raise concerns that the contaminants which the landscape contains can mean that people producing food for sale to others may fall foul of municipal food hygiene regulations, leading to wider outbreaks of food poisoning. Indeed, looking at a number of countries, including India, Tipple (2005b) noted the high proportion of home based enterprises in informal settlements producing food for sale, both in the settlement and on the streets of the main city or town.

Stability of food security

The 'stability' pillar of the food security framework is concerned with the risk of environmental, economic or other shocks which might impact on food security. From the cases discussed above we can identify several ways in which the landscape, urban change, and poverty relationship undermines the stability of food security.

The transformation of shrimp farming in Bangladesh into a major global food industry is causing irreparable damage to the delicate landscape along the Bay of Bengal (Ahmed, 2003; Hoq, 2007; Iftekhar and Takama, 2008). The environmental damage to the landscape is becoming an increasing threat to the land's productive capacity and the wider region's longer-term sustainability. It is directly threatening the sustainability of the food supply, through making the delicate costal area more vulnerable to climatic crisis and cyclical events, such as cyclones, which can destroy crops over a wide area (Paul and Vogl, 2011).

This threat is not limited to the case of shrimp in Bangladesh but is seen in the trend for industrialisation of agriculture more generally. Large-scale farms may be more productive than smaller ones in an industrial sense considering labour intensity, with higher yield through intensification and the use of agrochemicals. However, they may well be destructive to both the ecological and social environments in the longer term (Chapter 16). If that is the case, there will be even more rural migrants in urban areas.

In the development of major urban economic zones, like that of Gurgaon, cast in once rural, productive agricultural landscapes, we see the removal of productive land which could provide food for the local urban population. This is being repeated across India as urbanisation and economic development demands urban expansion into rural agricultural land. As consumption is concentrated in ever larger urban areas, pressure mounts on food transport networks to deliver ever greater amounts of food to growing urban populations. Any increase in oil or transport costs, or a breakdown in the supply chain networks, could be problematic, especially in times of stress. Already grain supply chains in India are coming under scrutiny, amongst fears that their failure is causing spoilage and increasing prices (Tsao et al., 2010).

Rural migrants to urban areas are generally very poorly paid and their employment insecure. Food comprises a considerable proportion of their budget and they struggle to maintain an adequate diet, even in times of stability. However, in the face of failing supply chains or other crises increasing prices, their food security strategies are constrained by inadequate, small housing in densely populated settlements with very limited open space for growing and restrictions on what any space may be used for (Speak, 2011).

As we see in the case of Bhalaswa, small housing with such limited space means food cannot be stored safely to preserve it from the environment or vermin. Thus, households cannot benefit from buying and storing cheap seasonal foods to support them through times of economic shock or poor harvest elsewhere, as the 'sustainability' pillar of food sovereignty might suggest. Nor is there any land to use as productive resource to support a household in times of stress.

Conclusions

This chapter has painted a picture of smallholders in Bangladesh and farmers in India being forced to leave their productive landscapes, and of migrants being rehoused in inhospitable landscapes. In doing so it sought to describe the way in which urbanisation and its drivers are impacting the food security and food sovereignty of the poor. However, this picture is not unique to those countries, nor is it uncommon across the developing world. The problems arise as a result of policy responses to economic growth which prioritise international trade and profit over the needs of the poor, and sacrifice the traditional relationships between the poor, the landscape, food security and food sovereignty.

Urbanisation itself is not the problem. Moreover, urbanisation may well be necessary in some contexts if the rural landscape is to be adequately productive to feed the world's growing population. This may be particularly true in India where the average farm size across the country is a mere two acres. In many areas, such small landholdings cannot support a household, let alone contribute to the nation's food reserves. It seems likely, therefore, that these smallholdings will eventually be appropriated and agglomerated into larger farms in line with a shift to industrial agriculture which is gaining weight in India (Niyogi et al., 2010).

The problems of the urban poor are exacerbated by urban planning policies. Too often these policies seek first to locate poor people into barren, isolated and, in some cases, toxic urban landscapes. These landscapes reduce food availability by limiting people's access to productive land. They limit access by placing people out of the reach of markets and supply networks. Through poor housing and lack of services, they also impact negatively on people's ability to store, cook and use food safely.

In a move to sanitise and commercialise the urban landscape, planners also overlay residential areas with rigid 'one size fits all' categories of acceptable use and appearance, which they impose on affluent and, when they 'upgrade' them, poor areas alike (Speak, 2012). While it might be perfectly logical and workable to zone more affluent areas in this way, the landscapes of the poor need to be more flexible and multi-purpose, allowing residential/domestic and productive uses, including food production and keeping of small livestock, to co-exist. In this way, the urban poor would be less dependent on insecure paid employment and formal food supply chains for their food security and might withstand economic, environmental and political shocks more easily. Many of the urban poor have come as migrants from rural agricultural backgrounds; given a little more land and support, they may also contribute to the urban area's food availability more generally. Through small scale commercial urban agriculture, they may help to support both the availability and sustainability pillars. What is needed is a willingness on the part of urban planning and policy making to break the urban/rural dichotomy and view the urban landscape in a more flexible way, and to acknowledge that it needs to perform different functions for different urban residents.

Key texts

Ahmed, A.U., Hill, R.V., Smith, L.C., Wiesmann D.M., & Frankenberger, T. (2008). 'The world's most deprived: Characteristics and causes of extreme poverty and hunger', *2020 Vision for Food, Agriculture, and the Environment Discussion Paper* No. 43, International Food Policy Research Institute, Washington, DC.

Barrett, C. B (2010) "Measuring food insecurity". *Science 327* (5967), 825–828

Hernandez-Garcia, J. (2016). Sustainability citizenship: Lessons and challenges from informal settlements. In: Horne, R., Fien, J., Beza, B.B., & Nelson, A. (eds.) *Sustainability Citizenship in Cities: Theory and practice*, Australia: Routledge.

Howe, J., & Wheeler, P. (1999). Urban food growing: The experience of two UK cities. *Sustainable Development*, 7(1), 13.

Paul, B.G., & Vogl, C.R. (2011). Impacts of shrimp farming in Bangladesh: Challenges and alternatives. *Ocean & Coastal Management*, 54(3), 201–211.

Pimbert, M. (2009). *Towards food sovereignty*. London: International Institute for Environment and Development.

Speak, S. (2014) Desperation, deviance or delight: conflicting cultural landscape of the urban poor in the developing world. In Roe, M., & Taylor, K. (eds.), *New cultural landscapes*. London: Routledge.

Further reading

Ahmed, N. (2003) Environmental impacts of freshwater prawn farming in Bangladesh. *Shellfish News*, 15, 25–28.

Ahmed, N., Demaine, H., & Muir, J.F. (2008) Freshwater prawn farming in Bangladesh: History, pres status and future prospects. *Aquaculture Research*, 39(8), 806–819.

Ahmed, R., Haggblade, S., & Chowdhury, T.E.E. (eds.) (2000) Out of the shadow of famine: Evolving food markets and food policy in Bangladesh. *International Food Policy Research Institute*.

Auyero, J., & Moran, T.P. (2007) The dynamics of collective violence: Dissecting food riots in contemporary Argentina. *Social Forces*, 85(3), 1341–1367.

Cohen, M., & Garrett, J. (2010) The food price crisis and urban food (in)security. *Environment and Urbanization*, 22, 467–482.

De Haan, A. (1999) Livelihoods and poverty: The role of migration-A critical review of the migration literature. *Journal of Development Studies*, 36(2), 1–47.

De Haan, A. (2002) *Migrants, livelihoods, and rights: the relevance of migration in development policies*. Department for International Development, Social Development Department, London. Social Development Working Paper No. 4.

De Haan, A., & Yaqub, S. (2010) Migration and poverty: Linkages, knowledge gaps and policy implications. In: *South-South Migration*. London: Palgrave Macmillan, pp. 190–219.

Deb, A.K. (1998) Fake blue revolution: Environmental and socio-economic impacts of shrimp culture in the coastal areas of Bangladesh. *Ocean & Coastal Management*, 41(1), 63–88.

Delhi Development Authority (2010) *Masterplan for Delhi 2012*. New Delhi: Delhi Development Authority.

Duncan, J.S. (1989) The power of place in Kandy: 1780–1980. In: J.A. Agnew, J.A., & Duncan, J.S. (eds.), *The Power of Place: Bringing together Geographical and Sociological Imaginations*. Winchester, MA: Unwin, pp. 185–201.

Food and Agriculture Organization of the United Nations (2005) *Voluntary guidelines to support the progressive realization of the right to adequate food in the context of national food security*. Rome: FAO.

Hoq, M.E. (2007) An analysis of fisheries exploitation and management practices in Sundarbans mangrove ecosystem, Bangladesh. *Ocean & Coastal Management*, 50(5), 411–427.

Iftekhar, M.S., & Takama, T. (2008) Perceptions of biodiversity, environmental services, and conservation of planted mangroves: A case study on Nijhum Dwip Island, Bangladesh. *Wetlands Ecology and Management*, 16(2), 119–137.

Islam, M.S., Milstein, A., Wahab, M.A., Kamal, A.H., & Dewan, S. (2005) Production and economic return of shrimp aquaculture in coastal ponds of different sizes and with different management regimes. *Aquaculture International*, 13(6), 489–500.

Ito, S. (2002) From rice to prawns: Economic transformation and agrarian structure in rural Bangladesh. *The Journal of Peasant Studies*, 29(2), 47–70.

Jhamnani, B., & Singh, S.K. (2009) Groundwater contamination due to Bhalaswa landfill site in New Delhi. *International Journal of Environmental Science and Engineering*, 1(3), 121–125.

Jonsson, M., Buckley, H.L., Case, B.S., Wratten, S.D., Hale, R.J., & Didham, R.K. (2012) Agricultural intensification drives landscape-context effects on host – parasitoid interactions in agroecosystems. *Journal of Applied Ecology*, 49(3), 706–714.

Maart, S., Eide, A., Jelsma, J., Loeb, M., & Ka, M. (2007) Environmental barriers experienced by urban and rural disabled people in South Africa. *Disability & Society*, 22(4), 357–369.

Mills, A.D. (2010) *A dictionary of London place-names*. Oxford: Oxford University Press.

Narain, V. (2009) Growing city, shrinking hinterland: Land acquisition, transition and conflict in peri-urban Gurgaon, India. *Environment and Urbanization*, 21(2), 501–512.

Nema, V., Agrawal, R., Kamboj, D.V., Goel, A.K., & Singh, L. (2007) Isolation and characterization of heat resistant enterotoxigenic Staphylococcus aureus from a food poisoning outbreak in Indian subcontinent. *International Journal of Food Microbiology*, 117(1), 29–35.

Niyogi, D., Kishtawal, C., Tripathi, S., & Govindaraju, R.S. (2010) Observational evidence that agricultural intensification and land use change may be reducing the Indian summer monsoon rainfall. *Water Resources Research*, 46(3).

Pauchard, A., Aguayo, M., Peña, E., & Urrutia, R. (2006) Multiple effects of urbanization on the biodiversity of developing countries: The case of a fast-growing metropolitan area (Concepción, Chile). *Biological conservation*, 127(3), 272–281.

Perez, I. (2013) Climate change and rising food prices heightened Arab Spring. *Scientific American*, 4 March.

Primavera, J.H. (1997) Socio-economic impacts of shrimp culture. *Aquaculture Research*, *28*(10), 815–827.

Rajagopalan, S., & Tabarrok, A.T. (2014) "Lessons from Gurgaon, India's Private City," *George Mason University Working Paper in Economics*, pp. 14–32.

Rane, S. (2011) Street vended food in developing world: Hazard analyses. *Indian Journal of Microbiology*, 51(1), 100–106.

Roy, A., & Ong, A. (eds.) (2011) *Worlding cities: Asian experiments and the art of being global*, Vol. 42. Oxford: John Wiley & Sons.

Ruel, M.T., Garrett, J.L., Hawkes, C., & Cohen, M.J. (2010) The food, fuel, and financial crises affect the urban and rural poor disproportionately: A review of the evidence. *The Journal of Nutrition*, 140(1), 170S–176S.

Sidhu, B.S., Sharma, D., Tuteja, T., Gupta, S., & Kumar, A. (2015) Human health risk assessment of heavy metals from Bhalaswa Landfill, New Delhi, India. In: *Management of natural resources in a changing environment*. New Delhi: Springer International Publishing, pp. 215–223.

Speak, S. (2011) Point of no return: Exploring the issues which trap migrants in a state of homelessness in the city. *International Development Planning Review*, 32(3–4), 225–243.

Speak, S. (2012) Planning for the needs of urban poor in the Global South: The value of a feminist approach. *Planning Theory*, 11(4), 343–360.

Speak, S. (2014) Desperation, deviance or delight: Conflicting cultural landscape of the urban poor in the developing world. In: Roe, M., & Taylor, K. (eds.), *New cultural landscapes*. London: Routledge.

Speak, S., & Kumar, A. (2016) Fit and misfit – ordinary urban injustice in a global context. In: Davoudi, S., & Bell, D. (eds.), *Justice and fairness in the city*. Bristol: University of Bristol, Policy Press.

Steel, C. (2008) *Hungry city: How food shapes our lives*. London: Chatto and Windus.

Tipple, G. (2005a). The place of home-based enterprises in the informal sector: Evidence from Cochabamba, New Delhi, Surabaya and Pretoria. *Urban studies*, 42(4), 611–632.

Tipple, G. (2005b). Pollution and waste production in home-based enterprises in developing countries: Perceptions and realities. *Journal of Environmental Planning and Management*, 48(2), 275–299.

Tipple, G., & Speak, S. (2009) *The hidden millions: Homeless in developing countries*. London: Routledge.

Tsao, H.S.J., Parikh, S., Ghosh, A.S., Pal, R., Ranalkar, M., Tarapore, H., & Venkatsubramanyan, S. (2010) Streamlining grain supply chains of India: Cloud computing and distributed hubbing for wholesale-retail logistics. In: *Service Operations and Logistics and Informatics (SOLI), 2010 IEEE International Conference on* (pp. 252–257). IEEE.

Vitello, D., & Brinkley, C. (2014) The hidden history of food system planning. *Journal of Planning History*, 13, 911–912.

Vlahov, D., Freudenberg, N., Proietti, F., Ompad, D., Quinn, A., Nandi, V., & Galea, S. (2007) Urban as a determinant of health. *Journal of Urban Health*, 84(1), 16–26.

Wittman, H., Desmarais, A., & Wiebe, N. (2010) The origins and potential of food sovereignty. *Food Sovereignty: Reconnecting Food, Nature and Community*, 1–14.

21

Connecting landscapes and food in Africa

Case studies from Ethiopia and Uganda

Max Kelly and Ruth Jackson

Introduction

The interrelationships between landscapes and food in Africa are deeply complex. The physical geography of Africa encompasses incredible diversity with eight regions across the continent: Sahara, Sahel, Ethiopian Highlands, Savanna, Swahili Coast, rain forest, African Great Lakes, and Southern Africa (National Geographic Society, 2016). Overlying this physical diversity that occupies just over 20% of the world's landmass, are 54 nation states (Western Sahara is currently disputed), and over one billion people speaking hundreds of native languages. Africa is the "cradle of humankind" (Wayman, 2011), the home of great Kingdoms, and vastly diverse cultures. It is also the site of the one of the largest land grabs in history, through colonialism, and the "scramble for Africa" (Pakenham, 1992). The continent is widely touted as one of the richest areas in the world for its natural and mineral resources and is collectively one of the best performing economies globally, behind only East Asia in 2015 (ADB et al., 2016). Despite impressive economic growth, which saw the percentage of the continent's population in extreme poverty decline from 57% in 1990 to 43% in 2012 (Beegle et al., 2016), the impact of population growth saw absolute numbers in extreme poverty grow from 280 to 330 million people from 1990 to 2012 (ibid.). In relation to hunger, the prevalence of hunger from 1990–92 to 2015 declined by 31% (FAO, 2015). At the other end of the scale, the number of high net worth individuals grew from 107,000 to 148,000 from 2009 to 2014 (CapGemini, 2015). These statistics hide startling differences, with, for example, less than 18% of Ethiopians living in cities compared to 70% of Algerians (NEPAD, 2013). A recent World Bank report detailing poverty across the continent notes that within-country inequality was increasing but African inequality was balanced by those countries where inequality is decreasing (Beegle et al., 2016). There is a robust relationship between geography and inequality with southern Africa showing the highest levels of inequality, West Africa the lowest, and mixed results in East Africa. This aside, there are no other significant patterns (ibid.).

Food, development, and landscapes

Food security relies on both availability of and access to food, stability, and utilisation (World Food Summit, 1996; Chapters 4, 20, 25 and 31). Luan et al. (2013) analysed food availability across Africa measuring the 'self sufficiency ratio' (SSR), essentially the ratio of production to domestic consumption or use. Their key finding is that overall, Africa's food self-sufficiency has declined over the period 1961 to 2007 with a drop in the SSR from almost 1 (0.908) back to 0.871. Shortfalls in food have been made up of increasing imports, and also food aid (Luan et al., 2013); a trend which they note is likely to continue.

Economic performance, especially in agriculture, is sharply influenced by commodity prices. Sub-Saharan Africa (48 countries south of the Sahara), has the highest prevalence of food insecure people (one in four) (World Bank, n.d.). The share of agriculture to value added as share of GDP (2009) was 14%. Agriculture is a key factor in rural areas, employing over half of Africa's labour force and accounting for 17.5% of GDP (World Bank, n.d.a). However, 80% of all farms in sub-Saharan Africa are less than two hectares (NEPAD, 2013). In the 10 years leading up to 2013, 20 million hectares of African farmland was subject to investment contracts (NEPAD, 2013). Some of this is foreign direct investment in agriculture through market seeking. Much is classed as 'land grabbing', or the purchase or long-term lease of land by international private and government investors, which is frequently associated with dispossession of rural communities, and privatisation of previously communal or public land (FAO, 2009; Chapters 31 and 32). The focus of commercial agricultural interests and a move to a more industrialised agricultural model raises concern, as there is widespread acknowledgement that "family farming is best placed to optimise labour use and to enable greater land use [. . .] with the greatest potential for increasing broad based growth and sustainable wealth creation" (NEPAD, 2013: 8).

Increases in agricultural production are currently based on increasing the area of land utilised for agriculture, rather than any significant overall increases in productivity. Land brought into production for agriculture, of course, encroaches onto land that may be less suitable for agricultural production or of significance for other purposes. This has implications for forest and other habitat destruction, and for environmental degradation of land, as well as climate change. Pretty et al. (2011) argue that sustainable agricultural intensification of existing land used for agriculture can increase productivity and associated social and economic benefits, but reduces the negative impacts of expansion of agriculture into new lands, and environmental and social costs of many forms of intensive agriculture. Intensification is not so simple however, with associated social and environmental costs of intensive agriculture (see Chapters 8, 9, 10 and 16). Despite the complex interactions between agriculture, food, and the environment, place-based policy and programme interventions traditionally tend to have a sectoral focus, for example, in food security, agriculture, energy, economic development, ecosystem conservation, climate change, and risk management. The implications of this sectoral thinking on strategic planning and management at scale are problematic.

The future challenges for Africa are to: feed a growing population (estimated at 1.5 billion by 2030 and 2 billion by 2050), address poverty, reverse environmental degradation, address climate change, cater to increasing urban populations, and reduce inequality and vulnerability (NEPAD, 2013). To do this, future policy options advocated across the continent include: increasing investment in agriculture, increasing agricultural production sustainably, climate-smart agriculture, and an increasing focus on landscape scale management and governance. The tone of the debate is influenced by social and environmental factors given the extremely high proportion of smallholder farmers, and the heavy reliance on many of the complicating factors outlined earlier, but also exacerbated by climatic and other factors such as limited irrigation across the continent. According to NEPAD (2013), only 6% of cultivated land is irrigated, and most of this is in just five countries, predominantly South Africa, Egypt, Madagascar, Morocco, and Sudan. The mantra of sustainable intensification of production is embedded in discourse around Africa, development, economic growth, food security, and poverty alleviation (cf. FAO, 2016). Sustainable intensification is driven by some key players, including the work of Conway and his colleagues (The Montpellier Panel, 2013) and Pretty et al (2011). The underlying message of the sustainable intensification 'paradigm' (The Montpellier Panel, 2013) is one of intensification (a growth paradigm), with a long-term view of larger farms as a given (a highly questionable assumption, given resource and environmental issues from a longer-term perspective), but driven by a pragmatic understanding that in the short-to-medium term, smallholders must provide food for increasing urban populations "for some years to come" (ibid.: 7). This runs counter to arguments promoting ecological agriculture, small scale family farming, and debates more focused on ecological sustainability (Chapters 10, 16 and 24). The implications of current and future land use decisions, in particular around food production, have intense implications not just on livelihoods and food security, but on economic growth, climate change, social systems, cultural heritage, and environmental issues more broadly including biodiversity, watersheds, and soil health.

There are multiple competing visions of the future (see Chapter 40), yet many of these visions are grounded in a political economy context that promotes an agroindustrial complex driven by food as commodity, with intensification of agriculture a central tenet (Poole et al., 2013). This issue is discussed elsewhere in this volume, and is not a core focus of this chapter (see Chapters 10, 16, 31 and 32). This chapter briefly explores emerging debates on moving beyond a sectoral focus on aspects of rural landscapes, to nascent initiatives in integrated landscape scale initiatives.

Landscape scale approaches are defined as "approaches which use landscape management practices to address multiple objectives and provide multiple benefits" (Hart et al., 2015: 90). The notion of multifunctionality is central to these approaches. Multifunctionality is used in relation to both agriculture, and landscapes. Multifunctional agriculture recognises the "inescapable interconnectedness of agriculture's different roles and functions i.e., agriculture is a multi-output activity producing not only commodities, but also non-commodity outputs such as environmental services, landscape amenities and cultural heritages" (IAASTD, 2008: 1). A multifunctional landscape is somewhat more of an ambiguous concept, as is frequently noted; all landscapes are, by nature, multifunctional. Vejre et al. (2007) argue that land use is the common denominator. And it is both to land use, and the multifunctional nature of land, landscapes, and agriculture that we now turn.

Agriculture is traditionally located at a farm scale. At a very simplified level, in a productive landscape, farming involves the modification of various landscape elements, from the choice of crop, to hedge and tree maintenance, drainage, fences, and so on (Kristensen et al., 2001). However, decision making related to these changes are predominantly private decisions, and in many cases rooted in traditional intergenerational knowledge and experience (Chapters 1, 2 and 13). As noted above, increasing recognition of the complex and interrelated nature of a sustainable, biodiverse, productive landscape requires a broader scale of engagement, and a broad scale of both planning and management, bringing a transdisciplinary perspective to bear (O'Farrell and Anderson, 2010). Calls for agriculture to be more sustainable, whether through sustainable intensification, agroecological approaches, or other variations on ecological, social, cultural and economically sustainable agriculture, focus on more holistic farm level solutions. Details on what constitute sustainable agriculture and agroecological approaches are addressed elsewhere in this volume (Chapters 10, 24 and 33). However, this is necessary, but unlikely to be sufficient, given that key ecosystem services that underpin all human development activities and outcomes operate at a larger scale, including landscape, watershed, and sub-national (Milder et al., 2014; Chapters 16 and 17).

Landscape scale focus and initiatives have gained momentum in the last decade. After surveying 191 landscape scale initiatives mostly in Latin America and Africa, Hart et al. (2015: 98) concluded that "outcomes can be achieved simultaneously in domains that at times have been thought to be incompatible (e.g. agriculture and conservation)" and that "a move toward multi-objective management can lead to the achievement of a broad set of outcomes, rather than a dilution of initiatives' effectiveness, particularly by reducing tradeoffs through cross-sector cooperation and enabling access to the resources and energy of multiple stakeholders." The emergence of the Global Landscapes Forum and the Landscapes for People, Food and Nature Initiative show the scale of engagement with landscapes. Milder et al. (2014) note that landscape scale management is not entirely a new phenomenon in Africa, primarily from the conservation sector, but recent (past decade) initiatives demonstrate a greater engagement with agriculture and livelihoods. Milder et al. (2014) completed a region wide assessment of integrated landscape initiatives to ascertain the scale of the emergence of integrated multi-objective management initiatives in Africa, finding that integrated landscape management is growing rapidly, albeit from a very low base, and at the current time initiatives are neither as widespread or not as well developed as elsewhere, particularly Latin America. As noted earlier, it is beyond the scope of this chapter to detail initiatives across the continent (see for example Milder et al., 2014; Landscapes for People, Food and Nature, 2014).

To explore landscape, it is important to locate that landscape, however briefly, in history and culture, because the present and the future are linked to the past. But which landscape? Whose history? Whose culture? We draw on the authors' experiences from two countries (Ethiopia and Uganda) to ground the discourse of 'African' landscapes and food. These are not designed to be exhaustive studies of landscape scale initiatives in either case, but to give depth to the analysis of the potential of landscape scale initiatives to promote sustainable outcomes in food and landscape management.

Ethiopia

As one of the least urbanised countries in the world, Ethiopia has one of the fastest growing populations. The population is projected to grow from 73,750,932 in 2007 to 92,206,005 in 2016, with urbanisation increasing from 11,862,821 to an estimated 18,256,000 in the same period (Central Statistical Agency [Ethiopia], 2007, 2013). Agriculture, the mainstay of Ethiopia's economy, accounts for 42% of GDP, 85% of total employment, and 70% of export earnings (Central Statistical Agency, 2015; Food and Agriculture Organization (FAO) of the United Nations, 2014). Cash crops such as coffee, sesame, and spices and livestock contribute to agricultural GDP – all exports where world demand is increasing (World Bank Group, 2014).

Ethiopia's agricultural systems have been in place for over two millennia, yet are undergoing significant change. In 1965, Ethiopia was described as an agricultural nation relying on exports of coffee, skins, and oilseeds (Levine, 1965), and then 20 to 30 years later as a nation burdened with drought and famine. McCann (1995) describes how the rural landscape has changed, the effect of population growth and other factors such as urbanisation, the effects of perennial tree crops and the introduction of maize, contrasting these factors with past stories of fertility and productivity. More recently, photographers such as Chester Higgins, (formerly of the *New York Times*), researching and photographing monuments and other antiquity sites along the River Nile from the 6,000-foot high mountains of Kush (modern-day Ethiopia) through Nubia (Sudan) down to the ancient land of Kemet (Egypt); documentary makers such as those filming 'The Hottest Place on Earth' (BBC), or Dutch investors visiting the export-oriented cut flower industry, present visual images to counteract the images of drought and famine. Although agriculture's share of GDP has declined over the past decade as the service and industrial sectors expand, mainly driven by the expansion in tourism and hotels, construction, mining, and manufacturing, agriculture is still dominated by small subsistence farms. Coffee, skins, and oilseeds are still among the largest exports along with pulses, gold, chat, and potential electricity exports when the Great Ethiopian Renaissance Dam is completed (Wondifraw et al., 2015).

The Ethiopian Government's notion of 'landscape' has shifted from one whereby agricultural productivity is judged to be low to that of one with many "untapped opportunities to increase production by promoting modern farming practices" (ibid.: 4). This includes plans to commercialise smallholder agriculture, promote high-value export crops, and develop large-scale commercial agriculture by adjusting these interventions to address the needs of the various agro-ecological zones. Agricultural Development Led Industrialization (ADLI), the Government's key policy response to Ethiopia's food security and agricultural productivity challenges, aims to reduce poverty and to reduce dependence on food aid through economic growth based on a free market economic system (Federal Democratic Republic of Ethiopia Ministry of Finance and Economic Development, IMF and World Bank, 2002).

Agriculture is practiced by 93% of rural and 42% of small town households. Average household land holdings are less than 1.5 ha (Central Statistical Agency and World Bank, 2013) or 1.22 ha (World Bank, 2015). Yet farm households can have 12 fields with an average field size of 0.15 ha per field (Central Statistical Agency and World Bank, 2013). Ethiopian farmers have developed a form of agrobiodiversity based on land fragmentation; crops have been developed and adapted to moisture and temperature regions so each farmer grows cereals, pulses, and oil crops as a drought risk-averting mechanism and possibly to reduce erosion (Dejene, 1990). In Menz, 300 km north of Addis Ababa (the capital city) for example, land is subdivided into as many as 13 plots; people prefer their land "to remain fragmented, because of the benefits to be derived from microecological variations in soils and climates" (Pankhurst, 1992: 81).

New and emerging changes to the landscape have also come from government leases of 'under-used' or 'uncultivated' land. According to the Ethiopian government, 45% of Ethiopia's landmass is arable but only about 10,556 thousand ha are presently under cultivation (Ethiopian government portal, 2016). Yet pastoralists or nomadic/semi-nomadic people live in much of this area (around 60% of the total land mass of Ethiopia) (Federal Ministry of Health, WHO and UNICEF, 2011). For example, Afar Region in the northeast, ranges from the Danakil depression which is largely desert scrubland with shallow salty lakes and long chains of volcanoes, to the Awash River in the south which flows into the northern lakes. Around 90% of the population base their livelihood on livestock rearing – cattle, camels, goats, sheep, and donkeys. This provides subsistence for Afar pastoralist households including household consumption (milk, meat, butter, hides, skins, and so on), and a source of cash from the market. Climate change is adding "to the burden of those who are already poor and vulnerable by affecting their livelihood pattern and strategies and triggering food, feed, water and social insecurity" (Afar National Regional State, 2010: ii).

There are numerous international reports and media stories about land grabs, land clearing for dam construction (Abbink, 2012), sugar plantations, and commercial agriculture which displace hundreds of thousands of indigenous peoples, creating conflict between different ethnic groups. Additionally, villagisation – or the clustering of agro-pastoralists "and/or shifting cultivator populations into more permanent, sedentary settlements" (Human Rights Watch, 2012: 11), has a long history in Ethiopia. Villagisation is often implemented with threat of force to rationalise land use; "conserve resources; strengthen security; and provide access to clean water, health and education infrastructure" with new villagers being forced to provide labour for "government projects – whether for road construction, agricultural production, or other infrastructure development" (Human Rights Watch, 2012: 12). More recently, forest-dwellers and pastoralists have been relocated to overcome "perceived problems of water and pasture scarcity in semi-arid areas" (Abbink et al., 2014: 10), often alongside large-scale acquisition of land by states and investors.

Ethiopia is also recognised as having one of the world's 34 biodiversity hotspots and one of the world's most species rich countries. In 2010, UNESCO accepted the Kafa Biosphere Reserve in Ethiopia's southwest, an area covering 760,000 hectares, into the World Network of Biosphere Reserves (Nature and Biodiversity Conservation Union (NABU), 2011). Home to approximately one million people, Kafa Zone is also the centre of the genetic diversity of *Coffea arabica* and provides a habitat for various endangered animal and plant species.

Food, culture, and landscape in Kafa Zone, Ethiopia

Before 1897, Kafa was a large and powerful kingdom with its own monarch exacting tribute from many of the smaller neighbouring kingdoms trading "gold, ivory, hides, honey, coffee and slaves" (Lange, 1982 cited in Pankhurst and Freeman, 2003: 70) with the northern Ethiopian empire. Coffee emerged as the 'mainstay' of the Ethiopian economy motivated by higher prices, improved transportation and low taxation as the export of ivory diminished (McLellan, 1980, 1986). While this altered the role of Ethiopia's colonial subjects, coffee "more so than ivory, could better serve the needs of a developing nation since its production could be sustained over an indefinite period of time" (McLellan, 1980: 78). Coffee remains the focal point of the Ethiopian economy to this day – not only is it closely intertwined with history, culture, and economy but it also still plays a significant role in daily life for most Ethiopians (Stellmacher, 2006). Preparing coffee is a daily activity frequently performed by women, and one that enables the building of good relationships with neighbours and the community. The coffee ceremony also gives women a time to sit, relax and gossip especially if the men are not around; it is impossible to separate the daily socialising from consumption (Pankhurst, 1992).

Kafa Zone's "mosaic of different land-use types" (Schmitt, 2006: 25) ensures most people are engaged in agriculture, with the average land holding per family between 1.25 and 2.0 ha. Of the total land area, only 28% of agricultural land is intensively cultivated. Much of Kafa Zone is still classified as forest, wetland, grassland, or bamboo forest, around 1% of the total area is tea, coffee, or timber plantation, and 1% is classified as urban. Despite this diversity of land use across the Zone some characteristics may be generalised. The livelihood of the original population, the Kafa and Manjo tribes, was based on dependence on the forest complemented with animal husbandry and typical crops planted around homesteads including enset (*Ensete ventricosum*), a large single-stemmed banana-like plant that has a variety of uses including food and fibre (Brandt et al., 1997).

As continuous population growth and migration from the over-populated highlands has resulted in higher population density and gradual changes of the farming systems and accelerated deforestation in Kafa Zone, settlers cleared the forests to grow more cereals and pulses, thus pushing back the forest fringe. Non-timber forest products such as coffee, honey, false cardamom (*Aframomum corrorima*), and wild pepper (*Piper capense*) are also important means of income. The forests are an important source for fuelwood, charcoal and timber for house construction, particularly for the Manjo who make up 5 to 10% of the total population and are the largest minority class still facing prejudice and discrimination (Pankhurst and Freeman, 2003).

Despite the Afromontane forests being recognised as an invaluable genetic source of coffee (*Coffea arabica*), they are greatly endangered – an estimated 43% of the wild coffee forests were lost due to resettlement and migration, timber extraction, forest grazing, and conversion to agricultural land between 1988 and 2008 (FAO and Swiss Agency for Development and Cooperation, 2011). One response was the establishment of the 760,000 ha Kafa Biosphere Reserve as one of Ethiopia's first biosphere reserves and the first United Nations Educational, Scientific and Cultural Organization (UNESCO) coffee biosphere reserve in the world (NABU, 2011).

Further changes to the landscape are due to deforestation, agriculture expansion, population pressures, resettlement, concessions for coffee plantations, and the Road Sector Development Program (RSDP). Ethiopia's RSDP has restored and expanded the non-urban road network from 26,550 km in 1997 to 85,966 km in 2014 (World Bank, 2014) to enhance interconnectivity and to bring the rural population to within two km from an all-weather road. One aim is to improve accessibility for women and their children to social, education, and medical facilities; to reduce the time women spend transporting products to market; and to improve opportunities for small-scale road-side commercial operations (World Bank, 2014). However, the construction of roads in rural communities can create conflicts between different stakeholders such as those advocating to protect the Kafa Biosphere Reserve and those building local roads, some of which have resulted in deforestation, surface run-off and soil erosion in areas nominated as reserve.

The preceding picture demonstrates aspects of the complexity embedded in exploring sustainable food systems that provide a sufficient, safe, and nutritious food supply that meets that meets the dietary needs and food preferences for an active and healthy life (FAO, 2106). The somewhat economic focus of both much of the government policy, and the livelihoods conceptualisation for food production is challenged by the social, cultural historical and climatic factors that influence how land is used, why, and for what purpose.

Kafa Biosphere Reserve

NABU's project 'Biodiversity under Climate Change: Community-Based Conservation, Management and Development Concepts for the Wild Coffee Forests' (2014–2017) is part of the International Climate Initiative funded by the German Federal Ministry for the Environment, Nature Conservation, Building and Nuclear Safety. It builds on earlier projects that aimed to make a significant contribution to the preservation of biological diversity, particularly that of *Coffea arabica* (NABU, 2011, 2015, 2017). In particular, the formation of the Kafa Biosphere Reserve was based on landscape multifunctionality that recognises the need to "reconcile people's needs with conservation pressures" (NABU, 2017: 331). As one of the few areas in Ethiopia that still has substantial forest cover, the project is one that brings together various stakeholders to preserve the remaining forest.

A study of small farmers, local NABU staff involved in nature conservation, and scientists participating in a biodiversity assessment examined perceptions of biodiversity; risks and threats to Kafa biodiversity; and what measures could be taken to protect diversity (NABU, 2017). The three groups had substantially different understandings of biodiversity based on their level of education and individual interests. Farmers' perception of biodiversity was equated with ecosystems or forests which they valued for use of wild species for fuel and construction. Other uses were for medicine, food, craft materials, and animal husbandry. Most of the scientists and 20% of the local staff gave the standard definition of biodiversity (genetics, species, and ecosystems). Despite the different understandings of biodiversity across the three groups, there was common understanding of the drivers of loss of biodiversity. Agricultural expansion was seen as the biggest threat to biodiversity. Small-scale logging was the second biggest threat. However, '[p]opulation pressure was unanimously seen as the biggest underlying risk for the loss of biodiversity' (NABU, 2017: 348) and the future of the Biosphere.

Though the biggest challenges overlapped with the drivers of biodiversity loss and the risk of population pressure, the challenge most frequently mentioned by local staff was the lack of benefits to the local communities. One-third of scientists supported this view. There was partial agreement between the three groups over the best ways to protect the biodiversity of the Biosphere. These included diverse home gardens, planting or maintaining trees on their properties, and occasionally planting flowering plants for bees (farmers). Finally, although both scientists and local staff suggested that Kafa's people have a unique culture based around protecting nature and the majority of farmers and scientists argued that the community was responsible for protecting biodiversity, half of the farmers thought this responsibility lay with the government. Both personnel and scientists thought that government involvement was essential.

Uganda, agriculture, intensification, and landscape scale change

Uganda sits further down the Rift Valley in the heart of the Great Lakes region. Uganda's ecosystems and landscapes are predominantly composed of a large plateau, occupying two-thirds of the country, bounded to the east and west by mountains. The soils are generally very fertile, with those in the vicinity of Lake Victoria being some of the most fertile in the world. Agriculture is a core part of the Ugandan economy and land use, with around 75% of the country available for agriculture (Mwebaze, n.d.). Approximately 21% of the land is cultivated and 45% is woodland and grassland, some of which has been cleared for roads, settlements, and farmland in the south. Some 13% of the land is set aside as national parks, forests, and game reserves.

Uganda emerged from British colonisation in 1963. The country has been dominated by 'big man' politics, with Obote and Amin ruling over a series of rebellions, human rights abuses, political oppression, ethnic persecution, civil wars, and intrastate conflicts. The current president, Mueseveni, has been in power since 1983. Although his reign has been controversial, he has overseen strong economic performance and stability in much of the country. Twenty years of conflict associated primarily with the Lord's Resistance Army in the north, and the ongoing tensions in the Karamoja (north east) region, dominated by nomadic pastoralists, have severely constrained development in these regions.

Of the total population of over 35 million (2014 Census), 80% of the population is involved in agriculture, which contributes 23% of GDP (Anderson et al., 2016). Uganda has one of the world's fastest growing populations with three quarters of the population under 30 years of age, and significant unemployment and underemployment of youth (Ahaibwe and Mbowa, 2014). Uganda's economy is growing quickly, driven by growth in the agricultural sector and investment, both foreign and domestic. Average landholding size is

1.1 ha, ranging from 1.6 ha in the northern region to 0.8 ha in the western region (Anderson et al., 2016). Smallholder farms (75% of land holdings) average area is 0.7 ha, and larger farms (25% of land holdings) average 2.4 ha. Despite the very high agricultural potential of Uganda, productivity is very low, and in many cases declining, due to a number of factors including lack of access to inputs (fertiliser), little uptake of 'modern' farming techniques, dependency on rainfed agriculture, and land tenure challenges (National Planning Authority, 2014). The very small landholdings, lack of productivity, increasing population, deforestation, land degradation, and land tenure conflicts arising from historical land use and land tenure policy in overview form give some sense of the complexity of policy and practice in agricultural and land use decisions.

The Uganda 2040 vision (National Planning Authority, 2014) focuses on making Uganda a modern and prosperous country. Aspects of the strategy relevant to land use and agriculture include targeting irrigation, land reform, and reversing land fragmentation to "secure land for mechanisation" (ibid.: 47). Land reform to support commercialisation is central to the strategy. Land disputes in Uganda are considerable, affecting an estimated 7% of agricultural holdings (2005/2006, UNHS). Key exports include coffee (Africa's second largest exporter after Ethiopia, and 22% of Ugandan export value), tea, cotton, copper, oil, and fish. The government is pushing a strong investment agenda, with a focus on agricultural led growth; Foreign Direct Investment (FDI) in Uganda grew from around US$4 million in 1980 to US$800 million in 2009 (Ogalo, 2011). However, only 20% of this is in agriculture, with floriculture being a favoured subsector. Other areas that have attracted FDI include improved seeds, and cotton and associated opportunities (Ogalo, 2011). Areas that the Government of Uganda are promoting as potential investments include dairying, meat and meat products, fisheries, and particularly aquaculture in the lower lying wetland and swamp areas, cotton and textiles, fruit and vegetables, floriculture, tea, coffee, and cocoa (Mwesigye, 2010).

Global land grabs have also received attention in relation to Uganda, as in much of Africa. Some examples include reports of 20,000 people evicted in Kiboga and Mubende districts for a UK forestry company land lease (Oxfam, 2011), or the impact of land leased for Palm Oil plantations, including removal of landholders, destruction of forest resources, and resulting large scale monocropping (Friends of the Earth, 2012). Increasing interest in foreign purchase of land has been touted as everything from an enormous development opportunity to a 'land grab' by countries desperate to increase their food supplies (Cotula et al., 2009). Although data on the land based investment is patchy, Cotula et al. (2009) found that there is a distinct upward trend in land acquisitions, and there are both risks and opportunities associated with land deals. Although they note that large scale land deals are only a small proportion of total land, it does put pressure on higher value lands, and land is already often under claim by local people, with potentially negative impacts. The impact of non-food agricultural products (biofuels and other commodities) is a key driver as well as food security concerns of investors and investment opportunities provided by agriculture in the 21st century.

The specific intent at national level to promote commercialised agriculture, with a focus on increasing productivity, farm size and output is a clear message. Alongside this, there are a number of initiatives that operate on a larger than individual landholding scale. Five of the six Landscape Scale Initiatives (LSIs) identified by Milder et al. (2014) relevant to Uganda focus on ecosystem maintenance and services, with all six initiatives linking with economic development and livelihood initiatives. The sixth initiative, the Consortium for Improving Agriculture-based Livelihoods in Central Africa (CIALCA), appears to have a physical office in Kampala rather than any specific activities in Uganda, focusing on Burundi, Rwanda, and The Democratic Republic of the Congo (DRC). Two of the initiatives are directly linked to a conservation area, namely Virunga (home of the mountain gorillas), and Mount Elgon (national park and fragile mountain ecosystem) (UNDP, 2015). A third is in an area of landscape significance (Murchison Smelikki, including Murchison Falls National Park), although the project is a fee for ecosystem services project to protect forests on private land, providing essential wildlife corridors, linking protected and public forests (WCS Uganda, 2017). A very recent initiative is the identification of priority landscapes in Uganda for restoration of forest cover, which is currently at 9% (Ministry of Water and Environment, 2016). The high levels of population growth, very fragmented landholdings and land tenure issues, and resulting soil and land degradation have prompted a partnership between the Government of Uganda, the International Union on the Conservation of Nature (IUCN), and the World Resources Institute (WRI) adopting a restoration of forest landscapes approach to increase forest cover to a goal of 24%, through identification and targeting of priority landscapes. A move from conservation landscapes (national parks and protected areas, and surrounding areas) to broader productive, yet degraded landscapes, is clear in these initiatives.

Land, food, and people in Africa: a case for landscape scale thinking?

Although these are compressed snapshots of food and landscapes in modern Africa, they provide a sense of the challenges faced in both in Ethiopia and Uganda. In many senses, they echo challenges and opportunities faced in many places across the continent, whether planted crops, pastoralism, nomadism, or land framed as 'unproductive' providing habitat, wild spaces, ecosystem services, or other land dynamics. Africa is a landscape driven by human encroachment and high levels of population growth, leading to ever greater pressures on land and landscapes. Farm level innovations can help increase productivity and therefore food availability, and, if drawing on agroecological premises, can contribute to long-term maintenance of multifunctional landscapes. However, individual decision making on private (or customary tenured; see Chapter 28) land is unlikely to provide strategic landscape scale impact. In an era of agricultural intensification, increasing pressure to respond to complex issues of degradation, social and economic sustainability, and changing climatic conditions, there is an imperative to explore how and what land and landscapes can provide the multifunctionality required of them. This brings us back to the multiple visions of future directions. In the context of the Ethiopian and Ugandan landscape and culture within a globalised world economy where agriculture must fit in with the dominant neoliberal paradigm (see for example Wolf, 2014), what actions will contribute to a sustainable future?

Sustainable intensification does engage with agroecology at a conceptual level (otherwise how can it be sustainable?). However, beyond the conceptual level, how can this be applied in practice? Responding to a 'Sustainable Intensification' report by Garnett and Godfrey (2012), Tudge (2012: 12) noted that intensification can entail one of two approaches:

> The first is the skills-intensive, artisanal, peasant route – to produce more food from the same space through more intricate husbandry (intercropping, rotations, individual care of plants and animals, etc.); and this requires us to build up the agrarian workforce and all that goes with it. The second is the high-tech, high capital, industrial route – to produce more food from the same space by more effective fertilizers, feeds, or growth promoters and (with or without GM) by precise manipulation of the animals' or plants' genomes, all with as little labour as possible.

Reflecting back to earlier debate, even 'sustainable' intensification is likely to increase farm size, with a resultant decrease in crop diversity, and a reduction in both the complexity, and by association the exposure to risk and climatic features, leading to further long-term changes to the 'mosaic of different land use types' as noted earlier. Larger scale commercial interests in agriculture are more difficult to engage in either landscape scale initiatives or agroecology more generally, as incentives for large scale commercial stakeholders may be substantially different (Hart et al., 2015).

The implications for smallholder agriculture and all that this entails in terms of people, social structures, cultural and human environment interactions are potentially immense. Despite strong rhetoric supporting smallholders – 2014 was the UN International Year of the Family Farm – it can be difficult to reconcile the competing discourses between commercialisation and retaining family farms. Collier and Dercon (2014) argue that African Agriculture will have to "change beyond recognition" (ibid.: 92). Their argument is that a focus on smallholders as the preferred mode of production is not likely to be optimal for growth. Although noting that "mega farm" deals (landgrabbing) (ibid.: 98) is also not a viable option, they do argue for economies of scale inherent in pursuing opportunities for what they refer to as new forms of commercialisation, in essence large (private) farms with potential for hybrid versions whereby smallholders 'interact' (ibid.: 99) with larger farmers. In another response to the commercialisation of African smallholders' narrative, Poole et al. (2013) argue strongly against the notion of a 'metanarrative' approach to development, stressing the importance of a context specific approach. They highlight the myriad of challenges that smallholders can face on the path to commercialisation, which is the only pathway promoted by development institutions (FAO, 2016; World Bank, 2008), and the potential for exclusion of those farmers least able to respond to market demands. Poole et al. (2013: 163) conclude that "the local context and farmer characteristics and attitudes need to be much better understood in order address the strengths and weaknesses of the sector participants and the opportunities and threats of the external environment." They call for a much more local narrative, including in respect to rural livelihoods conception and property rights (particularly those that are insecure) and a much more complex understanding of human capital. They argue that this is vital when assessing external factors (for example, migration, climate change, ecosystem sustainability, political, economic, and social power relations, weak institutions, and so on).

Within this debate lies one potential pathway; that of greater engagement with landscape scale perspectives, as previously introduced. Recent Ugandan initiatives in forest restoration demonstrate how environmental considerations can be integrated at a scale that is far broader than individual landholdings. The potential benefits in terms of environmental services are bolstered by better soil health, and therefore a more productive landscape, better livelihood outcomes, ecosystem services, and better social, economic, and environmental outcomes. However, there is little detail on the implementation of these landscape scale initiatives and how they link with the communities that inhabit priority areas. Although the focus has moved from protected landscapes to a broader social, economic, as well as environmental focus, the link from global and national priorities through to local scale action for landscape scale change is complex. The Ethiopian case study demonstrates both shortcomings and potential for future research. One of the shortcomings of the NABU study is that women's interaction with the Kafa Biosphere has not been explored in detail. Women in rural Ethiopia live and work around the radius of loosely scattered family households embedded in domestic agriculture and industry. Production, reproduction and consumption are oriented to the household unit of husband, wife and their children. While work takes place in and around the household with appropriate tasks allocated by age and gender (Jackson, 2010; Pankhurst, 1992; Poluha, 2002), women may have regular interaction in the forest to collect firewood, water, and wild coffee. To date, NABU projects have targeted women by supplying 11,200 energy saving stoves; training women in regional crafts such as pottery; and actively involving women in reforestation activities (NABU, 2015). One of the aims of the current NABU project is to create alternative income opportunities for at least 300 indigenous women. Other projects support women with projects they can run from home such as beekeeping or raffia weaving (FARM Africa, n.d.).

A study by Gobeze et al. (2009) indicates that there is potential for participatory forest management (PFM) to promote sustainable forest management, particularly as the extraction of forest honey and coffee has increased since the introduction of PFM. While this success had redirected "the income sources of the local community from destructive forest use to sustainable forest production system or other complementary activities" (ibid.: 354), and the project had helped address the challenges of gender and ethnic inequity by women and the Manja in decision making, benefit sharing and providing a role in forest management, none of these projects will reduce gender inequality or change gender norms as they reinforce the traditional gendered world of rural women in Ethiopia.

Identifying *Coffea arabica* as one of the flagship species for the Kafa Biosphere was a good choice as it can be seen in light of its contribution to national or global diversity, as well as the value farmers place on it as they directly profit from it (NABU, 2017). The implication from the study for NABU, the Kafa Zonal government, farmers, and scientists is that the future of the Biosphere depends on multifunctionality. However, the objectives of external donors who are often led by short-term project goals may not sit well in the complex process of forest devolution and protection, let alone PFM which aims to offer local communities "multiple social, economic and ecological roles" to generate "sufficient and sustainable livelihoods to take them out of poverty" (Ameha et al., 2014: 839).

As noted by O'Hara and Hesseldon (2015), the two funding approaches (the climate finance mechanism **r**educe **e**missions from **d**eforestation and forest **d**egradation (REDD+) and PFM) in Ethiopia's southwest can create confusion and undermine trust in local communities as:

> REDD+ is based on a premise that maintaining and managing the forest is an opportunity cost for communities, whereas PFM is predicated on a premise that under a conducive governance environment forest value itself can induce forest maintenance and management. [. . .] To put it simply, PFM assumes that communities see the forest as a benefit. REDD+ assumes communities see the forest as a burden. Merging both potentially conflicting views will need to be negotiated carefully.
>
> *(O'Hara et al., 2015)*

Paying guards to ensure protection of the Biosphere Reserve means that communities are less likely to be involved and do not 'own' Biosphere initiatives. Although there are no government institutional frameworks to properly manage them, the Biosphere approach could work with 'sufficient economic development' in areas where "there is less direct dependence by local people on forests" (Tolera et al., 2015: 6). However, PFM is more likely to offer "important opportunities for enhancing biodiversity conservation in forests because it is empowering and economically motivating to communities" (ibid.: 6). The ultimate aim must be to find the right balance between the protection of biodiversity and improving the forest based livelihoods of the local people.

Discussion

Both Ethiopia and Uganda provide immense potential for additional food production and agricultural development. They also provide an exceptional diversity of landscapes and ecosystems that are impacted by human activities. Both countries comprise a high number of small fragmented landholdings, supporting a vast array of livelihood opportunities, ecological systems, and cultural diversity. Many of the key policy and program level interventions in both countries focus on the industrialised model of agriculture that dominates globalised food systems and human development (see Chapters 9 and 16). Commercial agriculture currently covers a small proportion of land, and improved inputs (seeds, fertilisers, mechanisation, and so on) are much lower in Africa than other areas of the world. African agriculture is intensifying, however, this intensification is not inducing the level of investment in inputs or irrigation sufficient to maintain soil health and increase productivity (Binswanger-Mkhize and Savastano, 2016). The direction of agricultural development in many parts of Africa will impact directly on food security and environmental sustainability. This chapter highlights one mode of thinking around landscape multifunctionality that can bring together some of the diverse requirements of natural resources and human endeavour on the continent.

Although landscape scale changes can create new challenges and conflicts, our aim is to recognise the complexities and to suggest that as the landscape has evolved over centuries it will continue to do so as we seek ways to balance climate change adaptation and biodiversity protection with regional development and the improvement of people's living conditions, rather than to paint a simplistic picture of cause and effect. However, it is clear that mainstream development economics, promoting economic growth, as demonstrated through investment in large scale commercialised agricultural development projects, is a very narrow perspective within which to operate. There needs to be a conceptual space for alternative paradigms. We earlier explored landscape scale initiatives and identified the increasing recognition of the complex and interrelated nature of a sustainable, biodiverse, and productive landscapes, at global and national levels. The broader scale of engagement, and scale of both planning and management, requires a transdisciplinary perspective (O'Farrell and Anderson, 2010), yet has huge potential to try to reconcile some of the contradictions in visions of the agricultural future of the continent. Large scale farming, monocropped landscapes, intensive animal husbandry, genetically modified crops are all part of one vision. However it is far from the only vision. The role smallholders play now and into the future, and the role of government and the private sector in determining land use is vital, particularly with the increased risks associated with farming complex diverse risk prone food landscapes in a changing climate. The case for long-term strategic collaboration between key stakeholders, but particularly communities, is strong.

From the growing focus on landscape scale initiatives an 'African Landscapes Action Plan' (LPFN, 2014) identifies key themes where action is required. These are in the areas of governance, policy formulation, provision of enabling environment for private sector investment, the identification and support of capacity required to fully realise landscape scale management and planning, and finally the support of integrated, multi-disciplinary, and Africa-focused scientific research to better understand the complexity of African agricultural landscapes and make them more productive, sustainable, and inclusive (Negra, 2014).

Drawing on some of the previous discussions, arguably the most important issue to arise is the acknowledgement of the importance of, and requirement to build on, understandings of the multifunctionality of agriculture and landscapes, and linking communities with government, private sector, and other actors to ensure policy and practice is based on strategic, transdisciplinary, multisectoral understandings of how to work at a larger and more effective scale. This requires the debunking of the development metanarrative based on economic growth and agriculture, and food as commodity, rather than that of spaces of environmental, social, cultural, (and economic) engagement.

Conclusion

The two case studies demonstrate a fundamental disconnect of the various discourses on the future of agriculture, food, and human development. At one end of a broad spectrum there is a metanarrative of economic growth and development, focusing on commercialisation of agriculture, particularly in the smallholder sector, and increasing investment in agriculture, either by foreign entities, or domestic. Against this view of the future, embedded in government policy and national development planning, there is the narrative of support for smallholder agriculture that embeds social and environmental concerns. As both case studies demonstrate, smallholders are not necessarily well positioned to commercialise. Decades of agricultural development have not necessarily resulted in increased productivity, on average, from the smallholder sector. There is a significant disconnect between discourses of intensification and commercialisation on one hand, and the importance of family farming on the other. This chapter demonstrates the vital engagement with both agriculture and landscapes as multifunctional. Each unique landscape supports a range of products services, and systems, economic, social, and cultural environmental. Landscapes have evolved through human intervention over centuries and are diverse, complex, and provide the basis of planetary health through ecosystem services and livelihoods. The recent emergence of both discourse and initiatives in practice on landscape scale governance and management results from ecological and social issues, but also in response to increasing focus on agriculture as a force for economic growth and development, as well as providing food for a growing global population.

It is indeed unequivocal that African landscapes, particularly those where food production is optimal, or indeed, even possible, will change dramatically over the coming decades. Requiring commercialisation of smallholders is a far more complex undertaking in many contexts. Landscape scale thinking, encompassing multifunctionality, and ensuring context specificity, as core concepts, are critical.

Further reading

Abbink, J. (2012) Dam controversies: Contested governance and developmental discourse on the Ethiopian Omo River dam, *Social Anthropology*, 20(2), 125–144.

Abbink, J., et al. (2014) *Lands of the future: Transforming pastoral lands and livelihoods in eastern Africa*. Max Planck Institute for Social Anthropology. www.eth.mpg.de/pubs/wps/pdf/mpi-eth-working-paper-0154 (accessed 26 May 2016).

ADB, OECD, UNDP. (2016) *African economic outlook 2016*. www.africaneconomicoutlook.org/en/african-economic-outlook%202016 (accessed 26 May 2016).

Afar National Regional State (2010) *Programme of plan on adaptation to climate change*, Environmental Protection Authority of the Federal Democratic Republic of Ethiopia, Semera www.epa.gov.et/Download/Climate/Regional%20Climate%20Change%20Adaptation%20Programmes/Afar%20National%20Regional%20State%20%20Climate%20Change%20Adaptation%20program.pdf (accessed 26 May 2016).

Ahaibwe, G., & Mbowa, S. (2014) *Youth unemployment challenge in Uganda and the role of employment policies in jobs creation*. www.brookings.edu/blogs/africa-in-focus/posts/2014/08/26-youth-unemployment-uganda-ahaibwe-mbowa (accessed 21 June 2016).

Ameha, A., Larsen, H.O., & Lemenih, M. (2014) Participatory forest management in Ethiopia: Learning from pilot projects. *Environmental Management*, 53(4), 838–854.

Anderson, J., Learch, C., & Gardner, S.T. (2016) *National survey and segmentation of smallholder households in Uganda: Understanding their demand for financial, agricultural, and digital solutions*. Washington, DC: CGAP.

Beegle, K., Christiaensen, L., Dabalen, A., & Gaddis, I. (2016) *Poverty in a rising Africa*. Washington, DC: World Bank. Available from: DOI: 10.1596/978-1-4648-0723-7.

Binswanger-Mkhize, H.P., & Savastano, S. (2016) Agricultural intensification: The status in six African countries. *World Bank Policy Research Working Paper No. 7116*. https://papers.ssrn.com/sol3/papers2.cfm?abstract_id=2526573 (accessed 21 June 2016).

Brandt, S.A., et al. (1997) *The tree against hunger: Enset based agricultural systems in Ethiopia*. http://users.clas.ufl.edu/aspring/publications/enset.pdf (accessed 10 February 2017).

CapGemini (2015) *World wealth report*. CapGemini, www.worldwealthreport.com.

Central Statistical Agency (2013) *Population projection of Ethiopia for all regions at wereda level from 2014–2017*. www.csa.gov.et/images/general/news/pop_pro_wer_2014-2017_final.

Central Statistical Agency (2015) *Key findings of the 2014/2015 (2007 E.C.) agricultural sample Surveys, Addis Ababa*. www.csa.gov.et/images/general/news/agss_2007_report.

Central Statistical Agency and World Bank (2013) *Ethiopia rural socioeconomic survey (ERSS) survey report*. http://siteresources.worldbank.org/INTLSMS/Resources/3358986-1233781970982/5800988-1367841456879/9170025-1367841502220/ERSS_Survey_Report.pdf.

Central Statistical Agency [Ethiopia], (2007) *Population and housing census atlas of Ethiopia*. www.csa.gov.et/images/documents/pdf_files/regional/CountryLevel.pdf.

Collier, P., & Dercon, S. (2014) African agriculture in 50 years: Smallholders in a rapidly changing world? *World Development*, 63, Supplement C, 92–101.

Cotula, L., Vermeulen, S., Leonard, R., & Keeley, J. (2009) *Land grab or development opportunity? Agricultural investment and international land deals in Africa*. London and Rome: IIED/FAO/IFAD. Avilable from: ISBN: 978-1-84369-741-1.

Dejene, A. (1990) *Environment, famine, and politics in Ethiopia: A view from the village*. Colorado: Lynne Rienner Publishers Inc.

Ethiopian government portal. (2016) www.ethiopia.gov.et/web/Pages/Economy (accessed 4 June 2017).

FAO (2009) *Foreign direct investment – win-win or land grab?* Rome: Food and Agticulture Organisation of the United Nations. ftp://ftp.fao.org/docrep/fao/meeting/018/k6358e.pdf.

FAO (2015) *Regional overview of food insecurity: African food security prospects brighter than ever*. Accra: FAO.

FAO (2016) *Food security statistics*. www.fao.org/economic/ess/ess-fs/en/ (accessed 15 June, 2016).

FAO (2016) *Sustainable production intensification and value chain development in Africa*. www.fao.org/africa/perspectives/agricultural-landscapes-africa/en/ (accessed 26 May 2016).

FAO and Swiss Agency for Development and Cooperation (2011) *Mountain forests in a changing world: Realizing values, addressing challenges*,.www.fao.org/3/a-i2481e.pdf.

FARM Africa. (n.d.). www.farmafrica.org/what-we-do-1/women-in-the-field (accessed 4 June 2017).

Federal Ministry of Health, World Health Organization and UNICEF (2011) Report on assessment of healthcare provision among pastoralist communities in Ethiopia (Final Report), Addis Ababa, Ethiopia.
Federal Democratic Republic of Ethiopia, Ministry of Finance and Economic Development, International Monetary Fund and World Bank (2002). *Ethiopia: Sustainable development and poverty reduction program*, Addis Ababa www.imf.org/External/NP/prsp/2002/eth/01/073102.pdf (accessed 5 May 2010).
Food and Agriculture Organization (FAO) of the United Nations (2014) *Country fact sheet on food and agriculture policy trends: Ethiopia*. www.fao.org/3/a-i4181e.pdf.
Friends of the Earth UK. (2012) *Landgrabs, forests and finance: Issue brief #5*. http://libcloud.s3.amazonaws.com/93/7b/3/3078/Issue_Brief_5_-_Wilmar_in_Uganda.pdf.
Garnett, T., & Godfray, C. (2012) *Sustainable intensification in agriculture: Navigating a course through competing food system priorities*, Food Climate Research Network and the Oxford Martin Programme on the Future of Food. Oxford: University of Oxford.
Gobeze, T., Bekele, M., Lemenih, M., & Kassa, H. (2009) Participatory forest management and its impacts on livelihoods and forest status: The case of Bonga forest in Ethiopia, *International Forestry Review*, 11(3), 346–358.
Hart, A.K., Milder, J.C., Estrada-Carmona, N., DeClerck, F.A.J., Harvey, C.A., & Dobie, P. (2015) Integrated landscape initiatives in practice: Assessing experiences from 191 landscapes in Africa and Latin America. In: Minang, P.A., van Noordwijk, M., Freeman, O.E., Mbow, C., de Leeuw, J., & Catacutan, D. (eds.), *ClimateSmart landscapes: Multifunctionality in practice*. Nairobi: World Agroforestry Centre (ICRAF), pp. 89–101.
Human Rights Watch (2012) *"Waiting here for death": Forced displacement and "Villagization" in Ethiopia's Gambella Region*. www.hrw.org/sites/default/files/reports/ethiopia0112webwcover_0.pdf.
IAASTD (2008). Towards multifunctional agriculture for social, environmental and economic sustainability, Issues in Brief, International Assessment of Agricultural Knowledge, Science and Technology for Development, Island Press.
Jackson, R. (2010) *(Un)safe routes: Maternal mortality and Ethiopia's development agenda*. PhD thesis, Deakin University.
Kristensen, S.P., Thenail, C., & Kristensen, L. (2001) Farmers' involvement in landscape activities: An analysis of the relationship between farm location, farm characteristics and landscape changes in two study areas in Jutland, Denmark. *Journal of Environmental Management*, 61(4), 301–318. Available from: DOI: 10.1006/jema.2000.0409.
Landscapes for People Food and Nature (2014) The African landscapes action plan, landscapes for people, food and nature in Africa, conference, 1–3 July, 2014, Nairobi, Kenya. http://peoplefoodandnature.org/wp-content/uploads/2014/09/African-Landscapes-Action-Plan-FINAL-FINALwNEPAD-Dec15.pdf.
Levine, D.N. (1965) *Wax and gold: Tradition and innovation in Ethiopian culture*. Chicago: University of Chicago Press.
LPFN, (2014) The African landscapes action plan, landscapes for people, food and nature, in Africa conference, 1–3 July, 2014, Nairobi, Kenya.
Luan, Y., Cui, X., & Ferrat, M. (2013) Historical trends of food self sufficiency. *African Food Security*, 5, 393. doi:10.1007/s12571-013-0260-1
McCann, J. (1995) *People of the plow: An agricultural history of Ethiopia, 1800–1990*. Madison, WI: University of Wisconsin Press.
McLellan, C.W. (1980) Land, labor, and coffee: The South's role in Ethiopian self-reliance, 1889–1935. *African Economic History*, 9, 69–83.
McLellan, C.W. (1986) Coffee in centre-periphery relations: Gedeo in the early twentieth century. In: Donham, D., & James, W. (eds.), *The Southern marches of imperial Ethiopia: Essays in history and social anthropology*. Cambridge: Cambridge University Press, pp. 175–195.
Milder, J.C., Hart, A.K., Dobie, P., Minai, J., Zaleski, C. (2014) Integrated landscape initiatives for African agriculture, development, and conservation: A region-wide assessment. *World Development*, 54, 68–80. http://dx.doi.org/10.1016/j.worlddev.2013.07.006.
Ministry of Water and Environment (2016) *Forest landscape restoration opportunity assessment report for Uganda 2016*. Ministry of Water and Environment – Uganda; IUCN. x + 42pp.
The Montpellier Panel (2013) *Sustainable intensification: A new paradigm for African agriculture*. London: The Montpelier Panel.
Mwebaze, S.M.N. (nd). *Uganda*. www.fao.org/ag/AGP/AGPC/doc/counprof/uganda.htm (accessed 18 June 2016).
Mwesigye, Hon. H.R. (2010) Investment opportunities in the Ugandan agriculture sector. *EMRC 2010 Agribusiness Forum*. Speke Reort, Munyonyo, Kampala. www.emrc.be/documents/document/20101125153530-session_viii_-_hope_mwesigye.pdf (accessed 21 June 2016).

NABU. (2015) Conservation and sustainable use of the last wild coffee forests of Ethiopia: A project for the preservation of biodiversity under climate change, http://imperia.verbandsnetz.nabu.de//imperia/md/content/nabude/international/nabu_biodiversity_project_ethiopia.pdf.

NABU (2017) *NABU's biodiversity assessment at the Kafa biosphere reserve.* Berlin and Addis Ababa: The Nature and Biodiversity Conservation Union (NABU).

National Geographic Society (2016). *Africa: Physical geography.* http://nationalgeographic.org/encyclopedia/africa-physical-geography/ (accessed 26 May 2016).

National Planning Authority (2014) *Uganda vision 2040, National Planning Authority of the Government of Uganda.* http://npa.ug/wp-content/themes/npatheme/documents/vision2040.pdf.

Nature and Biodiversity Conservation Union (NABU). (2011) *Protecting the last cloud forests of Ethiopia.* www.nabu.de/downloads/international/eng/Projektblatt_IKI_%C3%84thiopien_eng_06122011.pdf.

Negra, C., Kinuthia, W., Sola, P., Harvey, C.A., Jaquez, C., & Zeleke, G. (2014) *Priorities for research in African landscapes, landscapes for people, food and nature in Africa and TerrAfrica. Landscapes for people, food, and nature in Africa.* http://peoplefoodandnature.org/wp-content/uploads/2014/09/PrioritiesforResearchinAfricanLandscapes_Final.pdf (accessed 25 February 2017).

NEPAD. (2013) *African agriculture, transformation and outlook.* www.un.org/en/africa/osaa/pdf/pubs/2013africanagricultures.pdf (accessed 26 May 2016).

O'Farrell, P.J., & Anderson, P.M.L. (2010) Sustainable multifunctional landscapes: A review to implementation. *Current Opinion in Environmental Sustainability*, 2(1–2), 59–65. http://dx.doi.org/10.1016/j.cosust.2010.02.005.

Ogalo, V. (2011) *Foreign investment in agriculture in eastern Africa: A general overview of trends and issues.* www.fes-kenya.org/media/publications/FDI/overview-%20FDI%20in%20Agriculture%20in%20Eastern%20Africa_FES_VOO.pdf (accessed 21 June 2016).

O'Hara, P., & Hesselden, F. (2015) Reflections on participatory forest management in South West Ethiopia – beyond a project to a process of devolved forest management. Paper presented to 14th World Forestry Congress 2015, 7–11 September, Durban, South Africa. http://foris.fao.org/wfc2015/api/file/553f81d9aad66a0115ccca86/contents/deb1e9e5-09b5-4e4f-92ac-a7b927c4df09.pdf.

Oxfam (2011) *Land and power, the growing scandal surrounding the new wave of investments in land.* Oxfam Briefing Paper No 151, Oxford, Oxfam GB.

Pakenham, T. (1992) *The scramble for Africa: White man's conquest of the dark continent from 1876 to 1912.* New York: Harper Collins.

Pankhurst, A., & Freeman, D. (2003) Part II the North-West. In: Freeman, D., & Pankhurst, A. (eds.), *Peripheral people: The excluded minorities of Ethiopia.* London: Hurst and Co, pp. 70–79.

Pankhurst, H. (1992) *Gender, development and identity: An Ethiopian study.* London and New Jersey: Zed Books.

Poluha, E. (2002) Learning political behaviour: Peasant-state relations in Ethiopia. In: Poluha, E., & Rosendahl, M. (eds.), *Contesting 'Good' governance: Crosscultural perspectives on representation, accountability and public space.* London and New York: RoutledgeCurzon, pp. 101–136.

Poole, N.D., Chitundu, M., & Msoni, R. (2013) Commercialisation: A meta-approach for agricultural development among smallholder farmers in Africa? *Food Policy*, August, 155–165. Available from: http://dx.doi.org/10.1016/j.foodpol.2013.05.010.

Pretty, J., Toulmin, C., & Williams, S. (2011) Sustainable intensification in African agriculture. *International Journal of Agricultural Sustainability*, 9(1), 5–24.

Ros-Tonen, M.A.F., Van Leynseele, Y.P.B., Laven, A., & Sunderland, T. (2015) Landscapes of social inclusion: Inclusive value-chain collaboration through the lenses of food sovereignty and landscape governance. *European Journal of Development Research*, 27(4), 523–540. Available frm: DOI: 10.1057/ejdr.2015.50.

Schmitt, C. (2006) *Montane rainforest with wild Coffea arabica in the Bonga region (SW Ethiopia): Plant diversity, wild coffee management and implications for conservation.* PhD thesis, Center for Development Research, University of Bonn, Bonn.

Stellmacher, T. (2006) *Governing the Ethiopian coffee forests: A local level institutional analysis in Kaffa and Bale mountains.* PhD, Institut für Lebensmittel- und Ressourcenökonomik (ILR), University of Bonn, Bonn.

Tolera, M., Lemenih, M., O'Hara, P., & Wood, A (2015) *Insitu conservation of wild forest coffee exploring the potential of participatory forest management in south west Ethiopia.* Paper presented to WFC2015 – XIV World Forestry Congress, 7–11 September, Durban, South Africa. http://eprints.hud.ac.uk/26132/.

Tudge, C. (2012) *Sustainable intensification in agriculture navigating a course through competing food system priorities Stakeholders comments on the food climate research network/Oxford Martin Programme on the future of food 2012 report*. www.futureoffood.ox.ac.uk/sites/futureoffood.ox.ac.uk/files/Sustainable%20Intensification%20Comments.pdf (accessed 15 June 2016).
UNDP (2015) *Mountain ecosystem-based adaptation in Uganda*. www.adaptation-undp.org/projects/mountain-eba-uganda (accessed 25 February 2017).
Vejre, H., Abildtrup, J., Andersen, E., Andersen, P.S., Brandt, J., Busck, A., Dalgaard, T., Hasler, B., Huusom, H., Kristensen, L.S., Kristensen, S.P., & Præstholm, S. (2007) Multifunctional agriculture and multifunctional landscapes – land use as an interface. In: Mander, U., Wiggering, H., & Helming, K. (eds.), *Multifunctional land use*. Berlin and Heidelberg: Springer-Verlag, pp. 93–104.
Wayman, E. (2011) How Africa became the cradle of humankind. *Smithsonian*, October 17, 2011, Smithsonian Institution, Washington DC. www.smithsonianmag.com/science-nature/how-africa-became-the-cradle-of-humankind-108875040/ (accessed 24 February 2017).
Wildlife Conservation Society (WCS) Uganda (2017) *Murchison-Semliki*. https://uganda.wcs.org/Wild-Places/Murchison-Semliki.aspx (accessed 25 February, 2017).
Wolf, S. (2014) *The neoliberal regime in the agri-food sector: Crisis, resilience and restructuring*. New York: Routledge.
Wondifraw, A.Z., Kibret, H., & Wakaiga, J. (2015) *Ethiopia 2015*. African Development Bank, OECD Development Centre, UNDP. www.africaneconomicoutlook.org/fileadmin/uploads/aeo/2015/CN_data/CN_Long_EN/Ethiopia_GB_2015.pdf (accessed 8 March 2016).
World Bank (2008) *Africa – fact sheet: The World Bank and agriculture in Africa*. http://web.worldbank.org/WBSITE/EXTERNAL/COUNTRIES/AFRICAEXT/0,,contentMDK:21935583~pagePK:146736~piPK:146830~theSitePK:258644,00.html (accessed 26 May 2016).
World Bank (2014) *International development association project appraisal document on a proposed credit in the amount of SDR 208.5 million (US$320 Million Equivalent) to the Federal Democratic Republic of Ethiopia for a road sector support project, World Bank, Addis Ababa*. http://www-wds.worldbank.org/external/default/WDSContentServer/WDSP/IB/2014/02/03/000333037_20140203121700/Rendered/INDEX/PAD7000P131118010Box382121B00OUO090.txt.
World Bank (2015) *Agricultural Growth Program (AGP-II) social assessment report (Final), Addis Ababa*. http://www-wds.worldbank.org/external/default/WDSContentServer/WDSP/IB/2015/02/10/000442464_20150210153642/Rendered/PDF/IPP7690Social000Box385433B00PUBLIC0.pdf.
World Bank Group (2014) *3rd Ethiopia economic update strengthening export performance through improved competitiveness*. http://www-wds.worldbank.org/external/default/WDSContentServer/WDSP/IB/2014/11/12/000470435_20141112125152/Rendered/PDF/895480REVISED00c0Update0v0final0web.pdf.
World Bank (n.da). *Poverty data*. http://povertydata.worldbank.org/poverty/region/SSA.
World Bank (n.d.b). *World Bank data*. http://data.worldbank.org/indicator/NV.AGR.TOTL.ZS?locations=ZG.
World Food Summit. (1996) Rome declaration on world food security. 13–17 November, 1996, Rome, Italy, www.fao.org/wfs/index_en.htm.

22

Urban Agriculture in Bogotá's informal settlements

Open space transformation towards productive urban landscapes

Jaime Hernández-García and Sandra Caquimbo-Salazar

Introduction

In Colombia, informal settlements constitute an important part of cities. Their configuration possesses strong links to agricultural practices. These urban spaces have been the life alternative for many rural migrants in first, second, or third generations who have had to abandon the countryside because of the extensive period of violence experienced in the country and/or those seeking better economic opportunities in urban conglomerates (Gómez Rodríguez, 2014; Barriga Valencia and Leal Celis, 2011; Zaar, 2011; Cantor, 2010). People who move with their knowledge of the land are accustomed to a particular relationship with landscape – these people arrive in the city with nothing more than their culture and hope. People have organised their lives in the cities 'informally', due to lack of plans and opportunities provided by the government (Hernández-García, 2013b). They demonstrate self-initiated practices, bringing rural traditions and initiatives to urban areas which are often far from official, approved, and standard customs and procedures. In Bogotá, more than 50% of the city has been developed with informal practices (Martin Molano, 2000; Rueda García, 2000), allocating currently nearly four million people, half of the current population. Bogotá, beyond its condition of being the country's capital, is a city where urban informality has made a major impact in the socio-spatial conditions of many (Torres Tovar, 2009).

Urban Agriculture (UA) has been developed in informal settlements; on the one hand from the residents' point of view, as a way to connect to rural traditions (Zaar, 2011), and on the other hand, from the local government´s perspective as a way to contribute to nutrition of low income populations usually located in these areas. The government project's aim has been to strengthen the socio-economic, cultural, and environmental processes of food production through urban farming (Prieto Dueñas, 2005; Lara Sánchez, 2008; Gómez Rodríguez, 2014). UA practices can be traced in Colombian cities, such as Medellín and Bogotá, since the 1960s when major migration processes from rural to urban areas started (Sánchez, 2008; Marulanda Tabares, 2007). UA has been a private and in-house activity developed in courtyards, balconies, and terraces, and it is still that way (see Marulanda Tabares, 2007; Millán and Granados, 2006; Méndez et al., 2005). In the last decade, however, UA is gradually expanding from both private spaces to include open spaces and from nutritional concerns to a range of social motivations such as environmental, artistic, and political. This chapter aims to contribute to this emerging practice and discourse (see Figure 22.1).

Figure 22.1 Compostela's community garden, Usme, Bogotá. The tyres read "The Garden. Environmental Group. Compostela barrio".

Source: Sandra Caquimbo Salazar, 22 July 2016

Accordingly, this chapter provides a general perspective regarding UA in informal settlements and considers links to social practices and needs as well as the growing of food and its spatial dynamics. The aim of the chapter is to showcase Bogotá's UA practices and to explore transformation and appropriation of urban spaces in the *barrios* through food growing initiatives. Following this introduction, the themes of the chapter in relation to UA are explained. Then, an overview of agricultural practices in informal settlements in the city is presented, highlighting social characteristics and spatial particularities. From here, different UA initiatives in Bogotá are discussed and are grouped into two categories: firstly, those promoted directly by the people corresponding to their rural traditions, and secondly, those established by public programmes and academic and institutional projects, corresponding to social and food policies. All of this allows us to discuss the challenges of UA in informal settlements as well as the need for social, educational, economic, and spatial policies. The chapter finally argues for the possibility of perceiving open spaces in *barrios* as potential productive landscapes, in which spatial transformations may contribute to social, cultural, and nutritional practices.

Informal settlements, open spaces, and urban agriculture

Urban expansion in Latin America in the last five decades has contributed to the growth of informal settlements. In 2001 more than 75% of the population of the region lived in cities and over 30% (128 million people) were estimated to be living in conditions defined by United Nations Human Settlements as slums or informal settlements (UNCHS, 2003: 14). Informal settlements are usually located on the outskirts of cities and are described by UN-Habitat (UNCHS, 2003) by what they lack: (1) economic capital, (2) urban resources (e.g. food markets), (3) infrastructure, (4) 'proper' housing, and (5) social services. They typically include high levels of poverty, marginalisation and exclusion from social, economic and urban opportunities (DANE-SDP, 2007). Informal areas, however, also possess positive characteristics especially in terms of solidarity, social organization, and self-development (Hernández-García, 2016); regularly being able to accommodate most of their housing and urban needs. It can be argued that informal settlements are what people make of them through their own possibilities, initiative and imagination (Beza and Hernández-García, 2015). In this sense, informal settlements not only make up a large portion of Latin American cities, but are also a dynamic part of their physical, social and cultural characteristics. Fiori and Brandao (2010: 188) argue that: "urban informality is inexorably interwoven with the city as a whole – at all scales and levels – and has to be seen as another way of being in the city and constructing it."

Informal settlements are part of the urban environment of Latin America; they are not a marginal component but a central one, as different authors have suggested (Brillembourg et al., 2005; Hernández and Kellett, 2010; Lima and Pallamin, 2010). Informal practices of urbanisation and urban life coexist with formal and regulated urban development in cities of the region, having "contrasting physical, social and legal conditions" (Lima and Pallamin, 2010: 39) but connected to such a point that it is not easy to identify the informal or the formal origins (Bromley, 1978; Moser, 1994; Kellett, 1995; Santos, 2000; Ward, 2004). Furthermore, some urban informality "emerges under a paradigm of liberalization" (AlSayyad, 2004: 26), demonstrating an alternative way of thinking and performing with "urban informality not only as a political economy but also as a way of life" (AlSayyad, 2004: 27). Informality also applies to social and cultural practices, containing non-mainstream city expressions such as rural traditions. In this context, open spaces and UA are part of those "informal" responses to both the transformation and appropriation of the urban space in the *barrios* (Hernández-García, 2013a).

Open spaces in informal settlements play an important role in physical and social dynamics, and, as in the case of housing and urban facilities, they are largely self-developed by local people (Hernández-García, 2013b). They are highly symbolic places, with many telling the story of struggle and patience of people to fulfil their needs (Hernández-García, 2013b). Open spaces are the most important social places in the *barrio*; they are also places for cultural exchange and building values (Niño and Chaparro, 1997; Viviescas, 1997; Segovia and Oviedo, 2000; Hernández Bonilla, 2001), including food growing initiatives that are starting gradually to appear (Millán and Granados, 2006). These places are a mixture of sports pitches, playgrounds, children's facilities (Riaño, 1990; Beardsley and Werthmann, 2008) and some green spaces, in which community gardens and UA are beginning to become evident, as discussed in the following section (Figure 22.2).

Green spaces and vegetation can be seen as controversial in informal settlements as on the one hand, they are not a prominent spatial element in *barrios*, leading to the assumption that they do not matter to

Figure 22.2 Urban agriculture in an open space in Potosí, Bogotá. A fence has been recently installed to help the growing of the plants; however, the gates are always open.

Source: Sandra Caquimbo Salazar, 25 January 2017

inhabitants. As Monteiro and De Matos (2006: 2) found in Brazil: "vegetation is lacking or only in the form of sparse ground cover" and when observed, such spaces are not always well maintained. On the other hand, when asked about green elements in Bogotá and elsewhere, people tend to demonstrate enthusiasm: "[a]sked to rate specific positive design elements the survey population cited in sequence: tree-lined street, [a] very visible house, high front fence, cemented lot, big tree on lot, two-storey house and verandas" (Kowaltowski, 1998: 309). UA, as part of green spaces and vegetation initiatives in the *barrios*, is apparently seen as unimportant, but through careful observation and when talking to inhabitants, its importance becomes clearer (Millán and Granados, 2006).

In terms of landscape, the European Landscape Convention (2000, article 1) defines it as "an area, as perceived by people, whose character is the result of the action and interaction of natural and/or human factors". The characteristics of the landscape, therefore, is based on the close relationship between the people and the place they inhabit. This "represents the formative role of landscape in building identity, both collective and individual, in response to the basic human need to belong" (Egoz, 2013: 272). In this sense, everyday activities in the settlements constitute identity building experiences that "allows for the identification of space as a unit with its own character, and they can assign a symbolic level to specific landscape qualities and features" (Lee in Ramos et al., 2016: 38). It can be inferred that communities with strong rural traditions, such as those found in informal settlements, reflect their cultural values through agricultural practices and by doing so, they build identity towards living in the city.

Regarding basic nutrition for vast populations, the Food and Agriculture Organization of the United Nations (FAO) and other international institutions reported around 842 million people are underfed in the world (FAO et al., 2013). Most of this population is in low income countries, and within them, in informal settlements (FAO, 1999, 2014). This population shows food nutrition below the minimum energy requirements, causing loss of body mass and undernourishment. A strategy to mitigate the problem is family farming, a way to ensure the permanent physical, social and economic access of people to healthy and nutritious food to satisfy their basic needs. From this perspective, UA has become a tool in Latin American cities to mitigate food insecurity (FAO, 2014); in a similar sense, Speak (2014) suggests that agriculture in urban spaces is linked with food safety, poverty reduction, climate change adaptation and health strategies for cities. However, Badami and Ramankutty (2015) challenge this arguing that there is a limited role for food growing spaces to alleviate poverty and provide food access for low income communities; showing that the debate is ongoing.

According to FAO (1999), there is food security when three characteristics are met: access, availability, and health (see Chapters 20 and 25). Access to resources to get food (economic and transport), availability of sufficient food, and health in terms of nutrition to live an active and healthy life. In this sense, UA in the *barrios* appears, in particular small-scale practices, as a way to support the biological and psychological survival of informal urban settlers. Parker (2010) points out that most of the literature regarding UA in low income settlements is about survival strategies, food access and economic benefits. Little is said about leisure, community building, environmental amelioration, beautification, and political protest (Crouch and Ward, 1997; Certomà, 2011; Adams and Hardman, 2014). In this context it is interesting to examine which characteristics are evident in informal settlements in Colombia and Bogotá. The next section explores UA experiences in Bogotá discussing with empirical cases the topics that this section has presented.

Food growing experiences

Urban Agriculture in Latin America is widespread through multiple activities located in small spaces (FAO, 2014). The main forms of this food production are school gardens and backyard family horticulture. Most farmers are low-income people and the role of women in this practice is particularly high in Bolivia, Colombia, Ecuador, Honduras, Nicaragua, and in the Caribbean, as noted by the FAO (ibid.). UA practices in the region have been seen as a way to improve nutrition of low income population by increasing the consumption of fruits and vegetables. The challenge, however, is the lack of space, the poor quality of soils and problems with water access (FAO, 2014). Two types of policies have been implemented in the region. Firstly, countries like Cuba and the Caribbean Islands have established national policies promoting UA. Secondly, others countries like Peru, Ecuador and Colombia, programmes and projects promoted by local governments are in place due to the lack of a national policy (FAO, 2014).

UA in Colombia is a heterogeneous practice with varying motivations, strategies and places where it is observed. Although it is found in most urban areas, it can be argued that it is mainly practiced in informal settlements linked to low income populations, as explained in the first section of the chapter. UA in Colombia is associated with different things, especially with food security of low income groups and cultural identity of informal settlers with strong rural traditions, but also with open space appropriation and ideological and artistic representations, as later discussed. UA in the country demonstrates what Marulanda Tabares (2014: 37) argues: "urban agriculture reflects a rural identity of people struggling to survive under the conditions imposed by city 'life.'" Particularly aimed to improve families' livelihoods, different organisations such as the FAO, CORPOICA (Colombian Agricultural Research Corporation), GIAUP (Colombian Inter-institutional Group to Support Urban and Peri-urban Agriculture) and Águila Network (Latin American Research Network in Urban Agriculture), have been providing support to UA in Colombia (Leandro, 2013). Although there is not accurate data about farmers or the activity itself (Leandro, 2013), there has been a range of initiatives around the country, especially evident in the experiences in Antioquia Province, particularly in Medellín and in the capital of Bogotá.

In Medellín, UA programmes started in the 1990s to address city violence with environmental initiatives as a way to offer different alternatives to conflicts (Gómez Rodriguez, 2014). In 2005, UA programmes were focused more on nutritional contribution to low income populations than as an environmental activity (Brand and Muñoz, 2007). Brand and Muñoz (2007) explain how the most important programmes of that time were: *Solares Ecológicos* (Ecological Courtyards), *Agricultura Urbana para una población en condiciones de desplazamiento* (Urban Agriculture for Displaced Population), and *Ciudades Cultivando para el Futuro* (Growing Cities for the Future). In the last few years, the *Ecohuertas Urbanas* (Urban Gardens) programme has been the main UA initiative driven by the local government of Medellín in association with FAO-MANA (Gómez Rodriguez, 2014). It is a city agricultural, environmental and educational project oriented to enhance the social fabric with urban agriculture initiatives (Alcaldía de Medellín, n.d.).

Cultivation practices in Bogotá

Previous sections briefly presented the context of UA in informal settlements in Latin America and Colombia, with the focus of this chapter now shifting to Bogotá (Figure 22.3). The city is the first host of internal migrants from rural areas (CODHES, 2012), and informal settlements are where many of them first arrive and establish their lives (CODHES, 2014). Rural families coming to Bogotá know about agriculture, and some see the practice as a way to either provide some food and/or money, or 'to connect' with their rural roots (Linares, 2007; Leandro, 2013; Marulanda Tabares, 2014). For them, cultivation has symbolic value, because agricultural activity in the city incorporates value in the landscape, embodying tranquility for farmers and becoming a space for individual resistance to the difficulties they found when moving to the city (Cantor, 2010). For those coming to the city due to forced displacement, this symbolic value is stronger, it represents dignity and autonomy, a way to take root in their current place and reconstruct their life project. They can often have nostalgic memories of the countryside, of a better and quieter life, and the contact with the land allows them to feel useful through their own knowledge and identity, establishing bonds in their new habitat (Cantor, 2010).

Food growing spaces in Bogotá´s *barrios* show the same heterogeneity as the practices that take place in them. Some people, within their houses, have made use of unbuilt areas, roofs, patios and front gardens, and several make use of their facades and windows (see Figures 22.4 and 22.5). Although instances are

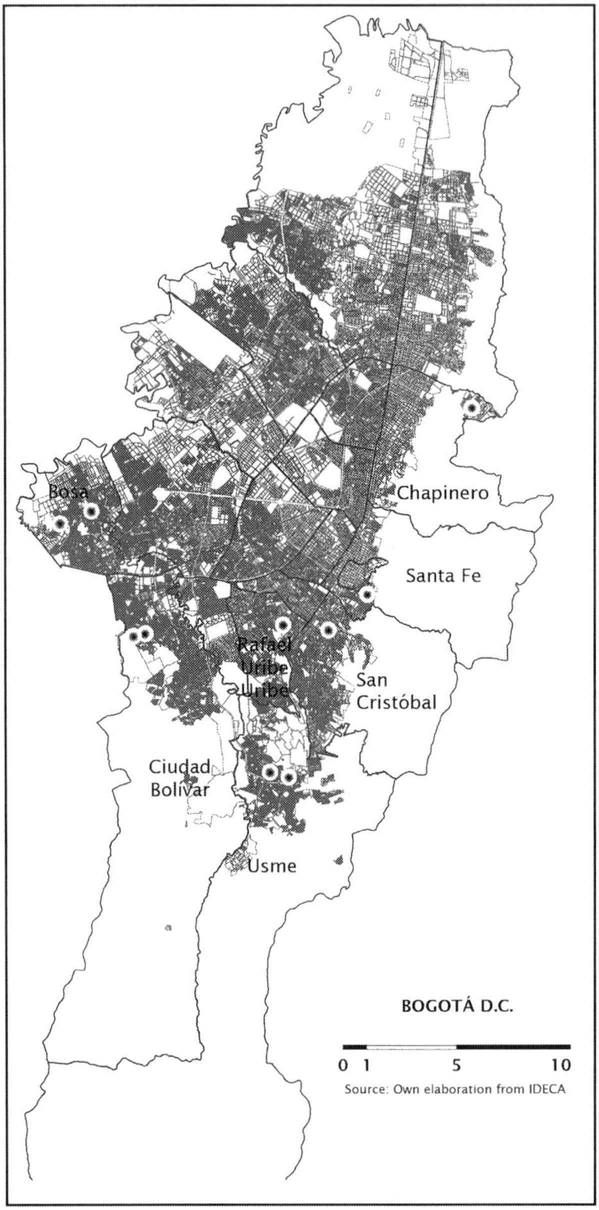

Figure 22.3 A map of Bogotá with location of the case studies. Most are in the southeast and west of the city, coinciding with the informal settlements.

Source: The base map is from IDECA (www.ideca.gov.co/)

few to date, UA is gradually occupying some streets and green open spaces within informal settlements. Green open spaces constitute a key aspect for people's sense of well-being and their connection to urban agriculture and community gardens (Caputo, Schwab and Tsiambaos, 2016). In informal settlements, of 57 cases studied over ten years in Bogotá, Hernández-García (2013b) found that 20% have no green areas

Figure 22.4 Garden and cultivation initiatives in informal settlements in Usme, Bogotá.
Source: Sandra Caquimbo Salazar, 22 July 2016

Figure 22.5 Garden and cultivation initiatives in informal settlements in Usme, Bogotá.
Source: Sandra Caquimbo Salazar, 22 July 2016

or trees at all. Of the remaining 80%, half (23 cases) have some degree of green within the open space, and the other half had a considerable amount of green space. The establishment of these areas starts with grass-roots initiative, and can sometimes be later formalised by local governments. The quality and level of care of the green spaces greatly varies, some being very well looked after, while others seemed abandoned. In Villa Sonia's park, for example, there is no green space at all; and there is an evident lack of green areas and trees in the settlement generally. This is partly explained by the arid conditions in this area of the city; the relatively small size of the park and the absence of spaces where trees could eventually be planted. By contrast, Los Cerezos' park contains the most green spaces, vegetation and trees in the study, appearing to work particularly well in terms of use and general care. La Andrea's phase three park, or el 'ocho' also offers a good example: the park does not have as much green open space as Los Cerezos, but its few green areas, trees and flower beds are very important to the people and present an identifiable picture of the open space which makes it unique. The flowerbeds are important and were developed between the community and the municipality.

Social and cultural initiatives

According to Marulanda Tabares (2014), Urban Agriculture in Bogotá is different from other Colombian cities because of the diverse origins of its population. While in other Colombian cities, UA tends to be more a food productive practice, in Bogotá it is mainly seen as a way of recovering rural traditions and a way to seek an identity in the city. Urban agriculture turns out to be not only a work alternative but also represents an aspiration towards both individual and collective wellness with high symbolic value. The new urban inhabitants attempt to recreate their rural landscapes and recover their customs and their culture. This is maybe the reason why in Bogotá UA initiatives have flourished more than in other Colombian cities. People organise themselves and look for options, in the same sense as the urban and housing development, by means of mutual aid.

The discussion about urban agriculture in Bogotá is driven by grassroots organisations different social groups and organizations such as one called *Red de Agricultura Urbana de Bogotá* (UA network of Bogotá). It was established as an exchange of ideas and discussion of experiences related with UA, the same as strategies to improve and promote the practice. From there, the city government developed an understanding of the diversity of processes that had been taking place for some time in the city. For example, starting in 2004, the institutional initiative called *Mesa Distrital de Agricultura Urbana* (Bogotá permanent panel for UA) commenced, which some time later gave rise to Bogotá's Urban Agriculture project (Lara Sánchez, 2008).

Among those initiatives developed within these community and institutional efforts, one interesting instance is the *Arte Productivo* (productive art) initiative. It is a community garden placed at the informal developed area called *Rafael Uribe Uribe* (historic figure). The project started with a temporary union of five people and this has been growing, the majority of which have rural origins. They received support from a training agreement between the city government and *Uniminuto* University where they learned to grow plants for cold climates (Bogotá is in the mountains at an elevation of 2,640 meters and experiences yearly temperatures between 5 and 15 degrees centigrade). Although this training has not been provided for all of the rural migrants, Bogotá's government has been consistently implementing the programmes. In the first four years of the agreement there were around 700 people of all ages trained. The garden is organically cultivated, with medicinal and ornamental plants, and about 150 varieties of food species. The original soil in the garden was not suitable for agriculture, so Community Gardeners used compost made with the raw food waste of restaurants and supermarkets with the addition of Californian red worms (*Eisenia fetida*) to create humus. The production of the garden is mainly for self-consumption. When surplus is available it is sold or used for *Ollas Comunitarias*, (community picnics), where gardeners cook – usually a soup called *sancocho*, which is shared with everyone who has participated in the cultivation (Lara Sánchez, 2008).

Another case is found in the informal settlement called *Bosa*, with the *Cabildo Indígena* of *San Bernardino* (indigenous association of San Bernardino). It is an indigenous community developing urban agriculture in empty plots set out for future social housing projects. This farming in soft patches (conventional agriculture) is in contrast to with the experience of the NGO called *Cuaremántima*, which have interesting work in hard areas of the city (roof terraces in houses) (Lara Sánchez, 2008). In the same sense, Luz Helena Hernández, who worked in the project (cited in: Lara Sánchez, 2008), explains that in *Bosa*, despite people not having land for gardens, they use part of their flat roofs and balconies for cultivation. In *San Cristóbal*, another informal settlement, there are self-consumption and *trueque* (exchange) experiences with farming production from rooftop growing. Similar initiatives have been observed in other barrios of the city, like *Bolonia*, *Guavio* and *San Luis*, which confirms that although UA is still a marginal activity in terms of size and impact, it has been gradually gaining interest, especially with informal settlers and public and private institutions working in these areas.

Community participation and NGO involvement

Another informal area within Bogotá that includes several UA experiences is *Ciudad Bolívar*. Here we find Potosi *barrio*, where some urban gardens can be observed in open spaces, yet these gardens are still at a very basic level. UA practices are also in community walled courtyards and attached to façades with ornamental plants seeded in recycled plastic bottles (Figure 22.6). Agricultural practices in this *barrio* have their origins in the people who have cultivated land for many years (Gómez Pérez et al., 2014). Based on their rural traditions and working with different institutions, they have been able to improve their life conditions through urban farming, giving new uses and significance to their open space (Millán Guzmán and Granados Ortiz, 2006).

In 2002, the *Instituto Cerros del Sur* (a community educational institution) with the community put together a project called *Bioseguridad Alimentaria con Participación Comunitaria* (Food Bio-safety with Community Participation). The project was about a group of single mothers who worked together for the nutritional improvement of their households, putting into practice years of farming experience in their front and back yards. They organised and helped others to cultivate in *Mingas* (an old word meaning a gathering of people who work together) and *Ollas Comunitarias*. They were able to grow more than 46 species of plants, and they also implemented water reuse, compost production with worms, and applied homemade techniques of allopathy and biological preparations. This initiative also allowed people to exchange ideas

Figure 22.6 Food growing in recycled bottles, Potosí, Bogotá.

Source: Sandra Caquimbo Salazar, 12 August 2016

and knowledge about how to cultivate and manage seeds, in addition to the development of organisational initiatives such as school meals, seed networks, community markets, and group collection and recycling of organic waste (Millán Guzmán and Granados Ortiz, 2006).

The NGO *Planeta Paz* (Peace Planet) undertook another initiative in Potosi in 2004. They organised several workshops with the objective of making visible agricultural practices within homes and consolidating a team that would work on empty plots (Millán Guzmán and Granados Ortiz, 2006). The workshops also encouraged community reflection about food policy and the convenience of a market area in the neighbourhood, in order to connect rural producers with buyers. They promoted the idea of collectivism above individuality, to increase impact and obtain better UA results (Cantor, 2010).

The most recent initiative of urban agriculture in Potosi was at the end of 2015. The *Junta de Accion Comunal* (Community Action Group – JAC) organised two food growing areas in a square called *Cocinol*, in front of the *barrio* community room (Figure 22.7). In terms of results in food production, the square shows a transformation and also, another meaning, through the connection with rural traditions. This change for most of the residents is positive (but not for all, as some wish to 'claim back' the formerly 'empty' space).

Policies and programmes

From the government side, UA in Bogotá started formally in 2004, with Major Luis Eduardo Garzón and the policy *Bogotá sin Hambre* (Bogotá Hunger Free). With this policy, the UA project called *Agricultura Urbana: Sostenibilidad ambiental sin indiferencia para Bogotá* (Urban Agriculture: Environmental Sustainability without indifference for Bogotá), led by Bogotá's Botanical Garden, UA initiatives started to be promoted and supported by the municipality, and helped communities in the *barrios* who had already began their UA endeavours several years ago (Lara Sánchez, 2008; Gómez Rodríguez, 2014). The project's aim was to strengthen the socio-economic, cultural and environmental processes of food production through urban farming (Prieto Dueñas, 2005). In this project, they developed a master plan for food supply and food

Figure 22.7 Food growing area in Cocinol, Potosí, Bogotá.

Source: Sandra Caquimbo Salazar, 25 January 2017

security for Bogotá through the law 315 of 2006. This law adopts a master plan which aims to regulate the food supply in Bogotá. It seeks to guarantee the adequate and permanent availability and accessibility of quality food. It controls prices and strengthens economic urban and rural circuits. (Alcaldía Mayor de Bogotá, D.C., 2006). Also a community dining project was established. This is a social project aiming to provide, in dining rooms located in poor *barrios* of the city, at least one daily serving of balanced food for residents. (Giraldo Rodríguez, 2007). At the time of writing, the community dining initiative is still operating and the law is still in force.

The Botanical Garden of Bogotá is the municipality office in charge of promoting and supporting UA activities in the city. It focuses efforts in the knowledge transfer about farming alimentary plants through clean technologies to improve nutritional conditions of communities, especially in low-income settlements (Lara Sánchez, 2008). The project was implemented as an educational and a social strategy, with about 148 training initiatives around the city. To reinforce the strategy, an alliance with local municipalities was developed in order to apply for funding to reach more communities with the training, but also to develop research. Nowadays, the Botanical Garden continues investigating organic urban agriculture, and giving support to urban farmers for the use of organic technologies and management processes (Lara Sánchez, 2008).

After these first policies and programme initiatives, the subsequent Municipality Mayors continued them and further developed them. One of these was the programme *Bogotá bien Alimentada* (Bogotá is well fed), and currently *Bogotá Te Nutre* (Bogotá Nourishes You); which have been interpretations of the above law following the same aim. In 2015, the Bogotá Council approved the 605 Agreement, which refers to guidelines for an urban and peri-urban agro-ecological agriculture programme and its implementation in Bogotá. This agreement defines urban and peri-urban agro-ecological agriculture as a food production model located in urban and peri-urban spaces, which allows the organisation of neighbouring communities to implement agricultural systems (see Chapters 12, 14 and 15). The agreement's strategy is to promote community organisation through agricultural practices structured in systems to take advantage of waste, optimise resources and respect the environment, according to available technologies. Likewise, it indicates that the programme should be sustainable, low-cost, oriented to climate change adaptation and to healthy food, through agro-ecological practices for the strengthening of social networks and the establishment of sustainable farming.

Conclusion: challenges for productive landscapes in informal settlements

Although "sufficient land for cultivation and practical knowledge of how to grow food" are known as two essential prerequisites for urban food production (Barthel and Isendahl, 2013), in Colombia in general, and in informal settlements in particular, there are several other fundamental aspects of urban agriculture. The first is that UA tends to be associated with informal settlements. This situation may change in the future, but so far urban agriculture is considered a marginal activity, and, of course, informal settlements themselves are marginal. UA in informal settlements is seen mainly as an economic and nutritional activity (Parker, 2010; Cantor, 2010). Evidence, however, from the case studies presented above (and elsewhere, as well) challenges these categorisations. First of all, urban agriculture and community gardens are not so marginal in terms of the number of initiatives observed in informal settlements (the same is true of informal settlements: they are not marginal in size compared with the city). However, they are marginal in terms of organisation, impact, and productivity. Second, UA is still not a significant economic activity and apparently it is also insignificant as provider of a nutritional component of families' diets (more research is needed, but from other studies and this research, the nutritional contribution appears to be low). This will perhaps also be true in the future, but at the moment what is observed is that UA is more about traditional, cultural, and symbolic activity – a way to transform and appropriate open space, a community learning and empowerment experience, an artistic representation, and even an ideological statement. These aspects confirm the social dimension of urban agriculture, in which a population may benefit in terms of food or additional income, but in particular communities gain strength and urban and spatial improvements may occur, as a result of and in the context of a self-produced settlement. Regardless of people's reasons for migrating from the countryside, such as escaping violence or looking for better opportunities, through urban agriculture they establish supportive relationships and make bonds of reciprocity and trust (Cantor, 2010).

In this sense and as the evidence from the discussed cases in this chapter show, urban agriculture helps to create meaningful experiences associated with a place, and it contributes to establishing symbolic links with the urban condition, and to the wider landscape (Egoz, 2013). It is interesting that citizens in informal settlements use rural traditions to help them connect with urbanity. This dimension of place that endures in the memory is their landscape: a social construction linked to a mental image with collective recognition, developed from the experience of the space (Caquimbo et al., 2010: 198). It is through the everyday use of the territory, especially when it is spatially transformed as with UA, that a place is perceived and valued. Connected to this, UA practices in informal settlements help in the establishment of networks that allow a sense of belonging to a group. According to Enríquez (2000), social networking configuration depends on, among other things, mutual exchange and trust building between individuals who share values and norms. These networks are one of the most successful strategies for survival in contexts of economic poverty. The cases presented here show that people possess knowledge from their rural past of how to cultivate land. In this sense, urban agriculture is found especially in informal settlements where, as indicated, people mostly come from rural areas. However, growing food in rural areas is different from growing food in the city.

In this urban context, the challenge for public policy is to improve capacities of cities to meet social and environmental conditions for urban agriculture. Likewise the challenge is in the promotion of waste management, environmental and food sustainability in informal settlements, and generating engines for community development by creating employment and economic chains of production, distribution and marketing on small and medium scales. At the same time, as some authors suggest (Dabas, 1993; Abello et al., 1997), the construction of new networks is necessary – networks which are based not only in the exchange of goods and services, but also in emotional and supportive links between neighbours. This can allow people to sustain their farming activities over time, without relying exclusively on external support (such as authorities or charities). According to Marulanda Tabares (2007), this requires selection criteria to be applied for people who wish to become involved in the programmes of urban agriculture, to select those with a true interest in the activity, and who see a future in the practice. In the same regard, Cantor (2010) argues it is important that the activity mean something to people, in terms of the economy but also in the social and cultural significance – without this investment, UA initiatives risk becoming merely temporary.

Finally, there is the challenge to improve the economic and nutritional impacts of UA, because, as Schwab (2015) suggest for the case of Medellín, so far the contribution in these regards is very small. It is not clear if urban agriculture can contribute to ameliorating negative economic impacts as, for example, in India, where to some extent, agriculture is "a logical livelihood activity" (Speak, 2014: 144). Also in Cuba, where the practice of UA has contributed substantially to the nutrition of the people (FAO, 2014). In Colombia, and in Bogotá in particular, urban agriculture is a relatively small but growing phenomenon with great potential. Its potential lies in these areas, but is also about how landscapes can be transformed and made productive in every way.

Further reading

Abello, R., Madariaga, C., & Hoyos, O. (1997) Redes sociales como mecanismo de supervivencia: Un estudio de casos en sectores de extrema pobreza. *Revista Latinoamericana de Psicología*, 29, 115–137.

Adams, D., & Hardman, M. (2014) Gardening, observing guerrillas in the wild: Reinterpreting practices of urban guerrillas. *Urban Studies*, 51(6), 1103–1119.

Alcaldía Mayor de Bogotá, D.C. (2006) *Decreto 315*. www.alcaldiabogota.gov.co/sisjur/normas/Norma1.jsp?i=21063 (accessed 18 June 2016).

Alcaldía Mayor de Bogotá, D.C. (2015) *Acuerdo 605*. www.alcaldiabogota.gov.co/sisjur/normas/Norma1.jsp?i=62903 (accessed 11 February 2017).

Alcaldía de Medellín (n.d.) *Ecohuertas urbanas. Por una ciudad sostenible*. https://issuu.com/adri.ana/docs/cartilla_ecohuertas (accessed 11 February 2017).

AlSayyad, N. (2004) Urban informailty as a 'New' way of life. In: Roy, A., & AlSayyad, N. (eds.), *Urban informality: Transnational perspectives from the Middle East, Latin America, and South Asia*. Oxford: Lexington Books.

Badami, M., & Ramankutty, N. (2015) Urban agriculture and food security: A critique based on an assessment of urban land constraints. *Global Food Security*, 4, 8–15.

Barriga Valencia, L.M., & Leal Celis, D.C. (2011) *Agricultura Urbana en Bogotá. Una evaluación externa-participativa. X Congreso Nacional de Sociología, 2–4 Nov 2011, Cali, Colombia*. www.icesi.edu.co/congreso_sociologia/mesas_de_trabajo_y_ponencias/mesa_9_estado_y_politicas_publicas.php (accessed 18 June 2016).

Barthel, S., & Isendahl, C. (2013) Urban gardens, agriculture, and water management: Sources of resilience for long-term food security in cities. *Ecological Economics*, 86, 224–234.

Beardsley, J., & Werthmann, C. (2008) Improving informal settlements: Ideas from Latin America. *Harvard Desing Magazine*, 28(Spring/Summer), 31–35.

Beza, B.B. (2012) Place-making. In: James, P., Soguk, N., Fien, J., & Postma, M. (eds.), *Global cities annual review*. Melbourne: Global Cities Institute, RMIT University.

Beza, B.B., & Hernández-García, J. (2015) *Placemaking in informal settlements*, XXXIII International Congress of the Latin American Studies Associatsion, San Juan, Puerto Rico, May 27–30.

Brand, P., & Muñoz, E. (2007) Cultivando ciudadanos: Agricultura urbana desde una perspectiva política. *Cuadernos IPPUR*, XXI(1), 47–70.

Brillembourg, A., Feireiss, A., & Klumpner, H. (2005) *Informal city: Caracas case*. Munich: Prestel.

Bromley, R. (1978) The urban informal sector: Why is it worth discussing? *World Development*, 6(9/10), 1033–1039.

Cantor, K.M. (2010) Agricultura urbana: Elementos valorativos sobre su sostenibilidad. *Cuadernos de Desarrollo Rural*, 7(65), 59–84.

Caputo, S., Schwab, E., & Tsiambaos, K. (2016) Emergent approaches to urban gardening. In: Bell, S., Fox-Kämper, R., Keshavarz, N., Benson, M., Caputo, S., Noori, S., & Voigt, A. (eds.), *Urban allotment gardens in Europe*. Abingdon: Routledge.

Caquimbo Salazar, S.L., & Devoto Magofke, C. (2010) Variables de sustentabilidad para la construcción de paisaje habitacional. *Cuadernos de Vivienda y Urbanismo*, 3(6), 196–219.

Certomà, C. (2011) Critical urban gardening as a post-environmentalist practice. *Local Environment*, 16(10), 977–987.

CODHES (2012) Estadísticas Municipales. In: *Estadísticas Históricas de Desplazamiento*. www.codhes.org/index.php?option=com_si&type=1 (accessed 18 June 2016).

CODHES (2014) Caracterización del desplazamiento forzado intraurbano en Bogotá. In CODHES, *Desplazamiento forzado intraurbano y soluciones duraderas. Vol. II Bogotá, Cúcuta y Quibdó*, Bogotá.

Consejo de Bogotá, D.C. (2015) *Acuerdo 605*. www.alcaldiabogota.gov.co/sisjur/normas/Norma1.jsp?i=62903 (accessed 18 June 2016).

Council of Europe (2000) *European landscape convention*. Treaty. www.coe.int/en/web/conventions/full-list/-/conventions/rms/0900001680080621 (accessed 11 February 2017).

Crouch, D., & Ward, C. (1997) *The allotment – its landscape and culture*. Nottingham, Five Leaves.

Dabas, E. (1993) *Red de redes. Las prácticas de la intervención en redes sociales*. Buenos Aires: Paidós.

DANE-SDP (2007) *Encuesta de Calidad de Vida*. Bogotá, DANE.

Egoz, S. (2013) Landscape and identity: Beyond a geography of one place. In: Howard, P., Thompson, I., & Waterton, E. (eds.), *The Routledge companion to landscape studies*. New York: Routledge, pp. 272–285.

Enríquez, R. (2000) Redes sociales y pobreza: Mitos y realidades. *Revista de Estudios de Género. La Ventana*, 11, 36–72.

FAO (1999) *The state of food insecurity in the world*. Rome: FAO.

FAO (2014) *Growing greener cities in Latin America and the Caribbean*. Rome: FAO.

FAO (2016) *Peace and food security. Investing in resilience to sustain rural livelihoods amid conflict*. Rome: FAO.

FAO (n.d.) *Urban agriculture*. www.fao.org/urban-agriculture/en/ (accessed 11 February 2017).

FAO, FIDA and PMA (2013) *El estado de la inseguridad alimentaria en el mundo 2013. Las múltiples dimensiones de la seguridad alimentaria*. Roma: FAO.

Fiori, J., & Brandao, Z. (2010) Spatial strategies and urban policy: Urbanism and poverty reduction in the favelas of Rio de Janeiro. In: Hernandez, F., Kellett, P., & Allen, L. (eds.), *Rethinking the informal city: Critical perspectives from Latin America*. Oxford: Berghahn Books, pp. 181–206.

Gilbert, A. (1998) *The Latin American city*. London: Latin America Bureau.

Giraldo Rodríguez, J.A. (2007) *Los Comedores Comunitarios como Estrategia de Inclusión Social en Bogotá, Localidad Mártires, Barrio La Favorita, 2006–2007*, Degree Work at Universidad De La Salle, Bogotá.

Gómez Pérez, N., Benavides Acosta, O., & Robayo, Y. (2014) *Partir de lo que somos. Ciudad Bolívar, tierra, agua y luchas*. Bogotá, Alcaldía mayor de Bogotá.

Gómez Rodríguez, J.N. (2014) *Agricultura Urbana en América Latina y Colombia: Perspectivas y Elementos Agronómicos Diferenciadores*, Degree Work at the Universidad Nacional de Colombia Abierta y a Distancia de Medellín.

Hernández Bonilla, M. (2001) Public spaces in the urban popular habitat of Mexico: The case of Xalapa, Veracruz. *Forum Journal*, 5(1), 27–35.

Hernández, F., & Kellett, P. (2010) Re-imagining the informal settlements in Latin America. In: Hernandez, F., Kellett, P., & Allen, L. (eds.), *Rethinking the informal city: Critical perspectives from Latin America*. Oxford: Berghahn Books, pp. 1–19.

Hernández-García, J. (2013a) Construcción Social de Espacio Público en Barrios Populares de Bogotá. *Revista INVI*, 78(28), 143–178.

Hernández-García, J. (2013b) *Public space in informal settlements: The Barrios of Bogotá*. Cambridge: Cambridge Scholars Publishing.

Hernández-García, J. (2016) Sustainability citizenship: Lessons and challenges from informal settlements. In: Horne, R., Fien, J., Beza, B.B., & Nelson, A. (eds.), *Sustainability citizenship in cities: Theory and practice*. Australia: Routledge.

Holston, J., & Caldeira, T. (2008) Urban peripheries and the invention of citizenship. *Harvard Design Magazin*, 28, 18–23.

Kellett, P. (1995) *Constructing home: The production and consumption of popular housing in Northern Colombia*. Unpublished PhD Thesis, University of Newcastle Upon Tyne, Newcastle Upon Tyne.

Kowaltowski, D. (1998) Aesthetics and self-built houses: An analysis of a Brazilian setting. *Habitat International*, 22(3), 299–312.

Krasny, E. (2012) Hands-on urbanism. In: Krasny, E. (ed.), *Hands-on urbanism*. Vienna: Thuria+Kant.

Lara Sánchez, A.J. (2008) *Agricultura Urbana en Bogotá: Implicaciones en la Construcción de una Ciudad Sustentable*, Degree Work at the Pontificia Universidad Javeriana, Bogotá.

Leandro Hernández, A.V. (2013) *La Agricultura Urbana en Bogotá: Cómo llegar a tener un Modelo de Negocio*. Degree Work at Universidad Escuela de Administración de Negocios, Bogotá.

Liévano-Fiesco, M., & Tobar-Vargas, L. (2008) Caracterización de la seguridad alimentaria y nutricional de la población del Verjón Bajo, localidad de Chapinero, Bogotá, D.C. *Universitas Scientiarum*, 13(3), 298–305.

Lima, Z., & Pallamin, V. (2010) Informal practices in the formal city: Housing, disagreement and recognition in downtown Sao Paulo. In: Hernandez, F., Kellett, P., & Allen, L. (eds.), *Rethinking the informal city: Critical perspectives from Latin America*. Oxford: Berghahn Books, pp. 39–52.

Linares, P.F. (2007) La Agricultura Urbana en el barrio Ciudad Londres, localidad de San Cristóbal, Bogotá. Aproximación a su análisis bajo el prisma de los Medios de Vida Sostenibles, Degree Work at Pontificia Universidad Javeriana, Bogotá.

Low, S.M. (1996) Spatializing culture: The social production and social construction of public space in costa Rica. *American Ethnologist*, 23(4), 861–879.

Martin Molano, J. (2000) *Formación y Consolidación de la Ciudad Espontánea en Santafé de Bogotá: el Caso de Altos de la Estancia en Ciudad Bolívar*. CIDER (Centro Interdisciplinario de Estudios regionales), Bogotá, Universidad de Los Andes.

Marulanda Tabares, C.H. (2007) *Microhuertas familiares: Cultivo de esperanzas con rendimientos de paz*. Armenia: FAO.

Marulanda Tabares, C.H. (2014) Bogotá. In: FAO (ed.), *La Agricultura Urbana y Periurbana en América Latina y el Caribe: Compendio de estudios de casos*. s.l., FAO, pp. 37–44. www.fao.org/ag/agp/greenercities/pdf/Compendium.pdf (accessed 1 November 2017).

McFarlane, C., & Waibel, M. (2012) *Urban informalities: Reflections on the formal and informal*. London: Routledge.

Méndez, M., Ramírez, L., & Alzáte, A. (2005) La Práctica de la agricultura urbana como expresión de emergencia de nuevas ruralidades: reflexiones en torno a la evidencia empírica, *Cuadernos de Desarrollo Rural* (julio-diciembre), pp. 51–70.

Millán Guzmán, J., & Granados Ortiz, S. (2006) Evaluación ecológica, agronómica y socio-cultural de la agricultura urbana en el sector de Potosí. In: Planeta Paz (ed.), *Bogotá: autonomía agroalimentaria. Diálogos y controversias*. Bogotá: Planeta Paz, pp. 117–144.

Monteiro, E., & De Mattos, M. (2006) A sense of comfort in open space around owner-built houses: The case of Campinas, Brazil. *The 23rd Conference on Passive and Low Energy Architecture*, Geneva.

Moser, C. (1994) The informal sector debate, part 1: 1970–1983. In: Rakowski, C. (ed.), *Contrapunto: The informal sector debate in Latin America*. Albany, NY: University of New York Press, pp. 11–29.

Niño, C., & Chaparro, J. (1997) El Espacio Público en Algunos Barrios Populares de la Bogotá Actual. *Barrio Taller. Serie Ciudad y Habitat*, 4, 1–22.

Ontiveros, T., & De Freitas, J. (2006) Hacia la Comprensión del Uso de los Espacios Públicos – Privados en los Territorios Populares Contemporáneos. *Cuaderno Urbano*, 5(17).

Palma, D. (1988) *La Informalidad, Lo Popular y el Cambio Social*. Lima: DESCO.

Parker, J. (2010) *More than subsistence: Small-scale urban agriculture and household livelihoods in Havana, Cuba*. Halifax: Degree Work at the Dalhousie University.

Pearce, J. (1990) *Colombia: Inside the Labyrinth*. London: Latin American Bureau.

Prieto Dueñas, C.S. (ed.) (2005) *Análisis a la Política de Alimentación y Nutrición en el Distrito Capital*. Bogotá, Contraloría de Bogotá, D.C. www.contraloriabogota.gov.co/destino.asp?solicitud=informes.htm (accessed 7 June 2017).

Pugh, C. (2000) Squatter settlements. Their sustainability, architectural contribution, and socio-economic roles. *Cities*, 17(5), 325–337.

Ramos, I.L., Bernardo, F., Carvalho Ribeiro, S., & Van Eetvelde, V. (2016) Landscape identity: Implications for policy making. *Land Use Policy*, 53(2016), 36–43.

Riaño, Y. (1990) Understanding the cultural dynamics of popular habitats: From spatial activity patterns to local identity in the Barrios of Bogotá, Colombia. In: Bassand, M., & Bolay, J. (eds.), *International colloquium: Creatiive habitat, culture and participation*. Laussane: Institut de Recherche sur l'Envrionment Construit, Comission Nationale Suisse pour l'UNESCO.

Rueda García, N. (2000) *La Ciudad Que No Conocemos, Cien Años de Arquitectura en Colombia, XVII Bienal de Arquitectura*. Bogotá: Sociedad Colombiana de Arquitectos.

Sánchez, A.J. (2008) *Agricultura urbana en Bogotá: Implicaciones en la construcción de una ciudad sostenible*. Bogotá: Pontificia Universidad Javeriana.

Santos, M. (2000) *La Naturaleza del Espacio*. Barcelona: Ariel Geografia.

Schwab, E. (2015) *Urban promises? Spatial justice in public space based upgrading programmes of popular settlements in Latin America*. Unpublished PhD thesis, University of Life Science and Natural Resources, Vienna.

Segovia, O., & Oviedo, E. (2000) Espacios Públicos en la Ciudad y el Barrio. In: Segovia, O., & Dascal, G. (eds.), *Espacio Público, Participación y Ciudadanía*. Santiago de Chile: Ediciones Sur, pp. 51–69.

Speak, S. (2014) Desperation, delight or deviance: Conflicting cultural landscapes of the urban poor in developing countries. In: Roe, M., & Taylor, K. (eds.), *New cultural landscapes*. New York: Routledge, pp. 136–152.

Torres Tovar, C.A. (ed.) (2009) *Ciudad informal colombiana: barrios construidos por la gente*. Bogotá: Universidad Nacional de Colombia.

UNCHS (2003) *The challenge of slums: Global report on human settlements 2003*. London: Earthscan.

Vargas, S.B. (2009) Ruralidades Emergentes y Dinámicas Territoriales: Nuevas Percepciones y Medios de Vida. *Eleuthera*, 3, 194–205.

Viviescas, F. (1997) Espacio Público, Imaginación y Planeación Urbana. In Taller, B. (ed.), *La Calle: Lo Ajeno, Lo Público y Lo Imaginado*. Bogotá: Barrio Taller.

Ward, P. (2004) Introduction and overview: Marginality then and now. *Latin American Research Overview*, 39(1), 183–187.

Zaar, M.H. (2011) Agricultura urbana: algunas reflexiones sobre su origen e importancia actual. *Revista Bibliográfica de Geografía y Ciencias Sociales*, XVI (944). http://www.ub.es/geocrit/b3w-944.htm (accessed 1 November 2017).

23

Rural-urban food and nutrient dynamics and nutrient recovery from waste in developing countries

Munir A. Hanjra, Mary Lydecker, Pay Drechsel, and Johannes Paul

Introduction

Food production is frequently limited by under-supply of nitrogen (N) and phosphorous (P) in freshwater, marine, and terrestrial landscapes (Elser et al., 2007; Mayer et al., 2016). On the other hand, elevated levels of these nutrients have been implicated worldwide for massive changes in biodiversity and ecosystem health (Billen et al., 2009; Liu et al., 2014). Global cycles of N and P have been amplified by 400% and 100%, respectively, by post-industrial human activities (Elser et al., 2007; Cordell and White, 2015). Globally, human-generated waste streams are the major source of urban nutrient discharge to both surface water and agricultural recycling (Morée et al., 2013). Sustainable nutrient management requires an understanding of if, where, and how these key nutrients flow in food supplies across the rural-urban landscape (Heimersson et al., 2016; Nesme and Withers, 2016) as well as policy and investment options for recovering nutrients from municipal solid waste, fecal sludge, urine, and wastewater for reuse in food production in urban/rural areas (Montangero et al., 2007; Metson et al., 2015; Drechsel and Hanjra, 2016). This approach aims to close the nutrient loop locally while addressing nutrient limitations and sustainable food production across territories, establishing landscape resilience at all scales.

This chapter illustrates the opportunities and challenges for implementing resource recovery for reuse to promote nutrient-balanced landscapes and sustainable food production systems in developing countries, with in depth analysis from West Africa. Within the sanitation-agriculture interface, essential plant food nutrients are the focus which can be derived, for example, from municipal solid waste, fecal sludge, urine, and wastewater. While resource recovery and reuse can take place in household backyards, adopting a broader landscape perspective allows analysis of the relationship between people and place, at the scale of territories rather than localities. This chapter makes the landscape connection on the most fundamental level, through nutrient waste processes and human actions such that the circular flows discussed are of central importance to contemporary landscape concerns as well as public policy. The landscape perspective reveals how the success of the nutrient-food-waste loop relies on its meaningful integration into the economic, sociocultural, and ecological life of rural-urban areas (Lerner and Eakin, 2011; de Zeeuw and Drechsel, 2015; Verhougstraete et al., 2015). A sustainable landscape is greater than the physical realm as it supports how a society functions collectively, reinforcing shared values and providing essential needs, while protecting environmental and cultural resources for future generations (Sayer et al., 2013). This loop is an opportunity to establish new human behaviours at the local level that collectively influence processes and systems at the regional scale.

If present trends continue, world population is projected to increase between 9.6 and 12.3 billion by 2100 (Gerland et al., 2014; UN, 2015; World Bank, 2016). By 2050, about 66% of world population will likely be living in urban areas (54% today). Although urbanisation will mostly happen in medium-size cities and urban peripheries around the world (UN, 2015), the number of megacities – each with more than 10 million people – is increasing. Today there are 28 megacities, 16 located in Asia, and there are likely to be 41 megacities by 2030 (UN, 2015). Rural population proportions are expected to decrease as urban populations continue to grow. These trends pose complex challenges for meeting the needs of urban communities – including water, food, and energy security (see Chapters 16 and 19), as well as sanitation and waste management. Provision of nutrients for agriculture is a further challenge, for ensuring food security for urban populations and protecting peri-urban and rural landscapes (Villarroel Walker et al., 2012; Drechsel and Hanjra, 2016; Duchin, 2016; Chapters 14 and 15). Planning is needed to strengthen infrastructure and built form to facilitate robust rural-urban linkages, establishing functional and productive dynamics across the landscape (see Chapter 26). Policy and investment options are needed to support this planning, promoting sustainable integration between rural and urban areas for improved wellbeing and prosperity.

Aside from the often described water security-food-energy nexus (Scott et al., 2011; Jalilov et al., 2013; Biggs et al., 2015; Sadoff et al., 2015; Damerau et al., 2016; Yillia, 2016) the linkages between food

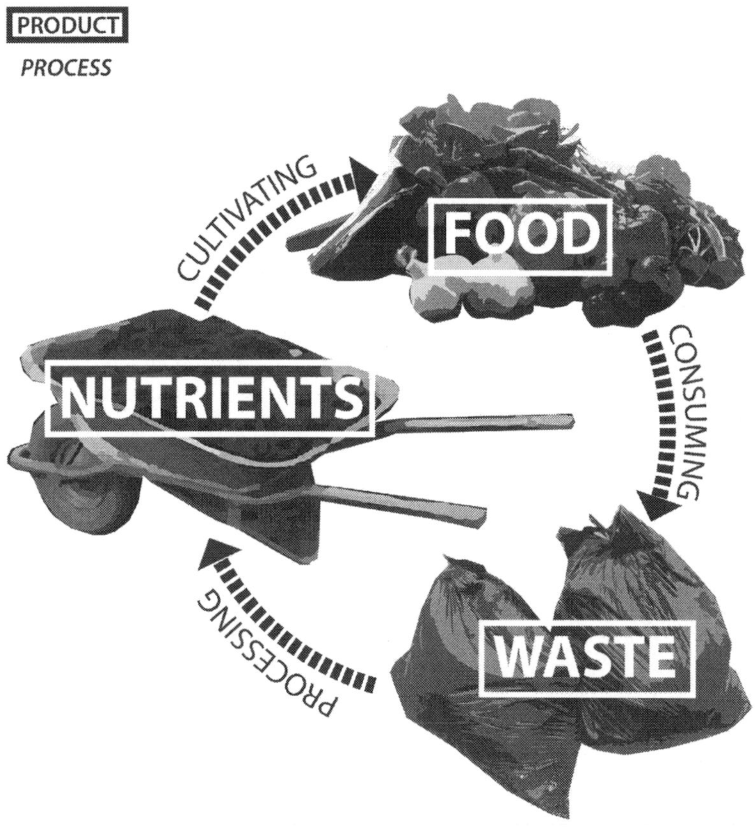

Figure 23.1 The 'nutrient-food-waste' loop includes the already strong 'food-waste' link, but underdeveloped 'waste-nutrient' and 'waste-nutrient-food' links. Closing this cycle offers the potential to simultaneously address other urban goals, such as sanitation and food security.

Source: Mary Lydecker

production, consumption, waste, and sanitation are increasingly recognised (Drechsel and Hanjra, 2016; Hutton and Varughese, 2016) (Figure 23.1). The United Nations Sustainable Development Goals (SDGs), for example, including SDG 6 and SDG 11 and 12, call for more sustainable consumption, enhanced rural-urban linkages and resource recovery for reuse (ODI, 2015; Hanjra et al., 2017). To protect our rural-urban landscapes, they must be as *functional and productive* as possible. These two characteristics we consider the most important, and they are highly interdependent. Therefore, the objective of this chapter is to highlight opportunities for promoting closed loop processes – where resources are used again and again instead of being irretrievably lost to the environment, and re-visioning urban waste and sanitation challenges as opportunities to promote food security through the recovery of nutrients for reuse in agriculture (Lazarova et al., 2012; Mo and Zhang, 2013; Drechsel and Hanjra, 2016). This chapter focuses on the relationship of these resource and waste systems in developing countries with in depth examples from West Africa.

The urban metabolism and its impact on rural-urban resources

Urban areas depend on peri-urban and rural landscapes. Reciprocal processes impact rural-urban linkages – flows of goods, services, people, information, finance, water, food, nutrients, and waste (Muzondi, 2014; Capps et al., 2016; Zhang et al., 2016). Cultural and social dynamics overlay each of these systems, influencing the complex interrelationships of key economic sectors including agriculture, services, and industry. Urbanisation is much more than just a demographic change involving population growth in urban areas, supported by rural-urban migration. Urban areas are extending further into peri-urban areas and rural hinterlands (see Chapters 14 and 15). The physical growth of cities takes arable land out of food production (Molden, 2007; Satterthwaite et al., 2010), transforming the character and function of these rural and peri-urban landscapes which are often situated on agricultural lands that are more productive than the global average (Bren d'Amour et al., 2016). Urbanisation also impacts the dynamics of trade, such as the location, quantity, and structural mix of food demanded for urban areas corresponding to income, cultural diversity, lifestyle, and diets (see Chapter 20). All of these factors significantly influence agricultural production and food supply chains (Schmidt et al., 2015). Urbanisation thus has clear linkages with agriculture, which are, however, largely ignored in agricultural policies or urban planning and implementation. In an ongoing study by International Water Management Institute/Food and Agriculture Organisation of the United Nations (IWMI/FAO) on city region food systems in Colombo, Sri Lanka, for example, the only municipal department touching on agriculture or food supply was the one in charge of public health (Fernando et al., 2016). It is worth noting that this is also the case even in western cities/economies, where the relationship of agricultural policy to urban planning is often weak (Chapter 26). Meanwhile, the urban metabolism is not only hungry, it is also thirsty. Similar to urban food supply, growing cities are in dire need of reliable water which is increasingly affecting aquifers and surface water resources (Chapter 19) in peri-urban areas and much further afield. Again, this rural-urban resource imbalance and unsustainable dependence is not exclusive to low-income countries, as demonstrated by the case of California's water supply (Binz et al., 2016).

Urbanisation also means solid waste generation, including food waste, and in most low-income countries waste management has been outpaced by population growth (Yoada et al., 2014; Gould et al., 2016). Waste production has grown tenfold in the past century and will likely double again by 2025 up from 3.5 million tons (in 2010) to 6 million tons per day (Hoornweg et al., 2013). Accordingly, solid waste management is one of the greatest costs to municipal budgets (Parthan et al., 2012; Lohri et al., 2014) and this highlights the need for cities to implement systems-based approaches to waste management. Curitiba's system in Brazil, for example, recycles an estimated 70% of its solid waste, thus integrating fundamental urban objectives of environmental, social, and financial sustainability (IFC, 2015; Põldnurk, 2015). Population and waste will continue to rise in fast-growing cities of South Asia and Africa, presenting prime opportunities for introducing new paradigms of nutrient cycling, from food consumption to sanitation systems and back to agriculture, promoting landscapes that support clean living and food security (Hoornweg et al., 2013; Zabaleta and Rodic, 2015).

Coincident challenges apply to wastewater and sanitation services, including the safe removal of human excreta, which is the ultimate waste product of the food chain and a significant nutrient and pathogen source in the urban environment. Where wastewater and fecal matter can be collected from households without interference of industrial sources and related waste flows, as is the case in on-site sanitation systems, 'humanure' can be collected and treated safely for crops (Nikiema et al., 2014). Safety measures, as those suggested by the World Health Organisation (WHO, 2006) are in particular important where wastewater collection and treatment remain limited, while farmers rely on increasingly polluted streams and rivers for irrigation. Around four out of every five cities surveyed in developing countries use irrigation water for food crops that is polluted by urban wastewater (Raschid-Sally and Jayakody, 2008).

The modern use of the urban metabolism concept was pioneered by Wolman (1965) in his study on "The Metabolism of Cities" for an imaginary city of one million people, where he examined total resource inputs into the city and waste outputs from the city (cited in Kennedy et al., 2015; Bai, 2016). Urban metabolism is defined as the sum total of the technical and socioeconomic processes that occur in cities, resulting in growth, production of energy, and elimination of waste (Kennedy et al., 2015). During recent decades, urban metabolism research has witnessed a significant shift in focus (Bai, 2016). This involves a shift from direct flows and hidden flows from elsewhere that produce goods and services consumed by urban areas; to cross boundary to intra city distribution; from single city resource budgets to cross city patterns for comparison and benchmarking of resource use; from static snapshots to dynamics of change over time and space; from quantitative analysis to regulating processes and drivers such as consumption; from resource and environmental impacts to policy and planning implications (Kennedy et al., 2011; Zhang, 2013; Dinarès, 2014). Benchmarking these flows per population unit can help visualise emerging issues in rural-urban dynamics to promote, for example, nutrient recovery from waste to ultimately develop effective strategies for recycling these nutrients back into agriculture to enhance food security and sustainability (Zabaleta and Rodic, 2015; Drechsel and Hanjra, 2016).

There are many related concepts assessing, for example, the water and energy footprint in catchments (Khan and Hanjra, 2009; Khan et al., 2009; Hoornweg et al., 2016) or the resilience of urban foodsheds and city-region food systems (Karg et al., 2015), which link the challenge of sustainable landscapes with climate change and social conflicts affecting rural-urban supply chains and business relations. The application of 'closed loop' nutrient cycle across a landscape is both generic and specific. While this chapter outlines generic features of a sustainable nutrient system, the specific characteristics of a given site – physical, political, social – deeply shape implementation, and are guided by public policy for its implications in spatial urban form and socioeconomic conditions.

Rural-urban nutrient flows

Cities use materials and services that occur outside their boundaries such as water supply, food supply, energy, and wastewater treatment. Cities also generate vast quantities of nutrient rich wastes, which are only to a minor extent, mostly via wastewater irrigation, reused for food production in peri urban areas of developing countries. With the majority of food coming from rural areas, the 'City Region Food System' concept is helpful for analysing the contributions urban centres receive from rural, peri-urban, and urban agriculture, including the nutrients embedded in imported food (Karg et al., 2015). In Ghana, about 1,000,000 tons of food enters the cities of Accra and Kumasi each year, of which one third (Accra) to one half (Kumasi) might leave the urban trade hubs again (Drechsel et al., 2007). There are clear commodity specific foodsheds – geographically defined source areas of particular commodities supplied to the urban area – which vary in scale and extent between lean and bumper seasons (Karg et al., 2015). To meet the growing urban calorie demand, cities in Ghana and across West Africa rely heavily on rural production, especially for staple crops like cereals and tubers, but also increasingly on large-scale, long-distance transportation of food, like rice and poultry, often suppressing local production with competitive prices (Karg et al., 2015). Figure 23.2 shows an example of Tamale, the capital city of the northern region of Ghana with a population of 3.7 million.

Converting all food flows into nutrients including nitrogen and phosphorous, the quantity of N that flows annually as food to the city of Kumasi, Ghana, is – to visualise the dimension – more than the total amount of all annually imported chemical N fertiliser in the whole of Ghana over several years (Drechsel

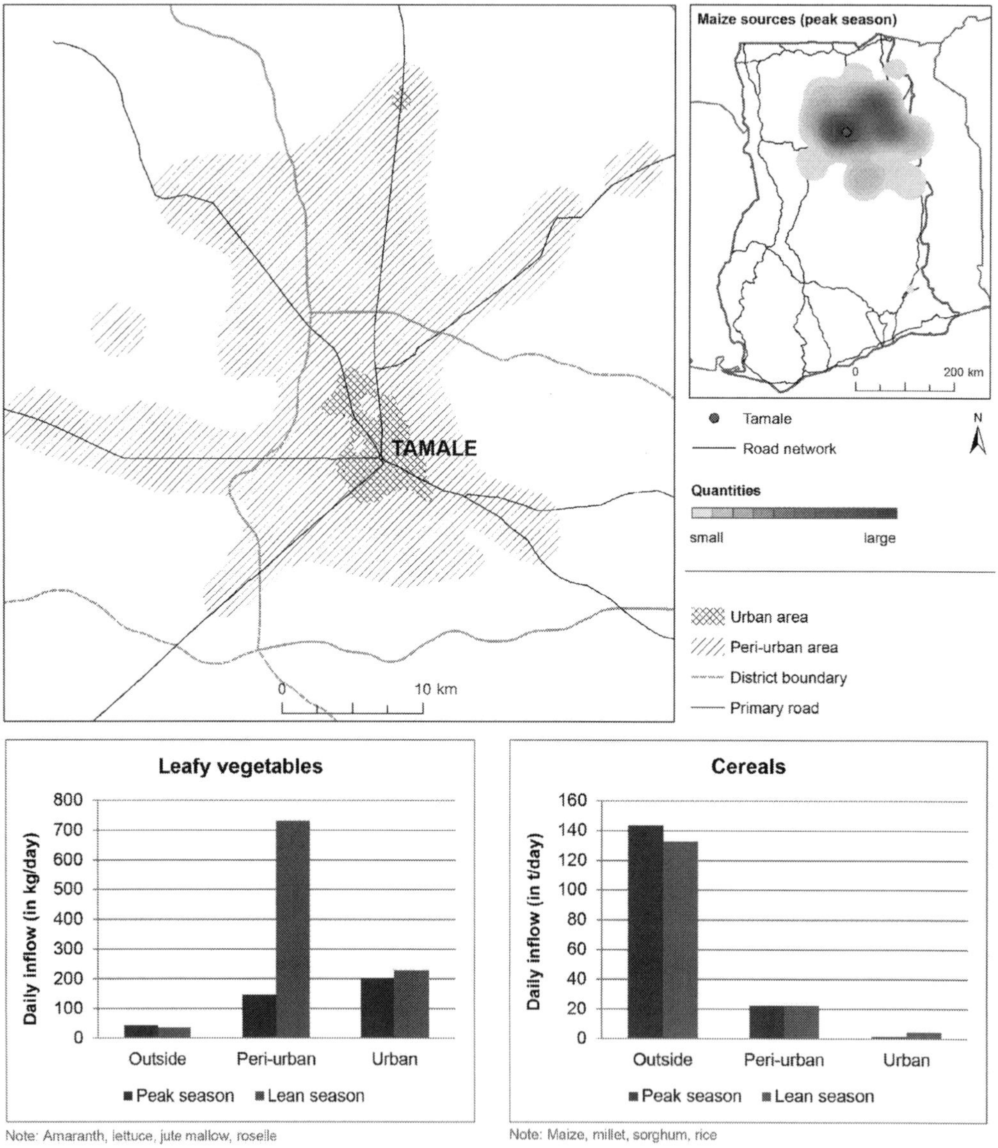

Figure 23.2 The contrasting role of different regions (urban, peri-urban, and rural/import) for Tamale's supply with cereals (local rice, maize, sorghum, millet, groundnut) and leafy vegetables (amaranth, roselle, jute, and lettuce). The inlay is Tamale's overall maize foodshed in 2014.

Source: Karg et al., 2016, modified

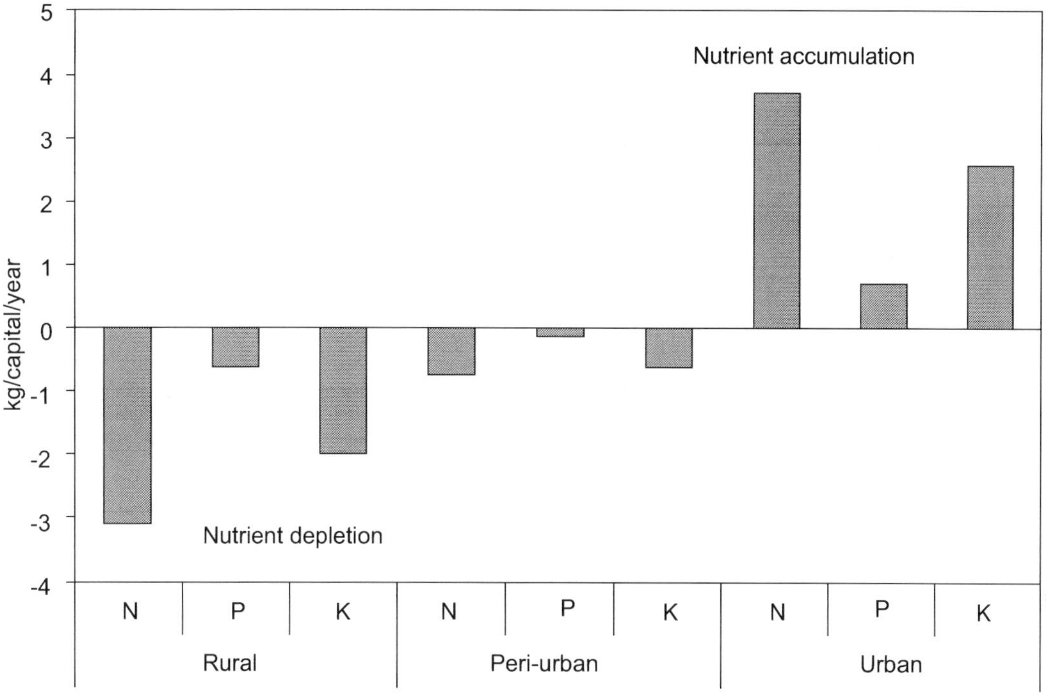

Figure 23.3 Nutrient shift along the rural-urban landscape in kilograms/year based on data from four West African cities, standardised per capita.

Source: Drechsel et al., 2007

et al., 2007). As both chemical N fertiliser use and urban waste recycling activities are negligible in Ghana, soils in agricultural production areas are continuously mined for plant food nutrients, resulting in declining fertility. Meanwhile, the bulk of food waste and other nutrient-rich wastes end either in landfills, street drains, or the environment, making it evident that such urban centres are indeed nutrient 'sinks' (Keeling, 2005; Drechsel et al., 2007) (Figure 23.3). Nutrient loading in the urban landscape has implications for surface and groundwater pollution, environmental degradation and public health (Craswell et al., 2004). The approach of analysing nutrient budgets has been used widely and at different scales from farm plots to landscapes, districts, and national levels (Roy et al., 2003).

Significant amounts of 'wasted' nutrients are in human excreta – primarily urine. There are some examples that are seeking to close this loop (Ganesapillai et al., 2015), including historic hot bed horticulture, for example, as was once used in Europe and Paris in particular. In the city of Kumasi, material flow analysis combining solid and liquid flows shows that before investment in a new fecal sludge treatment facility, between 70% and 80% of the food-based N and P consumed in Kumasi eventually polluted the urban environment. As a result, the amount of N and P leaving the city via different streams was 10 times greater than the amount that enters the city upstream and through precipitation (Erni et al., 2010). Smallholders cultivating vegetables on about 10,000 ha downstream of Kumasi use the untreated/mixed wastewater from these streams. While dilution is undermining the fertiliser benefit of the water, the dilution is not strong enough to eliminate potential pathogenic health risks (Erni et al., 2011). Depending on the water volume of the receiving rivers and the wastewater inflow, some smaller stream can carry high nutrient loads while the dilution in larger ones would not impact farmers' need to apply additional fertiliser (Erni et al., 2011).

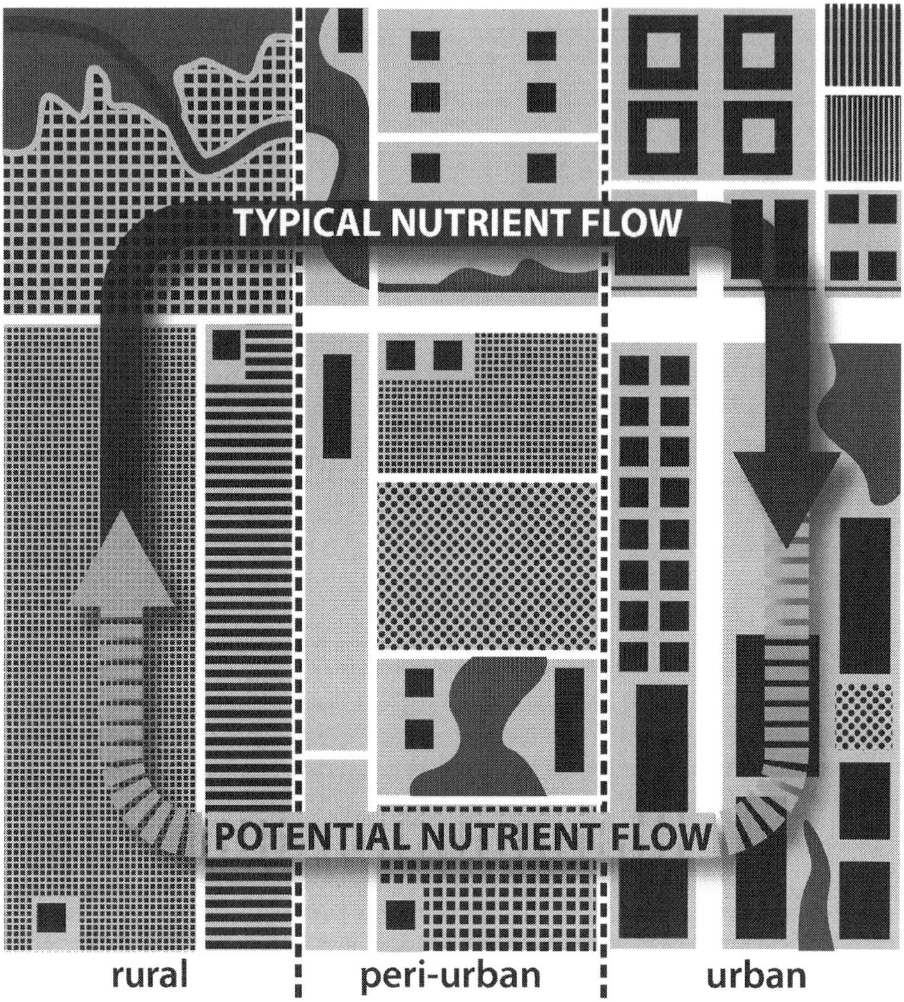

Figure 23.4 Opportunity to alter the typical nutrient flow from the rural to urban landscape with a counterflow back to agricultural use, with positive impacts on the landscape.

Source: Mary Lydecker

From a technical perspective, a planned resource recovery system does not require high capital investments. There is much that can be done simply in landscape management, land use planning, and agricultural practices that require education more than expensive physical infrastructure, which is particularly useful for those who are designing urban and peri-urban spaces. Significant parts of the food waste from, for example, markets and food outlets can be recycled through composting or co-composting with nutrient-rich fecal sludge. Considering the existing waste collection capacity of Kumasi, the entire N and P demand of urban farming could be recovered through solid waste recycling, as well as 18% of the N and 25% of the P needs of peri-urban agriculture within a 40 km radius around Kumasi (Belevi, 2002). Figure 23.4 illustrates the typical rural-urban nutrient flow, as well as the opportunity for urban-rural nutrient flow for reuse, which is currently poorly leveraged as a driver of resilient landscapes, with potential to address multiple objectives such as economic vitality, sustainable resource management and public health. With low fertiliser use in agriculture in most African countries, if this nutrient loop could be closed with education, urban design, landscape management and policy, it could altogether cut out chemical fertilisers, averting their potential negative impacts on the environment and human health.

Transforming challenges to opportunities at scale

Urbanisation poses challenges for sustainable resource management, but also offers powerful economic opportunities such as market proximity, shorter transport distances, higher purchasing power, export hubs, and economies of scale which can attract private capital (Hanjra et al., 2015b). All of these factors give urban areas key advantages over rural areas. Small private sector operator participation is welcomed in some sectors, for example waste management, which is traditionally reliant upon public subsidies and struggles to meet required capital investment and operational costs, resulting in poor service delivery (Wang et al., 2015). Amid concerns about promoting privatisation (Chapters 10, 16, 20, 24, 25, 31, 32 and 33) and dominance by foreign-based companies (Chapter 20), if human excreta is seen as a resource, with a corresponding market value, there may be potential opportunities for small, local start-up businesses to become involved in this market (Diener et al., 2014; Nikiema et al., 2014; Tilmans et al., 2014; Rao et al., 2017). Indeed, local businesses and communities have the advantage of direct knowledge of local resource and economic dynamics, which could be synergistically integrated in a waste recovery and reuse initiative. This means a new planning and design paradigm to involve and empower locals and the urban poor to vision and create their own functional, productive, and safe landscape (Guest et al., 2009; Cornejo et al., 2013) without requiring involvement and/or substantial capital investment from global corporations.

Nutrient recovery from food waste is particularly compelling for a private operator at the community level, as it allows additional revenue streams beyond typical tipping fees (Amir et al., 2015). It also applies to more sophisticated phosphorous recovery through wastewater treatment (Mayer et al., 2016) as well as more basic solid waste composting (Alvarenga et al., 2015; Otoo et al., 2015) based on well-known technologies for nutrient and organic matter recovery. However, despite the win-win in terms of cost savings and organic nutrient supply for waste management and agriculture, there are hardly any compost facilities at the scale of serving towns or cities (Drechsel et al., 2010b; Viaene et al., 2016). One of the identified key reasons for failure is over-reliance on technical solutions without sound market analysis, assessment of transport costs, business models, or strategies for collaboration with public sector services, municipalities, and farmers (Truc et al., 2012; Hanjra et al., 2015b; Maroušek et al., 2016).

The situation in developing countries is, however, changing. Success stories driven by private entrepreneurs are emerging, ranging from community-level initiatives to large-scale composting operations (Otoo and Drechsel, 2018). A well-known example is 'Waste Concern' which, founded in 1995, collects and composts more than 100,000 tons of waste in Dhaka city, Bangladesh, tapping into carbon credits as an additional revenue stream. Between 2001 and 2006, it produced compost in the larger Bangladesh area worth more than US$1 million (www.wasteconcern.org). Waste Concern and other such enterprises are shaping landscapes and landscape practices in important ways, including improved pit design to suit local physical conditions, on-demand and safer pit emptying services, nutrient recycling, and compost supply to local farmers, all supporting cleaner and healthier environments across the rural-urban transect (Matter et al., 2015). Successful enterprises demonstrate a mix of strategies from physical design of equipment, thoughtful integration into existing urban landscapes, and efficient operational/behavioural practices – all conceived and implemented based on the studied appreciation of local socio-cultural characteristics and economies. Companies are increasingly focusing on nutrient-rich fecal sludge from non-sewered on-site sanitation facilities, which have very limited risks of chemical contamination compared with sewage sludge derived from sewered industrial-cum-residential areas (Nikiema et al., 2014). Such on-site facilities (septic tanks, pit latrines) are the most common sanitation system in developing countries, poised for new design

Figure 23.5 Strategies to promote 'closed loop' nutrient management range from site scale interventions, such as composting toilets (left) or community composting of solid waste (middle tumbler) and composting bins (middle bottom), to regional scale operations, such as septage good practices/fecal sludge collection and reuse (right, top and bottom). Long-term resilient solutions will require an inter-weaving of operational approaches and technologies at multiple scales.

Source: David Omick/Pearl Mast, Urban Ore, Smith Family Garden, Michigan Department of Environmental Quality, USA respectively

approaches that blend local decentralised strategies with regional infrastructures and larger landscape systems (Cairns-Smith et al., 2014) (Figure 23.5).

Data from Ghana, Senegal and Uganda show a high market potential for fecal sludge, not only as a fertiliser but also when considering other value propositions – dry sludge fuel, biogas, sludge in building materials, and protein derived from sludge as animal feed (protein valued at US$22–32 per ton of dry sludge) (Diener et al., 2014). While fecal sludge as soil conditioner (US$7–16 per ton of sludge) remains the most common end use of treated sludge, related businesses are usually financed through the double value proposition of safeguarding public health (which warrants public subsidies) and revenues from resource recovery. The exploration of the reuse market can be supported by innovative policy and finance instruments and mechanisms (Box 23.1).

To make waste-based compost an asset, that is, to tailor it to farmers' needs and better compete with other soil inputs, compost quality can be 'boosted' through, for example, (1) co-composting organic solid waste with dewatered but nutrient-rich urban fecal sludge or other nutrient-rich waste products; (2) further enriching the compost with inorganic fertiliser, rock-phosphate, or urine to create a 'fortified' organo-mineral material; and/or (3) pelletising the compost to reduce its bulkiness, generating a product similar in its appearance and handling to an inorganic fertiliser (Nikiema et al., 2014). These options can also be combined with due care that any related increase in production costs is matched by the targeted customer segments, and remains competitive to alternative fertilisers. Composting also eliminates the mining process and associated environmental impacts of chemical fertiliser production and methane emission from landfill. In terms of energy, composting will always win due to less methane emissions from landfills to landscape.

Box 23.1 Public policies to promote the acceptance of city compost.

Some countries fail to provide laws or protections on composting. The Indian Cabinet approved in early 2016 a policy on the Promotion of City Compost. The Ministry of Urban Development in consultation with the Ministry of Chemicals and Fertilisers agreed to subsidise compost sale with INR 1,500 (US$ 22.5) per ton of city compost. This market development assistance will be paid to fertiliser companies with the expectation of co-marketing city compost together with chemical fertilisers. The co-marketing details will be decided by the Department of Fertilisers depending on supply and demand. Earlier suggestions for co-marketing were to sell, for example, one bag of municipal compost with every two bags of chemical fertiliser, or that only a co-marketing arrangement gives access to the subsidy on chemical fertiliser. Such a directive would urge fertiliser companies to seek compost from compost stations; turning the common situation around where compost plants have to seek out customers (Otoo and Drechsel, 2018). Yet, the policy has no explicit objective to wean farmers off unsustainable chemical fertilisers (see Chapter 16) and eventually to replace them entirely with compost. While the effort is notable, there are not yet typically better outcomes than in western contexts.

In a similar effort, the Ghanaian Ministry of Food and Agriculture decided in 2016 to expand its fertiliser subsidies to large scale compost production. This move was supported by long-term field research on safe co-composting of municipal solid waste and fecal sludge by the International Water Management Institute (IWMI) in Ghana.

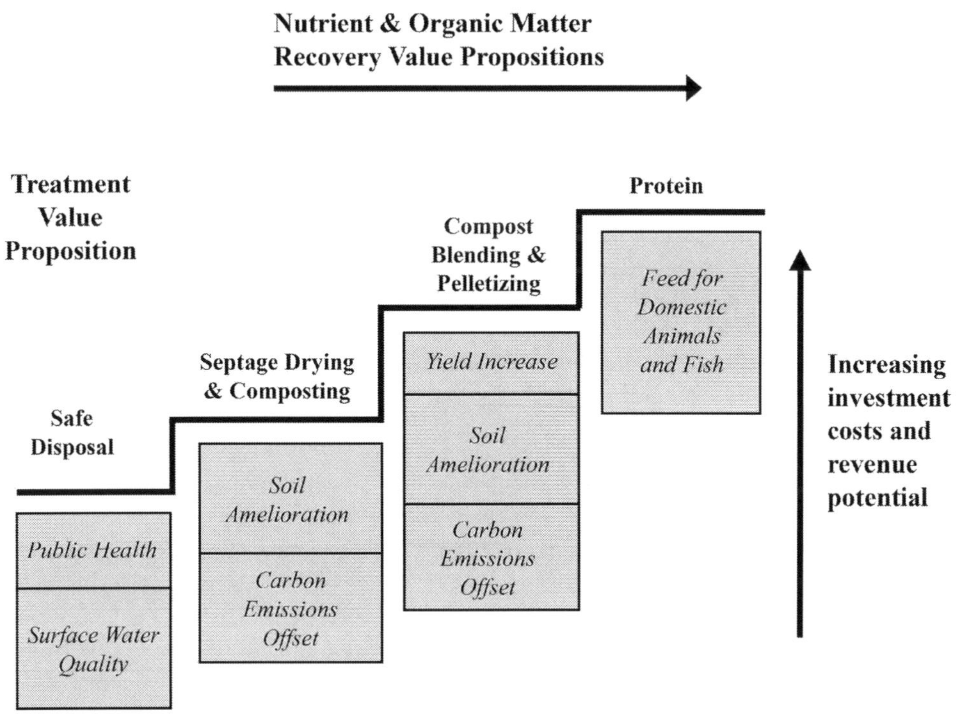

Figure 23.6 Value propositions for nutrient and organic matter recovery and reuse from septage from household-based sanitation systems.

Source: Otoo et al., 2015, modified

Composting is also a low energy process, primarily involving sun drying and binders made of organic waste materials sourced from agriculture (no chemicals); external energy inputs are minimal compared to fertilisers, except for the optional process of pelletising, which requires high energy to press the pellets and gamma radiation to sanitise the product, if demanded by customers. Composting compares far better in terms of sustainability as, in the least, there is no mining involved.

There are emerging possibilities for sanitation systems that separate urine and reuse it as the primary humanure in Burkina Faso, Uganda, and other landscapes (Andersson, 2015; Makaya et al., 2015) but regional-scale reuse programs are rare due to high cost and virus survival risk, and fundamental design challenge posed by overlaying human activities with resource recycling. The global initiative to 'Reinvent the Toilet' illustrates in particular the deep link between design, social behaviours, and public health. Options for increasing the value of organic waste are shown in Figure 23.6 including the use of Black Soldier fly larvae (*Hermetia illucens*) which feeds on organic matter, such as fecal sludge and organic wastes, and leapfrogs the nutrient extraction via crops by directly generating high-value protein and fat, which can be marketed for poultry, duck, and fish feed (Diener et al., 2014).

Another important pathway of nutrient recycling in urbanised landscapes is wastewater irrigation (see Chapter 19). In contrast to the efforts related to nutrient recovery from organic waste, most wastewater reuse in developing economies happens without any support from government. Due to poor sanitation in these contexts, nearly all surface water sources in city vicinities are contaminated, and unplanned reuse of untreated but usually diluted wastewater is taking place on an area globally at least ten times larger than any planned reuse with treated wastewater (Scott et al., 2010). For example, the Musi River, which flows through Hyderabad in India, carries – despite increasing treatment efforts – significant amounts of nutrients to downstream farms where they are informally recycled (Rao et al., 2015; Starkl et al., 2015). Other authors describe similar scenarios for East and West Africa where wastewater treatment capacities are even smaller (Drechsel et al., 2007).

Key direct benefits of wastewater use in food production systems include the nutrients in the water, especially in raw wastewater, as well as the year-round reliable and low/no cost supply of water itself (as opposed to groundwater or detained water, or when freshwater is not available) (Hanjra et al., 2012). Undiluted wastewater contains nutrients that can significantly contribute to crop growth and improve soil fertility. It is estimated that 1000 m^3 of municipal wastewater (enough for irrigating one hectare) can contribute 16–62 kg total nitrogen, 4–24 kg phosphorus, 2–69 kg potassium, 18–208 kg calcium, 9–110 kg magnesium, and 27–182 kg sodium (Qadir et al., 2007) – sufficient nutrients needed for one hectare of commonly grown food crops and pastures. In Mexico's Mezquital (Tula) Valley, wastewater irrigation provides 2,400 kg of organic matter, 195 kg of nitrogen and 81 kg of phosphorus/ha/year, contributing significantly to crop yields (Jiménez, 2005). Wastewater not only adds nutrients to soil, but can also amend soils through its organic matter content (biosolids or biochar) (Christie et al., 2001). Compared to freshwater, there is a significant body of literature showing production advantages for wastewater irrigation, although many comparative assessments have to be read with caution if different soils are compared (Hanjra et al., 2015a; Minhas et al., 2015; Tal, 2016).

While wastewater reuse offers significant opportunities for closed loop processes across the rural-urban trajectory, its common unplanned nature poses significant challenges for safeguarding public health (Drechsel et al., 2010a; Hanjra et al., 2012; Fuhrimann et al., 2016; Rietveld et al., 2016). Municipal wastewater treatment must protect and enhance local reuse practices, and avoid undermining their efficacy. In Guanajuato, Mexico, for example, the estimated cost for farmers to replace the nitrogen and phosphorus loss through wastewater treatment was estimated at US$900/ha (Scott et al., 2000) and thus is prohibitively expensive. Landscapes are rarely isolated but instead are deeply interdependent through physical, sociocultural, and economic systems. This case demonstrates the potential synergistic benefits for linking local agricultural practices and policies with the planning and design of future regional water treatment plants.

Managing trade-offs across landscapes

Escalating food demand under rapid urbanisation can drive farmers to excessive fertilisation, leading to large scale reactive nitrogen enrichment, river eutrophication, and potential human and environmental health issues across the rural-urban range (Billen et al., 2009; Xian et al., 2016) Data for Yongan River in eastern China for 1980–2010 show that agricultural pollution consistently dominated the annual riverine phosphorous load (55–93%) due to run-off, soil erosion and improved drainage of agricultural fields (Chen et al., 2015). Basin scale temporal analysis linking P dynamics with agricultural intensification for food security as well as the sanitation sector in rural/urban contexts are, however, rare (Box 23.2). In many places, people still defecate directly in the fields, which leads to direct fertilisation, but also to non-point source pollution in runoff and the possible spread of disease.

Nutrient management as part of wastewater treatment is a challenging task. While on one hand, nutrients are valuable for irrigated farming, and farming helps to absorb and recycle these resources, excessive nutrient concentrations, in particular in undiluted wastewater, can severely harm soils and water bodies (Reznik et al., 2017). In the case of macronutrients such as N and P, there are three possible impact pathways (Qadir et al., 2007): (1) excess N applied through wastewater may lead to excessive vegetative growth (green biomass), delay in maturity, lodging, and low economic yield (in cereals for example), while the yields of leafy vegetables will be enhanced; (2) excess N and P in wastewater can cause eutrophication of natural water bodies and in irrigation systems, undesirable growth of algae and weeds; and (3) leaching of N can cause groundwater pollution, and in the case of drinking N-rich groundwater (particularly high levels of nitrates) effect human health and children in particular, including blue baby syndrome. Many other nutrients can lead to soil salinisation if applied in high concentration, particularly in arid and semi-arid regions (Qadir et al., 2007). Aside from appropriate wastewater treatment there is also a range of on-farm options which exist to manage excessive nutrient and salt levels (Simmons et al., 2010). The same applies to pathogens where untreated wastewater or fecal sludge is used (Drechsel et al., 2010b). Such farm management practices can be coupled with water-sensitive urban design elements, such as biofiltration and constructed wetlands, as relatively low cost initiatives to improve water quality, provide ecological benefits (see Chapter 17) and/or a range of co-benefits for communities through their integration with public open spaces.

In peri-urban Tamale, Ghana, for example, drivers of trucks that empty household septic tanks discharge their fecal load on farm fields on demand by local farmers who pay for the 'human manure' (Cofie

Box 23.2 Phosphorous dynamics at a rural-urban watershed scale

Analysis of P cycling in Chaohu watershed located in the southeast of China during 1978–2012 (Jiang and Yuan, 2015) shows:

- *A five-fold increase in annual P inputs to expand and intensify agriculture.* Annually, most P inputs (75%) were stored within the watershed causing buildup of legacy P in cultivated land (42–69%), uncultivated land (8–20%), and surface water (22–38%) – with animal husbandry contributing two-thirds of total P inputs to surface waters.
- *Animal food-P increased greatly for urban residents* (7–24%) but less for rural residents (1–8%). Yet, per capita food-P consumption decreased for urban residents (from 0.71 to 0.46 kg) and for rural residents (from 1.1 to 0.69 kg) – denoting changes in diets.
- *P recycling rate from waste decreased* sharply in growing urban areas (46 to 4%) compared to a slight decline in rural areas (82 to 78%) as it is still popular in rural areas to use excreta as organic fertiliser (Jiang and Yuan, 2015).

et al., 2005). The practice is very common and less risky from a health perspective than it might appear as the sludge is usually sun-dried over several months before cereals are grown. An interesting aspect is the reversed cash flow of the model. While normally the drivers have to pay for desludging their trucks, farmers instead pay the drivers (Otoo and Drechsel, 2018). The safe use of humanure, especially by those marketing composting toilets, raises the question of how to incorporate latrines and cesspits into this cycle, since they often leach into groundwater (Anand and Apul, 2011). Additionally, many toilets can open directly into water bodies (ADB, 2014) – not just in the developing world but as evidenced by examples in Amsterdam and for much of the UK's passenger rail system. Indeed, the regional-scale nutrient loop is deeply influenced by the design and implementation of site-scale infrastructure – such as the typical, replicable toilet system, reverberating impacts across landscape systems.

Other innovations for wastewater recycling include 'living machines', that is, ecological sewage treatment within building complexes through constructed wetlands, or ecological sanitation such as at GIZ headquarters in Bonn, Germany, where urine is separated for possible reuse within its building (Otoo et al., 2015). Although such decentralised treatment and resource recovery options are appealing for developing countries as an alternative to conventional systems, regulations and perceptions in support of reuse have to be in place, while technical expertise to operate and maintain this kind of system might limit its potential as a model for broad application (Kuttuva et al., 2017). However, the nutrient recycling demonstrated through 'living machines' at the building and site-scale is exactly the kind of closed loop which can easily be extended within watersheds to the landscape scale, making use of grey or green infrastructure (WRI, 2012).

Policyscape

Achieving closed-loop resource flows and resilient landscape systems at scale requires supportive polices, regulations, and investments. The concept of a 'policyscape' recognises a policy mix with actual spatial overlap of instruments, functional overlap of instruments in a cost-effectiveness space, and complementary spatial targeting of instruments as determined by business models (Barton et al., 2013), which can be evaluated and targeted for adjustment in order to enhance the sustainable and productive use of the landscape. Nutrient management takes place amid a policyscape which is often densely laden with public policies created in the past that are functionally weak, or instruments that are out of sync with contemporary nutrient management and thus inhibit implementation of new approaches (Carey et al., 2016; Selvaratnam et al., 2016). Policy instruments and financial incentives like subsidies or legislation to couple inorganic and organic soil ameliorants as described above for India and Ghana show emerging options for supporting a circular economy model within the waste sector through supportive policy. This section also presents examples from the developed world which have future application to the developing world. Available evidence shows that nutrient recovery from wastewater or reuse of organic waste and sludge for composting, the most common end use of treated sludge, are seldom profitable and hence require supportive regulations and public subsidies at par with bioenergy economy and chemical fertilisers (Hanjra et al., 2015b; Mininni et al., 2015) to recognise the environmental and public health value of such reuse (Grant et al., 2012; Hanjra et al., 2015a).

Nutrient recovery and reuse must transition from informal to formal, allowing planned reuse directed by supportive policy. Wastewater irrigation, for example, is globally the primary resource recovery channel. But it most frequently takes place without sufficient (or any) treatment within the informal agricultural/irrigation sector. To sustain the benefits of this practice, it is important to safeguard public health. This will only be possible if the sector receives official recognition and capacity development, such as low-cost options for safe irrigation as promoted by the World Health Organisation (WHO, 2006) and its recent efforts for sanitation safety planning (SSP Manual) (Gordon et al., 2015). In fact, there are few policy instruments in the nutrient space. Similarly, formal quality standards for waste-based compost are needed to certify the safety of such products to enhance public confidence in reuse.

Urbanising landscapes are facing significant resource relocation processes linked to centralised food flows and limited resource recovery for reuse. To boost nutrient recovery from municipal solid waste (organic waste or food waste), technical solutions must be combined with market analysis and business models (Ganesapillai et al., 2015; Hanjra et al., 2015b). The same applies to nutrient recovery from wastewater – although wastewater treatment was and is first of all a 'social business model', with a strong economic justification and returns on investments through safeguarding public health and the environment. A second (reuse-based) value proposition in support of cost recovery can however be a significant advantage from the financial perspective, also to encourage private sector engagement.

There are, for example, increasing numbers of new technologies targeting phosphorous recovery and business approaches for market uptake at scale in the developed world (Mayer et al., 2016). At the current phosphorous price in the global market, the recovered phosphorous offers only a minor revenue stream next to the service of removing it. In fact, many resource recovery business models are hybrids with revenue streams comprising sales of technology/patent or operational service charges; with subsidies based on social or environmental benefits, building on the total value of P recovery (Mayer et al., 2016; Roy, 2017).

The synergistic integration of urban infrastructural networks with the physical, ecological, socio-cultural, and economic systems of cities has emerged as a powerful strategy for innovative urban design and planning in developed countries. Where cities have traditionally landfilled wastes, they are now exploring opportunities to position sanitation and wastewater systems as drivers of urban form, recreational and cultural expression, and economic vitality for greater social value (Figure 23.7). Thus, there is a push in 'landscape' to make infrastructures multifunctional rather than be isolated or inaccessible monoliths (Zeunert, 2017).

Above all, cities seek to transform sanitation and wastewater management from systems of subsidised civic management to significant cultural resources that echo stewardship through recognised qualitative *and* quantitative amenity value (Murray and Buckley, 2010; Hanjra et al., 2015b). In developing countries, where these systems are already often exposed and conventional approaches to waste management

Figure 23.7 The planning and design of cities is evolving to re-vision traditionally 'invisible' infrastructure into cultural resources. Buffalo Bayou Park in Houston, Texas, USA (top left) converts riverbanks below highways into recreational trails, which improves public access to the water front and creates spatial benefits like better habitat and amenity value. The Brightwater Treatment Plant in Washington (bottom left) integrates wastewater treatment with a regional public open space amenity. Also, Qunli Stormwater Park in China (right) demonstrates how 'new' cities are positioning urban stormwater management as a driver of urban form. New design thinking is needed for innovating how the food-waste-nutrient loop can be leveraged to enhance the productive and functional resilience of landscapes.

Source: Tom Fox, Hargreaves Associates USA, Kongjian Yu respectively

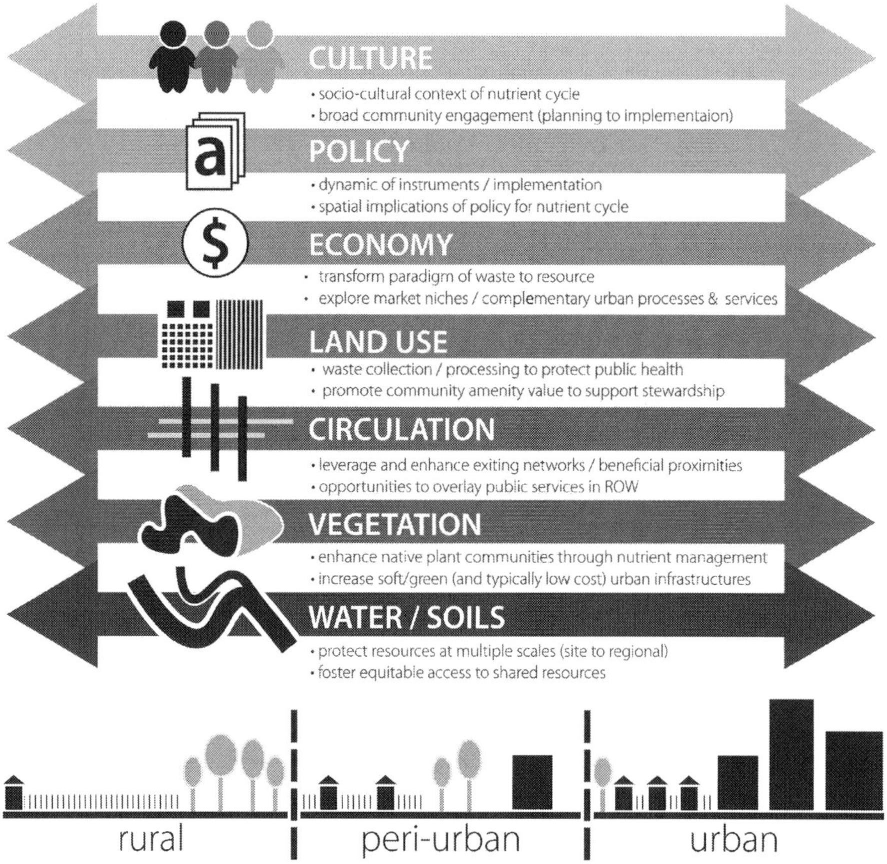

Figure 23.8 Successfully closing the nutrient management loop will require engaging a broad range of systems, as well as an understanding of their relationship to landscape from across the rural/peri-urban/urban transect.

Source: Mary Lydecker

are challenged, there is enormous opportunity for design that promotes closed loop systems, at both the regional and local scale. Re-visioning municipal waste services as process-driven resource streams aims at preserving and enhancing the inherent value, inspiration, and beauty of indigenous lands and landscape systems, while integrating with the functional and routine activities of communities. Figure 23.8 illustrates the complex relationship of these systems – from physical resources to the policies and cultural features that shape those resources. The successful shift towards 'closed loop' nutrient management systems will require engagement with all of these forces, appreciating their collective power to shape the landscape.

Potential landscape planning, design, and management strategies in support of closed-loop waste recovery for nutrient reuse include:

- **Agri-food systems**: Move beyond landscape retention systems for nutrient management, and design *site-specific*, multi-species food production systems that can completely utilise a wide array of nutrients in wastewater, fecal sludge, and organic waste (de Zeeuw and Drechsel, 2015; Minhas et al., 2015).

- **Spatial planning tools**: Develop specific spatial planning tools for linking urban areas with peri-urban and rural landscapes for efficient resource recycling aimed at sustainable development and a better quality of life in rural areas, by supporting integration of land use, economic, agricultural, and landscape planning (Tassinari et al., 2013).
- **Design**: Promote sustainable solid waste and wastewater reuse systems, including improved site-scale infrastructure such as toilets, city-scale sanitation systems to reuse nutrients and wastewater, and regional-scale planning and design strategies that close resource loops and transition towards a circular economy (Guest et al., 2009).
- **Construction**: Explore opportunities for green/soft infrastructure instead of technical/hard infrastructure. Soft systems engage the local ecology, require less capital investment and are less likely to break down or require high-skilled operation. Collectively, these characteristics establish landscapes that are resilient in the context of evolving site and global conditions (Chiew et al., 2015; Sussams et al., 2015).
- **Design for reuse**: Build reuse-oriented sanitation infrastructure, designing sustainable wastewater reuse systems (Murray and Buckley, 2010), build buffer zones around drains carrying wastewater to safeguard against floods and serve as sites for nutrient recovery and reuse in local vegetable production, to cut emissions and food miles (Hanjra et al., 2015).
- **Education**: Invest in the education of graduates charged with designing landscape futures, and capacity development for farmers engaged in wastewater reuse (Otoo and Drechsel, 2018). Engage communities and opinion leaders/teachers in sanitation promotion throughout the design and implementation process (Das, 2015; Crocker et al., 2016). A sustainable landscape is achieved only through the commitment and long-term stewardship of the local and regional users.
- **Resource policy**: Invest in multiple element, and multiple resource approach to resource recovery and reuse at scale (Mayer et al., 2016) for supporting both environmental objectives such as eutrophication control and social goals such as food security (Hanjra et al., 2017).

Acknowledgements

This chapter was financially supported by the International Water Management Institute (IWMI), funded through the CGIAR Research Program on Water, Land and Ecosystems (WLE). We appreciate the helpful comments provided by several reviewers.

Key texts

de Zeeuw, H., & Drechsel, P (eds.) (2015) *Cities and agriculture: Developing resilient urban food systems*. Oxon, UK: Routledge – Earthscan.

Hanjra, M.A., Blackwell, J., Carr, G., Zhang, F., & Jackson, T.M. (2012) Wastewater irrigation and environmental health: Implications for water governance and public policy. *International Journal of Hygiene and Environmental Health*, 215, 255–269.

Hoornweg, D., Bhada-Tata, P., & Kennedy, C. (2013) Waste production must peak this century. *Nature*, 502, 615–617.

Kennedy, C., Pincetl, S., & Bunje, P. (2011) The study of urban metabolism and its applications to urban planning and design. *Environmental Pollution*, 159, 1965–1973.

Mayer, B.K., Baker, L.A., Boyer, T.H., Drechsel, P., Gifford, M., Hanjra, M.A., Parameswaran, P., Stoltzfus, J., Westerhoff, P., & Rittmann, B.E. (2016) Total value of phosphorus recovery. *Environmental Science and Technology*, 10.1021/acs.est.6b01239.

Murray, A., & Buckley, C. (2010) Designing reuse-oriented sanitation infrastructure: the design for service planning approach: Earthscan, London. In: Drechsel, P., Scott, C.A., Raschid-Sally, L., Redwood, M., & Bahri, A. (eds.),

Wastewater irrigation and health: Assessing and mitigation risks in low-income countries. London, UK: Earthscan-IDRC-IWMI, pp. 303–318.
Otoo, M., & Drechsel, P. (eds.) (2018) *Resource recovery from waste: Business models for energy, nutrients and water reuse in low- and middle-income countries*: Routledge, USA.
Redwood, M. (ed.) (2011) *Agriculture in urban planning: Generating livelihoods and food security*. International Development Research Centre (IDRC) Ottawa, Canada and Earthscan UK and USA.
Roy, R.N., Misra, R.V., Lesschen, J.P., & Smaling, E.M. (2003) *Assessment of soil nutrient balance: Approaches and methodologies*. Rome: FAO.
Zabaleta, I., & Rodic, L. (2015) Recovery of essential nutrients from municipal solid waste – Impact of waste management infrastructure and governance aspects. *Waste Management*, 44, 178–187.

Further reading

ADB. 2014. *From toilets to rivers: Experiences, new opportunities and innovative solutions*. Manilla, The Philippines: Asian Development Bank.
Alvarenga, P., Mourinha, C., Farto, M., Santos, T., Palma, P., Sengo, J., Morais, M.-C., & Cunha-Queda, C. (2015) Sewage sludge, compost and other representative organic wastes as agricultural soil amendments: Benefits versus limiting factors. *Waste Management*, 40, 44–52.
Amir, E., Hophmayer-Tokich, S., & Kurnani, T. (2015) Socio-economic considerations of converting food waste into biogas on a household level in Indonesia: The case of the city of Bandung. *Recycling*, 1, 61.
Anand, C., & Apul, D.S. (2011) Economic and environmental analysis of standard, high efficiency, rainwater flushed, and composting toilets. *Journal of Environmental Management*, 92, 419–428.
Andersson, E. (2015) Turning waste into value: Using human urine to enrich soils for sustainable food production in Uganda. *Journal of Cleaner Production*, 96, 290–298.
Bai, X. (2016) Eight energy and material flow characteristics of urban ecosystems. *Ambio*, 1–12.
Barton, D.N., Blumentrath, S., & Rusch, G. (2013) Policyscape – a spatially explicit evaluation of voluntary conservation in a policy mix for biodiversity conservation in Norway. *Society and Natural Resources*, 26, 1185–1201.
Belevi, H. (2002) Material flow analysis as a strategic planning tool for regional waste water and solid waste management. Proceedings of the workshop "Globale Zukunft: Kreislaufwirtschaftskonzepte im kommunalen Abwasser- und Fäkalienmanagement". Munich, 13–15 May, 2002.
Biggs, E.M., Bruce, E., Boruff, B., Duncan, J.M.A., Horsley, J., Pauli, N., McNeill, K., Neef, A., Van Ogtrop, F., Curnow, J., Haworth, B., Duce, S., & Imanari, Y. (2015) Sustainable development and the water – energy – food nexus: A perspective on livelihoods. *Environmental Science and Policy*, 54, 389–397.
Billen, G., Barles, S., Garnier, J., Rouillard, J., & Benoit, P. (2009) The food-print of Paris: Long-term reconstruction of the nitrogen flows imported into the city from its rural hinterland. *Regional Environmental Change*, 9, 13–24.
Binz, C., Harris-Lovett, S., Kiparsky, M., Sedlak, D.L., & Truffer, B. (2016) The thorny road to technology legitimation – institutional work for potable water reuse in California. *Technological Forecasting and Social Change*, 103, 249–263.
Bren d'Amour, C., Reitsma, F., Baiocchi, G., Barthel, S., Güneralp, B., Erb, K.-H., Haberl, H., Creutzig, F., & Seto, K.C. (2017) Future urban land expansion and implications for global croplands. *Proceedings of the National Academy of Sciences*, 114, 8939–8944.
Cairns-Smith, S., Hill, H., & Nazarenko, E. (2014) *Urban sanitation: Why a portfolio of solutions is needed*. Working Paper. The Boston Consulting Group. www.bcg.com/documents/file178928.pdf (accessed 30 November 2016).
Capps, K.A., Bentsen, C.N., & Ramírez, A. (2016) Poverty, urbanization, and environmental degradation: Urban streams in the developing world. *Freshwater Science*, 35, 429–435.
Carey, D.E., Yang, Y., McNamara, P.J., & Mayer, B.K. (2016) Recovery of agricultural nutrients from biorefineries. *Bioresource Technology*, 215, 186–198.
Chen, D., Hu, M., Guo, Y., & Dahlgren, R.A. (2015) Reconstructing historical changes in phosphorus inputs to rivers from point and nonpoint sources in a rapidly developing watershed in eastern China, 1980–2010. *Science of The Total Environment*, 533, 196–204.

Chiew, Y.L., Spångberg, J., Baky, A., Hansson, P.-A., & Jönsson, H. (2015) Environmental impact of recycling digested food waste as a fertilizer in agriculture – a case study. *Resources, Conservation and Recycling*, 95, 1–14.

Christie, P., Easson, D.L., Picton, J.R., & Love, S.C.P. (2001) Agronomic value of alkaline-stabilized sewage biosolids for spring barley. *Agronomy Journal*, 93, 144–151.

Cofie, O., Kranjac-Berisavljevic, G., & Drechsel, P. (2005) The use of human waste for peri-urban agriculture in Northern Ghana *Renewable Agriculture and Food Systems*, 20, 73–80.

Cordell, D., & White, S. (2015) Tracking phosphorus security: Indicators of phosphorus vulnerability in the global food system. *Food Security*, 7, 337–350.

Cornejo, P.K., Zhang, Q., & Mihelcic, J.R. (2013) Quantifying benefits of resource recovery from sanitation provision in a developing world setting. *Journal of Environmental Management*, 131, 7–15.

Craswell, E., Grote, U., Henao, J., & Vlek, P. (2004) *Nutrient flows in agricultural production and international trade: Ecological and policy issues*. ZEF Discussion Papers on Development Policy No. 78. Bonn, Germany: ZEF (Center for Development Research, University of Bonn). 62 pp.

Crocker, J., Geremew, A., Atalie, F., Yetie, M., & Bartram, J. (2016) Teachers and sanitation promotion: An assessment of community-led total sanitation in Ethiopia. *Environmental Science and Technology*, 50, 6517–6525.

Damerau, K., Patt, A.G., & van Vliet, O.P.R. (2016) Water saving potentials and possible trade-offs for future food and energy supply. *Global Environmental Change*, 39, 15–25.

Das, P. (2015) The urban sanitation conundrum: What can community-managed programmes in India unravel? *Environment and Urbanization*, 27, 505–524.

Diener, S., Semiyaga, S., Niwagaba, C.B., Muspratt, A.M., Gning, J.B., Mbéguéré, M., Ennin, J.E., Zurbrugg, C., & Strande, L. (2014) A value proposition: Resource recovery from faecal sludge – can it be the driver for improved sanitation? *Resources, Conservation and Recycling*, 88, 32–38.

Dinarès, M. (2014) Urban metabolism: A review of recent literature on the subject. *Inici – Documents d'Anàlisi Geogràfica*, 60, 551–571.

Drechsel, P., Graefe, S., & Fink, M. (2007) *Rural-urban food, nutrient and virtual water flows in selected West African cities*. Colombo, Sri Lanka: International Water Management Institute (IWMI) 29p. (IWMI Research Report 115).

Drechsel, P., & Hanjra, M.A. (2016) Green opportunities for urban sanitation challenges through energy, water and nutrient recovery. In: Dodds, F., & Bartram, J. (eds.), *The water, food, energy and climate nexus: Challenges and an agenda for action*. London and New York: Earthscan from Routledge.

Drechsel, P., Cofie, O., & Danso, G. (2010) Closing the rural-urban food and nutrient loops in West Africa: A reality check. *Urban Agriculture Magazine*, 23, 8–10.

Drechsel, P., Scott, C.A., Raschid-Sally, Redwood, M., & Bahri, A. (eds.) (2010) *Wastewater irrigation and health: Assessing and mitigation risks in low-income countries. Earthscan-IDRC-IWMI, UK.*

Duchin, F. (2016) A global case-study framework applied to water supply and sanitation. *Journal of Industrial Ecology*, 20, 387–395.

Elser, J.J., Bracken, M.E.S., Cleland, E.E., Gruner, D.S., Harpole, W.S., Hillebrand, H., Ngai, J.T., Seabloom, E.W., Shurin, J.B., & Smith, J.E. (2007) Global analysis of nitrogen and phosphorus limitation of primary producers in freshwater, marine and terrestrial ecosystems. *Ecology Letters*, 10, 1135–1142.

Erni, M., Bader, H.P., Drechsel, P., Scheidegger, R., Zurbrügg, C., & Kipfer, R. (2011) Urban water and nutrient flows in Kumasi, Ghana. *Urban Water Journal*, 8, 135–153.

Erni, M., Drechsel, P., Bader, H.P., Scheidegger, R., Zurbruegg, C., & Kipfer, R. (2010) Bad for the environment, good for the farmer? Urban sanitation and nutrient flows. *Irrigation and Drainage Systems*, 24, 113–125.

Fernando, S., Semasinghe, C., Jayathilake, N., Wijayamunie, R., Wickramasinghe, N., & Dissanayake, S. (2016) *City region food system situational analysis*. Colombo: FAO – Food for the Cities Programme. Available from: www.fao.org/3/a-bl821e.pdf?galog=no [Accessed 7 February 2017].

Fuhrimann, S., Pham-Duc, P., Cissé, G., Tram, N.T., Thu Ha, H., Dung, D.T., Ngoc, P., Nguyen-Viet, H., Anh Vuong, T., Utzinger, J., Schindler, C., & Winkler, M.S. (2016) Microbial contamination along the main open wastewater and storm water channel of Hanoi, Vietnam, and potential health risks for urban farmers. *Science of The Total Environment*, 566–567, 1014–1022.

Ganesapillai, M., Simha, P., & Zabaniotou, A. (2015) Closed-loop fertility cycle: Realizing sustainability in sanitation and agricultural production through the design and implementation of nutrient recovery systems for human urine. *Sustainable Production and Consumption*, 4, 36–46.

Gerland, P., Raftery, A.E., Ševčíková, H., Li, N., Gu, D., Spoorenberg, T., Alkema, L., Fosdick, B.K., Chunn, J., Lalic, N., Bay, G., Buettner, T., Heilig, G.K., & Wilmoth, J. (2014) World population stabilization unlikely this century. *Science*, 346, 234–237.

Gordon, B., Jackson, D., Medlicott, K., Winkler, M., & Stenström, T. (2015) *Sanitation safety planning: Manual for safe use and disposal of wastewater, greywater and excreta*. Geneva: World Health Organization.

Gould, R.K., Ardoin, N.M., Biggar, M., Cravens, A.E., & Wojcik, D. (2016) Environmental behavior's dirty secret: The prevalence of waste management in discussions of environmental concern and action. *Environmental Management*, 58, 268–282.

Grant, S.B., Saphores, J.-D., Feldman, D.L., Hamilton, A.J., Fletcher, T.D., Cook, P.L.M., Stewardson, M., Sanders, B.F., Levin, L.A., Ambrose, R.F., Deletic, A., Brown, R., Jiang, S.C., Rosso, D., Cooper, W.J., & Marusic, I. (2012) Taking the "Waste" out of "Wastewater" for human water security and ecosystem sustainability. *Science*, 337, 681–686.

Guest, J.S., Skerlos, S.J., Barnard, J.L., Beck, M.B., Daigger, G.T., Hilger, H., Jackson, S.J., Karvazy, K., Kelly, L., Macpherson, L., Mihelcic, J.R., Pramanik, A., Raskin, L., Van Loosdrecht, M.C.M., Yeh, D., & Love, N.G. (2009) A new planning and design paradigm to achieve sustainable resource recovery from wastewater. *Environmental Science and Technology*, 43, 6126–6130.

Hanjra, M.A., Drechsel, P., Mateo-Sagasta, J., Otoo, M., & Hernandez-Sancho, F. (2015a) Assessing the finance and economics of resource recovery and reuse solutions across scales, chapter 7. In: Drechsel, P., Qadir, M., & Wichelns, D. (eds.), *Wastewater: economic asset in an urbanizing world*. Dordrecht, The Netherlands, Heidelberg, New York and London: Springer, on behalf of IWMI/ CGIAR/ UNU NWEH.

Hanjra, M.A., Noble, A., Langan, S., & Lautze, J. (2017) Feeding the 10 billion within the sustainable development goals framework, chapter 2. In: Gordon, I.J., Prins, H.H.T., & Squire, G.R. (eds.), *Food production and nature conservation: Conflicts and solutions*. London and New York: Earthscan, Routledge.

Hanjra, M.A., Drechsel, P., Wichelns, D., & Qadir, M. (2015b) Transforming urban wastewater into an economic asset: Opportunities and challenges, chapter 14. In: Drechsel, P., Qadir, M., & Wichelns, D. (eds.), *Wastewater: Economic asset in an urbanizing world*. Dordrecht, The Netherlands, Heidelberg, New York and London: Springer, on behalf of IWMI/ CGIAR/ UNU INWEH.

Heimersson, S., Svanström, M., Laera, G., & Peters, G. (2016) Life cycle inventory practices for major nitrogen, phosphorus and carbon flows in wastewater and sludge management systems. *The International Journal of Life Cycle Assessment*, 21, 1197–1212.

Hoornweg, D., Hosseini, M., Kennedy, C., & Behdadi, A. (2016) An urban approach to planetary boundaries. *Ambio*, 45, 567–580.

Hutton, G., & Varughese, M. (2016) *The costs of meeting the 2030 sustainable development goal targets on drinking water, sanitation, and hygiene*. Washington, DC: World Bank/ Water and Sanitation Program (WSP).

IFC. (2015) *Curitiba solid waste management*. Washington, DC: International Finance Corporation (IFC).

Jalilov, S.-M., Amer, S., & Ward, F. (2013) Water, food, and energy security: An elusive search for balance in Central Asia. *Water Resources Management*, 27, 3959–3979.

Jiang, S., & Yuan, Z. (2015) Phosphorus flow patterns in the Chaohu Watershed from 1978 to 2012. *Environmental Science and Technology*, 49, 13973–13982.

Jiménez, B. (2005) Treatment technology and standards for agricultural wastewater reuse: A case study in Mexico. *Irrigation and Drainage*, 54, S23–S33.

Karg, H., Akoto-Danso, E., & Drechsel, P. (2015) Examining food sources in the city of Tamale, Ghana. *Urban Agriculture Magazine*, 29, 38–40.

Karg, H., Drechsel, P., Akoto-Danso, E.K., Glaser, R., Nyarko, G., & Buerkert, A. (2016) Foodsheds and city region food systems in two West African cities. *Sustainability*, 8, 1175.

Keeling, A. (2005) Urban waste sinks as a natural resource: The case of the Fraser river. *Urban History Review*, 34, 58–70.

Kennedy, C.A., Stewart, I., Facchini, A., Cersosimo, I., Mele, R., Chen, B., Uda, M., Kansal, A., Chiu, A., Kim, K.-G., Dubeux, C., Lebre La Rovere, E., Cunha, B., Pincetl, S., Keirstead, J., Barles, S., et al. (2015) Energy and material flows of megacities. *Proceedings of the National Academy of Sciences*, 112, 5985–5990.

Khan, S., & Hanjra, M.A. (2009) Footprints of water and energy inputs in food production – Global perspectives. *Food Policy*, 34, 130–140.

Khan, S., Khan, M.A., Hanjra, M.A., & Mu, J. (2009) Pathways to reduce the environmental footprints of water and energy inputs in food production. *Food Policy*, 34, 141–149.

Kuttuva, P., Lele, S., & Mendez, G.V. (2017) Decentralized wastewater systems in Bengaluru, India: Success or failure? *Water Economics and Policy*, 1650043.

Lazarova, V., Choo, K.-H., & Cornel, P. (2012) Meeting the challenges of the water-energy nexus: The role of reuse and wastewater treatment. *Water21*, 14(2), 2–17.

Lerner, A.M., & Eakin, H. (2011) An obsolete dichotomy? Rethinking the rural – urban interface in terms of food security and production in the global south. *The Geographical Journal*, 177, 311–320.

Liu, C., Zou, C., Wang, Q., Hayashi, Y., & Yasunari, T. (2014) Impact assessment of human diet changes with rapid urbanization on regional nitrogen and phosphorus flows – a case study of the megacity Shanghai. *Environmental Science and Pollution Research*, 21, 1905–1914.

Lohri, C.R., Camenzind, E.J., & Zurbrügg, C. (2014) Financial sustainability in municipal solid waste management – Costs and revenues in Bahir Dar, Ethiopia. *Waste Management*, 34, 542–552.

Makaya, J.M., Kaplon, J., Fremy, C., Barro, N., Aho, S., Pothier, P., Belliot, G., & Traoré, A.S. (2015) Norovirus and rotavirus survival in urine collected from a public ecological sanitation system in Ouagadougou, Burkina Faso. *Food and Environmental Virology*, 7, 41–48.

Maroušek, J., Hašková, S., Zeman, R., Žák, J., Vaníčková, R., Maroušková, A., Váchal, J., & Myšková, K. (2016) Polemics on ethical aspects in the compost business. *Science and Engineering Ethics*, 22, 581–590.

Matter, A., Ahsan, M., Marbach, M., & Zurbrügg, C. (2015) Impacts of policy and market incentives for solid waste recycling in Dhaka, Bangladesh. *Waste Management*, 39, 321–328.

Metson, G.S., Iwaniec, D.M., Baker, L.A., Bennett, E.M., Childers, D.L., Cordell, D., Grimm, N.B., Grove, J.M., Nidzgorski, D.A., & White, S. (2015) Urban phosphorus sustainability: Systemically incorporating social, ecological, and technological factors into phosphorus flow analysis. *Environmental Science and Policy*, 47, 1–11.

Minhas, P.S., Khajanchi, L., Yadav, R.K., Dubey, S.K., & Chaturvedi, R.K. (2015) Long term impact of waste water irrigation and nutrient rates: I. Performance, sustainability and produce quality of peri urban cropping systems. *Agricultural Water Management*, 156, 100–109.

Mininni, G., Blanch, A.R., Lucena, F., & Berselli, S. (2015) EU policy on sewage sludge utilization and perspectives on new approaches of sludge management. *Environmental Science and Pollution Research International*, 22, 7361–7374.

Mo, W., & Zhang, Q. (2013) Energy – nutrients – water nexus: Integrated resource recovery in municipal wastewater treatment plants. *Journal of Environmental Management*, 127, 255–267.

Molden, D. (2007) Water responses to urbanization. *Paddy and Water Environment*, 5, 207–209.

Montangero, A., Cau, L.N., Anh, N.V., Tuan, V.D., Nga, P.T., & Belevi, H. (2007) Optimising water and phosphorus management in the urban environmental sanitation system of Hanoi, Vietnam. *Science of The Total Environment*, 384, 55–66.

Morée, A.L., Beusen, A.H.W., Bouwman, A.F., & Willems, W.J. (2013) Exploring global nitrogen and phosphorus flows in urban wastes during the twentieth century. *Global Biogeochemical Cycles*, 27, 836–846.

Muzondi, L. (2014) Urbanization and service delivery planning: Analysis of water and sanitation management systems in the city of Harare, Zimbabwe. *Mediterranean Journal of Social Sciences*, 5, 2909–2915.

Nesme, T., & Withers, P.J.A. (2016) Sustainable strategies towards a phosphorus circular economy. *Nutrient Cycling in Agroecosystems*, 104, 259–264.

Nikiema, J., Cofie, O., & Impraim, R. (2014) *Technological options for safe resource recovery from fecal sludge*. Colombo, Sri Lanka: IWMI/WLE, 47p. (Resource Recovery and Reuse Series 2). CGIAR Research Program on Water, Land and Ecosystems (WLE). Online at www.iwmi.cgiar.org/Publications/wle/rrr/resource_recovery_and_reuse-series_2.pdf (accessed 11 August 2016)

ODI (2015) *Projecting progress reaching the SDGs by 2030*. London: Overseas Development Institute (ODI).

Otoo, M., Drechsel, P., & Hanjra, M.A. (2015) Nutrient recovery: Business models and economic approaches for nutrient recovery from wastewater and faecal sludge, chapter 13. In: Drechsel, P., Qadir, M., & Wichelns, D. (eds.), *Wastewater: Economic asset in an urbanizing world*. Dordrecht, The Netherlands, Heidelberg, New York and London: Springer, on behalf of IWMI/ CGIAR/ UNU INWEH.

Parthan, S.R., Milke, M.W., Wilson, D.C., & Cocks, J.H. (2012) Cost estimation for solid waste management in industrialising regions – precedents, problems and prospects. *Waste Management*, 32, 584–594.
Põldnurk, J. (2015) Optimisation of the economic, environmental and administrative efficiency of the municipal waste management model in rural areas. *Resources, Conservation and Recycling*, 97, 55–65.
Qadir, M., Sharma, B.R., Bruggeman, A., Choukr-Allah, R., & Karajeh, F. (2007) Non-conventional water resources and opportunities for water augmentation to achieve food security in water scarce countries. *Agricultural Water Management*, 87, 2–22.
Rao, K., Hanjra, M.A., Drechsel, P., & Danso, G. (2015) Business models and economic approaches supporting water reuse. In: Drechsel, P., Qadir, M., & Wichelns, D. (eds.), *Wastewater: Economic asset in an urbanizing world*. Dordrecht, The Netherlands, Heidelberg, New York and London: Springer, on behalf of IWMI/ CGIAR/ UNU-INWEH.
Rao, K.C., Otoo, M., Drechsel, P., & Hanjra, M.A. (2017) Resource recovery and reuse as an incentive for a more viable sanitation service chain *Water Alternatives (Themed Issue: Water for Sustainable Growth)*, 10, 493–512.
Raschid-Sally, L., & Jayakody, P. (2008) *Drivers and characteristics of wastewater agriculture in developing countries: results from a global assessment*. Colombo, Sri Lanka: International Water Management Institute (IWMI). 29p. (IWMI Research Report 127).
Reznik, A., Feinerman, E., Finkelshtain, I., Fisher, F., Huber-Lee, A., Joyce, B., & Kan, I. (2017) Economic implications of agricultural reuse of treated wastewater in Israel: A statewide long-term perspective. *Ecological Economics*, 135, 222–233.
Rietveld, L.C., Siri, J.G., Chakravarty, I., Arsénio, A.M., Biswas, R., & Chatterjee, A. (2016) Improving health in cities through systems approaches for urban water management. *Environmental Health*, 15, 151–160.
Roy, E.D. (2017) Phosphorus recovery and recycling with ecological engineering: A review. *Ecological Engineering*, 98, 213–227.
Roy, R.N., Misra, R.V., Lesschen, J.P., & Smaling, E.M. (2003) *Assessment of soil nutrient balance: Approaches and methodologies*. Rome: FAO.
Sadoff, C.W., Hall, J.W., Grey, D., Aerts, J.C.J.H., Ait-Kadi, M., Brown, C., Cox, A., Dadson, S., Garrick, D., Kelman, J., McCornick, P., Ringler, C., Rosegrant, M., Whittington, D., & Wiberg, D. (2015) *Securing water, sustaining growth: Report of the GWP/OECD task force on water security and sustainable growth*. Oxford: University of Oxford, 180pp.
Satterthwaite, D., McGranahan, G., & Tacoli, C. (2010) Urbanization and its implications for food and farming. *Philosophical Transactions of the Royal Society of London B: Biological Sciences*, 365, 2809–2820.
Sayer, J., Sunderland, T., Ghazoul, J., Pfund, J.-L., Sheil, D., Meijaard, E., Venter, M., Boedhihartono, A.K., Day, M., Garcia, C., van Oosten, C., & Buck, L.E. (2013) Ten principles for a landscape approach to reconciling agriculture, conservation, and other competing land uses. *Proceedings of the National Academy of Sciences*, 110, 8349–8356.
Schmidt, S., Magigi, W., & Godfrey, B. (2015) The organization of urban agriculture: Farmer associations and urbanization in Tanzania. *Cities*, 42, Part B, 153–159.
Scott, C.A., Pierce, S.A., Pasqualetti, M.J., Jones, A.L., Montz, B.E., & Hoover, J.H. (2011) Policy and institutional dimensions of the water-energy nexus. *Energy Policy*, 39, 6622–6630.
Scott, C., Drechsel, P., Raschid-Sally, L., Bahri, A., Mara, D., Redwood, M., & Jimenez, B. (2010) Wastewater irrigation and health: Challenges and outlook for mitigating risks in low-income countries. In: Drechsel, P. Scott, C.A., Raschid-Sally, L., Redwood, M., & Bahri, A. (eds.), *Wastewater irrigation and health: Assessing and mitigating risk in low-income countries*. London: Earthscan.
Scott, C.A., Zarazua, J.A., & Levine, G. (2000) *Urban wastewater reuse for crop production in the water-short Guanajuato River Basin, Mexico*, Research Report 41. Colombo: International Water Management Institute.
Selvaratnam, T., Henkanatte-Gedera, S.M., Muppaneni, T., Nirmalakhandan, N., Deng, S., & Lammers, P.J. (2016) Maximizing recovery of energy and nutrients from urban wastewaters. *Energy*, 104, 16–23.
Simmons, R., Qadir, M., & Drechsel, P. (2010) Farm-based measures for reducing human and environmental health risks from chemical constituents in wastewater. In: Drechsel, P. Scott, C.A., Raschid-Sally, L., Redwood, M., & Bahri, A. (eds.), *Wastewater irrigation and health: Assessing and mitigating risk in low-income countries*. London: Earthscan.
Starkl, M., Brunner, N., Amerasinghe, P., Mahesh, J., Kumar, D., Asolekar, S.R., Sonkamble, S., Ahmed, S., Wajihuddin, M., Pratyusha, A., & Sarah, S. (2015) Stakeholder views, financing and policy implications for reuse of wastewater for irrigation: A case from Hyderabad, India. *Water (Switzerland)*, 7, 300–328.
Sussams, L.W., Sheate, W.R., & Eales, R.P. (2015) Green infrastructure as a climate change adaptation policy intervention: Muddying the waters or clearing a path to a more secure future? *Journal of Environmental Management*, 147, 184–193.

Tal, A. (2016) Rethinking the sustainability of Israel's irrigation practices in the Drylands. *Water Research*, 90, 387–394.

Tassinari, P., Torreggiani, D., & Benni, S. (2013) Dealing with agriculture, environment and landscape in spatial planning: A discussion about the Italian case study. *Land Use Policy*, 30, 739–747.

Tilmans, S., Diaz-Hernandez, A., Nyman, E., & Davis, J. (2014) The potential for financing small-scale wastewater treatment through resource recovery: Experience from bocas del Toro, Panama. *Journal of Water Sanitation and Hygiene for Development*, 4, 449–459.

Truc, N.T.T., Sumalde, Z.M., Espaldon, M.V.O., Pacardo, E.P., Rapera, C.L., & Palis, F.G. (2012) Farmers' awareness and factors affecting adoption of rapid composting in Mekong Delta, Vietnam and central Luzon, Philippines. *Journal of Environmental Science and Management*, 15, 59–73.

UN. (2015) World population prospects: The 2015 revision, key findings and advance tables. ESA/P/WP.242. United Nations, Department of Economic and Social Affairs, Population Division. https://esa.un.org/unpd/wpp/Publications/Files/Key_Findings_WPP_2015.pdf (accessed 17 July 2016).

Verhougstraete, M.P., Martin, S.L., Kendall, A.D., Hyndman, D.W., & Rose, J.B. (2015) Linking fecal bacteria in rivers to landscape, geochemical, and hydrologic factors and sources at the basin scale. *Proceedings of the National Academy of Sciences*, 112, 10419–10424.

Viaene, J., Van Lancker, J., Vandecasteele, B., Willekens, K., Bijttebier, J., Ruysschaert, G., De Neve, S., & Reubens, B. (2016) Opportunities and barriers to on-farm composting and compost application: A case study from northwestern Europe. *Waste Management*, 48, 181–192.

Villarroel Walker, R., Beck, M.B., & Hall, J.W. (2012) Water – and nutrient and energy – systems in urbanizing watersheds. *Frontiers of Environmental Science and Engineering*, 6, 596–611.

Wang, X., McCarty, P.L., Liu, J., Ren, N.-Q., Lee, D.-J., Yu, H.-Q., Qian, Y., & Qu, J. (2015) Probabilistic evaluation of integrating resource recovery into wastewater treatment to improve environmental sustainability. *Proceedings of the National Academy of Sciences*, 112, 1630–1635.

WHO. 2006. *Guidelines for the sea use of wastewater, excreta and grey Water: Wastewater use in agriculture (Volume 2)*. [Online]. Geneva, Switzerland: World Health Organization.

Wolman, A. (1965) The metabolism of cities. *Scientific American*, 213, 178–190.

World Bank (2016) *Urban development. Urban development in Africa*. Washington, DC: World Bank. Available from: www.worldbank.org/en/topic/urbandevelopment/x/afr [Accessed 23 June 2016].

WRI. (2012) *Green vs. Gray infrastructure: When nature is better than concrete*. Washington, DC: World Resources Institute (WRI). www.wri.org/blog/2012/06/green-vs-gray-infrastructure-when-nature-better-concrete (accessed 20 March 2017).

Xian, C., Ouyang, Z., Lu, F., Xiao, Y., & Li, Y. (2016) Quantitative evaluation of reactive nitrogen emissions with urbanization: A case study in Beijing megacity, China. *Environmental Science and Pollution Research*, 1–13.

Yillia, P.T. (2016) Water-energy-food nexus: Framing the opportunities, challenges and synergies for implementing the SDGs. *Österreichische Wasser- und Abfallwirtschaft*, 68, 86–98.

Yoada, R.M., Chirawurah, D., & Adongo, P.B. (2014) Domestic waste disposal practice and perceptions of private sector waste management in urban Accra. *BMC Public Health*, 14, 1–10.

Zeunert, J. (2017) *Landscape architecture and environmental sustainability: Creating positive change through design*. London and New York: Bloomsbury.

Zhang, W., Li, X., Swaney, D.P., & Du, X. (2016) Does food demand and rapid urbanization growth accelerate regional nitrogen inputs? *Journal of Cleaner Production*, 112, Part 2, 1401–1409.

Zhang, Y. (2013) Urban metabolism: A review of research methodologies. *Environmental Pollution*, 178, 463–473.

Part V

Intellectual, political, and economic realms

24
Ethics of agricultural landscapes and food production

Isis Brook

Introduction

The most obvious area for ethical thinking about landscape and food is agriculture. Here, the ethical issues have typically revolved around food being equitably shared or at least relieving famine, being safe and accurately described, produced without oppression or harm to those working in its production, sensitive to land rights and ownership, and, latterly, protecting animal welfare and produced without harm to the environment – or at least sustainably, where harm is minimised and future potential production protected (Zimdahl, 2012). These are important issues for food ethics and will be covered in general terms. Concerns regarding agriculture and landscape as scenographic or place redolent, as opposed to simply the site of agricultural production, have featured in discussions of aesthetics (Brady, 2006; Carlson, 1985; Brook, 2013). However, an ethics of landscape and food suggests that producing food is also about producing meaningful landscapes, ecologies, and human communities. With the growing appreciation of *terroir* (see Chapter 36), local sourcing, slow food, *terra madre*, local distinctiveness (see Chapter 35), and landscape character as concepts, rather than just lived experience, there is a need to articulate an ethics of agroecological landscapes that could guide action and enhance appreciation.

Normative ethical frameworks

Normative ethics, the branch of ethics that is about how we should live our lives and the kind of things it is right or wrong to do, has traditionally been focused on placing constraints on our behaviour towards other humans (Sterba, 2001). The three major ethical theories of *virtue ethics*, *deontological ethics*, and *consequentialism* place a reasoning human being at their centre and, particularly the latter two, have extolled core principles as the means of adjudicating right from wrong (Mizzoni, 2009; Benn, 1997).

The most significant form of deontological ethics, Kantianism, has the categorical imperative: 'act only in such a way as you could will your act to be a universal law.' It focuses on one's duty to behave in particular ways, such as truth telling, and is connected to the idea of humans as bearers of rights that should never be overridden. The most significant form of consequentialism, utilitarianism, has the core principle: 'act always such that your actions bring about the greatest happiness (or diminishment of suffering) of the greatest number.' Its focus is on the outcomes of actions and bringing about the best consequences without necessarily having regard for particular individuals. Both of these approaches require the decision maker to adopt the position of a detached onlooker to see their action in a wider perspective and ensure that it is unbiased by, for example, personal attachment. With virtue ethics the focus is more on the character of the individual and the kinds of qualities they need to have to be a good person, for example: courageous, honest, and caring. Which qualities are most valued depends, to some extent, on the particular context of that person and the situation of their wider community.

We can see how these forms of ethics fare in a key food-related ethical problem situation: inequality and famine.

Feeding the world viewed from three ethical frameworks

A major ethical issue of food is its fair distribution. We could see the idea that everyone should have enough to eat to sustain life as a basic premise of a rights based view (Chapter 31). This is enshrined in Article 25 of the Universal Declaration of Human Rights which was proclaimed by the United Nations General Assembly in 1948. This states that:

> Everyone has the right to a standard of living adequate for the health and well-being of himself and of his family, including food, clothing, housing and medical care and necessary social services, and the right to security in the event of unemployment, sickness, disability, widowhood, old age or other lack of livelihood in circumstances beyond his control.

Part two of this article also speaks of mothers and children needing additional care. The gendered language is of course part of the norm for a document crafted in 1948 and the key driving force of the whole document (just after the horrors of the second world war) was the equality of peoples of different nations and creeds and respecting the dignity and freedom of the individual. Stating a right does not mean that it will be respected – as continuing poverty and recurring famines in many parts of the world shows (see Chapters 20, 21 and 22) – but it does place the signatories to the declaration and those they represented (that is, most of us) as having a duty to one another such that it would be ethically wrong to contribute to another's hunger or poverty. The assertion of a right implies a corresponding duty but for most deontological ethicists this can be interpreted in a strong or weak form. For example, I have a strong duty not to cause harm to another but a rather weaker duty (if I have one at all) to prevent everyone from being harmed. The focus of duty ethics is on whether my own act is right or wrong as judged by whether I could will that such an act should always be done.

Utilitarianism as an ethical theory does not recognise the concept of rights as such and places its moral weight on the principle of utility. The situation with famine would be that we all should act to bring about the diminishment of that suffering. In a seminal paper by the controversial philosopher Peter Singer 'Famine, Affluence, and Morality' (1971), he sets out the moral case for helping to prevent bad things happening if we can do so without sacrificing "anything of comparable moral importance" (2007: 614). Singer questions the status quo of public opinion; where it is a good thing to give to charity but no moral blame attaches to those who do not. For him, if we can prevent suffering we should do so and in affluent countries there are many things that we could easily sacrifice in order to help those in a famine situation. In the paper he also clearly articulates that, for the utilitarian, distance is morally irrelevant. We would not, he says, ignore the plight of a neighbour's drowning child when we could easily save them, so we should not ignore the plight of a child starving in another part of the world when we could so easily save them. Thus distance is irrelevant as is whether our action is one where we are the only person who can help or where we are just one of many who, together, can help. How much one should contribute will depend on what is needed and how many contribute, but a situation could arise where one should make sacrifices in order to contribute to famine relief to what Singer calls "the point of marginal utility" (2007: 615). This is where to give more would put us in a situation that would be worse than the situation of those we are trying to help.

This is a striking demand when compared to our usual understanding of ethics, but it does follow logically from the principle of utility.

Can virtue theory help us here? Can it present a way of thinking that does not make our actions bad if we do not make sacrifices to the point of marginal utility? One difference is that with virtue theory motives matter. What matters with regard to giving is that we demonstrate the virtue of generosity not out of a moral duty or a moral principle but just because it comes naturally to us, it is part of who we are. What matters with regard to the suffering of others is that we feel empathy for them and our actions arise from that quality of character. Virtue theory is also generally regarded as not imposing a universalising principle on us (Hursthouse, 1999; Darwell, 2003; Sandler, 2007). What matters is our familial and local connections and contributing to our own communities first and foremost. Also, it does not expect the same of everyone; the virtues are demonstrated by actions within our purview. A small child might demonstrate caring on the family farm by remembering to take a snack out to older siblings who are helping in the lambing shed, whereas the farmer might demonstrate caring by going out all night in a storm to bring a few lost sheep in to safety. As we will see later, with the examination of ecofeminism and the ethics of care, doing the right thing does not always come about through reasoning alone; feelings and emotional bonds can be relevant ethical guides rather than hindrances to ethical behaviour. Does this mean that virtue ethics has nothing to say about wider world issues such as famine? It does not seem as immediately applicable as utilitarianism but it is applicable in the sense of developing qualities of character in people and in communities; qualities such as a love of fairness and anger at injustice, gratitude, and avoidance of greed. In this sense virtue ethics is about transforming and developing the inclinations and attitudes of everyone such that inequalities would not be tolerated.

Is agriculture inherently ethically good?

One of the reasons that agriculture ethics, as a philosophical area, is relatively new is the sense – common to farmers – that their task is inherently good. Providing food for people is often seen by agriculturalists as an undeniably good thing and thus no further ethical theorising about it is necessary (Dundon, 2003: 427). When we dig beneath the surface of that 'good thing' there is a long history of agricultural work seen as ennobling: the labour of land and animal husbandry is seen as developing traits such as patience, humility, and general strength of character. This view is often linked with Thomas Jefferson (1743–1826) as a proponent of the morally improving nature of working the land (Holowchak, 2011) although his emphasis on the ethical, as opposed to the democratic and pragmatic value of agrarianism, can be questioned (Thompson and Kutach, 1990: 135). The view struck a chord – agrarianism instilled the virtues and farming communities were seen as peaceful and incorruptible in comparison with urbanites. Jefferson's appreciation of the rural life as morally improving builds on the same view from George Washington (1732–1799) and his view draws on a similar one from the Ancient Romans (Morrison, 2009). An eloquent proponent of this view in contemporary writing is Wendell Berry whose poetic/spiritual writings are steeped in the sense that virtues such as humility and respect for nature are instilled by working within the limits of, and communing with, the land (Berry, 1997, 2009). Such writing, and indeed most of what could be called agricultural ethics, is not explicitly engaging with specific normative theories. They are articulating the values of social and landscape structures of traditional farming that are in danger of being lost (Dundon, 2003: 435).

It seems clear that a form of virtue ethics is at play here. The critique of focusing on feeding your local community is usually made from a utilitarian perspective of needing to feed the world. From the mid 20th century 'Green Revolution', which also introduced industrialised and mechanised forms of agricultural production into developing countries (see Chapter 9), to the current controversies with genetically modified (GM or bioengineered) crops, the technological trump card has always been – *this will feed people who are currently starving and to do otherwise would be wrong*. However, we are now becoming aware of the problems that Western industrialised agricultural models are creating (Glaeser, 2011; Kimbell, 2002; Shiva, 2016; Chapters 16, 25, 31, 32 and 33).

Expanding ethical considerations to the land itself

A key problem with industrialised forms of modern agriculture is the mind-set that sees food production as something to be extracted from the land almost as if one is mining what is in the ground. This is mitigated or solved in some agroecological approaches (see Chapter 10). Agroecology attempts to encompass the scientific discipline of researching food production within an ecologically aware framework. It puts this together with a politically participative approach that honours indigenous knowledges and tackles practical, real-world, problems for the benefit of people not multinational corporations (Shiva, 2016; Gliessman, 2015; Levidow et al., 2014; Altieri, 1995). The extractive industrial approach is most extreme in ecosystem clearance, for example rainforests, where complex ecosystems are cleared and the land used for pasture until depleted, then more forests need to be destroyed (see Chapter 17). The 'destruct and move on' strategy is a different approach from making sensitive changes to a landscape to allow for crop production. In rural areas dominated by agribusiness the picture is less extreme than rainforest destruction, but the soil is treated as if it is an inert growing medium; synthetic fossil fuel based fertilisers are introduced for each round of cropping, pesticides and herbicides protect the crop at the expense of the wider environment, multiple passes with heavy machinery compact the soil and what was a complex living system eventually does become a lifeless growing medium. From an agroecology perspective what is needed is a shift in our thinking about land and to return to, or to newly develop, a sense of partnership with the landscape. New ways of thinking about ethics towards the land itself were developed in the 20th century, but, with one exception, they did not initially have much to say about farming.

With the rise of environmental ethics in the 1970s how we should behave was re-examined and in some cases ethical obligations broadened to include living entities and/or communities such as ecosystems or species (animal ethics will be addressed in a later section) (Callicot, 1989; Devall and Sessions, 1985; Rolston, 1989). This marked a radical shift away from humans – now viewed as the problem – towards nature and valuing nature particularly in its most natural state – the wilderness. The philosophical work was often focused on finding reasons to preserve wilderness areas that did not rely on any of the numerous use-value reasons we might have. It was seen as necessary to forge an ecocentric position in order to redress the problems caused by anthropocentric views: that see the human being as the only entity of importance. Growing awareness of how much damage humans had done to the Earth led to the message being framed as protecting nature from humans, particularly nature that had any kind of pedigree of pristineness (Elliot, 1982).

As the field developed, that radical stance was tempered by other thinking, such as William Cronon's influential 1995 paper 'The Trouble with Wilderness; or, Getting Back to the Wrong Kind of Nature'. Here he questioned the idea of wilderness being 'untouched'. These landscapes were changed by their indigenous populations so this Western environmental wilderness discourse was in fact engaged in a quasi-colonisation of this land to serve its own Romantic imaginings. The focus on 'wilderness' had also left other landscapes unprotected and outside of environmental concern. It didn't begin to address the issues of the environments where we do live, where we do source our food, and so on (Zeunert, 2013). The wilderness discourse was seen as driving us too far from being able to address how humans could live in a responsible way that took account of the environment. Early philosophical environmental thinking was often from the 'new world' of North America and Australia where the wilderness discourse made more sense, but in the 'old world' of European cultural landscapes, the world did not look quite like that stark division of wilderness and built environments (Artzen and Brady, 2008). Cultural landscapes were already valued for their historical synthesis of human and land relationships. Moreover, there was a long cultural history of nature writing such as found in the works of Wordsworth, Ruskin, Rousseau, and Clare that extolled the value of these farmed landscapes and the human work of their shaping and maintenance.

A precursor of the new philosophical environmental thinking about valuing the environment in a different way was that of the North American ecologist Aldo Leopold (1887–1948). His essay 'The Land Ethic', part of a collection of nature writings called *A Sand County Almanac*, was published in 1949. Leopold was writing for an audience that included farmers and those shaping the land within their own homesteads and he brings to this his background in wildlife ecology and conservation. He wants to see a broadening of the sympathies that we hold for other humans with the addition of the biotic community to our ethical considerations. He presents this as a code which says: "A thing is right when it tends to preserve the integrity, stability, and beauty of the biotic community. It is wrong when it tends otherwise" (Leopold, 1981: 224–225). In his writings he is drawing the reader's attention to the interlocking nature of the ecological community and the way we should be, as he says, not a conqueror, but a plain member and citizen of that community (ibid.: 204). What emerges here – radical for its time and place – is a critique of the use value or economic conception of land as purely a resource for humans. He certainly pointed out the long-term self-interest in behaving differently, and in many ways he anticipates ideas of sustainability and even regenerative design (see Chapter 16). However, in couching his approach as a land *ethic* he was entreating us to undergo a shift in attitude and way of being toward the land, to, as he says, develop an "ecological conscience" (ibid.: 207). He makes the case that we should extend the love we hold for our nearest and dearest to the biotic community. In this way the farmsteader who pollutes the river or causes erosion through aggressively logging the slopes or eradicates wolves, is seen as morally flawed rather than as an upstanding local citizen.

What is interesting about Leopold is not just how early he was making these pronouncements, but how he was blending a respect for wild nature with a practical farmsteading outlook. He talks of respectful management of land as well as preserving some land in a natural state.

Human environment synthesis and the local

The productionist paradigm of industrial, and particularly intensive, agriculture separates humans, as the consumers, from land as the provider. Traditional ethics posited that human needs and the reduction of human suffering were paramount and so could be seen as supportive of an exploitative approach to land use. Getting the maximum from it, by any means, became the right thing to do. Three major developments have called this approach into question. One is pragmatic and linked to better agricultural science recognising effects such as: compacted, degraded, and eroding soils; inefficiencies and declining yields; and pest problems inherent in agri-business style monocultures (Nawaz et al., 2013). The second is the political shift from assuming a benign market economy working with nation states to provide food, to seeing big business providing for shareholders and the globalisation of an agribusiness model as the problem (McMichael, 2006, 2015; Pretty, 2002; Shiva, 2016; Chapters 16, 31 and 32). The third is the growing public perception of nature as something to care about for its own sake, regardless of any benefit to humans. Thus we could see, for example, the protection of, or return of, hedgerows in the English landscape as serving all three impulses. A reduction in field size (and thereby large soil compacting machinery) brings about an increase in crop diversity and farmer autonomy; the hedgerows and field margins increase biodiversity with birds and small mammals assisting in pest control. By making a space for nature, richer synergistic relationships begin to build. The understanding in biodynamics of the farm as an organism was a very early expression of that working with nature from the local to the cosmic scale (Hurter, 2014; Chapter 10).

Those synergistic relationships include the farming community and those who live within these landscapes. A focus on local conditions and growing what works well in the area, both in terms of climate and cultural needs, creates landscapes with meaning and unique qualities. This is what helps to make one place different from another. The concept of place is connected to a large body of literature with its underlying values based argument of preserving or creating the characterful nature of places and resisting the forces of homogenisation that result in placelessness (Relph, 1976; Tuan, 1996; Beatley and Manning, 1997; Hough, 1990; Brook, 2012; Seddon, 1997; Cresswell, 2004). Large-scale agribusiness is seen as one of the drivers of placelessness and thus a return to the smaller scale local farm is seen as a way of protecting and building biodiversity and cultural diversity as well as employment opportunity and thus communities (Berry, 2009). The attractiveness of this can be seen in the rise of interest in local food, from the community supported agriculture schemes (Janssen, 2010), the 100-mile diet (Smith and MacKinnon, 2007) and rise of locavores (Gray, 2013) to the growing attention to and gourmand interest in *terroir* (Chapter 36), specialist food producers, and the Slow Food movement (Chapters 5, 6, 7, 31 and 35). This brings together an appreciation for local, caringly produced food and savouring the experience of eating as a socially, culturally, and politically nourishing activity (Lotti, 2010). The local goes beyond the landscape of traditional farms as can be seen by the rise of demand for allotments, community orchards and gardens (Chapter 30), and urban agriculture in a surprising range of spaces, including rooftop market gardens and integrated fish and vegetable aquaponic systems (Ladner, 2011; Chapters 12, 13 and 22). These developments point to a desire for engagement in growing, in landscapes, and with preparing and relishing our food (Chapters 37 and 39). This indicates caring about where things come from and how they are produced and consumed (Chapters 35 and 36). There is an increasing understanding that every decision and particularly every purchase is a political act where we endorse the production methods of the article we buy (Bray et al., 2011; Vermeir and Verbeke, 2006; Chapter 16).

The political and ethical import of an emphasis on the local and the shift from agribusiness is seen in a radical form in the 2007 Nyéléni declaration on food sovereignty (Chapter 31), which states:

> Food sovereignty is the right of peoples to healthy and culturally appropriate food produced through ecologically sound and sustainable methods, and their right to define their own food and agriculture systems [. . .] It offers a strategy to resist and dismantle the current corporate trade and food regime, and directions for food, farming, pastoral and fisheries systems determined by local producers and users. Food sovereignty prioritises local and national economies and markets and empowers peasant and family farmer-driven agriculture, artisanal-fishing, pastoralist-led grazing, and food production, distribution and consumption based on environmental, social and economic sustainability.

Embedded in the Nyéléni declaration is a rejection of the kind of bio-technological fixes that agribusiness proposes as a response to the perceived need to increase yields (or at least yield per unit of labour) (Shiva, 2000, 2016) and the trade deals that serve governments and business but not farmers or the Earth. In the rights language we can see a drive for fairness for all as a form of universalising, but one that attempts to be place specific by endorsing cultural diversity as well as ecological biodiversity.

New ethical approaches

Many of the ethical questions that involve landscape and food can over-stretch the applicability of traditional philosophical ethical approaches. Particularly the reorientation of human beings towards the Earth, as something of intrinsic value that we should care about, calls for a new type of ethical relationship. Following are four developments in ethics that could provide food for thought and even food for people.

Animal ethics

The focus of this collection on landscape and food precludes an exhaustive treatment of the well developed and complex area of animal ethics. However, an overview is necessary to note some changes that have occurred within and beyond philosophical ethics. Along with the rise of concern about environmental issues the 1960s–70s also marked a watershed in thinking about our relationships with non-human animals. The burgeoning intensive farming industry sparked concern about the treatment of these animals. In 1965 Ruth Harrison published her book *Animal Machines*, which brought to wider attention the conditions in factory farms. This text contributed to the 1976 European Convention for the Protection of Animals Kept for Farming Purposes. The general thrust of this convention has been developed further into what are called the five freedoms that guide contemporary animal welfare (Webster, 2008: 6). These are:

1. Freedom from hunger and thirst, by ready access to water and a diet to maintain health and vigour.
2. Freedom from discomfort, by providing an appropriate environment.
3. Freedom from pain, injury and disease, by prevention or rapid diagnosis and treatment.
4. Freedom to express normal behaviour, by providing sufficient space, proper facilities and appropriate company of the animal's own kind.
5. Freedom from fear and distress, by ensuring conditions and treatment, which avoid mental suffering. (FAWC, 2009: 2)

For the development of foundational ethical arguments in this area a seminal text was Peter Singer's, 1975 book *Animal Liberation*. Here Singer applied the principle of utility to all entities capable of experiencing pleasure or pain. Quantities of pleasure or pain are the ethically relevant aspect so not to include non-human animals in the calculation would be, as Singer says, speciesist. Thus it is unjustifiable to prioritise the wishes of human beings. The production of food from animals inevitably involves some suffering. For Singer, pleasure should never trump the alleviation of pain. Thus, for Singer, the times when consumption of animals is justifiable are very rare.

The most influential approach to animal ethics from a deontological perspective was Tom Regan's, 1983 book *The Case for Animal Rights*. For him, some non-human animals are rights bearers based on them being 'subjects-of-a-life', as evidenced, according to Regan, by their possession of a psychophysical identity over time (Regan, 1983: 243). Although the kinds of animals included would be fewer than the utilitarian criterion of 'ability to feel pain' such animals as a rights bearer should *never* be used for food. To deny their freedom and life is to infringe their rights and this would be as ethically wrong as we generally consider it to be to do this to a human being.

Both the utilitarian and the rights positions have certain logical problems, not least what we as ethical agents should do about natural predation (Sagoff, 1984: 299) or the fact that farm animals often have a less painful life and death in the context of a farm than they would in the wild (ibid.: 303). However, the work of philosophical campaigners such as Singer, Regan, and Richard Ryder (who coined the term speciesism) were important in constructing supporting arguments to bring about changes in attitudes and practical changes via legislation in the welfare of animals reared for food. This can be seen in the growing popularity, in the west, of vegetarianism and veganism (Morgan, 2016). For some this choice is based on an animal ethics position and for others an environmentalist position. Animal production is seen as a wasteful use of agricultural land that could produce much more plant-based food, and farm animals are a significant contributory factor in the production of greenhouses gas emissions (Bland, 2012).

Permaculture ethics

It is interesting that ethical language is central to one of the more recent innovations in growing food; recent under the name *permaculture* (Mollison and Holmgren, 1978; also see Chapter 10), but of course linking back to and using many traditional techniques. Likewise, the ethical ideas claim to be grounded in those of traditional cultures with a respect for the Earth as a living being. The tripartite mantra of permaculture is Earth care, people care, and fair shares (Holmgren, 2011). Placing the Earth at the beginning clearly denotes that this is not about enlightened self-interest but a placing of the planet, all life forms, and particularly the living soil at the centre of the ethic. People care is what ethics has a long history of promoting but here it particularly emphasises caring for one's self, friends, and neighbours and building resilient communities (Macnamara, 2012). Fair shares is about sharing the abundance of nature and not hoarding food or taking more than we need, and ensuring we return nutrients to the Earth. The examples often given create the sense of a homespun philosophy of being a good person in one's local situation and endorse a positive outlook and the idea of living in balance with nature. Indeed they are said by one of the founders of permaculture to be "simple and relatively unquestioned ethical foundations for permaculture design within the movement and within the wider "global nation" of like-minded people" (Holmgren, 2013: 7).

The twelve principles of permaculture (Holmgren, 2002) appeared before the ethics had been articulated, but those principles could be seen as emerging out of an implicit ethos that later became articulated. For example, the first principle is about working with nature not against it. The principles are closest to a virtues approach as can be seen by the idea of doing the best you can and through that improving as well as accepting personal responsibility rather than blaming others. The focus on those nearest to you and your community also sits well with a virtues approach. The idea of global justice appears in permaculture but it is seen as contributed to by living lightly on the land and moderating your own desires. One of the critiques of permaculture is about the problem of scaling it up to agricultural size projects (Ferguson and Lovell, 2014). However, this could be seen as a lack of vision about how it could be done, or indeed, whether it should be done. Yield, the holy grail of agribusiness, can be very high with the dense planting and synergistic multi cropping of permaculture. Moreover, the crops produced are – or should be – desired. This does not fit the business model of mainstream agriculture: it calls for a completely different understanding of markets and supply chains, but it could feed people and protect the environment.

Ecofeminist ethics

Since its beginning in the mid-1970s *ecofeminism* has become a large but diverse movement, encompassing thinkers from a wide range of perspectives from within the feminist movement and the environmental movement. Although ecofeminism is diverse, there does seem to be one core shared premise: that there is a link between what is understood as the domination of nature and the domination of women. Sometimes this is expressed as the view that oppression of women and the natural world are 'twin' oppressions, stemming from the same cause.

Ecofeminists reject ethical approaches that rely solely on or emphasise reason or that work on abstract principles such as justice, with the implication that they can apply to all people everywhere (Warren, 2000; Gruen, 1993). The problem they see with these approaches comes from the claims of universalisability; whether in the form of a universal law such as Kant's categorical imperative or the idea of an ideal detached observer or applying utilitarian calculations that would be the same no matter who calculated them. It is the impersonal or detached aspect of this that the ecofeminists see as working against how we actually are and how we should behave ethically in the world. An engaged ethics is one that arises out of the personal, the emotional, and is exemplified when we care for another or a place or nature. Thus an ethics of care is seen as recognising our interdependence and that some people are affected by our decisions. It brings ethical thinking closer to home, and makes it more contextual and nuanced. As with permaculture this ethics of care approach is closest to, or a development of, virtue ethics. Here we can see that caring is not an abstract principle that must be applied in the appropriate circumstances. As Nell Noddings explains:

> One might suggest as a basic principle: always act so as to establish, maintain or enhance caring relations. A carer, however, does not refer to this principle when she responds to a person who addresses her. The "principle" is *descriptive*, not prescriptive. The behaviour of carers is well described by this principle, but their motivation arises either spontaneously (in natural caring) or through deliberate reflection on an ideal of caring that has become part of their character.
>
> *(2002: 30; emphasis in the original)*

Similarly, the farmer with a caring approach to land and animal husbandry does not care because a rule or maxim dictates it, they care because they are in a *relationship of care* with the land and the animals.

Responsive cohesion

Responsive cohesion is an approach developed by Warwick Fox in answer to the basic ethical question: 'how should we live?' Unusually for an ethical theory, responsive cohesion is based on the qualities of structure or form of organisation. Things, anything at all, can be organised or disorganised – in Fox's terms, have *cohesion* or be *discohesive*. If something is organised then it can be organised by having a rigid structure or by the way its various parts respond to each another; in Fox's terms have *fixed cohesion* or *responsive cohesion* (Fox, 2006). Which is better: discohesion, fixed cohesion or responsive cohesion? We already know which we value the most. As Fox points out, if we take any realm of endeavour or any entity and ask considered judges what makes a good 'X' the answer, regardless of how it is couched, will be directly translatable into 'the one with the most responsive cohesion'. For example, we could ask about the best political system, would it be the discohesive lawlessness, the rigid dictatorship, or the responsive democracy that we value? When we investigate why a politically informed person would prefer democracy it is going to have something to do with how the parts have to respond to one another. Democracy only works if there is an answering to and an accommodation between the parts (the people, the parliament, the judiciary, a free press, access to education, and so on) rather than a rigid form. The democracy hangs together by virtue of that flexible but structured answering to each other. Fox points out we can run the same 'what's the best?' question with any human endeavour and get the same answer. Let us take farming and ask what the best farm would be. The answer will always be one that is not a rigid fixed pattern, for example, a vast monoculture with minimal relationship to the local community, employing few people, and selling to a single client who could go elsewhere. Nor will it be a farm that is disorganised, for example, a farm often left untended such that cattle could break in to crops and harm both the plants and themselves, or weeds left to seed and rain water wasted rather than diverted to where it is needed, where no long-term, but flexible, planning takes place, and so on. A good farm will be one where the various parts and activities will answer to each other. The following points just serve as examples. If there is livestock the balance of animals to arable land will be such that animal manure is a boon rather than a hazardous waste product. Crops can be rotated to ensure the best replenishment of the soil. Food is produced that suits the soil and climatic conditions and local markets found for it. There is a diversity of produce to cope with difficult weather conditions or fluctuations in demand and the ecological affordances of synergistic relationships are enhanced. Local people are employed or co-create the farm and it plays a role in the local community such as farm visits for schools (see Chapter 33).

Responsive cohesion is offered as an ethical gauge. For any ethical problem situation we just need to work out what will bring about or preserve the maximum amount responsive cohesion and this will inevitably bring about the most valuable situation. Obviously there is rather more to this theory than set out so far. For our purposes, of reflecting on our ethical obligations regarding food and landscape, a particularly important element is the theory of contexts and the requirement to prioritise responsive cohesion in the widest context (Fox, 2006). Fox outlines three broad contexts: (1) the biophysical realm (ecosystems, Gaia), (2) the socio/cultural realm, and (3) the human constructed realm. These are nested one within the other and it should be clear (3) is dependent on (2) and (1), and (2) is dependent on (1). That is, you don't get the rice paddy fertilised by food fish stocks without the socio/cultural realm of human culture and ingenuity and you don't get human ingenuity in shaping the landscape without a landscape to be shaped and, moreover, a planetary situation that has brought about and supported life in the first place.

Thus in any situation to do with cultural landscapes, or indeed most things, we need to be attentive to responsive cohesion in the broadest context, then the next broadest to ensure that the 'direction of fit' for our creation or preservation of responsive cohesion is right. For example, a problem of insect damage to crops can be solved with a pesticide, but this ignores aspects such as the role that the insect plays in the local ecosystem and the follow on effects of pesticide residue in the crops to be eaten and the impact on rivers and downstream ecosystems. Being mindful of the wider systems, and seeing all of these as part of the situation we need to consider, means that the most immediately effective solution to the crop damage is not necessarily the right one, as it could destroy more responsive cohesion than it brings about.

This could suggest that farming – or any changes to naturally occurring ecosystems – is wrong, but this is not the case. The moral imperative that arises from the theory of responsive cohesion is that "in living *your* life" (Fox, 2006: 302) you protect or create more responsive cohesion and, of course, the wider the context the more responsive cohesion you create. The "in living *your* life" formulation allows for the fact that we need to build shelter and source food. Also it allows for the fact that we are not psychologically capable of caring for everyone, let alone every aspect of nature impartially. A normative ethical theory should not impose something that is impossible to do. We, human beings, as living organisms are examples of responsive cohesion. Our socio/cultural systems and our material creations can also exhibit responsive cohesion. And by living in dialogue with nature we can bring ourselves into a more responsively cohesive relationship with the wider biosphere.

Conclusion

It is clear that thinking about ethics, landscape and food presents some highly complex issues that need to be brought to the forefront of agriculture as an academic discipline and food production as a practical endeavour. Of the three main approaches in traditional ethics – virtues, deontology, and utilitarianism – we saw that it was virtue ethics that possibly allowed for a nuanced way of dealing with that complexity. However, it does not issue in clear directives or guide us by generating rules to follow. The partiality of responding to our nearest and dearest is a more comfortable place to be, but how do we, as a society, nurture those virtues that will stretch our empathy to encompass others and the biotic community?

The social and ecological critique of industrial methods and agribusiness market models has alerted us to a need to rethink the simplistic idea that food production *per se* is a 'good thing' and therefore ethical. The development of agroecology as an approach has been helpful in creating a more holistic way of thinking about food production. Aspects of agroecology, with its respect for local knowledge and ecological sensitivity, could be seen as a reinvigoration of the traditional idea of husbandry of the land as a noble and morally improving activity.

Early developments in environmental ethics paved the way for thinking beyond seeing the Earth as there to provide services for humans. The four new developments outlined – animal ethics, permaculture ethics, ecofeminist ethics and responsive cohesion – explore the implications of that ethical thinking and help to make sense of care and relationality as well as practical applicability (also see Chapter 34). We flourish in relationship with our communities and through engagement with the world around us. Environmentally sensitive food production is one of the ways in which this flourishing is seen at its best.

Further reading

Alroe, H.F., Byrne, J., & Glover, L. (2006) Organic agriculture and ecological justice: Ethics and practice. In: Halberg, N., Alroe, H.F., Knudsen, M.T., & Kristensen, E.S. (eds.), *Global development of organic agriculture; challenges and prospects*. CAB International: Wallingford, pp. 75–112.

Altieri, M. (1995) *Agroecology: The science of sustainable agriculture*. Boulder, CO: Westview Press.

Artzen, S., & Brady, E. (eds.) (2008) *Humans in the land: The ethics and aesthetics of the cultural landscape*. Oslo: Unipu.

Beatley, T., & Manning, K. (1997) *The ecology of place*. Washington, DC: Island Press.

Benn, P. (1997) *Ethics*. London: Routledge.

Berry, W. (2009) *Bringing it to the table: On farming and food*. Berkeley: Counterpoint.

Berry, W. (2015) *The unsettling of America: Culture and agriculture*, reprint ed. Berkeley, CA: Counterpoint Publishers.

Bland, A. (2012) Is the livestock industry destroying the planet. *Smithsonian.com* www.smithsonianmag.com/travel/is-the-livestock-industry-destroying-the-planet-11308007/ (accessed 21 May 2017).

Brady, E. (2006) The aesthetics of agricultural landscapes and the relationship between humans and nature. *Ethics, Place and Environment*, 9(1), 1–19.
Bray, J., Jones, N., & Kilburn, D. (2011) An exploratory study into the factors impeding ethical consumption. *Journal of Business Ethics*, 98(4), 597–608.
Brook, I. (2008) Wildness in the English Garden tradition: A reassessment of the picturesque from environmental philosophy. *Ethics and the Environment*, 13(1), 105–120.
Brook, I. (2012) Make do and mend: Solving placelessness through embodied environmental engagement. In: Brady, E., & Phemister (eds.), *Human-environment relations: Transformative values in theory and practice*. BV, Springer, pp. 109–133.
Brook, I. (2013) Aesthetic appreciation of landscape, chapter 9. In: Howard, P., Thompson, I., & Waterton, E. (eds.), *The Routledge companion to landscape studies*. London: Routledge, pp. 108–118.
Callicott, J.B. (1989) *In defense of the land ethic: Essays on environmental philosophy*. New York: SUNY.
Calson, A. (1985) On appreciating agricultural landscapes. Journal of Aesthetics and Art Criticism, 43(3), 301–312.
Cresswell, T. (2004) *Place: A short introduction*. London: Blackwell.
Cronon, W. (1995) The trouble with wilderness; or, getting back to the wrong nature. In: Cronon, W. (ed.), *Uncommon ground*. New York: Norton and Company Ltd, pp. 69–90.
Darwell, S. (2003) *Virtue ethics*. London: Blackwell.
DeLind, L.B. (2011) Are local food and the local food movement taking us where we want to go? Or are we hitching our wagons to the wrong stars? *Agriculture and Human Values*, 28(2), 273–283.
Devall, B., & Sessions, G. (1985) *Deep ecology*. Leyton, UT: Gibbs Smith.
Dundon, S. (2003) Agricultural ethics and multifunctionality are unavoidable. *Plant Physiology*, 133, pp. 427–437.
Elliot, R. (1982) Faking nature. *Enquiry*, 25, 81–93.
European Convention for the protection of animals kept for farming purposes. https://ec.europa.eu/food/sites/food/files/animals/docs/aw_european_convention_protection_animals_en.pdf (accessed 21 May 2017).
FAWC (2009) *Farm animal welfare in Britain: Past present and future*. London: Farm Animal Welfare Council.
Ferguson, R., & Lovell, S. (2014) Permaculture for agroecology: Design, movement, practice and worldview: A review. *Agronomy for Sustainable Development*, 34(2), 251–274.
Fox, W. (2006) *A theory of general ethics: Human relationships, nature, and the built environment*. Cambridge, MA: The MIT Press.
Gamborg, C., Millar, K., Shortall, O., & Sandøe, P. (2012) Bioenergy and land use: Framing the ethical debate. *Journal of Agricultural and Environmental Ethics*, 25(6), 909–925.
Glaeser, B. (ed.) (2011) *The green revolution revisited*. London: Routledge.
Gliessman, S. (2015) *Agroecology*, 3rd ed. Boca Rabon: CRC Press.
Gray, M. (2013) *Labor and the locavore: The making of a comprehensive food ethic*. Berkeley, CA: University of California Press.
Gruen, L. (1993) Ecofeminism: Toward global justice and planetary health. *Society and Nature* 2(1), 1–35.
Harrison, R. (2013 [1964]) *Animal machines*. Wallingford: CABI.
Holmgren, D. (2002) *Permaculture principles and pathways beyond sustainability*. Hepburn, Vic: Holmgren Design Services.
Holmgren, D. (2011) *Permaculture principles and pathways beyond sustainability*. 2nd ed. East Meon: Permanent Publications.
Holmgren, D. (2013) *Essence of permaculture*. https://holmgren.com.au/downloads/Essence_of_Pc_EN.pdf (accessed 16 August 2016).
Holowchak, M. (2011) Jefferson's moral agrarianism: Poetic fiction or normative vision? *Agriculture and Human Values*, 28(4), 497–506.
Hopkins, R. (2008) *The transition handbook*. Totnes: Green Books.
Hough, M. (1990) *Out of place: Restoring identity to the regional landscape*. New Haven, CT: Yale University Press.
Hursthouse, R. (1999) *On virtue ethics*. Oxford: Oxford University Press.
Hurter, U. (ed.) (2014) *Agriculture for the future: Biodynamic agriculture today*. Basal: Verlag am Goetheanum.
Jansson, B. (2010) Local food, local engagement: Community-supported agriculture in Eastern Iowa. *Journal of Culture and Agriculture*, 32(1), 4–16.
Kimbell, A. (ed.) (2002) *The fatal harvest: The tragedy of industrial agriculture*. Washington, DC: Island Press.
Kirschenmann, F. (2010) *Cultivating an ecological conscience: Essays from a farmer philosopher*. Berkeley, CA: Counterpoint Press.
Ladner, P. (2011) *The urban food revolution: Changing the way we feed cities*. BC, Canada: New Society Publishers.

Lapping, M. (2005) Toward the recovery of the local in the globalizing food system: The Role of alternative agricultural and food models in the US. *Ethics, Place and Environment*, 7(3), 141–150.
Lehman, H. (1995) *Rationality and ethics in agriculture*. Caldwell, ID: Caxton Press.
Leopold, A. (1981 [1949]) *A sand county Almanac*. Oxford: Oxford University Press.
Levidow, L., Pimbert, M., & Vanloqueren, G. (2014) Agroecological research: Conforming or transforming the dominant agro-food regime? *Agroecology and Sustainable Food Systems*, 38, 1127–1155.
Lewis, T. (ed.) (2010) *Ethical consumption: A critical introduction*. London: Routledge.
Lotti, A. (2010) The commoditization of products and taste: Slow food and the conservation of biodiversity. *Agriculture and Human Values*, 27, 71–83.
Macnamara, L. (2012) *People and permaculture*. East Meon: Permanent Publications.
McMichael, P. (2006) Peasant prospects in a Neoliberal Age. *New Political Economy*, 11(3), 407–418.
McMichael, P. (2015) The land question in the food sovereignty project. *Globalizations* 12(4), 434–451.
Mizzoni, J. (2009) *Ethics: The basics*. London: Wiley.
Mollison, B., & Holmgren, D. (1978) *Permaculture one: A perennial agriculture for human settlements*. Melbourne: Transworld.
Morgan, R. (2016) *Roy Morgan research*. www.roymorgan.com/findings/vegetarianisms-slow-but-steady-rise-in-australia-201608151105 (accessed 21 May 2017).
Morrison, J. (2009) *The political philosophy of George Washington*. Baltimore, MD: John Hopkins University Press.
Nawaz, M. Bourrié, G., & Trolard, F. (2013) Soil compaction impact and modelling: a review. *Agronomy for Sustainable Development*, 33, 291–309.
Noddings, N. (2002) *Starting at home: Caring and social policy*. Berkeley, CA: University of California Press.
Orr, D. (2001) The urban Agrarian mind. In: Freyfoggle, E. (ed.), *The new Agrarianism: Land, culture, and the community of life*. Washington, DC: Island Press, pp. 93–110.
Pollan, M. (2006) *The omnivore's dilemma*. London: Bloomsbury.
Pretty, J. (2002) *Agri-culture: Reconnecting people land and nature*. London: Earthscan.
Regan, T. (1983) *The case for animal rights*. Berkeley, CA: University of California Press.
Relph, E. (1976) *Place and placelessness*. London: Pion Ltd.
Rolston, H. III (1989) *Philosophy gone wild: Environmental ethics*. New York: Prometheus Books.
Sagoff, M. (1984) Animal liberation and environmental ethics: Bad marriage, quick divorce. *Osgoode Hall Law Journal*, 22(2), 297–307.
Sandler, R. (2007) *Character and environment*. New York: Columbia University Press.
Sanford, A. (2011) Ethics, narrative, and agriculture: Transforming agricultural practice through ecological imagination. *Journal of Agricultural and Environmental Ethics*, 24(3), 283–303.
Seddon, G. (1997) *Landprints: Reflections on place and landscape*. Cambridge: Cambridge University Press.
Shiva, V. (2000) *Stolen Harvest: The Hijacking of the global food supply*. Cambridge: South End Press.
Shiva, V. (2016) *Who really feeds the world? The failures of agribusiness and the promise of agroecology*. Berkeley: North Atlantic Books.
Singer, P. (1975) *Animal liberation*. New York: Harper Collins.
Singer, P. (2007) Famine, affluence, and morality. In: La Follette, H. (ed.), *Ethics in practice*, 3rd ed. London: Blackwell Publishing.
Smith, A., & MacKinnon, J. (2007) *The 100 mile diet*. Toronto: Vintage House.
Sterba, J. (2001) *Three challenges to ethics: Environmentalism, feminism and multiculturalism*. Oxford: Oxford University Press.
Sumberg, J., Thompson, J., & Woodhouse, P. (2012) Why agronomy in the developing world has become contentious. *Agriculture and Human Values*, 30(1), 71–83.
Thompson, P. (1994) *The spirit of the soil: Agriculture and environmental ethics,* Environmental Philosophies Series. New York: Routledge.
Thompson, P. (2010) *The Agrarian vision: Sustainability and environmental ethics*. Knoxville, KY: University of Kentucky Press.
Thompson, P., & Kutach, D. (1990) Agricultural ethics in rural education. *Peabody Journal of Education*, 67(4), 131–153.
Tuan, Y. (1996) *Cosmos and hearth*. Minneapolis, MN. University of Minnesota Press.
United Nations General Assembly, (1948) *Universal declaration of human rights*. www.ohchr.org/EN/UDHR/Documents/UDHR_Translations/eng.pdf (accessed 24 August 2016).

Vermeir, I., & Verbeke, W. (2006) Sustainable food consumption: Exploring the consumer "Attitude- Behavioural Identity" gap. *Journal of Agricultural and Environmental Ethics*, 19(2), 169–194.
Warren, K. (2000) *Ecofeminist philosophy*. London: Rowman and Littlefield.
Webster, J. (2008) *Animal welfare: Limping toward Eden*. London: John Wiley and Sons.
Zeunert, J. (2013) Challenging assumptions in urban restoration ecology. *Landscape Journal*, 32(2), 231–242.
Zimdahl, R. (2012) *Agriculture's ethical horizon*, 2nd ed. Amsterdam: Academic Press.

Websites

Biodynamic Land Trust www.biodynamiclandtrust.org.uk/
Biodynamic Association www.biodynamics.com/content/overview-biodynamic-research
Community Supported Agriculture www.communitysupportedagriculture.org.uk/
Compassion in World Farming www.ciwf.org.uk/
Food Research Collaboration www.foodresearch.org.uk
Open Food Network www.facebook.com/openfoodnetaus/
The Orchard Project www.theorchardproject.org.uk/
Permaculture Association www.permaculture.org.uk
Philosophy of food project www.food.unt.edu/
The Food Ethics Council www.foodethicscouncil.org
Food and Agriculture Organisation of the United Nations www.fao.org/home/en/
Declaration of Nyeleni on global food sovereignty https://nyeleni.org/IMG/pdf/DeclNyeleni-en.pdf

25

The new food insecurity

Damian Maye

Price volatility and the 'new food insecurity'

Food security is a multi-dimensional concept. For example, it can be analysed at different scales, from the individual and household level to a region, nation state or globally (Lee, 2007). The UN FAO's definition of food security is the most commonly cited (Lang and Barling, 2012). It suggests that "[food security] exists when all people at all times have physical, social and economic access to sufficient, safe and nutritious food to meet dietary needs and food preferences for an active and healthy life" (FAO, 2009: 8). Food security is therefore about more than just producing more food (i.e. availability); it is also about access, utilisation and social agency (Ericksen, 2008). Moreover, food security involves two-way interactions between food systems *activities* (producing food, processing and packaging food, retailing and distributing food, consuming food) that give rise to food security *outcomes* (food availability, access to food, food utilisation) (Ingram, 2011). In essence, "[t]he attainment of food security is primarily about the social, economic, cultural and political circumstances that either enable or restrict the provisioning of food to needy populations" (Pritchard, 2016: 162).

This chapter focuses on a specific aspect of the food security debate, which I term the 'new food insecurity'. This relates to relatively recent national and global food security discourses and specifically the re-emergence of the term in global food politics since the infamous 2007–08 price spikes (see Tomlinson, 2013). Before 2008 food security was viewed in developed market economies as a backward-looking form of agricultural fundamentalism, because it framed rural land use as a mono-functional space used primarily for food production (Lobley and Winter, 2009; Maye, 2013). After the 2007–08 price spikes food security had renewed geopolitical status and meaning (Maye and Kirwan, 2013; Hopma and Woods, 2014), with politicians more aware of the potential threats to food security and the "political and social importance of affordable food" (Ambler-Edwards et al., 2009: 5). Food security was once again an issue of global and national urgency, symbolised by international meetings, such as the Food and Agriculture Organization of the United Nations (FAO) Rome Summit on World Food Security in June 2008, global food security research programmes, such as Foresight's (2011) *The Future of Food and Farming: Challenges and Choices for Global Sustainability*, and a number of academic publications and special issues (see, for example, Ambler-Edwards et al., 2009; McDonald, 2010; Foresight, 2011; Maye and Kirwan, 2013; Hinrichs, 2013; Marsden, 2013; Hopma and Woods, 2014; Candel et al., 2014; Sonnino et al., 2016). From a global politics and food discourse perspective, the June 2008 FAO Rome Summit was particularly important because it introduced two projections that became central pillars of the new global food security consensus frame (Tomlinson,

2013): first, that food production needed to increase by 50% by 2030 to meet rising demand (Ki-moon, 2008); and second, that food production needed to double by 2050 to feed a world population of 9 billion (Diouf, 2008). The projections have since been questioned; however, meetings like this were crucial in helping to establish a consensus that the 2007–08 price spike was not merely a blip.

The 2007–08 price spike event itself is now well documented. In summary, global food prices increased sharply during the period (by 83% over three years up to February 2008; see World Bank, 2008; Ghosh, 2010; Tomlinson, 2013). Prices had started to increase in 2006 but were initially steady. They rose sharply in the latter part of 2007 and the first half of 2008. Figure 25.1 uses FAO food price data from 2002–16 to show price changes in 2007–08, contextualising that particular spike event in terms of price changes before and after. The FAO food price index is derived from the price of five main commodity groups (meat, dairy, cereals, oils and sugar) and is therefore a reliable general indicator of commodity price changes. Figure 25.1 also shows individual prices for meat, dairy and cereals. Very sharp increases were evident from 2006 for dairy and cereals and for all four in 2008, although less dramatic for meat. Prices rose very sharply in the first half of 2008 then peaked around the middle of that year and then fell back again. We have seen price spikes again since this time in 2011 and prices remained high until 2014 before dropping again. Price spikes and volatility are nothing new for primary agricultural commodities (Ghosh, 2010; Gardner, 2013) but the sharp nature of the spikes is historically unprecedented. The 2007–08 spikes were also distinct in that they applied to most food and feed commodities (Evans, 2008).

The feature that captured most political and media attention in relation to price spikes was the public reaction, with riots and demonstrations in Africa, Asia, South America and the Caribbean (Maye and

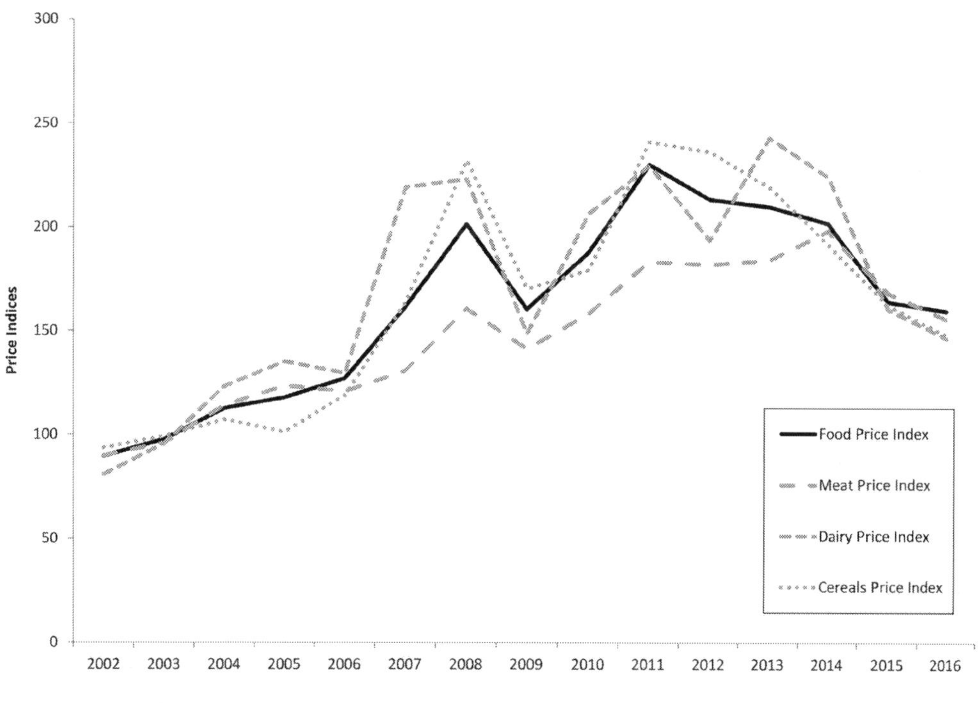

Figure 25.1 Annual food price indices, 2002–2016.

Source: www.fao.org/worldfoodsituation/foodpricesindex/en/; accessed: 03/06/2017

Note: All indices have been deflated using the World Bank Manufactures Unit Value Index (MUV) rebased from 2010=100 to 2002–2004=100

Kirwan, 2013). Addressing high levels of volatility in global markets is therefore critical "because of the adverse effects they have on consumers and producers, because of the disruption they cause to the global food system, and, when particularly severe, because of the general economic and political instability that can occur" (Foresight, 2011, 22, exec. summary). Impacts on agricultural landscapes and land use are not mentioned here, but they are potentially very significant, if we think, for instance, in terms of how food security discourse influences and promotes particular forms of agricultural land use practice and/or potentially how land is used (to grow food rather than other land uses, scales of agricultural land use, etc.).

There is now significant debate about the factors causing volatility in global food prices, including the extent to which financial forces influenced and exacerbated the 2007–08 crisis (Clapp, 2012; Gardner, 2013; House of Commons, 2013). Understanding this debate and related analysis of relations between food systems and finance – the 'middle space' of the world food economy (Clapp, 2012) – is a key focus of this chapter. It is important in a handbook about landscape and food because it points to a fundamental but often forgotten relationship in contemporary foodscapes between finance markets and agriculture (Clapp et al., 2017), a relationship which can determine how farms and agricultural landscapes are valued and used (also see Chapters 9, 19 and 32).

To develop this argument the rest of the chapter reviews the food security discourse and links between food systems and finance, including debates about how best to respond and how to make future agri-food systems and agricultural landscapes more resilient (also see Chapter 16). The first part of the chapter examines supply and demand arguments that were used initially to explain price volatility. Although initially convincing, such arguments oversimplify food system complexity and relations, linking the food crisis to a resource supply argument (Deutsche Bank, 2009; Foresight, 2011). The food system is connected and related to wider global environmental change and socio-economic drivers (Ingram, 2011). To explain the food crisis – and in turn to fully comprehend relations between food systems, agricultural landscapes and land use – a systems perspective is needed that goes beyond agriculture and resource inputs. Key work from agricultural economics is reviewed that examines the role of finance markets and their links with the food system. Capturing the perceptions and discursive framings of this new food security discourse is also critical, as these mechanisms *produce social (and spatial) realities* and determine agri-food governance and land uses (Nally, 2014). The rest of the chapter thus looks at governance responses to the 'new food insecurity', comparing science-driven bio-economy approaches with critical political economy, eco-economy and place-based approaches. Examining these debates (Foresight, 2011; McMichael, 2009; De Schutter, 2011; Sonnino et al., 2016) reveals the complex nature of global agri-foodscapes and associated modes of governance and the contested nature and geography of food security politics, including some tentative suggestions about what this means in relation to agriculutral landscapes and land use (Chapter 32).

Supply and demand arguments to explain the global food crisis

The new food security debate concerns the character and origins of the crisis and also what needs to be done in the name of food security. A number of reports, discussion papers and statements are now available from public and private institutions to document and explain the 2007–08 price spikes and their consequences (Rosin et al., 2012; Maye and Kirwan, 2013). Early analysis of the 2008 food crisis and volatility in food prices pointed to "problems with the market fundamentals of supply and demand" (Clapp, 2012, 126). On the demand side of the equation you have the emerging demand for grain to feed biofuel production, rapid population growth, and escalating middle-class demands for dairy and meat products in India and China (Rosin et al., 2012; Gardner, 2013). On the supply side, factors include short-term extreme weather events in Australia, Russia and elsewhere, which reduced the harvest of staple grains, and longer-term factors such as global climate change concerns that will disrupt food production in some parts of the world (see Chapter 18). This market fundamentals perspective was described as a "silent tsunami" (Economist, 2008) and a "perfect storm" (Lacey, 2008); in other words, the food price crisis was not caused by one single disrupting factor but was instead a "result of the compounding effects of coincidental factors" (Rosin et al., 2012: 4). A number of reports were published (e.g. Evans, 2008; Ambler-Edwards et al., 2009; Deutsche Bank, 2009; House of Commons, 2013) which recognised the impact played by various *short-term factors*, especially a series of extreme weather events, low global stock levels, the use of food crops for biofuels, rising energy prices, export bans and increased financial speculation.

Studies also recognised the influence of *longer-term trends* and recognition that the food crisis was also a consequence of structural problems rooted in global resource limits (Goodman et al., 2012), including changing global patterns in diet and world population pressures. The Foresight (2011) report, for example, identified a list of 'key drivers' of change in the food chain, noting population growth, changes in food demand, both in terms of size and the nature of demand (especially for meat and fish), climate change impacts (both to the biophysical environment and how policy responds) and competition for key resources (namely, land, energy and water). Foresight also identifies how national and international food system governance develops as a key driver. This includes globalisation, production subsidies, trade restrictions, and how and whether governments work together in the future to solve food problems. Changes in the values and ethical stances of consumers is listed as a final driver (see Chapter 24). Other reports use different language but usually point to similar landscape conditions and often present them in similarly pervasive, uncontested terms. For example, Ambler-Edwards et al. (2009: 11–18) describe seven "*new fundamentals*" which will affect the global food system: (1) a rapidly rising world population, (2) the effect of the 'nutrition transition', (3) the cost of energy and dependency on energy, (4) land resource issues, (5) global stresses on available water for consumption and food production, (6) climate change and (7) labour in both developed and developing world countries.

Looking for explanations beyond the food system: the 'middle space' of finance markets

Initial analysis of the 2007–08 price spikes is open to critique. A key weakness in the supply and demand analysis presented is the failure to recognise links between food and finance markets. Some fundamental pressures may be pushing prices upwards over the long-term, but *volatility* (i.e. rapid fluctuation in prices both upwards and downwards), evident through the huge fluctuations in prices over a short period of time, suggests a strong link to the "expansion of the intermingling of food and financial markets" (Clapp, 2012: 126; Ghosh, 2010; Gardner, 2013; House of Commons, 2013). In other words, fluctuations in financial markets may have played a significant role in recent periods of food price volatility. Crucially, supply and demand arguments can only explain some of the underlying pressures that created price volatility in 2008. They do not fully explain the extent of the crisis. The demand for food and 'nutrition transition' processes taking place in India and China, for example, has been known about for a while, making it an unlikely trigger for sudden food price rises, especially as both are largely self-sufficient in food. The short supply argument is therefore open to question. Furthermore, the export restrictions that many suggested caused volatility were in many cases a *response* to already rising prices (Clapp, 2012).

Price volatility then is not explainable by supply and demand for food arguments alone. As Ghosh (2010: 77) argued, such wild swings "cannot be explained by short-term supply and demand factors or any other 'real economy' tendencies . . . these acute price movements are clearly the result of speculative activity in these markets". Financial markets provided "an additional ingredient to the perform storm" (Clapp, 2012: 134). Early accounts of the food crisis tended to downplay financial markets' role. This view is now changing, prompted by the global financial crisis which led to a significant dropping of food prices at that time and alerted analysts to the role finance played in the food crisis and food pricing. Financial actors occupy a new 'middle space' in the world food economy (Clapp, 2012), with food and changing farming conditions used increasingly by financial institutions to make money.

This 'middle space' of the world food economy – that is, the financial dimensions of food pricing – is often *hidden* from view, with greater attention on the material aspects of supply and demand. The financial transactions discussed in this 'middle space' of the world economy are nothing new. Agricultural futures markets have been around for hundreds of years. 'Hedging', for example, was traditionally viewed as beneficial to producers and users of the commodity (Gardner, 2013: 57), with speculation allowing both parties to see what market conditions would be in the future and therefore helping them to plan for volatility. However, because traditional markets, including food commodities, are now absorbed into a much wider market for financial instruments (Clapp et al., 2017) some suggest this acted as a major price-inflating factor during the 2007–08 food price spikes and subsequently (De Schutter, 2010).

A key product sold to investors by banks was *commodity index funds (CIFs)*. "The root of speculative pressure on food prices in 2006–2008, according to the critics, came from [CIFs]" (Gardner, 2013: 58). CIFs bundled commodities, including oil, minerals and agricultural commodities, into a single financial instrument based on a commodity price index (Clapp, 2012: 140). Crucially, CIFs are a further step removed from the physical commodity because "investors are able to bet on market price movements, rather than the purchase of the commodity itself" (ibid.). Investors do not need detailed knowledge of futures exchanges or even the nature of the commodities themselves. They have only to deal directly with the bank selling them, which tracks a specific group of commodities via indexes like the Dow Jones. Financial investment in commodity futures markets has increased significantly since the early 2000s due to the economic climate, with a weak US dollar and rising commodity prices, but also because of deregulation of agricultural commodity markets in the US which allowed banks to offer CIFs to investors (Clapp, 2012). The upshot of this deregulation of agricultural derivatives is that it attracted a new set of investors (both small and large) and the trade increased sharply, in turn exposing agricultural trade to global financial markets.

As a result of the increased investment in agricultural derivatives the link between food and finance is now much stronger. "[F]ood and agriculture had effectively been financialized, with agricultural derivatives widely seen as simply a financial investment, interchangeable with any other financial asset" (Clapp, 2012: 143; see also Clapp et al., 2017). The risks that this new exposure to global financial markets posed to global food security were not understood until after the 2007–08 crisis. As prices rose sharply in late 2007 financial speculation was a key factor causing price volatility, with speculators using CIFs as a way to protect themselves from financial instability, which was driving up food prices. De Schutter (2010), the former UN's special rapporteur on the right to food, argued, for example, that the price spike in wheat and other agricultural commodities was mainly a consequence of a 'speculative boom', which he asserts is further justified in the subsequent price rises evident in 2010–11. During 2010–11 food stocks had been rebuilt, demand was reduced as a result of the global recession and biofuel production was also static. Consequently, "the main cause of price rises is a speculative bubble created by the massive growth in financial instruments linked to food commodities" (De Schutter, quoted in Gardner, 2013: 62).

De Schutter is by no means alone in his critique of speculation as a contributing factor to food price rises in 2006–08 and beyond (see Ghosh, 2010, for example). For some speculation was a central factor causing the food price rises. Other analysis rejects the argument that the financialisation of food and agricultural markets was a major factor leading to price volatility and food price increases. These analysts are usually pro free trade, in contrast to consumer groups, aid charities and humanitarian agencies (e.g. the World Development Movement, the United Nations). The World Bank argues, for example, that supply and demand fundamentals and policy decisions are the real factors, not speculation. A major OECD study of 2006–08 food prices by Irwin and Sanders (2010) viewed speculation as a significant force that *stabilised markets* because speculators provide an immediate connection between buyers and sellers (Gardner, 2013; see also Foresight, 2011, 22, executive summary).

Governance responses to global food insecurity and links to landscape

Debates about the role of financial speculation are therefore hotly contested. Most analysts now accept that "financial speculation on futures markets could be responsible for either causing or exacerbating food price volatility" (Clapp, 2012: 146). This 'middle space' of the world food economy was initially ignored.

Some distinct ideological positions divide opinion about the role of speculation (e.g. human rights vs. market liberalisation). From a human rights perspective, the financialisation of food markets and speculation in food futures have tangible landscape impacts beyond materially what is grown on land by further separating people from land access in terms of food production and recreation. Policy failings also play a part here (Pinstrup-Andersen, 2013; Candel et al., 2014). Price spikes can be caused by a variety of factors but *poor policymaking* by affected countries and regions can exacerbate the problem. In Pinstrup-Andersen's (2013) study, changes in world market prices were transmitted differently to national markets. They varied depending on the degree of openness the trade policies followed and the nature of domestic markets and infrastructure. In this next part of the chapter I turn to examine responses to the food crisis and the 'new food insecurity'. This is also contested and political, with different framings shaping policy responses and landscape connections.

Techno-innovation and sustainable intensification

The UK Foresight (2011) programme on global food security was commissioned in response to the 2007–08 food price spikes. The report identified five key challenges for future food security and some associated actions:

- Challenge A: Balancing future demand and supply sustainably

  Five classes of action are identified in the report to address this challenge:

  - *The better use of existing knowledge*: the application of existing knowledge and technology could increase average yields two- to threefold in many parts of Africa, and twofold in the Russian Federation. However, in determining where and how much to invest in producing more food, policy-makers will need to consider a range of criteria rather than increases in production alone.
  - *Capitalising on new science and technology*: this is necessary, it is argued, to raise the limits of 'sustainable production' and address new threats. The need for new technology not just to raise productivity and manage threats such as pests, but also to ensure ecosystems sustainability.
  - *Reducing waste*: from post-harvest (especially in poor countries) through to consumers.
  - *Improving governance* (of the food system, fisheries and corporate governance): self-sufficiency as a goal is rejected. The need to end rich world production subsidies and to reduce trade restrictions at times of crisis is emphasised.
  - *Influencing demand*: various approaches are identified, including economic measures, choice editing and consumer awareness.

- Challenge B: Addressing the threat of future volatility in the food system

  Determining acceptable levels of volatility in food prices is regarded as a political judgement that needs to consider the negative effects of volatility, but also the costs of intervention.

- Challenge C: Ending (world) hunger

  Producing enough food in the world so that everyone can *potentially* be fed is not the same thing as ensuring food security for all (Garnett, 2012). The report points out that "For many governments, the purpose of agriculture is seen primarily as food production," and argues for "a repositioning of agriculture as a profession dedicated to multiple ends, of which hunger and poverty reduction are central. *Food production is the means, not the end*" (Foresight, 2011: 25, emphasis added).

- Challenge D: Meeting the challenges of a low emissions world

  A strong case is made for "substantially integrating and improving considerations of agriculture and food production in negotiations on global emissions reductions" (ibid.: 28), although it is noted that the special features of the sector must be taken into account.

- Challenge E: Maintaining biodiversity and ecosystem services while feeding the world

  Foresight points out that "policies in conservation and in food security were largely developed in isolation. However, increasingly and rightly, they are being pursued together, driven by a growing realisation of their interdependence." It goes on to argue that "the global food supply will need to increase without the use of substantially more land and with diminishing impact on the environment: sustainable intensification is a necessity. Pursuit of this agenda requires a much better understanding of how different policy options, both within and outside the food system, affect biodiversity and ecosystem services" (ibid.: 31; see also Chapters 19 and 23).

A key point of debate to emerge from Foresight (2011) is the recommendation that the best way to achieve global food security is through the delivery of *sustainable intensification* (see also Royal Society, 2009). This suggests that we need to produce more food from less land, resources, energy and water using a mix of 'eco-efficiency' approaches such as genetic modification, nanotechnology, genomics and computerisation (Garnett, 2012; Garnett and Godfray, 2012; Lang and Barling, 2012). The relationship to landscape is in terms of how technologies can make land and food growing productive but eco-efficient. Crucially, solutions to food security are not just about food production but also how we use resources and agricultural landscapes to maintain biodiversity and ecosystem services and how the food system interacts with other systems (Chapter 17). As noted under Challenge C, food production is the means not the end and issues such as volatility, sustainability, climate change and hunger are *not* just food system issues but should be developed in conjunction with policies about energy, water supply, land use and biodiversity.

Food rights and food sovereignty

Despite its techno-scientific bias, Foresight (2011) is broadly in keeping with a food systems-focused 'emerging' discourse that accepts the need to address a complex range of problems, not just food production (Lang and Barling, 2012). In this regard, it suggests that landscape needs to be framed as a *relational concept* with landscape relationships viewed in holistic ways. This also raises questions about the financialisation of food markets that continue to homogenise and centralise agricultural practice in the hands of big business. In a system discourse what is required is a greater diversity and range of scales of agricultural land use, including urban, peri-urban and suburban land. However, when the oil and agricultural commodity prices spiked in 2007–08, this emerging complex analysis was often side-lined. What we have seen instead was a renewed international focus on primary production and the needs of low income countries that highlighted production and demography as the key factors (i.e. neo-Malthusian; Jarosz, 2009; Pritchard, 2016).

In response to this neo-Malthusian framing, other responses to the crisis have emerged that place greater emphasis on food sovereignty (Hopma and Woods, 2014; Sonnino et al., 2016). The most significant is in relation to 'food rights' (see Chapters 31 and 32). Leading food policy thinkers, notably De Schutter (2011), actively challenge neo-productivist strategies to food security on the basis that such an approach may not enhance the *human rights* to adequate food. What we see through this human rights approach is a shift in scale from a rather abstract and complex global food system problem to consider more concretely the needs and rights of individuals, local communities and specific agricultural landscapes (Sonnino et al., 2016), much more in keeping with the entitlement arguments that underpinned early definitions of food security and hunger (see Pritchard, 2016). The largely productivist strategy associated with the exhortation to double food output at a global scale is therefore inappropriate, it is argued, because of structural problems linked to the food system (Sage, 2013), and because the industrial food system is strongly resistant to alternative framings and may consequently actually increase vulnerability and food insecurity for many.

International political economy and ecology

Work in international political economy and ecology conceptualises the problem of food crises and hunger as:

> rooted in the social relations of agricultural production and the political, cultural, and economic relations of food access, distribution and consumption . . . [which are] . . . socially and spatially constructed through long-term historical processes across the globe and within specific regions and places.
>
> *(Jarosz, 2009: 2066)*

The 2007–08 food crisis (and other subsequent shock events) are not an unexpected 'perfect storm' but "an entirely predictable outcome of an oil-dependent feedgrain-livestock complex" (ibid.: 2067) that supplies a meat-centric diet to those who can afford to buy it. This links international patterns of food production and consumption to the development of the capitalist system (Jarosz, 2009; McMichael, 2009; Sage, 2013). The crisis is a consequence of industrial agriculture's long-term over-dependence on fossil fuel (Chapter 16), combined with the inflation-producing effects of biofuels offsets, financial speculation activities, and the concentration and centralisation practices of agribusiness capital. The food crises therefore emerged out of the conditions of the globalised food system, which is designed to produce large quantities of cheap meat using food grains that are increasingly commodified under neoliberal food security policy. It also views the food crisis in its relational position to other crises. The 'feedgrain-livestock' complex significantly contributes to climate change and the food crisis is linked to much wider energy, climate change and financial crises. Thus this more critically draws attention to the structural conditions of the neoliberal food system that have caused the problem and the need to view landscape and agri-food relationships in holistic and relational terms.

Multi-level and place-based reflexive governance

A multi-level and place-based reflexive governance approach emerges out of a critique of various existing conceptual frameworks (productivism, food sovereignty, livelihood security, the right to food, food democracy, food citizenship, community food security) which have a tendency to be 'locked into' fixed scales and represent oppositional assumptions (Sonnino et al., 2016). It calls for new forms of food governance as mechanisms to respond to food insecurity issues. The thinking behind this approach is connected to wider work examining transitions to sustainability (Kirwan et al., 2017). In the Netherlands, for example, Hendriks and Grin (2007: 345) suggest that "steering for sustainability can be understood as reflexive governance – a process of fundamentally reconsidering the way our socio-technical systems are structured, practised and most significantly governed". They distinguish between first- and second-order reflexivity. First-order reflexivity is largely an unconscious process that does not necessarily result in substantive change to the existing order of things. It is more about adapting to external pressures that may have been created by the unintended consequences of the actions of a particular system (Sonnino et al., 2014). In contrast, second-order reflexivity is much more dynamic and intent on criticising the processes associated with modernity.

Dialogue and the development of collective action and understanding through inclusivity in that dialogue is critical here (Sonnino et al., 2014); moreover, these practices require particular spaces for their enaction. Governance mechanisms are needed that encompass a wider range of perspectives that include state, private and civil sectors, who are all recognised as having a valid perspective (Pereira and Ruysenaar, 2012), and the different actors may be positioned at different scales (Kirwan et al., 2017). In practical terms this requires *place-based (or landscape-based) integrated approaches to food governance* (for example, bringing together different interests and voices rather than working in silos). A place-based approach "engages with the complex multi-actor, multilevel and reflexive political and social structures that support the emergence of distinct food security trajectories in a highly contested and unequal landscape" (ibid.: 486).

The key to a reflexive governance framework is that it is flexible and dynamic, as well as providing adequate spaces for deliberation (Pereira and Ruysenaar, 2012; Kirwan et al., 2017). These fora give voice to a variety of discourses and interests. Examples of such deliberative spaces include commodity fora that multinationals have activated to local level for a such as Solidarity Purchasing Groups, Community Supported Agriculture (Renting et al., 2012) and food councils (Pothukuchi and Kaufman, 1999). Food policy councils, for example, many of which are city-based, bring together public, private and civic actors and in theory, encourage the sort of reflexive policy thinking previously described. The extent to which reflexive governance can challenge and transform the perspective of the dominant food paradigm will vary, dependent on the scale involved, the context and the changes demanded of the normative framings of what is considered to be acceptable practice (Marsden, 2013).

Conclusion

This chapter has focused on a specific aspect of the food security debate, the 'new food insecurity', describing the re-emergence of the term in global geopolitics since the 2007–08 price spikes. Metaphors such as 'perfect storm' and related short-term explanations "identify the world food crisis as a onetime, contingent event that will end when food prices decline" (Jarosz, 2009: 2066). The implication is that "the global food system functions well under normal conditions and that the real failure during the crisis was the unfortunate and ill-timed simultaneous pressures on food commodity supplies and prices" (Rosin et al., 2012: 4). Analysis of the 2007–08 price spikes points towards longer-term structural problems associated with the food system. The chapter has revealed the importance of price volatility not price rises per se. Until recently the role of financial markets in determining food prices has been underplayed.

This 'middle space' of the world food economy is therefore an important part of the food security crisis narrative. Understanding finance and food economy relations is important, particularly in relation to agricultural landscapes and related land use debates. Sustainable agricultural land use will require two things: first, it will be necessary to break the link between financialisation and food systems (trends suggest financial speculation will maintain homogenised and centralised agricultural practices when greater diversity and a range of scales of agricultural land uses is required, with landscape relationships viewed in integrated and relational ways); second, if market-led approaches dominate, bad agricultural landscape practices will likely persist, unless corporate social responsibility of finance markets is made more transparent in terms of impacts on food system governance.

The different food security perspectives reviewed in this chapter have helped to develop this critique. Food security as we have seen is a concept that has *discursive fungibility* in response to global political economy changes, national pressures, social movements and a wider neoliberal agenda (Hinrichs, 2013). It can be used to support different arguments from different perspectives. Perceptions and discursive framings produce social realities (Nally, 2014). In this regard, techno-scientific approaches were reviewed (esp. Foresight, 2011) that view sustainable intensification as one important solution to the food crisis. Foresight (2011) also recognises that we need to view food security challenges as part of a more holistic mode of governance. Other more critical social science perspectives suggest structural conditions of the food system and neoliberal responses to it need to be challenged and transformed to foster more sustainable food systems and land uses. This includes critiques by transnational social movements which frame food security in terms of food rights and democracy and discourse-based critiques which concern the way food is framed as a market-based commodity and less as a public good. Place-based approaches to food security, including new forms of multi-level reflexive governance (Sonnino et al., 2016; Kirwan et al., 2017), are arguably the most progressive. They have the potential to allow landscape and food relationships to be viewed holistically, in the sense that it is using place/landscape as a focus for discussion which can deal with more complexity and diversity than a single food security narrative. Reflexive governance can thus foster social action at multi-levels and change the perspective of the dominant food paradigm in more transformative ways to fully address new food security challenges.

Further reading

Ambler-Edwards, S., Bailey, K., Kiff, A., Lang, T., Lee, R., Marsden, T., Simons, D., & Tibbs, H. (2009) *Food futures: Rethinking UK strategy*. The Royal Institute of International Affairs, Chatham House. [online] [cited 02.02.2009] www.chathamhouse.org.uk/publications/papers/view/-/id/695/.

Candel, J.J.L., Breeman, G.E., Stiller, S.J., & Termeer, C.J.A.M. (2014) Disentangling the consensus frame of food security: The case of the EU Common Agricultural Policy reform debate. *Food Policy*, 44, 47–58.

Clapp, J. (2012) *Food*. Cambridge: Polity Press.

Clapp, J., Isakson, S.R., & Visser, O. (2017) The complex dynamics of agriculture as a financial asset: Introduction to symposium. *Agriculture and Human Values*, 34(1), 179–183.

De Schutter, O. (2010) Food commodities speculation and food price crises. Briefing Note 2 by the UN special rapporteur on the right to food (September 23, 2010) www.srfood.org/index.php/en/component/content/article/894-food-commodities-speculation-and-food-price-crises (accessed 7 November 2013).

De Schutter, O. (2011) The right of everyone to enjoy the benefits of scientific progress and the right to food: From conflict to complementarity. *Human Rights Quarterly*, 33, 304–350.

Deutsche Bank Group. (2009) *Investing in agriculture: Far-reaching challenge, significant opportunity. An asset management perspective*. Frankfurt: Deutsche Bank Group.

Diouf, J. (2008) *Address by Jacques Diouf, Director-General of the Food and Agriculture Organization of the United Nations, at the FAO High-Level Conference on World Food Security: The challenges of climate change and bioenergy*. Rome, June 3, 2008. www.fao.org/fileadmin/user-upload/foodclimate/statements/fao-diouf-e.pdf (accessed 24 February 2017).

Economist (2008) The silent tsunami. *The Economist*, April 17. www.economist.com/opinion/displaystory.cfm?story_id=11050146 (accessed 17 July 2013).

Ericksen, P. (2008) Conceptualizing food systems for global environmental change research. *Global Environmental Change*, 18, 234–245.

Evans, A. (2008) *Rising food prices: Drivers and implications for development*. Briefing Paper, Chatham House, London.

FAO (2009) *State of food insecurity 2009*. Rome: Food and Agriculture Organization.

Foresight (2011) *The future of food and farming: Challenges and Choices for Global Sustainability*. London: The Government Office for Science.

Gardner, B. (2013) *Global food futures: Feeding the world in 2050*. London: Bloomsbury.

Garnett, T. (2012) *Summary of foresight study: The future of farming and food*. Unpublished paper prepared for the Food Climate Research Network.

Garnett, T., & Godfray, C. (2012) *Sustainable intensification in agriculture. Navigating a course through competing food system priorities*, Food Climate Research Network and the Oxford Martin Programme on the Future of Food. Oxford: University of Oxford.

Ghosh, J. (2010) The unnatural coupling: Food and global finance. *Journal of Agrarian Change*, 10(1), 72–86.

Goodman, D., DuPuis, E., & Goodman, M. (2012) *Alternative food networks: Knowledge, practice, and politics*. Abingdon: Routledge.

Hendriks, C.M., & Grin, J. (2007) Contextualizing reflexive governance: The politics of Dutch transitions to sustainability. *Journal of Environmental Policy and Planning*, 9(3–4), 333–350.

Hinrichs, C. (2013) Regionalizing food security? Imperatives, intersections and contestations in a post-9/11 world. *Journal of Rural Studies*, 29, 7–18.

Hopma, J., & Woods, M. (2014) Political geographies of 'foodsecurity' and 'food sovereignty'. *Geography Compass*, 8, 773–784.

House of Commons (2013) *Global food security. First report of session 2013–14*. London: The Stationery Office Limited.

Ingram, J. (2011) A food systems approach to researching food security and its interactions with global environmental change. *Food Security*, 3, 417–431.

Irwin, S.H., & Sanders, D.R. (2010) *The impact of index and swap funds on commodity futures markets: Preliminary results*. OECD Food, Agriculture and Fisheries Working Paper No. 27, OECD Publishing, Paris.

Jarosz, L. (2009) Energy, climate change, meat, and markets: Mapping the coordinates of the current world food crisis. *Geography Compass*, 3/6, 2065–2083.

Ki-moon, B. (2008) Address by Ban Ki-moon, Secretary-General of the United Nations, at the FAO High-Level Conference on World Food Security: The challenges of climate change and bioenergy. Rome, June 3, 2008www.fao.org/fileadmin/ user-upload/foodclimate/statements/un-ki-moon.pdf (accessed 24 February 2017).

Kirwan, J., Maye, D., & Brunori, G. (2017) Reflexive governance, incorporating ethics and changing understandings of food chain performance. *Sociologia Ruralis*, 57(3), 357–377.

Lacey, M. (2008) 'Perfect storm' food crisis grips globe. *Scotland on Sunday*, April 20. http://scotlandonsunday.scotsman.com/world/39Perfect-storm39-food-crisis-grips.4000872.jp (accessed 17 July 2013).

Lang, T., & Barling, D. (2012) Food security and food sustainability: Reformulating the debate. *The Geographical Journal*, 178, 313–326.

Lee, R. (2007) *Food security and food sovereignty*. Centre for Rural Economy Discussion Paper Series No. 11, Newcastle.

Lobley, M., & Winter, M. (2009) Introduction: Knowing the land. In: Winter, M., & Lobley, M. (eds.), *What is land for? The food, fuel and climate change debate*. London: Earthscan, pp. 1–20.

Marsden, T. (2013) From post-productionism to reflexive governance: Contested transitions in securing more sustainable food futures. *Journal of Rural Studies*, 29, 123–134.

Maye, D. (2013) Reassessing agricultural multifunctionality in the context of food security, climate change and the new land use debate. In: Frantal, B., & Martinat, S. (eds.), *New rural spaces: Towards renewable energies, multifunctional farming, and sustainable tourism*. Bruo: Institute of Geonics, Academy of Sciences of the Czech Republic, pp. 60–71.

Maye, D., & Kirwan, J. (2013) Food security: A fractured consensus. *Journal of Rural Studies*, 29(1), 1–6.

McDonald, B.L. (2010) *Food security*. Cambridge: Polity Press.

McMichael, P. (2009) A food regime analysis of the 'world food crisis'. *Agriculture and Human Values*, 26, 281–295.

Nally, D. (2014) Governing precarious lives: Land grabs, geopolitics, and 'food security'. *The Geographical Journal*, 181, 340–349.

Pereira, L.M., & Ruysenaar, S. (2012) Moving from traditional government to new adaptive governance: The changing face of food security responses in South Africa. *Food Security*, 4(1), 41–58.

Pinstrup-Andersen, P. (2013) *The political economy of food price policy: An overview*. Policy Brief No 1, March 2013, Cornell University, UNUWIDER and the University of Copenhagen.

Pothukuchi, K., & Kaufman, J.L. (1999) Placing the food system on the urban agenda: The role of municipal institutions in food systems planning. *Agriculture and Human Values*, 16(2), 213–224.

Pritchard, W. (2016) Food security. In: Daniels, P., Bradshaw, M., Shaw, D., & Sidaway, J. (eds.), *An introduction to human geography*. Essex: Pearson Education, pp. 156–169.

Renting, H., Schermer, M., & Rossi, A. (2012) Building food democracy: Exploring civic food networks and newly emerging forms of food citizenship. *International Journal of Society of Agriculture and Food*, 19(3), 289–307.

Rosin, C., Stock, P., & Campbell, H. (eds.) (2012) *Food systems failure: The global food crisis and the future of agriculture*. London: Taylor and Francis.

Royal Society (2009) *Reaping the benefits: Science and the sustainable intensification of global agriculture*. London: The Royal Society.

Sage, C. (2013) The interconnected challenges for food security from a food regimes perspective: Energy, climate and malconsumption. *Journal of Rural Studies*, 29, 71–80.

Sonnino, R., Marsden, T., & Moragues-Faus, A. (2016) Relationalities and convergences in food security narratives: Towards a place-based approach. *Transactions of the Institute of British Geographers*, 41, 477–489.

Sonnino, R., Lozano Torres, C., & Schneider, S. (2014) Reflexive governance for food security: The example of school feeding in Brazil. *Journal of Rural Studies*, 36, 1–12.

Tomlinson, I. (2013) Doubling food production to feed the 9 billion: A critical perspective on a key discourse of food security in the UK. *Journal of Rural Studies*, 29, 81–90.

World Bank (2008) *Rising food prices: Policy options and World Bank response*. Washington, DC: World Bank.

26

Food-sensitive urban planning

Australian perspectives

David S. Jones and Beau B. Beza

Australian planning and landscape architecture literature and discourse

Academic works in Australian planning and landscape architecture in the last 30 years have largely focused upon approaches to spatial patterns, social and economic contexts, and governance associated with the urban footprint of city centres. Several publications have resulted from these works and ventured into edited essays that profile and discuss a raft of contemporary issues prevalent in Australia (Brunner and Glasson, 2015; Byrne et al., 2014; Gleeson and Low, 2000; Gurran, 2011; Hutchings and Bunker, 1986; McLoughlin and Huxley, 1986; McLoughlin, 1992; Thompson and Maginn, 2012; Wright, 2001) including thematic focuses upon suburbia (Gleeson, 2006), green infrastructure (Beatley with Newman, 2009; Downton, 2009; Hopkins and Goodwin, 2011), infrastructure and transport disconnects (Gleeson and Beza, 2014; Mees, 2000, 2010), and histories (Freestone, 2009, 2010; Sulman, 1921). Increasingly, several scholars are discussing issues involved in peri-urban planning (Kennedy et al., 2016; Martin and Budge, 2011), environmental sustainability (Byrne et al., 2014; Gleeson, 2014; Zeunert, 2017), ethics (Cook and Sarkissian, 2000; Sarkissian, 1996), Indigenous planning (Pieris et al., 2014; Porter and Barry, 2016), landscape architecture projects (Aitken and Looker, 2002; Bull, 2002; Raxworthy and Blood, 2004; Ware and Raxworthy, 2011), and healthy cities (Barton et al., 2015; Townsend and Weerasuriaya, 2010). Food security and planning, however, has not been in these predominant publication discourses, except by a few and often in a generalised discussion (Downton, 2009; James, 2016; Mobbs, 2012; Slade and Wardell-Johnson, 2016; Weller, 2009), and more recent discrete focused discussions linked to human health, community gardens (Calverley, 2017), and urban agriculture (Donovan et al., 2011; Downton et al., 2017; Food Alliance and NHFA, 2012; Gaynor, 2006; NHFA, 2004; Lyons et al., 2013; Nettle, 2014; Pires, 2011). What is clear form this research in and about Australia, is that notions of food planning are in their infancy and formative stages. With this in mind, the authors have identified eight threads that underpin alternate and future food planning discourse in the Australian urban context and these are discussed in more detail herein.

Food-sensitive urban planning in Australia

In Australia, human-urban relationship issues with food are linked to health, energy (Gaballa and Abraham, 2008), land depletion (James, 2016), cost inequity, and inefficiency. Food quality is increasingly linked to rising levels of disease including cardiovascular and cancer, as well as obesity, cholesterol levels, hypertension, anxiety, depression, loneliness and disproportionate consumption of total kilojoules (energy), saturated fats and salt, low fruit and vegetable intake, and an addictive passion for fast foods (AIHW, 2006; Donovan et al., 2011; VHPF, 2010). Social factors of income, transport, physical access, and socio-economic standing compound disproportionate consumption(s) as they are individually or collectively linked to physical and economic access to food. This has been resulting in 'food deserts' in suburbia, rationalisation, and loss of peri-urban food venues due to operational costs thereby increasing travel distances, and the stealth-like strategic generic-brand replacement of food-related products on small/medium/large sized supermarket shelves (ABC, 2011; Chapters 27 and 32). Low density living and spatial arrangements, particularly prevalent in suburban, peri-urban areas, and rural/regional Australia, are heightening accessibility hardship resulting in physical and psychological health deprivation and disadvantage (Buxton et al., 2003; Buxton et al., 2011).

To address this growing health deprivation and disadvantage a range of research and applied fields (e.g. behavioural epidemiology, environmental psychology, landscape architecture) offer solutions in reversing the above effects on people living in cities (e.g. Beza et al., 2015; Townsend and Weerasuriya, 2010), but little attention is being placed on Australia's rural and regional areas (Buxton et al., 2011; Thompson and Maginn, 2012). Within this context, urban planning has largely been left behind in regards to proscribing corrective measures that address issues of public health related to food and its access by communities (Australia, 2013). Current urban planning is market driven by land development interests and stakeholders (e.g, Buxton et al., 2016; Thompson and Maginn, 2012) and is 'tinkering around the edges' in relation to food and the realisation of new or redevelopment or regeneration of existing settings. Urban planning documents, such as *Plan Melbourne Refresh* (2015) are attempting to promote city growth and renewal with community servicing being a goal through policy aspirations supporting a service destination to dwelling 20-minute walking distance (Plan Melbourne Refresh, 2015). These facilities include, "schools, shops, meeting places, open space, cafés, doctors, childcare and access to public transport" (Plan Melbourne Refresh, 2015: 22). Unfortunately, the provision of food-human relationships in this document are primarily discussed in terms of non-urban and peri-urban areas. Food production in the urban setting is inferred, via reference to 'community gardens' and productive streetscapes, as parodied in the Australian Broadcasting Corporation's *Utopia* (2016), but provides no provision of what these community food services may entail or how they may be achieved within contemporary communities.

Threads of food in urban planning in Australia

In colonial Anglophone Australia, food production has been integral to survival and allied with industry, but predicated upon a primary production and export-oriented economic agenda. Unfortunately, colonising cultures overlooked some 60,000 years of Indigenous intergenerational knowledge about terrestrial and aquatic food sources on and in Australia: which also included 'foodways' (Clarke, 2007; Kerwin, 2010; Newton, 2016; Chapters 3 and 34).

When the original colonists arrived in the late 18th century there were no facilities such as conventional 'supermarkets' (until 1946 in the USA) to draw upon, less the larger town markets, and they had to sow/plant, cultivate, harvest, import, and prepare their produce to feed individuals, families, and communities. They also had to navigate sourcing and management of potable water for drinking and food production, and to store it in isolation for increased pollution risk in the absence of conventional sewerage and septic systems (Jones, 2005; Jones and Payne, 1998). During this establishment period of Anglophone Australia there were devastating environmental effects upon the country's arable areas and extensive seed, root-stock, and (stud and noxious) animal invasion assimilation actions in the colonies. The imports formed the basis of initial domestic survival gardens, plant nurseries, farms, and agricultural plots (Gammage, 2011; Jones, 1993; Rolls, 1980). At the same time land was cleared, appropriated, and organised into paddocks, burnt and fenced, and farrowed and sown. Also, accompanying these imported food stocks and practices was the mass transmission of noxious flora and fauna and the imposition of a northern hemisphere-inspired land management system and accompanying (and ongoing) social/cultural expectations for European and British foods/diets that are largely inappropriate for the Australian landscape and environment (Gammage, 2011; Jones, 1993; Jones and Payne, 1998; Pascoe, 2014; Rolls, 1981).

In the urban setting from the end of the Second World War up until the 1970s the yards of many working-class and middle-class families, that owned or were buying properties, accommodated livestock (e.g. chickens) and became small vegetable patches dotted with citrus and nut trees as well as fruit bearing vines (Gaynor, 2006). Even the newly realised estates and public open spaces were planted with food producing flora; which in some areas of Australia and its suburbs are still harvested (Eltahir, 1992; Stallard, 2017). Greek and Italian migrants brought, in particular, agendas of change for domestic gardens, albeit many of their dense and innovative productions were obscured from view in inner city and suburban backyards (Timms, 2006). As people shifted physically out of the colonial inner 'city' the more they relied upon these types of gardens; irrespective of family size or land allotment (Jones and Payne, 1998). The highly desired 'quarter acre lot' in Australia, that provided space for both the garden and home, was born out of the survival imperative (Timms, 2006). While progressively the yard merged into a formal or gardenesque assemblage of floral and aesthetic displays, historically it retained its food supplement role (Gaynor, 2006; Jones and Payne, 1998; Timms, 2006).

With waves of migrations coming to Australia in the 1860s, 1880s, 1910s, 1930s, and 1950s, successive layers of migrants re-appropriated their front and back yards; which were informed by their cultural background (e.g. Chinese, French, German, Turkish) – with many elderly families in Australia today still maintaining the integrity of their highly prosperous food gardens (Aitken and Looker, 2002; Gaynor, 2006; Jones and Payne, 1998; Timms, 2006). In this sense Gaynor (2006: 3) has observed that:

> For these people, home production has long been a source of food valued for its freshness, purity and health-giving qualities. Perhaps it is not so easy, then to neatly assign fowl coops, fruit trees and veggie patches to the category of working-class and rural functionality.

Contemporary Australian society has increasingly, however, turned its back upon the typical colonial sustenance backyard (vegetable garden orchard, vines, chicken coops, fruit trees) in favour of gardenesque representations informed by period garden design fashions and associated literature (Aitken and Looker, 2002; Weller, 2009). Such transformations met current aesthetic trends, the availability of plants and water resources, and also witnessed a shift towards markedly smaller housing lot sizes, larger houses, and the advent of Australian suburbia as an aspiration and practice (Gaynor, 2006; Timms, 2006; Weller, 2009). Thus, progressively over the last 100 years, the Australian suburban 'lot' has decreased from the 'quarter acre' (1011.7 m^2) to a smaller allotment (typically 350 m^2).

Current food production in Australia has largely departed from the supplemental 'backyard' in favour of corporate supermarket solutions resulting from industrial food production methods (see Chapter 16). In this regard, food production at such large spatial scales directly impacts upon our environment through excessive potable water consumption, household greenhouse emissions, inefficient food residual waste, landfills and non-recycling, soil quality depletion and non-renourishment, inappropriate herbicide/fertiliser dispersal and post-farm transmission (Buxton et al., 2011; Donovan et al., 2011; Food Alliance and NHFA, 2012). Many of these variables in the transformation to large scale agricultural production were addressed in the acclaimed *Silent Spring* (Carson, 1962) wherein Carson critiqued the impact of inappropriate chemical usage, dispersal and transmission through soil, water and air; including their long-term impacts directly and indirectly upon human health and wellbeing (see also Barton et al., 2015; Buxton et al., 2011; Townsend and Weerasuriya, 2010). Townsend and Weerasuriya (2010), for example, have discussed contemporary examples of this effect in Melbourne. Yet, a more apt analytical review is contained in the Marxist-inspired analysis offered by White et al. (1978) and in Lintern (2015), where recent archaeo-hydraulic investigations of the post-colonisation suburban and industrial pollution of Melbourne's Yarra River is discussed and analysed;

both conclude that our wasteful 'colonisation' and treatment of the physical landscape of Melbourne has disregarded quality productive soils with building development and polluted soils with industrial product and by-product waste solids and liquids echoing the same concerns of Carson (1962).

Research on food planning in Australia

The above material provides a brief historical overview of informal food planning practices in Australia. Unfortunately, research examining contemporary interactions, perceptions and understanding about food related issues by urban planners in Australia is limited. Much contemporary discourse about food and city planning, for example, has lamented the Australian inability to foresee, plan, and commit to sensitive land use planning (for example MTPC, 1929; Plan Melbourne, 2014). Such concerns include the prevalence of car and road dominance, fossil fuel depletion, carbon emissions and climate change, consumption of quality land for suburbia, and the panacea of multi-energy consumption demands. Responses to these concerns, and research thereto, tends to focus on the city fringe rather than looking within the inner and middle ring suburbs, and few proactively seek to offer innovative interventions to tackle these issues (James, 2016). Threads of community gardens, urban agriculture, biophilic designed cities, permaculture, vertical gardens, roof gardens, green urban infrastructure, temporary gardens, urban forests, and the forgotten 'backyard' offer possibilities and innovations from which Australian cities could benefit (see: Downton et al., 2017; VEIL, 2017). Many of these 'threads', however, are ill understood by the community, municipal officials, and practitioners. Allen (2004), for example, in an Australian state-based investigation, examined the present and potential influence of permaculture on landscape architecture in South Australia. Her research concluded that the present exposure and understanding of permaculture by Australian landscape architects is varied; their overall interaction with, and influence of permaculture concepts is very limited, and where present, unpublicised. Thus, the key issues contributing to this low level of interaction and influence were determined as current community and client priorities, differences in approaches to aesthetics and the role of animals, and establishment and maintenance timeframes.

In North America Pothukuchi and Kaufman (2000: 113) raised similar topics, concluding that "the food system [. . .] is notable by its absence from the writing of planning scholars, from the plans prepared by planning practitioners, and from the classrooms in which planning students are taught" of which the same conclusion was implicitly drawn about Australian planning programs and their ethical relationship to human and land sustainability (Cook and Sarkissian, 2000; Sarkissian, 1996). For other international perspectives on food and urban planning, see: Cabannes and Dubbeling (2000), APA (2007), Hodgson (2010) LA and GLA (2010), Viljoen and Wiskerke (2012), Morgan (2013), Roggema (2016), and Ilieva (2016). Yet, in some contemporary Australian institutions the topic of food and planning has gained formative currency and is now taught and researched in some urban planning programs (e.g. Deakin and UNSW universities).

In terms of planning agencies, Pothukuchi and Kaufman's (2000) survey of 22 USA organisations in 1997–1998 identified that the top ten issues confronting planners in North America as it related to the food system mainly reflected consideration of product/consumer relationships rather than what Nasr and Komisar (2012: 47) conclude as "the disconnection between [agricultural] food systems and city dwellers." In this context, there is an evident disconnection between agricultural food systems and city dwellers, but urban foodways are complex, intricate, and meaningful. Therefore, there is a problem that 'food systems' is taken just to mean production and distribution. Such is, in its own right, a little discussed issue confronting planning and design practitioners/academics generally, despite a coalescence under the 'healthy cities' epithet (Barton et al., 2015; Thompson and Maginn, 2012). Contrasting this, in the late 1900s in North America and Australia the issues practicing and academic land use planners (and to a lesser extent health practitioners and associated academics) identified were activities that involved what Pothukuchi and Kaufman (2000: 115–116) summarised as:

mapping of food sector activities, sustainable agriculture, community supported agriculture, food cooperatives and distribution alternatives, nutrition and health education, food safety, food impacts on environment and solid waste, community food security, and other – an open category.

Additionally, Barton et al. (2015), Donovan et al. (2011), and Thompson and Maginn (2012) have also observed these issues in their respective settings.

To address the preceding problems both planners and politicians need to have greater awareness of issues related to food as well as the need to initiate a policy shift in their approach to land use and food planning. Land use and food planning therefore have the potential to be a central tenet of sustainable development and new community benefit. Within this potential, however, one must reconcile and accommodate the perceptual difference between food producing agricultural land adjacent to and increasingly encroached upon by cities, and the urban agriculture possibilities of producing food on building rooftops, in a city's open spaces and unused transport corridors, and within community gardens, 'ecovillages', and school grounds (see Chapter 27). Thus, rethinking the linear associations of the overall food production/consumption journey, including applying chain metaphors (APA, 2007), to understanding and applying larger conceptual metaphors that characterise and account for the various scales of interdependence and interaction humans, the environment and the food we have with one another (Morton, 2007, 2010). A major obstacle in the rethinking new land use/food planning directions is that it depends upon a legislative governance commitment by political representatives to this realm of change and as to whether or not any concern indeed exists by planning decision-makers and their action then following.

In the State of Victoria, for example, the *Planning and Environment Act 1987* (Victoria, 1987: 1) is empowered "to establish a framework for planning [...] in the present and long term interests of all Victorians"; the latter element being a potential key feature in the protection of current and long-term agricultural land as it can guard against urban encroachment. The Victorian State Planning Policy Framework (Victoria, 1987: clause 10.02), within the *Act*, echoes the *Act's* purpose requiring "appropriate land use and development policies and practices which integrate relevant environmental, social and economic factors in the interests of net community benefit and sustainable development".

Similar land use planning imperatives appear in comparable state and territory settings such as Western Australia, South Australia, Tasmania, New South Wales, and Queensland, but each state has different land use planning policies, governance, and political agendas, and peri-urban/agricultural land interpretation approaches. State/Territory action also depends upon the natural topographical barriers that envelop Australia's respective metropolitan areas like the Darling Hills in Perth (Western Australia), The Hills Face Zone in Adelaide (South Australia), and the River Derwent corridor bisecting Hobart (Tasmania).

Any such rethinking and/or repositioning of the *Act*, in an urban planning sense, would include shifting perceptions that food production is an integral part of planning and requires political commitment. Food production as an integral part of planning requires:

- Positioning long-term food production into relevant upper level planning policy instruments;
- Identifying that food production must be a mandatory consideration of development approvals, and its demonstrable inclusion as a percentage of horizontal/vertical area;
- Prioritizing food as an integral component of sustainability (performance) and evaluation; and
- Increasing planning practitioner, politician, and community comprehension of the topic and its implications.

While legislative policy change at these 'upper levels' of planning explicitly requires due concern for food security and the conservation of lands that enable quality produce to be grown and harvested, what transpires on the 'ground' level is ill-informed by those responsible (e.g. private practitioner and local government planners) for respecting and implementing this *Act's* aim and framework aspirations. In terms of the latter, aspirations pertain to the origins of the urban planning profession, often influenced by Ebenezer Howard's Garden City ideals that urban centres need to be logically ordered so as to permit efficiencies of movement, quality human lifestyles, access to rural landscapes and their foodbowls, and an order of development (Freestone, 1989, 2009, 2010; Chapters 27 and 40). Paralleling these aspirations municipal planners in, for example, Victoria, mostly act upon 'right of use' and whether or not (re)development conforms or not to the local planning scheme (or development plan). Many of these planners and the legislative bodies (e.g. Victorian Civil and Administrative Tribunal) commonly overlook the *spirit of intent* that lead to the development and application of the planning framework.

Food and planning

To go back a step, in terms of food and planning, the overall food system involves what the American Planning Association (APA, 2007) defines, as a "chain of activities [. . .] [including] the production of food [. . .] [its] processing, distributing, wholesaling, retailing, and consumption of food and eventually the disposal of waste". In essence, the food system comprises four key phases: producing food; processing and transporting food; access and utilisation; and waste, re-use, and post-use management (APA, 2007).

In the conventional planning sense, this aforementioned 'chain' assumes commercial production and resourcing of food and, as previously mentioned, follows a linear interpretation of human/environment/food relationships. The food metaphor is very linear, food systems, their interactions, and foodways are ecological and systematic in their construction; the web or mesh is often used as a preferred metaphor, though of course it is also imperfect (Ingold, 2011; Morton, 2016, 2017; Chapter 34). Conventional planning in Australia, as the authors observe, does not generally think alternatively, or, in the spirit of this discussion, make allowances for the alternate communal, urban domestic garden, or 'farm gate' possibilities of growing and/or acquiring food. Thus, conventional state and municipal related planning does not account for the micro in scale or human communal rich 'industries' and contributions that are prevalent in today's society. This results in planning failing to facilitate and adequately accommodate human scale and complex systems in its forward planning deliberations and ideations. Nor does conventional planning engage with Indigenous (Chapter 3), or pre-colonisation modes of Australian food systems (Newton, 2016; Pascoe, 2014), or indeed comprehend that humans are a segment within a biota in which wildlife equally engages in food systems production and management; often in a synergistic and co-dependent relationship with human food production.

At Australia's national level, food and planning also reflect a linear producer/consumer relationship where documents such as the *National Food Plan* (2013: 14) focus on:

> ensuring that Australia has a sustainable, globally competitive and resilient food supply that supports access to nutritious and affordable food.

A noble endeavour, yet in terms of land use and its relationship with food in Australia this consideration is 'farmed out' to other national urban policy documents, such as *Our City, Our Future* (Australia, 2011). Yet, in this document land is generally considered as a commodity in the provision of housing for Australia's growing urban populations, with planning related considerations delegated down to the country's States and Territories by directing reform, amongst other considerations, to "improve planning and land release processes" and "to improve housing supply and affordability" (Australia, 2011: 61).

In contrast, Dutch land use planners Broekhof and van der Valk (2012: 393) argue that "food relates to transport planning, planning green spaces, commercial strips and retail markets, conservation of fertile land, mixed land uses, localising urban farms, and much more." Instead, conventional land use planning in Australia remains human-centric, devoid of ecological systems accountability, macro-scale in its municipal bound (re)development approaches and interventions when it comes to food planning and security. In one sense this approach to land use planning in Australia supports what Broekhof and van der Valk (2012: 394) refer to as de-territorialisation and its "consequences of the global-industrial food system [...] such as monoculture, spatial homogeneity, uniformity in the landscape".

Where Australian national and state level planning fails to address the food/planning nexus, organisations (such as Urban Agriculture Australia, Australian City Farms, Sustain: The Australian Food Network, and Community Gardens Network) have consolidated various urban agricultural activities throughout Australia under operational umbrellas. This collective activity is, in a sense, 'informal' or outside the formal regulatory system in Australia, self-governing, and being initiated to help disseminate urban agricultural knowledge and practices throughout the country.

Lateral to 'formal or legislated' food/planning considerations, the Heart Foundation of Australia (Donovan et al., 2011) has taken a particular interest in the role and impacts of the built environment upon the health of the Australian population. The Foundation's pioneering impetus over the past 20 years has heralded several investigations that have explored, scaffolded evidence and put forward recommendations to change the fabric of the built environment (new and existing) to improve the health of Australia citizens (Donovan et al., 2011). This includes raising the concept of 'food sensitive planning and urban design' (FSPUD) as a political agenda (Donovan et al., 2011; Larsen, 2008; Victoria, 2010).

In this sense Donovan et al. (2011: 5) have explained that:

> FSPUD is an approach to planning and urban design that explicitly addresses the way food is produced, moved, processed and consumed, to create places that make it easy for people to meet their food needs. FSPUD also considers how meeting food needs can contribute to other objectives of planning and urban design, classified in this resource as: health and fairness; sustainability and resilience; livelihoods and opportunity; [and] community and amenity.

Returning to the state level of planning considerations in Australia, where the Heart Foundation's framework could be implemented (Donovan et al., 2011), the Victorian State Planning Policy (SPP) (Victoria, 2016) articulates a state vision downwards into respective Local Planning Policy Frameworks per municipality with, generally, no explicitly stated provision or accountability for urban agriculture in either the State or local planning frameworks. The SPP informs planning accountability across the state, requiring land use "issues to be determined and [the need to] balance conflicting objectives in favour of net community benefit and sustainable development" (Victoria, 2010: 2) and accountability discretion.

At the local level, however, provision for urban agriculture does exist, to a degree, and can be found in the various municipal related literature pertaining to community gardens (e.g. CERES, 2017; City of Melbourne, 2013; City of Yarra, 2014; City of Port Phillip, 2016, 2017; City of Glen Eira, 2017) where policy, guidelines, and/or management plans help direct the development and use of these by community members. Such places and their respective municipal policies, guide the development and use of potential or existing gardens on municipal-owned land but not the individually-owned household parcel. What municipality-specific gardening material and the previously mentioned gardening networks identify is that urban agriculture in Australia is largely an uncoordinated food growing endeavour. Thus, to have the greatest effect in use, production and accessibility to the community, urban food planning as a topic should be embedded at the State level and as an element in this SPP. In this sense a Municipal Local Planning Policy Framework then would take its lead from the SPP and be expressed at a micro-level in planning schemes (or development plans) via strategies, zones, overlays, and policies.

Additionally, and in terms of supporting such a micro-level application, municipalities have many statutory and/or strategic instruments that they can draw from to improve urban agriculture in their respective setting. These include rural land strategies, municipal public health plans established under the Victorian *Public Health and Wellbeing Act 2008*, open space requirements and recreational provision strategies, subdivision guidelines and standards, retail planning policies, structure plans and growth strategies, and urban design/urban landscape strategies, of which comparable legislation exists in other Australian States and Territories. Each instrument or avenue offers opportunities for planners to address and request food planning concerns and measures. However, to exercise these instruments proactively and for the benefit of the community, municipalities must demonstrate leadership, establish a need, and establish commitment to the growing of agriculture in urban settings.

Innovations and opportunities

Stemming from the material discussed in this chapter, the authors identify eight threads that frame and underpin the food planning discourse within Australia: (1) the emergence of food planning and production (i.e. urban agriculture that is seeking commercial viability and a community/network scale beyond a domestic scale); (2) the nature of backyard and community garden yield; (3) the potential of 'food hubs' or community-based nodes that enable barter/swap/purchase of farm and allied produce; (4) *biophilia* and biophilic design (i.e. the intertwining of urban development and nature within existing/proposed built settings; (5) the invisible urban landscape (i.e. urban spaces that may allow for the production of food); (6) a lack of uncertainty and capacity as to Indigenous Australian native plant food sources; (7) the role of permaculture in urban food production facilitation and yield; and (8) the larger discussion about urban forests (i.e. a tree-dominated urban ecosystem).

The first thread to be discussed relates to the emergence of food planning and production. Innovation in Australian food planning over the last ten years has been largely absent, but small threads of inquiry have paralleled the emergence of urban agriculture and its application in urban Australia. Pires (2011: 3) concludes, and as alluded to earlier, that while there are "hundreds of potential urban agriculture related policies, strategic plans and regulations" in Australia, none of the municipalities he surveyed had "a substantial document holistically addressing urban agriculture and all of its facets." Additionally, Pires (2011) also concludes that the majority of these documents rotated around animal keeping, poultry keeping, beekeeping, community gardens composting, green roofs, markets, and streetscapes. In this sense Australian urban agriculture may be acknowledged as being in its infancy, has yet to realise its maturation, and has not yet been considered a major element to be accounted for in planning documentation. However, urban agriculture is well established at the grassroots, community, and social level; which is an urban agriculture dimension described in Chapter 12.

A second thread of discourse in food planning in Australia is the lack of data on what is transpiring in household backyards and community garden settings, including, for example: how much food is being produced for individual/family consumption in a domestic or community setting; the volume of food production; whether or not excess produce is being sold (or traded), transported, and consumed off-site; and what are the net individual/community benefits and/or disadvantages of urban agriculture (VEIL, 2017). Zainuddin (2013) has concluded, from a survey of 20 households in the Coburg-Preston regions in Melbourne that individual garden plots of between 7–250 m^2 were evident. Of these the highest volume yield was 50 kg arising from an 80 m^2 plot but that yields were extremely variable accordingly to land holder passion. Further, Zainuddin recorded a collective yield of 389 kg from 15 backyards totalling 1,096 m^2 over a 12-month period (Zainuddin, 2013: 37, 49). Similar data is coming through the work of the Brookman's family permaculture farm and education centre in Gawler, South Australia (Food Forest, 2017).

In the last five years in Australia there has been discussion about and movement towards the notion of 'food hubs' (AFHN, 2017; Deakin University, 2016; VEIL, 2017), the third thread. The Australian Food Hubs Network (2017) define Hubs as providing:

> the 'missing local middleman' in local food infrastructure. [They] [. . .] work directly with farmers to assist in the marketing and distribution of their products, making it easier for local businesses and communities to access fresh local food [. . .] and usually focus first on coordinated marketing and

> distribution of local fresh produce, [...] range in scale from volunteer-run buying groups using temporary spaces for receipt and packing of goods (like community or school halls, churches and garages) to permanent and well-established Hubs providing a variety of business, educational and/or food access services. Some have no staff and are purely run by volunteers or coop members, others are significant employers.

Barham et al. (2012: 4) also define food hubs as "a business or organisation that actively manages the aggregation, distribution and marketing of source identified food production primarily from local and regional produces to strengthen their ability to satisfy wholesale, retail and institutional demand". Thus, hubs are distillation and exchange nodes but there is uncertainty as to the economic feasibility and viability of such hubs.

A fourth thread is *biophilia* (Wilson, 2016) and biophilic design (Beatley, 2017), which should not be confused with Tuan's *Topophilia* concept which literally means 'love of place' (Tuan, 1972). In Australia there is growing concern that nature is increasingly disappearing in cities through various urban (re)development means and, in particular, 'nature', 'green', and food planning are integral elements of human attachment to place that are being lost. Yet interest in this concept is in part aided by Beatley's and Newman's advocacy of this concept, but their engagement of food in *biophilia* is cursory (see Beatley, 2010, 2017; Newman, 2012). Thus, the real possibilities of food-plant informed *biophilia* urban transformations is overlooked. Further, integrating nature and 'green' along with food planning into our cities scaffolds our wellbeing (Donovan et al., 2011) and are sub-themes in the larger healthy cities planning inquiry (Beatley, 2010, 2017; Beza et al., 2015; Downton et al., 2017; Newman, 2012; Söderlund, 2015). Thus, as urban populations increase and denser cities become more prevalent, biophilic design is emerging as a critical reaction to the hardscape built environment generally created in the urban setting.

The fifth thread, mentioned earlier, refers to the invisible urban landscape that exists within our Australian private and public settings and refers to the notion of flora as a supplemental source of food production in the city and scaffold to support one's wellbeing, rather than the volume of yield discussed in the second thread (Anon., 2017). This invisible realm includes the food-producing (and ornamental) plant life in the front and backyards of residents, in nature strips, parklands, roadsides, freeway corridors, and so on, that were consciously planted in these tracts of land by acts of design or vernacular activities; some of which are formally and or informally cultivated and harvested (Eltahir, 1992; Stallard, 2017). Unfortunately, there is little research on what is transpiring in this realm, but it is evident in the above historic related food discussion that the invisible urban landscapes, that maybe taken for granted, are actually productive in some manner. Such invisible landscapes are either formal (e.g. planted gardens, street tree plantings) or informal (e.g. pop-up gardens, ad hoc, and/or infrequently planted street trees). Such plantings, in terms of the latter, are invariably hard to locate and map, and most locations of these plantings are known by oral traditions or knowledge. Increasingly, however, they are being noticed as part of municipal open space GIS-positioned asset management inventories.

A complementary element to the above discussion is the sixth thread; which revolves around the Australian interest in the previously ignored Indigenous food sources, and their commercial harvesting possibilities. As mentioned previously in the historic related food discussion, when Europeans ventured onto the Australian landscape they experienced a new terrestrial and aquatic plant and wildlife regime – thus, a new food resource regime for which there were no guide books on how to use. Jones and Clarke discuss in detail this context, and its contemporary discourses in Chapter 3 of this book. However, in essence and in terms of this current chapter Australian planning and landscape architecture academics, practitioners, and policy makers have little addressed the alignment and potential of Australian 'bush tukka' plants in their plans and research aspirations (see Brunner and Glasson, 2015; Bull, 2002; Byrne et al., 2014; Kennedy et al., 2016; Raxworthy and Blood, 2004; Thompson and Maginn, 2012; Ware and Raxworthy, 2011; Weller, 2009). This is a remarkable oversight, as this food source sustained human life in Australia for 60,000 years.

A seventh thread is the role of permaculture in urban food production facilitation and yield (Chapters 10, 16 and 24). Permaculture is a system of social and agricultural design principles focused upon enhancing or directly using natural ecosystem patterns and features. Permaculture co-originator Mollison (1991: 1) has said:

> Permaculture is a philosophy of working with, rather than against nature; of protracted and thoughtful observation rather than protracted and thoughtless labour; and of looking at plants and animals in all their functions, rather than treating any area as a single product system.

Food security is a basic tenet of permaculture theory and practice. While permaculture is clearly acknowledged in literature and can be positioned in any number of the previously mentioned threads, as briefly discussed in the second, the discipline still today lacks academic credibility despite the many adherents inside and outside academia (Holmgren, 2002; Mollison, 1979, 1988, 1991; Mollison and Holmgren, 1978). What influence permaculture has had upon Australian society is equally unclear. Holmgren (2002: 271) expresses permaculture as:

> a dynamic interplay between two phases: on one hand, sustaining life within the cycle of the seasons, and on the other, conceptual abstraction and emotional intensity of creativity and design [. . .] the relationship between these two [is] like the pulsing relationship between stability and change. It is the steady, cyclical and humble engagement with nature that provides the sustenance for the spark of insight and integration (integrity), which, in turn, informs and transforms the practice. The first is harmonious and enduring; the second is episodic and powerful. The joyful asymmetric balance between the two expresses our humanity.

The eighth thread is the larger urban forest discourse prevalent in cities like Melbourne (City of Melbourne, c.2013, 2012), Sydney (City of Sydney, 2013, 2017), Geelong (City of Greater Geelong, 2015, 2017), and Adelaide (Jones, 2007; South Australia, 2013) where the emphasis has been upon urban forest security policies and actions informed by tree deterioration, carbon sequestration and addressing the urban heat island affect. Significantly the urban forest movement in Australia, until very recently, altogether excluded the idea of these trees being productive and edible (Eltahir, 1992; Zeunert, 2011; Stallard, 2017). Urban forests are notionally artificially (human-created/cultural landscapes) creating tree-dominated ecosystems that are occupied by a different regime of wildlife. This is different from a natural ecosystem whereby nature creates the vegetation and wildlife succession unaided by substantive human intervention (Hobbs and Higgs, 2013; Jonnes, 2016).

Adelaide, by virtue of historical chance, has enveloped sections of its national heritage listed Park Lands (Australia, 2017) in productive olive (*Olea europaea*) groves. These groves unconsciously achieve an urban forest setting at the same time as historically replanting a barren expanse that was deforested in the initial years of European colonisation. This original open woodland provided employment, and an economic fuel resource in the city's first instance that later became a food resource (Jones, 2005, 2007). The latter is quite instrumental because it is an essential part of the Adelaide Park Lands, and sections of land in metropolitan Adelaide are food forests of olive, nut, and fruit and not just urban ornamental forests (Eltahir, 1992; Jones, 2005, 2007; Peter, 2008). As a note, olive (*Olea europaea*) plants are legally deemed a noxious weed in South Australia under Section 174 of their *Natural Resource Management Act 2004* except where "not planted and maintained for domestic or commercial use" (South Australia, 2017a, b).

Cultivating change and innovation

A key aspect and challenge of Australian food planning is heightening its acceptance and primacy as an integral component of urban sustainability. Hence, to effectively move forward on food planning in Australia, we would conclude that urban planning in Australia and, particularly, at the state level needs to:

- Incorporate food planning and security as a measurable outcome of sustainability in state strategic polices and associated legislation;
- Encourage decision makers and design practitioners and academics, across Australia, to develop an understanding of urban agriculture systems and relationships for its application in Australian cities;
- Encourage by way of planning/design incentives the insertion of food conducive landscapes into both the private and public domains of council and Crown land jurisdictions (e.g. nature strips, orchards and groves, green corridors, common land, public green spaces, community gardens, and village gardens); and
- To re-imagine the productive potential of the Australian 'backyard' aligned to its historical antecedent as a survival landscape but to also support people's wellbeing through interaction with greenery.

In terms of the Australian citizen advancing the use of food planning in the state legislative framework we encourage individuals and community groups to continue: to self-organise and promote achievements in urban agriculture; to engage with municipal councils and Crown land managers to demonstrate the benefits of urban agriculture; to work with design practitioners and municipal council officers to develop a planning incorporated urban agriculture vision and plan to achieve it; to addresses site specific applications of urban agriculture through to coordination of the 'chain of activities'; and to draw from one another and work collectively to lobby other councils where urban agriculture is not 'seen' as a positive attribute in the urban setting.

Further reading

Aiken, R., & Looker, M. (eds.) (2002) *Oxford companion to Australian gardens*. Melbourne: Oxford University Press in association with the Australian Garden History Society.

Allen, S. (2004) *The influence of permaculture upon landscape architecture in Australia: Present and potential*. Unpublished BLArch (Hons) Thesis, University of Adelaide, Adelaide.

American Planning Association (APA). (2007) *Policy guide on community and regional food planning*. https://www.planning.org/policy/guides/adopted/food.htm (accessed 1 November 2017).

Anon (2017) *Urban food street*. www.urbanfoodstreet.com/ (accessed 1 January 2017).

Australia (2011) *Our cities, our future – a national urban policy for a productive, sustainable and liveable future, 2011*. Canberra: Infrastructure Australia.

Australia (2013) *National food plan, our food future*. Canberra: Department of Agriculture, Fisheries and Forestry.

Australia. (2017) *National heritage places: Adelaide park lands and city layout*. www.environment.gov.au/heritage/places/national/adelaide-parklands (accessed 1 February 2017).

Australian Broadcasting Commission (ABC). (2011) *Hungry beast: Coles and woolies*, www.youtube.com/watch?v=lnQMOyGeP_I (accessed 24 May 2011).

Australian Broadcasting Commission (ABC). (2016) *Utopia*. www.abc.net.au/tv/programs/utopia/ (accessed 1 February 2017).

Australian Food Hubs Network (AFHN). (2017) *What are food hubs?* www.foodhubs.org.au/ (accessed 1 February 2017).

Australian Institute of Health and Welfare (AIHW). (2006) *Australia's health 2006*. Cat. No. AUS 73. Canberra: Australian Institute of Health and Welfare.

Barham, J., Tropp, D., Enterline, K., Farbman, J., Fisk, J., & Kiraly, S. (2012) *Regional food hub resource guide*. U.S. Department of Agriculture, Agricultural Marketing Service. http://ngfn.org/resources/ngfn-database/knowledge/FoodHubResourceGuide.pdf/view (accessed 1 February 2017).

Barton, H., Thompson, S., Burgess, S., & Grant, M. (eds.) (2015) *The Routledge handbook of planning for health and well-being: Shaping a sustainable and healthy future*. Oxford: Routledge.

Beatley, T. (2010) *Biophilic cities: Integrating nature into urban design and planning*. Washington, DC: Island Press.

Beatley, T. (2017) *Handbook of biophilic city planning and design*. Washington, DC: Island Press.
Beatley, T., & Newman, P. (2009) *Green urbanism down under: Learning from sustainable communities in Australia*. Washington, DC: Island Press.
Beza, B.B., Veitch, J., & Hanson, F. (2015) Improving the health of Australians by applying evidence from behavioural epidemiology to urban design projects. In: Beza, B.B., & Jones, D. (eds.), *Book of proceedings: 8th international urban design conference 2015: Empowering change: Transformative innovations and projects* (pp. 1–18). Nerang: AST Management Pty Limited.
Broekhof, S.M., & van der Valk, A.J.J. (2012) Planning and quest for sustainable food systems: Explorations of unknown territory in planning research. In: Viljoen, A., & Wiskerke, J.S.C. (eds.), *Sustainable food planning: Evolving theory and practice*. Wageningen, The Netherlands: Wageningen Academic Publishers, pp. 393–404.
Brunner, J., & Glasson, J. (eds.) (2015) *Contemporary issues in Australia urban and regional planning*. Oxford: Routledge.
Bull, C.J. (2002) *New conversations with an old landscape – landscape architecture in contemporary Australia*. Melbourne: Images Publishing Group.
Buxton, M, Alvarez, A., Butt, A., Farrell, S. Densley, L., Pelikan, M., & O'Neill, D. (2011) *Scenario planning for Melbourne's Peri-urban region*. Melbourne: RMIT University.
Buxton, M., Goodman, R., & Budge, T. (2003) *A failed experiment: The performance of the Victoria planning provisions and the new format planning schemes*. Melbourne: School of Social Science and Planning, RMIT University.
Buxton, M., Goodman, R., & Moloney, S. (2016) *Planning Melbourne: Lessons for a sustainable city*. Melbourne: CSIRO Publishing.
Byrne, J, Sipe, N., & Dodson, J. (eds.) (2014) *Australian environmental planning: Challenges and future prospects*. Oxford: Routledge.
Cabannes, Y., & Dubbeling, M. (2000) Urban agriculture and urban planning: What should be taken into consideration to plan the city of the 21st century. Presented at *Urban Agriculture and Horticulture, the Linkage with Urban Planning*. July 7–9, 2000. Berlin, Germany.
Calverley, J. (2017) *The urban farmer: How to create a productive garden in any space*. Brunswick East: Centre for Education and Research in Environmental Strategies (CERES).
Carson, R. (1962) *Silent spring*. Boston, MA: Houghton Mifflin; Cambridge, MA: Riverside Press.
Centre for Education and Research in Environmental Strategies (CERES). (2017) *CERES*. http://ceres.org.au/ (accessed 1 January 2017).
City of Glen Eira. (2017) *Neighbourhood sustainable gardening program*. www.gleneira.vic.gov.au/Community-services/Environmental-sustainability/Neighbourhood-Sustainable-Gardening-Program (accessed 1 February 2017).
City of Greater Geelong (2015) *Urban forest strategy 2015–2025*. Geelong: City of Greater Geelong Council.
City of Greater Geelong. (2017) *Urban forest strategy 2015–2025*. www.geelongaustralia.com.au/urbanforest/documents/item/8d30153dfee2a6c.aspx (accessed 1 February 2017).
City of Melbourne. (2012) *Urban forest strategy: Making a great city greener 2012–2013*. www.melbourne.vic.gov.au/SiteCollectionDocuments/urban-forest-strategy.pdf (accessed 1 February 2017).
City of Melbourne. (c.2013) *Community gardens policy*. www.melbourne.vic.gov.au/ . . . /city_melbourne_community_garden_policy.pdf (accessed 1 February 2017).
City of Port Phillip. (2016) *City of Port Phillip urban agriculture policy snapshot*. www.circlesoffood.org/2016/10/17/port-phillip-urban-agriculture-policy-snapshot/ (accessed 1 February 2017).
City of Port Phillip. (2017) *Parks, playgrounds and trees: Community gardens*. www.portphillip.vic.gov.au/gardens.htm (accessed 1 February 2017).
City of Sydney. (2013) *Urban forest strategy*. www.cityofsydney.nsw.gov.au/live/trees/urban-forest (accessed 1 February 2017).
City of Sydney. (2017) *Urban forest strategy*. www.cityofsydney.nsw.gov.au/live/trees/urban-forest (accessed 1 February 2017).
City of Yarra. (2014) *Urban agriculture in Yarra: Growing and sharing food where you live – draft urban agriculture strategy 2014–2018*. www.yarracity.vic.gov.au/DownloadDocument.ashx?DocumentID=11156 (accessed 1 February 2017).
Clarke, P.A. (2007) *Aboriginal people and their plants*. Dural Delivery Centre: Rosenberg Publishing.
Cook, A., & Sarkissian, W. (2000) Who cares? Australian planners and ethics. In: Bishop, P., & Preston, N. (eds.), *Local government, public enterprise and ethics*. Sydney: The Federation Press, pp. 112–137.

Deakin University (2016) *Geelong food hub feasibility study*. Geelong: School of Architecture and Built Environment, Deakin University.

Donovan, J., Larsen, K., & McWhinnie, J.A. (2011) *Food-sensitive planning and urban design: A conceptual framework for achieving a sustainable and health food system*. Melbourne: David Lock and Associates, The University of Melbourne and National Heart Foundation of Australia.

Downton, P.F. (2009) *Ecopolis: Architecture and cities for a changing climate*. Cham: Springer.

Downton, P.F., Jones, D.S., Zeunert, J., & Roös, P.B. (2017) *Creating healthy places*. Melbourne: Melbourne Metropolitan Metro Authority.

Eltahir, S. (1992) *The management of food-producing trees in the public places of metropolitan adelaide*. Unpublished MEnvSt Thesis, Mawson Graduate Centre for Environmental Studies, University of Adelaide, Adelaide.

Food Alliance and National Heart Foundation of Australia (NHFA) (Victorian Division) (2012) *Planning for food: Towards a prosperous, resilient and healthy food system through Victoria's metropolitan planning strategy*. Melbourne: Food Alliance and National Heart Foundation of Australia (Vic Division).

Food Forest. (2017) *The food forest*. www.foodforest.com.au/ (accessed 1 February 2017).

Freestone, R. (1989) *Model communities: The garden city movement in Australia*. Melbourne: Nelson.

Freestone, R. (ed.) (2009) *Cities, citizens and environmental reform: Histories of Australian town planning associations*. Sydney: Sydney University Press.

Freestone, R. (2010) *Urban nation: Australia's planning heritage*. Melbourne: CSIRO Publishing.

Gaballa, S., & Abraham, A. (2008) *Food miles in Australia: A preliminary study of Melbourne, Victoria*. Melbourne: Centre for Education and Research in Environmental Strategies.

Gammage, B. (2011) *The biggest estate on Earth: How Aborigines made Australia*. Sydney: Allen and Unwin.

Gaynor, A. (2006) *Harvest of the suburbs – an environmental history of growing food in Australian cities*. Crawley: University of Western Australia Press.

Gleeson, B. (2006) *Australia's heartlands: Making space for hope in the suburbs*. Sydney: Allen and Unwin.

Gleeson, B. (2014) *Questioning cities: The urban condition*. Oxford: Routledge.

Gleeson, B., & Beza, B.B. (eds.) (2014) *The public city: Essays in honour of Paul Mees*. Carlton: Melbourne University Press.

Gleeson, B., & Low, N. (2000) *Australian urban planning: New challenges, new agendas*. Sydney: Allen and Unwin.

Gurran, N. (2011) *Australian land use planning: Principles, systems and practice*, 2nd ed. Sydney: Sydney University Press.

Hobbs, R.J., & Higgs, E.S. (2013) *Novel ecosystems: Intervening in the new ecological world order*. New York: John Wiley and Son.

Hodgson, K. (2010) *Urban agriculture growing healthy, sustainable communities*. Chicago: APA Planners Press.

Holmgren, D. (2002) *Permaculture: Principles and pathways beyond sustainability*, Hepburn Springs: Holmgren Design Services.

Hopkins, G., & Goodwin, C. (2011) *Living architecture: Green roofs and walls*. Melbourne: CSIRO Publishing.

Hutchings, A., & Bunker, R. (eds.) (1986) *With conscious purpose: A history of town planning in South Australia*. Adelaide: Wakefield Press in association with the Royal Australian Planning Institute (South Australian Division).

Ilieva, R. (2016) *Urban food planning: Seeds of transition in the global north*. Oxon: Routledge.

Ingold, T. (2011) *Being alive: Essays on movement, knowledge and description*. Oxford: Routledge.

James, S. (2016) *Farming on the fringe: Peri-urban agriculture, cultural diversity and sustainability in Sydney*. Cham: Springer International Publishing.

Jones, D.S. (1993) Traces in the country of the white cockatoo (Chinna junnak cha knæk grugidj): A quest for landscape meaning in the Western District, Victoria, Australia. Unpublished PhD, University of Pennsylvania, Philadelphia PA.

Jones, D.S. (1998) Planning a place in the Sun: Monarto, South Australia. *Urban Design Studies*, 4, 71–88.

Jones, D.S. (2005) The ecological history of Adelaide 3: The historical evolution of the present landscape. In: Daniels, C.B., & Tait, C.J. (eds.), *Adelaide – nature of a city: The ecology of a dynamic city from 1836 to 2036*. Adelaide: BioCity, pp. 56–74.

Jones, D.S. (2007) *Adelaide park lands and squares cultural heritage landscape assessment for the Corporation of the City of Adelaide Council*. Adelaide: School of Architecture, Landscape Architecture and Urban Design, The University of Adelaide.

Jones, D.S., & Payne, P. (1998) *Gardens in South Australia 1840–1940: Guidelines for design and conservation*. Adelaide: Heritage SA.

Jonnes, J. (2016) *Urban forests: A natural history of trees and people in the American cityscape*. New York: Viking Books.

Kennedy, M., Butt, A., & Amati, M. (eds.) (2016) *Conflict and change in Australia's Peri-urban landscapes*. Oxford: Routledge.

Kerwin, D. (2010) *Aboriginal dreaming paths and trading routes: The colonisation of the Australian economic landscape*. Eastbourne, East Sussex: Sussex Academic Press.

LA and Greater London Authority (GLA) (2010) *Cultivating the capital: Food growing and the planning system in London*. London: Greater London Authority.

Larsen, K. (2008) *Food sensitive urban design*. Presentation. Sydney: University of Sydney, Agri-Food XV conference, 2008: 25–29 November. www.ecoinnovationlab.com/uploads/attachments/123_VEIL%20FSUD%20FINAL%20small.pdf (accessed 1 September 2016).

Lintern, A.M. (2015) *Using sediment cores to reconstruct historical pollen records: Digging up the Yarra's dirty past*. Unpublished PhD thesis, Department of Civil Engineering, Monash University, Clayton.

Lyons, K., Richards, C., Desfours, L., & Amati, M. (2013) Food in the city: Urban food movements and the (re)- imagining of urban spaces. *Australian Planner*, 50(2), 157–163.

Martin, J., & Budge, T. (eds.) (2011) *The sustainability of Australia's country towns: Renewal, renaissance, resilience*. Ballarat: Victorian Universities Regional Research Network Press.

McLoughlin, J.B. (1992) *Shaping Melbourne's future? Town planning, the state and civil society*. Cambridge: Cambridge University Press.

McLoughlin, J.B., & Huxley, M. (eds.) (1986) *Urban planning in Australia: Critical readings*. South Melbourne: Longman Cheshire.

Mees, P. (2000) *A very public solution: Transport in the dispersed city*. Melbourne: Melbourne University Press.

Mees, P. (2010) *Transport for suburbia: Beyond the automobile age*. London: Earthscan.

Metropolitan Town Planning Commission (MTPC) (1929) *Plan for general development*. Melbourne: Metropolitan Town Planning Commission.

Mobbs, M. (2012) *Sustainable food*. Sydney: NewSouth Publishing.

Mollison, B. (1979) *Permaculture two*. Sisters Creek: Tagari Publishing.

Mollison, B. (1988) *Permaculture: A designer's manual*. Sisters Creek: Tagari Publishing.

Mollison, B. (1991) *Introduction to permaculture*. Sisters Creek: Tagari Publishing.

Mollison, B., & Holmgren, D. (1978) *Permaculture one*. London: Transworld Publishers.

Morgan, K. (2013) Feeding the city: The challenge of urban food planning. In: Miazzo, F., & Minkjan, M. (eds.), *Farming the city: Food as a tool for today's urbanisation*. Haarlem: Valiz/Trancity, pp. 18–21.

Morton, T. (2007) *Ecology without nature: Rethinking environmental aesthetics*. Cambridge, MA: Harvard University Press.

Morton, T. (2010) *The ecological thought*. Cambridge, MA: Harvard University Press.

Morton, T. (2016) *Dark ecology: For a logic of future coexistence*. New York: Columbia University Press.

Morton, T. (2017) *Ecology without nature*. http://ecologywithoutnature.blogspot.com.au/2009/05/mesh.html (accessed 1 February 2017).

Nasr, J.L., & Komisar, J.D. (2012) The integration of food and agriculture into urban planning and design practices. In: Viljoen, A., & Wiskerke, J.S.C. (eds.), *Sustainable food planning: Evolving theory and practice*. Wageningen, The Netherlands: Wageningen Academic Publishers, pp. 47–58.

National Food Plan (2013) *National food plan: Our cities our future*. Canberra: Commonwealth of Australia.

National Heart Foundation of Australia. (NHFA) (2004) *Healthy by design: A planners' guide to environments for active living*. www.heartfoundation.org.au/SiteCollectionDocuments/Healthy%20by%20Design.pdf (accessed 1 September 2016).

Nettle, C. (2014) *Community gardening as social action*. London: Ashgate.

Newman, P. (2012) *Can biophilic urbanism deliver strong economic and social benefits in cities?* Perth: National Research Centre for Sustainable Built Environment.

Newton, J. (2016) *The oldest foods on Earth: A history of Australian native foods, with recipes*. Sydney: NewSouth Publishing.

Pascoe, B. (2014) *Dark emu: Black seeds agriculture or accident?* Broome: Magabala Books.

Pearlmutter, D., Calfapietra, C., Samson, R., O'Brien, L., Ostoić, S.K., Sanesi, G., & del Amo, R.A. (eds.) (2017) *The urban forest 2017: Cultivating green infrastructure for people and the environment*. Cham: Springer International Publishing AG.

Peter, D. (2008) *Tree succession planning: Modelling tree longevity in Tuttangga/Park 17, the Adelaide Park lands*. Unpublished PhD thesis, School of Architecture, Landscape Architecture and Urban Design, University of Adelaide, Adelaide.

Pieris, A., Tootell, N., Johnson, F., McGaw, J., & Berg, R. (2014) *Indigenous place: contemporary buildings, landmarks and places of significance in South East Australia and beyond*. Melbourne: Melbourne School of Design, The University of Melbourne.

Pires, V. (2011) *Planning for urban agriculture planning in Australian cities*. Presented at State of Australian Cities Conference (SOAC) held in Melbourne from 29 November – 2 December 2011. www.soac.fbe.unsw.edu.au/2011/papers/SOAC2011_0055_final(1).pdf (accessed 1 November 2016).

Plan Melbourne (2014) *Plan Melbourne: Metropolitan planning strategy*. www.planmelbourne.vic.gov.au/__data/assets/pdf_file/0016/131362/Plan-Melbourne-May-2014.pdf (accessed 1 November 2016).

Plan Melbourne Refresh (2015) *Plan Melbourne: Metropolitan planning strategy*. Melbourne: State Government of Victoria.

Porter, E., & Barry, J. (2016) *Planning for coexistence: Recognizing indigenous rights through land-use planning in Canada and Australia*. Oxford: Routledge.

Pothukuchi, K., & Kaufman, J.L. (2000) The food system: A stranger to the planning field, *Journal of the American Planning Associates*, 66(2), 113–124.

Raxworthy, J., & Blood, J. (eds.) (2004) *The MESH book: Landscape/infrastructure*. Melbourne: RMIT Publishing.

Roggema, R. (ed.) (2016) *Sustainable urban agriculture and food planning*. Oxford: Routledge.

Rolls, E.C. (1981) *A million wild acres*. Melbourne: Nelson.

Sarkissian, W. (1996) *With a whole heart: Nurturing an ethic of caring for nature in the education of Australian planners*. Unpublished PhD thesis, Murdoch University, Perth, WA.

Slade, C., & Wardell-Johnson, A. (2016) Beyond governance of food systems: Planning for resilience in the Sunshine Coast, Queensland, Australia. In: Kennedy, M., Butt, A., & Amati, M. (eds.), *Conflict and change in Australia's Peri-urban landscapes*. Oxford: Routledge, pp. 221–240.

Söderlund, J.C. (2015) *Biophilic design: A social movement journey*. Unpublished PhD thesis, Curtin University, Perth.

South Australia (2013) *SA urban forests million trees program: Report and case studies of selected targeted project sites*. Adelaide: Urban Biodiversity Unit, Natural Resources Adelaide and Mt Lofty Ranges, Department of Environment, Water and Natural Resources.

South Australia (2017a) *Consolidated list of declarations of plants and animals*, www.pir.sa.gov.au/__data/assets/pdf_file/0003/231924/Declaration_of_Animals_and_Plants_Feb_2016.pdf (accessed 1 January 2017).

South Australia (2017b) *Weeds in South Australia*, www.pir.sa.gov.au/biosecurity/weeds_and_pest_animals/weeds_in_sa#toc0 (accessed 1 January 2017).

Stallard, K-E. (2017) *Can existing urban forestry strategies be scaled and replicated to overcome the challenges faced by Townsville, North Queensland?* Unpublished MLArch thesis, School of Architecture and Built Environment, Deakin University, Geelong, VIC.

Sulman, J. (1921) *An introduction to the study of town planning in Australia* (2007 reproduction). Sydney, NSW: William Applegate Gullick, Government. Printer 1921; Observatory Hill: National Trust of Australia (NSW), 2007.

Thompson, S., & Maginn, P.J. (eds.) (2012) *Planning Australia: An overview of urban and regional planning*, 2nd ed. Port Melbourne: Cambridge University Press.

Timms, P. (2006) *Australia's quarter acre: The story of the ordinary suburban garden*. Carlton: Miegunyah Press.

Townsend, M., & Weerasuriya, R. (2010) *Beyond blue to green: The benefits of contact with nature for mental health and well-being*. Melbourne: Beyond Blue Ltd.

Tuan, Y-F. (1972) *Topophilia: A study of environmental perception, attitudes and values*. Minneapolis, MN: Dept. of Geography, University of Minnesota.

Victoria. (1987) *Planning and Environment Act 1987*, No. 45 of 1987, Version No. 093. www.legislation.vic.gov.au/ (accessed 1 October 2016).

Victoria. (2010) *Development Committee of Parliament. Inquiry into sustainable development of agribusiness in outer suburban Melbourne, 2010*. www.parliament.vic.gov.au/osisdc/inquiries/agribusiness (accessed 1 September 2016).

Victoria. (2016) *State planning policy framework*. http://planningschemes.dpcd.vic.gov.au/schemes/combined-ordinances/VPPs_All_Clauses.pdf (accessed 1 August 2016).

Victorian Eco Innovation Laboratory (VEIL). (2017) *Current projects*. www.ecoinnovationlab.com/ (accessed 1 February 2017).

Victorian Health Promotion Foundation (VHPF) (2010) *Food Security – What is food insecurity?* www.vichealth.vic.gov.au/en/Programs-and-Projects/Healthy-Eating/Food-Security.aspx (accessed 1 November 2016).

Viljoen, A., & Wiskerke, J. (eds.) (2012) *Sustainable food planning: Evolving theory and practice*. Wageningen, The Netherlands: Wageningen Academic Publishers.

Ware, S.A., & Raxworthy, J. (2011) *Sunburnt: Landscape architecture in Australia*. Melbourne: SUN architecture.

Weller, R. (2009) *Boom town: Scenarios for a rapidly growing city*. Crawley: University of Western Australia Publishing.

White, D., Sutton, P., Pears, A., Mardon, C., Dick, J., & Crow, M. (1978) *Seeds for change: Creatively confronting the energy future*. Melbourne: Patchwork Press/Conservation Council of Victoria.

Wilson, E.O. (2016) *Half Earth: Our plant's fight for life*. New York: Liveright Publishing.

Wright, B. (2001) *Expectations of a better world: Planning Australian communities*. Canberra: Royal Australian Planning Institute.

Zainuddin, Z. (2013) *Melbourne urban backyard food production: How much and how?* Unpublished MSocSci thesis, RMIT University, Melbourne.

Zeunert, J. (2011) *Eating the landscape: Aesthetic foodscape design and its role in Australian landscape architecture*, Proceedings of AILA National Conference: Transform, Brisbane, 12–13 August 2011.

Zeunert, J. (2017) *Landscape architecture and environmental sustainability: Creating positive change through design*. London: Bloomsbury Publishing.

27
Food, landscape, and urban design

Susan Parham and Jacques Abelman

Introduction

Aim of the chapter

In this chapter we explore urban design in the context of food as a critical social, economic, and cultural factor in urban landscape formation and change. Focusing predominantly on the developed world, we see this as a dynamic interplay, in which food both influences the design of place and thus the development of urban landscapes, and is, in turn, influenced by it. The focus on food from an urban design perspective requires us to deal with a range of spaces that are easily recognised as food related – allotments, community gardens, or orchards for example – as well as others in which food makes a contribution to place design but in ways that are not necessarily fully understood. We have taken what constitutes a food space to be broadly conceived, arguing that food's interplay with place design relates to a series of stages in the food chain from production to waste (see Figure 27.1), and at the same time encompasses food loci in diverse settlement settings (see Figure 27.2). These include but are not limited to streetscapes, town squares, food market spaces, restaurants and dining districts, shopping centres, malls and food courts, urban gardens, orchards and allotments, public green spaces, and peri-urban farming of various kinds.

As the chapter elucidates the approach to urban design and food gives weight to food production, processing, distribution, retailing, consumption, and waste space rather than just one or two elements along this chain. This is because each part of the food chain intertwines with aspects of the urban 'armature' in complex design-related ways. 'Armature' is used here more broadly that in Shane's (2008: 267) sense, following Foucault, to denote physical elements that support his argued "heterotopias of illusion": places in cities which now spatially, culturally and economically meld consumption, leisure and commodified art and architecture. Food spaces are important even in Shane's (2008) very particular use of the term 'armature', with examples of the destruction of traditional food markets, such as Covent Garden in London and Les Halles in Paris, as elements of a formerly coherent urban armature undermined by the insertion of simulacra based urban forms.

Armature in this chapter is used to define in urban design terms "the principal elements . . . that will have a significant influence on the form, structure or use of the city" (Erikson, 2014: 21). Erikson (2014: 21) suggests (to paraphrase) that the armature of a city would include its main public spaces, squares, main roads and places where people come together, including community and religious buildings, shopping areas, and recreation spaces, as well as identity forming monuments, landmarks, and cultural institutions around

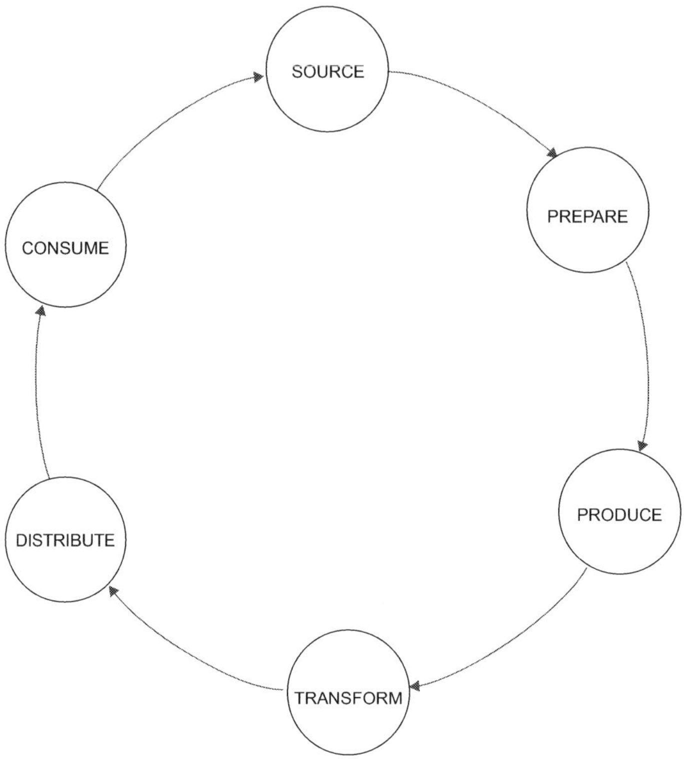

Figure 27.1 Urban food production cycle.
Source: J. Abelman

which public activity revolves. "Urban armature . . . describes[s] this core which can be thought of as a series of physical objects, buildings, spaces, roads, trees, monuments etc., with a particular configuration or arrangement" (Erikson, 2014: 22).

The chapter includes consideration of potential synergies between a food-centred armature and sustainable urban design for the future. It therefore starts by framing the discussion with reference to relevant contexts including sustainability and divergent urban design approaches. It then focuses in summary on how design relates to food in urban space and why this matters. It sketches out the design principles that underpin the analysis connecting urban design and food. It explores some applied approaches from designers who have applied food and urban design in practice for making new and retrofitting existing places. Finally the chapter summarises conclusions reached about the interplay with urban space and design in food terms.

Framing the urban design and food discussion – designers' perspectives

We explore ways that urban design theory and practice deal with spatial requirements for food at nested scales from the building, garden, and street to the city as a whole; within a sustainability focused design approach (Beatley, 2011; Wu and Wu, 2013, Zeunert, 2017). In later sections we situate the work against a background of landscape character appraisal, with its origins in Geddes' 'landscape section' and more recent evolution in gastronomic mapping (Parham, 1992), convivial green space (Parham, 1993), a food based transect (Duany and DPZ, 2011) and continuous productive urban landscapes (Viljoen et al., 2005; Bohn and Viljoen, 2011; Viljoen and Bohn, 2014) approaches. There is a focus on the theorised role of design in productive urban

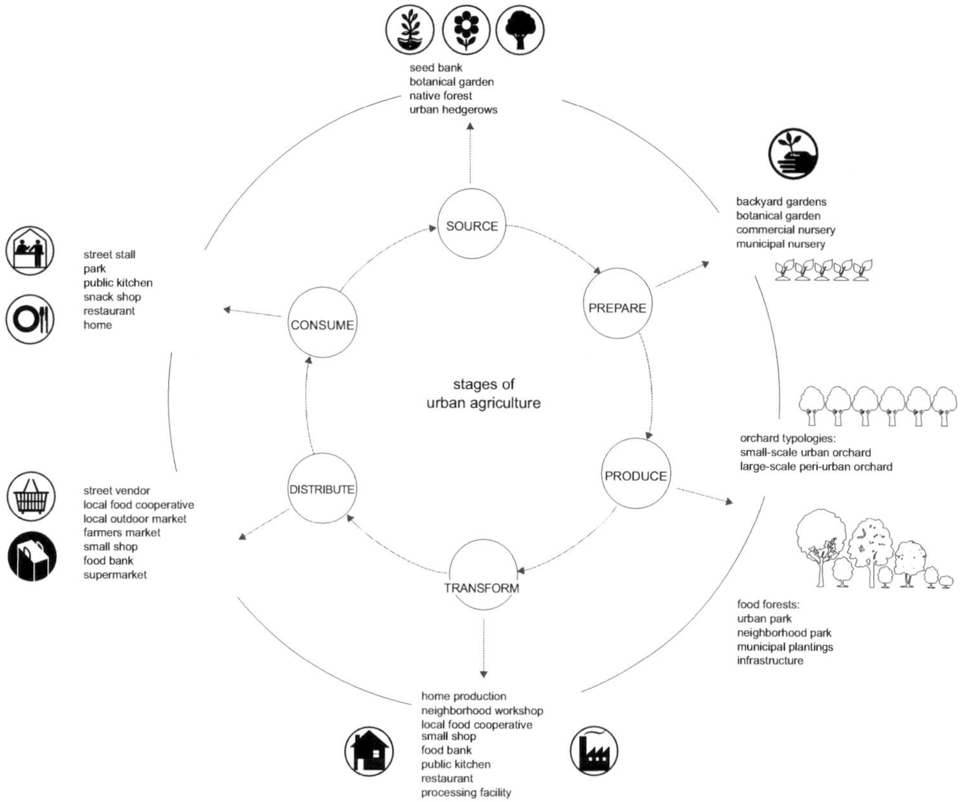

Figure 27.2 Spatialising the urban food cycle.
Source: J. Abelman

landscapes. It is argued that these landscapes can take on a more central role within overall green infrastructure, itself defined as "the network of natural and semi-natural features, green spaces, rivers and lakes that intersperse and connect villages, towns and cities" (Landscape Institute, 2013: 1; see also Austin, 2014).

Urban design approaches to urban space vary. Contextualist designers seeking to maintain long-term urban shaping principles seem to offer a sympathetic framework – explicit or implied – for treating food as an essential element in developing or supporting a coherent and well-functioning urban space focusing on the public realm (Peattie, 1998; Urban Design Compendium, 2000; Barton, Grant and Guise, 2002; Knox, 2005; Carmona, 2009; Nordahl, 2009; Zeunert, 2011; Philips, 2013). A review of design approaches relating to food (Parham, 2012a) demonstrates this point with regard to primary research on food markets in London. As work by Abelman (2012, 2015) shows, this does not mean that urban typologies with food potential all have to look or function in the same way (see Figure 27.3) but we argue that they do need to reflect implied design principles which we delve into below.

In the mid to late twentieth century, the idea of design to support and renew public space primacy in which food is a central element was undercut both theoretically and in practice before a more recent renewal of interest in urban agriculture dealt with in full in Chapter 12. The most obvious interplay between urban design and food in recent years has clearly related to urban agriculture (see, from a widening field, Viljoen, 2005; Nordahl, 2009; Gorgolewski et al., 2011; Viljoen and Bohn, 2014). However, particularly at

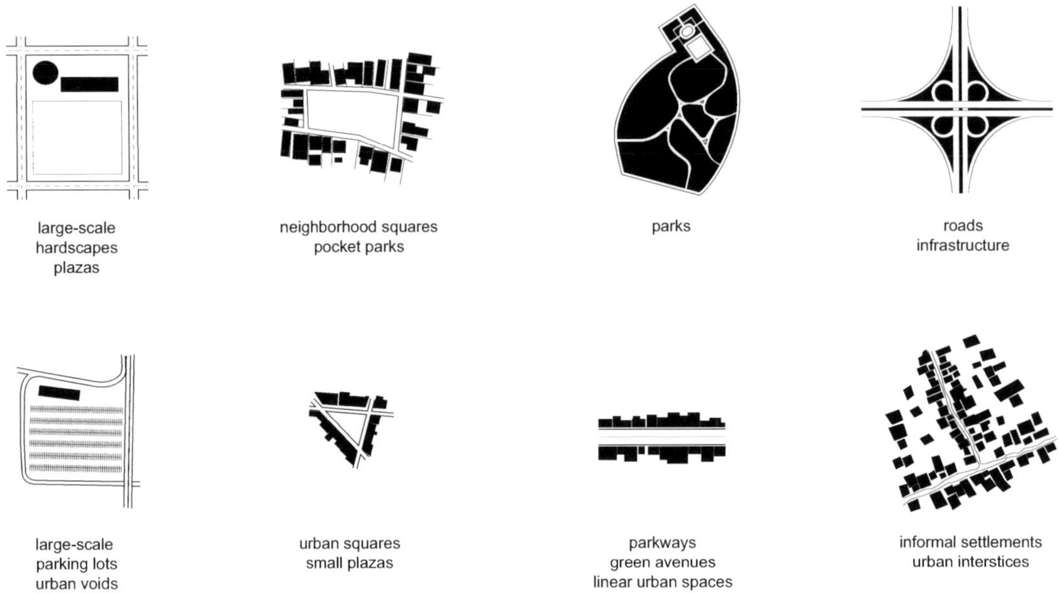

Figure 27.3 Urban typologies with food space potential.

Source: J. Abelman

the distribution and consumption stages of the food system, references to food as an urban design topic of major importance to place are difficult to find (outliers include Barton et al. (2002) and Donovan et al. (2011) as useful examples of applied urban design guidance that does focus on food). Food may be seen as something to 'add in' to the urban design of place; giving 'vibrancy' and 'colour' to public (and privately owned and managed) spaces once aspects of design deemed more important have been completed.

If food has not featured largely in an explicit way in parts of the urban design literature beyond urban agriculture (an issue explored in Parham, 2015a) this may reflect the urban form circumstances many designers see around them at home and on their travels. In many western cities, food may be a taken-for-granted category for designers or one with specialist characteristics. Food-conscious urban design may be seen as a kind of niche or elitist activity to cater for segments of the urban bourgeoisie and place gentrifiers (Bridge and Dowling, 2001; Zukin, 2008; Gonzalez and Waley, 2013). Food centred design can equally be treated as primarily an element of identity relating to immigrant and minority ethnic communities (see from a wide literature Law, 2001; Baker, 2004; Saldivar-Tanaka and Krasny, 2004). It can also be situated within a design intervention paradigm dealing with those in food poverty and food deserts (for example Gordon et al., 2011; Smith, 2012).

As has occurred in relation to the cognate discipline of town planning (Brinkley, 2013), there is increasing discussion in the urban design literature about designing more sustainable urban forms than has been symptomatic in twentieth-century practice (Dempsey, 2005: Farr, 2011; Beatley, 2011; Larice and Macdonald, 2013; Ritchie and Thomas, 2013; Pickett, Cadenasso and McGrath, 2013). There remains scope in both urban design guidance documents and more academic design literature to further explore the array, diversity, and complexity of food's connections to urban design at a range of scales, and consider current sustainability implications and sustainable design potential in food and design terms.

In exploring urban design and food in landscape terms in this chapter we seek to show how various food spaces which may appear to have surface similarities in design terms, can be analysed in a more subtle way based on urban design principles, and why this matters for sustainable social and cultural life in cities.

We argue that a better basis of knowledge about what constitute effective urban food spaces in design terms is needed. This allows a clearer potential to protect such spaces that exist now, retrofit food elements back into existing fabric where they have disappeared from the urban armature (or been actively removed), and ensure food-centred spaces form the heart of new urban areas.

Urban design, food, and landscape – some framing principles

As suggested in our introduction, the urban design literature discussed here presents a mixed picture in relation to food's interplay with spatial design. Many of its overarching themes and values are implicitly supportive of food-centred urban design even if there are relatively fewer mentions of food directly. At the level of principles, those for design across a range of thematic areas also tend to work for food. Thus we concentrate the rest of this sub-section on key urban design concepts from a selection of influential design approaches insofar as these have a bearing on designing urban space in ways that recognise the centrality of food.

Townscape and public space primacy

We start with the notion of 'townscape', which Powers (2012) points out has been an important design concept in the twentieth century for understanding place complexity; presaging the work of Jane Jacobs, and postwar design thinking more generally (Laurence, 2015). As influentially argued by the urban designer, Gordon Cullen (1961: 8) in his book of the same name, townscape requires taking "all the elements that go to create the environment: buildings, trees, nature, water, traffic, advertisements and so on, and to weave them together in such a way that drama is released".

Ideas about townscape are today in the 2010s associated with neo-traditional urbanism (Hebbert, 2003: Talen, 2005) including remaking main street food spaces as part of shopping mall retrofitting (Southworth, 2005a). More broadly such townscape inflected analysis and practice offers an urban design counterpoint to an understanding of place design in which landscape urbanism has become a more dominant (Steiner, 2011; Thompson, 2012; Waldheim, 2012) but contested design perspective (Duany and Talen, 2013). We argue townscape's conceptual stress on the interrelatedness and complexity in urban form (and how that plays out in practice) remains a useful jumping off point for exploring food and urban design in contemporary practice; with designers such as Carmona (2009) putting significant stress on the holistic nature of urban design.

Specifically, thinking about townscape as encompassing the holistic, dynamics of place may be a way into exploring the nature of public space. This is critical in food terms because urban design stress on the primacy of public space and the 'outdoor rooms' of streets and squares we argue tends to support well functioning urban food spaces. Over a reasonably long term, design theorists have undertaken detailed fieldwork to better understand the physical design dimensions, qualities or elements that would inform such principles. Carmona (2009), for example, has noted the morphological, perceptual, social, visual, functional, and temporal dimensions of urban design. Sitte (1965), Bacon (1982), Ashihara (1983), Trancik (1986), Broadbent (1990), Lozano (1990), Moughtin (2003), and Punter (2007) among others, have explored concepts of positive and negative space, and openness and enclosure to help define good city form. City form with a strong solid-to-void ratio is favourably compared with the weak solid-to-void arrangements of many cities shaped with reference to twentieth century design principles (Trancik, 1986).

As the modern movement gained ascendancy in the twentieth century, there was a loss of the design tradition of unbuilt space-built space duality in which built form enclosed urban spaces – town squares and streets where people interconnected with each other. Instead built form came to be treated as an independent element – often in the form of object buildings – that was understood as a non-spatial vacuum (Lozano, 1990: 40, 42). The approach to food moved from one in which urban armature space (and socio-spatial practices) relating to food – street markets, small food shops, cafés, restaurants, hot food providers – was carved out of strongly enclosed block structures which created a sense of place, to one in which separate buildings were scattered within a spatial void.

The loss of public space primacy and eventually of public space itself (Sennett, 1987, 1996; Valentine, 1998; Smiley, 2002; Minton, 2006, 2012) has had powerful results in urban design and food terms. Formerly, food was integrated as part of the collective qualities of the whole city, which Lozano (1990) argues were more important than any individual building or space. At the consumption end of the food system, all sorts of people would mix at the local food market, and might meet at the shops found in each neighbourhood: the baker, the grocer, the greengrocer, the fishmonger, and the butcher, among others. At the production end of the food system, many people would grow vegetables in their own gardens (Kostof, 1991). Some grew fruit, vegetables, and herbs in family plots close to home such as the *jardins ouvriers* of France (Cabedoce and Pierson, 1996) and the *kleingärten* of Germany (Groening, 2005), while from the nineteenth century onwards a number would run an allotment (Crouch and Ward, 1994). These were essentially public spaces, although allotments can become semi-private and excluding as Crouch and Wiltshire (2005: 127) have noted. In the latter part of the twentieth century much consumption moved to privately owned interiorised spaces of shopping centres and malls while urban food growing (prior to a recent resurgence) declined massively as did the public spaces to support it.

Urban space was also subject to hierarchy (Kostof, 1992; Broadbent, 1990; Trancik, 1986). Private spaces, domestic, economic, and institutional such as kitchens, gardens, and dining rooms were important, but public food space, as core public space elements in the city's 'design vocabulary', reflected the spatial primacy of the public realm (Trancik, 1986: 11). Moreover, this public space-focused design syntax supported social and economic vibrancy reflected through highly complex, mixed land uses (Jacobs, 1961).

In functionalist urban design, food spaces such as small food shops and food markets which had once acted spatially as critical townscape elements in the armature, knitting together the fabric of the city, no longer found a coherent place in that fabric (Parham, 1992). Instead, segregation of land uses, the privatisation of public space, and the rise of interiorised, private circulation and meeting space, drained food-related activity and vitality from the street (Kunstler, 1994; Jackson, 1996; Morrison, 2003; Zukin, 2004; Stanilov, 2004; Voyce, 2006). In contrast to the conviviality made possible by the design of public space focused food spaces, functions such as food shopping trips became instrumental, isolated, and isolating activities (Crawford, 1992; Cohen, 1996; Jackson, 1996; Zukin, 2004).

Different kinds of urban food spaces that began to emerge in the mid to late twentieth century such as shopping centres and malls, atriums, festival markets and gallerias (Knox, 1992; Goss, 1996; Robertson, 1997), were not solely caused by functionalist approaches to place – the causes were much more complicated – but were enabled by functionalist urban design expressions. What these approaches shared was a loss of the public space primacy that in traditional cities had been extremely important in providing the interface between, and maintaining the health of, private and civil society (Rowe, 1991; Gehl, 1996). In the latter part of the twentieth century these private spaces started to be reframed and claimed as public spaces by some owners and designers despite exclusionary practices (Mitchell, 1995; Kayden, 2000; Németh, 2009; Schmidt et al., 2011), further blurring their position in design and food space terms; a point we will return to later in the chapter.

Rowe (1991) points to the Sienese *Palio* in that city's *campo* as an example of the importance of maintaining public interface for the social and cultural health of the place. Field (1996) has explored the *Palio* as a process in which sharing food in public space through *contrade* dinners is a critical element in cementing social bonds. More recent examples also suggest that design should underpin socially rich food spaces to help provide the spatial conditions for so-called convivial ecologies to emerge (Shorthose, 2004; Bell, 2007). Such ecologies can revolve around special events like the *Palio* but also be based on ordinary day-to-day food activities in particular public spaces that give rise to rich forms of sociability, as in research in Manchester (Bell and Binnie, 2005) and London (Maitland, 2007). To underpin sociability, designers should work on the creation and maintenance of outdoor rooms based on a range of solid-to-void typologies some of which are shown in Figure 27.4.

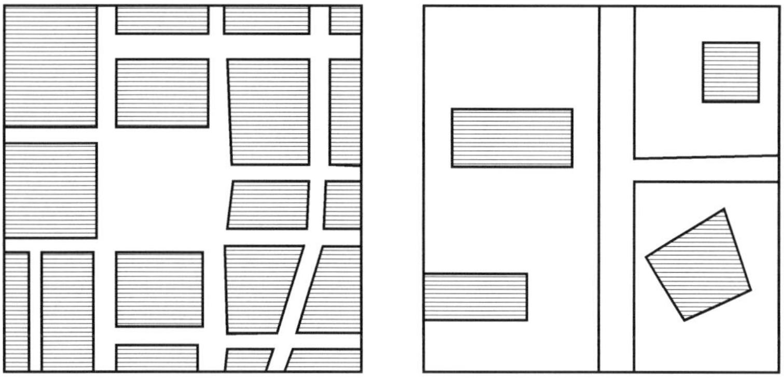

Figure 27.4 Solid to void typologies.
Source: S. Parham, J. Abelman

How urban designers have engaged with food as an urban element

Relating food conscious design to the interplay of land use and movement

The movement economy – the ways that people make their way around a city (Hillier, 1996), as well as land use patterns, has an important design bearing on urban food spaces. Design ideas and resultant place design techniques that stress movement as part of food friendly design include the notion of local walkable catchments for various services and facilities (Urban Design Compendium, 2000). Designers may seek to build in or protect existing walkable catchments (sometimes known as walkability radii or 'pedsheds' as in the Calgary Greenfield Toolbox (undated) whereby food land uses such as food shops, cafés, community gardens, and allotments can be easily accessed on foot. Where the urban design is configured appropriately (see Figure 27.5), walkability radii of between 400–600 metres, or around five to ten minutes in time terms, are taken as a rule of thumb for the distance people can conveniently walk to local food shops and other food-related land uses rather than being tipped into using car based transport for their food consumption trips (Lee and Moudon, 2008; Forsyth and Southworth, 2008). Such walkability can be built in to local place design and place retrofitting.

At a wider scale a strong strand in urban design theory and practice is on highly connected, mixed use approaches defined as transport oriented developments or 'TODs' (Cervero et al., 2002; Southworth, 2005b; Newman and Kenworthy, 2006; Dittmar and Ohland, 2012). TODs use mixed use place design approaches similar to those existing in traditional dense, compact cities but focusing on transport stops and nodes, walking, and cycling (see Figure 27.6). Such TOD influenced design is now used in a range of urban contexts and shares a number of elements including walkable street networks directly focusing in on a dense public space oriented node of activity (Kamruzzaman et al., 2014; Singh et al., 2014). In food terms this could contain a range of food-focused land uses such as markets at the transport stop or node for rail, light rail, buses, and bicycles. Sonnino (2016: 196) reports on such strategies from the cities of Philadelphia and New York. These approaches can be knitted together to provide form-giving properties at nested scales, from the neighbourhood to the region.

Urban design elements and qualities in relation to food spaces

Urban design approaches sympathetic to food space qualities are often grounded in a set of principles, elements or 'qualities including legibility, visual appropriateness, richness, robustness, permeability, and

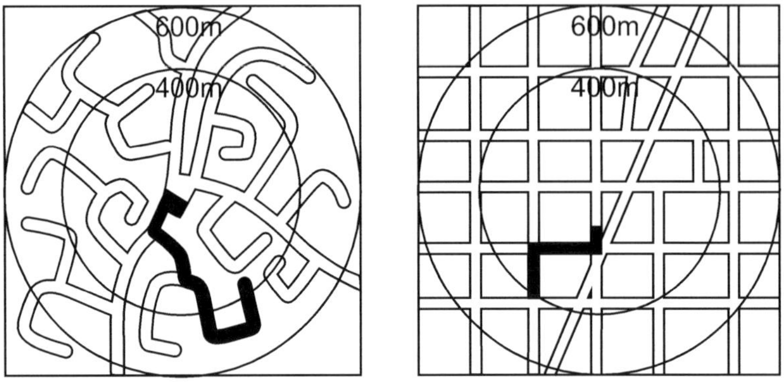

Figure 27.5 Diagram of walkability mapping showing conventional suburban vs. connected grid design.
Source: S. Parham, J. Abelman

Figure 27.6 Transport oriented development (TOD) diagram.
Source: S. Parham, J. Abelman

personalisation which feature in both theoretical work and applied design guidance documents (Greed and Roberts, 2014; Punter and Carmona, 1996; Bentley et al., 1985). Such principles, elements and qualities-based urban design can underpin food space to do specific things in design terms. These will include achieving an appropriate level of enclosure to make outdoor rooms, through the balance of positive-negative space and the right range of height-to-width ratios in public space, to support both vibrant and quieter, more reflective food related land uses and activities of all kinds.

While a detailed focus on food is often absent in such urban design approaches, some designers, including Bernard Rudofsky (1964, 1980) and William H. Whyte (1980) in the latter part of the twentieth century, and Gabrielle Esperdy (2002) in the early twenty-first have been notably interested in the way that urban design can influence food spaces in positive terms. Rudofsky noted the importance of city design to food while Whyte's (1980) famous work from New York City was based on his long-term observations from the late 1960s onwards about how people create the social life of small urban spaces. His fieldwork demonstrated how important being able to change and adapt aspects of a space (so-called personalisation) is in design terms. In Whyte's (1980) research, this in turn contributed to the rich sociability of a small informal space; showing how its design qualities underpinned its use as a food spaces.

In design terms the inference that can be made more broadly is that traditional food-centred spaces – markets, high streets with small food shops, and townscape that offers space for gardens and allotments – tend to achieve the qualities that support vitality in food spaces through highly mixed, fine-grained urban form. In relation to everyday life, such food space design could mean for instance that a person might easily visit their local food shops and food market on foot. They could walk to meet friends and share a meal at a local café, pub or bar; to visit their allotment; to forage fruit and herbs from local streets; and to cook at home using fresh, seasonal, and often local ingredients. Although such a convivial food-centred urban design context can be seen as no more than a nostalgic fantasy, fieldwork examples demonstrate that numerous such places exist today (Deener, 2007; Parham, 2008, 2012a, 2015a; Nordahl, 2009).

An 'urban patterns' food space design example

Food has been an explicit and holistic focus for place design in the era of planned settlements (Parham, 2015a: 137, 212), with Ebenezer Howard, for example, at the end of the nineteenth century giving significant attention to a food in his proposals for garden city settlements (Howard, 1902). Howard proposed an holistic design, social and economic vision of healthful food spaces stretching from a 'Crystal Palace' in the town centre given over to food and other shopping space; through co-operative kitchens and dining rooms; shared and private vegetable gardens; peri-urban edge allotments and dairying; and a productive farming and forestry greenbelt that returned both food and farm rental income to the town (Howard, 1902: 54). Land not yet in use for building could have fruit trees planted or temporary dairies set up so that urban land did not lie unproductive in food terms (Miller, in Parsons and Schuyler, 2002: 106).

These food ideas were played out in widely different ways in design terms including most famously through the development of Letchworth Garden City in the United Kingdom; through the work of landscape architects such as Leberecht Migge in the German Garden City Association (Haney, 2010) and in the United States in Frank Lloyd Wright's proposed 1930s plan for the semi-rural Broadacre City (Lloyd Wright, 1935: 348). We return to garden cities later in the chapter.

Some designers working at around the same time that Whyte was collecting his observational data, have taken food space extremely seriously in design terms. Best known of these design approaches, and strongly informed by countercultural perspectives, remains the work of Alexander et al. (1977) who were concerned with developing holistic urban 'patterns' from the scale of the kitchen and dining table outward. They developed the most detailed and integrated proposals relating food and design of any of the twentieth century urban designers broadly understood as contextual in their approach. Through *A Pattern Language*, Alexander et al. (1977) presented designs for everything from the quality of the eating space in domestic architecture (covering aspects including the table, the lighting, and the eating atmosphere), through to vegetable gardens, orchards, compost, food markets, cafés, pubs, beer halls, and more. Their 'patterns' approach was based on their own long-term observations and evidence from others' urban practice and made explicit design linkages between many of the patterns they developed.

The dominant urban design paradigm for urban food space today

Food and notions of modernism in design

It is clear that the dominance of modernism – the twentieth-century movement in politics, economics, the arts and humanities – in defining how to shape urban food spaces in western cities has been uneven, and we have retained or revived to some degree the kinds of food spaces predominantly discussed to date in this chapter – food markets, small shops, vegetable gardens, forageable streets and urban edges, allotments, and orchards among others. The central paradigm of modernism in urban design terms however has been rather different from the fine-grained, human-scaled mixed-use arrangements that went before and which contextualist designers of varying backgrounds – vernacular inspired, critical regionalist, or neo-traditional urbanist in perspective – tend to advocate. Much urban design of new space of large-scale shopping centres and malls, airports, high rise apartment buildings, business and office parks in the latter twentieth century and now in the twenty-first century has occurred at the edges of cities: first in suburbia, then in post-suburbia, and now in fractured, disparate megalopolitan regions (Graham and Marvin, 2001). As Marshall (2009: 243) argues, this is predicated on an "ad hoc, laissez-faire urbanism. The result is a mix of stuff all over the place . . . in other words, left to its own devices this is what a modern, car-oriented, telecommunications-rich society would naturally generate."

Specifically in relation to food spaces this has often produced significantly lower density, single land use based places than were prevalent previously. So, for example, over the twentieth century high streets of small food shops had their economic base destroyed by larger scale shopping centres spaced further apart, while similarly numerous local food markets were driven out of business by fewer, larger supermarkets owned by a few large corporations. Such place design change was predicated on and responded to car ownership and use. 'Big box' shopping centres and shopping malls of increasing size (with similarly large carparks) became the norm in the suburbs of western cities in the post second world war era (Humphery, 1998; Crawford, 2002; Stanilov, 2004; Voyce, 2006; Ellickson and Grieco, 2013). Food consumption itself was redesigned and made ostensibly more convenient as a car based weekly shop at a supermarket. Only poorer areas in inner cities were 'red-lined' out of such supermarket provision (Eisenhauer, 2001).

More recently, having driven traditional small food shops – grocers, greengrocers, bakers, butchers, and fishmongers – out of business, large operators have started to insert smaller food shops in inner urban areas. However the food available in these shops now reflects the way consumption has been further reshaped, increasingly towards ready prepared foods rather than on everyday, pedestrian based purchase of raw foodstuffs to be cooked at home (Lang and Caraher, 2001; Cromley, 2010). Additionally, a coarse-grained supermarket based spatial model continues to be retrofitted into some inner and traditional suburban city areas where operators are able to achieve this.

The design of suburbia developed alongside, as well as exploited and increasingly gave design centrality to, mass car ownership while dismantling the place design and movement alternatives that had underpinned previous food spaces. Food space loci increasingly reflected the way that the scale of food production, processing, distribution and consumption grew ever larger in spatial terms while becoming increasingly economically integrated and monopolised. Today, in 2017, this is represented in different ways along a food chain that spatially and economically reflects the workings of the modern food system.

For instance at the consumption end of the food system, the process of 'Walmartisation' (Senauer and Seltzer, 2010), super-regional malls and 'category killer' hypermarkets have further increased the scale of food buying. The destruction of the viability of smaller malls has led to dead malls and often-unsuccessful 'mallovers' in which declining malls are repurposed in makeovers intended to bring back customers. Food spaces such as food courts and cafés are often a prominent element in such attempts. Coarse-grained food spaces meanwhile are strongly correlated with creating the design conditions for obesity – through so-called obesogenic environments (Lake and Townshend, 2006; Lake et al., 2010; Courtemanche and Carden, 2011).

At the distribution stage of the food system we have seen the growth of enormous food delivery hubs and so-called dark stores where online food orders are packed and dispatched (Benedictus, 2014; Baraniuk, 2015). These are predicated on just-in-time logistics and large-scale road transport. Distribution landscapes are a feature of the urban design of sprawling areas around the edges of cities (Harvey, 2005; Hackworth, 2007; Leitner et al., 2007; Allmendinger, 2016), sometimes on land that may be better used for food production (see Chapters 14 and 15). At the production end of the food system is industrialised agriculture, vertically integrated with this distribution landscape in an economic sense, but in design terms often physically disconnected from it (Chapters 12 and 16).

These radical changes represent enormous shifts in food's political economy (Chapters 25, 31, 32 and 33) but also in the spatial design of urban landscapes related to food, and the transformations in food spaces have not only occurred in new urban edge spaces. The coarse-grained urban design norms of a suburban and post-suburban food space model have also intruded into finer-grained existing urban spaces, in what Dovey (2016: 163) describes as a form of metastasizing. Thus we see large floor-plate supermarkets with attendant carparks retrofitted into many inner urban shopping streets and town centre areas, undercutting smaller-scale, individually-owned food shops and public space focused conviviality, with consequent blight effects.

At a time when there is significant and growing criticism of these effects in western countries (Harvey, 2005; Hackworth, 2007; Leitner et al., 2007; Allmendinger, 2016), they are being introduced using similar spatial design and economic techniques in developing countries (Reardon and Hopkins, 2006; Reardon and Berdegué, 2008; Amine and Lazzaoui, 2011). Such supermarket and supermall based food consumption has become a sign of modernity in these contexts, where car ownership too has come to be seen as a hallmark of individual and societal success. Ironically, what might now be seen as significant urban design mistakes in food terms are thus being played out in this expanding urban territory (Roy and Ong, 2011).

Urban design and food into the future

Design enquiry and practice – transects, stitches, and landscape patterns

How then might we approach urban food spaces today as a design enquiry and practice? A range of approaches to such landscapes offer interesting design possibilities in relation to food. In this penultimate section we briefly touch on work from practice that reflects the insights explored earlier in the chapter. It can be argued that particularly on the food production side, applied food-related design stems from ideas broadly encompassed in the concept of 'foodsheds' pioneered in work by Hedden (1929). Peters et al. (2009: 2) have defined the foodshed as "the geographic area from which a population derives its food supply". Reflecting on insights from Kloppenburg et al. (1996: 33) and Peters et al. (2005), the urban foodshed can act as both a conceptual and methodological unit of analysis for understanding not only the way that food growing around an urban area is spatially organised but how it can be better aligned to needs for food resilience (Feagan, 2007; Peters et al., 2009; Chapter 16) and conviviality (Parham, 2015b).

On the landscape side of the equation, work by Patrick Geddes on the 'landscape section' (Meller, 2005) has been widely influential on applied design work. The integration of 'green belts', 'green wedges', and 'green fingers' approaches to urban design of city regions have been a feature of settlements in the twentieth century, famously in places including Copenhagen, Helsinki, Stockholm, and Christchurch in New Zealand (Ignatieva et al., 2011: 21). Landscape character has informed influential work such as Ian McHarg and Mumford's (1969) theories about designing with nature; and ideas about ecological networks in cities have become increasingly mainstream conceptually and as design frameworks (Ignatieva et al., 2011). Ignatieva et al.'s (2008) own work on low impact urban design and development (LIUDD) is a case in point.

Ecological notions about landscape have acted to underpin the notion that food spaces should be consciously 'designed in' as critical aspects of place (Hough, 1984, 1990; Morgan, 2009; Kennedy et al., 2011); and conceptualised in ways including as convivial green space within broader gastronomic strategies for cities (Parham, 1993). In this urban design framework, specific food proposals are placed at the centre of shaping and retrofitting settlement space from agricultural edges through private and public vegetable gardens and orchards in urban space; integrated with food processing and distribution spaces and built form; on into the townscape of markets, food shops, restaurants, and cafés; then into the domestic spaces of home kitchens and dining spaces; and finally into ways of dealing with food waste, including composting and anaerobic digestion.

Landscape characterisation has similarly been central to the design and form giving notion of the 'transect'. Duany points out that the transect as a landscape tool harks back as at least as far as the work of Alexander von Humbolt in the eighteenth century (www.dpz.com/Initiatives/Transect) but his design development of the transect conceptually and in practice is particularly relevant here because of the conscious focus on food. Duany (2002: 255) defines the transect as having:

> heretofore been understood as an ordering system deploying a geographic gradient to arrange the sequence of natural habitats. This conception proved to be extensible to the human habitat, as every component of urbanism also finds a place within a continuous rural-to-urban gradient. . . . Beyond being a system of classification, the transect has the potential to become an instrument of design. The correlation of the various specialized components by a common rural-to-urban continuum provides the basis for a new system of zoning, one that creates complex, contextually resonant natural and human environments.

More recently Duany and DPZ (2011: 8) have usefully developed the transect into a food-focused design schema as the basis for what they have termed 'agrarian urbanism'. Duany and DPZ (2011) posit four related models of food-based urbanism – agricultural retention, urban agriculture, agricultural urbanism, and agrarian urbanism. The last of these refers to "settlements where the entire society is involved with food in all its aspects: organising, growing, processing, distributing, cooking and eating it" (Duany and DPZ, 2011: 8). A key urban design aspect in food and landscape terms is the way that both convivial green space and transect approaches integrate with broader urbanist approaches to retrofitting sprawl, to which we will return.

From a landscape architectural perspective, a number of theorists have explored the capacity for urban green to support food land uses (see Lim and Liu, 2010; Gorgolewski, Komisar and Nasr, 2011; Miazzo and Minkjan, 2013; Phillips, 2013; Zeunert, 2016). Abelman (2015, 2012) has explored the idea of multifunctional green productive infrastructures stitched in as a landscape layer through pre-existing urban fabric. The first phase in the potential recolonisation of urban spaces with the productive capacities of food-producing landscape systems relies on the thorough cataloguing of urban spatial typologies and the development of matrices for food space integration (see Figure 27.7). Working through the typologies of these spaces entails examining their suitability for spatially integrating productive landscape elements such as agroforestry units. These units, such as orchards, tree rows, or groupings of structured food forest plantings, act as individual stitches which, over a larger scale, form a productive green fabric.

Productivity in this context is not limited to food but extends to environments that offer recreational, economic and socially supportive opportunities (Viljoen, 2005). The variety of specific forms reflect the designer's analysis of what could be spatially and socially possible. This creates a touchstone for further deliberation and negotiation of terms between top-down and bottom-up processes in the city's metabolism. Who plants? Who maintains? Who profits or is fed? Each landscape stitch has its own context-dependent identity inscribed in social and political structures. Examples range from highly maintained public plazas whose function is to educate and inform civic identity to very productive peri-urban agricultural projects run by local farmers (see Figure 27.8).

As noted in the introduction and also from an architectural perspective, are the so-called Continuous Productive Urban Landscapes (given the acronym CPULs by their originators) focused on food production related design interventions (Viljoen, 2005; Viljoen and Howe, 2012; Viljoen and Bohn, 2014). These have been conceived for – and in some instances applied to – real urban spaces. In work that is headlined as about 'the edible city', Bohn and Viljoen (2011: 150) argue that the CPUL is:

urban spatial typology	agroforestry type	top down vs bottom up	plant sources	project type/goal	maintenance	harvesting	products	food system endpoints
urban plazas large-scale hardscapes	forest/orchard	intitiated by municipality	municipal greenhouse + seedling center	changing identity of the city + work program	municipal work program	municipal workers + volunteers	fresh fruit and fruits/products	public market
large-scale parking lots urban voids	forest/orchard	intitiated by municipality	botanical garden + municipal greenhouse	intensive food plots	favela inhabitants + municipal worker support	favela inha - bitants with municipal worker support	fresh food for immediate consumption or resale	kitchen table + street stall/ public market
squares small plazas	forest/orchard	intitiated by municipality with local residents	botanical garden + municipal greenhouse	city image + improving urban neigh - borhoods	municipal workers/ locals/ gardening clubs/ volunteers	local residents + gardening clubs	fresh food for immediate consumption	kitchen table + local markets
streets green avenues	orchard	intitiated by municipality	botanical garden + municipal greenhouse	improving urban quality + soft transport portential	municipal workers	municipal workers	fruit for processing	food banks food coop/ processing plants
neighborhood squares pocket parks	forest	intitiated by local residents with municipal support	botanical garden + families	improving neighborhood greenspaces and social networks	local residents + support from municipal workers	local residents	fresh fruit and artisinal products	kitchen table/ exchange + neighborhood markets
large scale urban parks	forest	intitiated by municipality	botanical garden + municipal greenhouse	multifunctional park space/ education/ recreation	municipal workers	municipal workers	fresh fruit/ juices	sale in park + food banks
informal settlements urban periphery	orchard	intitiated by residents in partnership with entrepreneurs	university + community nursery	intensive commercial production`	entrepreneurs and paid workers	paid workers from favelas	fresh fruit for processing	commercial networks processing plants
roadsides infrastructure	orchard	intitiated by municipality	botanical garden + municipal greenhouse	improving urban quality + soft transport portential	municipal workers	municipal workers	fresh fruit for processing	commercial networks processing plants

Figure 27.7 Matrix of urban typologies for food spaces.

Source: J. Abelman

Figure 27.8 Matrix of potential functions of food spaces in an urban context.

Source: J. Abelman

a design concept advocating the coherent introduction of interlinked productive landscapes into cities as an essential element of sustainable urban infrastructure. Central to the CPUL concept is the creation of multi-functional open urban space networks that complement and support the built environment.

Edible urbanism and food retrofitting

In fact the idea of – and practical activities focused on – 'edible urbanism', 'edible landscapes', and 'edible cities' have come from a variety of directions since the early 2000s (Esperdy, 2002; Vitiello, 2008; Paull, 2011). Various initiatives have become internationally recognised including 'Incredible Edible Todmorden' located in the small Yorkshire town of that name (Paull, 2011). In the Upper Calder Valley in the UK, local residents started to informally take over public and semi-public spaces for productive plantings. This has opened up both gardening and foraging opportunities, bringing individual and societal benefit, as demonstrated elsewhere by Nordahl, (2009). From an urban design perspective it might well be argued that the traditional urban design structure of Todmorden's townscape made such actions work with the grain of the existing urbanism rather than against it. At Beacon Hills in Seattle, meanwhile, the Beacon Food Forest's goal has been to "plant and grow an edible urban forest garden that inspires our community to gather together, grow our own food and rehabilitate our local ecosystem" (http://beaconfoodforest.org/: undated). Numerous other examples now exist as referenced in Zeunert (2017: 166–195).

Famously influencing urban design at a whole settlement level, individual residents, communities and urban food groups have undertaken a range of food-based retrofitting projects in the face of Detroit's massive economic decline, equally profound spatial shrinkage and burgeoning array of lost landscapes (Bonfiglio, 2009; Giorda, 2012; Cockrall-King, 2012; Colasanti et al., 2012). These have among other things responded though design (as well as in process terms) to 'food desert' characteristics particularly affecting black residents and undercutting their food justice (Hashim, 2015). Detroit's experience of responding to catastrophic urban change with particularly serious consequences for black communities, has shown the utility of food space-focused urban design and landscape renewal. Thus, despite some contestation about effectiveness, food space retrofitting does appear to show how a series of design initiatives can create broader linkages to social, environmental, and economic improvements in urban landscapes at a large scale.

A new town food retrofitting case study

Urban agricultural work in Detroit makes explicit the connection of food to conditions of urban shrinkage and this is also reflected in an urban design-based series of retrofitting scenarios explored in recent work on food space retrofitting (Parham and McCabe, 2016a, 2016b) within a 'sprawl repair' paradigm (Dunham-Jones and Williamson, 2009; Tachieva, 2010). In Hatfield, Hertfordshire, just north of London, the context is a complicated one of both growth in academic and service industries in a New Town university campus and business park setting, and decline in traditional jobs, social engagement, and food spaces. The place is experiencing both urban sprawl conditions on the one hand and suffering continued economic and social difficulties stemming from de-industrialisation in the 1980s on the other.

A series of urban design scenarios have been developed for food space renewal (see Figure 27.9), including semi-rural hamlet *potager* gardens; urban edge and neighbourhood community orchards; community vegetable gardens in leftover spaces resulting from postwar new town landscaping norms; revival of moribund midblock allotments; public space focused townscape designs of the town centre marketplace and local shopping strips; edible urbanism planting schemes for a new village square; and a highly walkable urban 'foodway' between two university campuses (Parham and McCabe, 2016a, 2016b). Notably, in urban design terms, the work encompasses the economically successful but environmentally and socially dysfunctional and sterile business park landscape situated as a 'rich' food desert (Parham and McCabe, 2016a). A series of transitional retrofittings of vegetable gardens and orchards is proposed prior to the area being

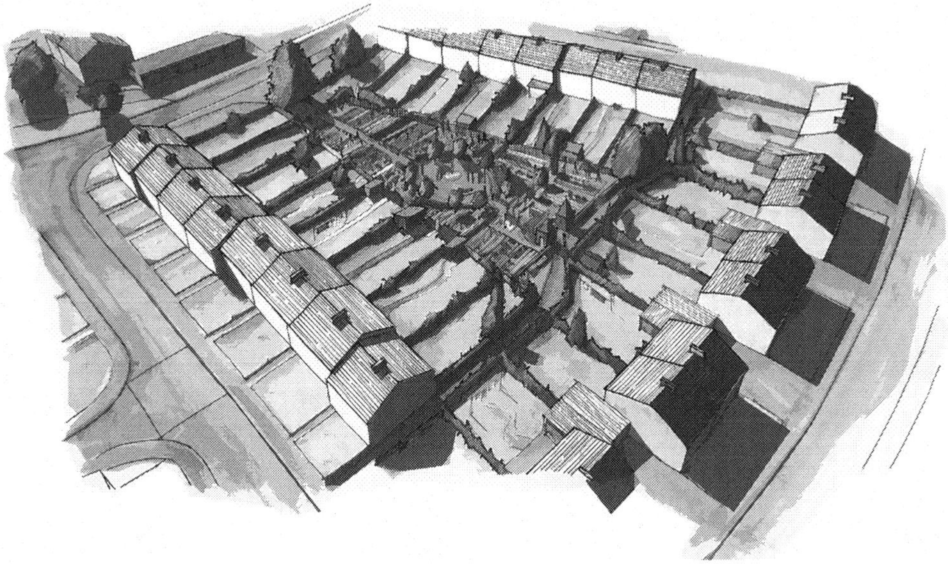

Figure 27.9 Scenario for New Town food space retrofitting design.

Source: S. Parham and B. McCabe

redeveloped more fully according to sprawl repair principles to produce more adequate urbanism both in food space terms and more broadly (Parham and McCabe, 2016a, 2016b).

Garden cities and food

In case it is thought that food-centred urban design is only a feature of the last few decades, it is worth noting that there have been planned settlement frameworks giving food a central place in spatial design terms since at least classical times (Kostof, 1991, 1992; Parham, 2015a, 2016). As noted earlier in the chapter, Ebenezer Howard's ideas and principles for garden cities are particularly notable and today a series of new garden cities, garden towns, and garden villages are in development or planned in the United Kingdom. At the proposed Symondshyde Garden Village and the Stanboroughbury neighbourhood, both on the north-west edge of Hatfield in Hertfordshire (mentioned earlier in relation to New Town food retrofitting), the masterplanners intend to include a range of food spaces in the design. These encompass local food shops, cafés and pubs, vegetable garden spaces, orchards, and allotments, as well as support for continued farming in the surrounding greenbelt. All of these food related urban armature elements are shown in the masterplanning documentation for the proposed new places and demonstrate highly contextual, mixed use, walkable, and food-focused settlements in the making (see Figure 27.10).

Conclusions

To briefly summarise key points from the chapter, we are arguing that urban design for food has been present as a theme in place design for millennia. In recent and contemporary practice we suggest that food could be better represented in both thinking about and shaping urban space design. Food loci are not of peripheral or marginal importance as landscapes even though that may be the way they have sometimes

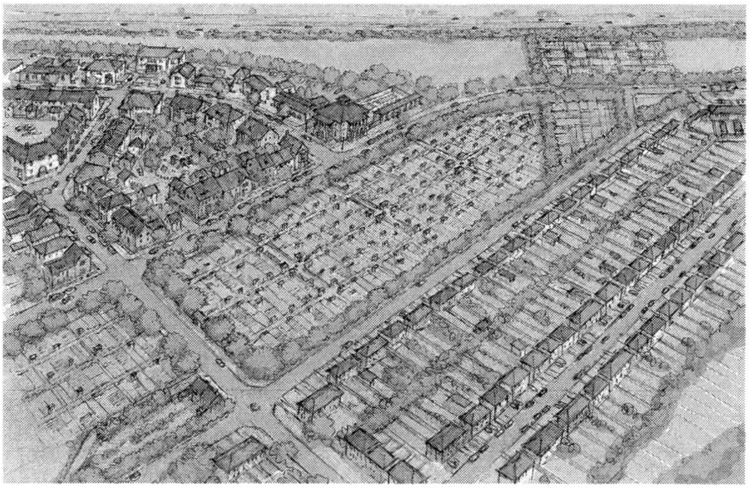

Figure 27.10 Masterplan detail for Stanboroughbury, Hatfield, UK, illustrating new allotment space.
Source: Illustration prepared by DPZ, reproduced with permission from Gascoyne Cecil Estates

been treated in urban design theory and practice. That understanding is increasingly recognised in urban design as well as town planning and landscape architecture, its sister disciplines.

Taking urban sustainability as a framing device it can be seen that environmental aspects of urban environments are informing urban design ideas and methods to an increasing extent. As a result, the range of urban food spaces is in turn starting to be better recognised and represented, at least in environmental sustainability terms. For instance, as the chapter explores, landscape design for urban agriculture has seen a renaissance of interest, although this is not yet matched by the same degree of spatial design focus on other parts of the food system such as food shopping and consumption spaces. Despite pioneering work by a number of designers to explore more broadly the landscape characteristics of food growing, distribution, retailing, consumption, and clean up loci, it is clear that urban design for food, and its sustainable representation in urban landscapes, has substantially further to go.

We can demonstrate how closely aligned food conscious design is to a landscape focused understanding of the urban including landscape character analysis. Yet it is clear that there are different and sometime competing approaches to the way that urban design principles, elements, and patterns are conceived and applied to food spaces. We situate and compare broadly contextual and modernist worldviews and urban design techniques for food spaces stemming from these different perspectives. We identify a kind of urban design circularity here starting from traditional city shaping, in which food spaces were largely the vernacular results of local constraints and opportunities, to a more technocratic phase especially in the latter part of the twentieth century, and now again a move toward more contextual approaches to urban design for food which are consciously reviving and employing principles previously rejected in favour of a supposed modernity.

Of course, in places with poor urban populations, especially in developing world settings (see Chapters 20, 21 and 22), productive food spaces have always been a resilience feature. This is especially obvious in relation to street food and food growing in informal and leftover spaces (Freeman, 1991; Tinker, 1997; Etzold, 2016; Patel et al., 2014). Even in western cities, urban food growing has been critical to resilience when formal food systems collapse (Barthel, Parker and Ernstson, 2015). More recently, with an upsurge in street food, 'disobedient' food activities in western cities can be considered part of the temporary and tactical urbanism that has been developing in relation to food in such urban space (Valentine, 1998: Newman and Burnett, 2013; Mould, 2014; Lydon and Garcia, 2015).

As well as the applied design examples discussed here, front yard farms, edible estates (Haeg et al., 2008), guerilla gardening (Tracey, 2007; Hardman and Larkham, 2014), and the 'tactical urbanism' of food trucks (Shouse, 2011; Tanenbaum, 2012; Mould, 2014), and a number of other food initiatives, have burgeoned in a diversity of urban settings. What does seem to link these is the kind of physical design approach that provides a positive context for such activities. As we demonstrate here, access-controlled shopping malls and privately managed public spaces (Schmidt et al., 2011) do not offer convincing design opportunities for sustainable food spaces to develop as convivial ecologies, whereas streets and other urban armature and fabric elements which retain a public face tend to do so.

We have explored a series of specific examples of food-focused urban design proposals from a variety of urban settings and scales, including deindustrialised urban regions, new towns, and garden cities, that show some approaches that we deem successful in urban design terms in principle. We ask how far such urban design techniques – stemming from ideas about urban food space typologies, convivial green space, food transects, sprawl repair, retrofitting edible cities, and continuous productive urban landscapes among others – can inform the design of settlements today? We suggest in answering that question that food-conscious urban design is centrally important to maintaining and protecting existing food related armature elements, renewing and retrofitting dysfunctional spaces and making new places in which food is central to sustainable urban landscapes in future.

Further reading

Abelman, J. (2012) *From vision to means: Articulating landscape architectural practices for urban agriculture – a design proposal for connected Brazilian foodscapes*. Amsterdam: Academy of Architecture.

Abelman, J. (2015) Cultivating the city: Infrastructures of abundance in urban Brazil. *Future of Food: Journal on Food, Agriculture and Society*, 3(1), 101–109.

Alexander, C., Ishikawa, S., & Silverstein, M., (1977) *A pattern language: Towns, buildings, construction*. New York: Oxford University Press.

Allmendinger, P. (2016) *Neoliberal spatial governance*. New York: Routledge.

Amine, A., & Lazzaoui, N. (2011) Shoppers' reactions to modern food retailing systems in an emerging country: The case of Morocco. *International Journal of Retail and Distribution Management*, 39(8), 562–581.

Ashihara, Y. (1983) *The aesthetic townscape*. London and Cambridge, MA: MIT Press.

Austin, G. (2014) *Green infrastructure for landscape planning: Integrating human and natural systems*. London: Routledge.

Bacon, E. (1982) *Design of cities*. London: Thames and Hudson.

Baker, L.E. (2004) Tending cultural landscapes and food citizenship in Toronto's community gardens. *Geographical Review*, 94(3), 305–325.

Baraniuk, C. (2015) Shop, but don't enter: The strange world of dark stores. *BBC Future*. www.bbc.com/future/story/20150804-shop-but-dont-enter-the-strange-world-of-dark-stores (accessed 15 January 2017).

Barthel, S., Parker, J., & Ernstson, H. (2015) Food and green space in cities: A resilience lens on gardens and urban environmental movements. *Urban Studies*, 52(7), 1321–1338.

Barton, H., Grant, M., & Guise, R. (2002) *Shaping neighbourhoods: Health, sustainability, vitality*. London: Routledge.

Beatley, T. (2011) *Biophilic cities: Integrating nature into urban design and planning*. Washington, DC, Covelo and London: Island Press.

Bell, D. (2007) The hospitable city: Social relations in commercial spaces. *Progress in Human Geography*, 31(1), 7–22.

Bell, D., & Binnie, J. (2005) What's eating Manchester? Gastro-culture and urban regeneration. In: Franks, K. (ed.), Food and the city, *Architectural Design*, 75(3), 78–85.

Benedictus, L. (2014) Inside the supermarkets' dark stores. *The Guardian*, 7 January 2014 (accessed 15 January 2017).

Bentley, I., Alcock, A., Murrain, P., McGlynn, S., & Smith, G. (1985) *Responsive environments: A manual for designers*. London: Routledge.

Bohn, K., & Viljoen, A. (2011) The edible city: Envisioning the continuous productive urban landscape (CPUL). *FIELD*, 4(1), 149–161.

Bonfiglio, O. (2009) Delicious in Detroit-Vacant land can be an asset, as Olga Bonfiglio reports. Sidebars on Montreal and on LA's vertical gardens. *Planning*, 75(8), 32.

Bridge, G., & Dowling, R. (2001) Microgeographies of retailing and gentrification. *Australian Geographer*, 32(1), 93–107.

Brinkley, C. (2013) Avenues into food planning: A review of scholarly food system research. *International Planning Studies*, 18(2), 243–266.

Broadbent, G. (1990) *Emerging concepts in urban space design*. London and New York: Van Nostrand Reinhold.

Cabedoce, B., & Pierson, P. (eds.) (1996) *Cent ans d'histoire des jardins ouvriers, 1896–1996: la Ligue française du coin de terre et du foyer Grâne*. Créaphis.

Carmona, M. (2009) Sustainable urban design: Principles to practice. *International Journal of Sustainable Development*, 12(1), 48–77.

Carmona, M., Heath, T., Oc, T., & Teisdell, S. (2003) *Public places, urban spaces: The dimensions of urban design*. Amsterdam: Architectural Press.

Cervero, R., Ferrell, C., & Murphy, S. (2002) Transit-oriented development and joint development in the United States: A literature review. *TCRP Research Results Digest* (52).

Cockrall-King, J. (2012) *Food and the city: Urban agriculture and the new food revolution*. Amherst, NY: Prometheus Books.

Cohen, L. (1996) From town center to shopping center: The reconfiguration of community marketplaces in postwar America. *The American Historical Review*, 101(4), 1050–1081.

Colasanti, K.J., Hamm, M.W., & Litjens, C.M. (2012) The city as an 'Agricultural Powerhouse'? Perspectives on expanding urban agriculture from Detroit, Michigan. *Urban Geography*, 33(3), 348–369.

Courtemanche, C., & Carden, A. (2011) Supersizing supercenters? The impact of Walmart Supercenters on body mass index and obesity. *Journal of Urban Economics*, 69(2), 165–181.

Crawford, M. (2002) Suburban life and public space. In: Smiley, D.J. (ed.) *Sprawl and public space: Redressing the mall*. Washington, DC: National Endowment for the Arts, pp. 21–30.

Cromley, E.C. (2010) *The food axis: Cooking, eating, and the architecture of American houses*. Virginia: University of Virginia Press.

Crouch, D., & Ward, C. (1994) *The allotment: Its landscape and culture*. Nottingham: Mushroom.

Crouch, D., & Wiltshire, R. (2005) Designs on the plot: The future for allotments in urban landscapes. In: Viljoen, A. (ed.), *Continuous Productive Urban Landscapes (CPULs): Designing urban agriculture for sustainable cities*. Oxford: Architectural Press, pp. 124–130.

Cullen, G. (1961) *The concise townscape*. London: Architectural Press.

Deener, A. (2007) Commerce as the structure and symbol of neighborhood life: Reshaping the meaning of community in Venice, California. *City and Community*, 6(4), 291–314.

Dempsey, N. (2005) *Future forms and design for sustainable cities*. London: Routledge.

Dittmar, H., & Ohland, G. (eds.) (2012) *The new transit town: Best practices in transit-oriented development*. Washington, DC: Island Press.

Donovan, J., Larsen, K., & McWhinnie, J. (2011) Food-sensitive planning and urban design: A conceptual framework for achieving a sustainable and healthy food system. *Melbourne: Report commissioned by the National Heart Foundation of Australia (Victorian Division)*.

Dovey, K. (2016) *Urban design thinking: A conceptual toolkit*. London: Bloomsbury Publishing.

Duany, A. (2002) Introduction to the special issue: The transect. *Journal of Urban Design*, 7(3), 251–260.

Duany, A., & DPZ (2011) *Garden cities: Theory and practice of agrarian urbanism*. Duany Plater-Zyberk and Company and The Prince's Foundation.

Duany, A., & Talen, E. (2013) Looking backward: Notes on a cultural episode. In: *Landscape urbanism and its discontents: Dissimulating the sustainable city*. Gabriola Island, Canada: New Society Publishers, pp. 1–16.

Dunham-Jones, E., & Williamson, J. (2009) *Retrofitting suburbia: Urban design solutions for redesigning suburbs*. Hoboken, NJ: John Wiley and Sons.

Eisenhauer, E. (2001) In poor health: Supermarket redlining and urban nutrition. *Geojournal*, 53, 125–133.

Ellickson, P.B., & Grieco, P.L. (2013) Wal-Mart and the geography of grocery retailing. *Journal of Urban Economics*, 75, 1–14.

Erikson, B. (2014) The 'armature' and 'fabric' as a model for understanding spatial organisation. In: Roberts, M., & Greed, C. (eds.), *Approaching urban design: The design process*, Vol. 5. New York and London: Routledge.

Esperdy, G. (2002) Edible urbanism. *Architectural Design*, 72, 44–50.
Etzold, B. (2016) Migration, informal labour and (trans) local productions of urban space – the case of Dhaka's street food vendors. *Population, Space and Place*, 22(2), 170–184.
Farr, D. (2011) *Sustainable urbanism: Urban design with nature*. Hoboken, NJ: John Wiley and Sons.
Feagan, R. (2007) The place of food: Mapping out the 'local' in local food systems. *Progress in Human Geography*, 31(1), 23–42.
Field, C. (1997) *Celebrating Italy: Tastes and traditions of Italy as revealed through its feasts, festivals and sumptuous foods*. New York: William Morrow and Company.
Forsyth, A., & Southworth, M. (2008) Cities afoot – Pedestrians, walkability and urban design. *Journal of Urban Design*, 13(1), 1–3.
Gehl, J. (1996) *Life between buildings: Using public space*. Copenhagen: Arkitektens Forlag.
Giorda, E. (2012) Farming in Motown: Competing narratives for urban development and urban agriculture in Detroit. In: Viljoen, A., & Johannes, S.C.W. (eds.), *Sustainable food planning: Evolving theory and practice*. Wageningen, The Netherlands: Wageningen Academic Publishers, pp. 271–281.
González, S., & Waley, P. (2013) Traditional retail markets: The new gentrification frontier? *Antipode*, 45(4), 965–983.
Gordon, C., Purciel-Hill, M., Ghai, N.R., Kaufman, L., Graham, R., & Van Wye, G. (2011) Measuring food deserts in New York City's low-income neighborhoods. *Health and Place*, 17(2), 696–700.
Gorgolewski, M., Komisar, J., & Nasr, J. (2011) *Carrot city*. New York: Monacelli.
Goss, J. (1996) Disquiet on the waterfront: Reflections on nostalgia and utopia in the urban archetypes of festival marketplaces. *Urban Geography*, 17(3), 221–247.
Graham, S., & Marvin, S. (2001) *Splintering urbanism: Networked infrastructures, technological mobilities and the urban condition*. London: Routledge.
Greed, C., & Roberts, M. (2014) *Introducing urban design: Interventions and responses*. London: Routledge.
Greenfield Tool Box Walkability Catchment (nd) *Calgary regional partnership*. http://greenfield.calgaryregion.ca/tools/greenfield_process_walkabilityCatchment.pdf (accessed 23 May 2017).
Groening, (2005) The world of small urban gardens. *Chronica Horticulturae*, 45(2), 22–25.
Hackworth, J. (2007) *The neoliberal city: Governance, ideology, and development in American urbanism*. Ithaca, NY: Cornell University Press.
Haeg, F., Farms, W., Riley, B., Fountain, F., Rinaldo, K., & Youngs, A. (2008) *Edible estates*. New York: Metropolis Books.
Haney, D. (2010) *When modern was green: Life and work of landscape architect Leberecht Migge*. London: Routledge.
Hardman, M., & Larkham, P. (2014) *Informal urban agriculture: The secret lives of Guerrilla Gardeners*. Cham: Springer International Publishing.
Harvey, D. (2005) *A brief history of neoliberalism*. Oxford: Oxford University Press.
Hashim, N. (2015) Reversing food desertification: Examining urban farming in Louisville, Chicago and Detroit. *Local Environment*, 20(6), 611–636.
Hebbert, M. (2003) New urbanism – the movement in context. *Built Environment*, 29(3), 193–209.
Hedden, W.P. (1929) *How great cities are fed*. Boston, MA: D.C. Heath and Company.
Hillier, B. (1996) Cities as movement economies. *Urban Design International*, 1(1), 41–60.
Hough, M. (1984) *City form and natural process. Towards a new urban vernacular*. London and New York: Routledge.
Hough, M. (1990) *Out of place, restoring identity to the regional landscape*. New Haven, CT and London: Yale University Press.
Howard, E. (1902) *Garden cities of tomorrow*, Facsimile of 2nd ed. London: Dodo Press.
Humphery, K. (1998) *Shelf life: Supermarkets and the changing cultures of consumption*. Cambridge: Cambridge University Press.
Ignatieva, M., Meurk, C.D., & Stewart, G.H. (2008) Low Impact Urban Design and Development (LIUDD): Matching urban design and urban ecology. *Landscape Review*, 12(2), 61–73.
Ignatieva, M., Stewart, G.H., & Meurk, C. (2011) Planning and design of ecological networks in urban areas. *Landscape and Ecological Engineering*, 7(1), 17–25.
Jackson, K.T. (1996) All the world's a mall: Reflections on the social and economic consequences of the American shopping center. *The American Historical Review*, 101(4), 1111–1121.
Jacobs, J. (1961) *The death and life of great American cities*. New York: Random House.

Kamruzzaman, M., Baker, D., Washington, S., & Turrell, G. (2014) Advance transit oriented development typology: Case study in Brisbane, Australia. *Journal of Transport Geography*, 34, 54–70.
Kayden, J.S. (2000) *Privately owned public space: The New York city experience*. New York and Chichester: John Wiley and Sons.
Kennedy, C., Pincetl, S., & Bunje, P. (2011) The study of urban metabolism and its applications to urban planning and design. *Environmental Pollution*, 159(8), 1965–1973.
Kloppenburg, Jr, J., Hendrickson, J., & Stevenson, G.W. (1996) Coming in to the foodshed. *Agriculture and Human Values*, 13(3), 33–42.
Knox, P. (1992) The packaged landscapes of post-suburban America. In: *Urban landscapes: International perspectives*, London: Routledge, pp. 207–226.
Knox, P.L. (2005) Creating ordinary places: Slow cities in a fast world. *Journal of Urban Design*, 10(1), 1–11.
Kostof, S. (1991) *The city shaped: Urban patterns and meanings through history*. London: Thames and Hudson.
Kostof, S. (1992) *The city assembled: The elements of urban form through history*. London: Thames and Hudson.
Kunstler, J. H. (1994) *Geography of nowhere: The rise and decline of America's man-made landscape*. New York: Simon and Schuster.
Lake, A.A., Townshend, T.G., & Alvanides, S. (eds.) (2010) *Obesogenic environments: complexities, perceptions and objective measures*. Chichester, West Sussex: Wiley-Blackwell.
Lake, A.A., & Townshend, T. G (2006) Obesogenic environments: Exploring the built and food environments. *The Journal of the Royal Society for the Promotion of Health*, 126, 262–267.
Landscape Institute (2013) *Green infrastructure. An integrated approach to land use*, Landscape Institute Position Statement, March. London: Landscape Institute.
Lang, T., & Caraher, M. (2001) Is there a culinary skills transition? Data and debate from the UK about changes in cooking culture. *Journal of the HEIA*, 8(2), 2–14.
Larice, M., & Macdonald, E. (eds.). (2013) *The urban design reader*. London: Routledge.
Laurence, P. L. (2015). Jane Jacobs, the Townscape Movement, and the Emergence of Critical Urban Design, in Pendlebury, John, Ertem, Erdem and J. Larkham, Peter (eds.), *Alternative Visions of Post-War Reconstruction: Creating the Modern Townscape,* London: Routledge, 2015, 216-226.Law, L. (2001) Home cooking: Filipino women and geographies of the senses in Hong Kong. *Ecumene*, 8(3), 264–283.
Lee, C., & Moudon, A.V. (2008) Neighbourhood design and physical activity. *Building Research and Information*, 36(5), 395–411.
Leitner, H., Peck, J., & Sheppard, E. (2007) *Contesting neoliberalism: Urban frontiers*. New York: The Guilford Press.
Lim, C.J., & Liu, E. (2010) *Smartcities and eco-warriors*. London: Routledge.
Lozano, E.E. (1990) *Community design and the culture of cities: The crossroad and the wall*. Cambridge: Cambridge University Press.
Lydon, M., & Garcia, A. (2015) A tactical urbanism how-to. In: *Tactical urbanism*. Washington, DC: Island Press/Center for Resource Economics, pp. 171–208.
Lynch, K. (1961) *The image of the city*. Cambridge, MA: MIT Press and Harvard University Press.
Lynch, K. (1985) *A theory of good city form*. Cambridge, MA and London: MIT Press.
Maitland, R. (2007) Conviviality and everyday life: The appeal of new areas of London for visitors. *International Journal of Tourism Research*, 10, 15–25.
Marshall, S. (2009) *Cities, design and evolution*. Abingdon: Routledge.
McHarg, I.L., & Mumford, L. (1969) *Design with nature*. New York: American Museum of Natural History.
Meller, H. (2005) *Patrick Geddes: Social evolutionist and city planner*. Abingdon: Routledge.
Miazzo, F., & Minkjan, M. (eds.). (2013) *Farming the city: Food as a tool for today's urbanisation*. CITIES trancity-valiz.
Miller, M. (2004) British and American community design: Clarence Stein's Manhattan transfer, 1924–1974. In: Parsons, K.C., & Schuyler, D. (eds.), *From garden city to green city: The legacy of Ebenezer Howard*. Baltimore, MD and London: The Johns Hopkins University Press.
Minton, A. (2006) *The privatisation of public space*. London: The Royal Institution of Chartered Surveyors.
Minton, A. (2012) *Ground control: Fear and happiness in the twenty-first-century city*. London: Penguin.
Mitchell, D. (1995) The end of public space? People's Park, definitions of the public, and democracy. *Annals of the Association of American Geographers*, 85(1), 108–133.
Morgan, K. (2009) Feeding the city: The challenge of urban food planning. *International Planning Studies*, 4(4).

Morrison, K.A. (2003) *English shops and shopping: An architectural history*. New Haven, CT and London: Yale University Press, English Heritage.

Moughtin, C. (2003) *Urban design: Street and square*. Amsterdam and London: Architectural Press.

Mould, O. (2014) Tactical urbanism: The new vernacular of the creative city. *Geography Compass*, 8(8), 529–539.

Németh, J. (2009) Defining a public: The management of privately owned public space. *Urban Studies*, 46(11), 2463–2490.

Newman, L.L., & Burnett, K. (2013) Street food and vibrant urban spaces: Lessons from Portland, Oregon. *Local Environment*, 18(2), 233–248.

Newman, P., & Kenworthy, J. (2006) Urban design to reduce automobile dependence. *Opolis*, 2(1).

Nordahl, D. (2009) *Public produce: The new urban agriculture*. Washington, DC: Island Press.

Parham, S. (1990) The table in space: A planning perspective. *Meanjin*, 49(2), Melbourne, Winter.

Parham, S. (1992) Gastronomic strategies for Australian cities. *Urban Futures*, 2(2), Canberra.

Parham, S. (1993) Convivial green space. *Proceedings*. Canberra: Seventh Australian Symposium of Gastronomy.

Parham, S. (1996) Productive land-use on the urban fringe: A comparative study in planning for regional economic development in languedoc and tuscany, Report, South Australia: Department of Housing and Urban Development.

Parham, S. (1996) *Gastronomic architecture: The cafe and beyond*. *Architecture Bulletin*, October. Sydney: RAIA.

Parham, S. (2001) *Developing market typologies: An urban design analysis*. Master's thesis, London. University of Westminster.

Parham, S. (2005) Designing the gastronomic quarter. In Franck, K. (ed.), *Food and the City Architectural Design*, 75(3) May/June, Wiley Academic.

Parham, S. (2008) The relationship between approaches to conservation and the idea of nostalgia: Looking at food-centred spaces within cities. In: Hardy, M. (ed.), *The Venice Charter revisited: Modernism, conservation and tradition in the 21st century world*. Newcastle upon Tyne: Cambridge Scholars Publishing.

Parham, S. (2011) Theory and practice of agrarian urbanism. Review, *Journal of Urbanism: International Research on Placemaking and Urban Sustainability*, 4(2), 193–194.

Parham, S. (2012a) *Market place: Food quarters, design and urban renewal in London*. Newcastle upon Tyne: Cambridge Scholars Publishing.

Parham, S. (2012b) Retrofitting for food – edible urbanism in inmidtown. Report. London: Unpublished.

Parham, S. (2015a) *Food and urbanism: The convivial city and a sustainable future*. London: Bloomsbury.

Parham, S. (2015b) The productive periphery: food space and urbanism on the edge. In: *Localizing urban food strategies. Farming cities and performing rurality,* 7th International Aesop Sustainable Food Planning Conference Proceedings, Torino, 7–9 October 2015, (eds.) Giuseppe Cinà and Egidio Dansero, Torino, Politecnico di Torino, 2015, 118–130.

Parham, S. (2016) Food and garden cities in principle and practice. *Letchworth: International Garden Cities Institute*. July 2016 www.gardencitiesinstitute.com/think-piece/food-and-garden-cities-in-principle-and-practice (accessed 4 April 2017).

Parham, S., & McCabe, B. (2016a) Shrinking cities and food: Place-making for sustainable renewal, reuse and retrofit. In: Neill, W.J., & Schlappa, H. (eds.), *Future directions for the European shrinking city*, 1. New York and London: Routledge.

Parham, S., & McCabe, B. (2016b) Making space for food, Research Report, Hatfield: University of Hertfordshire Press.

Patel, K., Guenther, D., Wiebe, K., & Seburn, R.A. (2014) Promoting food security and livelihoods for urban poor through the informal sector: A case study of street food vendors in Madurai, Tamil Nadu, India. *Food Security*, 6(6), 861–878.

Paull, J. (2011) Incredible edible todmorden: Eating the street. *Farming Matters*, 27(3), 28–29.

Peattie, L (1998) Convivial cities. In: Douglass, M., & Friedmann, J. (eds.), *Cities for citizens, planning and the rise of civil society in a Global Age*. Chichester. New York: John Wiley and Sons, pp. 247–253.

Pendlebury, J., & Erten, E. (2014) *Alternative visions of post-war reconstruction: Creating the modern townscape*. London: Routledge.

Peters, C.J., Bills, N.L., Wilkins, J.L., & Fick, G.W. (2009) Foodshed analysis and its relevance to sustainability. *Renewable Agriculture and Food Systems*, 24(1), 1–7.

Peters, C.J., Lembo, A.J., & Fick, G.W. (2005) A tale of two foodsheds: Mapping local food production capacity relative to local food requirements. http://crops.confex.com/crops/viewHandout.cgi?uploadid=226.

Philips, A. (2013) *Designing urban agriculture: A complete guide to the planning, design, construction, maintenance and management of edible landscapes*. Hoboken, NJ: John Wiley and Sons.

Pickett, S.T., Cadenasso, M.L., & McGrath, B. (eds.). (2013) *Resilience in ecology and urban design: Linking theory and practice for sustainable cities* (Vol. 3). Springer Science and Business Media.

Powers, A. (2012) Townscape as a model of organised complexity. *The Journal of Architecture*, 17(5), 691–702.
Punter, J. (2007) Developing urban design as public policy: Best practice principles for design review and development management. *Journal of Urban Design*, 12(2), 167–120.
Punter, J.V., & Carmona, M. (1996) Design policies in English local plans: The search for substantive principles and appropriate procedures. *Urban Design International*, 1(2), 125–143.
Reardon, T., & Hopkins, R. (2006) The supermarket revolution in developing countries: Policies to address emerging tensions among supermarkets, suppliers and traditional retailers. *The European Journal of Development Research*, 18(4), 522–545.
Reardon, T., & Berdegué, J.A. (2008) The rapid rise of supermarkets in Latin America: Challenges and opportunities for development. *Development Policy Review*, 20(4), 371–388.
Ritchie, A., & Thomas, R. (2013) *Sustainable urban design: An environmental approach*. Abingdon, NY: Taylor and Francis.
Robertson, K.A. (1997) Downtown retail revitalization: A review of American development strategies. *Planning Perspectives*, 12(4), 383–401.
Rowe, P. (1991) *Making a middle landscape*. Cambridge, MA and London: MIT Press.
Roy, A., & Ong, A. (eds.). (2011) *Worlding cities: Asian experiments and the art of being global* (Vol. 42). Chichester, UK: Wiley-Blackwell.
Rudofsky, B. (1964) *Architecture without architects: A short introduction to non-pedigreed architecture*. Albuquerque: University of New Mexico Press.
Rudofsky, B. (1980) *Now I lay me down to eat: Notes and footnotes on the lost art of living*. Garden City, NY: Anchor Press and Doubleday.
Saldivar-Tanaka, L., & Krasny, M.E. (2004) Culturing community development, neighborhood open space, and civic agriculture: The case of Latino community gardens in New York City. *Agriculture and Human Values*, 21(4), 399–412.
Schmidt, S., Nemeth, J., & Botsford, E. (2011) The evolution of privately owned public spaces in New York City. *Urban Design International*, 16(4), 270–284.
Senauer, B., & Seltzer, J. (2010) The changing face of food retailing. *Choices*, 25(4).
Sennett, R. (1987) The public domain. In: Glazer, N., & Lilla, M. (eds.), *The public face of architecture: Civic culture and public spaces*. New York: The Free Press, pp. 26–47.
Sennett, R. (1996) *Flesh and stone: The body and the city in Western civilization*. New York and London: WW Norton and Company.
Shane, D.G. (2008) Heterotopias of illusion: From Beauborg to Bilbao and beyond. In: Dehaene, M., & De Cauter, L. (eds.), *Heterotopia and the city: Public space in a postcivil society*. London: Routledge.
Shorthose, J. (2004) Nottingham's de facto cultural quarter: The lace market, independents and a convivial ecology. In: Jayne, M., & Bell, D. (eds.), *City of quarters: Urban villages in the contemporary city*. London: Ashgate, pp. 149–162.
Shouse, H. (2011) *Food trucks: Stories and recipes from America's best kitchens on wheels*. Berkeley, CA: Ten Speed Press.
Singh, Y.J., Fard, P., Zuidgeest, M., Brussel, M., & van Maarseveen, M. (2014) Measuring transit oriented development: A spatial multi criteria assessment approach for the City Region Arnhem and Nijmegen. *Journal of Transport Geography*, 35, 130–143.
Sitte, C. (1965) *City planning according to artistic principles*, translated by George R. Collins and Christiane Crasemann Collins, London: Phaidon Press.
Smiley, D.J. (2002) *Sprawl and public space: Redressing the mall*. Washington, DC: National Endowment for the Arts.
Smith, J. (2012) Encouraging the growth of urban agricultural in Trenton and Newark through amendments to the zoning codes: A proven approach to addressing the persistence of food deserts. *Vt. J. Envtl. L.*, 14, 71.
Sonnino, R. (2016) The new geography of food security: Exploring the potential of urban food strategies. *The Geographical Journal*, 182(2), 190–200.
Southworth, M. (2005a) Reinventing main street: From mall to townscape mall. *Journal of Urban Design*, 10(2), 151–170.
Southworth, M. (2005b) Designing the walkable city. *Journal of Urban Planning and Development*, 131(4), 246–257.
Stanilov, K. (2004) Planning for sprawl – the evolution of a regional shopping center. In: Scheer, B.C., & Stanilov, K. (eds.), *Suburban form: An international perspective*. New York and London: Routledge.
Steiner, F. (2011) Landscape ecological urbanism: Origins and trajectories. *Landscape and Urban Planning*, 100(4), 333–337.
Tachieva, G. (2010) *Sprawl repair manual*. Washington, DC: Island Press.
Talen, E. (2005) New urbanism and American planning: The conflict of cultures. London: Psychology Press.
Tanenbaum, J. (2012) *Regulating mobile food vending in Greenville, SC*. Diss. Clemson University.

Thompson, I.H. (2012) Ten tenets and six questions for landscape urbanism. *Landscape Research*, 37(1), 7–26.

Tinker, I. (1997) *Street foods: Urban food and employment in developing countries*. Oxford: Oxford University Press.

Tracey, D. (2007) *Guerrilla gardening: A manualfesto*. Gabriola Island, Canada: New Society.

Trancik, R. (1986) *Finding lost space: Theories of urban design*. New York: Wiley.

Valentine, G. (1998) Food and the production of the civilised street. In: Fyfe, N.R. (ed.), *Images of the street: Planning, identity, and control in public space*. London and New York: Routledge.

Viljoen, A. (ed.) (2005) *Continuous productive urban landscapes (CPULs): Designing urban agriculture for sustainable cities*. Oxford: Architectural Press.

Viljoen, A., & Bohn, K. (2014) *Second nature. Urban agriculture: Designing productive cities*. New York: Routledge.

Viljoen, A., & Howe, J. (eds.) (2012) *Continuous productive urban landscapes*. London: Routledge.

Viljoen, A., & Wiskerke, J.S.C. (eds.) (2012) *Sustainable food planning: Evolving theory and practice*. Wageningen, The Netherlands: Wageningen Academic Publishers.

Vitiello, D. (2008) Growing edible cities. In: Birch, E., & Watcher, S. (eds.), *Growing greener cities: Urban sustainability in the twenty-first century*. Philadelphia, PA: University of Pennsylvania Press, pp. 259–280.

Voyce, M. (2006) Shopping malls in Australia. The end of public space and the rise of 'consumerist citizenship'? *Journal of Sociology*, 42(3), 269–286.

Voyce, M. (2007) Shopping malls in India: New social 'Dividing Practices'. *Economic and Political Weekly*, 2055–2062.

Waldheim, C. (ed.) (2012) *The landscape urbanism reader*. New York: Princeton Architectural Press.

Wheeler, S.M., & Beatley, T. (2014) *Sustainable urban development reader*. London: Routledge.

Whyte, W.H. (1980) *The social life of small urban spaces*. Washington, DC: The Conservation Foundation.

Wright, F.L. (1935) *Broadacre city: A new community plan*. New York: Architectural Record Publishing Company.

Wu, J., & Wu, T. (2013) Ecological resilience as a foundation for urban design and sustainability. In: *Resilience in ecology and urban design*. Netherlands: Springer, pp. 211–229.

Zeunert, J. (2011) Eating the landscape: Aesthetic foodscape design and its role in Australian landscape architecture. Proceedings of AILA National Conference: Transform, Brisbane, 12–13 August 2011.

Zeunert, J. (2016) Urban agriculture up-scaled: Economically and socially productive public green space. In: Roggema, R. (ed.), *Sustainable urban agriculture and food planning*. Abingdon, UK: Routledge, pp. 107–125.

Zeunert, J. (2017) *Landscape architecture and environmental sustainability: Creating positive Change through design*. London and New York: Bloomsbury.

Zukin, S. (2004) *Point of purchase: How shopping changed American culture*. New York and London: Routledge.

Zukin, S. (2008) Consuming authenticity: From outposts of difference to means of exclusion. *Cultural Studies*, 22(5), 724–748.

Part VI
Social practices and meanings

28

Eating the commons landscape

Sacrificial food for thought concerning the meaning of landscape

Kenneth R. Olwig

Introduction

Food provides a means of understanding and evaluating both the social and the physical character of landscape as place. What is the connection between the eating of particular kinds of food, the physical character of the lands where that food is produced, and people's identification with landscape as place and community? This chapter examines this question, using philological analysis by looking at the parallels between the sacred role of sacrificial food in religion and the apparently profane, festive consumption of food. Food, it will be argued, provides a key to understanding the way that people and landscape are symbolically ingested and thereby incorporated within public community life.

Invocation

> "Behold the Lamb of God who takes away the sins of the world"
>
> John 1:29 (King James Bible, 1952 (orig. 1611))

> "The Lord Jesus on the night when he was betrayed took bread, and when he had given thanks, he broke it, and said, 'This is my body which is for you. Do this in remembrance of me.' In the same way he took the cup also, after supper, saying, 'This cup is the new covenant in my blood. Do this, as often as you drink it, in remembrance of me'"
>
> Paul 1 Corinthians 11:23–25 (King James Bible, 1952 (orig. 1611))

> Almighty and ever living God, we most heartily thank thee for that thou dost feed us, in these holy mysteries, with the spiritual food of the most precious Body and Blood of thy Son our Savior Jesus Christ; and . . . that we are very members incorporate in the mystical body of thy Son, the blessed company of all faithful people. . . . And we humbly beseech thee, O heavenly Father, so to assist us with thy grace, that we may continue in that holy fellowship, and do all such good works as thou hast prepared for us to walk in; through Jesus Christ . . . Amen.
>
> (Text from the celebration of the Communion Eucharist in The Episcopal *Book of Common Prayer (Church, 2007: 337–339)*)

The lamb and turkey of 'God' and community

The preceding texts from the Bible and the Episcopal (Anglican) Church's *Book of Common Prayer*, describe Christ as a sacrificial lamb whose body and blood is symbolically consumed in the celebration of the Communion Eucharist. It is by this means that the people assembled are incorporated together as a religious fellowship in communion with the holy spirit that is expressed in communal good works. These texts build upon age old sacrificial practices in many cultures, long preceding Christianity, where real animals, not the least lambs, were killed and eaten by an assembly of people who thereby were incorporated as a living and breathing community with common spiritual values – the word 'spirit' deriving from 'breath' (Hubert and Mauss, 1964 (orig. 1898); Buttimer, 1982). The *Book of Common Prayer* thus also refers to Jewish tradition, writing that Jesus "is the true Paschal Lamb" as celebrated at Passover (Church, 2007: 287). The Eucharist, like the Jewish Passover Seder, is a meal involving the symbolic eating of a sacrificial lamb, as well as other foods (such as bread and wine). The purpose of the sacred sacrificial ritual of the Eucharist is that this "spiritual food" enables individuals to become "members incorporate" in the "holy fellowship" that is the church's community through eating in communion.

At first glance the religious text above seems quite exotic with its talk of "members incorporate" in a "mystical body," but actually these metaphors are also very everyday in contemporary unreflected discourse. Our bodies literally have "members," e.g. arms and legs, but we also are metaphorically "members" of corporate bodies (the *corp-* in *corp*orate coming from the Latin *corpus* – as in *corpus Christi* – meaning body) and we are thereby incorporated, or initiated, into various bodies, including the body politic constituting a landscape as place and polity. Human beings thus still tend to do their thinking through these metaphors, they have just forgotten that the symbol refers back to real bodies! In the case of the Christian church Christians are incorporated into the church body, not the least through the celebration of the Eucharist, which literally means 'thanksgiving' (NOAD, 2005: Eucharist). Here too, then, this elevated ritual act actually has a meaning similar to a more down-to-earth harvest festivity such as the American *Thanksgiving* holiday with the carving of its (sacrificial?) turkey. The word holiday derives from holy day and festivity comes from feast, so even if the only religious element of the American Thanksgiving for many families may be the saying of 'grace' before the meal, there is a certain ritual character lurking in etymological recesses of these words that indicates some tie to a ceremony like that of the Eucharist. Eucharist derives from the Greek *eukharistos* meaning 'grateful'. The prefix *eu* means 'well' and the suffix *kharizesthai*, means 'offer graciously' (from *kharis* 'grace') and, when spelled *eukharistia* it means 'thanksgiving' (NOAD, 2005: Eucharist). So it turns out that the saying of 'grace' before digging into the turkey actually indicates that the turkey is being offered graciously, much as a sacrificed sacred animal might be offered graciously.

The American *Thanksgiving* belongs to a long line of harvest festival customs with time-out of-mind roots in Europe, and probably amongst the native Americans. But because the United States is a relatively new nation, going only back to the 17th-century colonial settlement of the continent, it has been consciously shaped as a national holiday, and this means that its key narratives and myths give a comprehensible picture of how food, 'sacrificial' ritual and landscape can be connected in contemporary society (Sifton, 2016). The narrative of *Thanksgiving*'s origins at the time of the nation's settlement in New England is taught in the schools throughout the United States (with a sibling version celebrated in Canada), so that every school child has been exposed to its historical imaginary and its symbolic iconography. The focus of this story is upon the first celebration of *Thanksgiving* by the Pilgrims which is believed to have taken place at Plymouth plantation in 1621, and its subsequent celebrations in what is now the Commonwealth of Massachusetts. A key element of the narrative as written from the perspective of the colonial settlers, and as later adopted more generally by the population of the United States of America, is the fact that a considerable number of Native Americans participated in the meal, and that the cultivation and gathering of much of the food, such as maize, squash and beans, was learned from the Native Americans, or provided by them, and many of these foods continue to be a traditional part of the feast as celebrated by many, though not necessarily all, human beings living in the United States of America. Though the event had religious origins it quickly also became a secular affair, supported by the Massachusetts governor. Key elements of the feast were the consumption of native foods, including wild turkeys, the inclusion of both Native Americans and settlers, and the identification of the harvest feast with Plymouth and other New England townships, ranging from villages to cities like Boston, that are all characterised by a landscape famous for its direct democracy as symbolised by the village or urban commons and the meeting house, often under the church, used for a representative assembly (Meinig, 1979). The event, thus, is very much about the incorporation of community and polity through the ingestion of foods that are germane to the land and made available through commons practices. It is easy, of course, to be dubious with regard to the irony of the colonists' subsequent ill treatment of the Native Americans who had fed them and kept them from starvation, but the point here is rather to exemplify a symbolic ideal image of how food and landscape, understood as the place of a polity and its milieu, are inextricably linked in a major holiday which seeks to communicate an exemplary message of community, multi-ethnic fellowship and landscape as place (even if the ideal is rarely achieved). This, however, is not *place* understood as a bounded excluding territory – the territory of particular white Anglo-Saxon New England villages – but a symbolic landscape place ideally belonging to a communion of everyone anywhere, of any faith or ethnicity, who is celebrating *Thanksgiving* – also in Canada. It is, in fact, more than an expression of American ideals because this New England tradition points backwards to the old England, and the English settlers' customs that they brought with them, and which were closely entwined with both religious and sociopolitical existence (Olwig, 2002). Many of the settlers were fleeing religious and political oppression and sought to rescue their customs by giving them new life in a new context and with new foods, like the turkey, the cranberries, squash and maize – the latter being a food of religious importance to the Native Americans. Against this complex religious and sociopolitical background it can thus be useful to return to the origins and landscapes of such seasonal festivals, where lamb is commonly eaten, in Europe, the Middle East and beyond, starting with the spring seasonal festivals, like Easter, Passover and equivalent Moslem, Hindu, Buddhist and pagan ritual meals, like those celebrated in the works of Homer (Hubert and Mauss, 1964 [orig., 1898]).

The pastoral landscape commons

Sheep spend much of the year grazing pastures in marginal areas that are not well suited to intensive cultivation, and which are often grazed in common. In Christian symbolism Christ is not only the sacrificial lamb, but the 'good shepherd' of the flock of people who make up his pastorate. The word pastor, which comes from Latin and means 'shepherd,' derives from *past-* meaning 'fed' or 'grazed' on the grazed land of a pasture. Food (a word related to fodder), also in the form of pasturage, is thus key to this key religious metaphor. The importance of pasturage is a theme that precedes the New Testament of the Bible, as for example in the Old Testament's Psalm 23: "The LORD is my shepherd; I shall not want. He maketh me to lie down in green pastures: he leadeth me beside the still waters" (King James Bible, 1952 [orig., 1611]: Psalm 23). The figure of the shepherd is clearly of religious importance, but also political, since the Israelites formed a nation as well as a religion. This political element is also of key significance to Platonic discourse, which influenced Christian discourse, as found most notably in Plato's *Statesman* (1961). This suggests the significance of the relation between shepherd, the flock and pasturage to the body politic (Foucault, 2007). On lands of the commons type there are basically no fences (Olwig, 2016). The animals tend to form flocks of varying size that are made up of interrelated animals, bred by generations of herding families (Galaty and Johnson, 1990). One flock shares a particular pasture, whereas other flocks tend to share other pastures, so that no fences are needed to separate the flocks. The attachment to these differing pastures is taught by the mother animal to their offspring through many generations in a process that is termed 'hefting' (spelled variously) in Britain and Scandinavia. In this way generations of shepherds, in turn, have hefted the rights of their families to certain unfenced pastures within a commons, thereby manifesting their own identification with the land

(Olwig, 2008a, 2016). This notion of hefting thus extends from the behaviour of sheep to that of people, to their sense of rights and justice (Olwig, 2008a). This suggests that there is more to the grazing of sheep on a commons, and their sacrifice to help incorporate a common community, than one might imagine if one's thought and imagination were a narrow-minded product of a 'modern' society in which enclosed private property has been constitutionally enshrined as the foundation of society. In the following I will examine the role of sheep, and by extension, other grazing animals as socially and symbolically foundational both to the landscapes surrounding the Mediterranean, and to those bordering the North Atlantic and Baltic.

The term *heft* is found particularly in areas of Britain settled by the Norse, such as the Lake District of Northwest England, where the flock is hefted to land to which the herder has use rights to the commons, or rights of 'usufruct,' meaning the right to use and enjoy the fruit of the land (Olwig, 2016). Such a use right is dependent upon its maintenance and to this end the possession of a hefted flock is vital because the even grazing of flexibly defined areas of the commons is necessary to maintaining both the sustainability of the pasturage and the functioning of the commons as a socio-pastoral entity. Fencing leads to a ridged grazing pattern with uneven grazing and increased erosion because fences hinder the flock's movement. The commons, as historically regulated by common law, facilitate improved movement patterns and reduced landscape impact (see Olwig, 2016). The combination of separately hefted flocks on a common pasture allows for the existence of separate farming entities while simultaneously requiring 'neighbourliness' because of the need to cooperate with regard to the overall management of the commons, and to prevent the conflicts that can easily arise when fenceless common areas are shared (Gray, 2000; Hart, 2004; Rodgers et al., 2011). The term 'heft' is not only used literally to refer to the flock behaviour of sheep and the pastures grazed. In a more abstract way 'heft' also captures the affective and customary legal attachment people develop to a particular place and landscape (Gray, 2000), and in Scandinavia where the term originates, it refers to a 'hefted right' (or a *hævdvunden ret* in Danish), or legal use right, or right of usufruct, gained under customary law through precedence (Olwig, 2008a). In this way abstract legal principles merge with the practices and affective attachments of both people and sheep in relation to living plant communities that in turn have an impact on the soil (Burton et al., 2008). The concept of 'heft' might thus be regarded as being foundational to the basic principles of customary law upon which British common law is based – and thereby, by extension, the law of its former colonies and commonwealth nations. Sheep grazing, as the above Biblical passages suggest, also has a social, moral and legal importance that goes deep in the landscapes surrounding the Mediterranean. It is little wonder, thus, that the English concept of "holy *fellowship*" mentioned above in the Episcopal *Book of Common Prayer*, derives from the Old Norse *fēlagi* where it referred originally to a group of shepherds who as a body share the shepherding of a common pasture (NOAD, 2005: fellow).

According to Michel Foucault the Pythagoreans, who played a foundational role in ancient Greek philosophy, derived *gnomes*, the law (*nomos*), etymologically "from *nomeus*, that is to say, the shepherd" (Foucault, 2007: 188). This fits well with the dictionary definition of *nomos*, which sees the concept of *nomos*, like hefting, as deriving from pastoralists practice of allotting pastures (Liddell and Scott, 1940: nomos):

Nomo: νέμω

Of herdsmen, pasture, graze their flocks, drive to pasture.

Nomos: νόμος

A. *that which is in habitual practice, use or possession,*

I. usage, custom, : "hence, law, ordinance."

II. melody, strain.

The nouns nómos and nomós both derive from the verb némō, to dispense or to allot, with, nomós being the result of allotment and nómos being the manner of allotment or dispensing. (justice)

The distinction between notions of law based upon the enclosed territorial space of property and those rooted in the practice of herdsmen is brought out by Foucault when he writes: "The shepherd's power is not exercised over a territory but, by definition over a flock, and more exactly, over the flock in its movement from one place to another" (Foucault, 2007: 171). This means that "in contrast with the power exercised on the unity of a territory, pastoral power is exercised on a multiplicity on the move" (ibid.). A spatial territorial approach to law, as Stuart Elden points out, would have been difficult in ancient Greece because: "to have *demes* [an ancient Greek township] that were divided . . . would have required an extensive land survey and mapping [but] there is 'little evidence of systematic land surveying,' [or that the land was] generally divided" (Elden, 2013: 14).

The exact etymological origin of the word land is complex and it varies somewhat from dictionary to dictionary, but the standard Danish unabridged dictionary gives a definition to the effect that land is "a portion of the earth's surface taken into use by humans; especially of that portion of the earth's surface used for, or considered in relation to, its use for agriculture or gardening" (trans. K.R:O., O.D.S. 1931: land). This fits well with the idea of flocks moving through and maintaining the land's pastures physically, but thereby also maintaining the shepherd's right to graze the land. The notion of use is key to this understanding of land, as it is to the definition of customary use rights. You have a moral obligation to use and maintain the land responsibly, and if you do not you can lose your use right (i.e. usufruct) to the land according to customary law. Sheep and other grazing animals thus do not graze pre-existing pastures, but create them through the process of pasturage, sometimes with the help of shepherds, who might clear or prune trees, or burn the land, much as suburbanites maintain a lawn with a mower (Vera, 2000). It is, as noted, the use and maintenance of the pasture that gives a 'hefted' customary use right to continue using it. The same applies, of course, to the practices of farmers and gardeners, whose land historically was created in partnership with pastoralism in which the animals grazed the fallow land and commons and provided vital fertiliser, milk, wool and even, in the case of cattle and horses, traction. When land is enclosed and mapped as the space of territory or property it is 'naturalised' according to the absolute 'natural' laws of geometry and becomes defined by the physical nature of whatever is enclosed within its boundaries, including raw and desolate abandoned wastelands (Olwig, 2016). When land is defined as individualised property corresponding to the space of the map, then common custom and use ceases to define legal rights to land, and the property right to land becomes, in principle, absolute, regardless of use or misuse (Olwig, 2016). The importance of custom's foundational relationship to the idea of law as inherited from ancient Greece is brought out by C.C.W. Taylor who argues that:

> Usage renders *nomos*, from the verb *nomizein*, to have a usage or custom. *Nomos* is *ho nomizetai*, what is customary or enshrined in usage. The term is standardly contrasted with *phusis*, literally 'nature,' which in this general contrast is simply the abstract noun for how things are independent of human thought or belief. A related sense of *nomos* is 'norm,' or more specifically 'law,' derived from the basic sense via the normativity of custom and usage.
>
> *(Taylor, 2007: 1)*

The connection between customary 'norm' and law is important because the norm also implies a moral imperative. The word moral, in fact, derives from the Latin *moralis*, meaning custom, as has been illustrated by the customary right to use the land being determined by the maintenance of land through use, as opposed to the abuse, of the land. This moral dimension is also apparent in the ties between shepherding and religion, exemplified at the outset, as manifested in the customary performance of the Eucharist. This, then, is the kind of law that is grounded in time out of mind customary morality, rather than in the law of property (Olwig, 2016).

What makes, one may wonder, the slaughtering of the sacrificial lamb a sacrifice if the animal is then roasted and eaten as part of a holy holiday feast? The answer lies in the core importance of the grazing animal in a society where pastoralism is of central importance. The animal is thus not only a source of meat, but also of milk, wool, manure and the genetic continuity of a stock bred over generations of animals and people. The "hefted" animals, the places where they heft and people are thus, in a certain sense, "related." The offering of these animals to a common shared festivity thereby becomes a sacrifice by those who make the offering, but it is a sacrifice that helps transform individuals into a community whole, which can be regarded as a kind of sacred miracle in which social health, wholeness and holiness are conjoined (health, whole and holiness being etymologically related) (Gudeman, 2001; Hubert and Mauss, 1964 [orig., 1898]). It is very important to note here, however, that health, wholeness and holiness are not pre-given by nature, by *phusis*, and hence scientific certainty, but by a ceremonial, ritual offering of individual wealth to the commonwealth, that is rooted in culture and faith. Neither community nor landscape, as the expression of community in the land, are thus given by nature, but by culture, and they are therefore the product of the laws of people, not the laws of nature. The idea that nature mandates, or should mandate, human law provides a recipe for tyranny by those who claim to speak for the laws of nature (Olwig, 2017). Historically the idea that law has a 'natural' foundation derives from the legal tradition of the Roman empire, where law was propounded by the state on the basis of what was believed to be eternal unchanging principles (like those of geometry). Customary law, which is the foundation of English common law, is based on the idea that law is based on the precedence of established practice as remembered for 'time out of mind', which means that if a custom is no longer practiced, and ceases to be remembered, it no longer applies (Olwig, 2002). The notion of sacrifice, as expressed in the Eucharist communion, means that the resultant incorporation of individuals as a (religious) community is an expression of belief, not nature – though one can believe in the existence of (secular) community without belonging to an organised religion.

Sharing the common wealth

When society is founded on the pre-enclosure idea that society is a commonwealth where the land is not private property, but the common wealth of a political community to which people have customary use rights, then the idea of sacrificing private wealth to the common good makes a certain sense. The paintings of Peter Breughel (c.1525–69) and other artists of his period helped establish a genre of landscape painting that paid close attention to landscape, be it rural or urban, as the substantive place of polity and community (Olwig, 2005). Among the activities recorded in these paintings is the eating of communal meals on festive occasions. The Russian literary theorist Mikhail Bakhtin has described the role of festive occasions, notably carnival (*carn-* being Latin for meat or flesh), in creating a sense of community through a form of 'belly' humour that emphasizes the way the body, with its bodily functions and fluids ('humors'), such as eating (and what follows thereafter), acts as a common denominator bringing people together as a body (Bakhtin, 1984).

Not all scholarly knowledge is gathered from texts. 'Participant-observation' is an important method employed by cultural anthropologists and cultural geographers to research phenomena that must be experienced to be understood. Carnival is such a phenomenon because its meaning comes through the body, not the least the belly, and key to this form of knowledge is that it does not derive from individual thought, but from common experience. Many years ago I thus, on a field excursion with a cultural geographer, was able to participate in the Cajun *Mardi Gras* carnival celebration held in the little town of Iota, the name of which means little, in Acadia Parish on the Louisiana bayou. This carnival feast began when the 'Courir de Mardi Gras,' a band of masked men on horseback and led by a whip-wielding 'Capitaine' 'raided' surrounding farms, 'stealing' a chicken at each farm, and 'begging' for foodstuffs (a somewhat frightening form of 'friendly' extortion). The raiders then returned to town where the chickens were 'sacrificed and cooked as a 'filé gumbo' in a big pot with local ingredients including filé powder (originating with the local the Choctaw Native American tribe) and sometimes okra (of African American origin), local spices, and the three main ingredients in Cajun French gumbo: 'holy trinity' of onions, celery and bell pepper. This was thus a literal melting pot including ingredients stemming from the area's multi-multi-ethnic population. The food was then served to the people of the community to the tune of lively Cajun music. Here we were very far from the stern Northern European protestant pilgrims of Massachusetts, but much of the cultural 'flavour' was the same, as was the fact that the tradition could be traced back to Europe, especially France (and Spain), as well as to the Native Americans of the region, and even the African American descendants of former slaves. As was the case with the Thanksgiving originating in New England, the central point here is that the food played a key role in incorporating a landscape community that was also able to assimilate visitors like myself. These Cajuns, like the Pilgrims were, metaphorically speaking, eating and drinking the body and blood of their landscape – to the point that this largely Roman Catholic population terms, in the spirit of Carnival, the main ingredients of the Cajun gumbo a 'holy trinity'.

Even though the busy medieval church calendar of holiday feast days has long since been replaced, even in Catholic countries, by Protestantism's and Capitalism's nose-to-the-grindstone workaday ethic (and holiday-starved calendar), one nevertheless does not have to travel very far today to see that the people of different European landscape communities are still gathering to feast together on various occasions that are often related to the seasonal cycle of land use and food production. These festivities are often named for a particular cherished food, for example chestnuts or truffles, though in Catholic communities the name of the area's patron saint is often also identified with the celebration. In Italy and France, for example, my family has eaten its way through food festivals where the local families have produced everything imaginable from chestnuts or fungi to truffles. In the case of truffles, the whole town can reek of their earthy raunchy smell. The most exotic local community festival we have attended, however, was undoubtedly the 15th annual 'Blue Food Festival' in 2013 that was held above Bloody Bay on the Caribbean island of Tobago. The festival featured everything imaginable and unimaginable (for example ice cream, pizza and wine) that could be made from the root of the dasheen plant (taro – *Colocasia esculenta*), which when cooked tends to turn a light blue colour. The festival, as might be expected given the colour of the food, attracted only a few tourists (though the few that came were welcome), but the Tobagonians flocked to the festival by foot, car, chartered bus and, in the case of migrants, airplane in order to dine on dasheen and listen to calypso music. Dasheen is typical of the kind of food African-Americans affectionately call 'soul food', which is based on the inexpensive staples of the poor that the wealthier classes find undesirable. In this case dasheen is one of the worlds oldest cultigens, brought with the slaves from Africa and cultivated on the 'provision grounds' that the slaves were allowed to develop on the marginal common lands outside the plantation fields where commercial crops were produced. Dasheen was thus a low status food identified with slavery, but the 'Blue Food Festival', through the humorous abundance of unlikely, and nominally high status, dasheen products worked in carnivalesque fashion, through the belly, to establish the food as a staple of Tobagonian landscape identity in the sense of belonging to the community of the slave's descendants and encompassing the island's flora and fauna.

What emerges when comparing these festivities is a realisation that a commons landscape is not simply an area defined by the presence of commonly used subsistence resources, but also the site of the common resources, notably foods, that when shared provide a basis for a group of individuals' incorporation as a community and polity. As the anthropologist Stephen Gudeman puts it: "The commons is a shared interest or value. It is the patrimony or legacy of a community and refers to anything that contributes to the material and social sustenance of a people with a shared identity: land, buildings, seed stock, knowledge of practices, a transportation network, an educational system, or rituals," and to this might be added the foods produced from that seed stock, and the genetic stock of the animals that graze the land. "The commons," according to Gudeman, is thus "the material thing or knowledge a people have in common, what they share, so that what happens to a commons is not a physical incident but a social event." It is for this reason that: "Taking away the commons destroys community, and destroying a complex of relationships demolishes a commons" (Gudeman, 2001: 27).

Eating the commons landscape today

Though the absolute, individual, naturalised rights of private property dominate in modern society, customary rights are still important in areas of the earth surface that are difficult or unprofitable to enclose for intensive agriculture, and therefore remain in something like the state of a commons. In such areas one often finds laws like the Scandinavian *allemansrätt* that give people a customary right to roam and gather treasured subsistence foods like berries and mushrooms. In the Catholic Mediterranean lands regulated by shared use rights are not only characterised by their marginality and their heritage of commons, but also by their more direct religious associations, sometimes by pilgrimages to holy sites. In the more culturally Protestant areas of Atlantic Europe, the practice of hiking, camping and food gathering in such areas is also often of significant importance to landscape identity and feeling of community, even if these activities are not directly religious. In both cases importance is attached to the eating of foods identified with these lands – their honeys, cheeses, milk, spirits, and cured meats, and the feeling of community that arises when hiking or making a pilgrimage through these lands together with others.

A 'commons' is defined not by boundaries on a map, or by a fixed 'scenic' arrangement of soil, flora, fauna, and so on, but by the nature of the resources to which different segments of society have access, and the places where they are located. Thus, if people have grazing rights to uncultivated village lands, they have rights to those places where grazing resources like grass and heather are available. Though the enclosed space of modernity has come to dominate the perception of landscape, a subaltern non-modern notion of landscape as a common good has persisted, as can be seen in the continued existence of common rights of way and access and the role that national parks and other forms of parkland continue to play as symbols of political and community identity (Olwig, 2008a). A substantive element fostering the maintaining of pre-enclosure rights of commons, and the sense of landscape as polity and place that they foster, is the fact that not all land is well suited to enclosed intensified agriculture. This means that much of the marginal land that historically was used as common land could persist under forms of land use reminiscent of historical commoning, which is to say grazing and subsistence gathering (i.e. everything from berries and mushrooms to the firewood to cook the food). It is the food from these places that plays a key role in the constitution of place and community landscape identities.

One could argue that the importance of eating the commons landscape is as marginal as the lands in which they take place, and this would also be the case if one treated these experiences and foods in quantitative terms. Most of the food sold comes from agribusiness's mass production of uniform foods. If, however, one looks at these marginal phenomena qualitatively, and from a temporal rather than a spatial perspective, one might argue that modern society's weekdays where people work boxed in as individuals in offices and factories are comparable to the spatially enclosed privatised lands where industrial crops are intensively produced and intensively marketed and consumed. The marginal lands with their heritage of commons, however, are more comparable to modern society's weekends. It is during the weekend that people 'recreate', by *recreating* themselves in a physical bodily sense, as well as socially, by doing things together with others, thereby (re)creating a sense of community. It is the tension of the temporal dialectic between the weekend and the weekday, and the spatial dialectic between centre and periphery, I would venture, that allows society to encompass both exploitive and alienating individualisation and the restorative needs of community and morality.

The economically marginal, but culturally central, landscape areas have importance to cultural identity in not only symbolic terms, but also literally. The word culture thus derives from the Latin *colere* meaning 'tend, or cultivate'. Originally applied to the tending and cultivation of the flora and fauna of the land it was extended to include care for the micro-organisms that are present in a body of land, and which are vital to the health of the land and its flora and fauna, much as with regard to a human body. These micro-organisms, in turn, are vital in the creation of products of the land such as yogurt 'culture' or the cultures used to create cheese, wine, cider and beer. It was only in the early 16th century that culture came to be applied in the modern sense of the cultivation of the mind (Williams, 1976: culture; NOAD, 2005: culture; Paxson, 2008: 12). It is in the areas, with small scale production maintained by generations of farmers that, for example, distinctive cheese cultures are developed as a product of a particular agricultural culture and community that uses the land in particular ways and breeds particular breeds of the cows, sheep or goats that eat the pastures that, together with various micro-organisms, produce a particular cheese or meat flavour (Paxson, 2008). This flavour, in turn, reflects the complex 'cultural' biodiversity of the place. Such products can often command high prices because of their perceived unique taste, and because they can only be produced on a small scale in limited landscape areas. It is difficult for large scale agribusiness to duplicate and compete with these producers (Paxson, 2008: 18, 32). These foods thus raise questions concerning not simply rarified issues of identity, community and heritage, but also the economic and ecological health of our landscapes. One can thus see that in many marginal areas the small scale production of high quality meats, cheeses, grains and vegetables that people may buy as much for symbolic as for sustenance reasons, can be highly viable economically and, simultaneously, keep alive forms of agriculture that can guide necessary reforms of more exploitive agricultural mass production. Likewise, when a population learns that the eating of inexpensive, environmentally friendly and healthy, but low prestige products like dasheen, is a mark of cultural identity, this is advantageous both to the economy of the agriculturalists who produce these marginal products and to the economy and health of those who consume them.

Defining landscape

As Denis Cosgrove has shown, the modern meaning of landscape was defined through the visual cartographic and perspectival representation of space – perspective essentially being an illusion of spatial depth created by the reconfiguration of the Euclidean geometric space of the map (Cosgrove, 1984). When this mode of spatial representation is applied to places on the earth it makes possible the definition of landscape according to boundaries drawn on the uniform space of map, and visualised as spatial scenery. This approach to landscape was well suited to the 18th-century era of enclosure when common land was redefined as private property, and estate owners sought to memorialise the process with idealised artistic images and gardens designed to glorify the private individual owner and the ideology of individually owned property (Barrell, 1972; Olwig, 2002). Enclosure also meant a simplification of agriculture whereby landowners, through the use of drainage and other methods, sought to create uniform fields with the intensive specialised production of uniform, standardised crops to fit the new spatialised structure of land use and a growing money economy. It was also well suited to the needs of the expanding top-down state bureaucracies of the time because they provided a means of registering information for taxation and control within cartographic regional boundaries (Olwig, 2016). This approach to landscape, however, with its Euclidean lines of fixed, absolute cartographic space, is ill suited to the topological and temporal complexity of the earth and its life. An approach based on pre-enclosed, or non-enclosed, use of common land, for example by grazing animals like sheep, or for the gathering of subsistence food and other resources, is more suited to this complexity. Likewise, as good as scenic landscape was at capturing the ethos of a modern social and political identity as defined by the ownership of private property – which, for example, determined the political rights of early liberal democracy – it was not a useful measure of the place identities of commoners in the political communities formed prior to enclosure. A commoner's rights and identity were thus largely defined in relation to the commoner's rights of common, which is the right of access to common resources usually located on 'the commons' (Barrell, 1972). If landscape is conceptualised as a substantive phenomenon, rather than as a form of scenery, it will be perceived as the common(s) social and material place of a political community (Olwig, 1996; Olwig, 2002). Then 'sacrificial' holy and holiday meals, where the wealth of the individual is sacrificed to incorporate, through a feast, the needs of a community common wealth, come to provide a vital key to understanding the sacred and the profane meaning of both landscape, and the role of meals in generating that meaning.

Further reading

Bakhtin, M. (1984) *Rabelais and his world*. Bloomington, IN: Indiana University Press.

Barrell, J. (1972) *The idea of landscape and the sense of place, 1730–1840: An approach to the poetry of John Clare*. Cambridge: Cambridge University Press.

Burton, R.J.F., Schwarz, G., Brown, K.M., Convery, I.T., & Mansfield, L. (2008) The future of public goods provision in upland regions: Learning from hefted commons in the Lake District, UK. In: Bonn, A., Hubacek, K., & Stewart, J. (eds.), *Drivers of environmental change in uplands*. London and New York: Routledge, pp. 629–657.

Buttimer, A. (1982) Musing on helicon: Root metaphors and geography. *Geografiska Annaler*, 64 B(2), 89–96.

Church, T.E. (2007) *The book of common prayer*. New York: Church Publishing Inc.

Cosgrove, D. (1984) *Social formation and symbolic landscape*. London: Croom Helm.

Elden, S. (2013) *The birth of territory*. Chicago: The University of Chicago Press.

Foucault, M. (2007) *Security, territory, population: Lectures at the Collège de France, 1977–78*. New York: Palgrave Macmillan.

Galaty, J.G., & Johnson, D.L. (eds.). (1990) *The world of pastoralism: Herding systems in comparative perspective*. New York: The Guilford Press.

Gray, J.N. (2000) *At home in the hills: Sense of place in the Scottish borders*. New York: Berghahn.

Gudeman, S. (2001) *The anthropology of economy: Community, market and culture*. Oxford: Blackwell.

Hart, E. (2004) *The practice of hefting*. Ludlow: The Farmers Fund.

Hubert, H., & Mauss, M. (1964 [1898]) *Sacrifice: Its nature and function*. Chicago: The University of Chicago Press.

King James Bible (1952 [1611]) *Holy Bible: Authorized King James version*. New York: Oxford University Press.

Liddell, H.G., & Scott, R. (1940) *A Greek-English Lexicon*. Oxford: Clarendon Press.

Meinig, D.W. (1979) Symbolic landscapes: Some idealizations of American communities. In: Meinig, D.W. (ed.), *The interpretation of ordinary landscapes: Geographical essays*. New York: Oxford University Press, pp. 164–192.

NOAD (2005) *New Oxford American dictionary*, E. McKean (ed.). Oxford: Oxford University Press.

O.D.S. (1931) *Ordbog over Det Danske Sprog*. Copenhagen: Gyldendal.

Olwig, K.R. (1996) Recovering the substantive nature of landscape. *Annals of the Association of American Geographers*, 86(4), 630–653.

Olwig, K.R. (2002) *Landscape, nature and the body politic: From Britain's renaissance to America's new world*. Madison, WI: University of Wisconsin Press.

Olwig, K.R. (2005) Liminality, seasonality and landscape. *Landscape Research*, 30(2), 259–271.

Olwig, K.R. (2008a). Performing on the landscape versus doing landscape: Perambulatory practice, sight and the sense of belonging. In: Ingold, T., & Vergunst, J.L. (eds.), *Ways of walking: Ethnography and practice on foot*. Aldershot: Ashgate, pp. 81–91.

Olwig, K.R. (2008b) Has "Geography" always been modern? Choros, (non)representation, performance, and the landscape. *Environment and Planning A*, 40, 1843–1861.

Olwig, K.R. (2016) Virtual enclosure, ecosystem services, landscape's character and the "Rewilding" of the commons: The "Lake District" case. *Landscape Research*, 41(2), 253–264.

Olwig, K.R. (2017) Geese, elves and the duplicitous, "Diabolical" landscaped space of reactionary-modernism: The cases of Holgersson, Hägerstrand, and Lorenz. *GeoHumanities*, 3(1), 41–64.

Paxson, H. (2008) Post-Pasteurian cultures: The microbiopolitics of raw-milk cheese in the United States. *Cultural Anthropology*, 23(1), 15–47.

Plato (1961) Statesman. *The collected dialogues of Plato*, Hamilton, E., & Cairns, H. (eds.). New York: Pantheon, pp. 1018–1085.

Rodgers, C.P., Straughton, E.A., Winchester, A.J.L., & Pieraccini, M. (2011) *Contested common land: Environmental governance past and present*. London: Earthscan.

Sifton, S. (2016) The American thanksgiving. *The New York Times*, 1–45.

Taylor, C.C.W. (2007) Nomos and phusis in democritus and Plato. *Social Philosophy and Policy*, 24(2), 1–20.

Vera, F.W.M. (2000) *Grazing ecology and forest history*. New York: CABI Publishing.

Williams, R. (1976) *Keywords: A vocabulary of culture and society*. London: Fontana.

29

From the agora to the modern marketplace

Food markets as landscapes of business and pleasure

Susan Parham

Defining and framing today's food markets

Food markets present critical and increasingly studied elements in the landscapes of food. As food itself has gained in legitimacy for scholarship in the social sciences (Atkins and Bowler, 2001), its spatiality has also gained prominence (Bell and Valentine, 1997). Within a range of disciplines, individual scholars have now written about the role of food markets as *places* (Parham, 1990, 1992, 2005, 2012, 2015; Esperdy, 2002; Franck, 2005; Watson, 2009). Yet there is still much to understand about the place-based, material aspects of market landscapes as dynamic contributors to everyday social, economic, and environmental practices and the broader landscapes and townscapes of human settlement.

It is important to be clear about the terms that are central to such research. The term 'landscape' is used to mean 'an area, as perceived by people, whose character is the result of the action and interaction of natural and/or human factors', consistent with the European Landscape Convention (Council of Europe, undated). 'Townscape', meanwhile, stems conceptually from the work of urban morphologists, including M.R.G. Conzen (1960), whose groundbreaking exploration of land use change over time in Alnwick modelled and prefigured close attention to the material, spatial aspects of urban space transformation, including in relation to changing food marketplaces (Cullen, 1961; Ashihara, 1983; Southworth, 2005; Parham, 1992, 2012). Gordon Cullen (1961: 8) has argued that townscape is based on "all the elements that go to create the environment: buildings, trees, nature, water, traffic, advertisements and so on, and to weave them together in such a way that drama is released". Ideas of character and sense of place have become enmeshed in the notion of unique townscapes, so that,

> in the course of time the landscape, whether that of a large region like a country or of a small locality like a market town, acquires its specific genius loci, its culture- and history-conditioned character which commonly reflects not only the work and aspirations of the society at present in occupancy but also that of its precursors in the area.
>
> *(Conzen, 1966: 57, in Jive'n and Larkham, 2003)*

The focus of this chapter is not primarily about 'food markets' in the sense of resource allocation systems understood as economic constructs. Nor are 'markets' conceived as a placeless economic interplay between actors in a food system assumed to be universally similar whatever the global context. Within the food markets literature, often the words 'food market', 'marketplace', and 'food landscapes', although implying some embodied food-centred site, are used in an imprecise and aspatial way to explore immaterial aspects of food-related networks and connections, both economic and social. As previously argued by Parham (2012, 2015), and as teased out by Stobard and Van Damme (2015), this immateriality is theoretically inadequate where the focus is on the way food markets and cities intersect. There are steps towards a *material*, place-based approach to food markets in work on Alternative Food Networks (such as in Renting et al., 2003; Jarosz, 2008; Goodman and Goodman, 2009) with farmers markets picked out as an important typological category. Markets receive some limited attention in commodity biography research, in ethical consumption critiques (such as Coles and Crang, 2011) and where there is a focus on so-called niche food markets (Ilbery and Kneafsey, 1999; Pearson et al., 2011). However despite references to spatiality in these research areas there is more to be done to explore the place aspects in depth.

In keeping with the centrality of landscape as the organising principle for this chapter, here instead the attention is firmly on specific places and their meanings. These material marketplaces are, it is argued, *physical* as well as social, environmental, and economic constructs that are produced by, respond to, and themselves may alter the unique and particular landscapes of which they are a part. The chapter therefore avoids the largely aspatial and universalising way 'markets', 'food markets', and 'marketplaces' are often discussed within urban geography, economics, planning, and related disciplines, and asserts the material connections to and identification of issues related to place in landscape terms.

Sources for place-based food market analysis in this chapter

In part as a response to the inadequacies discussed here, interest in the significance of food markets as *places* has emerged in recent years. This can be seen within and across a range of disciplines, as part of a wider recognition of the importance of food's spatiality in relation to human settlement. Integrating research *foci*, like those of the study of food and urbanism, is helping scholars to take a holistic view and draw together concerns about the roles of food markets within a paradigm of sustainable placemaking (Parham, 2005, 2012, 2015; Hinrichs and Lyson, 2007; Heller and Adams, 2009; Marsden, 2013).

In exploring this theme, the chapter refers to previous primary research work on this across a wide range of topics, including in relation to city design, ordinary places (Amin and Graham, 1997; Hall, 2012; Blake, 2013); everyday life (Jarvis, Pratt and Cheng-Chong Wu, 2001; Maitland, 2007; Pink, 2012); convivial ecologies (Bell and Binnie, 2005) and hospitable spaces (Bell, 2007); gastronomic travel and tourism (Hjalager and Richards, 2002; Sims, 2010; Hall and Gössling, 2013); and place consumption (Thrift and Glennie, 1993; Warde, 1997; Banerjee, 2001; Raki and Chambers, 2012). Theoretical constructs and frameworks relevant here include Bourdieu's (1984) notion of the *habitus*; the interplay of socio-spatial practices in constructing new forms of sociability, and the impact of gentrification on food space (Deener, 2012; Duruz and Khoo, 2014; Aiello, 2014). The chapter makes use of the author's own work as well as others in arguing for place-based gastronomic strategies and food quarters (Parham, 1992), through research into food markets as physical spaces (Carr et al., 1992; Parham, 2005, 2012; Smith, 2011; Smith et al., 2014), and by way of a more global research reach in spatial terms in relation to markets in *Food and Urbanism* (Parham, 2015). There is a particular stress on markets as aspects of an urban landscape approach to combatting and mitigating climate change and in enhancing human sociability.

First, the chapter contains a brief historical review of roles markets have played in societies over the course of urban settlement. Second, it then traces some of the ways food markets have developed into the contemporary urban period. Third, it provides critical readings of market landscapes in current urban practice. Fourth and finally, it draws conclusions about the likely and potential roles for food markets in future.

Markets in historical landscape review

Markets have been a critical element in urban space throughout its history; central to the landscapes of many human settlements (Kostof, 1991, 1992; Parham, 1990, 1992, 2005, 2012, 2015). Archeological and other historical sources provide evidence about the existence of markets that also references their ubiquity (Dahlin et al., 2007; Chase and Chase, 2014). From Mayan, Greek and Roman settlements, for example, quite a lot is known about food markets from various archeological, morphological, historical and literary sources (Bray, 2003; Wilkins and Hill, 2006; Dahlin et al., 2010). Three main types of markets seem to have been present in the European classical world: daily markets, usually in larger towns and cities, occasional markets tied to fairs and festivals, and periodic markets repeated regularly in the same location (Dalby, 2003: 208). Markets often started on an informal basis, as urban landscapes themselves developed, but over time became regularised and incorporated into the civic functions of the town under some control by urban authorities. Rome's urban markets, for instance, became big enough to justify permanent oversight of prices and quantities (Corbier, 1999: 130).

Classically, the marketplace was the centre for civic life and political engagement as well as the location for responding to the material needs of the body through food buying: "in ancient cities from early times, the marketplace, or *agora* in Greek, *forum* in Latin, was the focus for political gatherings, and also for the exchange or purchase of goods, including food" (Wilkins and Hill, 2006: 64). In Rome, the range of food on offer, often 'the fruits of empire' was a sign of prosperity, while specialist markets for different foodstuffs including meat, fish and grain also developed (Wilkins and Hill, 2006: 203). Examples from Greek city-states and later from Rome demonstrate how important these sites were for acquiring cooked dishes as well as raw ingredients. Rome had a *Forum Cuppedinis* where stallholders sold cooked foods, foreshadowing today's 'slow food food courts' (Parham, 2015: 75). Also in Rome, substantial market buildings were constructed, such as the *marcella* initiated by Trajan in the early second century CE that served as a model for other Roman cities and towns (Alcock, 2006: 24).

Natural as well as built landscape aspects were evidently important; influencing market locations close to production sites and relying on characteristics such as the confluence of rivers and roads. Much of the trade on which urban markets were reliant was local, with markets established in towns, and markets and fairs at "nodal points in the countryside", including religious sites, to offer the opportunity to trade, feast and profit from food surpluses (Alcock, 2006: 24). However, some foodstuffs came from further afield. Alcock (2006: 12) points out that in Greek classical times, trade helped define where markets might be established, noting the example of impromptu fish markets set up where fish for Athenians was first landed, before it was conveyed to the regular urban market.

Already evident was the close interconnection between market spaces, other buildings and streets and food related practices. Public and civic dining – both daily and honorific – occurred in buildings close to the market, while in Rome there is evidence that meat derived from ritual sacrifice found its way to the *forum boarium* market for sale (Dupont, 1999: 21). Markets acted as anchors for a range of other food-related land uses, reflecting the range and integration of everyday food practices in a largely walkable townscape. While well-off families could employ cooks (who sometimes hired out their services from food markets) and thus eat decently at home, many poorer residents did not have facilities to cook domestically and relied on walking to market stalls, cook shops, and snack bars, as well as visiting bars and taverns in the market's shadow (Parham, 2015: 98).

It is worth noting similar urban market landscapes are evident from Chinese cities, from the Islamic world and from Byzantium. Constantinople, like Rome, encompassed a huge and complex city and hinterland. Food markets there sold a wide range of foodstuffs, and it is thought that surplus food from urban farms made its way to local urban food markets (Barthel and Isendahl, 2013: 8). As in Rome, many people ate in nearby taverns rather than at home due to the cost of cooking fuel and lack of facilities. Similarly, evidence from food scholars in relation to the medieval period, suggests that with population growth, as urban settlements and rural production began to renew and expand in the early Middle Ages, fairs and markets played important roles in urban and regional economies and food provisioning systems as they had done in Classical times.

In Europe, by the end of the thirteenth century, the feudal settlement landscape of villages, castles, and cities was connected by land and river routes (Sarasúa and Scholliers, 2005): markets were "where the peasants disposed of their small agricultural surpluses and acquired a few manufactured articles sold by merchants from cities" (Riera-Melis, 1999: 257). As in Greece and Rome, especially in the later Medieval period, although there was an interplay of urban food growing with the European built townscape, most urban dwellers needed to buy at least some food at the market. Some large cities had everyday food markets while others hosted markets on a few regular days a week (Nicolas, 1996: 44). Many larger cities had built central market halls as well as specialist markets for imports by the thirteenth century, while smaller towns did so by the end of the fourteenth (Nicolas, 1996).

As Nicolas (1996) points out, buildings inserted into formerly open market space could sometimes be a technique for dealing with food shortages, with granaries and other large urban storage facilities developed to respond to such situations. In Florence, for example, the *Orsanmichele* was constructed in 1336, as a *loge* "open on all sides that served as a market and granary. This meant enclosing the market square, and the grain market was thus moved behind the city hall" (Nicolas, 1996: 44). The developing ensemble of the marketplace and market buildings was closely connected as an urban landscape to food shops such as butchers and bakers, and other food-centred land uses including inns and taverns. Gies and Gies (1969) show how around 1,250 food shops clustered by the food market in the French town of Troyes, and as elsewhere, their frontages worked as food stalls with shutters that could be let down to provide horizontal counters across which food could be sold. Clark (2000: 64) reports on similar frontages in English towns from the eleventh century while Sennett (1996: 193) reports on food shops with drop downcounters in Paris of the early 1100s onwards.

By looking at rural and urban diets in the Middle Ages, Cortenesi (1999) is able to also trace areas of spatial continuity and change related to food markets in European medieval cities and towns. Trade at various scales is one area of continuity. For example, a wide range of foodstuffs was being transported locally and regionally to town markets, including wheat, oil, and chestnuts, while others went considerably further: citrus fruits from Sicily to Rome, and onions from Brittany and Picardy were being sold in London markets by the thirteenth century (Cortenesi, 1999: 271). Food guilds began to develop and new food related trades started to emerge in the fourteenth and fifteenth centuries to join the bakers, butchers, and tavern keepers. In Provence, for example, pork butchery stalls were given a special place in the market (Desportes, 1999: 282).

Evidence from the fourth century BCE onward shows rich spatial and social relationships between the marketplace, market buildings, and these other food-related land uses. It suggests a need to think about food markets not just as a discrete element in the urban armature (Chapter 27) but connected to other parts of the townscape and to wider landscapes. Markets were usually connected to local and regional agriculture, food trade and distribution; the purchasing of food and cooking services; public and private feasting and dining; buying take away food; and eating and drinking out. The practices that occurred in and around the market landscape were embedded at a range of scales from place to street; from quarter to urban, suburban, and peri-urban space; and again to rural, regional, and even more global space. The food landscapes of which the market was a part were thus both intensely localised physical spaces, and the locus of a much broader spatial and economic sphere of economic and social relations, connected through trade routes on road, river, and sea.

It is also worth noting that a theme throughout the history of food markets has been paradoxically their connections to rural and urban poverty and hunger. Food producers generally had to send their best produce off to market while they themselves might go hungry. As Wilkins and Hill (2006) argue,

> In many ways, the Greeks and Romans resembled pre-modern societies in experiencing recurring food shortages (particularly in the spring) and disease. For example, shortages were frequent in Britain

> until the twentieth century; and France, for all its modern sophistication and the richness of its markets paraded weekly throughout the country from the smallest villages to the streets of Paris, also struggled to feed itself until the nineteenth century.
>
> *(ibid.: 15)*

Thus, in tracing the evolution of this critical urban landscape it is important to not romanticise the relationship between the urban food market, its supply chain and food consumption. This was a matter of practical necessity as it remains for many today.

Markets in the industrial era

It would be wrong to assume a sharp break point between traditional market forms and functions and the changes to retailing ushered in by the effects of maritime and later forms of capitalism at a global scale. Undoubtedly, food markets *as places* were affected and transformed by forces of urbanisation and the growth of large cities, although as noted earlier, discussion of place design in relation to the nature of such food markets tends to be aspatial in practice (Kearns and Philo, 1993). There are also strong aspects of continuity in market-based retailing and consumption practices well into the twentieth and twenty-first centuries. In the early modern period of urban development, markets retained their dominance in the townscape, and the United Kingdom, for example, saw a number of consolidating developments of market space including the widespread building of market halls (Schmeichen and Carls, 2000). In large cities like London there was growth of substantial wholesale markets to deal with massive population increases. Even though there remained a large presence of urban agriculture there was also a growing population of those who were spatially disconnected from food production, and who relied on markets and street venders for food sourced from an expanding foodshed.

Market halls became an increasingly ubiquitous feature of townscape, within explicitly defined 'market' towns and other urban settings. In the European tradition, urban neighbourhoods were market-centred, with the *marché de quartier* an essential element of the urban landscape in many places. A series of different market space typologies could be identified, and work using observations and morphological analysis demonstrates their regional and local variations within an overarching place shaping structure based on principles of urbanism derived from classical and medieval models (Sitte, 1986; Trancik, 1986; Calabi, 2004; Atkins et al., 2007; Parham, 2005, 2012).

As in classical and medieval times, in early modern towns and cities, food markets retained their close physical and social connections to other food spaces including food shops, bars, inns and taverns. The role of food providers with close associations to markets continued to be emphasised, as in the guilds of *boulangeurs* and *traiteurs* in Paris and smaller French cities and towns. New food-centred land uses, including the cafés and the *bouillon* began to emerge, as did restaurants as we might recognise them, at around the time of and following the French Revolution of 1789–1799 (Spang, 2000). Wholesale and retail food markets continued to connect urban populations closely to local and regional food production and distribution chains and offered an important local social and cultural space for everyday food practices of consumption and identity formation.

The spatial configuration and interplay between markets and other parts of the townscape did not change massively until the late nineteenth and again in the early and middle part of the twentieth century when structural forces of various kinds – spatial, economic, and social – began to shift consumption practices and the way space was organised in towns. For example, in nineteenth-century London the coming of the railways not only disrupted central city space, thus dislodging markets; it altered the scale and nature of wholesaling and retail market space. Food markets changed as the urban commuter population developed to a much larger physical extent than it had before, and food production and consumption patterns were transformed (Parham, 2012).

The fate of many market halls in Britain over the nineteenth and twentieth centuries demonstrates that markets were affected by the profound changes to settlement shaping and these in turn were allied to alterations in food production and consumption (Schmeichen and Carls, 2000). Markets, they argue, in the United Kingdom reached a kind of apogee in the late nineteenth century as a desirable urban space for both working and middle class consumers, and notably places women could legitimately be in the public domain (ibid.). By the mid to latter part of the twentieth century, though, markets had been overtaken by broader events in many developed countries, not only Britain. Most prominent of these were an increasing physical scale for food consumption made possible by mass car ownership (Humphery, 1998; Alexander et al., 2009). That shift also ushered in forms of retailing that challenged and then killed off many markets. Of course such changes were not confined to the United Kingdom but began earlier in North America as work from Zukin (2004) and others demonstrates. The rise of supermarket-based retailing in the latter part of the twentieth century is well documented elsewhere, but the key point in relation to this chapter is that many markets declined or disappeared in the countries where modernist spatial forms and the related economic practices of the modern food system became dominant (Parham, 2015).

In the post–Second World War era and into the later part of the twentieth century, food markets were broadly eclipsed by supermarket-based shopping as this form of retailing generated both massive profits and reshaped urban space toward car-centred consumption as a norm. Of course some continued to go to markets and buy food everyday rather than just in a weekly shop, so arguments that such practices disappeared entirely are not supported by evidence from the field (Parham, 2005, 2012; Watson, 2009). While the dominant tropes about how food is managed spatially in a modern food system undoubtedly saw food markets largely overshadowed or abandoned in some places, they also point to the resurgence in the role of food markets (as actual physical places) in the late twentieth century and the early part of the twenty-first. Those changes are explored later in the chapter.

Markets in contemporary theory and practice

Food markets as everyday places attract many accounts of people's experience of them. While such descriptions can be fascinating and useful in the sense that they give an indication of lived experience in relation to food markets, they are less helpful when employed to draw broad conclusions about markets *per se*. The risk is a tendency to generalise from the individual, to form universalising judgments that are not evidentially grounded in any wider reality. And of course there are also narrators with an axe to grind: something to sell or value to extract. Among these are the civic boosters and place branders sometimes associated with area regeneration, the marketing spiels of commercial developers, and the glowing accounts in real estate pages of newspapers. Within economic systems that commodify space, these tend to treat resurgent food markets as spatial markers of the potential for value extraction: of house price rises to be captured, local lifestyle benefits to be enjoyed, or externalities to be managed (Bloomfield, 2016).

Fortunately, beyond these perspectives, and a large body of speculative arguments and assertions about what markets mean, a certain amount of more useful work comes from researchers employing a wide range of primary techniques to explore market space in an evidential way. These include field observations (written and drawn), food mapping, food diaries, structured and semi-structured interviewing, photographic essays, morphological analysis, design retrofit scenario building, and more. Perhaps the most useful approaches triangulate methods, drawing from across different disciplines to increase the explanatory power of the research in understanding aspects of these complex urban landscape phenomena. Thus, in this section, the emergence of some market forms and related aspects of the urban landscape are explored with reference to some of this work.

Contemporary food markets offer a potentially bewildering array of physical forms, everyday social practices, and connections to broader structural elements including the political economy that underpins their operation in the townscape. As noted earlier, in some places the traditional food quarter has continued to function with the market as a core urban landscape, despite the spatial, food cultural, and economic incursions of late capitalism (Parham, 2005; 2012; 2015). In poorer countries food markets and street food systems have continued through sheer need as more people live on the edge of or in food deprivation (Tinker, 1997; Chapters 20 and 21), and because of the more limited (albeit rapidly growing) penetration of global retailing giants, industrialised agriculture, a fully modern food system, and car-based spatiality (Reardon et al., 2003; Parham, 2015; Chapter 16).

In both developed and developing countries, meanwhile, places which retain a strong sense of localised, regional gastronomy – as well as those with urban working class and minority ethnic communities – have sometimes been able to resist or mitigate the dominant move to supermarket-based consumption and production systems. For such communities shopping for food at markets remains important. In many other circumstances, with the United States as a front-runner in these changes, food markets were diminished or destroyed as first suburbanisation, and then post-urban development of sprawling urbanised regions occurred, and car-based consumption at supermarkets increasingly dominated (Parham, 2015). In fact, one of the overarching themes of the recent history of food markets in western countries is a proportional decline in such spaces and related consumption practices in the postwar and latter twentieth-century period – with a range of prescriptions offered to tackle these problematic changes (Pothukuchi and Kaufman, 1999; Frumkin et al., 2004; Vallianatos et al., 2004; Lopez, 2004; Lake and Townshend, 2006; Paül and McKenzie, F.2013).

It was in the late 1970s and early 1980s that urban landscapes began to shift again and witnessed the beginnings of a renaissance in food markets in new townscape configurations and hybrid urban landscape forms. In some places, food market renewal was part of a conscious strategy to deal with the effects of de-industrialisation and the economically deprived and physically dilapidated and decaying landscapes it left behind. The form of the so-called festival market plaza, often on a former industrial waterfront or using historic buildings of some architectural note, became the physical setting for a common regeneration technique in which a food (and wider) market with some food court elements was deployed as a flagship place attraction for inward tourists (Martin, 1984). Boston's Faneuil Hall, opened in 1976, is a famously successful example in financial terms; others like South Street Seaport in New York were judged as failures (Metzger, 2001). Food in the festival market plaza was one of a range of attractors, rather than the main point of the regeneration exercise, and was offered within an overall place branding and place marketing approach (also see Chapters 35 and 36).

Western cities also saw the emergence of other new or reconfigured kinds of markets more predominantly focused on food, and there is now a considerable literature dealing with such places as part of the urban landscape (Knox, 1992; Latrobe, 2001; Kirwan, 2006; Vecchio, 2009; Markowitz, 2010). As the shape of the urban world itself shifted in the late twentieth century, farmers markets sprang up in all kinds of urban circumstances: in conurbations, peri-urban edges, suburban centres, and inner urban and central city locations. For some researchers these remain a problematised form of foodspace: thought to model elitism and to attract proponents and visitors who lack reflexivity on issues of class and race inequality, and complicit in exclusion in relation to food (Alkon, 2008). In terms of townscape it is also of interest that farmers markets have perched in a wide range of non-traditional food locations, rather than solely inhabiting spaces that were historically market areas. Often these marginal or repurposed urban landscapes – car parks, schoolyards, vacant and leftover spaces – seem very unpromising foodscapes in physical design terms (a situation discussed in more detail in the handbook's chapter on food landscapes and urban design).

Other emerging food markets took more hybrid forms. This led to attempts to conceptualise and characterise such transforming spaces based on evidence about their material forms and socio-spatial practices, including the notion of the 'gastronomic quarter' or 'food quarter' (Parham, 2005, 2012). Resurgent market spaces were seen to have rather paradoxical qualities: often found in ordinary places yet seen as special by those who frequented them (Whitelegg, 2002; Parham, 2005, 2012). They provided hospitable spaces while also being largely based on an instrumental exchange of food goods and services (Bell, 2007). They were sites for convivial ecologies in Bell and Binnie's terms (2005), but also for individuals playing out an exclusive habitus connected to gentrification (Parham, 2005). Theorists of gentrification have shown how certain neighbourhoods and urban quarters have become the sites for identity formation in which food plays an important part as a habitus 'field' in establishing individual taste and distinctiveness (Butler, 2002; Butler and Robson, 2001; Bridge and Dowling, 2001; Butler and Lees, 2006). Bridge (2006: 1966) argues that, stemming from Butler's metropolitan habitus, "particular mixes of social, cultural, and economic capital produces a geography of and the development of 'mini' habituses with different neighbourhoods having distinct neighbourhood milieu".

One of the fascinating aspects of the development of such hybrid market areas is the often close material relationship between retrofitted pieces of traditional urban fabric and new socio-spatial practices in relation to food markets, as documented in work at Borough, Broadway, and Exmouth Markets in London in 2005–2008 (Parham, 2012). Here, a series of markets were revived and reconfigured from the bottom up by local individuals with food interests, restaurateurs, or charitable market trusts, as is the case at Borough Market in London. In some instances these renewal processes have grown to become large retail food markets of a hybrid kind, sharing some characteristics with farmers markets and attracting a similar discussion about their argued role as sites of food-led gentrification (Zukin, 2008; Quastel, 2009; González and Waley, 2013). In others they have developed more as 'slow food' food court market spaces, emphasising prepared foods to eat on the spot, as at Maltby Street, Whitecross Street, and Exmouth Market, again in inner London.

Such revitalisation processes have often prefigured and then interconnected with the more widespread emergence of food carts, trucks, street stalls, and night market style food courts. Recent food quarter type developments, which locate in or actively engage with the fine-grained and human-scaled spatiality of traditional urbanism, do so largely by way of growing socio-spatial food practices of street-based food consumption (Parham, 2015). Food-centred developments with a 'market' cast – night market food courts, street food markets, and the like – can be found in numerous places, especially in larger cities and are sometimes referred to as an aspect of tactical urbanism (Tanenbaum, 2012). Very often they are able to make use of leftover spaces such as railway arches, former industrial buildings or human-scaled streets that reflect townscapes developed up until the nineteenth century. Architects have started to engage with such design-led examples of food-focused developments, including Rotterdam's much publicised new *Markthal*, with its integrated housing in a giant arch over the food hall, Copenhagen's *Torvehallerne*, or Turin's *Eataly* at Lingotto with its gastronomically focused market hall and 'slow' food court located in renovated industrial buildings near the former FIAT factory famous for its roof top car testing track.

Salient characteristics of these transforming food market landscapes, both hybrid and farmers market based, are the often explicitly made connections to shorter food supply chains and alternative food networks than in the conventional modem food system (Chapters 12, 31, 32 and 33). Some fieldwork suggests buyers and sellers value artisanal methods, pay more conscious attention to ethical production and consumption, and to taste as a basis for eating (Parham, 2005; Guthrie et al., 2006; McEachern et al., 2010; Chapters 34, 35 and 36). In one such study it was argued that the market-based consumption reflected gastronomically informed food practices that valorised season, region and locality – while attracting substantial criticism on the basis that concern for food quality is an elitist preoccupation (Parham, 2012).

What can be seen is the interconnection between ideas about what constitutes good food, and the landscape expressions of elements thought to be essential for that designation. This is not just in the urban marketplace itself but operates spatially back along the food chain into urban peripheries and the countryside. Such thinking acknowledges that spatial elements of landscape from the marketplace outwards need to be in a positive interplay with each other if settlement is to be sustainable (Hough, 1984, 1990). The marketplace's direct spatial relationship with wider scales of place is encapsulated in the notion of the 'gastronomic landscape', whereby urban, peripheral and rural foodspace are understood as closely interconnected as a food system with particular spatial features which can be highly supportive of producing good food (Hardy, 1993; Parham, 2015). Markers would be likely to include a localised scale of production to limit food miles, a stress on regional and seasonal produce, explicit connections to *terroir*, food produced, processed and distributed in ways that support the wider food environment and the physical landscape, and localised consumption (Blue, 2009; Chapter 36). Of course, as well as supporting marketplaces from highly urban to very rural locations, it is perhaps unsurprising that spaces which demonstrate these characteristics are also often the preferred landscapes of urban and rural food tourism (Chapter 35), reflecting both admired foodways (Chapters 34) and pleasurable, human-scaled space (Hjalager and Richards, 2002; Sidali, Spiller and Schulze, 2011).

These expressions of food market landscapes are not the only ways that the character of food is developing in townscape terms. It is worth remembering that markets as forms sit within an overarching spatial framework globally of conurbation development, often of a splintered, sprawling character (Graham and Marvin, 2001). This has spawned a wide variety of food chain expressions but is predominantly, in developed countries, associated with a coarse-grained, car-based supermarket consumption pattern in the post Second World War era (Parham, 2015). Some have situated this largely market-free food landscape as a 'natural' expression of freely chosen individual food production, retail and consumption preferences (Bruegmann, 2005). However, in food terms this is untenable given the evidence of profound structural, economic and social forces shaping food landscapes (Sonnino, 2009; Morgan, 2015). One of the negative outcomes associated with urban landscapes where markets have been excised is a correlation with higher levels of obesity, obesogenic environments, food poverty, and poor health outcomes (Lake and Townshend, 2010; Chapter 32). This in turn, it is argued, is closely connected to the spatial design of megalopolises in which food consumption is at a large, incoherent and generally car-based spatial scale, rather than a walkable, everyday, food-market focused one (Parham, 1995, 2015).

So what has replaced traditional markets? Contemporary urbanism is associated with particular foodscape forms that are sometimes described as 'markets' (as in Westfield Stratford's 'Great Eastern Market' in east London) but are actually forms of food court within shopping malls. Such food 'markets' in so-called superregional malls, for example, are a part of highly sophisticated food court spaces that continue to fit well with Knox's (1992) arguments about these artful settings for lifestyles based on consumption. They are also associated with increasing 'dwell times' and impulse spending which makes them very popular adjuncts to malls in an extremely competitive retailing environment (Haynes and Talpade, 1996).

The food court forms of shopping malls have both become much more sophisticated and have grown enormously in scale as that shopping form has itself changed over time since the middle of the twentieth century (Crawford, 2002; Voyce, 2006; Scharoun, 2012). To effectively capture shoppers such courts try to reflect the food language of markets because these are associated with conviviality, pleasure, and distinctiveness. It is therefore unsurprising that sometimes the interior design details of such 'markets' are modelled on traditional, farmers and hybrid markets, echoing stall structures, materials and finishes, as well as food display techniques for apparently artisanal and regional products.

However such courts are a kind of market simulacrum as they do not contain the essential marketplace elements discussed earlier in the chapter such as a public space focus; often short, localised supply chains; and ownership of at least part of the space by a number of small independent food shop operators. Even though now sometimes described as public space (Kärrholm, 2016) or as in what is now referred to as privately owned public space (Németh and Schmidt, 2011; Minton, 2012), in reality these 'food markets' are controlled and limited access spaces in private hands, unlike the public spaces of the traditional market square. While aping the look of the market, they are located inside privately owned buildings to which the owners or operators allow access within certain hours, patrol the spaces with guards, and in some cases establish visitor codes to control behaviour, as well as using other techniques to dissuade unwelcome visitors. Some operators have developed hybrid forms in which they allow small-scale businesses to take food retail space but these courts are still largely dominated economically and spatially by food and restaurant chains.

In developing countries, meanwhile, notions of modernity are often tied to shopping mall and supermarket-based consumption (Reardon et al., 2003; Reardon and Hopkins, 2006). This aids the penetration of multinational food firms by tying market rejecting food practices to social norms and aspirations, especially among those on higher incomes (Shrivastava, 2008). De facto it undercuts traditional food market spaces although they are still very important aspects of townscape in most conurbations, where operators of micro-businesses serve large populations teetering on the edge of food deprivation (Lyons and Snoxall, 2005). Given their continued centrality for food consumption, especially among poorer urban dwellers in an increasingly urbanising world, food market landscapes play a particularly critical role in such settings and are studied, even if somewhat peripherally. For instance, a number of food researchers studying food markets as *economic* constructs in both developing and developed countries' spatial contexts use commodity biographies to trace particular foodstuffs through material space including marketplaces: examples include French beans in Burkina Faso (Freidburg, 2004) and Atlantic bluefin tuna for sushi in Tokyo's fish market (Bestor, 2001). Even so there is often a disappointing lack of clarity about where markets as economic networks and structures intersect with *actual* marketplaces and food quarters. More work here would be welcome.

These are also more sinister connections to the physical space of the market as a primary urban food landscape. As we know, systemic issues of sharpening inequality and environmental degradation have already led to widespread problems of lack of food security and resilience globally (Ericksen, Ingram and Liverman, 2009; Chapters 20, 21, 25, 31 and 32). Feedback loops between food insecurity and violence can be identified (Hendrix and Brinkman, 2013). It is possible to speculate that food market sites as locations for convivial cosmopolitanism (Duruz, Luckman and Bishop, 2011) may be particularly subject to place-specific threats of terrorist attack specifically because of these symbolic qualities as well as offering soft targets with large numbers of people physically present. Of course the background global conditions of insecurity, inequality, and worsening conditions for many people, exacerbated by climate change, are fertile ground for those wishing to engender such responses for their own political ends. Food markets may be under attack not just because of the sheer numbers of people who congregate in such spaces, but, I would contend, in acknowledgement of their critical role in defining sense of place and supporting civility. It is these civic virtues and the convivial everyday practices that the market landscape celebrates that are being assailed. Again, more evidence-based research in this area would be useful to explore these contentions.

Markets and an urban future

So where to next for urban food markets? This concluding section of the chapter briefly considers roles that markets (and their spatial spheres of influence) might play in future, with reference to contemporary understandings of landscapes of food. An increasingly urbanised future seems likely, but what is less clear is how food markets will be configured within those spaces and influence them for the better. Here, therefore, are some positive suggestions about landscape roles that markets could play to help make the townscapes of this urbanised future more sustainable and convivial. To state the obvious, food markets should be a critical part of remaking the food system along more equitable, environmentally conscious, and sociable lines. One lesson from very long-term urban experience over thousands of years is that until the fairly recent past, in most settlements food markets brought together and nourished the physical and social aspects of civility in cities, towns, and villages (Chapter 4). While markets still play that role in some places they have been under threat or entirely eclipsed elsewhere, especially in the sprawling conurbations of post-industrial developed countries. Massive urbanised regions are more and more characterised by food landscapes that are unhealthful, give rise to obesogenic environments, challenge food resilience and sustainability and create food poverty and food deserts. Developing countries appear to be trying to catch up with some of the worst outcomes this has created in their race towards an imagined modernity.

Thus, in an urbanised future which threatens to be physically fractured, chaotic, and grossly unequal and unsustainable, the food market's material role and positive landscape character is increasingly important as an element of human scaled and focused townscape. This is not a new insight, as work by Alexander et al. (1977) makes clear. So-called gastronomic strategies for cities which argue for remaking urban space along more food-focused lines were proposed over twenty years ago (Parham, 1992), and more recent work on food-centred, food-productive urbanism reinforces those proposals (highlighted earlier in this chapter; Chapters 12 and 27; Zeunert, 2017: 77, 166–195, 235). Now it seems there is more widespread recognition and acceptance that food markets are a vital aspect of urban landscapes and need concentrated attention: burgeoning food strategies and policy frameworks in many places are hopeful signs of the recognition of this fact.

Another lesson for the future from this chapter is that renewed centrality for the food market as a landscape implies some reshaping of food spatiality and urban design along the food chain. In other words, markets do not exist in some kind of spatial vacuum. The scales at which such reshaping might be considered start from the very local food quarter elements the market underpins; shops, cafés, pubs, bars, restaurants, other food processing and distribution businesses, and the convivial green spaces of productive gardens and orchards that exist in synergy with or may even supply the market. Extending outward from the food quarter, these food landscapes expand in a series of nested scales: interacting with the town, city, peri-urban edge, sub-region, and rural region, and global connections in which foodspace is a conscious material consideration.

As well rehearsed elsewhere, including Chapter 27, food markets have profound food system implications both back towards distribution, processing and production and forwards to consumption, cooking, eating, clean up, and so-called waste (Parham, 2015). In recognition of these complex landscape interplays there are a number of urban designers working on scenario-based design retrofitting to re-engineer places towards more convivial and sustainable ends in which food markets and other foodscape elements reclaim their centrality (Parham, 2005, 2016; Viljoen, 2005; Viljoen and Bohn, 2005, 2014; Duany and DPZ, 2011). Like the commodity based proposals to reverse engineer *terroir* into places dominated by industrialised agriculture, as so well described by Paxson (2010), these place-based techniques are gaining policy and design traction.

A further lesson is to understand how nuanced food market landscapes can be, and to ensure the benefits of a more food-conscious life are extended as widely as possible both in and around food markets themselves and on into other urban, peri-urban, and rural landscapes and populations. This chapter has by implication explored the largely inaccurate way that markets are situated as a leftover, nostalgic or romanticised form. Yet the celebration or renewal of food-centred spaces can raise problems of gentrification and exclusion from the benefits of access to more localised, ethically produced and high quality food, and thus reflects serious issues of food justice (as explored in Parham, 2012, 2015; Chapters 31 and 32). However, surely the answer to the problem that not everyone currently enjoys access to food market landscapes is to seek ways to involve more people regardless of race, class, age, or gender? Positive landscape-based proposals and actions cited here are all ways to assist in this task.

For food markets in future, urban dwellers are likely to be in the position of experiencing these elements of townscape as at once convivial ecologies, sites for gentrification, and places under attack culturally, economically, and physically. The human-scaled and human-focused food landscapes of markets are suffering from profound systemic threats globally, including serious and worsening economic inequality and climate change; and a modern food system which is largely predicated on these unequal and unsustainable frameworks remaining in place. Thus markets as highly valued landscapes are at risk for that very reason: they offer an alternative vision of what the good life (Chapter 40) is about in townscape terms and this is a reason to nurture them as inclusive places that reject cultural appropriation, exploitative economic relations, environmental disaster or social nihilism. Maintaining, renewing and expanding the range of food markets as physical landscape entities that contribute to the gastronomic townscape (Parham, 2015) is not just a nostalgic trope or superficial gloss on the landscape of food. It is a critical resource for continued human life.

Key texts

Alexander, C., Ishikawa, S., & Silverstein, M. (1977) *A pattern language: Towns, buildings, construction*. New York: Oxford University Press.

Bell, D., & Binnie, J. (2005) What's eating Manchester? Gastro-culture and urban regeneration. *Architectural Design*, 75(3), 78–85. doi: 10.1002/ad.80 (accessed 2 August 2016).

Bell, D., & Valentine, G. (1997) *Consuming geographies: We are where we eat*. London: Routledge.

Kostof, S. (1992) *The city assembled: The elements of urban form through history*. London: Thames and Hudson.

Parham, S. (2012) *Market place: Food quarters, design and urban renewal in London*. Newcastle upon Tyne: Cambridge Scholars Publishing.

Parham, S. (2015) *Food and urbanism: The convivial city and a sustainable future*. Bloomsbury Publishing.

Scharoun, L. (2012) *America at the mall: The cultural role of a retail Utopia*. McFarland.

Schmeichen, J., & Carls, K. (2000) *The British market hall: A social and architectural history* London: Frank Cass.

Trancik, R. (1986) *Finding lost space: Theories of urban design*. Wiley.

Viljoen, A. (ed.) (2005) *Continuous Productive Urban Landscapes (CPULs): Designing urban agriculture for sustainable cities*. Oxford: Architectural Press.

Further reading

Aiello, D. (2014) *Vancouver's downtown eastside: An ethnography of restaurateurs and neighbourhood change*. Doctoral dissertation, Environment: Department of Geography.

Alcock, J.P. (2006) *Food in the ancient world*. Westport, CT and London: Greenwood.

Alexander, A., Nell, D., Bailey, A.R., & Shaw, G. (2009) The co-creation of a retail innovation: Shoppers and the early supermarket in Britain. *Enterprise and Society*, 10(3), 529–558. Available from: DOI: http://dx.doi.org/10.1017/S1467222700008156 [Accessed 2 August 2016].

Alexander, C., Ishikawa, S., & Silverstein, M. (1977) *A pattern language: Towns, buildings, construction*. New York: Oxford University Press.

Alkon, A. (2008) Paradise or pavement: The social constructions of the environment in two urban farmers' markets and their implications for environmental justice and sustainability. *Local Environment*, 13(3), 271–289. Available from: DOI: 10.1080/13549830701669039 [Accessed 2 August 2016].

Amin, A., & Graham, S. (1997) The ordinary city. *Transactions of the Institute of British Geographers*, 22(4), December, 411–429.

Ashihara, Y. (1983) *The aesthetic townscape*. Cambridge, MA: The MIT Press.

Atkins, P., & Bowler, I. (2001) *Food in society*. London: Arnold.

Atkins, P.J., Lummel, P., & Oddy, D.J. (2007) *Food and the city in Europe since 1800*. Aldershot: Ashgate.

Banerjee, T. (2001) The future of public space: Beyond invented streets and reinvented places. *Journal of the American Planning Association*, 67(1), 9–24.

Barthel, S., & Isendahl, C. (2013) Urban gardens, agriculture, and water management: Sources of resilience for long-term food security in cities. *Ecological Economics*, 86, 224–234.

Beardsworth, A., & Keil, T. (1997) *Sociology on the menu: An invitation to the study of food and society London*. New York: Routledge.

Bell, D. (2007) The hospitable city: Social relations in commercial spaces. *Progress in Human Geography*, 31(1), 7–22. Available from: DOI: 10.1177/0309132507073526 [Accessed 20 January 2015].

Bell, D., & Binnie, J. (2005) What's eating Manchester? Gastro-culture and urban regeneration. *Architectural Design*, 75(3), 78–85. Available from: DOI: 10.1002/ad.80 [Accessed 2 August 2016].

Bell, D., & Valentine, G. (1997) *Consuming geographies: We are where we eat*. London: Routledge.

Bestor, T.C. (2001) Supply-Side Sushi: Commodity, market, and the global city. *American Anthropologist*, 103(1), 76–95. Available from: DOI: 10.1525/aa.2001.103.1.76 [Accessed 2 August 2016].

Blake, M.K. (2013) Ordinary food spaces in a global city: Hong Kong. *Streetnotes*, 21(21)

Bloomfield, R. (2016) Where to buy a property in London: Homes near London markets can be cheaper by up to 15 per cent – but there's a catch. . . *Evening Standard Homes and Property*, June 15, 10 [Accessed 27 July 2016].

Blue, G. (2009) On the politics and possibilities of locavores: Situating food sovereignty in the turn from government to governance. *Politics and Culture*, 9, 68–79.

Bourdieu, P. (1984) *Distinction: A social critique of the judgement of taste*, Nice, R. (trans.). London: Routledge and Kegan Paul.

Bray, T.L. (ed.) (2003) *The archaeology and politics of food and feasting in early states and empires*. New York: Kluwer Academic/Plenum.

Bridge, G. (2006) It's not Just a Question of Taste: Gentrification, the Neighbourhood, and Cultural Capital. *Environment and Planning A*, Vol 38, Issue 10.

Bridge, G., & Dowling, R. (2001) Microgeographies of retailing and gentrification. *Australian Geographer*, 32(1), 93–107. Available from: DOI: 10.1080/00049180020036259 [Accessed 2 August 2016].

Bruegmann, R. (2005) *Sprawl: A compact history*. Chicago, IL and London: University of Chicago Press.

Butler, T. (2002) [Online] Thinking global but acting local: The middle classes in the city *Sociological Research Online*, 7. Available from: www.socresonline.org.uk/7/3/timbutler.html [Accessed 2 August 2016].

Butler, T., & Lees, L. (2006) Super-gentrification in Barnsbury, London: Globalization and gentrifying global elites at the neighbourhood level *Transactions of the Institute of British Geographers*, NS (31), 467–487. Available from: DOI: 10.1111/j.1475-5661.2006.00220.x [Accessed 2 August 2016].

Butler, T., & Robson, G. (2001) Social capital, gentrification and neighbourhood change in London: A comparison of three South London neighbourhoods. *Urban Studies*, 38(12), 2145–2162. Available from: DOI: 0.1080/00420980120087090 [Accessed 2 August 2016].

Calabi, D. (2004) *The market and the city: Square, street and architecture in early modern Europe*. Aldershot: Ashgate.

Carr, S., Francis, M., Rivlin, L.G., & Stone, A.M. (1992) *Public space*. Cambridge: Cambridge University Press.

Chase, D.Z., & Chase, A.F. (2014) Ancient Maya markets and the economic integration of Caracol, Belize. *Ancient Mesoamerica*, 25(1), 239–250. Available from: DOI: http://dx.doi.org/10.1017/S0956536114000145 [Accessed 2 August, 2016].

Clark, D. (2000) The shop within? An analysis of the architectural evidence for medieval shops. *Architectural History*, 43, 58–87.

Coles, B.F., & Crang, P. (2011) Placing alternative consumption: Commodity fetishism in Borough fine foods market, London. In: Lewis, T., & Potter, E. (eds.), *Ethical consumption: A critical introduction*. London: Routledge, pp. 87–102.

Conzen, M.R.G. (1960) Alnwick, Northumberland: A study in town-plan analysis. *Transactions and Papers (Institute of British Geographers)*, (27), iii-122.

Corbier, M. (1999) The broad bean and the moray: Social hierarchies and Food in Rome. In: Flandrin, J.L., & Montanari, M. (eds.), *Food: A culinary history*. Columbia: Columbia University Press, pp. 128–140.

Cortenesi, A. (1999) Self-sufficiency and the market: Rural and urban diet. In: Flandrin, J.L., & Montanari, M. (eds.), *Food: A culinary history*. Columbia: Columbia University Press, pp. 268–274.

Council of Europe (nd). *European landscape convention* www.coe.int/en/web/landscape/about-the-convention (accessed 5 June 2017).

Crawford, M. (2002) Suburban life and public space. In: Smiley, D.J. (ed.), *Sprawl and public space: Redressing the mall*. Washington, DC: National Endowment for the Arts, pp. 21–30.

Cullen, G. (1961) *The concise townscape*. London: Architectural Press.

Dahlin, B.H., Bair, D., Beach, T., Moriarty, M., & Terry, R. (2010) The dirt on food: Ancient feasts and markets among the lowland Maya. In: *Pre-Columbian foodways*. New York: Springer, pp. 191–232.

Dahlin, B.H., Jensen, C.T., Terry, R.E., Wright, D.R., & Beach, T. (2007) In search of an ancient Maya market. *Latin American Antiquity*, 363–384. Available from: DOI: 10.2307/25478193 [Accessed 2 August 2016].

Dalby, A. (2003) *Food in the ancient world from A to Z*. London: Routledge.

Deener, A. (2012) *Venice: A contested bohemia in Los Angeles*. Chicago: University of Chicago Press.

Desportes, F (1999) "Food Trades", In Flandrin, J-L & Montanari, M (eds) (1999) *Food: A Culinary History From Antiquity to the Present* New York. Columbia Universiy Press, pp.275-286.

Duany, A. & DPZ (2011) *Garden cities: Theory & practice of agrarian urbanism*. Duany Plater-Zyberk & Company and The Prince's Foundation.

Dupont, F. (1999) The grammar of Roman dining. In: Flandrin, J.L., & Montanari, M. (eds.), *Food: A culinary history*. Columbia: Columbia University Press, pp. 113–127.

Duruz, J., & Khoo, G.C. (2014) *Eating together: Food, space, and identity in Malaysia and Singapore*. Lanham: Rowman and Littlefield.

Duruz, J., Luckman, S., & Bishop, P. (2011) Bazaar encounters: Food, markets, belonging and citizenship in the cosmopolitan city. *Continuum*, 25(5), 599–604.

Ericksen, P.J., Ingram, J.S., & Liverman, D.M. (2009) Food security and global environmental change: Emerging challenges. *Environmental Science and Policy*, 12(4), 373–377.

Esperdy, G. (2002) Edible urbanism. *Architectural Design*, 72, 44–50.

Franck, K.A. (2005) The city as dining room, market and farm. *Architectural Design*, 75(3), 5–10. Available from: DOI: 10.1002/ad.70 [Accessed 2 August 2016].

Freidberg, Susanne (2004) *French beans and food scares: Culture and commerce in an anxious age*. Oxford. Oxford University Press.

Frumkin, H., Frank, L., & Jackson, R.J. (2004) *Urban sprawl and public health: Designing, planning, and building for healthy communities*. London: Island Press.

Gies, J., & Gies, F. (1969) *Life in a medieval city*. London: Arthur Baker.

González, S., & Waley, P. (2013) Traditional retail markets: The new gentrification frontier?. *Antipode*, 45(4), 965–983.

Goodman, D., & Goodman, M. (2009) Alternative food networks. *International Encyclopedia of Human Geography*, 3, 208–220.

Graham, S., & Marvin, S. (2001) *Splintered urbanism*. New York: Routledge.

Guthrie, J., Guthrie, A., Lawson, R., & Cameron, A. (2006) Farmers' markets: The small business counter-revolution in food production and retailing. *British Food Journal*, 108(7), 560–573.

Hall, S. (2012) *City, street and citizen: The measure of the ordinary*. Routledge advances in ethnography. London: Routledge.

Hall, C.M., & Gössling, S. (2013) *Sustainable culinary systems: Local foods, innovation, and tourism and hospitality*. London: Routledge.

Hardy, M. (1993) *The gastronomic landscape: Food production and the cultural value of the countryside*, Gastronomic symposiette series. Adelaide: University of Adelaide.

Haynes, J., & Talpade, S. (1996) Does entertainment draw shoppers. The effects of Entertainment Centers on Shopping Behaviour in Malls. *Journal of Shopping Center Research*. Available from: http://jrdelisle.com/JSCR/IndArticles/Haynes_N296.pdf [Accessed 1 August 2016].

Heller, A., & Adams, T. (2009) Creating healthy cities through socially sustainable placemaking. *Australian Planner*, 46(2), 18–21.

Hendrix, C., & Brinkman, H.J. (2013) Food insecurity and conflict dynamics: Causal linkages and complex feedbacks. *Stability: International Journal of Security and Development*, 2(2).

Hinrichs, C.C., & Lyson, T.A. (eds.) (2007) *Remaking the North American food system: Strategies for sustainability*. Nebraska: University of Nebraska Press.

Hjalager, A., & Richards, G. (eds.) (2002) *Tourism and gastronomy*. London: Psychology Press.

Hough, M. (1984) *City form and natural process. Towards a new urban vernacular*. London. New York: Routledge.

Hough, M. (1990) *Out of place, restoring identity to the regional landscape*. New Haven, CT and London: Yale University Press.

Ilbery, B., & Kneafsey, M. (1999) Niche markets and regional speciality food products in Europe: Towards a research agenda. *Environment and Planning A*, 31(12), 2207–2222.

Jarosz, L. (2008) The city in the country: Growing alternative food networks in Metropolitan areas. *Journal of Rural Studies*, 24(3), 231–244.

Jarvis, H., Pratt, A. C. & Wu, Cheng-Chong (2001) *The Secret Life of Cities: The Social Reproduction of Everyday Life*. Harlow, Essex. Prentice Hall.

Jive´n, G., & Larkham, P. (2003) Sense of place, authenticity and character: A commentary. *Journal of Urban Design*, 8(1), 67–81. Available from: DOI: 10.1080/1357480032000064773 [Accessed 2 August, 2016].

Kärrholm, M. (2016) *Retailising space: Architecture, retail and the territorialisation of public space*. London: Routledge.

Kearns, G., & Philo, C. (1993) Culture, history, capital: A critical introduction to the selling of places. *Selling Places: The City as Cultural Capital, Past and Present*, 1–32.

Kirwan, J. (2006) The interpersonal world of direct marketing: Examining conventions of quality at UK farmers' markets. *Journal of Rural Studies*, 22, 301–312. Available from: [Accessed 2 August, 2016].

Knox, P. (1992) The packaged landscapes of post-suburban America. In: Whitehand, J., & Larkham, P. (eds.), *Urban landscapes: International perspectives*. London: Routledge.

Kostof, S. (1991) *The city shaped, urban patterns and meanings through history*. London: Thames and Hudson.

Kostof, S. (1992) *The city assembled: The elements of urban form through history*. London: Thames and Hudson.

La Trobe, H. (2001) Farmers' markets: Consuming local rural produce. *International Journal of Consumer Studies*, 25(3), 181–192. Available from: DOI: 10.1046/j.1470-6431.2001.00171.x [Accessed 2 August 2016].

Lake, A.A., Townshend, T.G., & Alvanides, S. (eds.) (2010) *Obesogenic environments: Complexities, perceptions and objective measures*. Chichester, West Sussex: Wiley-Blackwell.

Lake, A., & Townshend, T. (2006) Obesogenic environments: Exploring the built and food environments. *The Journal of the Royal Society for the Promotion of Health*, 126(6), 262–267.

Lopez, R. (2004) Urban sprawl and risk for being overweight or obese. *American Journal of Public Health*, 94(9), 1574–1579.

Lyons, M., & Snoxell, S. (2005) Sustainable urban livelihoods and marketplace social capital: Crisis and strategy in petty trade. *Urban Studies*, 42(8), 1301–1320. Available from: DOI: 10.1080/00420980500150631 [Accessed 2 August 2016].

Maitland, R. (2007) Conviviality and everyday life: The appeal of new areas of London for visitors. *International Journal of Tourism Research*, 10, 15–25. Available from: DOI: 10.1002/jtr.621 [Accessed 2 August 2016].

Markowitz, L. (2010) Expanding access and alternatives: Building farmers' markets in low-income communities. *Food and Foodways*, 18(1–2), 66–80. Available from: DOI: 10.1080/07409711003708512 [Accessed 2 August 2016].

Marsden, T. (2013) Sustainable place-making for sustainability science: The contested case of agri-food and urban – rural relations. *Sustainability Science*, 8(2), 213–226.

Martin, J.A. (1984) If baseball can't save cities anymore, what can a festival market do?. *Journal of Cultural Geography*, 5(1), 33–46. Available from: DOI: 10.1080/08873638409478560 [Accessed 2 August 2016].

McEachern, M.G., Warnaby, G., Carrigan, M., & Szmigin, I. (2010) Thinking locally, acting locally? Conscious consumers and farmers' markets. *Journal of Marketing Management*, 26(5–6), 395–412.

Mennell, S. (1996) *All manners of food*. Urbana and Chicago, IL: University of Illinois Press.

Metzger, J.T. (2001) The failed promise of a festival marketplace: South Street Seaport in lower Manhattan. *Planning Perspectives*, 16(1), 25–46. Available from: DOI: 10.1080/02665430010000760 [Accessed 2 August 2016].

Minton, A. (2012) *Ground control: Fear and happiness in the twenty-first-century city*. London Penguin Books.

Morgan, K. (2015) Nourishing the city: The rise of the urban food question in the Global North. *Urban Studies*, 52(8), 1379–1394.

Mosupye, F.M., & von Holy, A. (2000) Microbiological hazard identification and exposure assessment of street food vending in Johannesburg, South Africa. *International Journal of Food Microbiology*, 61(2), 137–145. Available from: DOI: 10.1016/S0168-1605(00)00264-6 [Accessed 2 August 2016].

Németh, J., & Schmidt, S. (2011) The privatization of public space: Modeling and measuring publicness. *Environment and Planning B: Planning and Design*, 38(1), 5–23.

Nicolas, D. (1997) *The later Medieval city 1300–1500*. Harlow: Addison Wesley Longman.

Parham, S. (1990) The table in space: A planning perspective [online]. *Meanjin*, 49(2), Winter, 213–219. Available from: http://search.informit.com.au/documentSummary;dn=133325577156576;res=IELLCC ISSN: 0025-6293 [Accessed 2 August 2016].

Parham, S. (1992) Gastronomic strategies for Australian cities. *Urban Futures*, 2(2).

Parham, S. (1995) Megalopolis. *Arena*, 16, April/May, 38–41.

Parham, S. (1996) Productive land-use on the urban fringe: A comparative study in planning for regional economic development in Languedoc and Tuscany, department of housing and urban development, South Australia. Report.

Parham, S. (1998) Fat City. Why Bologna works. *Australian Financial Times*, Saturday Review Section, February.

Parham, S. (2001) *Developing market typologies: An urban design analysis*. MA Dissertation, University of Westminster, London.

Parham, S. (2005) Designing the gastronomic quarter. *Architectural Design*, 75(3), 86–95. Available from: DOI: 10.1002/ad.81 [Accessed 1 August 2016].

Parham, S. (2012) *Market place: Food quarters, design and urban renewal in London*. Newcastle upon Tyne: Cambridge Scholars Publishing.

Parham, S. (2015) *Food and urbanism: The convivial city and a sustainable future*. London: Bloomsbury Publishing.

Parham, S. (2016) "Shrinking Cities and Food: Place-making for Sustainable Renewal, Reuse and Retrofit", in Neill, W. J., & Schlappa, H. (eds.) *Future Directions for the European Shrinking City*, RTPI Library Series. New York and London: Routledge

Paül, V., & McKenzie, F.H. (2013) Peri-urban farmland conservation and development of alternative food networks: Insights from a case-study area in metropolitan Barcelona (Catalonia, Spain). *Land Use Policy*, 30(1), 94–105.

Paxson, H. (2010) Locating value in artisan cheese: Reverse engineering terroir for new world landscapes. *American Anthropologist*, 112(3), 444–457. Available from: DOI: 10.1111/j.1548-1433.2010.01251.x [Accessed 1 August 2016].

Pearson, D., Henryks, J., Trott, A., Jones, P., Parker, G., Dumaresq, D., & Dyball, R. (2011) Local food: Understanding consumer motivations in innovative retail formats. *British Food Journal*, 113(7), 886–899.

Pink, S. (2012) *Situating everyday life: Practices and places*. London: Sage Publications Limited.

Pothukuchi, K., & Kaufman, J.L. (1999) Placing the food system on the urban agenda: The role of municipal institutions in food systems planning. *Agriculture and Human Values*, 16(2), 213–224.

Quastel, N. (2009) Political ecologies of gentrification. *Urban Geography*, 30(7), 694–725.

Raki, T., & Chambers, D. (2012) Rethinking the consumption of places. *Annals of Tourism Research*, 39(3), 1612–1633.

Reardon, T., & Hopkins, R. (2006) The supermarket revolution in developing countries: Policies to address emerging tensions among supermarkets, suppliers and traditional retailers. *The European Journal of Development Research*, 18(4), 522–545. Available from: DOI: 10.1080/09578810601070613 [Accessed 1 August 2016].

Reardon, T., Timmer, C.P., Barrett, C.B., & Berdegué, J. (2003) The rise of supermarkets in Africa, Asia, and Latin America. *American Journal of Agricultural Economics*, 85(5), 1140–1146. Available from: DOI: 10.1111/j.0092-5853.2003.00520.x [Accessed 1 August 2016].

Renting, H., Marsden, T.K., & Banks, J. (2003) Understanding alternative food networks: Exploring the role of short food supply chains in rural development. *Environment and Planning A*, 35(3), 393–411.

Riera-Melis, A. (2013 [1999]) Society, food and feudalism. In: Flandrin, J.L., & Montanari, M. (eds.), *Food: A culinary history*. Columbia: Columbia University Press, pp. 251–267.

Sarasúa, C., & Scholliers, P. (2005) The rise of a food market in European history. In: Sarasúa, C., Scholliers, P., & Van Molle, L. (eds.), *Land, shops and kitchens. Technology and the food chain in twentieth-century Europe*. Turnhout: Brepols, pp. 13–29.

Scharoun, L. (2012) *America at the mall: The cultural role of a retail Utopia*. North Carolina, US: McFarland.

Schmeichen, J., & Carls, K. (2000) *The British market hall: A social and architectural history*. London: Frank Cass.

Sennett, R. (1996) *Flesh and stone: The body and the city in Western civilization*. London. Faber and Faber.

Sidali, K.L., Spiller, A., & Schulze, B. (eds.). (2011) *Food, Agriculture and tourism: Linking local gastronomy and rural tourism: Interdisciplinary perspectives*. London: Springer.

Sims, R. (2010) Putting place on the menu: The negotiation of locality in UK food tourism, from production to consumption. *Journal of Rural Studies*, 26(2), 105–115.

Sitte, C. (1986) *City planning according to artistic principles*. New York: Rizzoli.

Smith, J.K. (2011) *The everyday life of food: The cultural economy of the traditional food market in England*. Doctoral dissertation, University of Gloucestershire.

Smith, J., Maye, D., & Ilbery, B. (2014) The traditional food market and place: New insights into fresh food provisioning in England. *Area*, 46(2), 122–128.

Sonnino, R. (2009) Feeding the city: Towards a new research and planning agenda. *International Planning Studies*, 14(4), 425–435.

Southworth, M. (2005) Reinventing main street: From mall to townscape mall. *Journal of Urban Design*, 10(2), 151–170. Available from: DOI: 10.1080/13574800500087319 [Accessed 2 August 2016].

Spang, R.L. (2000) *The invention of the restaurant: Paris and modern gastronomic culture*. Cambridge, MA: Harvard University Press.

Srivastava, R.K. (2008) Changing retail scene in India. *International Journal of Retail and Distribution Management*, 36(9), 714–721. Available from: DOI: 10.1108/09590550810890957 [Accessed 1 August 2016].

Stobart, J., & Van Damme, I. (2015) Introduction: Markets in modernization: Transformations in urban market space and practice, c. 1800–c. 1970. *Urban History*, 1–14. Available from: DOI: http://dx.doi.org/10.1017/S0963926815000206 [Accessed 2 August 2016].

Tanenbaum, J. (2012) *Regulating mobile food vending in Greenville, SC*. Diss. Clemson University.

Thrift, N., & Glennie, P. (1993) Historical geographies of urban life and modern consumption. In: Kearns, G., & Philo, C. (eds.), *Selling places: The city as cultural capital. past and present*. Oxford: Pergamon, pp. 33–48.

Tinker, I. (1997) *Street foods: Urban food and employment in developing countries*. Oxford: Oxford University Press.

Trancik, R. (1986) *Finding lost space: Theories of urban design*. New York: Reinhold.

Vallianatos, M., Gottlieb, R., & Haase, M.A. (2004) Farm-to-school: Strategies for urban health, combating sprawl, and establishing a community food systems approach. *Journal of Planning Education and Research*, 23(4), 414–423.

Vecchio, R. (2009) *European and United States farmers' markets: similarities, differences and potential developments*. [Paper] 113th EAAE Seminar, A resilient European food industry and food chain in a challenging world, Chania, Crete, Greece, September 3–6.

Viljoen, A. (ed.) (2005) *Continuous Productive Urban Landscapes (CPULs): Designing urban agriculture for sustainable cities*. Oxford: Architectural Press.

Viljoen, A., & Bohn, K. (2005) Continuous productive urban landscapes: Urban agriculture as an essential infrastructure. *Urban Agriculture Magazine*, (15), 34–36.

Viljoen, A., & Bohn, K. (2014) *Second nature urban agriculture: Designing productive cities*. London: Routledge.

Voyce, M. (2006) Shopping malls in Australia. The end of public space and the rise of 'consumerist citizenship'? *Journal of Sociology*, 42(3), 269–286. Available from: DOI: 10.1177/1440783306066727 [Accessed 27 July 2016].

Warde, A. (1997) *Consumption, food and taste: Culinary antinomies and commodity culture*. London: Sage.

Watson, S. (2009) The magic of the marketplace: Sociality in a neglected public space. *Urban Studies*, 46(8), 1577–1591. Available from: DOI: 10.1177/0042098009105506 [Accessed 27 July 2016].

Whitelegg, D. (2002) From market stalls to restaurant row: The recent transformation of exmouth market. *London Journal*, 27(2), 1–11. Available from: DOI: 10.1179/ldn.2002.27.2.77 [Accessed 20 January 2015].

Wilkins, J.M., & Hill, S. (2006) *Food in the ancient world*. Hoboken, NJ: Blackwell.

Zeunert, J. (2017) *Landscape architecture and environmental sustainability: Creating positive change through design*. London and New York: Bloomsbury.

Zukin, S. (2004) *Point of Purchase: How shopping changed American culture*. New York and London: Routledge.

Zukin, S. (2008) Consuming authenticity: From outposts of difference to means of exclusion. *Cultural Studies*, 22(5), 724–748.

30

Allotments and community gardens

History, culture and landscape in Britain, North America and Australia

Claire Nettle and David Crouch

Introduction

Food production has long been a part of urban landscapes as well as rural ones. Allotments and community gardens are two specific instances in a long history of food being grown in public and in common in cities and towns around the world. This chapter examines the growth and development of allotments in Britain, and the community gardens in Australia, North America and Britain that evolved from them. Both allotments and community gardens model participatory landscapes that bring food production into view and cultivate connections between people and place. These landscapes are neither urban nor rural, but both: a hybrid with horticultural production shaped by the particular fabric of the city. Both allotments and community gardens have their roots in the enclosure of the British commons and the defence of common spaces. They are accessible landscapes experienced by passers-by; community spaces where people feel a strong sense of collective ownership; they are also often intimate, very personal places. Crucially, allotments and community gardens are spaces that are shared, that relate to particular attitudes towards land and nature, and that display creative reuse of materials. These threads emerge significantly in what has become their landscapes – human and cultural – in distinctive ways. One particular insight that this chapter delivers is the distinctive importance of the role that the culture of these sites plays in constituting their landscapes of food through shared practice and the combination of human labour and sociality and their diverse, multiple creativity.

What are allotments and community gardens?

British allotments are clusters of individual food-producing gardens generally managed by municipal authorities with plots leased on an annual basis. Individual allotment plots in Britain average 250 square metres in size (Mok et al., 2014) – plenty of room for gardeners to experience the joys of getting their hands in the soil and to harvest a substantial amount of produce. Recently, the governments of England and Scotland considered halving the size of allotment plots in response to a considerable increase in demand in the wake of the 2007–2009 financial crisis. Halving their size would have increased the number of available plots locally and nationally; however, despite some plotters finding existing plots too large to manage, the majority wanted the current size to remain standard. The governments withdrew this plan, but a number of councils across the UK are working on schemes to achieve the same result by making new members take half-size plots.

Allotments are spaces of community and conviviality, alongside the emphasis on the individual cultivation of distinct plots. In the UK a legislative right to an allotment has been established for over a century. Although this right does not exist in other European countries, similar schemes are common across the continent (Keshavarz and Bell, 2016). Allotments were established in Denmark in the 1700s and Germany's Schrebergartens proliferated from the 1800s. The first allotment site in the UK was established in around 1795 (Keshavarz and Bell, 2016). Between 1978 and 2006 the number of allotments in Britain reduced from half a million to 330,000 (Crouch, 2006). However, popular demand for allotments remains high. Since 2007 the number of households wanting to have a plot has become greater than at any time since the Second World War, with 100,000 people on waiting lists. A government inquiry in 1998 supported the continuing future of allotments and made it clear that there would be no reduction of councils' obligations to support allotments, and there has not been since. Today, most local councils acknowledge the value of allotments across a range of policies from amenity to health. Since the turn of the 21st century, allotment facilities and management have been improved by charitable funding through an 'Allotments Regeneration Initiative'. This has directly affected several hundred allotment sites, and has been supported by both allotment and community garden organisations nationally (FCFCG/NAS, 2007). This project has informed the enhancement of provision in many other allotment sites.

Community gardens are younger offshoots of allotments, having begun growing in their present form in the 1970s. Community gardens are collectively self-managed, and often emphasise community building and other social aims in addition to horticulture (Howe et al., 2012; Nettle, 2014). There are now around 1000 community gardens across the UK (Quale, 2008) – sometimes established within existing allotment sites – and community gardens grow, in various forms, around much of the world. There are more than 600 community gardens in Australia (Nettle, 2014), and more than 21,000 in the United States and Canada (ACGA, 2015).

Community gardens have been described and defined in a range of ways, with tensions existing among different emphases, and a degree of resistance to limiting and prescribing a diverse and evolving practice with too narrow a definition. There is considerable diversity among community gardens in terms of landscape, activities and management. The smallest community gardens occupy only a few square metres; larger ones span several hectares. Some community gardens are completely open to passers-by; others have high fences limiting access. Some community gardens host a wide range of activities, from playgroups to performances; some seek to address a range of social and environmental issues; others keep their focus on tending the soil. Community gardens are initiated and managed by local residents, grassroots organisations, NGOs, health and community welfare agencies and public housing organisations. They grow in reclaimed blight sites, public parks, community centres and churchyards, and in cities, suburbs and even rural areas. Some community gardens are cultivated and harvested communally; others comprise plots allocated to individuals and groups – many contain some combination of the two. Where there are allocated garden plots, they range in size from a single square metre to over 200 square metres, though many are around 10 square metres – diminutive when compared with Britain's allotments (Kurtz, 2001; Lawson, 2004; Nettle, 2014).

This diversity on the ground is reflected in the contrasting ways scholars have defined community gardens. Sometimes collectively cultivated ornamental or therapeutic gardens are considered community gardens; sometimes the term is reserved for sites focussing on food production. Some consider kitchen garden programs in prisons and schools, or commercial urban agricultural pursuits, to fall under the rubric of community gardening; others see community gardens as being defined by voluntarism and collectivism (Ferris et al., 2001; Nettle, 2014; Pudup, 2008).

Whatever their model, community gardens are landscapes that are often highly valued by their communities but also highly vulnerable. Most community gardens have little security of tenure, particularly when situated on once-derelict municipal or privately owned land (Barraclough, 2009; Schmelzkopf, 1995). Levels of government support for community gardens vary significantly. Australia has had no federal government programs aimed specifically at supporting community gardens, and a scattering of ad hoc programs at state and local levels. The national organisation of community gardens and city farms is unfunded. Their counterparts in the US have been supported by government-funded programs such as the USDA-sponsored Urban Gardening Program established in 1976, which promoted vegetable gardening and community gardens in 23 US cities (Lawson, 2005) and Philadelphia Green, founded in 1974, which in its first 20 years assisted in the development of over 2000 community gardens in low-income neighbourhoods (Hynes, 1996). Although US administrations and local authorities have championed and subsidised urban agricultural movements in periods of fiscal crisis, when crises end (at least for some), governments have generally withdrawn their support, turning to "focus instead on profitable real estate development on the former garden plots" (Schmelzkopf, 1995: 364). In the UK, allotments and community gardens are beginning to converge as allotment holders increasingly take part in the management and maintenance of their sites. The futures of allotments and community gardens in the UK are equally vulnerable, as many community gardens are on council-owned land, though each is buoyed up by the enthusiasm of their gardeners and often by other local bodies like civic societies, health and environmental organisations and charities.

Community gardens grow in an environment shaped by those who worked for the right to an allotment, and continue to view themselves in their lineage. The following section begins with a brief overview of the history of allotments in Britain, followed by a history of community gardens in North America, Britain and Australia. We then go on to explore the particular landscapes of allotments and community gardens, and the roles that landscape professionals can fruitfully play in their development.

The origins of allotments in Britain

Britain's pre-industrial villages were a patchwork of privately owned land, open fields, village commons and pastures, and 'waste lands' that were used for grazing animals and harvesting materials (Warner, 1987). The commons and waste lands supported an economy sustained by a complex system of customary rules and conventions that enabled the commoners to maintain and defend this land for centuries (Crouch and Ward, 2008; Linebaugh, 2014; Olwig, 2002; Chapter 28).

In the 1700s, new agricultural technologies made broad-acre farming viable. Owners of these new large-scale farms petitioned Parliament to divide undocumented common land into formal, privately owned titles. Because customary tenants could not prove any established title, most were denied a share of the newly enclosed land. Others received parcels too small to farm and sold these to the large landholders (Warner, 1987).

The enclosures created great disturbance in Britain. Commoners' livelihoods were destroyed, resulting in an "explosion of crime" (Endersby, 2000: 315). Many flocked to urban areas in search of employment, leading to overcrowding and pollution in the rapidly expanding cities. Others remained in the villages, working as seasonal labourers on the big farms or becoming dependent on local poor relief. For local authorities, this meant a sudden, substantial increase in welfare payments, which led to increasing property taxes for landholders (Warner, 1987).

Charitable organisations emerged to respond to the widespread poverty. William Wilberforce's charity, 'Society for Bettering the Condition and Increasing the Comforts of the Poor', was the first to propose that allotments large enough to produce a family's food (though not big enough to provide an income) be provided, a campaign taken up by other charities (Warner, 1987; Hoyles, 1991). Simultaneously, people living in the increasingly overcrowded cities worked together to create clusters of garden plots just outside of the city.

Campaigns for allotments were not universally welcomed. Farmers actively opposed them, fearing that their labourers would desert them to tend their own plots. Others saw them as rewarding the least ambitious and promoting laziness and overpopulation. By the late 18th century, a few wealthy estate owners had begun setting aside small parcels of land to rent out to villagers. They hoped the allotments would improve the behaviour of the landless commoners (that is, distract them from drunkenness and protest), and hoped it would add to their own benevolent image (Burchardt, 2002). Opposition to allotments meant that when they were established, they were accompanied by paternalistic and punitive regulations, forbidding visits during regular working hours, and making church attendance a condition of access.

In 1819, the practice of allocating small parcels of estates as allotments was reinforced by an Act of Parliament. In 1845 Parliament instructed that thereafter "fields of allotment gardens be established as a precondition attending all subsequent enclosures" (Warner, 1987: 10). Finally, in 1908, the *Smallholdings and Allotments Act* was introduced, calling on municipal governments to meet demands for allotments where they were not available on private land. It is this legislation and its successors that provide the basis for Britain's current-day allotment schemes.

Early allotments were established by a combination of charitable campaigns, philanthropy on the part of landholders, and the lobbying of self-help associations formed by landless commoners. Accounts vary in their emphases on each of these factors, portraying the allotments as either paternalistic and controlling, or as the result of collective social action. The philanthropic initiatives of industrialists and politicians may have been inspired by the City Beautiful Movement to improve urban living conditions (Keshavarz and Bell, 2016); however, Warner (1987: xiii) describes the establishment of allotments as an "unpleasant bit of charity", and a "philanthropic experiment" designed to reduce the costs of public welfare payments made to those deprived of sustenance by the enclosures. On the other hand, allotments have been associated with trade unions, friendly societies and the co-operative movement as examples of self-help and mutual aid (Crouch and Ward, 2008). The provision of allotments can be seen as the result of sustained agitation for improved conditions, mobilising tactics including squatting, mass trespass of enclosed common land and rioting (Crouch and Ward, 2008). Indeed, the main period of development of allotments was immediately following the Swing Riots. In the 1830s and 1840s, it was the Labourer's Friend Society's efforts that helped spread allotments across the country (Burchardt, 2002; see also Moran, 1990).

Though their origins were in part "palliative, preventing workers from political struggle" (Hoyles, 1991: 270), once established, allotments brought workers together and fostered mutual aid, agricultural trade unionism and Chartism (Burchardt, 2002). They brought practical benefits too, enabling people to generate an income by growing high-value cash crops as well as producing food for subsistence. Burchardt argues that allotment holders made more money from this than from wage labouring, and that allotments made them more independent and self-sufficient. Allotments also had symbolic value for landless commoners. With political rights enshrined in property, the landless had no rights (Davies, 2007). Allotments represented political franchise (Scott, 1985).

Local councils, parishes and districts were required to provide allotments for people living in their areas since the *Small Holdings and Allotments Act* (1908), a law still in place. The law does not state where, when or how soon people can have their plots. In the succeeding years the numbers of local council plots increased significantly, as did privately owned ones. In the century following the *Smallholdings and Allotments Act*, the demand for allotments rose and fell in response to economic cycles (Bassett, 1981) and cultural change (Acton, 2011). A great scale of change occurred in one sweep with the onset of World War I. Just beforehand there were an estimated half a million plots in the land. Within a year of its starting, there were nearly one and a half million, a threefold jump in 12 months. One in five families had a plot. Common land, parks and playing fields were subdivided and government inspectors were despatched to seek more land for allotments. Schools, hospitals and asylums were urged to provide more on their lands. Demand for allotments peaked again after the Dig for Victory campaigns of World War II. The 1960s saw a significant fall in interest in allotments. The national waiting lists fell to single figures and the number of plots available dropped by almost a third. The total number of plots in England and Wales declined from over one million plots in 1950 to 800,000 in 1960 to around half a million by 1970. In response to this decline, an enquiry into the condition of allotments was conducted. Its final report, published in 1969, became known as the Thorpe Report after its chair. Its concerns centred on the improvement of allotment sites to overcome their probable disincentive to take up a plot. An important focus was to provide or improve facilities like water supply and security fencing. The report recommended changing the image of allotments, to present a more contemporary association, from an appeal to *growing to live* to *gardening for leisure*. Although the report spurred some local councils to revive their allotment sites, there was no legislative change to implement the report's recommendations more widely. Interest in allotments rose once more in the 1970s in response to emerging environmental movements. Food and agricultural issues were key concerns of the environmental movement at this time and allotments were bolstered by food safety scares and new approaches to food ethics – increased interest in vegetarianism, organics, 'whole' and uncontaminated foods, a questioning of commercialisation and a desire for 'authenticity' (Belasco, 1993; Crouch and Ward, 2008; Howe et al., 2012; Pollan, 2006). Throughout the 1970s the numbers of people on waiting lists increased significantly but the numbers of plots declined very slowly, amid enormous demand for development and the land price boom at that time. Many campaigns to save allotment sites happened at this time.

The allotment story continues in Britain. The value of allotments is marked by their contributions to life in terms of building community, ecological quality and amenity, and improving mental and physical health. During the fertile period of the 1970s, allotmenteers were joined in their gardening pursuits by community gardeners.

The origins of community gardens in North America, Britain and Australia

Community gardens were created in urban, suburban and regional areas around the world in the 1970s. The first community gardens of this era were in the United States, where the term community garden had already been in use since World War I (Lawson, 2005). However, the present community gardening movement can be seen as originating with two events: the establishment of People's Park in Berkeley in April 1969, which included vegetable production as well as space for civic engagement and recreation (Carlsson, 2008; Mitchell, 2014), and the formation of the Green Guerillas in New York in 1973. The Green Guerillas were a group of activists who used 'seed green-aids' – balls of seeds and clay soil sometimes cased in Christmas tree ornaments – to beautify blight sites on New York's Lower East Side. They began taking over vacant lots; clearing out building rubble, abandoned cars and mountains of rubbish; and turning them into gardens. Despite initial opposition, the City of New York eventually issued leases to the Green Guerilla gardeners, and numerous other local residents became inspired to create community gardens in abandoned privately owned lots in their own neighbourhoods (Schmelzkopf, 1995; Smith and Kurtz, 2003).

Community gardens were created in Britain at around the same time, with the first established in 1972. The UK's Federation of City Farms and Community Gardens attributes the emergence of community gardens to the rise of community self-help initiatives, which inspired groups of neighbours to transform derelict land into community gardens that were run by communities to meet their own needs. The Federation of City Farms and Community Gardens also describes the growth of the community garden movement in the United States as a direct inspiration (FCFCG, 2006). Australia's first community gardens were established just a few years after these British and North American gardens, with the first including Nunawading Community Garden, established in Melbourne in 1977.

The British allotment system was a key inspiration for the instigation of community gardens in both North America and Australia. Many of the people who established community gardens in Australia in the late 1970s and early 1980s were inspired by visits to allotments in the United Kingdom. The initiator of the Nunawading Community Garden, for example, got the idea from reading a book about allotments at his local library (Brandenberg, 1978). Many early Australian community gardeners felt a stronger sense of connection with the British allotment tradition than with the other community gardens emerging in their own country (Phillips, 1996).

Community gardens have clearly been influenced by their predecessors, including Victory Gardens and school garden schemes as well as allotments. Indeed, people involved with gardens during World War II were central to the development of community gardens in the 1970s (Carlsson, 2008). However, the community gardens established in the 1970s can be seen as distinct from earlier forms of urban collective agriculture for several reasons. Unlike allotments, which are managed by municipal authorities, these new community gardens emphasised participation and collective self-management (Lawson, 2005; Bell, 2016). Unlike wartime Victory Gardens, the community gardens that originated in the 1970s were citizen initiatives with a broad range of concerns, not government-sponsored programs focused solely on food provisioning. Community activism was the driving force for these community gardens, with concerns of civil rights movements (Warner, 1987; Hynes, 1996) converging with an emerging environmental consciousness (Schukoske, 2000; Bartolomei et al., 2003) to address issues of urban decay and community self-help (Harvey, 2007).

Community gardening is often associated with the environmental justice movement (Ferris et al., 2001; Schukoske, 2000). Paul Milbourne (2012: 943) calls them "socio-ecological spaces of justice within the city". The first American community gardeners were experienced activists, who had been involved in anti-war movements, the National Association for the Advancement of Colored People and the Peace Corps (Warner, 1987). Accounts of community gardens as the result of collective social action are in contrast with those that see them primarily as means of producing food in times of economic crisis. They correspond more closely with the complex range of reasons that people cite for becoming and remaining involved with community gardening: access to green space for children, forming social networks, improving neighbourhood blight sites, enacting ecological values and accessing preferred foods (see for example research by Nemore, 1998; Armstrong, 2000; Blair et al., 1991; Holland, 2004; Lovell et al., 2014; Saldivar-Tanaka and Krasny, 2004).

Since those early gardens were created, community gardening has been taken up by a broad range of actors: health agencies seeking to increase vegetable consumption, schools wanting to increase food literacy, physical and psychological therapists seeking places of calm for their clients, public housing authorities seeking to address social isolation, local government authorities managing residents' demands for a place to garden, radical social movements seeking symbols of non-capitalist social and spatial relations. This plurality of actors has led to a concomitant diversity in the social and physical landscape of community gardens.

The landscape of allotments and community gardens

All gardens communicate something about those who design and tend them (Lewis, 1979; Francis and Hester, 1990). It's fair to say that in the past, allotments have been judged lacking in aesthetic virtues. However, allotment holders and community gardeners numerously express the delight of being on their plots: of finding calm there and something even spiritual. This can be felt in the aesthetic among plot holders. To dig as a way to freedom, to be active, to experience rhythm, to feed family.

> This may sound really simplistic, but there's a special magic in being able to plant a seed in the ground and water it and watch it grow into a plant – and then being able to eat it. It's magical. I guess if you look at lots of areas of people's lives, there's disempowerment, sadness. It's a real joy to be able to grow and harvest food.
>
> *(Community gardener in Nettle, 2014: 106)*

The social and cultural, as well as creative character of community and allotment gardens and gardening play a crucial role in the making of these food landscapes. Plot holders express the ways in which this character is practiced, and give insights into a unique diversity of their landscapes of cultivation. In short, personal and shared landscapes occur in both the material character of the ground and plants grown and in the imagination and feeling that they find in what they do. Crucially their landscape is multi-sensually experienced and formed; changing and as much formed through their feeling as their labour. The aesthetics of allotments is less in the view than in the doing. There is a complex and nuanced, fluid character to the ways in which allotments and community gardens occur as landscapes. The landscape is not fixed and certain, but fluid and full of surprises. It is in the closer involvement of cultivating a plot that other dimensions of an allotment or community garden aesthetic become clear. This is an aesthetic that is not merely visual, but derives from the ways in which gardeners use these spaces. Its aesthetic absorbs the senses. In the growing, the close-up intimacy of things brings us in to close contact with what is around us and the feeling is one of depth. Community gardens and allotments transcend the myriad details across the beds and give a kind of glow. The feeling is of mingling or commingling with vegetation, supporting twigs, smells and touch. The rhythm of cultivation is an important component of what emerges into landscape (Crouch, 2010; Meyer, 2008).

In sketching this distinctive way of creating landscape, it is necessary to let the plot holders themselves articulate the feeling of the making. Carol, living near Consett, County Durham, describes her experience of tending her plot:

> Working outdoors feels much better for you somehow . . . more vigorous than day to day housework, much more variety and stimulus. The air is always different and alerts the skin . . . unexpected scents are brought by breezes. Only when on your hands and knees do you notice insects and other small wonders. My allotment is a central part of my life. I feel strongly that everyone should have some access to land, to establish a close relationship with the earth, something increasingly missing in our society, but essential as our surroundings become more artificial.
>
> *(Crouch and Ward, 2008: 22)*

Carol describes what she does, and what her emerging feelings are, through how she grasps what is around her. Her body engages in an intimacy of space, significantly on her own terms, with movements in her encounters deploying all the senses. Carol touches, bends and kneels; she moves her body and spaces the gaps, and their objects, between vegetation, earth, insects, the air and herself. She finds and re-establishes her feeling of life; her values and attitudes through what her body does there. In this there is ambivalence between the content and process of her moment to moment of doing and she refers to her prefigured intentions and works those ideas through doing (Chapter 34). Amongst the intimacies of senses and mental connections, she refers that back to contexts of her life. There is a double checking or confirming of the close-up atmosphere and its lively resonance with prior context. She engages simultaneously an implicit acknowledgement of being surrounded by so much (Crouch, 2013).

The aesthetic may be considered in relation to its roots: caring, sharing, struggling, freedom. In the landscape is reflected the free creative action of what is done and felt. The landscape is an art product; a product of human action and care. There is a distinctive history of recycling materials: to build sheds, to create compost, to recycle bamboo sticks, and to save seed. Curiouser things are used, too, such as discarded CDs hanging loosely to scare away birds, and abandoned pieces of carpet laid down to suppress weeds. These are all part, with the growing plants and swelling fruits, of the material of the landscape. There is artistry in the handling and positioning of recycled materials, with care and not least ingenuity required for construction, and perhaps endless repair. DIY spills over into the gift relationship, where new uses can be discovered in old materials: seedlings find a home, to grow differently from where they came from. Considering this idea of space in which we cultivate as vital experience, as fundamental in the concept of landscape, ethically and aesthetically, is to realise allotment gardens as a space of the landscape that includes, in fact, the living experience, even when only contemplated (Crouch, 2010).

While community gardens may curve and sprawl in all directions, the rectilinear layout of allotment sites partly responds to the demands, from the earliest years, to provide as many plots as possible in the available space. The strict layout also speaks of a moral landscape, as sites were often also a means to regulate behaviour and the grid has long been associated with technologies of control. However, tended organic growth and the dynamism of the seasons provide variety and relief – naturalness – to site layouts and the often stiffness of row cultivation. Along with the vegetables themselves, the changing features of human structures provide a rich and varied texture to allotment sites (Crouch, 2003a).

The presence of allotments and community gardens in the urban landscape – and in the agri-food system – is a visible, tangible counter to the seeming inevitability of the industrial food system (see Chapters 16, 31 and 32). In ways that can be seen, touched, smelled and eaten, allotments and community gardens communicate that other ways of producing food are possible, that the industrial food system has not achieved complete saturation. Specific elements of garden design reinforce the distinctions between the possible food landscape of community gardens and the dominant industrial agri-food landscape. Allotments and community gardens are a distinctly human-scale form of food production, with little evidence of the use of machinery, and a general eschewing of agricultural chemicals. The diversity of species cultivated is counter to the narrow range of transportable, storable varieties produced in the mass monocultures of industrial farms and available on supermarket shelves. Diversity is valued for its ecological services (mitigating against crop failures, providing habitat for predatory insects), for its cultural value, for its beauty and for its utility – as ripe yellow tomatoes and purple and white striped aubergines are less tempting to unauthorised harvesters than their more familiar cousins. The allocation of space for play and socialising and the cultivation of flowers, even those gardens with the greatest demand for individual vegetable plots, conveys an interest in land that is not purely economic but is also concerned with beauty, connection to place and human connectedness (Nettle, 2014).

The space of allotments and community gardens provides a reprieve and engagement that town parks and other open spaces often cannot (Zeunert, 2016). The regular caring, working and replanting of the garden brings closer encounters and an extent of creativity that is otherwise more limited. Being creative occurs as we work the ground, nurture plants and make structures to support them. Allotment gardening produces a shared heritage of relationships with natural processes, commingling with cultivation and friendship. In public parks, the social relations that enabled their creation are largely hidden. In allotments and community gardens, the process of making the garden is ongoing. They are not simply to be observed or to provide space for passive leisure, but are created and re-created by their users (Francis, 1987; Linn, 1999).

Gardeners are producers. Not only of food, but also of culture. Their landscapes reflect this. Since the 1970s community gardens have become places of festivals and community celebrations; galleries for sculpture, murals and mosaic works; venues for concerts and performances. Community gardens are also repositories of traditional cultural knowledges from around the world, centres of expertise about increasingly rare plant varieties, methods of cultivation and the preparation of a myriad of foods. They are places where people attempt to develop and disseminate new cultural practices and values, where people seek to create and embed cultures of food and ecological literacy, connection to place, celebration of diversity and commitment to community and collaboration. Community gardeners produce culture not only through the arts, but through the creation of shared knowledges and ways of living and working: their organisational practices, decision-making processes, the ways they tell stories about what they do, the spaces they create and the ways they tend their gardens (Crouch, 2003b; Nettle 2014).

The role of landscape professionals

Landscape designers and thinkers have long been interested in the integration of the urban and the rural. Early writings on garden cities and suburbs (Howard, 1898 – see Hall and Ward, 1998; Migge, 1919 – see Haney, 2010) are echoed in recent policy-minded writing, suggesting the continued attraction of 'greening the city'. Despite this ongoing interest, allotments and community gardens have often proved difficult for designers and planners to locate in their scheme of things. These spaces developed for a long time before professional planners and landscape architects became involved. The continuing surges of interest and commitment to allotments and community gardens have made professionals rethink the role they play in greening environments and diversifying otherwise agricultural environments and closely developed cities and towns. Where once they were regarded as temporary, though beneficial, forms of land use (Lawson, 2004), increasingly, allotments and community garden sites are recognised in local and central government policy as important components of amenity, contributing to a variety of strategic aims (Crouch, 2006; Chapter 26).

In contrast with most public open space, community gardens and allotments are idiosyncratic, developed for the most part without the input of professional landscape architects and designers, though in the case of community gardens often influenced by design philosophies such as permaculture (Chapter 10). Parks and community gardens are models that represent fundamentally different ways of managing, designing and maintaining open spaces. Allotments and community gardens speak to the possibility of people working together at a neighbourhood level to imagine and create desired community spaces. In this way, the common land of community gardens is distinct not only from private land but also from public land, which is designed and managed by the state putatively for the benefit of an unidentified public (Chan and Sharma, 2007; Hardt and Negri, 2009).

Where landscape professionals have been involved at the level of site design the results have been mixed, particularly when conceptual and aesthetic considerations have been given precedence over functionality: for example, when raised garden beds have been designed with grand proportions that make it impossible for gardeners to reach the soil at their centres, or when plots for individual use are laid out in a visually pleasing mix of shapes and sizes that leads to competition for the largest beds and those with the best solar access. One site, Manor Gardens, in Stratford, East London, was erased for the London Olympics in 2012 (Crouch, 2014). A huge swathe of land along the city's Lea Valley became part of a landscaping scheme. It just happened that there was an allotment site, very well cultivated by its plot holders, within this larger area. The designers considered it to be in the way, perhaps because they feared its marks of individualism. It was cleared away and the remaining emptiness added into the wider grand scheme. They felt, therefore, that the site *had become landscaped*. The designers lacked a grasp of the liveliness and rich texture that the well-maintained site could offer their bland effort. A new allotment site was provided out of the way. With the site cleared, the plot holders had to leave. With their allotments gone, their years of collective effort and cultivation there were lost.

Where projects have taken seriously the social aspects of design, and where genuinely participatory and democratic processes have been embedded from the outset, landscape professionals have contributed to projects that are both beautiful and fully owned by their communities – though with perhaps a little less of the radical malleability of their more organically designed cousins (see for example NVA, 2013; Spirn and Marcucci, 1991; Gibbons and Gibbons, 2012; Linn, 2008).

However, it is at the strategic and policy levels that there is perhaps the greatest potential for landscape professionals to support and enable community food growing activities. Community gardens, whatever their participants' aims, have been seen by governments as a short-term emergency measure and this view of community gardens as temporary has led to their exclusion from planning and long-term support, particularly in times of relative prosperity (Francis, 1987; Lawson, 2004; McClintock, 2010). However, increasing recognition of the benefits of community food growing for wellbeing and food security has led to an interest in food sensitive urban design and planning (Chapter 26). Harris (2009) suggests that planning policy at local and state levels should, as a starting point, make explicit mention of community gardening as a desirable land use. Specific interventions, such as changing zoning regulations to allow and encourage urban agriculture in all zones, and to reserve land specifically for community gardens, could aid potential community gardeners and allotmenteers in accessing land and maintaining security of tenure. Designers and planners can also play an important role in helping community gardeners to navigate planning laws and processes and a range of stakeholders (Zammit and Suklje, 2016).

Conclusion

The continuing story of allotments and community gardens through the last hundred years is one of change and continuity; from early efforts to secure land to a contemporary concern to hold on to it. Change takes the form of intermittent advances and threats; enthusiasm and decline; extension, improvement and disruption. Ways of cultivation shift and return. Continuity expresses itself in the steady survival and inclusion of the increasing diversity of people who become gardeners. Cultivation remains at the centre of conviviality, friendships, things grown – and communities are cultivated too.

Moments of crisis sharpen the attitudes and values developed in gardening towards an overt kind of politics that speaks out in defence. The politics of values and attitudes, of ways of doing things, and regard for the community and cultivation reveal a slower, ongoing politics. This is one of freedom amid reciprocity, the gift relationship and simple, often very deep, friendship (Crouch and Ward, 2008).

In the UK the broadening of allotment and community gardening culture has brought a continuing wave of increasing numbers of skilled and professional people and increasing proportions of women, able to lend further strength to these kinds of cultivation and their defence. The newcomers mix with the continuing gardeners, each of whom has skills to bring to both gardening and efforts to hold on to the ground. Altogether there emerges a significant interest and commitment, political experience in communication, working with a wider community, and speaking out over their concerns. Allotments and their sisters, community gardens, have developed from being a significant cultural heritage into an increasingly complex and dynamic part of contemporary life and our relationship with nature.

There is a long history of gardeners growing mutual aid and solidarity along with their cabbages. With anxiety about climate change, peaking oil supplies and dwindling finite resources, crises in global agri-food systems and increasing attention to food security, vegetable gardening has taken on new significance. In 2009, there were successful campaigns for organic vegetable gardens to be dug at the White House and Buckingham Palace (Crouch, 2014; Nettle, 2014). In 2011 and 2012, 'Occupy' groups in the US, UK, Canada and Aotearoa/New Zealand established urban farms and food gardens. Far from being a retreat, growing vegetables in public spaces has become a widely adopted political statement. Community gardening in particular has had a resurgence in recent years and is increasingly associated with other local and alternative food initiatives (Bell, 2016; Nettle, 2014).

Although the more collectivist focus of many community gardens marks them as distinct from allotments, which were established with food production and independence at their core, both have roots in a commitment to the preservation of common land. The history of both allotments and community gardens can be traced directly to the enclosure of the British commons – and the ensuing social disorder (Carlsson, 2008; Crouch and Ward, 2008; Parker et al., 2007; Wilson, 1999). However, the enshrinement of allotments in British law and culture is not the only reason that the defence of commons has continuing resonances for community gardeners and allotmenteers. The commons, as a "way of thinking and operating in the world, a way of organising social relations and resources" (Eizenberg, 2012: 764), has become a powerful metaphor for the collective sphere in general, including not just land but political and social 'space' in all its forms (Klein, 2002; Wilson, 1999). The commons is also emblematic of an agrarian environmental ethic, in which stewardship is linked to cultivation (Judd, 1997). Commons are neither private property nor a free-for-all wilderness, but a form of corporate ownership (Judd, 1997). As actually existing neighbourhood commons, allotments and community gardens belong to the community and are managed collectively for mutual benefit (Eizenberg, 2012). The urban foodscapes they create speak to the desire for connection to the land, to community and to the source of our sustenance.

Key texts

Crouch, D., & Stott, M. (1997) *City fields and country gardens: Allotment essays of Michael Hyde*. Nottingham: Five Leaves.

Crouch, D., & Ward, C. (2008) *The allotment: Its landscape and culture*. 4th edition. Nottingham: Five Leaves (first edition London Faber and Faber 1988).

Lawson, L.J. (2005) *City bountiful: A century of community gardening in America*. Berkley: University of California Press.

Lovell, R., Husk, K., Bethel, A., & Garside R. (2014) What are the health and well-being impacts of community gardening for adults and children: A mixed method systematic review protocol. *Environmental Evidence*, 3(20), 1–13.

Nettle, C. (2014) *Community gardening as social action*. London and New York: Routledge.
Viljoen, A., & Bohn, K. (2014) *Second nature urban agriculture: Designing productive cities*. London: Routledge.
Warner, S.B. (1987) *To dwell is to garden: A history of Boston's community gardens*. Boston: Northeastern University Press.

Further reading

ACGA. (2015) *American Community Gardening Association Annual Report 2015*. College Park: ACGA.
Acton, L. (2011) Allotment gardens: A reflection of history, heritage, community and self. *Papers From the Institute of Archaeology*, 21, 46–58.
Armstrong, D. (2000) A survey of community gardens in upstate New York: Implications for health promotion and community development. *Health and Place*, 6(4), 319–327.
Bakker, N., Dubbeling, M., Guendel, S., Sabel-Koschella, U., & de Zeeuw, H. (eds.) (1999) *Growing cities, growing food: Urban agriculture on the policy agenda: A reader on urban agriculture*. Wielinger: Deutsche Stiftung für Internationale Entwicklung, Zentralstelle für Ernährung und Landwirtschaft.
Barraclough, L.R. (2009) South Central farmers and Shadow Hills homeowners: Land use policy and relational radicalization in Los Angeles. *The Professional Geographer*, 61(2), 164–186.
Bartolomei, L., Corkery, L., Judd, B., & Thompson, S. (2003) *A bountiful harvest: Community gardens and neigbourhood renewal in Waterloo*. Sydney, New South Wales Government – Department of Housing and University of New South Wales.
Bassett, T.J. (1981) Reaping on the margins: A century of community gardening in America. *Landscape*, 25(2), 1–8.
Belasco, W. (1993) *Appetite for change*. Ithaca, NY: Cornell University Press.
Bell, S. (2016) Introduction. In: Bell, S., Fox-Kamper, R., Keshavarz, N., Benson, M., Caputo, S., Noori, S., & Voigh, A. (eds.), *Urban allotment gardens in Europe*. Oxon: Routledge, pp. 1–7.
Blair, D., Giesecke, C.C., & Sherman, S. (1991) A dietary, social and economic evaluation of the Philadelphia Urban Gardening Project. *The Journal of Nutrition Education*, 23, 161–167.
Brandenberg, J.P. (1978) *Community participation with land use: A Nunawading example*. Paper presented to Land for Leisure, Annual Conference of the Royal Institute of Parks and Recreation, Burnie, Tasmania.
Burchardt, J. (2002) *The allotment movement in England, 1793–1873*. Cambridge: Boydell Press.
Carlsson, C. (2008) *Nowtopia: How pirate programmers, outlaw bicyclists, and vacant-lot gardeners are inventing the future today!* Oakland: AK Press.
Chan, G., & Sharma, N. (2007) Eating in public. In: Shukaitis, S., & Graeber, D. (eds.), *Constituent imagination: Militant investigations / collective theorization*. Oakland: AK Press, pp. 180–188.
Crouch, D. (1988) *Expert evidence, Report of the Government Enquiry into The Future of Allotments*. London: HMSO.
Crouch, D. (2003a) *The art of allotments: Culture and cultivation*. Nottingham: Five Leaves.
Crouch, D. (2003b) Spacing, performance and becoming: The tangle of the mundane. *Environment and Planning A*, 35, 1945–1960.
Crouch, D. (2006) *Report of survey of allotments and community gardens in England and Wales* University of Derby, H.M. Department of Local Government and Communities. (published in full by Government 2011: delayed as it had criticised the method required by Government.)
Crouch, D. (2010) Flirting with space: Thinking landscape relationally. *Cultural Studies*, 17(1), 5–18.
Crouch, D. (2013) Landscape, performance and performativity. In: Howard, P., & Waterton, E. (eds.), *The Routledge companion to landscape studies*. London: Routledge.
Crouch, D. (2014) The allotment and community garden. In: Viljoen, A., & Bohn, K. (eds.), *Second nature urban agriculture: Designing productive cities*. London: Routledge, p. 263.
Crouch, D., & Ward, C. (2008) *The allotment: Its landscape and culture*, 5th ed. Nottingham: Five Leaves Books (first edition London Faber and Faber 1988).
Crouch, D., & Wiltshier, R. (2005) Designs on the plot: The future of allotments in urban landscapes. In: Viljoen, A. (ed.), *Continuous productive urban landscapes*. London: Architectural Press, pp. 125–131.
Davies, M. (2007) *Property: Meanings, histories, theories*. Oxon: Routledge Cavendish.
Eizenberg, E. (2012) Actually existing commons: Three moments of space of community gardens in New York City. *Antipode*, 44, 764–782.

Endersby, J. (2000) A garden enclosed: Botanical barter in Sydney 1818–39. *British Journal for the History of Science*, 33(118), 313–334.

FCFCG/NAS 2007. *Allotments regeneration initiative*. www.farmgarden.org.uk.

FCFCG (2006) *A bit of history*. Bristol: Federation of City Farms and Community Gardens.

Ferris, J., Norman, C., & Sempik, J. (2001) People, land, and sustainability: Community gardens and the social dimension of sustainable development. *Social Policy and Administration*, 35(5), 559–568.

Francis, M. (1987) Some different meanings attached to a public park and community gardens. *Landscape Journal*, 6(2), 101–112.

Francis, M., & Hester, R.T. (1990) *The meaning of gardens: Idea, place and action*. Cambridge, MA: MIT Press.

Fryer, P., & Rigby, G. (1992) *Peaceable kingdoms*. Newcastle on Tyne Leisure Services.

Hall, P., & Ward, C. (1998) *Sociable cities: The legacy of Ebenezer Howard*. London: Wiley.

Haney, D.H. (2010) *When modern was green: Life and work of landscape architect Leberecht Migge*. Oxon: Routledge.

Hardt, M., & Negri, A. (2009) *Commonwealth*. Cambridge, MA: Harvard University Press.

Harris, E. (2009) The role of community gardens in creating healthy communities. *Australian Planner*, 46(2), 24–27.

Harvey, A. (2007) Food for thought: Southwark Queen Village Community Garden. *Pennsylvania Legacies: The Journal of the Historical Society of Pennsylvania*, 7(2).

Holland, L. (2004) Diversity and connections in community gardens: A contribution to local sustainability. *Local Environment*, 9(3), 285–305.

Howard, E. (1898) *To-morrow: A peaceful path to reform*. London: Routledge.

Howe, J., Bohn, K., & Viljoen, A. (2012) Food in time: The history of English urban space as a European example. In: Vilojen, A., & Howe, J. (eds.), *Continuous productive urban landscapes*. Oxford: Architectural Press, pp. 95–107.

Howe, J., & Wheeler, P. (1999) Urban food growing: The experience of two UK cities. *Sustainable Development*, 7(1), 13–24.

Hoyles, M. (1991) *The story of gardening*. London: Journeyman Press.

Hynes, H.P. (1996) *A patch of Eden: America's innercity gardeners*. Vermont: Chelsea Green.

J and L Gibbons Landscape Architects (2012) Dalton space making. *Landezine*. www.landezine.com/index.php/2012/01/making-space-in-dalston-by-j-l-gibbons-landscape-architects/ (accessed 4 July 2016).

Judd, R.W. (1997) *Common lands, common people: The origins of conservation in Northern New England*. Cambridge, MA: Harvard University Press.

Keshavarz, N., & Bell, S. (2016) A history of urban gardens in Europe. In: Bell, S., Fox- Kamper, R., Keshavarz, N., Benson, M., Caputo, S., Noori, S., & Voigh, A. (eds.), *Urban allotment gardens in Europe*. Oxon: Routledge, pp. 8–30.

Klein, N. (2002) *Fences and windows*. New York: Harper Collins.

Kurtz, H. (2001) Differentiating multiple meanings of garden and community. *Urban Geography*, 22(7), 656–670.

Lawson, L.J. (2004) The planner in the garden: A historical view into the relationship between planning and community gardens. *Journal of Planning History*, 3(2), 151–176.

Lawson, L.J. (2005) *City bountiful: A century of community gardening in America*. Berkley, CA: University of California Press.

Lewis, P.F. (1979) Axioms for reading the landscape: Some guides to the American scene. In: Meinig, D.W. (ed.), *The interpretation of ordinary landscapes*. New York: Oxford University Press.

Linebaugh, P. (2014) *Stop, thief! The commons, enclosures, and resistance*. Oakland: PM Press.

Linn, K. (1999) Reclaiming the sacred commons. *New Village*, 7(1), 42–49.

Linn, K. (2008) *Reclaiming commons and community*. New York: New Village Press.

McClintock, N. (2010) Why farm the city? Theorising urban agriculture through a lens of metabolic rift. *Cambridge Journal of Regions, Economy and Society*, 3, 191–207.

Meyer, E.K. (2008) Sustaining beauty. The performance of appearance: A manifesto in three parts. *Journal of Landscape Architecture*, Spring, 6–23.

Migge, L. (1919) *Jedermann Selbstversorger! (Everybody self-sufficient!)*. Jena: Diederichs.

Milbourne, P. (2012) Everyday (in)justices and ordinary environmentalisms: Community gardening in disadvantaged urban neighbourhoods. *Local Environment*, 17(9), 943–957.

Mitchell, D. (2014) *The right to the city: Social justice and the fight for public space*. New York: Guilford Publications.

Mok, H, Williamson, V.G., Grove, J.R., Burry, K., Barker, S.F., & Hamilton, A.J. (2014) Strawberry fields forever? Urban agriculture in developed countries: A review. *Agronomy for Sustainable Development*, 34(1), 21–43.

Moran, D.M. (1990) *The allotment movement in Britain*. New York: Lang.

Nemore, C. (1998) *Rooted in community community gardens in New York city: A report to the New York state senate*. New York: Senate Minority Office.

NVA Public Art (2013) Sage: Sew and grow everywhere. *NVA Public Art*. http://nva.org.uk/artwork/sage/ (accessed 4 July 2016).

Olwig, K.R. (2002) *Landscape, nature, and the body politic*. Madison, Wi: University of Wisconsin Press.

Parker, M., Fournier, V., & Reedy, P. (2007) *Dictionary of alternatives: Utopianism and organization*. New York: Zed Books.

Phillips, D. (1996) *Australian city farms, community gardens, and enterprize centres inventory*. Hobart: Symbion.

Pollan, M. (1996) *The omnivore's dilemma: A natural history of four meals*. London: Penguin.

Pudup, M.B. (2008) It takes a garden: Cultivating citizen-subjects in organized Garden Projects. *Geoforum*, 38(3), 1228–1240.

Quale, H. (2008) *The true value of community farms and gardens: Social, environmental, health and economic*. Bristol: Federation of City Farms and Community Gardens.

Saldivar-Tanaka, L., & Krasny, M.E. (2004) Culturing community development, neighbourhood open space, and civic agriculture: The case of Latino community gardens in New York City. *Agriculture and Human Values*, 21, 399–412.

Schmelzkopf, K. (1995) Urban community gardens as contested space. *Geographical Review*, 85(3), 364–381.

Schukoske, J.E. (2000) Community development through gardening: State and local policies transforming urban open space. *Legislation and Public Policy*, 3(35), 341–355.

Scott, J.C. (1985) *Weapons of the weak: Everyday forms of peasant resistance*. New Haven, CT: Yale University Press.

Smith, C.M., & Kurtz, H.E. (2003) Community gardens and politics of scale in New York city. *Geographical Review*, 93(2), 193–212.

Spirn, A.W., & Marcucci, D. (1991) Models of success: Landscape improvements and community development. In: *West Philadelphia landscape plan*. Philadelphia: Department of Landscape Architecture and Regional Planning, University of Pennsylvania.

Wilson, P.L. (1999) Avant gardening. In: Wilson, P.L., & Weinberg, B. (eds.), *Avant gardening: The ecological struggle in the city and the world*. New York: Autonomedia.

Zammit, A., & Erjavec, I.S. (2016) Allotment gardens as a challenge for an urban designer. In: Bell, S., Fox-Kamper, R., Keshavarz, N., Benson, M., Caputo, S., Noori, S., & Voigh, A. (eds.), *Urban allotment gardens in Europe*. Oxon: Routledge.

Zeunert, J. (2016) Urban agriculture upscaled: Can economic returns and social benefits co-exist in public green spaces? In Roggema, R. (ed.), *Sustainable urban agriculture and food planning*. London: Taylor & Francis, pp. 107–125.

31
Food sovereignty

Max Kelly

Introduction

Food sovereignty is a concept, a global movement, and an evolving project. It is also a rhetorical device and a political programme (Akram-Lodhi, 2015). At a very broad level, it is a radical proposal for substantive change in the existing food regime. As McMichael (2015: 434) notes, it is open to multiple interpretations; however, food sovereignty as an overarching term has struck a major chord with substantial numbers of individuals, farmers, consumers, social movements, and non-gonvermental organization (NGOs) around the world. It is a groundswell of ideas and actions driven from the grassroots, and has entered the highest levels of discourse on issues of food production and consumption, and was endorsed by the International Assessment of Agricultural Knowledge, Science and Technology for Development (McIntyre et al., 2009). However, despite the integration of food sovereignty into national legislation in a small number of countries, there has been limited impact in the corridors of power that has translated to significant fundamental change in agriculture and food policy (Hospes, 2014).

Food sovereignty is based on the right of peoples and countries to define their own agricultural and food policies. Food sovereignty has emerged from a complex series of related strands of thought, primarily ecological sustainability, agrarian citizenship, and alternatives to increasingly globalised and industrialised farming and food systems. Despite the significant attention it has garnered, Hospes (2014: 128) argues that contemporary debate on food sovereignty as "an alternative paradigm and counterframe to neoliberal thinking on food security" is in essence at either a deadlock, or a crossroads. Food security (see Chapters 20 and 25) focuses on availability of, access to, and utilisation of food. Food sovereignty is a fundamentally political project focusing on means and mode of production, social justice, and human rights.

Herein lies the crux of the debate on food production and consumption. Food sovereignty is, as Bernstein (2014: 1031) notes "a comprehensive attack on corporate industrialsied agriculture" and the exaltation of the peasant (or small scale farmer), and their "social and moral superiority [. . .] and [. . .] ecological superiority". Critics of this vast project point to: the issue of translating the conceptualisation of a new food system into a feasible project; the capacity of small-scale farmers to feed the world; the insular nature of the localisation discourse; issues of scale and scalability; and its location within the structures of capitalism, among other issues (Bernstein, 2014; Agarwal, 2014). Jansen (2014) explores the 'debate' (and its nuances) between McMichael as proponent and Bernstein as critic.

The principal aim of this chapter is an exploration of the conceptual underpinnings of the food sovereignty movement, from its peasant origins to the explosion of the concept globally, through food regime theory debates, and the relationship between food sovereignty, the 'right to food', and food security. There is an embedded localisation narrative underpinning food sovereignty (Iles and Montenegro De Wit, 2015), and an unequivocally clear relationship between land and food production, from a number of angles. Issues such as land reform and land grabs have received attention, with both of fundamental importance to small-scale farmers and the landless. Land reform is a change in the way land is held or owned, and can include land redistribution (to the landless) and change of tenure (for example from customary to freehold) (Peters, 2009). Land grabs are large-scale acquisitions of land for commercial and often foreign investors. This can result in the forced displacement of farmers and communities (McMichael, 2015; Borras Jr. et al., 2015). However, there is little specific debate on the interrelationships between food sovereignty and landscapes. This chapter draws on Australian and international experiences to explore critical debates and the realities of the food sovereignty paradigm for the maintenance and development of sustainable, productive, and diverse landscapes in complex socioeconomic and political contexts. The chapter first briefly unpacks food sovereignty through an analysis of its evolution as a concept and movement. The theoretical constructs embedded in food sovereignty are explored, with particular reference to links to food security and food regime theory. Contemporary challenges and opportunities in the food sovereignty 'project' are explored, and finally the relationship between food sovereignty and landscapes is critically analysed, calling for more focus on scale and interlinkages between politics, food systems, and landscapes.

Food sovereignty

The global project to feed the world is a vastly complex undertaking, and, as Schanbacher (2010) notes, sits somewhere between broader narratives concerning globalisation, human rights, food security, and poverty. Feeding a growing world population a safe, secure, and nutritious food supply is essential (Chapters 20 and 25). However, part of the broader agenda is, of course, social justice, ecological sustainability, and fighting poverty and inequality. This unquestionably requires engagement across a complex web of political, social, environmental, and cultural concerns, as well as the predominant focus on economics. Can the existing dominant food system meet future challenges, or is there a requirement for a fundamental rethink of food production and consumption, embracing what Charlton (2016: 3) refers to as 'non-traditional' approaches (see Chapters 4, 10 and 16). Given the acknowledgement of the role of small-scale and family farmers in feeding the growing world population (FAO, 2017) an understanding of the call for radical change as envisaged in the food sovereignty debate is essential. Existing discourse, policy, and practice on food production and consumption are primarily focused on a globalised, commoditised food system (see Chapters 4 and 32). There is significant focus on increasing productivity of small-scale producers and the commercialisation of the smallholder sector (Delgado, 1999; Chapter 21). However, this commercialisation metanarrative, via incremental change to existing food systems, is problematic, and there is significant debate as to the best ways to provide required increases in world food production (Poole et al., 2013; Schanbacher, 2010; Chapters 16 and 19).

Food security is the lens through which most food discussions occur. Food security is most commonly referred to as "when all people, at all times, have physical, social and economic access to sufficient, safe and nutritious food that meets their dietary needs and food preferences for an active and healthy life" (FAO, 2001: glossary). Food security is discussed elsewhere in this volume (Chapters 4, 19, 20 and 25), but a core critique of food security discourse is that food security is used by multilateral development institutions as a way of supporting market-based and monetary approaches to address a lack of food (Akram-Lodhi, 2015; Beddington, 2010; Sage, 2013; Schanbacher, 2010). La Via Campesina (1996) argues that food sovereignty is a precursor to food security. Patel (2009) makes explicit the link between the context of food security, or power relations that underpin how food security can be achieved, the politics, and associated rights and democracy. Food security *per se* has no embedded acknowledgement of or challenge to the entrenched structural, systemic, and inequitable food system that promotes food insecurity. Food sovereignty is frequently portrayed as the 'counter hegemonic' discourse to the discourse of food security (Hopma and Woods, 2014).

Food sovereignty entered global food discourse in 1996 by La Via Campesina as an alternative construct to the dominant food security discourse. It was launched at the World Food Summit as a strongly political statement that sought to present an alternative to dominant neoliberal modernised, agricultural production systems (see Chapter 16). The 'alternative' view of what a food system could or should be is best encapsulated in the 2007 Nyéléni Declaration extract below:

> Food sovereignty is the right of peoples to healthy and culturally appropriate food produced through ecologically sound and sustainable methods, and their right to define their own food and agriculture systems. It puts the aspirations and needs of those who produce, distribute and consume food at the heart of food systems and policies rather than the demands of markets and corporations. It defends the interests and inclusion of the next generation. It offers a strategy to resist and dismantle the current corporate trade and food regime, and directions for food, farming, pastoral and fisheries systems determined by local producers and users. Food sovereignty prioritises local and national economies and markets and empowers peasant and family farmer-driven agriculture, artisanal–fishing, pastoralist-led grazing, and food production, distribution and consumption based on environmental, social and economic sustainability. Food sovereignty promotes transparent trade that guarantees just incomes to all peoples as well as the rights of consumers to control their food and nutrition. It ensures that the rights to use and manage lands, territories, waters, seeds, livestock and biodiversity are in the hands of those of us who produce food. Food sovereignty implies new social relations free of oppression and inequality between men and women, peoples, racial groups, social and economic classes and generations.
>
> *(Declaration of Nyéléni, 2007)*

As later discussed, food sovereignty has evolved over time. At the 1996 Word Food Summit, national food sovereignty was the focus – and trade and the World Trade Organization (WTO) were the targets. The comprehensive definition just quoted shows the expansion of the concept. There are a number of core points: the primacy of citizens and communities rather than the market, the notion of intergenerational equity and sustainability, and a call for action against the dominant food system. It focuses on small-scale rather than industrialised models of agriculture, social justice and empowerment, and embraces food trading systems for the benefit of people and ecosystems. It also puts control of land decisions into the hands of the producers. There are many variations on definitions of food sovereignty, driven by the various contexts in which it exists. However, it is, as Alonso-Fradejas et al. (2015) note, ultimately a political challenge. La Via Campesina, the international peasant movement, is the originator of this concept. They are a grassroots, decentralised movement that asserts its representation of "millions of peasants, small and medium-size farmers, landless people, women farmers, indigenous people, migrants and agricultural workers from around the world" (La Via Campesina, no date).

Food regime theory and food sovereignty

Patel (2009) argues that it is difficult to clearly link changes to food sovereignty over time, due to the 'big tent' politics of the concept, which needs to incorporate a diversity of opinions. This, however, has, in particular, led to some inconsistency in the (re)location of sovereignty, in the issues of harmonising interests between farm workers and farm owners or the transnational corporations who may own the land.

Food sovereignty originally targeted a food system that was dominated by transnational corporations (TNCs), underpinned by World Trade Organization trade rules, and driven by corporate profitability and food commodification (Akram-Lodhi, 2015; McMichael, 2015). To place this in context, it is useful to draw on the food regime theories of Harriet Friedman and Phillip McMichael. Regime theory is detailed in McMichael (2009), but in brief there are three food regimes identified in this theoretical contribution. The first food regime (1870–1930s) was based on imports of goods from colonies and growing (monocultural) grain production in settler countries (USA, Canada, Australia) for Britain's outsourced food needs.

The second, more American-centric food regime (1930s–1970s) involved rerouting surplus production from the US back to newly postcolonial states, with emphasis on Cold War political manoeuvring (such as Vietnam, Korea, and the targeting of food aid to allies in Africa, such as Zaire DRC). Agribusiness transformed agriculture through increasingly globalised supply chains, and what McMichael (2009) terms agro-industrialisation. Holt-Giménez (2015) raises the spectre of a food system that has evolved in sync with 30 years of neoliberal globalisation, privatising public goods, and deregulating corporate capital (see also Wolf, 2014). This American-centric food regime prompted food 'dumping' in developing countries (benefitting US farmers), and the privileging of large-scale, increasingly intensive industrial agriculture, under the rhetoric of free trade and economic growth.

This is best encapsulated by the notion of the third food regime, the corporate food regime (Friedman and McMichael, 1989; McMichael, 2009, 2013, 2015). This system results in loss of country-owned food and agricultural policies, whereby the developing countries become increasingly food dependent on international, cheaply produced crops (primarily grain). This growing reliance by developing countries, both producers and consumers, helped catalyse evolution of the food sovereignty movement (McMichael, 2015). The 2008 food crisis saw record grain harvests juxtaposed with record levels of hunger (Chapter 25). This resulted in food riots and some 'rumblings of reform' (Holt-Giménez and Shattuck, 2011) in recognition of the negative impacts of the corporate food regime on the poor. However, these rumblings have not produced any appreciable changes to the dominant food regime (Chapter 25).

De Schutter (2015: para 2) notes that the original concept of food sovereignty was very much "a product of its times." As noted previously the concept has evolved and grown, and this evolution has in part responded to criticism of the food sovereignty approach, but in real terms, to the increasing globalisation of agriculture and value chains (Chapter 9). Part of the push for commercialisation of smallholder agriculture focuses on improving value chains. A value chain conceptualises the vertical and horizontal linkages between actors (farmers, suppliers, consumers, finance providers, and so on) in food production. The whole value chain approach to agricultural development is firmly rooted in commercialisation and industrialisation of practices (cf Altenburg, 2007).

Food sovereignty, food justice, or food security?

Food regime analysis gives fascinating insights into the relevance of food sovereignty globally, as it charts the role of agriculture and food in the emergence and development of the capitalist world economy, as well as the positioning of states within a global market; a political economy analysis of food (see Chapter 32). So if food sovereignty was a reaction *against* globalisation, economic liberalisation, and the corporate food regime, the question becomes: What does food sovereignty promote, what is it for, and where is it positioned in relation to other approaches to ensure a sustainable, equitable food supply globally?

Holt-Giménez and Shattuck (2011) provides a useful point of comparison across a spectrum of approaches to food in what he terms the food regime/food movement matrix. In essence, he tracks the predominant food discourses against their political underpinnings and orientation. This is summarised in Table 31.1.

Of course, this rather neat distinction does not fully capture the complexity of the discussion, particularly at the global level. There are some overlaps between the matrix shown here with 'a right to food', now part of high-level discourse, particularly within the Food and Agricultural Organization of the United Nations (FAO; 2017a). There are links between progressive and radical movements also, which are promoted by Holt-Giménez and Shattuck (2011) as desirable alliances in trying to force some regime change. However,

Table 31.1 Dominant trends in current food systems according to their politics and production models.

Politics	*Neoliberal*	*Reformist*	*Progressive*	*Radical*
Discourse	Food Enterprise	Food Security	Food Justice	Food Sovereignty
Orientation	Corporate	Development	Empowerment	Entitlement
Model	Overproduction; corporate concentration; unregulated markets and monopolies; monocultures (including organic); GMOs; agrofuels; mass global consumption of industrial food; phasing out of peasant and family agriculture and local retail	Mainstreaming/ certification of niche markets (e.g. organic, fair, local, sustainable); maintaining northern agricultural subsidies; 'sustainable' roundtables for agrofuels, soy, forest products, etc.; market-led land reform	Agroecologically produced local food; investment in underserved communities; new business models and community benefit packages for production, processing, and retail; better wages for agricultural workers; solidarity economies; land and food access	Dismantle corporate agri-foods monopoly power; parity; redistributive land reform; community rights to water and seed; regionally based food systems; democratisation of food system; sustainable livelihoods; protection from dumping/ overproduction; revival of agroecologically managed peasant agriculture to distribute wealth and cool the planet; regulated markets and supply
Approach to the Food Crisis	Increased industrial production; unregulated corporate monopolies; land grabs; expansion of GMOs; public-private partnerships; liberal markets; international sourced food aid	Same as neoliberal but with increased medium farmer production and some locally sourced food aid; more agricultural aid but tied to GMOs and 'bio-fortified/climate resistant' crops	Right to food; better safety nets; sustainably produced, locally sourced food; agroecologically based agricultural development	Human right to food sovereignty; locally sourced, sustainably produced, culturally appropriate, democratically controlled focus on UN/FAO negotiations

Source: Holt-Giménez and Shattuck, 2011: 117–118

it is worth noting the dissonance between a right to food, and a right of people to control their own food systems. The right to food (food justice – see Chapter 32) places responsibility on the state to provide an enabling environment for people to access adequate food and the resources necessary for the sustainable enjoyment of food security, but does not define the specific model or system to achieve this (OHCHR, 2010). Food sovereignty is the right of people to determine their food system, which requires engagement with the underlying political and economic systems that determine food production and consumption. The link between rights-based food networks that do not specifically embrace food sovereignty and food sovereignty itself is not so clear (Edelman et al., 2014). In essence then, at a high level, food sovereignty is a radical proposal for substantive change in the existing food regime. The matrix in Table 31.1 would suggest some clarity of concept, and generic agreement in both what the changed food regime may look like and how to achieve the desired change. However, this is highly problematic, and the following section, in brief, addresses some contemporary challenges to the food sovereignty 'project'.

Contemporary challenges to food sovereignty

The concept of food sovereignty provides a point of common ground and appealing alternative for many, including NGOs, social movements, academics, and food and agricultural professionals. It has provided a common cause for those who oppose corporate-driven agriculture, market-driven commodification of food, and who decry the failures of existing agriculture and food policy to provide a fair, equitable, or sustainable food system (Chapter 16). However, despite the 20 years that have elapsed since the concept was launched, there is limited agreement on the concept of sovereignty, and an equally limited adoption of a food sovereignty paradigm at a food or agricultural policy level by government or intergovernmental institutions. Hospes (2014) argues that this is driven by a lack of conceptual clarity and a lack of a common framework for debate, exacerbated by what Hospes (2014: 121) refers to as a "rather closed epistemic community". Alonso Fradejas et al. (2015: 432) identify the need to engage with external social forces (external to the collective food sovereignty movement), but also for "constant, internal negotiation within the emerging social forces constructing food sovereignty".

Hospes (2014) notes two core concerns. The first is a challenge to the existing political order, and the second is a challenge to the notion of food as a commodity, providing a rights-based context for a range of ideas, approaches, policies, and practices. These range from agroecology as a foundational concept; democracy; food as a human right; sustainability; social justice; localisation of food; a focus on food for people not profit; the support of the food providers, the women and men who grow, harvest, and process food; valuing small-scale producers; valuing local knowledge, seeds, technology, and skills. The initial framing of food sovereignty as agricultural and rural has changed radically. Food sovereignty now embraces a range of issues, as Alonso-Fradejas et al. (2015: 433) discuss in more detail: food politics, land reform, pastoralism, GMOs (also known as bioengineering), urban agriculture, and a wide range of other issues. Food sovereignty also encompasses non-agricultural, urban, production, consumption, trade, and, specifically within both developed and developing countries, class and other social attributes (Alonso-Fradejas et al., 2015). This is both a core seat of power and perhaps a limitation of food sovereignty. It appeals to many, but is an exceptionally complex concept to work with.

There has been widespread and critical debate recently on the potential contradictions and possible future directions of food sovereignty, focusing on its origins, what the practical and conceptual limitations may be, and whether and how it may be implemented across diverse contexts (Edelman et al., 2014). The depth of discussions are captured in three special editions of journals: *Journal of Peasant Studies* (2014, 41(6)), *Globalizations* (2015, 12(4)), and *Third World Quarterly* (2015, 36(3)). The following section summarises some of this discourse relevant to a landscape debate. Some of the key issues briefly addressed are: multiple understandings and positions; challenges of addressing change to a political order through policy; food sovereignty as an ontological alternative; and lastly the scale at which food sovereignty is conceived.

The evolution of the concept is perhaps the basis of many of the dilemmas associated with food sovereignty discourse. As de Schutter (2015) noted, it was very much a product of its time – the 1990s – and the key focus of solidarity, providing a widely accepted platform amongst farmers and peasants against the adversity provided by big business. However, there have been a number of significant changes both in the food system and responses to its shortcomings. The food system has shown increasing corporatisation and industrialisation, particularly of agricultural systems. Social movements promoting change in food systems are increasingly engaging urban consumers as well as producers across the globe. In response, food sovereignty now encompasses a vast array of initiatives and ideologies, and one of the drivers behind the energetic debate in the corridors of academia, as well as activists, is to address this ambiguity, which is both a strength and a core weakness. McMichael (2014) analysis of the evolution of food sovereignty incorporates much of the preceding in detail. De Schutter (2015) goes a little further and argues that we are actually in a second generation of food sovereignty characterised by five key attributes:

- Bridge building between local producers and consumers (localisation);
- Food democracy, through active citizenship;
- Focus on rebuilding social links;
- Resiliency rather than efficiency in food and social systems (although it must be noted that agroecological efficiency, rather than efficiency as a purely economic construct, is a desirable contributor to resiliency – see Chapter 16); and
- The intertwining of food sovereignty with agroecology (see Chapters 10, 17 and 24).

Whether De Schutter's analysis truly captures the diversity of debate and opinion, there has been a fairly dramatic level of critical engagement with food sovereignty as a concept, as a social movement, and as a process. This forms the basis of what should perhaps be a second generation of food sovereignty, although the form and nature of this is still evolving, both in response to the challenges of the task at hand, and to a more open debate about the possibilities and limitations of the previous thinking. What is clear is that there is no one-size-fits-all approach, no absolute consensus, and there are clearly embodied contradictions and ideologies that may never be reconciled. The limited application of food sovereignty in national policy does give some sense of the current potential, though. In terms of engaging public authorities with food sovereignty, the lack of conceptual clarity is a significant barrier to progressive discussion of and embedding of food sovereignty into policy (Hospes, 2014). Although the core premise is sound and as relevant as ever, the operationalisation of food sovereignty is a challenge.

As previously noted, there are a small number of countries that have started to redesign policy frameworks based around food sovereignty, despite the lack of consensus on what it is (OHCHR, 2010). The essence of the policy frameworks, as embodied through a handful of countries (Bolivia, the Dominican Republic, Ecuador, Mali, Nepal, Nicaragua, Senegal, and Venezuela according to Alonso-Fradejas et al., 2015), is the reclamation of local governance over food and productive resources (land, water, seeds). Food sovereignty policies in the regions just noted address privatisation of natural resources, the industrialisation of food production, and wealth accumulation embodied in a commoditised food system (Araujo, 2010). As part of this, Araujo identifies as the focus of food sovereignty policies which seek to rebuild food reserves, to strengthen small-scale agriculture, and to protect food staples from market speculation (and protect resource rights particularly of indigenous populations). So far so good – however, as Araujo (2010: 500) notes, despite the presence of some good policy initiatives there is still a fundamental issue in that "faced with the hardship of chronic poverty and hunger, the majority of nations have been unable to cut ties with trade-orientated development".

Agarwal (2014) raises a related issue – the notion of self-sufficiency. Initial food sovereignty discourse focused on national food self-sufficiency and diversity in agricultural production of livestock and crop species as a counter to monocropping. However this focus shifted to the rights of people, and a resulting shift to notions of local self-sufficiency, a truly debatable concept (Delind, 2011). Food sovereignty started with a very clear narrative of smallholders and local production systems, yet, the challenge of food sovereignty was a challenge directly to global food supply chains through confronting industrialisation of agricultural production and the associated global movement of food and supply chains. An example of this would be corporate control of seed and genetic resources versus local seed banks. This paradox is still heavily embedded in most discussions. The primacy of family farming in the 2007 Nyéléni definition discussed previously does little to address embodied inequities, particularly gender, where women's unpaid labour and household gender inequality provide a counter to a somewhat rose-tinted image of the family farm (Agarwal, 2014; Park et al., 2015). Similarly, the focus on family farming does little to address a potential disconnect between all producers (commercial, large-scale, small-scale, and so on) and even the landless (Agarwal, 2014; Patel, 2009).

Some of these previously raised challenges and contradictions demonstrate the range of critical discourse, but there are many, many more areas of critique. The breadth of the concept, and the scale of the movement(s) engaged in food sovereignty theory and practice leave plenty of room for analysis. However, as Shattuck et al. (2015) note, a relational approach to food sovereignty should focus on power more than food. In the context of this volume, therefore, there are a range of questions around the relationship between food sovereignty and landscapes.

Food sovereignty, land, and landscapes

The ecological basis of food sovereignty is again a complex area of debate given the wide variety of alternatives to industrialised, large-scale, and predominantly monocropped agriculture. However, as previously noted, the intertwining of food sovereignty, agroecology, and human rights demonstrates an essential issue – the ecological sustainability of agricultural production systems (Chapter 16). This section deals predominantly with productive and/or rural landscapes, whilst acknowledging the growing focus in food sovereignty on urban dwellers and consumers.

The scale at which food sovereignty is discussed is often vague at best. It defends small-scale sustainable agriculture as a way to promote social justice and dignity. It strongly opposes corporate-driven agriculture and transnational companies. The embedding of a 'local' narrative as opposition to the corporate food regime is again compelling, but issues of scale are problematic (Iles and Montenegro de Wit, 2015). However there is little explicit discussion that engages with landscapes and landscape scale policy and practice. McMichael (2015) argues that land use and land user rights are foundational to food sovereignty. Corporate-driven land grabs are one of the most explicit debates. Land grabbing – the large-scale acquisition of land by domestic or foreign corporate entities – involves the flow of financial capital into land resources, which, if the agricultural technologies used are unsustainable in terms of land, soil, or water quality, and particularly if the economic benefit and physical produce is sent offshore, is, in essence, a form of asset stripping.

Landscapes are undeniably human constructs, with productive, environmental, social, and cultural values. However, the economic value of land resources places the design of the landscape firmly in the hands of those who control the land. In the market-based, industrialised, large-scale food systems that form the basis of the corporate regime, power over landscape scale decisions rests with corporate entities. Landscapes reflect the values of the people who shape them. Unpacking conceptualisations of development demonstrates the urban-centric nature of development. There is an embedded discourse in international development literature where rural poverty is perceived as the absence of development (McMichael and Schneider, 2011; Chapter 21). This perspective sees agriculture as the engine of economic growth, and therefore as development. This provides a sharp contrast to the alternate (if we polarise the debate somewhat) view that sees agriculture as the foundation of social and ecological sustainability, with the economic component a contributor to rural livelihoods and subsistence – in other words, food sovereignty. Depopulation of rural areas, through land use intensification and industrialisation, and the resulting employment of rural dwellers in the urban industrial complex is a desired outcome in the first scenario (see Chapter 9). Economic efficiency, and the capacity to increasingly engage with markets through value chain development would be the overarching goal. However, the counterargument has much more resonance with a diverse, productive landscape encompassing social, environmental, cultural, and economic aspects. Some middle ground may be found in the agroecological debate (see for example United Nations, 2014; Chapter 17).

Access to and control over land has been the focus of debate within food sovereignty. As Edelman et al. (2014) identify, production is based on three intersections and essential aspects, namely democratic land control, ecologically sustainable practices, and food securing outcomes. Food sovereignty has potential to expand each of these as a coherent whole, however, this is not a common scenario, and some combination of the three aspects (or none) are more frequent. Land tenure and land redistributive land reform are widely accepted as central to the debate, with a specific requirement to address access to and control over land. Borras and Franco (2012a, 2012b) as well as McMichael (2015) explicitly contribute to the land debate.

Land grabbing can be both direct or less obvious and indirect. Direct land grabbing is seen in foreign or external investment in industrialised agriculture through land acquisition and production models. A less obvious aspect is the renewed focus by both the private sector and development institutions on linking farmers to markets – which McMichael (2015) argues is a further assault on land and land use practices of smallholders, through 'control grabbing' (McMichael, 2015: 438) of production decisions. Land grabbing helps refocus food sovereignty on producer rights, rather than just national food sovereignty (McMichael, 2015). These aspects are really the only explicit focus on land outside of a farming unit.

The link between food sovereignty discourse and landscape form and function is more of an implicit nature. A multifunctional, diverse, risk averse, sustainable, productive landscape that retains, rebuilds, or evolves environmental, social, and cultural values emerges from discourses on family farms, agroecology, localisation, and ethics (Chapters 10, 16 and 24). The evolution of agroecology to incorporate both geographical and system-wide aspects of food production (Edelman et al., 2014; Chapter 17) provides a positive move in relation to the food sovereignty/landscape relationship. The engagement of food sovereignty with both the means and mode of production is vital to attaining or regaining control over landscapes by producers and consumers. How this is achieved is, like most aspects of food sovereignty, contested.

Land sovereignty is a recently emerging discourse that specifically focuses food sovereignty thinking to land (Borras and Franco, 2012a). Land sovereignty is the "right of working peoples to have effective access to, use of, and control over land and the benefits of its use and occupation, where land is understood as resource, territory, and landscape" (Borras and Franco, 2012b: 1). Land sovereignty contributes to actions on property rights, mainly through land tenure and titling, with secure titling providing a logical means to protect vulnerable communities and individuals (Edelman et al., 2104). However, even this area is open to critique, as titling can both reinforce existing inequalities including patriarchal structures and provide a legal means by which to alienate, expropriate, or foreclose land and productive resources. Relationships between individual, communal, collective, or cooperative ownership arrangements, open access, or public property need to be contextualised within a food sovereignty discourse by forms of collective organisation, cultural relevance, economic viability, and ecological sustainability. Yet again, there appears a requirement for multiple and potentially contradictory visions.

The integration of discourse on land access and control is necessary but not sufficient to engage in landscape discussion. As previously noted, asserting control over land use and productive decisions requires both access (or control) of land, but also access to and control over land use, and resulting engagement with production and trade. Factors that contribute to landscape scale decision-making are currently land use policy and the value and ideals that shape those who make decisions over land use (landholders). There is no central body in a food sovereignty vision. Control is decentralised, and there are potentially competing sovereignties (Borras et al., 2015) with state and also non-state entities, including corporations and local communities, as sovereigns. This brings the debate back at least at some level to notions of scale. From soil health through healthy landscapes, there is a requirement for different levels of fora to engage with diverse issues, particularly at a transnational or watershed scale. This may require local, national, regional, continental, international forms of communication to ensure cooperation. Moves within food sovereignty towards land sovereignty or democratic land governance (Borras et al., 2015: 610) provide ample space for discussions of landscape, or potentially even landscape sovereignty. Landscape sovereignty is another potential expansion of the concept that may promote not just agrarian citizenship and the rights of people to determine their own food systems, but the right of people to determine landscapes. This is a debate which has not yet occurred explicitly, but a focus on landscapes in a dramatically changing political, economic, and environmental climate provides a scale of engagement that is vital.

Conclusion

Food sovereignty is an enormously broad idea, with extensive appeal as an alternative to increasingly globalised, commoditised agricultural production and consumption models that reproduce inequality and food insecurity. Food sovereignty is an indisputably political call for action, and a platform to address inequalities of power through social change. It is exceptionally widely supported by social movements across the globe.

Sovereignty can be, and is, critiqued and disputed. There are a range of inbuilt contradictions, or points where divergent views are entirely problematic. Social relations within the home that may reinforce patriarchy, between landholders and the landless, between the urban and rural, rich and poor, black and white, individuals, communities, and corporations, are all sites of potential power imbalances. Central are the links between local and global, and competing discourses on localisation and the championing of local food systems, versus the radical political change to state policy, international institutions, and the corporate control of food systems.

Food sovereignty is a process, not a fixed principle. From a landscape perspective, food sovereignty draws on agroecological perspectives to alter the human/nature relationship. Yet whether it truly is a counter-hegemonic discourse that can address the fundamental embeddedness of the capitalist economy, and embracing political structures, remains to be seen. Reflection on issues of scale are fundamental to the debate, from the micro (soil health) to the macro landscape scale. The integration of a landscape perspective in food sovereignty discourse is implicit, but is easily simplified to land reform and democratic land governance, or land sovereignty. Unless there is a clear engagement with a landscape scale discourse, it is likely that the framing of current and future landscapes will be by default rather than by design. Given the vast impact that food systems have on the Earth's landmass, it seems that food sovereignty may provide a means of engaging ecological process with social justice, citizenship, culture, and heritage in a food-securing, ecologically sound food system with democratic land and landscape control. It is, however, a work in progress.

Further reading

Agarwal, B. (2014) Food sovereignty, food security and democratic choice: Critical contradictions, difficult conciliations. *The Journal of Peasant Studies*, 41(6), 1247–1268. Available from: DOI: 10.1080/03066150.2013.876996.

Akram-Lodhi, H.A. (2013) *Hungry for change: Farmers, food justice and the agrarian question*. United States: Kumarian Press.

Akram-Lodhi, A.H. (2015) Accelerating towards food sovereignty. *Third World Quarterly*, 36(3).

Alonso Fradejas, A., Borras, S.M., Holmes, T., Holt-Giménez, E., & Robbins, M.J. (2015) Food sovereignty: Convergence and contradictions, conditions and challenges. *Third World Quarterly*, 36(3), 431–448. doi: 10.1080/01436597.2015.1023567.

Altenburg, T. (2007) Donor approaches to supporting pro-poor value chains*; report prepared for the donor committee for enterprise development working group on linkages and value chains*. www.eda.admin.ch/content/dam/deza/en/documents/themen/privatsektorentwicklung/162916-donor-appoaches-pro-poor-value-chains_EN.pdf (accessed 29 July 2016).

Araújo, S. (2010) The promise and challenges of food sovereignty policies in Latin America. *Yale Human Rights and Development Journal*, 13(2, Art 6), 493–506.

Beddington, J. (2010) Food security: Contributions from science to a new and greener revolution. *Philosophical Transactions of the Royal Society*, 365(2010), 61–71.

Bernstein, H. (2013) Food sovereignty: A sceptical view conference paper for discussion at: Food Sovereignty: A Critical Dialogue International Conference September 14–15, 2013.

Bernstein, H. (2014) Food sovereignty via the 'peasant way': A sceptical view. *Journal of Peasant Studies*, 41(6), 1031–1063. Available from: http://dx.doi.org/10.1080/03066150.2013.852082.

Borras, Jr., S.M., & Franco, J.C. (2012a) Global land grabbing and trajectories of agrarian change: A preliminary analysis. *Journal of Agrarian Change*, 12(1), 34–59. Available from: DOI: 10.1111/j.1471-0366.2011.00339.x.

Borras JR, S.M., & Franco, J.C. (2012b) A 'Land Sovereignty' alternative? Towards a peoples' counter-enclosure, TNI Agrarian Justice Programme Discussion Paper www.tni.org/files/a_land_sovereignty_alternative_.pdf.

Borras, S.M., Franco, J.C., & Suárez, S.M. (2015) Land and food sovereignty. *Third World Quarterly*, 36(3), 600–617. Available from: DOI: 10.1080/01436597.2015.1029225.

Charlton, K.E. (2016) Food security, food systems and food sovereignty in the 21st century: A new paradigm required to meet sustainable development goals. *Nutrition and Dietetics*, 73(1), 3–12. Available from: DOI: 10.1111/1747–0080.12264.

De Schutter, O. (2015) *Food democracy south and north: From food sovereignty to transition initiatives*. www.opendemocracy.net/olivier-de-schutter/food-democracy-south-and-north-from-food-sovereignty-to-transition-initiatives (accessed 16 April 2016).

Declaration of Nyéléni (2007) Declaration of Nyéléni, declaration of the forum for food sovereignty, Nyéléni Village, Sélingué, Mali, February, 27, 2007. https://nyéléni.org/spip.php?article290.

Delgado, C. (1999) Sources of growth in smallholder agriculture in SubSaharan Africa: the role of vertical integration of smallholders with processors and marketers of high-value items. Agrekon (special issue – May): 165–189.

DeLind, L. (2011) Are local food and the local food movement taking us where we want to go? Or are we hitching our wagons to the wrong stars? *Agriculture and Human Values*, 28, 273–283. Available from: DOI: 10.1007/s10460-010-9263-0.

Edelman, M., Weis, T., Baviskar, A., Borras, Jr, S.M., Holt-Giménez, E., Kandiyoti, D., & Wolford, W. (2014) Introduction: critical perspectives on food sovereignty. *The Journal of Peasant Studies*, 41(6), 911–931.

FAO (2001) *The state of food insecurity in the world*. Rome: Food and Agriculture Organization of the United Nations.

FAO (2017) *Family farming knowledge platform*. Rome: Food and Agriculture Organization of the United Nations, www.fao.org/family-farming/en/ (accessed 16 April 2016).

FAO (2017a) *The right to food knowledge platform*. Rome: Food and Agriculture Organization of the United Nations. www.fao.org/righttofood/right-to-food-home/en/ (accessed 16 April 2016).

Friedmann, H., & McMichael, P. (1989) Agriculture and the state system: The rise and decline of national agricultures, 1870 to the present. *Sociologia Ruralis*, 29(2), 93–117. Available from: DOI: 10.1111/j.1467–9523.1989.tb00360.x.

Globalizations (2015) Food sovereignty: Concept, practice and social movements, Special Edition, 12(4). www.tandfonline.com/toc/rglo20/12/4?nav=tocList.

Holt-Giménez, E. (2015) Racism and capitalism: Dual challenges for the food movement. *Journal of Agriculture, Food Systems, and Community Development*, 23–25. Available from: DOI: 10.5304/jafscd.2015.052.014.

Holt-Giménez, E., & Shattuck, A. (2011) Food crises, food regimes and food movements: Rumblings of reform or tides of transformation? *Journal of Peasant Studies*, 38(1), 109–144. Available from: DOI: 10.1080/03066150.2010.538578.

Hopma, J., & M. Woods (2014) Political geographies of 'Food Security' and 'Food Sovereignty'. *Geography Compass*, 8(11), 773–784.

Hospes, O. (2014) Food sovereignty: The debate, the deadlock, and a suggested detour. *Agriculture and Human Values* (2014) 31: 119. doi:10.1007/s10460-013-9449-3.

Iles, A., & Montenegro de Wit, M. (2015) Sovereignty at what scale? An inquiry into multiple dimensions of food sovereignty, *Globalizations*, 12(4), 481–497. Available from: DOI: 10.1080/14747731.2014.957587.

Jansen, K. (2015) The debate on food sovereignty theory: Agrarian capitalism, dispossession and agroecology, *The Journal of Peasant Studies*, 42(1), 213–232. Available from: DOI: 10.1080/03066150.2014.945166.

Journal of Peasant Studies. (2014) Global Agrarian transformations Volume 2: Critical perspectives on food sovereignty, 41(6).

La Via Campesina (nd) *The international peasant's voice*. http://viacampesina.org/en/index.php/organisation-main-menu-44 (accessed 9 March 2016).

La Via Campesina (1996) *The right to produce and access to land*. 11–17 November, 1996, Rome, Italy. Available from: http://www.acordinternational.org/silo/files/decfoodsov1996.pdf (accessed 1 November, 2017).

Lang, T., & Heasman, M. (2004) *Food wars: The global battle for mouths, minds and markets*. London: Earthscan Publications.

McIntyre, B.D., Herren, H.R., Wakhungu, J., & Watson, R.T. (eds.) (2009) *International assessment of agricultural knowledge, science and technology for development.* Washington, DC: Island Press.
McMichael, P. (2009) A food regime genealogy. *The Journal of Peasant Studies*, 36(1), 139–169.
McMichael, P. (2014) Historicizing food sovereignty. *The Journal of Peasant Studies*, 41(6), 933–957. Available from: DOI: 10.1080/03066150.2013.876999.
McMichael, P. (2015) The land question in the food sovereignty project. *Globalizations*, 12(4), 434–451. Available from: DOI: 10.1080/14747731.2014.971615.
McMichael, P., & Schneider, M. (2011) Food security politics and the millennium development goals. *Third World Quarterly*, 32(1), 119–139. Available from: DOI: 10.1080/01436597.2011.543818.
Nestle, M. (2007) *Food politics: How the food industry influences nutrition and health.* Berkeley, CA: University of California Press.
OHCHR (2010) *The right to adequate food, Fact Sheet No. 34.* www.ohchr.org/Documents/Publications/FactSheet34en.pdf (accessed 14 April 2016).
OHCHR (2016) *Sustainable development goals and human rights.* www.ohchr.org/EN/Issues/MDG/Pages/The2030Agenda.aspx (accessed 9 March 2016).
Park, C.M.Y., & Ben White & Julia (2015) We are not all the same: Taking gender seriously in food sovereignty discourse. *Third World Quarterly*, 36(3), 584–599. Available from: DOI: 10.1080/01436597.2015.1002988.
Patel, R. (2008) *Stuffed and starved: The hidden battle for the world food system*, 4th ed. United States: Melville House Publishing.
Patel, R. (2009) Food sovereignty. *Journal of Peasant Studies*, 36(3), 663–706.
Peters, P.E. (2009) Challenges in land tenure and land reform in Africa: Anthropological contributions. *World Development*, 37(8), 1317–1325. Available from: DOI: 10.1016/j.worlddev.2008.08.021.
Poole, N.D., Chitundu, M., & Msoni, R. (2013) Commercialisation: A meta-approach for agricultural development among smallholder farmers in Africa? *Food Policy*, 41, 155–165. Available from: DOI: 10.1016/j.foodpol.2013.05.010.
Sage, C. (2013) The interconnected challenges for food security from a food regimes perspective: Energy, climate and malconsumption. *Journal of Rural Studies*, 29(2013), 71–80.
Schanbacher, W.D. (2010) *The politics of food: The global conflict between food security and food sovereignty.* Westport, CT: Praeger.
Shattuck, A., Schiavoni, C.M., & VanGelder, Z. (2015) Translating the politics of food sovereignty: Digging into contradictions, uncovering new dimensions. *Globalizations*, 12(4), 421–433.
Sexsmith, K., & McMichael, P. (2015) Formulating the SDGs: Reproducing or reimagining state-centered development? *Globalizations*, 12(4), 581–596. Available from: DOI: 10.1080/14747731.2015.1038096.
Third World Quarterly (2015) Food sovereignty: Convergence and contradictions, condition and challenges, Special Edition, 36(3).
United Nations (2014) *Trade and environment review 2013: Wake up before it is too late – make agriculture truly sustainable now for food security in a changing climate.* United States: United Nations.
United Nations (2015) *Goal 2. :. Sustainable development knowledge platform.* Available at: https://sustainabledevelopment.un.org/sdg2 (accessed 9 March 2016)
Wittman, H., Desmarais, A., & Wiebe, N. (eds.) (2010) *Food sovereignty: Reconnecting food, nature and community.* Halifax, Winnipeg: Fernwood Publishing/Oakland: Food First Books/Oxford: Pambazuka Press.
Wolf, S. (2014) *The neoliberal regime in the agri-food sector: Crisis, resilience and restructuring.* New York: Routledge.

32

Landscape and the politics of food justice

Megan Blake

Introduction

Understanding how landscapes have evolved involves recognising the ways in which a landscape becomes the product of social, political, cultural, and material processes. As part of a landscape that is both urban and rural, food systems are not just shaped by the local physical and cultural conditions of a place; they are also shaped by the ways that social, political, and economic processes are enacted through that place. These intersections produce enabling and disabling conditions for those who participate in the practices that are intimately linked to that place. Schatzki (2010), for example, has discussed the relationship between social practices and place, and Massey (2010) has considered constructions of place, in the context of food systems.

Food is integral to the production of places, in that it does not just come from a place but helps to create places symbolically (Cook and Crang, 1996; also see Chapters 35 and 36). As such, food is enfolded into the lives of all people because we are all eaters. Food landscapes, or foodscapes, like the more general landscape, can be understood as a bricolage of smaller scale places and spaces and the materiality that composes those places and spaces (Dolphijn, 2004; Katz, 2008). These landscapes are designed and curated to serve not just the symbolic representation of a particular location but also the material practices of those who have the power to design and curate them in the first instance (Katz, 2008; Zukin, 1996). Further, as with all landscapes, foodscapes are ever-changing as a result of the processes that form and reform them. In the words of Katz (2008: 16), "every landscape can reveal sedimented and contentious histories of occupation; struggles over land use and clashes over meaning, rights of occupancy, and rights to resources." Finally, as with any landscape view, the foodscape view reveals or hides those ecologies that encompass all points from farm to fork and belly and beyond (Johnson et al., 2009). These processes that can contribute to producing landscapes of food injustice or what we might call *unjust foodscapes*.

The political responses to the inequalities in our food system that produce unjust foodscapes target areas of contention and conflict over who can wield the power over the production, control, regulation, inspection, distribution, and consumption of food; over what can become food and where and how that food becomes located in space; and for what purpose that power is wielded. *Food politics* broadly involves material interests and social values, including values that encompass human relations, human-environment relations, and human-animal relations (Chapter 24). Importantly, food justice, in its focus primarily on eaters, local social activism, and institutionalised disadvantage, is just one response to the alienation, hunger, and want that are endemic to our increasingly marketised and commodified food system. There is not as yet, however, a single counter-politics movement but rather a number of acknowledged partial alternatives that may broadly be categorised as *food sovereignty*, *right to food*, *food democracy*, and *food justice*.

- *Food sovereignty* is concerned primarily with reconfiguring how food is grown and produced with the aim of handing the power over the ownership of land and other inputs (seeds, for example) – and the decisions about what and how much is produced and marketed – back to those who do the work of producing food in the places where they live (Rosset, 2003; Chapter 31).
- *Right to food* approaches argue that food is a human right before it is a commodity and locates its power within legal frameworks that seek to enable those whose rights are obstructed and seeks avenues through which to hold governments accountable in order to provide avenues that give redress to violations of that right (De Schutter, 2014; Mechlem, 2004).
- *Food democracy* advocates for a commitment to addressing conflicts that arise out of the clashes of values that present themselves within the food system through democratic participation, negotiation, and compromise (Hassanein, 2003, 2008; Lang, 2003).
- *Food justice* recognises that everyone eats and that in our current food system some groups are able to access and afford sufficient and safe food in ways that recognise their cultural practices and preferences, while other groups are not (Gottlieb and Joshi, 2010). Food justice seeks to redress these imbalances by first making visible food injustices as they occur and are produced in place, and second through community-based activism and interventions aimed at overcoming institutionalised inequalities (Alkon, 2014a; Alkon and Agyeman, 2011; Cadieux and Slocum, 2015; Gottlieb and Joshi, 2010; Levkoe, 2011).

Of these, food justice is the more organic and bottom-up of the food inequality movements. Food justice has no overarching or coordinating organisation as is the case with food sovereignty's La Via Campesina, or single focus of responsibility for change as is the case with right to food, or a central decision-making mechanism as is the case with food democracy. Instead, food justice activism has been distinct in its focus on programmatic and bottom-up interventions that occur in place, though like the other movements there are efforts at wider consciousness-raising concerning systematic inequalities (see also Cadieux and Slocum, 2015).

The framing of food justice politics speaks directly to the ways in which landscapes of inequality and struggle are produced and reproduced within and through the institutionalised and ordinary spaces of everyday life but also shapes how inequality can be interrogated and challenged. The chapter first turns to an exploration of unjust foodscapes. It starts by examining the food desert concept (also see Chapter 26) and the interventions that have sought to redress this problem to illustrate why not all actions enact food justice and the dynamic nature of unjust foodscapes. The chapter then turns to other landscape processes, both physical and symbolic that contribute toward the materialisation of injustice to suggest that there are a multitude of landscape spaces that warrant examinations and interventions that seek to increase justice and equality in the foodscape. Following this discussion, the chapter then examines food justice as a social moment by considering in more detail how food justice fits within the wider field of food inequalities political movements. Food justice is then examined internally to identify areas where this relatively new, emerging, and as yet highly dispersed social movement contains some inherent contradictions and ambiguities.

Changing landscapes – food (in)justice

The links between social justice and landscapes of food access are at the heart of the food justice movement and a key component of this access is how food is made available in ways that are affordable, at a quality that is both safe and healthy, and in a manner that speaks to the cultural foodways of those who live in these environments (Chapters 20, 22 and 34). Food justice has arisen as response to and against those in agri-food science and industry who summon the argument that with growing population we need new food technologies and further industrialisation and its attendant concentrations of power to justify the ratcheting up of capitalism's control over our food system (see for example Agyeman and McEntee, 2014; Foley et al., 2011; Chapters 16 and 17). The idea of an unjust foodscape is implied in the food justice literature but has not been typically labelled as such (but see Blake, 2017). An *unjust foodscape* provides a landscape view that reveals not plenty but want, by looking for what is seen as well as what is absent (after Sobal and Wansink, 2007). Importantly, unjust foodscapes must be interrogated in the ways that they reveal or hide what is or is not encountered – and by whom and how the encountering is or is not enabled (after Dolphijn, 2004; see also Wenzer, 2015) – as well as who benefits or does not benefit from that encounter (Blake, 2017). The food desert concept provides the most fully researched example of how unjust foodscapes are produced and then reproduced differently through political-economic processes, discussed later. This discussion is followed by a consideration of further ways that landscape processes may also be considered through this justice lens, which can give rise to further domains for food justice research and activism.

Food deserts as unjust foodscapes

Concerns regarding access to food that is healthy and affordable in urban contexts continue to inspire an abundance of scholarly research seeking to understand and explain how, in lands of plenty, there can still be sub-populations who do not have access to food. In response to this, an extensive body of literature has emerged that identifies areas characterised by high economic deprivation, social exclusion, and low access to healthy and affordable food as *food deserts* (see for example Beaulac et al., 2009; Cummins, 2014; Larsen and Gilliland, 2008; Wrigley et al., 2003; Walker et al., 2010). The US Department of Agriculture (USDA, n.d.) has adopted a definition of a 'food desert' as a place where there is not "ready access to fresh, healthy, affordable food". The bulk of food desert research is quantitative in its methodological approach. The work typically focuses on the US and UK contexts (but see Battersby, 2012 for a discussion in relation to South Africa). The research argues that lack of access occurs because of a combination of urban development that has created an absence of supermarkets in inner-city spaces and transport linkages that render it difficult for low-income residents to access those suburban spaces where healthy food is identified as plentiful (Bedore, 2010; Shannon, 2016). To identify food deserts the USDA (n.d.), for example, produced a map of census tracts where at least a third of the population lives beyond a mile from a major supermarket as a tool for supporting food access (Shannon, 2016: 187; see also Donald, 2013). The proposed solution pursued most frequently by local and national state actors is to create a policy to encourage the re-population of these urban landscape spaces with supermarkets to provide better access. Examples of such policies are tax incentives and financial support to supermarket chains to encourage them to locate within what are considered low commercial value neighbourhoods (Donald, 2013; Pothukuchi, 2005; Shannon, 2016).

While the research on food deserts could be understood as showing concern for low-income consumers, there is a strong and multi-pronged argument that the approaches typically taken to ameliorate food deserts are not in tune with establishing greater food justice. There are a number of salient critiques as follows. Firstly, study and policy that focuses on supermarket access as measured by their presence or absence do little to interrogate the social inequalities that give rise to a lack of access to food in the first instance (Donald, 2013; Shannon, 2016). Secondly, such policy solutions are neoliberal in that they address a complex social problem by deferring to the market (Blake, 2017; Guthman, 2011; Wolf, 2014). Thirdly, the tax incentives and other financial supports amount to welfare for wealthy corporations and thus divert needed resources away from those who are most vulnerable within society (Donald, 2013). Fourthly, the presence of a supermarket in a neighbourhood does not necessarily ensure access to healthy, affordable, and culturally appropriate food (Drewnowski, 2009). Fifthly, the support of already wealthy private enterprises does nothing toward enabling communities to empower themselves (McClintock, 2011; Shannon, 2016). Sixthly, supermarkets can further exacerbate the problem of food injustice and insecurity in neighbourhoods by pushing out the smaller scale and independent stores who may already be providing access to healthy and affordable food (Donald, 2013; McClintock, 2011). Seventhly, the threat that large-scale supermarkets pose to the economic viability of smaller independent stores threatens household livelihood strategies for those groups who tend to concentrate in this occupational sector (Donald, 2013). Eighthly, the focus on American and British cities has failed to adequately theorise the food desert in relation to cities within other contexts that have different political economies and where urban residents have differently configured everyday lives (Battersby, 2012). Finally, food desert policy is a data-driven solution that hides as much as it reveals because not all supermarkets offer high-quality food at price points that are acceptable for those living in low-income families who may or may not live in low-income communities (Shannon, 2016). What these critiques also reveal is how a solution to a problem of absence within primarily urban landscapes may reshape these spaces without actually solving the problem in a way that is just or fair. Moreover, by populating the urban contexts with large supermarkets, the presence of an unjust foodscape is obscured by this new state-sponsored urban landscape of corporate-controlled food, which is also unjust but in different ways, compared to expressions of absence (Chapters 16 and 31).

While these critiques are damning of state-sponsored food desert solutions that operate under neoliberal principles that favour large corporations, the term is useful for revealing how unjust foodscapes are formed. McClintock (2011), for example, examines food deserts in the context of Oakland, California, highlighting the ways in which historical material processes within the city led to the capital devaluation of some neighbourhoods and coupled with racist redlining practices produce a racialised and unjust urban landscape. The subsequent trapping of a population of people of colour in this neighbourhood and the abandonment by major food retailers who had in an earlier period pushed out smaller, independent groceries has left the area underserved by food outlets that provide fresh and healthy food. Importantly, these abandoned areas also remained populated by low-quality, low-nutrition food outlets, areas which British public health policy circles now refer to as 'food swamps'. McClintock's (2011) account of Oakland's flatlands illustrates the ways in which political-economic landscape processes produce inequality (Chapters 25 and 31).

As a response to dispossession by abandonment, in some of these communities, residents have sought to create their own solutions by repopulating these urban spaces with community gardens, social eating activities, cooperative grocery stores, and the like. McClintock (2011), for example, points to bottom-up, community-based interventions that aim to create local opportunities such as people's supermarkets and community gardens that have begun to emerge in Oakland. While food justice activity in Oakland has been well researched (for example, in Alkon and Nortgard, 2009; McClintock, 2011; Sbicca, 2012), there is considerable evidence of similar activity occurring in other locations. For example. McKinney and Kato (2017) identify activity in New Orleans that seeks to support those living in food deserts. Likewise, White (2011) reveals the ways in which the Detroit Black Community Food Security Network has sought to work through food to introduce transformative resilience in communities shaped by racist policy and the abandonment of capital. Further examples can be found in the chapters in Alkon and Agyeman (2011), and several other examples are discussed in Cadieux and Slocum (2015). The chapter now turns to consider other spaces where the lens of the unjust foodscape could usefully be applied.

Locating other unjust foodscapes

The discussion of food deserts illustrates how political-economies produce and reproduce unjust foodscapes through the ways that they are designed, regulated, built, and abandoned. As with all landscapes, foodscapes are also produced through a myriad of means. In this section, the chapter briefly identifies further landscape processes that illustrate the ways in which unjust foodscapes can emerge. Specifically, the section focuses on environmental processes and presences or absences that create landscapes and then turns to symbolic and representational landscape production. Considering the ways these landscape formations are also unjust opens up possibilities for further research and intervention that seek to give greater access to food in all its domains to those who lack.

Work on environmental justice has shown how those who are the least powerful in social systems are also more likely to be exposed to those places where natural and environmental hazards will have the greatest impact (Fothergill and Peek, 2004; Gottlieb and Joshi, 2010; Kahn, 2005; Chapters 20–23). The creation of toxic landscapes, for example, has an influence on how specific populations experience their foodscapes. Toxic dumping and pollution also render spaces that might become urban gardens more problematic by limiting the possibilities for growing food or by creating caustic foodscapes (Armar-Klemesu, 2000; Harrison, 2008). Likewise, exposure to frequent flooding makes livelihood practices including food practices more precarious for those who are already living precarious existences (Maantay and Maroko, 2009). Food security interventions such as genetically modified organism (GMO) cropping have created landscapes populated by indebted farmers because of the privatisation of what was one part of the commons (for example seeds; see Shiva, 2004). At the same time, these interventions contribute to the loss of biodiversity in the landscape and produce a homogenised view (Agyeman, 2013; Heynen et al., 2012). While food sovereignty might and does claim some of the political ground related to the control of resources and decision-making in the production of food in these environments, there is overlap with and space for food justice interventions that consider how, for example, community seed banks also support the preservation of cultural culinary and dietary diversity, and enable the foodways of marginal groups (Schlosberg, 2004).

Symbolic acts of landscape production also play a major role in obfuscating the injustice within the food system. An example of this is the production of urban skylines and elevated landscape views often associated with global cities, such as those found in Hong Kong, Dubai, New York, and elsewhere (Sassen, 2001). Here, maid-employer relationships can exist within high-rise accommodation in places such as Singapore, Hong Kong, and Dubai (Yeoh and Huang, 1999; Law, 2001; Inhorn, 2012). These relationships not only limit the ability of the maids to practice their own foodways but also exploit their labour to facilitate the reproduction of dominant domestic cultures that include reproducing foodways not their own (for example being required to cook and eat foods prohibited by religious belief as is the case for some Muslim maids who serve non-Muslim families who eat pork). The building of these skylines has involved neoliberal policy interventions of privatisation leading to the loss of livelihoods and social spaces once catered for by public market spaces (Blake, 2013, 2017; Chapter 29). Similarly, present but obscured within this skyline are the low-waged and insecure restaurant workers who maintain the flourishing restaurant industry that services highly-paid and mainly expatriate communities who work in the global financial sector (Sassen, 2001). These skylines do much to distract and hide from our attention uneven and exploitative social relations that exist behind such urban spectacles and are produced in ways that deny food justice.

Foodscape views are also used on food labels, in advertisements, and to help story food for retailers (Kenner, 2008). Blake et al. (2010), for example, points to the sepia-toned images of white, British farmers in supermarket advertisements that are used to attract white, middle-class shoppers who wish to support locally based farming communities, reduce food miles as a form of environmental sustainability, and preserve place-based heritage (see also Guthman's (2007) critique of local designation food labels as an enactment of neoliberalism, and Chapter 36). The result is an obfuscation of labour practices, environmental destruction, and access issues that limit participation by and recognition of non-socially dominant groups in the food system more generally (Blake et al., 2010; Jackson et al., 2009; Johnston et al., 2009).

Unjust foodscapes are all around us, to such an extent that they have become naturalised as to what the landscapes that they help comprise *do* look like, rather than what they *should* look like. Certainly, expanding research in these areas seeks to identify, document, and redress the production of unjust foodscapes, and take a broad view of the origins of injustice. How this plays out in the constructions of landscape views is an area of important further consideration for food justice scholars and activists. Scholarship on the right to the city may be useful as it develops a deeper interrogation of urban space than has generally been incorporated into urban food justice scholarship thus far (for greater elaboration of the 'right to the city' as social justice see Brenner et al., 2012; Harvey, 2003; Lefebvre, 1996; Marcuse, 2009). Considering who produces these landscapes and for what purpose is integral to this, as is considering ways to support and make visible those that are made invisible in landscape processes. Dixon (2014: 176) provides pointers for how we might develop more nuanced skills of "seeing what is ethically salient about concrete particular details of the world" to "see food justice".

Food justice as food politics

Food justice is both social intervention and political activism. Of critical importance for this discussion is the way that the political is conceptualised because it helps to situate food justice within the wider context of alternative food movements and distinguishes what might be considered as food justice in relation to other forms of food fairness politics. Understanding how food justice is framed also helps to reveal the possibilities and limitations of the food justice approach. This section of the chapter firstly considers food justice and how it relates to other forms of food politics that are concerned with fairness in the food system. The discussion then turns to how endeavours that seek to enact food justice may be limited in their abilities to achieve their stated aims.

Food justice politics

Food justice seeks to empower socially marginal communities to feed themselves in a manner that is fair, healthy, safe, environmentally sensitive, and culturally sympathetic. As such, food justice tends to start with the eaters of food, rather than the producers of food (as opposed to food sovereignty that aims to empower peasants and small-scale farmers in food growing – see Cadieux and Slocum, 2015, for a further review), to address several aims. Firstly, food justice seeks food security by advocating change in the food system to reduce hunger, malnutrition, and diet-related illness that are endemic for certain groups. A second aim is to improve the long-term access to and availability of food that is healthy, high quality, and culturally appropriate for those living in underserved areas. Thirdly, food justice advocates aim to ensure that those who are working in international, national, and local scale food systems are fairly compensated for their effort. These three aims are based on the belief that food systems – as they are currently configured – de-skill eaters and alienate them from the food they eat (Hassanein, 2008; Jaffe and Gertler, 2006). Central to these aims is a politics that ultimately seeks the empowerment of eaters and may pragmatically recognise the need to engage with market-based mechanisms such as programmes that utilise microloans to support small business start-ups (Cadieux and Slocum, 2015) or promote food hubs that provide avenues through which people might support local food production (Blay-Palmer et al., 2013).

The most widely recognised and internationally coordinated bottom-up approach is food sovereignty. Holt-Giménez (2011) argues that food sovereignty encompasses a radical politics that demands a new system not enmeshed in the current global political-economic order while food justice adopts a more progressive stance (though Cadieux and Slocum (2015) argue food justice is also radical in its politics). The food justice stance seeks change first that enables better and fairer getting on (see Cadieux and Slocum, 2015). Other differences between food justice and food sovereignty include food justice's focus on food consumption and how to provide food to eaters, compared to food sovereignty's concern with land tenure and land claims. Chapter 31 in this volume outlines the approaches to food sovereignty in more detail (see also Holt-Giménez, 2011).

While both food justice and food sovereignty advocate handing power over decision-making and resources to marginal groups in order to challenge the dominant food system (albeit in importantly different ways), the right to food movement aims to engage existing political structures in order to change the way food is made available to those who are without food (Pritchard et al., 2016). Right to food does this by advocating for food as a basic and inalienable human right that must be guaranteed by national governments (De Schutter, 2014). The right to food approach, while it seeks a minimum access to safe and nutritious food for all, does not intrinsically seek to enable control over the food system for those who are marginal to that system, although there are some interventions that seek greater progressive reform. Instead, within this approach, problems of food and nutrition security are framed as outcomes of neoliberal food policy frameworks, and solutions advocate for state-led solutions and avenues for redress through international courts (Mechlem, 2004; Pritchard et al., 2016). Right to food interventions, for example, can include state provision or support of emergency food parcels, which, while they are needed by those who are hungry right now, in and of themselves do little to solve food insecurity in the long-term (Loopstra et al., 2015; Riches, 2002). Another approach involves ensuring the presence of minimum food stores (Pritchard et al., 2016). Right to food as a political movement focuses on a return to a state-supported social contract, unlike food justice, which seeks a more transformative form of resilience through primarily grass-roots interventions, a point I will elaborate in the next section.

Similar to the right to food, food democracy is concerned specifically with exposing and challenging the anti-democratic nature of our current food system, while at the same time demanding rights and acknowledging the responsibilities of citizens to participate in decision-making about what and how food enters into that system (Lang and Heasman, 2004). Food democracy has its roots in the environmental sustainability and environmental and social nutrition debates and is less concerned with identity politics compared to food justice and food sovereignty. The primary actors have tended to be, though not exclusively, from backgrounds that have historically had more political purchase in local, national, and global political arenas than those within food justice activism. The food democracy approach advocates for a commitment to what Hassanein (2003) has outlined as a 'strong democracy' (Johnston et al., 2009). Strong democracy considers that conflict is based on a clash of values and that the only way for these conflicts to be overcome is through democratic participation and compromise. Food democracy comes about because of the engagements of active citizens who are well informed about the issues at hand and are also skilled at engaging directly with existing political structures to change them from within. Importantly, while the democratic political process is an important aspect, this does not mean accepting the dominant political ideologies that may be in play, nor does it mean working from within those businesses that currently control the agri-food sector (Lang, 2003; Lang and Heasman, 2004). Food democracy, however, does not have a sufficient consideration for how groups who may be alienated from or suspicious of existing political structures can find a voice within those structures.

What this discussion has illustrated is that the political movements seeking to promote food access and fairness in the food system are diverse. Each approach takes a different point of entry and suggests different mechanisms through which the problems of an unfair food system might be addressed. Importantly, these categorisations are intended to add clarity to the terrain of food politics interventions by suggesting boundaries between and central characteristics of each. In practice, specific interventions may not fit easily into any single domain of food politics and alliances may and have formed between different groups (for example La Via Campesina has representation in the FAO framings around right to food). The chapter now focuses more closely on food justice to expand on how food justice interventions are intended to be more than coping strategies that are directed by and toward institutionally marginalised groups (as opposed to those who experience individual disadvantage or just lack of availability that is experienced by all within a specific context). Food justice strategies are those that aim to enable adapting and transformative resilience in communities that are vulnerable to environmental and political economic shock because of the ways in which they are systematically disadvantaged.

Food justice strategies

Food justice has emerged out of environmental justice activism and is a relatively new entrant into the domain of food fairness political activism (Alkon, 2014; Gottlieb and Joshi, 2013). Gottlieb and Joshi (2013:

ix) state that, as recently as 2007, "food justice was a relatively obscure term" with little clarity concerning how it should be defined and for whom it applies. As such, it remains an area of intervention that is characterised by a desire to address the immediacy of food insecurity that is faced by those who live in communities marginalised by mainstream political economies. However, food justice advocates recognise that to address problems in communities right now, what is needed are strategies that offer more than coping. Cadieux and Slocum (2015) suggest that food justice interventions should exhibit four interlinked characteristics. These characteristics include (1) an acknowledgement and confrontation of historical trauma, (2) a building of communal reliance and control, (3) a drive toward finding creative ways to produce new social ecologies that are outside of the speculative market, and (4) a commitment to pursue labour relations that guarantee livelihood that is not alienating and that recognises a need not to rely on the value of unpaid activity, particularly that of women. Importantly, the mechanisms seen as useful for mobilising around food justice include actions that incorporate "discourses of ownership, empowerment, and control [. . .] and involve the inclusion of business skills, equitable livelihoods and leadership roles in providing food for their communities" (Cadieux and Slocum, 2015: 5; see also Alkon and Agyeman, 2011). Food justice interventions tend to eschew corporate partnerships and are reluctant to engage directly with state mechanisms. As such, the pragmatism of food justice does not translate into advocacy of private sector equity investment and neoliberal governance regimes that continue to enhance the purses of the wealthy (Springer et al., 2016).

Food justice involves a reworking of systems to make them more useful for everyday life and may involve small-scale incremental changes that lead to the removal of constraints on free agency (Dixon, 2014) and the circumventing of inhibiting institutional structures (Smith, 2005). Food justice interventions, therefore, aim for forms of resilience that move beyond coping strategies. DeVerteuil (2016), in his account of the resilience of community service hubs within a context of neoliberalism, draws on Katz (2004) and others to define resilience as the ability to absorb disturbance (also see Chapter 16 for a discussion of resilience in relation to food production). Within the context of food interventions, many of these that may be mistaken for food justice are those that support or provide protective care that enables getting-by, recovery, or survival. These include interventions such as food banks. DeVerteuil goes on to note, however, that such coping activity can offer a pause, a resting place, or a place to be, and as such prevents displacement and erasure. Such a moment for repair can be the first step toward transformation, however, in and of themselves these interventions are not sufficient for moving beyond survival and such activity is not a long-term response that supports the aims of justice, addressing symptoms and not causes.

Keck and Sakdapolrak (2013) point out that coping is just one form of resilience. Two other forms of resilience include 'adapting resilience' and 'transformative resilience'. Adapting resilience enables incremental change and some degree of security. These forms of resilience activity reduce risk by preventing crises before they occur. Here, an example may be a community garden that helps residents gain some control over their urban spaces and some aspect of urban life and returns the land to the public sphere and as such expresses action toward greater urban social justice. An example of such an intervention is run by the Detroit Black Community Food Security Network (DBCFSN, n.d.) in the United States (see also Chapter 30). Adapting resilience tends to rely on resources that are immediately available within communities and as such offer security for communities but do not substantially transform the conditions for those who live within those communities in the longer term, as would be the case for transformative resilience. An example of transformative resilience may be the development of cooperatives that offer opportunities for livelihoods while at the same time offer present and future enhancement of place-based and community opportunities (Alkon, 2014b). It is important to remember, however, that resilience practices are enacted by both the powerful and the powerless; DeVerteuil (2016) notes, for example, that neoliberalism and elitism have demonstrated much resilience.

The bottom-up nature of food justice initiatives and the fragmented nature of food justice as a social movement means that there are some important limitations to the approach. Practical limitations of interventions, for example, can limit the security and sustainability of such programmes. McClintock (2011), for example, warns that without sufficient planning consent many of the interventions have uncertain futures as they become vulnerable to gentrifying forces (see also Guthman, 2008; Slocum, 2006, 2007; Alkon and Norgaard, 2009). Bradley and Herrera (2016) in their analysis of the Community Food Coalition in Central Los Angeles also point to the ways in which such interventions are vulnerable to becoming reinscribed with white, patriarchal systems of power and privilege. This inscription occurs through the ways that those who take leadership roles do not always live in, or share or understand the values of the communities that they seek to serve (Sbicca, 2012). Cadieux and Slocum (2015) also highlight how the term food justice has been adopted by organisations and advocates without a clear set of criteria for how what is advocated and implemented actually produces justice, and for whom.

There are also some disconnections between the practical doing of food justice and the critical aim of food justice to develop a food system that is fair for all. Sbicca (2012: 465) in an analysis of the People's Grocery in Oakland, California, points to the fact that while "critical of capitalist organisation of food production, People's Grocery is still largely reliant on capitalism's economic logic." Food justice with its emphasis on the cultural specificity of foodways (Guthman, 2011; Chapter 34), also means that the interventions have tended to advocate a politics of local food. Local food initiatives have been heavily criticised for having both a lack of engagement with how the short-hand of local food can lack a sufficiently reflexive stance and thereby reinforce existing centres of control (Born and Purcell, 2006), or be defensive and thereby replace one centre of control with another (Johnston et al., 2009). Critiques of local food solutions have also emphasised the lack of a sympathetic investigation into how eaters can become active and empowered citizens in historically produced constrained landscapes (Blake et al., 2010). Agyeman (2013) makes some important points about the need for food justice not to become locked into the local, including the point that an overly local and insular focus is easy for the wider agri-food system to ignore (Magdoff et al., 2000; Hassanein, 2003). As a result, food justice interventions, while supporting those who are hungry right now, can also be criticised for drawing on market logics and not actually changing the wider political-economic conditions that give rise to the problems in the first place. Sbicca (2012: 465) concludes with the statement: "inequality does not disappear with a spatial reordering of food production and consumption."

Much of the US activism within food justice is concerned with providing alternatives to the ways that the current system is institutionally racist. As a result, the majority of the research examining food justice has tended to focus on how food justice interventions seek to counter white privilege within the American food system (for example, Slocum, 2006). Interrogations of the racialisation of the food system have revealed much about the landscapes of food insecurity, but to date have not been sufficiently criticised with regard to the ways that such interventions might also undermine the intention of fairness for all. For example, there is room for further consideration of how might the emergence of particularly racialised and closed communities build bridges with other marginalised groups to secure fairness for all. Likewise, there is scope for greater consideration concerning how the prioritisation of racialised disadvantage undermines or obscures other dimensions of institutionalised and intersecting disadvantage concerning for example gender, sexuality, age, and so forth (Crenshaw, 1991; Nash, 2008). Food justice activism and research must consider how these interventions might be reworked to meet equity demands. Finally, the emphasis on white privilege does not leave space for considering how privilege can be experienced as a disadvantage, if for example you are part of the ethnic majority but are poor, as Blake (2017) illustrates in her discussion of food injustice and de-racialisation in Hong Kong (see also Barraclough, 2009, for a discussion of relational racialisation). Food justice, if it wants to become a globally recognised social movement aiming at achieving food for all, still has much work to do in order to smooth out the contradictions and disconnections that exist between food justice practice and food justice as a social goal.

Conclusion

This chapter has shown that, firstly, food justice is an excellent lens through which to identify what I have called *unjust foodscapes*. Food injustice is in and part of the landscape. Food justice scholarship and activism seek to unpick and undermine those processes of landscape formation that reproduce landscapes of inequality. This is done by making visible the processes and practices that physically and symbolically shape environments with the aim of enabling access to food that does not discriminate by race, class, gender, and other dimensions upon which institutionalised social inequalities are built. The relationships between food (in)justice and landscape production also demonstrate the value of a historical materialist understanding of how social, physical, cultural, political, and economic landscape processes produce racialised (and other) structures of systematic inequality that deny access to and participation in a food system that offers healthy, safe, affordable, and culturally appropriate food. We need, however, further research and methods that interrogate a wider array of landscape processes to understand how approaches to ways of seeing food injustice that is positioned within the lived experiences of those at the sharp end of that injustice. Participatory and narrative approaches offer promising ways to begin. Such accounts not only reveal how unjust foodscapes emerge but also brings into view that which the landscape view often seeks to obscure.

Food justice itself sits within a rapidly growing and complex political landscape. While there are a number of differently configured social movements that aim to address food insecurity in ways that are fair and which seek greater fairness overall, these movements are also overlapping, and internally may be incomplete and contradictory. Importantly, redressing the harms inherent in socially produced unjust food landscapes is not easily achieved and is certainly not achieved by rolling back the clock to a time bathed in sepia tones. As LaDonna Redmond, a food justice activist based in Detroit, said in her 2013 TEDxManhattan talk:

> There has never been, in the history of the United States, a fair or just food system.

Certainly, one would be hard-pressed to find a historical example of a fair or just food system in any location, not just within the USA. This is not to say that history is not important.

The chapter also illustrated that by bringing together an understanding of landscape processes with what food justice would consider as injustice can help with identification of how certain places facilitate or minimise engagements between people over food and between people and their food. These engagements are structured by absence and presence within landscapes. The lens of the unjust foodscape also helps to reveal that not all interventions in the food system that proclaim to support those who are food insecure can easily be enfolded into the concept of food justice. Food banks are one example and introducing multinational supermarket chains into the food landscape is another, due to their failure to move beyond coping resilience and their reliance on dominant political-economic conditions that give rise to the problems in the first instance. Certainly, there are some forms of food justice intervention and practice that aim to support currently marginalised groups to use the resources that they have to hand in order to improve the conditions for those living in their communities and which seek transformative resilience. While these have the potential to be internally contradictory, food justice advocates, however, are also pragmatic and progressive in their recognition that people are hungry now, and to pour all effort into larger and more radical structural change may be a case of waiting too long.

Key readings

Alkon, A.H. 2014a. Food justice: An overview. In Albala, K. (ed.), *Routledge international handbook of food studies*. London and New York: Routledge.

Alkon, A.H., and Agyeman, J. (2011) *Cultivating food justice: Race, class, and sustainability*. Boston, MA: MIT Press.

Bedore, M 2010. Just urban food systems: A new direction for food access and urban social justice. *Geography Compass*, 4(9), 1418–1432.

Cadieux, K.V., & Slocum, R. (2015) What does it mean to do food justice? *Journal of political ecology*, 22, 1.

Dixon, B.A. (2014) Learning to see food justice. *Agriculture and Human Values*, 31(2), 175–184.

Gottlieb, R., & Joshi, A. (2010) *Food justice*. Boston, MA: MIT Press.

Guthman, J. (2011) *Weighing in: Obesity, food justice and the limits of capitalism*. Berkeley and Los Angeles, CA: University of California Press.

Heynen, N., Kurtz, H.E., & Trauger, A. (2012) Food justice, hunger and the city. *Geography Compass*, 6(5), 304–311.

Levkoe, C.Z. (2006) Learning democracy through food justice movements. *Agriculture and Human Values*, 23(1), 89–98.

References

Agyeman, J. (2013) *Introducing just sustainabilities: Policy, planning, and practice*. London: Zed Books.

Agyeman, J., & McEntee, J., 2014. Moving the field of food justice forward through the lens of urban political ecology. *Geography Compass*, 8(3), 211–220.

Alkon, A.H. (2014b) Food justice and the challenge to neoliberalism. *Gastronomica: The Journal of Critical Food Studies*, 14(2), 27–40.

Alkon, A.H., & Norgaard, K.M. (2009) Breaking the food chains: An investigation of food justice activism. *Sociological Inquiry*, 79(3), 289–305.

Armar-Klemesu, M. (2000) Urban agriculture and food security, nutrition and health. In: *Growing cities, growing food. Urban agriculture on the policy agenda*, pp. 99–118. Available at http://www.futuredirections.org.au/

Barraclough, L.R. (2009) South Central farmers and Shadow Hills homeowners: Land use policy and relational racialization in Los Angeles. *The Professional Geographer*, 61(2), 164–168.

Battersby, J. (2012) Beyond the food desert: Finding wasy to speak about urban food security in South Africa. *Geographaska Annaler: Series B, Human Geography*, 94(2),141–159.

Beaulac, J., Kristjansson, E., & Cummins, S., 2009. A systematic review of food deserts, 1966–2007. *Prev Chronic Dis*, *6*(3), p.A105.

Blake, M.K. (2013) Ordinary food spaces in a global city: Hong Kong. *Streetnotes*, 21, pp1–12.

Blake, M.K. (2017) Building an unjust foodscape: shifting governance regimes, urban place making and the making of Chinese food as ordinary in Hong Kong. *Local Environment*, pp. 1–16. (e-first edition www.tandfonline.com/doi/abs/10.1080/13549839.2017.1328674).

Blake, M.K., Mellor, J., & Crane, L. (2010) Buying local food: Shopping practices, place, and consumption networks in defining food as "local". *Annals of the Association of American Geographers*, 100(2), 409–426.

Blay-Palmer, A., Landman, K., Knezevic, I., & Hayhurst, R. (2013) Constructing resilient, transformative communities through sustainable "food hubs". *Local Environment*, 18(5), 521–528.

Born, B., & Purcell, M. (2006) Avoiding the local trap scale and food systems in planning research. *Journal of Planning Education and Research*, 26(2), 195–207.

Bradley, K., & Herrera, H. (2016) Decolonizing food justice: Naming, resisting, and researching colonizing forces in the movement. *Antipode*, 48(1), 97–114.

Brenner, N., Marcuse, P., & Mayer, M. (eds.) (2012) *Cities for people, not for profit: Critical urban theory and the right to the city*. London and New York: Routledge.

Cook, I., & Crang, P. (1996) The world on a plate: Culinary culture, displacement and geographical knowledges. *Journal of Material Culture*, 1(2), 131–153.

Crenshaw, K. (1991) Mapping the margins: Intersectionality, identity politics, and violence against women of color. *Stanford Law Review*, 1241–1299.

Cummins, S. (2014) Food deserts. In: *The Wiley-Blackwell Encyclopaedia of Health, Illness, Behavior, and Society*, pp. 562-564. Hoboken, NJ: Wiley-Blackwell.

De Schutter, O. (2014) UN special rapporteur on the right to food. *Report on agroecology and the right to food*. UN, New York.

Detroit Black Community Food Security Network (DBCFSN) (nd) http://detroitblackfoodsecurity.org/ (accessed 5 December 2016).

DeVerteuil, G. (2016) *Resilience in the post-welfare inner city*. Bristol: Policy Press.

Dolphijn, R. (2004) *Foodscapes: Towards a Deleuzian ethics of consumption*. Delft: Eburon Publishers.

Donald, B. (2013) Food retail and access after the crash: Rethinking the food desert problem. *Journal of Economic Geography*, 13(2), 231–237.

Drewnowski, A. (2009). Obesity, diets, and social inequalities. *Nutrition Reviews*, 67(suppl 1), S36–S39.

Foley, J.A., et al. (2011) Solutions for a cultivated planet. *Nature*, 478(7369), 337–342.

Fothergill, A., & Peek, L.A., 2004. Poverty and disasters in the United States: A review of recent sociological findings. *Natural Hazards*, 32(1), 89–110.

Guthman, J., 2007. The Polanyian way? Voluntary food labels as neoliberal governance. *Antipode*, 39(3), 456–478.

Guthman, J. (2008). Bringing good food to others: Investigating the subjects of alternative food practice. *Cultural Geographies*, 15(4), 431–447.

Harrison, J. (2008) Abandoned bodies and spaces of sacrifice: Pesticide drift activism and the contestation of neoliberal environmental politics in California. *Geoforum*, 29(3), 1197–1214.

Harvey, D. (2003) The right to the city. *International Journal of Urban and Regional Research*, 27(4), 939–941.

Hassanein, N. (2003) Practicing food democracy: A pragmatic politics of transformation. *Journal of Rural Studies*, 19(1), 77–86.

Hassanein, N. (2008). Locating food democracy: Theoretical and practical ingredients. *Journal of Hunger & Environmental Nutrition*, 3(2–3), 286–308.

Holt-Giménez, E. (2011) Food security, food justice, or food sovereignty. Crises, food movements, and regime change. In: Alkon, A.H., & Agyeman, J. (eds.), *Cultivating food justice: Race, class, and sustainability*, pp. 309–330. Cambridge, MA: The MIT Press.

Inhorn, M.C. (2012) Reproductive exile in global Dubai: South Asian stories. *Cultural Politics*, 8(2), 283–306.

Jackson, P., Ward, N., & Russell, P., 2009. Moral economies of food and geographies of responsibility. *Transactions of the Institute of British Geographers*, 34(1), 12–24.

Jaffe, J., & Gertler, M. (2006) Victual vicissitudes: Consumer deskilling and the (gendered) transformation of food systems. *Agriculture and human values*, 23(2), 143–162.

Johnston, J., Biro, A., & MacKendrick, N. (2009) Lost in the supermarket: The corporate-organic foodscape and the struggle for food democracy. *Antipode*, 41(3), 509–532.

Kahn, M.E. (2005) The death toll from natural disasters: the role of income, geography, and institutions. *Review of Economics and Statistics*, 87(2), 271–284.

Katz, C. (2004) *Growing up global: Economic restructuring and children's everyday lives*. Minneapolis, MN: University of Minnesota Press.

Katz, C. (2008). Bad elements: Katrina and the scoured landscape of social reproduction. *Gender, Place and Culture*, 15(1), 15–29.

Keck, M., & Sakdapolrak, P. (2013). What is social resilience? Lessons learned and ways forward. *Erdkunde*, 5–19.

Kenner, R. (2008) *Food, Inc.* New York: Participant Media and River Road Entertainment, distributed by Magnolia Pictures.

Lang, T. (2003) Food industrialisation and food power: Implications for food governance. *Development Policy Review*, 21(5–6), 555–568.

Lang, T., & Heasman, M. (2004) *Food wars: The battle for mouths, minds and markets*. London: Earthscan.

Larsen, K., & Gilliland, J. (2008) Mapping the evolution of 'food deserts' in a Canadian city: Supermarket accessibility in London, Ontario, 1961–2005. *International Journal of Health Geographics*, 7(1), 16.

Law, L. (2001) Home cooking: Filipino women and geographies of the senses in Hong Kong. *Ecumene*, 8(3), 264–283.

Lefebvre, H. (1996) The right to the city. In: *Writings on cities*. New York: Wiley, pp. 63–181.

Levkoe, C. Z. (2011). Towards a transformative food politics. *Local Environment, 16*(7), 687-705.

Loopstra, R., Reeves, A., Taylor-Robinson, D., Barr, B., McKee, M., & Stuckler, D. (2015) Austerity, sanctions, and the rise of food banks in the UK. *British Medical Journal*, 350. Available from: www.bmj.com/content/350/bmj.h1775

Maantay, J., & Maroko, A. (2009) Mapping urban risk: Flood hazards, race, & environmental justice in New York. *Applied Geography*, 29(1), 111–124.

Magdoff, F., Foster, J.B., & Buttel, F.H. (eds.) (2000) *Hungry for profit: The agribusiness threat to farmers, food, and the environment*. New York: NYU Press.

Marcuse, P. (2009) From critical urban theory to the right to the city. *City*, 13(2–3), 185–197.

Massey, D. (2004) Geographies of responsibility. *Geografiska Annaler: Series B, Human Geography*, 86(1), 5–18.

Massey, D. (2010) *A global sense of place*. Aughty.Org. www.aughty.org/pdf/global_sense_place.pdf.

McClintock, N. (2011) From industrial garden to food desert: Demarcated devaluation in the flatlands of Oakland, California. In: Alkon, A.H., & Agyeman, J. (eds.), *Cultivating food justice: Race, class, and sustainability*. Boston, MA: MIT Press, pp. 89–120.

McKinney, L., & Kato, Y. (2017) Community context of food justice: reflections on a free local produce program in a New Orleans food desert. *AIMS Agriculture and Food*, 2(2),183–200.

Mechlem, K. (2004) Food security and the right to food in the discourse of the United Nations. *European Law Journal*, 10(5), 631–648.

Nash, J.C. (2008) Re-thinking intersectionality. *Feminist Review*, 89(1), 1–15.

Pothukuchi, K. (2005). Attracting supermarkets to inner-city neighborhoods: economic development outside the box. *Economic Development Quarterly*, 19(3), 232–244.

Pritchard, B., Dixon, J., Hull, E., & Choithani, C. (2016) 'Stepping back and moving in': The role of the state in the contemporary food regime. *The Journal of Peasant Studies*, 43(3), 693–710.

Redmond, L. (2013) *Food Justice*. New York: TEDxManhattan. Available from: https://youtu.be/ydZfSuz-Hu8 [Accessed 25 October 2016].

Riches, G. (2002) Food banks and food security: Welfare reform, human rights and social policy. Lessons from Canada? *Social Policy & Administration*, 36(6), 648–663.

Rosset, P. (2003) Food sovereignty: Global rallying cry of farmer movements. *Food First Backgrounder*, 9(4), 1.

Sassen, S. (2001) *The global city: New York, London, Tokyo*. Princeton, NJ: Princeton University Press.

Sbicca, J. (2012) Growing food justice by planting an anti-oppression foundation: Opportunities and obstacles for a budding social movement. *Agriculture and Human Values*, 29(4), 455–466.

Schatzki, T.R. (2010) *Site of the social: A philosophical account of the constitution of social life and change*. University Park, PA: Penn State University Press.

Schlosberg, D. (2004) Reconceiving environmental justice: Global movements and political theories. *Environmental Politics*, 13(3), 517–540.

Shannon, J. (2016) Beyond the supermarket solution: Linking food deserts, neighborhood context, and everyday mobility. *Annals of the American Association of Geographers*, 106(1), 186–202.

Shiva, V. (2004) The furture of food: countering globalisation and recolonization of Indian agriculture. *Futures*, 36(6), 715–732.

Slocum, R. (2006) Anti-racist practice and the work of community food organisations. *Antipode*, 38(2), 327–349.

Slocum, R. (2007) Whiteness, space and alternative food practice. *Geoforum*, 38(3), 520–533.

Smith, D.E. (2005) *Institutional ethnography: A sociology for people*. Lanham, MD: Rowman Altamira.

Sobal, J., & Wansink, B. (2007). Kitchenscapes, tablescapes, platescapes, and foodscapes: Influences of microscale built environments on food intake. *Environment and Behavior*, 39(1), 124–142.

Springer, S., Birch, K., & MacLeavy, J. (2016) *The handbook of neoliberalism*. New York: Routledge.

USDA (nd). Documentation. *Food Access Research Atlas*. United States Department of Agriculture Economic Research Service. www.ers.usda.gov/data-products/food-access-research-atlas/documentation/ (accessed 23 June 2017).

Walker, R.E., Keane, C.R., & Burke, J.G. (2010). Disparities and access to healthy food in the United States: A review of food deserts literature. *Health & Place*, 16(5), 876–884.

Wenzer, J. (2015) Foodscapes. In: Jackson, P., & The CONAX Group (eds.), *Food words: Essays in culinary culture*. London: Bloomsbury.

White, M.M. (2011). Environmental reviews & case studies: D-town farm: African American resistance to food insecurity and the transformation of detroit. *Environmental Practice*, 13(4), 406–417.

Wolf, S. (2014) *The neoliberal regime in the agri-food sector: Crisis, resilience and restructuring*. New York: Routledge.

Wrigley, N., Warm, D., & Margetts, B. (2003). Deprivation, diet, and food-retail access: Findings from the Leeds 'food deserts' study. *Environment and Planning A*, 35(1), 151–188.

Yeoh, B.S., & Huang, S. (1999). Spaces at the margins: Migrant domestic workers and the development of civil society in Singapore. *Environment and Planning A*, 31(7), 1149–1167.

Zukin, S. (1996). Space and symbols in an age of decline. In *Re-presenting the City* (pp. 43–59). London: Macmillan Education.

33

Grassroots activism, agroecology, and the food and farming movement

Ten years in Bristol's food story

Angela E. Raffle and Joy Carey

About the authors

We both live and work in Bristol, and have been active participants in Bristol's food and farming movement for the past ten years. Bringing experience from research and activism in public health and in local food, we inevitably give an account of events and projects as seen through our own experience. Complex living systems with multiple interacting variables (Boulton et al., 2015) are familiar territory for us. Simplistic linear models of cause and effect have limited power in this arena. Our practice uses the principles of systems change (Scharmer, 2017) and advocacy (Chapman, 2013) underpinned by continual evaluation of best available evidence.

Introduction

Crofts, farms, orchards, kitchen gardens, drove roads, waterways, abattoirs, warehouses, and food marketplaces have shaped our landscape throughout history. Cities arise from and are connected with the regional farmlands to which they belong (Steel, 2008). In this chapter we aim to show that cities can choose to use their significant influence in pursuit of a food and farming landscape that is good for people, planet, and place. We describe examples of food activism and policy change in one city in the southwest of England, during the last ten years. The changes achieved are small, but they are a beginning. There remains a hard road to travel to achieve truly sustainable food systems.

In Britain, during the past 2,000 years, the rights of ordinary people to live on and work the land for food have been steadily eroded (Fairlie, 2009). Land ownership is concentrated in the hands of a few. Emphasis is on land management primarily for making money rather than for sustainable production of wholesome food for local people (Linklater, 2014; Shrubshole, 2017; Chapter 16). Colonialism, cheap oil, and market driven policy have brought massive changes to farming and landscape in England from the 19th century, with a shift to cheap imports, mechanisation, monocrops, artificial fertilisers, pesticides, and market consolidation (Lang, 2004; Steel, 2008; Hawkes, 2012; Lawrence, 2013; Tudge, 2016). Our cities have sprawled and food is now globally transported, warehoused, processed, and sold in ways previously undreamed of. Farming has consistently viewed the species we share the Earth with as pests and vermin to be destroyed (Harvey, 1998). The 2016 State of Nature Report for the UK tells us that 75% of landscape is agricultural, farmland bird and butterfly species have declined substantially, and the UK is among the most nature-depleted countries in the world (Hayhow et al., 2016). Current national government policy in England treats food solely as a commodity, no reserves are held, and no national body is responsible (Lang, 2004; Carey, 2011). The supply system is dominated by a handful of multinational companies (Lang, 2004; Statistics Portal, 2017; Welch, 2011). Power and control are highly centralised with little remaining of production meeting local needs (Davis and Bollard, 1986) and infrastructure for supporting diverse independent producers, caterers, and retailers has all but disappeared (Carey, 2011).

Concerns about the impact of human activity on the natural world are not new. In England for example there were statutes for 'conservacie' of London's river Thames in the Middle Ages, and penalties for unlawful tree felling in the 7th century (Thomas, 1983). Awareness reached near universal levels in the 1960s, enhanced by images of planet Earth and first-hand accounts of the Apollo astronauts experience as captured for example in 'Overview' (Overview, 2012). The downsides of industrialised and globalised agriculture have become increasingly apparent, with concerns for the biosphere, for human health, for social justice, for human rights, and for fair employment (Morgan et al., 2006; Kessler, 2009; Headey and Fan, 2010; Wise, 2010; Zagema, 2011; Clutterbuck, 2013; Cooper and Dumpleton, 2013; Rappel, 2015). Yet humans are adept at distancing themselves from the consequences of their collective action. This is a key message of a massive open online course (MOOC), involving tens of thousands of participants worldwide and run from MIT Business School. The course provides evidence of this distancing mindset, and demonstrates ways of overcoming the collective insanity that arises when exchange of evermore money is put as a goal above all others (Scharmer, 2016, 2017).

In response to the negative impacts of the green revolution and industrialised agribusiness, the last 50 years has seen emergence of a food and farming movement pursuing a new direction. Grounded not on ideology but on biological and ecological principles, it is known as 'agroecology' (Altieri, 2015; ILEIA, 2016; Chapter 24). It is underpinned by a rationality that puts care of the Earth and social justice as primary. It aims to support the model of farming which currently generates food for some 70% of the world's population – produced from over 500 million small farms that mix crops, trees, and livestock; comprise low-impact rural settlement, short supply chains, decentralised infrastructure, efficient use of fuel and water; and that create jobs and livelihoods (IAASTD, 2009; UNCTAD, 2013; FAO, 2014; Graeub et al., 2016; Lowder et al., 2016; Willis, 2016; Tudge, 2016; Latham, 2016). *La Via Campesina*, representing some 200 million farmers, is a key campaigning group (La Via Campesina, 2017; Chapter 31). The UK has its own Landworkers' Alliance (Fernandes and Laughton, 2014), which is a full member of La Via Campesina.

A dominant narrative, promoted by multinational corporations, states that hunger can only be solved by industrialised agriculture. The claim is that so-called sustainable intensification (Cook et al., 2015) is essential and that 50% more food is needed by 2050 to solve global hunger (Chapter 19). An article in *Farming UK* quotes the Chair of Monsanto and claims "Agri-business aims to make 50% more food available and reduce emissions by 50% by 2030" (Farming UK, 2015). It also contains the arrogant and chilling statement that the World Business Council for Sustainable Developments (Pepsico, Walmart, Monsanto, and others) has "to reach, teach, and support [...] the millions of smallholders providing our daily coffee, cotton and chocolate fix" (ibid.: n.p.). It is important to look behind such claims. In 2014 the production of food worldwide was enough to feed 14 billion people, which is three to four billion more than the UN predict as the maximum future projected world population (Tudge, 2016). Global hunger is due to maldistribution, and not to food shortage (Roberts, 2008). It is agribusiness that consumes the majority of non-renewable inputs, despite producing a minority of the world's food (IAASTD, 2009). Unfortunately the UK government's own Future of Food and Farming report (Government Office for Science, 2011) produced in response to the 2009 IAASTD report, is firmly 'business as usual' despite a statement in the introduction that business as usual cannot continue.

Bristol context

Bristol lies on the River Avon, nine miles from the coast. There is evidence of human toolmaking since Paleolithic times, and iron age forts on the high banks of Avon Gorge (also see Chapter 1). The city became a busy port by the 1300s, trading wool, textiles, leather, and wine. In the 1600s and 1700s Bristol merchants grew rich through colonial exploitation, but as ships grew larger the Bristol docks declined, replaced eventually by a modern dock on the coast. The city today is home to creative and digital industries, aviation, finance, tobacco, and tourism (Bristol City Council, 2017). The population of 442,500 is ethnically diverse, with a marked divide in wealth between affluent and deprived parts of the city (Bristol City Council, 2016). Local government has a range of powers to raise taxes and provide services. Frameworks for land use planning in Great Britain take little or no account of food, and local government has limited direct powers over food and farming policy (Morgan, 2009). Nevertheless, the ways that cities can influence the food agenda are increasingly being explored (White and Natelson, 2011; Moragues-Faus and Morgan, 2015; Soto et al., 2016).

The beginnings of the current phase of food activism

Brownlee (2011) highlights the 1960s Bristol inner ring road plans as an important stimulus for strengthening Bristol's grassroots environmental activism. Diani and Rimbaldo (2007: 771) describe, on the basis of 134 study interviews performed in 2001, "a lively presence of environmental groups and activists". The Soil Association, which campaigns for healthy soils and oversees an organic farm certification scheme, moved its headquarters to Bristol in 1985 (Conford and Holden, 2007). A range of food-related projects became established from the 1970s and 80s, including three City Farms, and a number of buying co-operatives, community kitchens, and market gardens. In 1992 Bristol included food in its Agenda 21 sustainability commitments (Beavis, 2002), arising from the United Nations and Rio Earth Summit, thus making it easier for food projects to gain recognition. Permaculture networks were formed and continue to thrive in the city (Henfrey, 2017). Workers in local government (sustainability, allotments, food standards, health policy), workers in the National Health Service (public health, dietetics, catering), and local activists worked together through a project known as 'Food Links'. A central farmers market began in Corn Street in 1998 and is now one of seven regular farmers markets in the city. The Soil Association used Bristol and Bath to help pilot an innovative approach to transforming food in schools, now a successful nationwide programme called 'Food for Life' (Steering Group, 2006) and Bristol became a FairTrade city in 2005 and is one of only five cities in the world to hold continuous FairTrade status for over ten years.

Bristol's food activism 2007 to 2017

We now move to describing year by year, some of the activities and events affecting the food landscape in Bristol since 2007, and Table 33.1 provides a summary. We do not yet have hard outcome measures showing, for example, reduction in consumption of highly processed packaged food, shift in farming practices, or increased biodiversity on farmland, although this could form a worthwhile area of research for local universities. The last ten years have, however, seen new and stronger food activist networks, a substantial shift in attitudes, many new independent food businesses, more community food growing projects, some changes in local government policy, a rise in the number of restaurants which achieve Sustainable Restaurant Association or Food for Life standards, and a rise in sales of organically grown food (Carey, 2013; Morgan, 2014; Reed and Keech, 2015; Henfrey, 2017; Soil Association, 2017; Bristol Food Policy Council, 2016). The opening speech at the 2017 Oxford Real Farming Conference was given by Olivier De Schutter, Special Rapporteur to the United Nations on the right to food (De Schutter, 2017). De Schutter stressed that central government policy is highly susceptible to corporate hijack, and therefore cities play an essential role, citing Bristol's Food Policy Council as an example. He emphasised also the key role of innovative pioneering prototypes such as community farms, producer cooperatives, and the numerous forms of innovation arising across the world. These may at first seem too small to be useful, yet they represent the fissures through which new models and patterns can emerge (Scharmer, 2016).

Table 33.1 Key events in Bristol's food story.

1960s	Plans for an inner ring road that would disfigure the Avon gorge and its architectural heritage helps the city's environmental activism to manifest.
1970s	Buying cooperatives, local growing projects, three City Farms, and local Friends of the Earth and Greenpeace groups begin.
1980s	Soil Association locates its headquarters in the city. Food Links project begins.
1990s	Food is included in the Agenda 21 sustainability commitments. A central farmers market begins.
2005	FairTrade status begins.
2007	Transition Bristol runs a year-long programme of awareness raising, Green Capital Partnership forms, FareShare Southwest forms. Household food waste collections begin.
2008	'Building a Positive Future for Bristol after Peak Oil' report commissioned, Love Food Festivals begin.
2009	Bristol Food Network is formed. Officers in the City Council collaborate through an internal 'Food Interests Group'.
2010	'Feeding Bristol in the Future' conference. 'Sustainable Food Charter' adopted by City Council. UNICEF 'UK Baby Friendly City' standards achieved. Campaigns against Tesco in Stokes Croft and Sainsburys superstore in place of the football stadium.
2011	'Tesco' riots. 'Who Feeds Bristol' report published. 'Food Policy Council' is established. Inaugural conference of Sustainable Food Cities Network hosted. Two Community Supported Agriculture projects begin.
2012	Bristol Pound local currency launches. 'West of England Public Sector Food Procurement Group' begins. 'Bristol Good Food' definition launched by the Food Policy Council and immersion events begin.
2013	'A Good Food Plan for Bristol' launched, and Good Food Systems planning recognised within some local government policies.
2014	Peer Review of City Council work on food highlights reliance on junior staff, activists, and volunteers to lead the agenda, a message also confirmed by a published study of Bristol and Malmö. Bristol Food Producers forms, and baseline report published.
2015	Bristol's year as European Green Capital winner brings new short-term investment in food projects. Successful work on food by the Healthy Schools programme features strongly. 'Rising Up' group occupies trees in protest against a transport scheme that destroys some food growing land.
2016	Silver Award in the Sustainable Food Cities scheme is achieved. The food movement in the city continues to evolve and strengthen.

2007 – new partnerships and new dialogue in Bristol

It is hard to say why 2007 felt like an important year, but it seemed to mark a coming together of planetary concerns including climate change, biodiversity depletion, resource depletion, social injustice, violent conflict, limits to growth, and economic shock. In Bristol several initiatives helped strengthen the sustainability agenda. One was the formation in January 2007 of Transition Bristol, initiated by a group inspired by hearing Rob Hopkins speak at a conference (Hopkins, 2011, 2017). Hopkins is co-founder of what has now become a global movement of thousands of communities in 50 different countries coming together to reimagine and rebuild our world, working towards a healthier, happier, socially just future within the ecological boundaries of planet Earth. The Transition Bristol volunteers ran a program of talks, films, workshops, projects, and Open Space gatherings (Owen, 2017) including a 'Big Event' attended by some 400 people, including politicians and local government senior officers (Transition Bristol, 2017a). From a food and landscape perspective Transition Bristol effectively brought the principles of the permaculture movement into a prominent position (Henfrey, 2017), it helped trigger numerous projects, outlined in the rest of this chapter, and led to establishment of Shift Bristol (Shift Bristol, 2017) which runs highly regarded hands-on training.

Another key initiative was the Bristol Green Capital Partnership, launched in July 2007 by the then leader of Bristol City Council. It brought together local government, businesses, campaign groups, health sector, voluntary sector, and other bodies with the aim of making Bristol a low carbon city with a high quality of life for all (Bristol Green Capital, 2017). Subsequently the European Commission created its Green Capital Award and Bristol was the winning city for 2015. The Green Capital Partnership has now grown to a membership of 800 organisations and is beginning to turn welcome attention to the issue of inclusivity within the city's environmental activism (Cabot Institute, 2016).

In 2007 FareShare South West was established in Bristol. This organisation now involves around 100 volunteers each week in redistributing around 30–40 tons of quality surplus food to some 150 organisations working with vulnerable individuals in and around Bristol. FareShare has also created a catering arm called 'Surplus Supper Club' and a toolkit known as 'Eighth Plate' for use by festivals. In the same year Bristol City Council was one of the first English cities to introduce household collection of food waste for recycling (Bristol Food Policy Council, 2016).

2008 – Bristol's Peak Oil report

In 2008, partly influenced by a study session for senior local government officers with Daniel Lerch of the Post Carbon Institute (Hopkins, 2011), the Green Capital Partnership commissioned a report *Building a Positive Future for Bristol after Peak Oil* (Osborn, 2009). The report, which includes a chapter on food, is highly accessible, creative in its style, soundly evidence-based, and well referenced. It helped build positive commitment to creating local resilient systems for food and energy.

The year 2008 also saw the first of Bristol's 'Love Food Festivals', which added a celebratory and family-friendly addition to the organic food festivals and vegan food festivals already taking place in the city.

2009 – Bristol Food Network and the beginnings of new policy approaches

In 2009, the Bristol Food Network was created as natural successor to the Food Links Project (Bristol Food Network, 2017). The membership included individuals, campaign groups, and businesses united by their desire for a better food system. Powered mainly by volunteers, together with in-kind help and some funding from local government, the Network connected people, created local directories, published research, hosted events, and produced a high quality electronic newsletter, 'Local Food', now distributed to over 1,400 members. They conducted widespread community engagement to create a Sustainable Food Strategy for the city, many elements of which have informed subsequent actions and projects. Later, the Bristol Food Network played a key role in Bristol's year as European Green Capital, 2015, ensuring that food was an explicit theme and coordinating the development and oversight of Green Capital funded food projects.

Bristol City Council adopted a new waste strategy in 2009 including a commitment to tackle climate change by reducing food miles, maximising recycling, and reducing waste. A Sustainable Procurement Policy was also adopted although it contained limited detail on food procurement (Bristol Food Policy Council, 2016). Within the City Council the separation between healthy food policy and sustainable food policy began to lessen, with an internal 'Food Interests Group' established that brought middle managers from different departments to one table, helping all to see the roles they could play.

2010 – Feeding Bristol in the Future conference

A conference called 'Feeding Bristol in the Future' was hosted in City Hall in 2010 with the aim of connecting, inspiring, and informing, and was attended by nearly 200 people (Bristol Food Policy Council, 2016). It included presentations and workshops on many themes, and the adoption by Bristol City Council of a Sustainable Food Charter. One workshop was devoted to plans for the 'Who Feeds Bristol' project commissioned by National Health Service (NHS) Bristol and Bristol City Council, and supported by Bristol Green Capital Partnership.

The work of the Public Health team led to Bristol being the first UK city in England and Wales to achieve the UNICEF UK Baby Friendly Initiative best practice standards for breast and bottle-feeding (Bristol Food Policy Council, 2016).

Bristol Food Network held the first 'Get Growing Trail' event with the aim of making visible, and supporting and inspiring, the numerous food growing projects in the city. The event takes place over a weekend, using a colourful map and with each project hosting visitors, some providing refreshments. The event has now taken place for six years running, and 2017 saw a new format for the trail.

The year 2010 also saw growth in public concern, in Bristol and nationally, about rate of expansion and increasing monopoly by multinational supermarket chains and the consequent negative impact on local independent high street shops. During two years there had been 21 planning permissions granted in Greater Bristol for new supermarkets from the 'Big Four' supermarkets – Sainsbury's, Tesco, Asda, Morrisons – and the number of Big Four stores had risen to 76 compared with 19 in 2004 (Carey, 2011). A planning application for demolition of Bristol City Football Club's Ashton Gate stadium, with replacement by a Sainsbury's superstore, attracted strong opposition. Planning permission for the store to go ahead was initially refused then granted at appeal stage, but the delays associated with the linked permission for the new stadium build meant that the football club eventually opted instead for expansion of the existing stadium (Bristol City Council, 2010). In the same year a planning application for a Tesco Express on Stokes Croft, home to radical community action and The People's Republic of Stokes Croft (People's Republic of Stokes Croft, 2017), mobilised fierce opposition with marches, street demonstrations, application for judicial review, and occupation of the premises by protestors leading to an eviction that required the presence of 70 police officers (BBC News, 2010). On a hot bank holiday weekend the following April rioting ensued in Stokes Croft and the windows of the new Tesco store were broken, attracting major national media attention (Kemp, 2011).

A 'Bristol Independents' campaign was launched in 2010, to positively channel the energy from the supermarket campaigns. With the aim of 'Keeping Our High Streets Alive' it used a local marketing and branding tool to help highlight and strengthen Bristol's independent retail and café/restaurant sectors and local high streets.

2011 – Who Feeds Bristol report and formation of Bristol Food Policy Council

A 2011 food conference in City Hall fulfilled the combined launch of a newly formed entity called the 'Bristol Food Policy Council', and of the 'Who Feeds Bristol Report; Towards a Resilient Food Plan' (Carey, 2011). The full *Who Feeds Bristol* report was 135 pages, and a short summary version was distributed widely, with presentations made to local politicians and many other organisations. The report put a spotlight on the many producers and businesses that put food on our plates. It highlighted the fact that one in ten jobs in the West of England area are related to food and drink and showed the value – for local economy, for health, and for the planet – of revitalising our food system. It helped many people to recognise food and farming as a local as well as a global issue.

The idea for the Food Policy Council had been proposed at the 2010 conference. This was followed up by a survey of all Bristol Food Network and Bristol Green Capital Partnership members, a review of published experience in other nations (Williams, 2010) and political advice from Bristol City Council. These sources of intelligence all pointed to the need for a committed group of key stakeholders who, through their influence and networks, could put the food system onto the city's agenda and keep it there. Full meetings, usually involving a learning visit, take place four times a year. Working groups have delivered events, communications, input to procurement, to land use planning (see for example Bristol Food Policy Council, 2011a, 2011b), preparation of 'A Good Food Plan for Bristol' (Bristol Food Policy Council, 2013a), co-creation of the 2015 Action Plan (Bristol food Policy Council, 2015), and compilation and submission of the evidence for achieving the Sustainable Food Cities Silver Award in 2016 (Bristol Food Policy Council, 2016). The Food Policy Council helped host the inaugural conference of the 'Sustainable Food Cities Network' in October 2011 and its 2015 conference. It acted as steering group for Bristol's engagement with eight other European cities in the European Union URBACT project, comprising a shared learning and exchange program on sustainable food in urban communities. Membership of the Food Policy Council includes individuals from different elements of the food system and from key sectors, including health, business, grassroots, non-government organisations, education, and local government.

The Food Policy Council has from the outset adopted a systems approach (Meadows and Wright, 2009). An early task was to make explicit the desired aim for a food system. The Food Policy Council has articulated this aim, restated it at every opportunity, and embedded it in as many policies, and in as many institutions and organisations as possible. The aim of the food system is framed as follows:

> We want our food system to produce food that is tasty, healthy, affordable, good for nature, good for workers, good for local businesses and good for animal welfare. This is what we mean by Good Food.
> *(Bristol Food Policy Council, 2012b)*

By framing the underlying aim in this way, and by working to make it explicit, actors with divergent aims have increasingly worked together under a single banner of good food.

In 2011 Bristol City Council Planning Team signed a protocol with the Bristol Public Health Team whereby health comments would be sought on relevant planning policies and planning applications. This built on several years of joint working, with a key role played by a member of staff funded and managed by Public Health, jointly qualified in planning and public health, and working within the planning team (Grant et al., 2011). This joint working continued until 2014, by which time food systems planning was explicitly acknowledged within Bristol City Council's Development Management policies. This helped familiarise planners with the good food agenda, but the system and mindset barriers to achieving major change remain substantial.

In 2011 Bristol City Council made representations to Central Government, based on a motion from Full Council 6 September 2011, requesting that Local Government be granted powers to limit the uncontrolled expansion of multiple chain retail stores and specifically that a separate Use Class be created in planning law to allow local independent businesses to be distinguished from multiple chains so that the nature of a business could be used as a material factor in determining whether or not to grant planning permission. No such powers have yet been granted.

The year 2011 also saw the start of two community-supported agriculture projects, Sims Hill Shared Harvest (Sims Hill, 2017) and The Community Farm (The Community Farm, 2017). New and innovative growing projects have started in and around the city in every subsequent year, and many of these projects are now members of Bristol Food Producers, which is working to support and strengthen the network and infrastructure for ethical and sustainable food businesses.

2012 – launch of the Bristol Pound local currency and of the West of England food procurement group

The Bristol Pound local currency (£B) was launched in 2012, a project of Transition Bristol, Transition Network, and Bristol Credit Union. It has attracted media interest from all over the world. £B can be exchanged electronically as well as through paper notes and residents can use it to pay their Council Tax and bus fares. It supports local independent retailers by making them more visible, it strengthens the local economy by keeping more money circulating locally, and it helps link local producers with local retailers. At the time of compiling the 2016 Silver Award evidence, £B119,502 had been paid in business rates,

£B680,000 had been issued, and £B460,000 were in circulation. Around 800 locally owned businesses accept the currency with nearly a third of those (265) selling food and drink.

A 'Public Sector Food Procurement Group' involving 14 different organisations in the West of England was initiated in 2012. This group has been working together to share good practice and is achieving major shifts in how contracts are managed. Universities, colleges, schools, children's nurseries, local government-run cafés, and care homes are all part of this change towards local, sustainable, fairly traded, and healthier food. One measure of the progress is the attainment of bronze, silver, or gold on the 'Food for Life Catering Award'. By 2016 there were 114 primary schools serving 12,000 silver catering mark meals each day, 11 academies at both bronze and silver, serving around 2,000 catering mark meals daily, and 49 Early Years/ Nursery sites at both silver and gold serving around 1,800 catering mark meals daily (all of these figures refer to term time).

The Food Policy Council published its 'Bristol Good Food Charter' setting out the aim that should underpin our food system (described earlier) and set about disseminating this message through immersion events (see for example Bristol Food Policy Council, 2012a), film showings, summits, gatherings, and a short animated film (Bristol Food Policy Council, 2013b). The Council also published a report based on in-depth interviews with six key figures in the independent food business sector (Raabe and Raffle, 2011). Bristol City Council used the findings to build better support for the independent small business sector.

Avon Wildlife Trust, with the help of lottery funding, initiated the Feed Bristol urban growing project to engage more people with wildlife-friendly urban food growing. The project has supported 5,500 disadvantaged people, 4,000 school children from 38 schools, and 100 different community groups. Twenty-three thousand people have engaged on the site and 453 volunteers have given 22,000 hours of time (Bristol Food Policy Council, 2016).

In 2012, a mayoral election was held in Bristol for the first time, and George Ferguson, an independent candidate with a strong commitment to environmental issues was elected in November. He served until May 2016. The evidence submission for the European Green Capital was already assembled. Ferguson lead the panel presenting the Bristol bid in 2013 and he oversaw the Green Capital 2015 year helping to secure £7 million of central government investment and £1 million from Bristol, overseen by an arms length short life company. Ferguson's support undoubtedly assisted the good food agenda. Maintaining this support from local government is a constant struggle, and regime change can mean that incoming politicians deliberately distance themselves from issues championed by their predecessors.

2013 – launch of the Good Food Plan and inclusion of good food in local government strategies

The Food Policy Council published and disseminated a report on food poverty (Maslen et al., 2013) and co-hosted a Sustainable Food Summit in May 2013 held in the Eat Drink Bristol giant tipis in the beautiful location of Queen Square. This event attracted a new and diverse audience of chefs, caterers, and media experts.

In autumn 2013, the Food Policy Council launched its Good Food Plan for Bristol (Bristol Food Policy Council, 2013a). This summarised a vision and the system changes needed. The launch event, recorded in a three-minute film (Bristol Food Policy Council, 2013c), involved 120 people and used the Open Space meeting method (Owen, 2017).

Bristol City Council acknowledged the good food system aim of the Food Policy Council by including this within three statutory documents; the Health and Wellbeing Strategy, the Mayor's vision, and the Joint Strategic Needs Assessment (Bristol Food Policy Council, 2016). Representations made by the Food Policy Council led to acknowledgement of the local food system within the Strategic Economic Plan of the sub-regional business development body the Local Enterprise Partnership.

Also in 2013 North Bristol NHS Trust became the first hospital in England to achieve Silver in the Soil Association Food for Life Catering Award for its patient meals, and it has a Silver FairTrade Business Award. The Trust uses milk from a family-run pasture-fed dairy farm in Somerset. All meat is local and farm assured, with organic beef mince, eggs are free-range, all cheese and ice cream is local, and all fish is sustainably sourced and Marine Stewardship Council certified.

2014 – World Health Organisation collaborating centre peer review of Bristol City Council

In 2014 Bristol City Council invited an independent Food and Planning Developmental Review (WHO Collaborating Centre and SHINE, 2014). It asked the question "What could Bristol City Council do better to support health and health equity through a spatial planning approach to food governance at city and city-regions scale?" (ibid.: foreword). Fifteen interviews were conducted with staff whose responsibilities affected the food agenda, yet from widely divergent angles. The overall conclusion was strong commendation for the skills, expertise, and dedicated work of staff on the ground, but it was noted that at senior level there was ambivalence about the good food agenda and a lack of leadership. The "buoyant network of innovators and partners in the community and business sector" was reported as a positive driver by interviewees (ibid.: 1).

A similar theme was evident from a research study of Bristol and Malmö, summarised in this quote from a Bristol City Council interviewee describing Bristol's unpaid activists; "They are the primary community drivers, they've got endless energy, they've got fantastic ideas, great skills, and they are just pushing and pushing" (Moragues-Faus and Morgan, 2015: 1, 657). Crucially, the Bristol and Malmö study also highlighted:

> "This lack of resources and powers' is mainly due to a historical omission of food as a municipal mandate, further reinforced by the path-dependency of institutional structures and current squeeze of municipal budgets [. . .] there is an inordinate reliance on voluntarism, motivation and individual commitment, with the civil society actors having to invest the most time when they have the least resources."
>
> *(Moragues-Faus and Morgan, 2015: 1, 568)*

Considerable work took place to align the plans for the 2015 European Green Capital Year with the food system aims, and to ensure a robust process for collaborative development of candidate projects for Green Capital funding. The City Council approved an innovative 'Edible Parks Policy' to promote food growing in public green spaces. The charity Incredible Edible Bristol extended its reach to achieve Incredible Edible Schools and projects at railway stations and businesses. The inaugural meeting of Bristol Food Producers took place. In 2014 the Food Policy Council also published a Baseline Report summarising indicators of good food in the city (Webb et al., 2014).

2015 – a year as European Green Capital

The nine-day-long Food Connections Festival took place for the second time. The combined impact of 2014 and 2015 festivals was to engage over 100,000 people at over 300 events across the city, helping to create connections and make good food part of the city's culture.

The Green Capital year brought many new projects and activities into being, some aimed at awareness raising and celebration, and some that created new prototypes continuing beyond 2015. The Green Capital Citywide report (Bristol European Green Capital, 2015), the Bristol Method paper on food (Carey, 2015), and the Silver Award evidence (Bristol Food Policy Council, 2016) contain considerable detail.

The Healthy Schools Programme, run by the Director of Public Health's team within Bristol City Council, had been building in strength for ten years and featured strongly in the 2015 Green Capital projects. It streamlines all health-related work in Bristol schools and nurseries and has successfully embedded good food into all aspects of the school curriculum, reaching all cultures and all parts of the city.

In February 2015 protest against a new entry road in Stapleton for a Metrobus route, sited on 'best and most versatile agricultural land' came to a head (Bristol 24/7, 2015). The Rising Up group occupied the site for 42 days before being evicted, attracting considerable national publicity. The nationally funded plans for the transport scheme had been many years in preparation and designation of the site for transport use was long-standing (Ferguson, 2015). The protests highlighted the lack of consideration for agricultural land in English planning law, and led to a commitment by the outgoing Mayor George Ferguson of a 99-year lease to protect the remaining land in the area known as the Blue Finger. Securing this protection is not a foregone conclusion. Planning officers will have to make a case for the land having unique qualities in order for it to be designated in the adopted 'Local Plan' as not for development.

2016 – the second city in England to achieve Silver on the Sustainable Food Cities Award

In 2016, Bristol was awarded Silver status in the Sustainable Food Cities award (Bristol Food Policy Council, 2016). Only Brighton and Hove, on the South coast, with a long history of green activism, has silver already. The annual food event hosted by the Food Policy Council consisted of three nested celebratory events, each with distinct audiences. Much of the activity in 2016 was devoted to maintaining and strengthening existing pioneering projects. Incredible Edible Bristol created an 'Urban Food Growing Trail'; Sims Hill Shared Harvest ran a crowd-funding campaign for a new tractor; the Community Farm won a national award for its organic vegetable box scheme; the independent organic supermarket 'Better Food' opened a third shop; the successful 'Canteen' sustainable food restaurant opened a new venture in an old assembly rooms; and Bristol Food Producers secured funding to build its training, coordination, and business development work – to give just a few examples. All this is against a background of severe funding cuts, and loss through voluntary redundancy of many experienced staff, within Local Government.

Reflection on change-making in turbulent times

We are at an early stage of realising the changes needed in our food system, and there is much to learn. Whilst activism is evolving to take account of insights from modern psychology, equivalent evolution within statutory institutions seems elusive. A capacity for cynicism, for judging others, for selectively attending to evidence that fits our preconceptions, and for fearing to act lies deep in the human psyche (Cosmides and Tooby, 1999; Kurzban, 2012). Attention to our own inner world of 'locked-in' thought processes, alongside the outer world of action, is at the heart of the Transition Movement approach. This has informed practice within Bristol's food activism since 2007 and can be summarised by the following golden rules:

- Articulate the desired purpose of the food system and keep on saying it; get it written into the aims of all sustainability and health institutions.
- Show rather than tell; use facts and case stories creatively and non-judgmentally; enable people to see for themselves why producing food without destroying soil and habitats, without displacing people and cultures, without exploiting workers, and without polluting water-courses or creating greenhouse gas emissions, is an essential part of a sustainable and healthy existence; inspire people with a compelling description of what could be.
- Be open to working with anyone who shares your aims and values, and be comfortable with all kinds of difference. Be kind – most of us use supermarkets and convenience foods so take care not to make

people feel guilty or under attack for their actions; but don't compromise on your deep aims just because other agendas are disturbed or threatened by your values.

- Keep going. Don't expect overnight miracles. Do expect problems. Don't expect appreciation. Remind yourself why it matters, and stick with it.
- Be aware of advocacy techniques used by agribusiness and multinational retailers. These were perfected by the tobacco industry (Chapman, 2008). Key methods are confusion, spurious evidence, 'firecracker' arguments (designed to divide and antagonise), claims that they are solving the world's problems, dire economic warnings, use of 'front groups', ridicule, and portraying those who disagree with them as romantic fools, isolationists, or idiots who simply don't understand the economy. Become expert at calmly standing your ground, using evidence, and sticking to the deep purpose.
- Be alert to the capacity for community action to be appropriated, diluted, diverted, and dumbed down by incumbent systems locked into inevitable repetition of the patterns of the past.

Valuable though the work in Bristol is, there is so much more that needs to be done, locally, regionally, nationally, and globally. We hope that our case study will give encouragement, ideas, and insights to others engaged in similar work. Every place has its own unique circumstances, and there is no single way of doing things. There are nevertheless some universal themes.

Further reading

Altieri, M.A. (2015) Agroecology: Who will feed us in a planet in crisis. *Earth Talk Schumacher College*. Dartington Hall Devon England: Dartington Trust. www.youtube.com/watch?v=LKfiabQ-j0E (accessed 10 March 2017).

Anonymous. (2015) Agri-business aims to make 50% more food available and reduce emissions by 50% by 2030. *Farming UK*. www.farminguk.com/News/Agri-business-aims-to-make-50-more-food-available-and-reduce-emissions-by-50-by-2030_37917.html (accessed 10 March 2017).

BBC. (2010) Squatters at Tesco site in Stokes Croft are evicted. *BBC news online*, 16 March 2010. http://news.bbc.co.uk/1/hi/england/bristol/8570076.stm (accessed 10 March 2017).

Beavis, S. (2002) Sustainable development in Bristol. *The Guardian*.

Boulton, J.G., Allen, P.M., & Bowan, C. (2015) *Embracing complexity strategic perspectives for an age of turbulence*. Oxford: Oxford University Press.

Bristol 2015 European Green Capital. (2015) Citywide review. The story of Bristol's green capital year. http://bristol-review.s3-website-eu-west-1.amazonaws.com/mobile/index.html#p=1 (accessed 10 March 2017).

Bristol 24/7. (2015) Protest at Bristol MetroBus: Environmentalists tree protest at Feed Bristol Stapleton allotments. www.youtube.com/watch?v=i2a0FqkB4GE (accessed 10 March 2017).

Bristol City Council (2016) *Director of public health annual report city of Bristol*. Bristol: Bristol City Council.

Bristol City Council (2017) Economic information and analysis www.bristol.gov.uk/business-support-advice/economic-information-and-analysis (accessed 12 May 2017).

Bristol City Council Online Planning Portal. (2010) http://planningonline.bristol.gov.uk/online-applications/applicationDetails.do?activeTab=summary&keyVal=LB932UDN00C00 (accessed 10 March 2017).

Bristol Food Network. (2017) www.bristolfoodnetwork.org (accessed 10 March 2017).

Bristol Food Policy Council. (2011a) Planning comments for development management policy. http://bristolfoodpolicycouncil.org/planning-comments-for-development-management-policy/ (accessed 10 March 2017).

Bristol Food Policy Council. (2011b) Response to Sainsburys memorial ground application. http://bristolfoodpolicycouncil.org/wp-content/uploads/2013/06/BFPC-letter-14-01-2013.pdf (accessed 10 March 2017).

Bristol Food Policy Council (2012b) The Bristol good food charter http://bristolgoodfood.org/wp-content/uploads/2012/03/The-Bristol-good-food-charter.pdf (accessed 27 June 2017).

Bristol Food Policy Council. (2012a) Yeo Valley day report. http://bristolfoodpolicycouncil.org/wp-content/uploads/2012/10/Yeo-Valley-Day-final-Report.pdf (accessed 10 March 2017).

Bristol Food Policy Council. (2013a) A good food plan for Bristol. http://bristolfoodpolicycouncil.org/wp-content/uploads/2013/03/Bristol-Good-Food-Plan_lowres.pdf (accessed 10 March 2017).

Bristol Food Policy Council (2013b) *What does good food mean for Bristol?* https://vimeo.com/81582401 (accessed 10 March 2017).
Bristol Food Policy Council *(2013c) A good food plan for Bristol (film)* https://vimeo.com/81580858 (accessed 10 March 2017).
Bristol Food Policy Council. (2015) Bristol's good food action plan 2015–2018. http://bristolfoodpolicycouncil.org/bristols-good-food-action-plan-2015-2018-now-released/ (accessed 10 March 2017).
Bristol Food Policy Council. (2016) *Bristol sustainable food cities award 2016 silver.* Bristol: Bristol Food Policy Council. Available from: http://bristolfoodpolicycouncil.org/wp-content/uploads/2016/04/Bristol-SFC-2016-Silver1.pdf (accessed 10th March 2017).
Bristol Green Capital Partnership. (2017) http://bristolgreencapital.org. (accessed 10 March 2017).
Brownlee, E. (2011) *Bristol's green roots: The growth of environmental movement in the city*, Bristol, The Schumacher Centre.
Cabot Institute (2016) *Green and black ambassadors initiative.* University of Bristol. www.bristol.ac.uk/cabot/research/casestudies/2016/green-black-ambassadors.html (accessed 10 March 2017).
Carey, J. (2011) Who feeds Bristol? Towards a resilient food plan. Bristol: Bristol Green Capital, NHS Bristol, Bristol City Council. www.bristol.gov.uk/documents/20182/32619/Who-feeds-Bristol-report.pdf/6c305e1f-67f4-46bb-be4b-80192e8d64ea (accessed 10 March 2017).
Carey, J. (2013) Urban and community food strategies: The case of Bristol. *International Planning Studies*, 18, 111–128.
Carey, J. (2015) The Bristol method, how to become a more sustainable food city. Bristol 2015 European Green Capital. www.bristol2015.co.uk/media/filer_public/06/19/06197fd5-1f73-4573-bf49-15d7061c29cc/17_bristol_method_how_to_become_a_more_sustainable_food_city.pdf (accessed 10 March 2017).
Chapman, S. (2008) *Public health advocacy and tobacco control: Making smoking history*, Wiley Online Library, Wiley-Blackwell.
Chapman, S. (2013) Media advocacy for policy influence. In: Guest, C., Ricciardi, W., Kawachi, I., & Lang, I. (eds.), *Oxford handbook of public health practice*, 3rd ed. Oxford: Oxford University Press, pp. 302–307.
Clutterbuck, C. (2013) Worker exploitation in food production and service. In: Clark, P., & Ritson C. (eds.), *Practical ethics for food professionals: Ethics in research, education and the workplace.* Chichester: Wiley-Blackwell.
The Community Farm. (2017) www.thecommunityfarm.co.uk (accessed 10 March 2017).
Conford, P., & Holden, P. (2007) The soil association. In: Lockeretz, W. (ed.), *Organic farming. An international history.* Trowbridge: CAB International.
Cook, S. Silici, L. Adolph, B., & Walker, S. (2015) Sustainable intensification revisited. Issue Paper March 2015. International Institute for Environment and Development IIED Natural Resource Group.
Cooper, N., & Dumpleton, S. (2013) *Walking the breadline: The scandal of food poverty in 21st century Britain.* Manchester: Church Action on Poverty and Oxfam.
Cosmides, L., & Tooby, J. (1999) *Evolutionary psychology: An introduction.* London: Weidenfeld and Nicolson.
Davis, J., & Bollard, A. (1986) *As though people mattered – 1st edition: A prospect for Britain.* Rugby, UK: Practical Action Publishing Ltd.
De Schutter, O. (2017) Opening plenary for Oxford real farming conference. Oxford Real Farming Conference 2017, Oxford Town Hall. ORFC. http://orfc.org.uk/opening-plenary-3/ (accessed 10 March 2017).
Diani, M., & Rambaldo, E. (2007) Still the time of environmental movements? A local perspective. *Environmental Politics*, 16, 765–784.
Fairlie, S. (2009) A short history of enclosure in Britain. *The Land*, www.thelandmagazine.org.uk/articles/short-history-enclosure-britain (accessed 10 March 2017).
FAO Food and Agriculture Organisation of the United Nations. (2014) Innovation in family farming. *The State of Food and Agriculture.*
Farming UK (2015) *Agri-business aims to make 50% more food available and reduce emissions by 50% by 2030.* https://www.farminguk.com/news/Agri-business-aims-to-make-50-more-food-available-and-reduce-emissions-by-50-by-2030_37917.html (accessed 7 November 2017).
Ferguson, G. (2015) Mayors response to the metrobus open letter. www.kerrymccarthymp.org/news/bristol_news/news.aspx?p=1091217 (accessed 10 March 2017).
Fernandez, J., Laughton, R. (2014) Feeding the future. Landworkers' Alliance.
Government Office for Science. (2011) Foresight. The future of food and farming, Final Project Report. London: The Government Office for Science.

Graeub, B.E., Chappell, M.J., Wittman, H., et al. (2016) The state of family farms in the world. *World Development*, 87, 1–15.

Grant, M., Raffle, A., & Hewitt, S. (2011) Health triage in development management. *Project - Journal of the Department of Planning and Architecture. University of the West of England*, 3, 24–26.

Harvey, G. (1998) *The killing of the countryside*, Vintage.

Hawkes, C. (2012) Food policies for healthy populations and healthy economies. *British Medical Journal*, 344.

Hayhow, D.B., Burns, F., Eaton, M.A., et al.. (2016) State of nature 2016. The State of Nature partnership.

Headey, D., & Fan, S. (2010) Reflections on the global food crisis. How did it happen? How has it hurt? And how can we prevent the next one? *Research Monograph*. International Food Policy Research Institute.

Henfrey, T. (2017) Resilience and community action in Bristol. In: Henfrey, T., Maschkowski, G., & Penha-Lopes, G. (eds.), *Resilience, community action and societal transformation*. East Meon: Permanent Publications.

Hopkins, R. (2011) *The transition companion. Making your communities more resilient in uncertain times*, Totnes, Green Books.

Hopkins, R. (2017) 10 years of transition network: The early days. . . . https://transitionnetwork.org/news-and-blog/10years/ (accessed 10 March 2017).

IAASTD International Assessment of Agricultural Knowledge, Science and Technology for Development. (2009) Agriculture at a Crossroads: Global Report. Washington.

ILEIA. (2016) Making the case for agroecology. *Farming Matters*, 32.3.

Kemp, P. (2011) Tesco riot drives call for supermarket planning review. *BBC Radio 4 News*. BBC. www.bbc.co.uk/news/business-13431552 (accessed 10 March 2017).

Kessler, D. (2009) *The end of overeating: Taking control of the insatiable American appetite*, Rodale books.

Kurzban, R. (2012) *Why everyone (else) is a hypocrite: Evolution and the modular mind*. Princeton, NJ: Princeton University Press.

La Via Campesina. (2017) www.viacampesina.org/en/ (accessed 10 March 2017).

Lang, T. (2004) *Food industrialisation and food power: Implications for food governance*, The gatekeeper series. London: International Institute for Environment and Development.

Latham, J. (2016) Why the food movement is unstoppable. *Independent Science News Food Health and Agriculture Bioscience News*, www.independentsciencenews.org/health/why-the-food-movement-is-unstoppable/ (accessed 10 March 2017).

Lawrence, F. (2013) *Not on the label: What really goes into the food on your plate*, Penguin.

Linklater, A. (2014) *Owning the Earth: The transforming history of land ownership*, Bloomsbury.

Lowder, S.K., Skoet, J., & Raney, T. (2016) The number, size, and distribution of farms, smallholder farms, and family farms worldwide. *World Development*, 87, 16–19.

Maslen, C., Raffle, A., Marriott, S., & Smith, N. (2013) Food poverty. What does the evidence tell us? Bristol: Bristol City Council. http://bristolfoodpolicycouncil.org/wp-content/uploads/2013/08/Food-Poverty-Report-July-2013-for-publication.pdf (accessed 10 March 2017).

Meadows, D., & Wright, D. (2009) *Thinking in systems: A primer*. White River Junction, VT: Chelsea Green Publishing.

Moragues-Faus, A., & Morgan, K. (2015) Reframing the foodscape: The emergent world of urban food policy. *Environment and Planning*, 47, 1558–1573.

Morgan, K. (2009) Feeding the city: The challenge of urban food planning. *International Planning Studies*, 14, 429–436.

Morgan, K. (2013) The rise of urban food planning. *International Planning Studies*, 18, 1–4.

Morgan, K. (2014) The new urban foodscape: Planning politics and power. In: Viljoen, A., & Bohn, K. (eds.), *Second nature urban agriculture. Designing productive cities*. London and New York: Routledge.

Morgan, K. (2015) Nourishing the city: The rise of the urban food question in the Global North. *Urban Studies*, 52, 1379–1394.

Morgan, K., Marsden, T., & Murdoch, J. (2006) *Worlds of food; place, power, and provenance in the food chain*. Oxford: Oxford University Press.

Osborn, S. (2009) *Building a positive future for Bristol after peak oil*. Bristol: Bristol Local Strategic Partnership. Available from: http://transitionbristol.net/wp-content/uploads/2013/03/Peak-Oil-Report_01.pdf (accessed 10 March 2017).

Owen, H. (2017) *A brief user's guide to open space technology* www.openspaceworld.com/users_guide.htm (accessed 10 March 2017).

People's Republic of Stokes Croft website (2017) www.prsc.org.uk (accessed 10 March 2017).

Raabe, W., & Raffle, A. (2011) What problems do independent food businesses experience in Bristol? and how can a Bristol Independents campaign help? Bristol: NHS Bristol. http://bristolindependents.co.uk/wp/wp-content/uploads/2012/06/Bristol-independents-interviews-24Apr12.pdf (accessed 10 March 2017).

Rappel, I. (2015) Capitalism and species extinction. *International Socialism*, http://isj.org.uk/capitalism-and-species-extinction/ (accessed 10 March 2017).

Reed, M., & Keech, D. (2015) Building a Bristol food city region from the grass roots up: Food strategies, action plans and food policy councils. *Urban Agriculture Magazine*, 29.

Reid, G. (2012) *Overview*, (film) Planetary Collective. https://vimeo.com/55073825 (accessed 10 March 2017).

Roberts, P. (2008) *The end of food*. Houghton Mifflin Harcourt.

Scharmer, O. (2016) *Theory U: Leading from the future as it emerges*. Oakland, CA: Berrett-Koehler Publishers Inc.

Scharmer, O. (2017) *Awareness-based systems change with u.lab*. Cambridge, MA: Massachusetts Institute of Technology. Available from: www.edx.org/course/awareness-based-systems-change-u-lab-how-mitx-15-671-0x (accessed 10 March 2017).

Shift Bristol Website https://shiftbristol.org.uk (accessed 10 March 2017).

Shrubshole, G. (2017) Who owns England; what we know so far and where next. *Who Owns England* https://whoownsengland.org/2017/01/01/who-owns-england-what-we-know-so-far-where-next/ 1 January 2017 (accessed 10 March 2017).

Sims Hill Shared Harvest. (2017) https://simshill.co.uk (accessed 10 March 2017).

Soil Association (2017) *The organic market report 2017*. Bristol: Soil Association.

Soto, P., Lamb, D., Van Der Schans, J-W., Kneafsey, M., & Mcglynn, D. (2016) Smart and competitive food and drink supply chains. In: Thorpe, E. (ed.), *EU rural review*. Luxembourg: European Network for Rural Development.

Statistics Portal. (2017) *Market share of grocery stores in Great Britain from January 2015 to December 2016* www.statista.com/statistics/280208/grocery-market-share-in-the-united-kingdom-uk/ (accessed 6 March 2017).

Steel, C. (2008) *Hungry city: How food shapes our lives*, ProQuest LLC.

Steering Group Partnership for the food for Life Project. (2006) The food for life project in bath and North East Somerset; a report on the Pilot Project Feb 2005 – Feb 2006. Bath and NE Somerset Public Health Department. www.bathnes.gov.uk/sites/default/files/sitedocuments/Sport-Leisure-and-Parks/Healthandfitness/food_for_life_report.pdf (accessed 10 March 2017).

Thomas, K. (1983) *Man and the natural world. Changing attitudes in England 1500–1800*, Allen Lane.

Transition Bristol (2017b) http://transitionbristol.net (accessed 10 March 2017).

Transition Bristol Timeline (2017a) http://transitionbristol.net/transition-bristol-timeline/ (accessed 10 March 2017).

Tudge, C. (2016) *Six steps back to the land*. Cambridge: Green Books.

UNCTAD United Nations conference on Trade and Development (2013) *Wake up before it is too late trade and environment review*. Geneva: United Nations.

Webb, B. Bennett-Britton, B., & Lowman, C. (2014) What do we know about the state of good food in Bristol? A baseline report for Bristol Food Policy Council. http://bristolfoodpolicycouncil.org/wp-content/uploads/2015/06/Bristol-Good-Food-Baseline-Report.pdf (accessed 10 March 2017).

Welch, D. (2011) Endless choice or limitless damage? We name the greenest supermarkets. *Ethical Consumer*.

White, H., & Natelson, S. (2011) *Good planning for good food. How the planning system in England can support healthy and sustainable food*. London: Sustain; the alliance for better food and farming.

WHO Collaborating Centre for Healthy Urban Environments & SHINE (2014) *Food and planning developmental review. A report based on interviews with Bristol City Council staff about their work on food*: University of the West of England, Bristol Health Partners. Available from: http://bristolfoodpolicycouncil.org/wp-content/uploads/2014/07/WHO-Food-and-Planning-Review-BCC-040614.pdf (accessed 10 March 2017).

Williams, J. (2010) Briefing note on food policy councils. Bristol City Council.

Willis, G. (2016) New model farming: Resilience through diversity. *Food and Farming Foresight – Paper 1*. Campaign to Protect Rural England.

Wise, T. (2010) The true cost of cheap food. *Resurgence*.

Zagema, B. (2011) Land and power: The growing scandal surrounding the new wave of investments in land. *Oxfam Briefing Paper*. Oxford: Oxfam International.

Part VII

Food cultures and foodways

34

Taste, foodways, and everyday life

Tim Waterman

Introduction

The project of modernity and enlightenment reason has strenuously pursued objectivity in all the sciences, including the social, and the humanities. The separation of mind from body and the concomitant suppression of the sensual in favour of cold rationality and abstraction, along with objectivity led western humanity to the conclusion that building a better world begins with the act of erasure (Toulmin, 1990). Civilisation could thus be envisioned to be built upon a *tabula rasa*, logically and technologically constructed from a higher form of truth, the perfection of which would inevitably arise to finally answer the problem of living. The sciences, necessary as they are, can also necessarily be limiting when knowledge is overly compartmentalised. The rise of a deracinated science of economics would become a surrogate for religion, exciting a mystical fervor for the market; quasi-spiritual, but lacking grace. Such a mindframe makes difficult the comprehension of a triad of subjects that are essential to the construction of everyday life: taste, foodways (including the vaunted realms of gastronomy), and landscape. This chapter undertakes a broad survey of each in order to uncover, through ecological thinking – an ecology of mind – their interrelation and how together they allow for a fresh conception of everyday life for the future. It will necessarily comprise a whistle-stop tour through each subject, however this will allow a larger synthesis with the aim of introducing a relational framework that can be applied to the widest possible range of scales, places, and people. Practice theory, the study of the way that humans, as social beings, make themselves and their world, is introduced as useful for organising the study of landscape and food.

"[F]ood and landscape are mutually constituted," writes Matthew Potteiger (2013: 261). Kitchen and garden are inextricably linked, spatially, practically, and culturally, and the cuisines that arise in places link the kitchen to larger society and, consequentially, to civil society and the democratic public sphere. Both landscape and cuisine are collective works that pull the past into the present and which also of necessity address the future. Taste plays a decisive role in relating the practices and phenomena of landscape and food from the human perspective, and it motivates and describes our individual and collective ambitions. It is educated taste that undergirds the good life (Strong, 2011). Jean-Robert Pitte affirms this in a Gallic context: "[t]o remain an oasis of optimism and *joie de vivre* in the French imagination, to avoid becoming a vague memory of times gone by, the garden of delights requires jealous care and vigilance at every moment" (2002: xv). The 'garden of delights' – a vision of a good life – is an everyday utopia. It is an emblem of striving, both tasty in itself and the object of the efforts people put into acquiring taste and maintaining their homes, gardens, communities, and the larger body politic. "*Ubi panis, ibi patria*," wrote J. Hector St. John de Crèvecoeur (1997 [1782]: 54). Where there is bread, there is home (or homeland). Finally, the idea of homeland points not only to dwelling, but to sociopolitical ideals and aspirations, and this chapter does not shy from connecting taste, foodways, and everyday life to the conflictual realm of politics.

Taste

Over drinks after a talk I gave on taste and landscape, an architectural theorist who would later become a colleague confided in me that he disliked the term 'taste' and that he preferred 'judgement'. Perhaps this illustrates the primary problem with discussing taste: that the prevailing western common sense of modernity asks that we conceive of the workings of the human mind and our associations with the landscape around us as computational, objective, rational, and linear. Thus judgement may be passed without emotion or subjectivity, while taste is clouded by these factors and is therefore suspect. This common inability to see ourselves as sensate, sexed and sexuate (see Irigaray, 1993; Rawes, 2007), and emotional *as well as* reasonable – as part of the constitution of reason – refuses us access to the whole of the human decision-making apparatus, involving, as it does, complex multi-layered, often simultaneous assessments of psychosocial and cultural value (see Code, 2006; Lakoff and Johnson, 1999; Schatzki, 1996; Smith, 1990). It is a paradox, and one urgently needing further evaluation, that science, the humanities, and design are skittish around the question of taste, whereas practices of marketing and advertising have permeated our lifeworlds from the local to the global, extensively manipulating people's tastes, appetites, desires, and prejudices.

This is Taste with a capital 'T'. Philosopher Carolyn Korsmeyer (1999: 38) has employed the capitalisation in order to distinguish "literal, gustatory taste" (with a lower-case 't'), which describes the physical bodily sense – complicated as it is in its interdependence with the other senses – from its companion "analogic or metaphorical, aesthetic" Taste. For those undertaking more extensive explorations of taste it is a useful technique. It is necessary, though, to stress that 'aesthetic' must not be taken in its narrow sense, referring only to beauty and its appreciation, but in the larger sense of 'aisthesis' (αἴσθησις), which comes to us from Aristotle (see Ward Modrak, 2013) which describes all of sensate experience, extending to perception, cognition, and discernment. The analogic and metaphorical functions of aesthetic taste mirror the conception of human cognition as metaphorical (and ecological) as it has emerged in contemporary philosophy, particularly as it has been enriched by discoveries in cognitive neuroscience (Code, 2006; Lakoff and Johnson, 1999, 1980). This understanding has been further shaped by study into the relations between other primates, and the role of cultural evolution in the development of species (Tomasello, 1999). Understanding taste may well be a significant way that both scientists and philosophers will decode not just human interactions, but politics at the largest scales:

> The meritocracy of eating, of being able to employ one's senses in the service of identifying good food, is a function of the ideology of equality, while at the same time providing evidence for the assertion that equality does not mean that everyone is the same.
>
> *(Allen, 2012: 80)*

Taste, it may be seen, plays a role in politics and civil society. Václav Havel wrote in his *Summer Meditations* (1992) that good taste and manners in the populace are essential elements of democratic civil society. It is deeply unfashionable, but deeply necessary to talk about both. As Havel notes, these conversations are critical not just to human relations and human behaviour, but to:

> the quality of people's relationships to nature, to animals, to the atmosphere, to the landscape, to towns, to gardens, to their homes – the culture of housing and architecture, of public catering, of big

> business and small shops; the culture of work and advertising; the culture of fashion, behaviour, and entertainment.
>
> *(Havel, 1992: 14)*

Taste exists in a dialogue with appetite. Food tastes better if we have an appetite for it, naturally: hunger is the best sauce. Hunger and appetite for food involve the whole body. The nervous system, the gut, a variety of hormones, and the brain are all involved in telling us when it is time to feed and when to stop eating (Allen, 2012: 117–122). The brain's regulator for appetite, located in the hypothalamus, is called the *appestat*, and, to use the analogy of a car, it is the acceleration and brakes on our feeding. The appestat can fail, as can the body's other systems that control appetite. Faults may occur for a variety of reasons, and problems of appetite are part of problems with eating disorders or obesity (see Bordo, 2013; Albritton, 2013). Taste can be likened, using the same analogy, to the car's steering, driving us towards our desires and goals, and away from not just our fears and prejudices, but those foods or life choices that might actively be a danger. Life goals and life choices are often involved in the acquisition of taste: when our social and cultural appetites lead us to change our eating habits and/or our social habits in order to fit into a new social group. Matters are often confusing where taste is concerned, because subtle gradations are difficult to read, especially from outside a particular social set. Further, because gustatory taste and the other senses are so interdependent, and because all of our social and aesthetic tastes are determined in relation to one another and to particular times and settings, they must be evaluated always as both particular and general. Finally, to add another layer of complexity, taste and appetite are often conflated. When it is said someone 'has a taste for something', what is meant is actually an appetite (though of course this is an appetite for a specific flavour). It is precisely the interwoven character of taste and appetite, taste and the other senses, taste and sociocultural setting and landscape setting that makes taste such an important metaphor for (or perhaps simply descriptor of) an integrated, holistic, relational, ecological lifeworld (see Abram, 1996). Taste and appetite are also linked to curiosity, that rare and beautiful human trait which is our key to betterment as a species, and to ambition, which is seen, often simultaneously, as both wicked and monstrous and desirable and necessary. Indeed, it is often both, and curiosity, too, is both key to learning and invention, and that which kills the cat.

Curiosity and ambition require that we discuss acquired taste. Acquired taste is distinguished from discovered taste. Discovered tastes are those which may be considered innate, basic, or automatic, such as those tastes for bland, salty, or sweet "comfort" foods (see Capaldi, 1996). Acquired taste, rather, is not innate, but represents a taste for a new flavour or food which has been built up over time. My own contribution to the study of taste is to have identified two separate categories of acquired taste: accrued taste and educated taste. Accrued taste, I argue, "is acquired taste that is learned, constructed and additive, contextual and associative, but not necessarily intentional" (Waterman, 2011: 68). Educated taste, on the other hand, is "acquired taste derived from intentional taste explorations. [. . .] It is the taste equivalent of 'working out' or 'reading up'" (ibid.: 68). Gaining access to new social realms, acquiring what Pierre Bourdieu refers to as social capital and cultural capital, is the goal of such strivings (Peters, 2012; Wright, 2010; Bourdieu, 1979, 1977). This might be as commonplace as acquiring a taste for lager and football in order to enter a pub and benefit from camaraderie while watching a televised match, or it might be learning to appreciate abstract expressionism, twelve tone serialism, minimalist gardens, or particular Burgundy appellations and vintages. In all of these cases, gaining an aesthetic appreciation is supported by intellectual effort and a display and deployment of specific knowledges. Suffice to say, tastes and appetites are closely tied to desires, ambitions, hope, achievement, status, and the attainment of pleasure (Flammang, 2009; Korsmeyer, 1999). It is important to stress that educated taste is not a phenomenon peculiar to a particular class, but rather that the cultural transmission of how what is good, comforting, delicious, and mutually supporting is made in any place, class, culture, or time: "[w]hat people eat reflects not just who they are, but also who they want to be" (Peters, 2012: xiv).

Accrued acquired tastes are simpler than educated acquired tastes and are often locally inflected. It is not surprising, for example, to meet a Mediterranean child with a liking for olives and olive oil, which would be considered a complex, sophisticated, even disgusting taste for a child from a more northerly clime (reflecting Europe's famous butter/olive oil divide). The existence of accrued tastes points to a fundamental connection between acquired tastes (and prejudices) and landscapes, in the same way that the butter/olive oil divide describes a geography of taste.

One final complication in regarding taste is the contradictory and complementary coexistence of neophilia and neophobia. People both fear and love the new, the foreign, in individually differing but often simultaneous measures. The distress of disgust exists along the same continuum, and, depending on the individual, often occupying the same place as the eustress of the thrill of new sensations.

> Foreign bodies are not only the fly in the soup, but also everything that is unusual, forbidden, unfashionable, threatening. It was in the process of learning to eat that humans made their fear of foreign bodies a virtue and called it taste.
>
> *(Zeldin, 1998: 93)*

Those same forces and fears that inspire the worst in people – insularity and bigotry that may result in violent rejection – are also the same stimuli that reward the more adventurous with the rich mixing and mingling of cosmopolitanism. "An open mind about food, and about the tastes of foreigners, inevitably modifies one's attitudes to one's neighbours" (ibid.: 94).

The need to introduce sustenance into our bodies is one of our most fundamental yet complex encounters with the world. The consumption and tasting of food:

> takes place inside the open, biting, rending, chewing mouth, [and] is one of the most ancient, and most important objects of human thought and imagery. Here man tastes the world, introduces it into his body, makes it part of himself. Man's awakening consciousness could not but concentrate on this moment, could not help borrowing from it a number of substantial images determining its interrelation with the world.
>
> *(Bakhtin, 1984: 281)*

The biting, rending, chewing mouth, as the threshold between world and body, is a site where pleasure and danger (from a variety of threats, from food-borne to sexually transmitted illnesses, for example) coexist. The moment of this interaction is particularly powerful not just because it is the consummation of a relationship with the world, but because it represents a climactic moment in the busy sets of processes, forces, actions, interactions, decisions, and determinations involved in producing, procuring, preparing, consuming, sharing, and cleaning up after that are all part of feeding.

Foodways

The term 'foodway' has been used to describe the passage for introducing food into the body, beginning with the lips and mouth. More importantly, though, 'foodways,' in the plural, describe the sets of collective and individual practices associated with cuisine which "account for everything about eating, including what we consume, how we acquire it, who prepares it, and who is at the table" (Lawrance and de la Peña, 2012: 2). In short, what Bakhtin (1984) describes earlier as our interrelation with the world. This extends into agricultural practices as well ("Eating is an agricultural act," famously wrote Wendell Berry (2010:146)), and 'foodways' can refer to global and transnational food and landscape practices and cultures as much as it does to local, regional, and heritage practices. Carole Counihan, one of the foremost voices in the study of foodways, writes, "In every culture, foodways constitute an organized system, a language that – through its structure and components – conveys meaning and contributes to the organization of the natural and social world" (1999: 19). In many traditional cultures, foodways may be inseparably embedded in agricultural and horticultural practices (Lawrance and de la Peña, 2012; Gudeman, 2001). In such cultures, the practices and rituals of cuisine and commensality are full expressions of landscape and a people's relationship to it over generations. 'Foodways' is not a universal term, being used first and foremost in American cultural anthropology (again, Carole Counihan (2009, 2004, 1999) is a stalwart), sociology, and geography (see Harris et al., 2005) and particularly in the interdisciplinary journal *Food and Foodways*. There are many nuances and inflections that are impossible to cover here, but the French discourse, for example, is notably different, and is rooted in a different canon. Jean-Pierre Poulain, one of France's foremost food sociologists, uses the term *espace social alimentaire* – 'food social space', which follows from Georges Condominas (1980: 77) to define a "site of connection between nature and culture" (Poulain, 2017: 198). Thus it extends to describe much more than either nature or culture alone, or habitat, describing dialogues between "physical, climatological, and biological elements" in the natural world, and to "linguistic, technological, and imaginary dimensions" of culture. Further, "in the case of eating practices, the social space likewise enables connections to be made between the social, psychological, and physiological dimensions" (ibid.: 199). For my purposes here, I am gathering all these ideas under the heading of 'foodways', and while this practice could be criticised for its mixing of intellectual traditions, it allows me to quote, in the same paragraph, not just Counihan, but Roland Barthes, who writes, in the French tradition, that food (and by extension, though he would not have used the term, foodways) "is not only a collection of products that can be used for statistical or nutritional studies. It is also, and at the same time, a system of communication, a body of images, a protocol of usages, situations and behavior" (2008 [1961]: 29).

Understanding, or even glimpsing foodways and food systems *in toto* in the contemporary city can be difficult. Communities might be alienated from the sources of their nourishment, encountering meat not as whole animals to be raised for slaughter, but as cutlets, and vegetables cleaned of soil and roots and packaged with plastic trays and wrappers (see Belasco, 2008). One or the other experience – factory food vs. 'whole' foods – might be viewed as impoverished depending upon one's point of view (see Patel, 2007; Laudan, 2015) but the study of foodways is as important amongst subsistence agrarians or hunter-gatherers as it is in the streets and suburbs of the landscape of the developed world.

Carole Counihan provides further assistance in defining foodways:

> In examining the meaning of food, social scientists have studied *cuisine*, the food elements used and rules for their combination and preparation; *etiquette and food rules*, the customs governing what, with whom, when, and where one eats; *taboo*, the prohibitions and restrictions on the consumption of certain foods by certain people under certain conditions; and *symbolism*, the specific meanings attributed to foods in specific domains.
>
> *(2013: 19–20)*

Foodways must be studied as intricate sets of systems, processes, forces, and meanings that are not dialectic in their construction, but rather *dialogic*. They take the form of elements in conversation with one another, in a highly contingent, relational manner. Landscapes, too, must be understood in this way, and foodways have primary importance in how the forms and occupation of built environments come into being. Even in those global cities where buildings have become almost purely real-estate based financial instruments rather than vessels for daily life and daily practices (see Minton, 2017), foodways still have power to determine urban form. Even the faceless neoliberal precariat (Standing, 2011) must be fed:

> Because it concerns, at once, production and consumption, material and taste, object and self, food and drink offers scholars 'limitless grist' for exploring the personal and political, the material and symbolic, the self and other and the body and soul.
>
> *(Lawrance and de la Peña, 2012: 12)*

Foodways explicitly situate food practices as spatial, embodied (and sensual, in the full sense of *aisthesis*), and affective, and they are expressive of myth, rituals, attitudes, affections, identities, beliefs, and, importantly, ethnicities. Foodways are performed and enacted as processes in landscapes which are themselves processual (as opposed to static conceptions that see landscape as artefact or scenery).

> Foodways bind individuals together, define the limits of the group's outreach and identity, distinguish in-group from out-group, serve as a medium of inter-group communication, celebrate cultural cohesion, and provide a context for performance of group rituals. Although this process may be easier to discern in ethnic groups than in regional contexts, it is no less operative in the latter. Indeed, region and ethnicity are often intertwined."
>
> *(Brown and Mussell, 1984: 5)*

For the purposes here, read 'landscape' for 'region'. Taste, landscape, and foodways are all bound together by the study of everyday life.

Everyday life

Michael Gardiner describes the everyday as "where we develop our manifold capacities, both in an individual and collective sense, and become fully integrated and truly *human* persons" (2000: 2). In continental Europe, the quality of everyday life is a constant concern, and where British and American criticism has tended towards (often metaphysical) abstraction and calculated, rational analysis, theorists on the continent, particularly in France, have consistently worked to exercise philosophy in the realm of the quotidian and the common (see de Certeau, (1984); Lefebvre, 2000 [1968]). The discredited conception of the human brain's operation as purely computational and rational, which took hold in the western imagination increasingly throughout modernity, fuelled by the Cartesian separation of mind and body (see Johnson, 2007; Crane and Patterson, 2001), and bolstered by the increasing mechanisation of all factors of life and landscape from industrialisation to the digital, was never quite as fully accepted in a culture in which epicurean pleasure and emotional fulfilment were so completely embedded in existence. The extensive body of criticism of everyday life on the continent was obscured, particularly in the United States, by the acceptance of structuralism in American curricula, which more readily fit an academic culture accustomed to specialising and taxonomising (Kaplan and Ross, 1987). It was a milestone, then, at the turn of the millennium when an English translation was produced of Michel de Certeau, Luce Giard, and Pierre Mayol's collaborative (and dialogic) second volume of de Certeau's *The Practice of Everyday Life* (1998). This was subtitled 'Living and Cooking' and its content had previously been considered to be so alien to the Anglosphere that there was no reason to translate it (1998: ix).

The rhythmic continuity of everyday life possesses an in-built conservatism that is resistant to change, but that does not reject it outright. Total reinvention of urban processes or functions of everyday life was rare in history until modernity introduced the urge to write the spaces and processes afresh on a *tabula rasa* (see Toulmin, 1990). Jean-Robert Pitte helps to illuminate how food and foodways shed light on practices of everyday life, in a passage in which he pictures the "cook-inventor" as one who seeks to discard continuity: "They make themselves happy, they believe, in hatching some novel thing, not grown under natural sunlight but in the neon of their hodgepodge ideas" (2002: 171). He rejects this revolutionary compulsion. "Rather than privilege originality at any cost, the cook must attempt to capture what remains of a heritage, to continually question himself about culinary idiosyncrasies, to try to know the deep motivations behind them" (ibid.: 171). Cuisine, in this conception, is a collective work over time that may be influenced or modified in part, which may exhibit an openness to the other and to change in the future, but is, at its heart, customary. This connects with Henri Lefebvre's conception of landscape (city and countryside both) as *oeuvre*, and to include cuisine (and, by extension horticulture and agriculture) as part of that whole

(Lefebvre, 2008 [1947], 2002 [1961], 2005 [1981]; Waterman, 2015). The landscape is "slowly shaped by centuries of work, of patient, humble gestures. The result of these gestures, their totality, is what contains greatness" (Lefebvre, 2008 [1947]: 134). A failure to treat landscapes as works, as great totalities, as processes of incremental change that have an orientation to the future, is the root of the most spectacular disasters of contemporary development.

Everyday life is often referred to as only the most banal "repetitiveness as well as the mundane, prosaic, or ordinary" (Schatzki, 2010:12). Everyday life, however, is also guided by hopes and dreams (as evidenced by the existence of educated taste as a mode of social and cultural striving) and by the extraordinary, as the rhythmic, repetitive nature of everyday life invites interruption by the festal – the feast or the festival – which provides a welcome opposition to "the protective, timeless stability, the unchanging established order and ideology," and pronounces an emphasis on "the element of change and renewal" (Bakhtin, 1984: 81). This change is propelled not only by deed but by language: "words and gestures express an *action*, and not simply some ready-made 'internal reality'. When men speak they move forward along their line of action in a force field of possibilities" (Lefebvre, 2008: 134).

Michael Gardiner puts this well, as a processual and embedded vision of utopia – a sensibility, an urge, an emotion, a tendency, rather than an abstract diagram for the spatial relations of the future:

> This 'blueprint' paradigm of utopianism, which is a salient feature of modernity that has been excoriated (not without some justification) by conservatives and postmodernists alike (this is discussed in Sargisson, 2012), is not espoused by Lefebvre, Certeau, et al. Rather, the utopian moment is emblematic of a longing for a different, and better way of living, a reconciliation of thought and life, desire and the real, in a manner that critiques the *status quo* without projecting a full-blown image of what a future society should look like.
>
> *(2000: 17)*

"We are all utopians," says Henri Lefebvre, "so soon as we wish for something different" (2000 [1968]: 75). This wish for something different has transformative power. It is that which fuels meaningful change. Whether in political protest or in processes of design, our imaginations know "how to use the past in order to invent the future" (Lefebvre, 2002, 288). This positive drive, futurity, is part and parcel of the theory of everyday life (Chapter 40). Such a theory is "to elevate lived experience to the status of a critical concept – not merely in order to describe lived experience, but in order to change it" (Kaplan and Ross, 1987: 1). In the same way that the festal is necessary to everyday life, and is, in fact, produced by it, it is "in the midst of the utterly ordinary, in the space where the dominant relations of production are tirelessly and relentlessly re-produced, that we must look for utopian and political aspirations to crystallize" (Kaplan and Ross: 1987, 3).

Finally, it is important to understand that everyday life is constructed of and animated by collective practices that take place in physical spaces, either within buildings or in the landscape.

> Spaces play three roles in human life. They constitute, first, room-for-maneuver, open areas through which people can proceed and deploy their skills. Spaces, second, set bounds on movement, physically constraining what people do. These first two functions are material in nature, a matter of physical opportunities and restrictions. The third function of spaces is to serve as 'stage setting'.
>
> *(Schatzki, 2010: 18)*

The first two categories here are self-explanatory, but the 'stage setting' needs further elaboration. It is not merely a simple backdrop or static scene, but rather an active and three-dimensional inhabited background within which, and in interaction with, everyday life is played out. The 'stage setting' speaks not only of the acted-out, performative aspects of everyday life, but of the performativity of spaces themselves. Spaces are constructed of practices and practices are adapted to spaces. All practices are social practices, all practices are spatial practices, and all practices are acted and/or performed in time by actors, whether human, other species, or any of the various natural forces and processes which have a tangible physical influence on human lifeworlds.

Theories of practice

Practice theory, in the words of cultural anthropologist Sherry Ortner, "seeks to explain the relationship(s) that obtain between human action, on the one hand, and some global entity which we call 'the system' on the other" (1984: 148). To illustrate this embeddedness, or perhaps the fact that the relationship is often obscured in everyday life, consider the example of a punk crook in Alex Cox's 1984 cult film *Repo Man*, who waxes unexpectedly eloquent as he lies dying and bleeding on a convenience store floor after being shot in an attempted robbery, "I know a life of crime has led me to this sorry fate, and yet, *I blame society*." The line is funny for the archly Victorian phrasing of this deathbed contrition, but also because it is patently foolish to see 'society' as a single, overarching, totalising entity that acts so decisively upon the individual (Warde, 2016; Schatzki, 1996). When Margaret Thatcher said of society that "There is no such thing!" she was referring precisely to the concept of society as a totality (Thatcher, 1987). Her neoliberal conception was of a society composed of individuals, each responsible for his or her own actions and successes or failures. This conception has been referred to as competitive individualism (Gilbert, 2014) or idiotism (Curtis, 2013). In competitive individualism, "[i]ndividuals are not the product of social relations. Social relations are things that happen to individuals rather than things which actually define their identity and the co-ordinates of their existence" (Gilbert, 2014: 32). However, neoliberalism still expresses an underlying belief in another type of totality, that of a bounded and unified free market, which magically possesses properties that are equilibrating, just, and moral: an 'invisible hand' (ibid.). This spurious logic can be extended, blaming injustices or inequities not on social, moral, or political shortcomings, but rather on the insistence that the market – here describing not just financial markets but entire lifeworlds – simply is not yet free enough to fully self-regulate. The dilemma here is the casting of the societal and the individual as a binary, which forgets entirely to address the social, which is where individuals are bound into collective actions and identities that have unitary, if unstable, qualities (Barnes, 2001: 24–25). It is this social middle ground that is the realm of social practice, and these practices both shape and are shaped by places and landscapes.

The most useful guide to theories of practice as they relate to food is to be found in Alan Warde's recent *The Practice of Eating* (2016), an exhaustive and rigorous resource. It is also relatively straightforward for those engaged in landscape research to draw conclusions regarding landscape practices from his analysis of theories of food practices. Warde describes some of the basic characteristics of a diverse range of theories of practice:

> Existing theories exhibit family resemblance; all give precedence to practical activity as the means by which people secure their passages through the world thereby emphasizing doing over thinking, practical competence over strategic reasoning, mutual intelligibility over personal motivation, and body over mind.
>
> *(2016: 32)*

They are thus often directly counter to individualist models or explanations. To some readers, this set of qualities will be cause for alarm, raising the spectre of cultural and moral relativism or of postmodern post-factuality. There is, however, a body of theory in contemporary ethics (see Jonas, 2001 [1966]; Heller, 1999, 1990; Irigaray, 1993; Smith, 1990; Chapter 24) that grounds itself in a relational, not relativist, stance which allows for contingency and difference while recognising that there are universal measures of good and evil based in such fundamentals as planetary survival and human registers of pain and pleasure. These measures are also grounded in bodily and sensate experience, in situations, in emotions, and in physical places. Tim Cresswell (1996: 9) sees geography as constitutive of moral life: "[t]he geographical setting of actions plays a central role in defining our judgment of whether actions are good or bad". Andreas Reckwitz extends this further:

> It does not seem out of place to assume that practice theory encourages us to regard the ethical problem as the question of creating and taking care of social routines, not as a question of the just, but of the 'good' life as it is expressed in certain body/understanding/things complexes. Ethics thus does not refer only to the relation between subjects, but also to the relation to things (including nature) and to oneself (including the body, the motivations and emotions).
>
> *(2002: 259)*

Theories of practice, then, are grounded in customary social (and landscape) relations, where examples from the past may be recruited to provide examples of just and ethical situations and solutions. There is also an aspect of futurity to theories of practice, in that constructing or evolving current practice must also have an eye on a positive vision that is being worked towards. Practices, finally, are performed and enacted: they combine doing and thinking. There is no simple separation of theory and practice. Barry Barnes explains this with the example of an acupuncturist at work. In a view which see theory as prior to practice, the sticking-in of needles by the acupuncturist would be purely "the expression of a theory of the body" (2001: 19). He goes on to show that a simple inversion is also wrong, that acupuncture is "not the sticking-in of needles without thought. The practice should be treated as involving thought and action together, and in so far as this is the case, embodied theory, as it were, is a part of practice itself" (2001: 20). This is not dissimilar to what de Certeau et al. (1998) call "doing-cooking". This process is described well by Stephen Gudeman (2001), although without specifically using their term. He calls it 'reason-in-action' (Schatzki situates reason and action in 'activity timespace' (2010) – also worth comparing to Poulain's 'food social space' (2017)) and describes it as a process that is always in flux, responding to conditions and context:

> A woman cooking a pot of potatoes is in continual motion as she adjusts the fire, blows on it, selects a different type of wood for cooking, keeps an eye on the boiling water, adds potatoes or slices in half ones already cooking, fans the smoke out of the kitchen, and checks the flavor of the broth as she sprinkles in condiments or adds salt or water to suit the tastes of her household. I have watched farmers adjust their practices through the year as they try to judge the changing weather; household necessities; amount of rice they need; which plots to seed; when to seed; which seed to use (for seed strains have different times of maturity, resistance characteristics, and consumption features); as well as when, how often, and how well to weed. No meal preparation exactly resembles another; no farming year is precisely like the prior. Each requires adjustments, and as these are tried, they become part of a person's repertoire of skills or foundation.
>
> *(Gudeman, 2001: 40)*

Theories of practice suggest that narratives, moral tales, perhaps, are more useful for guiding good and just behaviour than absolute proscriptions, commandments, or prohibitions. The construction, too, of standards of good taste that are necessary for an ordered and considerate populace are formed out of dialogues and customs that have narrative qualities (Potteiger and Purinton, 1998). Landscapes, too, are formed of lived narratives from quotidian family and community stories to nationalistic imagery. There are also implications for the design of landscapes, as theories of practice aren't well accommodated by totalising masterplanning approaches – these clearly put theory ahead of practice – and suggest instead that codes for action might have greater currency. Most preferred, however, would be design processes that engage directly with landscape practices and, like the acupuncturist, combine thinking with doing, design with making. This same way of thinking about landscape and design can also be applied to landscape's relationship with architecture: both are buildings, and neither is prior to the other. The kitchen and garden, for example, have an inseparable and contingent relationship that must be conceived as hanging together.

Conclusion

Taste and appetite guide and impel human individuals through the complexities of the lived landscapes that are the setting for everyday life. We are constantly told we are individuals, each and own, but our lifeworlds are embedded in the social and societal as inseparably as theory and practice are interwoven. All human action and interaction is fundamentally relational and collective, shaped and influenced by natural and ecological processes and forces, and performed as work that takes place in space over time, in the past and in the future. What the sense of the collective in emplaced everyday life and an analysis of foodways prepares us for and celebrates is the ascendance of the idea of the commons (Chapter 28; Catlow and Waterman, 2015) in political theory (and, indeed, in law) as a primary consideration in all areas of life rather than a residual or historical notion still lingering from another era. People's agency in the shaping of their lives, foodways, and landscapes – and in the commons – has been eroded by various political processes such as bureaucratisation, agnotology, or authoritarianism (all tendencies of neoliberalism) and revolutionary, totalising reinvention in development. Also implicated in the erosion of widespread human agency is the overweening influence of the global agri-food industry and the irresistible concentration of power within it (Chapters 16, 31 and 32), aided and amplified by international policy, marketing, advertising, and the shaping of diets and practices of cooking, agriculture, and horticulture.

> It is apparent that, since the rise of the modern agri-food system after 1945 [. . .], there has been a relentless squeezing of the public into the category of 'consumers'. Under the cover of a continuous and heavy barrage of advertising, consumers have had their attention refocused onto individualised concerns for convenience and low price. This has not only led to the exclusion from the mainstream of other food quality criteria (its sensual attributes, nutritional value, production methods, sourcing and traceability), but has allowed corporate interests to dominate the food system.
>
> *(Sage, 2012: 6)*

Sage encourages a relocalisation of foodways which he terms 'food citizenship', an idea which reflects many of the politically and socially embedded approaches to food and landscape practices elaborated earlier. He says we need food that is "traceable and trustworthy, but also good to eat; food that is culturally, as well as physiologically, nourishing" (Sage, 2012: 7). Food citizenship asks that we not be consumers who are isolated particles in a system of competitive individualism, but rather that we engage in dialogues with each other and with our foodways as communities, in specific landscapes. Food citizenship in everyday life will require, to echo Václav Havel's words, a populace endowed with mutually constructed good taste, educated taste, produced by mutual striving: by design.

> Even the simplest societies, the ones that looked most stable, traditional, unreflective, and unquestioned, were products of continuous effort on a daily basis. Ordinary life was in fact the product of unrelenting efforts to make sure that catastrophic change, entropy, disenchantment, and weak attachment did not take the toll they so easily could. Thus, daily life in even the simplest societies must be seen as an outcome of design.
>
> *(Appadurai, 2013: 254)*

And design begins with research, with understanding what has come before, what was possible, and what might become possible. Research roots design in everyday life, and, as Henri Lefebvre puts it so astutely,

"To penetrate ever more deeply into the content of life, to seize it in its shifting reality, to be ever more lucid about the lessons it has to teach us – this is the essential precept of research" (Lefebvre, 2008: 182). To make our world, and to make it better, we must begin with an understanding of and appreciation for taste, foodways, and everyday life – and be prepared to make a collective effort.

References

Abram, D. (1996) *The spell of the sensuous: Perception and language in a more-than-human world.* New York: Vintage Books.

Albritton, R. (2013) Between obesity and hunger: The capitalist food industry. In: Counihan, C., & Van Esterik, P. (eds.), *Food and culture: A reader*, 3rd ed. New York and London: Routledge, pp. 342–354.

Allen, J.S. (2012) *The omnivorous mind: Our evolving relationship with food.* Cambridge, MA and London: Harvard University Press.

Appadurai, A. (2013) *The future as cultural fact: Essays on the global condition.* London and New York: Verso.

Bakhtin, M. (1984) *Rabelais and his world*, Iswolsky, H. (trans.). Bloomington, IN: Indiana University Press.

Barnes, B. (2001) Practice as collective action. In: Schatzki, T.R., Cetina, K.K., & von Savigny, E. (eds.), *The practice turn in contemporary theory.* London and New York: Routledge, pp. 1–14.

Barthes, R. (1961) Towards a psychosociology of contemporary food consumption. In: Counihan, C., & Van Esterik, P. (eds.), *Food and culture: A reader.* New York: Routledge.

Belasco, W. (2008) *Food: The key concepts.* Oxford and New York: Berg.

Berry, W. (2010) *What are people for?* Berkeley, CA: Counterpoint Press.

Bordo, S. (2013) 'Not just 'a White girl's thing': The changing face of food and body image problems. In: Counihan, C., & Van Esterik, P. (eds.), *Food and culture: A reader*, 3rd ed. New York and London: Routledge, pp. 265–275.

Bourdieu, P. (1977) *Outline of a theory of practice*, Nice, R. (trans.). Cambridge: Cambridge University Press.

Bourdieu, P. (1979) *Distinction: A social critique of the judgement of taste*, Nice, R. (trans.). New York and London: Routledge.

Brown, L.K., & Mussell, K. (eds.) (1984) Introduction. In: *Ethnic and regional foodways in the United States: The performance of group identity.* Knoxville, TN: University of Tennessee Press, pp. 3–15.

Capaldi, E.D. (ed.) (1996) *Why we eat what we eat: The psychology of eating.* Washington, DC: American Psychological Association.

Catlow, R., & Waterman, T. (2015) 'Situating the digital commons: A conversation between Tim Waterman and Ruth Catlow' *Furtherfield* website www.furtherfield.org/features/interviews/situating-digital-commons-conversation-between-ruth-catlow-and-tim-waterman (accessed 25 May 2016).

Code, L. (2006) *Ecological thinking: The politics of epistemic location.* Oxford and New York: Oxford University Press.

Condominas, G. (1980) *L'Espace social. A propos de l'Asie du Sud-Est.* Paris: Flammarion.

Counihan, C.M. (1999) *The anthropology of food and body: Gender, meaning, and power.* New York and London: Routledge.

Counihan, C. (2004) *Around the Tuscan Table: Food, family, and gender in twentieth-century Florence.* London and New York: Routledge.

Counihan, C. (2009) *A tortilla is like life: Food and culture in the San Luis Valley of Colorado.* Austin, TX: University of Texas Press.

Counihan, C., & Van Esterik, P. (2013) *Food and culture: A reader*, 3rd ed. London and New York: Routledge.

Crane, T., & Patterson, S. (eds.) (2001) *History of the mind-body problem.* New York and London: Routledge.

Cresswell, T. (1996) *In place/out of place: Geography, ideology, and transgression.* Minneapolis, MN: University of Minnesota Press.

Curtis, N. (2013) *Idiotism: Capitalism and the privatisation of life.* London: Pluto Press.

de Certeau, M. (1984) *The practice of everyday life*, Rendall, S. (trans.). Berkeley, CA: University of California Press.

de Certeau, M., Giard, L., & Mayol, P. (1998) *The practice of everyday life Vol. 2: Living and cooking*, Tomasik, T.J. (trans.). Minneapolis, MN: University of Minnesota Press.

de Crèvecoeur, J.H. St. J. (1997 [1782]) Manning, S. (ed.), *Letters from an American farmer.* New York: Oxford University Press.

Flammang, J.A. (2009) *The taste for civilization: Food, politics, and civil society.* Urbana and Chicago, IL: University of Illinois Press.

Gardiner, M.E. (2000) *Critiques of everyday life.* Abingdon and New York: Routledge.

Gilbert, J. (2014) *Common ground: Democracy and collectivity in an age of individualism*. London: Pluto Press.
Gudeman, S. (2001) *The anthropology of economy: Community, market, and culture*. Malden, MA and Oxford: Blackwell Publishers.
Harris, P., Lyon, D., & McLoughlin, S. (2005) *The meaning of food*. Guilford, CT: Globe Pequot Press.
Havel, V. (1992) *Summer meditations: On politics, morality and civility in a time of transition*, Wilson, P. (trans.). London and Boston: Faber and Faber.
Heller, A. (1990) *A philosophy of morals*. Oxford and Cambridge, MA: Basil Blackwell.
Heller, A. (1999) *A theory of modernity*. Malden, MA and Oxford: Blackwell Publishers.
Irigaray, L. (1993) *Je, Tu, Nous, Towards a culture of difference*, Martin, A. (trans.). London and New York: Routledge.
Johnson, M. (2007) *The meaning of the body: Aesthetics of human understanding*. Chicago and London: University of Chicago Press.
Jonas, H. (2001 [1966]) *The phenomenon of life: Toward a philosophical biology*. Evanston, IL: Northwestern University Press.
Kaplan, A., & Ross, K. (1987) Introduction. In: Kaplan, A.Y., & Ross, K. (eds.), *Yale French studies No. 73, Everyday life*. New Haven, CT: Yale University Press, pp. 1–4.
Kaplan, A.Y. (1987) Taste wars: American professions of French culture. In: Kaplan, A.Y., & Ross, K. (eds.), *Yale French studies no. 73, Everyday life*. New Haven, CT: Yale University Press, pp. 156–172.
Korsmeyer, C. (1999) *Making sense of taste: Food and philosophy*. Ithaca and London: Cornell University Press.
Lakoff, G., & Johnson, M. (1980) *Metaphors we live by*. Chicago: University of Chicago Press.
Lakoff, G., & Johnson, M. (1999) *Philosophy in the flesh*. New York: Basic Books.
Laudan, Rachel (2015) 'A plea for culinary modernism' in *Jacobin*, 22 May 2015, www.jacobinmag.com/2015/05/slow-food-artisanal-natural-preservatives/ (accessed 3 June 2017).
Lawrance, B.N., & de la Peña, C. (eds.) (2012) *Local foods meet global foodways: Tasting history*. London and New York: Routledge.
Lefebvre, H. (2000 [1968]) *Everyday life in the modern world*, Rabinovitch, S. (trans.). London: The Athlone Press.
Lefebvre, H. (2002 [1961]) *Critique of everyday life Volume 2: Foundations for a sociology of the everyday*, Moore, J. (trans.). London: Verso.
Lefebvre, H. (2005 [1981]) *Critique of everyday life Volume 3: From modernity to modernism*, Moore, J. (trans.). London: Verso.
Lefebvre, H. (2008 [1947]) *Critique of everyday life Volume 1*, Moore, J. (trans.). London: Verso.
Minton, A. (2017) *Big capital: Who is London for?* London: Penguin.
Ortner, Sherry (1984) "Theory and anthropology since the sixties" in *Comparative Studies in Society and History* 26(1): 126–66.
Patel, R. (2007) *Stuffed and starved: Markets, power, and the hidden battle for the world food system*. London: Portobello.
Peters, E.J. (2012) *Appetites and aspirations in Vietnam: Food and drink in the long nineteenth century*. Lanham, MD and Plymouth: Altamira Press.
Pitte, J-R. (2002) *French gastronomy: The history and geography of a passion*. Translated by Jody Gladding. New York: Columbia University Press.
Potteiger, M. (2013) Eating places: Food systems, narratives, networks, and spaces. *Landscape Journal*, 32–2. Madison, WI: University of Wisconsin, pp. 261–275.
Potteiger, M., & Purinton, J. (1998) *Landscape narratives: Design practices for telling stories*. New York and Chichester: John Wiley & Sons, Inc.
Poulain, J-P. (2017) *The sociology of food: Eating and the place of food in society*, Dörr, A>. London and New York: Bloomsbury Academic.
Rawes, P. (2007) *Irigaray for architects*. London and New York: Routledge.
Reckwitz, A. (2002) Toward a theory of social practices: A development in culturalist theorizing. *European Journal of Social Theory*, 5(2), 243–263.
Sage, C. (2012) *Environment and food*. London and New York: Routledge.
Sargisson, L. (2012) *Fool's gold? Utopianism in the twenty-first century*. Basingstoke and New York: Palgrave Macmillan.
Schatzki, T.R. (1996) *Social practices: A Wittgensteinian approach to human activity and the social*. Cambridge: Cambridge University Press.
Schatzki, T.R. (2010) *The timespace of human activity: On performance, society, and history as indeterminate teleological events*. Lanham, MD: Lexington Books.
Smith, D.E. (1990) *The conceptual practices of power: A feminist sociology of knowledge*. Toronto: University of Toronto Press.

Standing, G. (2011) 'The precariat – the new dangerous class' *Policy Network* 24 May 2011 www.policy-network.net/pno_detail.aspx?ID=4004&title=+The+Precariat+%e2%80%93+The+new+dangerous+class (accessed 3 June 2017).

Strong, J. (ed.) (2011) *Educated tastes: Food, drink, and connoisseur culture*. Lincoln, NE: University of Nebraska Press.

Thatcher, M. (1987) "Interview for *woman's own* ('no such thing as society')" *Margaret Thatcher Foundation* website. www.margaretthatcher.org/document/106689 (accessed 17 February 2017).

Tomasello, M. (1999) *The cultural origins of human cognition*. Cambridge, MA and London: Harvard University Press.

Toulmin, S. (1990) *Cosmopolis: The hidden agenda of modernity*. Chicago: University of Chicago Press.

Ward Modrak, D.K. (2013) Sensation and desire. In: Anagnostopoulos, G. (ed.), *A companion to Aristotle*. Chichester: Wiley Blackwell, pp. 310–321.

Warde, A. (2016) *The practice of eating*. Cambridge and Malden, MA: Polity Press.

Waterman, T. (2011) The flavor of the place: Eating and drinking in Payottenland. In: Strong, J. (ed.), *Educated tastes: Food, drink, and connoisseur culture*. Lincoln, NE: University of Nebraska Press, pp. 58–80.

Waterman, T. (2015) Critique of everyday life. Review of Henri Lefebvre's *Critique of everyday life*, Vols. 1–3. *Journal of Architecture*, 20(3). London: Routledge, pp. 566–573.

Wright, D. (2010) Cultural capital and tastes: The persistence of distinction. In: Hall, J.R., Grindstaff, L., & Lo, M-C. (eds.), *Handbook of cultural sociology*. London and New York: Routledge, pp. 275–284.

Zeldin, T. (1998) *An intimate history of humanity*. London: Vintage.

Further reading

Belasco, W. (2006) *Meals to come: A history of the future of food*. Berkeley, CA: University of California Press.

Brillat-Savarin, J.A. (2009 [1949]) *The physiology of taste, or meditations on transcendental gastronomy*, Fisher, M.F.K. (trans.). New York: Alfred A. Knopf, Everyman's Library.

Cosgrove, D. (1998 [1984]) *Social formation and symbolic landscape*. Madison, WI: The University of Wisconsin Press.

de Geus, M. (2009) Sustainable hedonism: The pleasures of living within environmental limits. In: Soper, K., Ryle, M., & Thomas, L. (eds.), *The politics and pleasures of consuming differently*. Basingstoke and New York: Palgrave Macmillan.

Douglas, M. (1972) Deciphering a meal. *Daedalus*, 101(1), 61–81.

Douglas, M. (1966) *Purity and danger: An analysis of concepts of pollution and taboo*. Baltimore, MD: Penguin Books.

Elias, N. (2000) *The civilizing process*, rev. ed., Jephcott, E. (trans.). Malden, MA: Blackwell Publishing.

Fernández-Armesto, F. (2002) *Near a thousand tables: A history of food*. New York and London: The Free Press.

Fisher, M.F.K. (2009 [1943]) *The gastronomical me*. New York: North Point Press.

Frascari, M. (2004) *Semiotica ab Edendo*: Taste in architecture. In: Horwitz, J., & Singley, P. (eds.), *Eating architecture*. Cambridge, MA and London: MIT Press.

Gilbert, S.M. (2014) *The culinary imagination: From myth to modernity*. New York and London: WW Norton & Co.

Goody, J. (1982) *Cooking, cuisine and class: A study in comparative sociology*. Cambridge and New York: Cambridge University Press.

Horwitz, J., & Singley, P. (eds.) (2004) *Eating architecture*. Cambridge, MA and London: MIT Press.

Johnson, P. (1999) *The philosophy of manners: A study of the 'Little Virtues'*. Bristol: Thoemmes Press.

Klein, J.A., & Murcott, A. (eds.) (2014) *Food consumption in global perspective: essays in the anthropology of food in honour of Jack Goody*. Basingstoke and New York: Palgrave Macmillan.

Laudan, R. (2013) *Cuisine and empire: Cooking in world history*. Berkeley, Los Angeles, CA and London: University of California Press.

Mennell, S. (2005) Of gastronomes and guides. In: Korsmeyer, C. (ed.), *The taste culture reader: Experiencing food and drink*. Oxford: Berg.

Ortner, S. (2006) *Anthropology and social theory: Culture, power, and the acting subject*. Durham, NC: Duke University Press.

Parham, S. (2015) *Food and urbanism: The convivial city and a sustainable future*. London and New York: Bloomsbury Academic.

Probyn, E. (2012) In the interests of taste and place: Economies of attachment. In: Pratt, G., & Rosner, V. (eds.), *The global and the intimate: Feminism in our time*. New York: Columbia University Press.

Rudofsky, B. (1980) *Now I lay me down to eat: Notes and footnotes on the lost art of living*. Garden City, NY: Anchor Books.

Sayer, A. (2011) *Why things matter to people: Social science, values and ethical life*. Cambridge: Cambridge University Press.

Schehr, L., & Weiss, A.S. (2001) *French food: On the table, on the page, and in French culture*. London and New York: Routledge.

Soper, K., Ryle, M., & Thomas, L. (eds.) (2009) *The politics and pleasures of consuming differently*. Basingstoke and New York: Palgrave Macmillan.

Steel, C. (2008) *Hungry city: How food shapes our lives*. London: Chatto and Windus.

Stohr, K. (2012) *On manners*. Abingdon and New York: Routledge.

Toussaint-Samat, M. (2009 [1987]) *A history of food*, 2nd ed., Bell, A (trans.). Malden, MA and Oxford: Wiley-Blackwell.

Visser, M. (2009) *The gift of thanks: The roots and rituals of gratitude*. Boston, MA and New York: Houghton Mifflin Harcourt.

Waterman, T. (2017) Publicity and propriety: Democracy and manners in Britain's public landscape. In: Wall, E., & Waterman, T. (eds.), *Landscape and agency: Critical essays*. London: Routledge.

Weiss, A.S. (2002) *Feast and folly: Cuisine, intoxication, and the poetics of the sublime*. Albany, New York: State University of New York Press.

Wrangham, R. (2009) *Catching fire: How cooking made us human*. London: Profile Books.

Wylie, J. (2007) *Landscape*. Abingdon and New York: Routledge.

35
Food and landscape tourism

Jo Russell-Clarke

Introduction: defining tourist interest in the mutual attractions of food and landscape

Food tourism includes a wide-ranging cluster of place-specific activities, increasingly recognised and participated in globally by diverse groups and individuals. It is also a new area of research, crossing and linking formal and informal studies about food tourism topics, as well as experiments through or within the range of practices that deliver food tourism experiences themselves. More than just eating excursions, food tourism can encompass the full spectrum of food identification, gathering, production, distribution, and preparation, as well as gastronomic consumption, acknowledging too the many overlaps and complementarities that are suggested in such comprehensiveness. From drinking wine at wineries; to jam-making classes at berry farms; hunting parties; and foraging excursions ending in picnics and banquets; and planting, pruning, or harvesting days rewarded with celebratory seasonal feasts, there are seemingly endless opportunities for observation as well as direct, visceral engagement with cultures through the landscapes and diversely edible produce that they simultaneously create and reflect.

Food tourism fundamentally connects food and place to the experience of a specific culture. This chapter is not much concerned with the historic migration of foodstuffs or the history of tourism, but rather focuses on how these are linked. It considers the development of a type of cultural tourism dependent on the demonstrable interaction of peoples with the landscapes that provide them with sustenance (Cros and McKercher, 2015 [2002]; Timothy, 2001).

Although literature across this field uses the terms indistinguishably, 'food tourism' is now more common than culinary tourism, gastronomic travel, or their variations. Recent shifts to the more general term 'food' appear to acknowledge broader classes of experience, as well as more ways to link the tourist with edible produce beyond the activity of eating. The Ontario Culinary Food Alliance describes culinary tourism as "any tourism experience in which one learns about, appreciates, and/or consumes food and drink that reflects the local, regional, or national cuisine, heritage, culture, tradition or culinary techniques" (CTA, n.d.-b). More simply, the Caribbean Tourism Organization declares: "Gastronomic tourism refers to trips made to destinations where the local food and beverages are the main motivating factors for travel" (CTO, n.d.). At the start of the millennium, Hall and Mitchell similarly concluded it was "visitation to primary and secondary food producers, food festivals, restaurants and specific locations for which food tasting and/

or experiencing the attributes of specialist food production regions are the primary motivating factor for travel" (Hall and Mitchell, 2001: 308).

With extensive publications on food tourism, Lucy Long explores definitions in *Culinary Tourism*, offering a comprehensive opening view:

> Culinary tourism is about food as a subject and a medium, destination and vehicle, for tourism. It is about individuals exploring foods new to them as well as using food to explore new cultures and ways of being. It is about groups using food to 'sell' their histories and to construct marketable and publicity attractive identities, and it is about individuals satisfying curiosity. Finally it is about the experiencing of food in a mode that is out of the ordinary, that steps outside the normal routine to notice difference and the power of food to represent and negotiate that difference.
>
> *(Long, 2003: 20)*

Almost a decade later Long again notes personal and commercial motivations, emphasising the cultural significance and power of food tourism's "adventurous eating, eating out of curiosity, exploring other cultures through food, intentionally participating in the foodways of an 'other', and developing food as tourist destination and attraction" (Long, 2012: 389). A key characteristic of contemporary food tourism is the conscious exploration of 'otherness', whether in place or time or both, exemplified by the term 'foodways' which captures the intersections of food cultures, traditions, and history through lived practices (see Jones et al., 1983; the academic journal *Food and Foodways*; Chapter 34). Experience and knowledge of the cultures of other places and periods are offered to the tourist through food – food that is located in a particular landscape setting (Urry and Larsen, 2005: 158).

Eating and travel, agriculture and terroir, produce and place have long been intertwined (see Chapters 34 and 36). The constantly evolving cultures of food and landscape clearly influence and enhance each other in ways which are fundamentally inextricable (Chapter 8). They both offer cultural immersion, encompassing an expansive array of activities and encounters affecting all aspects of individual and collective life on a daily basis as well as seasonal cycles. Food tourism explores and exploits opportunities for experiences across contexts from highly controlled human spaces to wildscapes, with potential to target any age, race, creed, and cost, in any geography, climate, or ecotype. Nevertheless, at present, the dominant demographic of those with a food interest is the same for those with interests in landscape, travel and cultural activities in general, tending to be "couples that have above-average income, are usually professionals and are aged 30 to 50" (CTO, n.d.).

The literature of food tourism crosses disciplinary boundaries (Montero, 2015: 308; Long, 2003: 23). Not simply a confluence of interests in food and interests in travel, it attracts investigations by specialists in the arts and sciences, including anthropologists; historians; cultural geographers; economists; social scientists; designers; and those with specialisms in botany, geology, chemistry, medicine, agronomy, and more. It encompasses more recently recognised areas of formal education including food studies (Albala, 2013; Pilcher, 2012; Ashley et al., 2004) and other books within the series Routledge Studies of Gastronomy, Food and Drink (Routledge, 2017a), tourism studies (Smith and Richards, 2013), and the extensive Routledge Series in Tourism (Routledge, 2017c), as well as revived landscape studies (Howard et al., 2013) and the Routledge Series in Landscape (Routledge, 2017b) with frequent cross-referencing chapters. Similarly, Elsevier has a series on Hospitality, Leisure and Tourism (Elsevier, 2017). Food tourism is enmeshed in studies that cut across key issues affecting contemporary scholarship and society.

History: an overview of food travels and the media communicating their experience

Food tourism can be argued to arise in any period, in any society, where consumption of food accrues more significance and permits more choice than eating for mere subsistence (Fussell, 2009 [1987]). On the other hand, it might be said that it is not possible to live by bread alone even when it is only bread we eat, as any food always carries with it more meaning and more forms of sustenance than that provided by its consumable materiality. Whether in pursuit of abundance, rarity, or simplicity and frugal good health, what we eat carries conscious and unconscious associations and aspirations for a way of life, and it is records of these experiences that convey and excite these motivations.

From the earliest targeted accounts of places and their special dishes, such as advice to medieval pilgrims, through the first published guides for mass travellers including those of Baedeker and Michelin, to television shows, multi-media accounts, and global online resources marketing ever-shifting itineraries of special events, the story and diversity of food tourism media reflects that of food tourism itself. Significantly, food-focused tourism:

> can take place without travel having actually occurred at all. [. . .] Culinary tourism activity has been regarded as 'travelling through food' and eating the food of another out of curiosity or to encounter a new culture using food as an interpretative vehicle.
>
> *(Wight, 2008: 154)*

"Whether you go to food or food comes to you, the nature of the encounter is what defines a food experience as culinary tourism" (Kirchenblatt-Gimblett, 2003: xi).

While many describe food tourism as a recent phenomenon, the idea of travelling for gastronomic experience is an old one. Lucy Long notes debates amongst food scholars as to when tourism started as an industry (Long, 2013: 343), citing the pilgrimages of the European Middle Ages and the religious or nomadic journeys of various cultures marked by some consideration of and diversions for food provision. The seventeenth century saw trips by the working class to healthier destinations away from urban industrial centres and upper-class youth began Grand Tours to Europe (Black, 2009 [1992]; Sweet, 2015), although food did not play a major part in these travel plans.

It was the late eighteenth century, Stephen Smith notes, that marked a shift from "the focus of those involved with food as a human and economic phenomenon [. . .] its production, preservation, distribution, pricing and other utilitarian concerns" toward interests in food as "more than just a life necessity" (Smith, 2015: xiii). Food choice has always served as a sign of wealth and status, through rarity and quality of ingredients, to complexity of preparation, opulence of presentation, and fame of the preparer. The eighteenth century marks a period of global colonisation and immigration that began with such expressions of power experimenting with new ingredients such as tomatoes, potatoes, and chilies. Recipes developed that would become everyday staples of modern cuisines, the recipe books themselves serving armchair food tourism by referencing special places and their histories. "Cookbooks and other culinary literature could perhaps be seen as the first virtual media for culinary tourism, offering readers a window into other people's food" (Long, 2012: 391). Contemplating William Sitwell's *A History of Food in 100 Recipes* (2013), Bee Wilson suggests that "being asked to read recipes for their own sake, rather than with a view to cooking, gives a clearer sense of how they stimulate our imaginations" (2013).

Nineteenth century industrialisation saw radical changes in farming practices and food production, as it did in publication for a growing urban middle class (Chapters 2 and 9). Guide books are a significant category within the genre of travel literature, which includes outdoor literature, nature writing, and travel memoir. The widespread and highly popular Baedeker travellers' guides were published from 1828. The guides offered information on hotels as well as cafés and culture, including agriculture – for example, agriculture, vegetation, and general farm produce receive attention in the 1898 edition on Egypt (Baedeker, 1898) – but did not draw attention to food and food culture as a key part of the tourist experience. It was the rise of personal transport in the twentieth century that initiated the influential publication of Michelin Guides from 1900.

One of the more famous restaurant critics and voluminous reviewer and author, Maurice Edmond Sailland, known by his pen-name Curnonsky, is considered the inventor of gastronomic motor tourism through his regular writings in several newspapers and journals for Michelin. Interestingly, landscape architect Beatrix Farrand gifted a 1913 Michelin Guide to the British Isles (Michelin Tyre Company Ltd., 1913) to the General Library of the University of California, Berkeley. Michelin launched in 1900 at the Paris World Fair (Murcott et al., 2013: 79) although the star rating for restaurants was not introduced until 1926 with a three-star system appearing in the 1930s. During this time Curnonsky and novelist Marcel Rouff produced 24 volumes of *Tour de France Gastronomique* and in 1933, with another collaborator, Curnonsky published *Tresor gastronomique de la France* (Abramson, 2007: 124). Michelin's guides encouraged more car travel and so the need for more frequent changes of tyres, using the pursuit of quality to boost the quantity of their product that was consumed.

The twentieth century continued the traditions of cookery books and travel guides, with food tourism boosted later in the century by the mass media platforms of television, the internet, and the beginnings of digital interactive technologies (Collins, 2015). Recipe books compete by offering far more than the recipes themselves, elaborating on ingredients of specific places and their histories, along with information about where they might be purchased or grown. In particular, a chef's personal and family history can provide an important reference for readers to appreciate dishes beyond their palatability. *Giorgio Locatelli* and his wife Plaxy run the Michelin-starred Locanda Locatelli in Mayfair, London. His 2006 book *Made in Italy: Food and Stories* (Locatelli, 2006) is more an autobiography than recipe book, full of food stories of his growing up in the northern Italian town of Corgeno by Lake Comabbio. It explicitly evokes a life and a culture and intends that this knowledge is necessary to fully appreciate the food.

A key part of modern travel guides is the ever-shifting 'where to eat' section. Exemplary or unusual produce, producers, and places of production are also often described as worthy excursion destinations, with details provided of opening hours and tours available. Lonely Planet – called the largest travel guide book publisher in the world (Fildes, 2007) – was first published in the seventies (Wheeler, 1973). It was aimed at budget travellers, backpackers and youth who viewed travel as part of countercultural experiment and education. Originally based on the overland trip from London to Australia through Europe and Asia, guides were subsequently developed for India (1981), South America, Africa, and Korea. Lonely Planet moved decisively to a food focus with *Food Lover's Guide to the World: Experience the Great Global Cuisines* (Bittman, 2012), followed by similar publications for spicy food and street food. The focus on younger travellers may represent a liberating and even democratising trend in self-education, however it also represents a new and growing consumer market.

Television emerging from the mid twentieth century can similarly be examined as a democratising educational platform, while clearly also marketing products and services through carefully targeted direct advertising, product placement, and sponsorship (Samuel, 2001). The success of Julia Child's 1963 daytime program *The French Chef* aimed at American housewives cemented television food program popularity, but it was the establishment of *The Food Network* in 1993 which "helped transform the perception of cooking from a domestic chore into a culinary art" (Long, 2012: 391), one that could be experienced through travel. In *Foodie Makeovers: Public Service Television and Lifestyle Guidance* Isabelle de Solier begins her discussion with the hugely popular 1997 launch of *The Food Lovers' Guide to Australia* from the multicultural television broadcaster, the Special Broadcaster Service (SBS) (Solier, 2012: 65). Importantly she notes the shift from mere cooking show – the presenters were not chefs – to lifestyle consumer program. This was a show about consuming rather than producing, deliberately aimed at food lovers or 'foodies'. The show covered stories not just about those who prepared food professionally, but also home cooks, primary producers and others involved in the introduction, continuation, or evolution of food practices. With an emphasis on location and diverse environments as well as diverse immigrant cultures, the show was as much travel show as food program.

The twentieth century consolidated earlier financial and other administrative, scientific, and technological advances that would bring mass production of many standardised food products to an international consumer market. A consequence of the long trajectory of western, first-world influence has been waves of countercultural resistance along with popular appetite for experiences of food and landscapes which that very affluence threatens. As Joni Seager notes: "Travel is now the world's largest industry. The irony of tourism is that it often destroys some of the very places tourists travel to see" (Seager, 1995: 35). Indeed, they are made more attractive and interesting for being under threat. In the twenty-first century, interest in different, unusual, or threatened landscapes and produce has resulted in the unprecedented growth of food tourism. It is recognised as the fastest growing area of overall tourism (Stanley and Stanley, 2015; UNWTO, 2012; WFTA, 2013; TourismAustralia, 2016b; Hall and Sharples, 2008b; Parmar, 2015), increasingly promoted by governments, industry groups and other NGOs. There is flexible, cross-sector enthusiasm for the potential of food tourism to make substantial contributions to the economic development of strategically identified areas.

This growth has been spurred by satellite and digital broadcasting which has greatly expanded the reach of food shows and also changed the tools available to producers. Interactivity encourages consumer participation. The Food Network "invites you to share your own everyday food adventures and escapades [. . .] share a photo on Instagram with #MyFoodonTV, or post to our Facebook page and it could appear on the channel!" (SBS, 2015). In another entrepreneurial development, food product producers can be consumers of the online service producers. Two other 'Food Channels' – one Australian and the other based in the USA – are essentially promotional or advertorial channels collecting and packaging information on food products and services (The Food Channel, 2008; Food Channel, 2016).

Food travel blogs (Powell, 2005), apps, and online material for the foodie tourist are legion. Innumerable freelance, sponsored, and government sites offer advice and information, attempting to capture and advertise ever-changing offerings. These include the growing range of festivals (Hall and Sharples, 2008a) and ephemeral opportunities including food truck locations, markets, and flash-mob cooking. While much material is simply the most current version of information potentially available in other formats, digital and portable sites have additional qualities. A new smartphone app wakes you up to the smell of bacon cooking utilising a small plug-in external device to emit a puff of scent while the phone itself plays sizzling sounds (Global News, 2014). Other smart phone smell-o-vision could extend the possibilities for armchair travel and foodie travel research (Duell, 2014). Virtual reality technologies offer the ultimate armchair experience, however the tourism industry sees this not as competition but as introductory encouragement to actual travel (Las Vegas, 2016). Advances in what is primarily a visual virtual reality may soon permit a fuller multi-sensory virtual experience involving the smell and taste of food, and with the possibility of its being located in multi-sensorial, rather than merely scenic, landscapes.

Current concerns: economics, ethics, and authenticity

In an environment of global travel, distribution and reach, associating food with place becomes a key tactic in the development of strategies for economic development (Hall and Gossling, 2016). Placemaking and the marketing of place – irrespective of physical visitation – drives recognition and competition between cities, regions, states, and nations. Food provides popular and recognisable place distinction. Evident as a significant market sector from the beginning of the millennium, food tourism has more recently been promoted as an area to be improved by competitive providers, rather than simply explored by consumers (Stanley and Stanley, 2015; Wolf, 2016). In a world of increasingly documented experiences, there is considerable help available about how to be a food tourist. However, there is less advice about how to be a food tourism operator, conceiving, creating, and advertising food experiences for profit. This is of increasing interest to governments as much as private investors (Hall et al., 2011 [2003]; PIRSA, 2016; Tourism Australia, 2016a; CTA, n.d.).

Considering future issues and trends C. Michael Hall and Liz Sharples have noted that:

> the growth of these [ethics-focused] movements parallel to that of the growth of food events is no accident and reflects broader societal concern in many developed countries not only about the qualities

> of the food we are consuming, but also how rural regions and communities, agricultural practices and life-styles and certain forms of small-scale retailing can continue to survive if not prosper.
>
> *(Hall and Sharples, 2008b: 334)*

In discussing 'Cultural Tourism as Alternative, Moral Tourism', Craig Wight recognises that "[f]ood and eating are firmly established components within the burgeoning mix of new ethical, culturally concerned tourism experiences in the West" (Wight, 2008: 153). In language that suggests a criticism in turn of the motivations behind critique of established tourism models, he goes on to characterise food tourism as "providing the increasingly thrill-seeking bourgeoisie tourist with a new type of moral, alternative tourism" (ibid.: 154). Food tourism then suggests a complex, nested set of concerns and counter-concerns.

The ethics of food tourism include the spectrum of ethical issues facing food production and provision in general (see Chapter 24). Indeed, food tourism advocates and providers typically explain their purpose as actively responding to these issues.

> Whether you're a local or a tourist, edible experiences will change the way you think about your food, beyond the plate. We believe in local. We believe in authentic [. . .] We believe commitment and collaboration are the avenues to success in this industry. Together, these fundamental ideas reinforce the economic, environmental and social health of our communities.
>
> *(CTA, n.d.-a)*

In this view, food tourism tackles practices that threaten triple bottom line sustainability (Elkington, 2004). It is the thesis of Priscilla Boniface's book *Tasting Tourism* that "a considerable counter-revolution against food industrialisation has appeared [that] is a main trigger to food and drink tourism emerging" (2003: 4). Concerned critiques look at longer-term threats, emphasising environmental and social values in the face of shorter-term financial gains which have prevailed to give us a globally organised food industry characterised by contemporary agribusiness, genetically modified (GM or bioengineered) cropping, and other technologies for yield enhancement of plants and animals, and shelf-life extension techniques such as irradiation, high processing, and additives. Essentially, food and food tourism ethics look to slower, smaller, more integrated production and consumption, reducing food miles, reversing loss of plant and animal species, and strengthening local farming, cooking, and eating practices traditionally suited to their local climates and ecologies. Food tourism requires and encourages transparency which feeds more ethical practices. Interestingly, food tourism and the concept of foodways has arisen with the concept of a quadruple bottom line of four Ps. To the usual three – planet (environmental), people (social), and profit (economic) – is added a more malleable, inclusive fourth – purpose or progress – recognising exactly the sort of integrated and evolving cultural condition and values of site-specific foods (Sawaf and Gabrielle, 2014 [2007]; Taback and Ramanan, 2013). It is the less precise but no less important measures of historical site-specificity that identify authenticity (Urry and Larsen, 2005; Gilmore and Pine, 2007; Wight, 2008; Willingham, 2014).

The concept of authenticity embraces several aspects of the ethical experience of food and food production that consider both 'local' landscape and 'local' people. Local landscape considerations encompass ideas of terroir and place-specific environmental influences on produce (Chapter 36). Considerations of local people recognise traditional or alternative processes of farming and identification of the cultural significance of ethnic or indigenous foods (Chapters 1, 7 and 10). Protection of those landscapes and traditional processes is sought through listing, registration, and certification, variously covering the cultivation and processing of endemic species, inherited or heirloom plant varieties and animal breeds (Ghione, 2017; Abramson, 2007). Legislation offers a range of protections for the consumer, but also for producers and local culture and landscape integrity. Amongst the most well-known and oldest certifications are France's *Appellation d'Origine Contrôlée* (AOC), Italy's *Denominazione di Origine Protetta* (DOP) and less strict *Indicazione*

Geografica Protetta (IGP). Other countries have labelling laws that require identification of a variety of aspects of a food's origin and processing.

This is not to say there is not conflict in the sustainability of traditions that may, for example, be thought cruel, such as force-feeding geese for foie gras, aspects of veal production, and other time-honoured practices that arguably cause stress or potential stress to animals. Nevertheless, it is mass-farming including stall-rearing, caged egg production, and other concentrated animal feeding operations (CAFOs) which is most targeted rather than old-tradition local practices (Gurian-Sherman, 2008; MacDonald and McBride, 2009). Impressive sites of large infrastructure once held a place, particularly in local itineraries for tourists, as popular attractions. Tea towels featuring dams and drainage projects such as the Snowy Mountains Hydro-Electric Scheme and Ord River Irrigation system, or salt and pepper shakers emblazoned with canning or food processing factories were once commonly available celebrating the size and grandeur of industrial agri-food operations, collected by tourists who marvelled in the spirit of scientific positivism that dominated the earlier half of the twentieth century.

Ever at the forefront of analysis of cultural trends, architect Rem Koolhaas has dispassionately examined radical changes to the 'countryside' throughout the world. He demonstrates that there has been an unprecedented recent expansion in the scale and rapid digitisation of agricultural practices through the use of programmed and driverless harvesters, drones and similar hardware, in combination with temperature, moisture and weather indication informing optimum decision-making and productivity.

> The countryside is now the frontline of transformation. A world formally dictated by the seasons and the organization of agriculture is now a toxic mix of genetic experiment, industrial nostalgia, seasonal immigration, territorial buying sprees, massive subsidies, incidental inhabitation, tax incentives, political turmoil, digital informers, flex farming, species homogenization [...] in other words more volatile than the most accelerated city.
>
> *(Koolhaas, 2015)*

The reality of massive agricultural change belies popular perception of the countryside and preferred food production methods as championed, for example, by the Slow Food movement (Slow Food, 2015). This is one of many related paradoxes, contradictions, and conflicts that riddle discussions within food, landscape, and tourism studies.

Both food products and landscapes are marketed in paradoxical ways where the local is promoted globally (Wight, 2008: 153), fads of simple ingredients and holiday experiences are the subject of sophisticated marketing (humble kale becomes a superfood, camping becomes *glamping*), the most accessible foods and travel itineraries are the most highly processed and packaged, and the popularity of cooking and travel shows, food and tour guide celebrities, exists at a time when fewer people can or do cook, working longer hours with less time for breaks (Hall and Mitchell, 2005: 73; Wight, 2008). Food tourism embraces, defines and re-presents 'high' and 'low' cultures to each other. On the one hand, lowly, simple food is packaged for consumption by the wealthy as health benefit or unique artisanal exotica. On the other, ways are briskly and profitably found to render expensive ingredients or processes available to the masses. In either case, it is the explicit or implied stories of food-landscape origin which lend it prestige or quaintness, piquing curiosity and framing the potential tourist attraction.

'Glocalisation' is a neologism from the late 1990s used to identify efforts to adapt global practices to local conditions, and to deproblematise tensions between local and global imperatives (Roudometof, 2016). Glocalisation "describes the naturalization of MacDonald's in diverse settings through the use of local food sources, the transformation of the beef patty through adaptation to local taste, and accommodations to local consumer practices" (Gabaccia, 2012: 318). A parallel effect has shaped the architecture of MacDonald's restaurants (as it has of other national and international corporate outlets and franchises, including banks), with attempts, albeit superficial, to respond to local settings and history, whether urban, suburban, or embedded in rural roadside petrol stations. Urban settings in particular are seen to contribute to local culture through adaptive reuse of languishing heritage buildings. But there are further waves of complexity in the marketing of site-responsive food places. In 2016, a long-standing, family-owned Adelaide supermarket – the Chapley's Frewville IGA – was awarded International Retailer of the Year (Changarathil, 2016) at a conference in the US. This popular 'local' food place has to compete with not only glocalising corporate adaptors for customers, however, but also newer 'hipster' locals focused even more rigorously on a market for ethically oriented produce and service (Boisvert, 2015).

The nature of food tourism appears often unproblematically to be both profitable and sensitive to environmental and social conditions with a view to sustaining and improving them (Everett and Aitchison, 2008). However, as in many other areas, the ethical ideals of food tourism and its links with a green economy are not without contests and contradictions. In *Black, White, and Green: Farmers Markets, Race, and the Green Economy*, Alison Alkon points out that "economic, environmental and social goals remain at odds" (Alkon, 2012: 124) and offers empathetic critique that recognises severe limitations as well as already well-articulated strengths.

Food tourism is marketed as a win-win arena where not-for-profit interest groups align with business interests and government support for both. Ontario's cross-sector *Culinary Tourism Alliance* declares: "It is our goal to #GrowFoodTourism globally by working with locals to create sustainable, authentic, and meaningful tourism experiences through food and drink" (CTA, n.d.-a). Their 'industry affiliations' include local, national, and international partners in government, business, and not-for-profit organisations. At a global scale, the potential benefits of food tourism have been aligned with the United Nations Sustainable Development Goals by the World Travel and Tourism Council (WTTC) and even argued to foster world peace:

> Travel and Tourism has often been recognised for its ability to drive peace, security and understanding. World leaders, from John F Kennedy to Bill Clinton to Tony Blair, have highlighted the importance of the sector. Now, for the first time, the World Travel and Tourism Council (WTTC) has partnered with the Institute of Economics and Peace (IEP) to explore the links between tourism and peace. The research shows that countries with a more open and sustainable tourism sector tend to be more peaceful.
>
> *(WTTC, 2016)*

The growing success of food tourism enterprises and food tourist enjoyment is assisted by bipartisan appeal to both conservative and radical agendas; to both advanced capitalism and its new markets of spectacle and sensation, and to a hopeful sense that consumers may somehow challenge rampant consumerism. Hall et al.'s first chapter of *Food Tourism Around the World* is titled "The Consumption of Experiences or The Experience of Consumption? An Introduction to The Tourism of Taste" (2011 [2003]: 1–24). Ideas about the commodification of immaterial sensation and experience are not new. Guy Debord's *Society of the Spectacle* (1977 [1967]) along with other Situationist works, thoroughly examined advanced capitalism's commodification of experience and perception, outlining "the principle of commodity fetishism, the domination of society by intangible as well as tangible things" (ibid.: clause 35).

Conclusion

The dynamic conditions of food tourism mirror core challenges for contemporary society. It may yet prove a utopian hope (Chapter 40) to suppose that the educative, broadening, and visceral experiences that food tourism can provide will help realise a future such as that outlined in the UN Sustainable Development Goals, and create global agreement on ethical solutions for delivering inclusive, fulfilling environments for ongoing human habitation. As many are empowered to pursue the rich opportunities which emerging food and place appreciations are revealing, it is a vital time to be alert to the layering and slippages of intention in these pursuits. Similar conflicts exist between nostalgia-driven protectionism and creativity-fuelled future heritage in decisions regarding our built environments and use of landscapes.

The rise of food tourism can represent both the greatest, inclusive achievements of human art, ingenuity, and generosity, simultaneously with its most pretentious, destructive conceits. The enthusiastic veneration of local cuisines, single dishes, site-specific ingredients, and distinctive traditions is an economic and educational blessing, broadly enriching lives, stimulating interests in and tolerance of the unfamiliar, and diversifying our diets. At the same time, it can generate popular standardisation or fetishism; threatening the loss of that distinction, impacting negatively on local landscapes, weakening personal judgement and responsibility, and potentially causing the collapse of specific food supplies through over-demand, or stagnation of cultures through the protectionism of consumer product guarantees (Chapter 36). How food tourism evolves to articulate and deal with the intellectual and ethical challenges facing it, will, no doubt, reflect how human cultures will strive to deal with the same consumption challenges facing the full spectrum of future human endeavour.

Key texts

Abramson, J. (2007) *Food culture in France.* Westport: Greenwood Press.

Albala, K. (ed.) (2013) *Routledge international handbook of food studies.* New York: Routledge.

Boniface, P. (2003) *Tasting tourism: Travelling for food and drink* Aldershot: Ashgate.

Gilmore, J., & Pine, J. (2007) *Authenticity: What consumers really Want.* Boston: Harvard Business Review Press.

Hall, C.M., & Gossling, S. (eds.) (2016) *Food tourism and regional development: Networks, products and trajectories.* London and New York: Routledge.

Hall, C.M., Sharples, L., Mitchell, R., Macionis, N., & Cambourne, B. (eds.) (2011 [2003]) *Food tourism around the world.* Abingdon: Routledge.

Helstosky, C. (ed.) (2015) *The Routledge history of food.* New York: Routledge.

Howard, P., Thompson, I., & Waterton, E. (eds.) (2013) *The Routledge companion to landscape studies.* New York: Routledge.

Long, L.M. (ed.) (2003) *Culinary tourism.* Lexington: The University Press of Kentucky.

Pilcher, J.M. (ed.) (2012) *The Oxford handbook of food history.* Oxford: Oxford University Press

Stanley, J., & Stanley, L. (2015) *Food tourism: A practical marketing guide.* Wallingford: CABI.

Further reading

Albala, K. (ed.) (2013) *Routledge international handbook of food studies.* New York: Routledge.

Alkon, A.H. (2012) *Black, white, and green: Farmers markets, race, and the green economy.* Athens: University of Georgia Press.

Ashley, B., Hollows, J., Jones, S., & Taylor, B. (eds.) (2004) *Food and cultural studies.* London: Routledge.

Baedeker, K. (1898) *Egypt handbook for travellers.* Karl Baedeker. Available from: https://archive.org/details/01329502.5457.emory.edu (accessed 6 May 2017).

Bittman, M. (2012), *Food lover's guide to the world: Experience the great global cuisines.* Footscray, Lonely Planet.

Black, J. (2009 [1992]) *The British abroad: The grand tour in the eighteenth century.* Stroud, Gloucestershire: The History Press.

Boisvert, E. (2015) New Glen Osmond Road Market upping the cool, hipster stakes against popular Frewville Foodland *Eastern Courier Messenger.* August 30, 2015 2:09pm, www.adelaidenow.com.au/messenger/east-hills/new-glen-osmond-road-market-upping-the-cool-hipster-stakes-against-popular-frewville-foodland/news-story/8e2c70e617fe64e7bdc5ccfcba21a2bf (accessed 19 June 2017).

Boniface, P. (2003) *Tasting tourism: Travelling for food and drink.* Aldershot: Ashgate.

Changarathil, V. (2016) Chapleys-owned Frewville Foodland supermarket named best in the world *The Advertiser.* March 2, 2016 9:39am.

Collins, K. (2015) Cooking class: The rise of the 'foodie' and the role of mass media. In: Helstosky, C. (ed.), *The Routledge history of food.* New York: Routledge, pp. 270–290.

Cros, H.D., & McKercher, B. (2015 [2002]) *Cultural tourism.* Abingdon and New York: Routledge.

CTA (n.d.-b) *Culinary tourism: A recipie for economic development success*. Culinary Tourism Alliance. https://ontarioculinary.com//wp-content/uploads/2012/06/OCTA-INFO.REBRAND.pdf (accessed 6 May 2017).
CTA (n.d.-a) About the Culinary Tourism Alliance. *Culinary Tourism Alliance*. www.growfoodtourism.com/about-us/ (accessed 6 May 2017).
CTO (n.d.) *Culinary (or Gastronomic) tourism*. Caribbean Tourism Organization. www.onecaribbean.org/content/files/CulinaryCaribbeanNicheMarkets.pdf (accessed 6 May 2017).
Debord, G. (1977 [1967)) *Society of the spectacle*. Detroit: Black & Red.
de Solier, I. (2012) Foodie makeovers: Public service television and lifestyle guidance. In: Palmer, G. (ed.), *Exposing lifestyle television: The big reveal*. Aldershot, UK: Ashgate Publishing, pp. 65–81.
Duell, M. (2014) Smell-O-vision for the 21st century: Phones able to send scented messages are among ten emerging technologies for 2015. *Daily Mail Australia*. www.dailymail.co.uk/sciencetech/article-2880142/Smell-O-Vision-21st-Century-Phones-able-send-scented-messages-ten-emerging-technologies-2015.html#ixzz4WSLundvv (accessed 6 May 2017).
Elkington, J. (2004) Enter the triple bottom line. In: Henriques, A., & Richardson, J. (eds.), *The triple bottom line, does it all add up? Assessing the sustainability of business and CSR*. London: Earthscan Publications Ltd, pp. 1–16.
Elsevier (2017) *Hospitality, leisure and tourism series*. Butterworth Heinemann. www.elsevier.com/books/book-series/hospitality,-leisure-and-tourism (accessed 6 May 2017).
Everett, S., & Aitchison, C. (2008) The role of food tourism in sustaining regional identity: A case study of Cornwall, South West England. *Journal of Sustainable Tourism*, 16(2), 150–167.
Fildes, N. (2007) BBC gives lonely planet guides a home in first major acquisition. *Independent*. www.independent.co.uk/news/media/bbc-gives-lonely-planet-guides-a-home-in-first-major-acquisition-395739.html (accessed 6 May 2017).
The Food Channel (2008) *The food channel*. www.foodchannel.com/about/ (accessed 6 May 2017).
FoodChannel (2016) *Food channel*. www.foodchannel.com.au/ (accessed 6 May 2017).
Fussell, B. (2009 [1987]) Foreward. In: Toussaint-Samat, M. (ed.), *A history of food*. Chichester: Blackwell Publishing, pp. xiii–xiv.
Gabaccia, D.R. (2012) Food, mobility and world history. In: Pilcher, J.M. (ed.), *The Oxford handbook of food history*. Oxford: Oxford University Press, pp. 305–323.
Ghione, J. (2017) Country of origin labeling: France at the finishing line, Italy still on the starting blocks. *Slow Food*. www.slowfood.com/country-origin-labeling-france-finishing-line-italy-still-starting-blocks/ (accessed 6 May 2017).
Gilmore, J., & Pine, J. (2007) *Authenticity: What consumers really want*. Boston, MA: Harvard Business Review Press.
Global News (2014) *New smartphone app wakes you up to the smell of bacon cooking*. Associated Press. Available from: http://globalnews.ca/news/1193058/new-smartphone-app-wakes-you-up-to-the-smell-of-bacon-cooking/ [Accessed 6 May 2017].
Gurian-Sherman, D. (2008) *CAFOs uncovered: The untold costs of confined animal feeding operations*, April. Cambridge, MA: Union of Concerned Scientists.
Hall, C.M., & Mitchell, R. (2001) Wine and food tourism. In: Douglas, N., & Derret, R. (eds.), *Special interest tourism: Context and cases*. Brisbane, John Wiley & Sons, Australia, pp. 307–329.
Hall, C.M., & S. Gossling (eds.) (2016) *Food tourism and regional development: Networks, products and trajectories*. London and New York: Routledge.
Hall, C.M., & Mitchell, R. (2005) Gastronomic tourism: Comparing food and wine tourism experiences. In: Novelli, M. (ed.) *Niche tourism: Contemporary issues, trends and cases*. Oxford: Taylor & Francis, pp. 73–88.
Hall, C.M., & Sharples, L. (eds.) (2008a) *Food and wine festivals and events around the world: Development, management and markets*. Oxford: Elsevier Butterworth-Heinemann.
Hall, C.M., & Sharples, L. (2008b) Future issues and trends: Food events, festivals and farmers' markets. In: Hall, C.M., & Sharples, L. (eds.), *Food and wine festivals and events around the world: Development, management and markets*. Oxford: Elsevier Butterworth-Heinemann, pp. 331–348.
Hall, C.M., & Sharples, L. (2011 [2003]) The consumption of experiences or the experience of consumption? An introduction to the tourism of taste. In: Hall, C.M., Sharples, L., Mitchell, R., Macionis, N., & Cambourne, B. (eds.) *Food tourism around the world*. Abingdon: Routledge, pp. 1–24.
Hall, C.M., Sharples, L., Mitchell, R., Macionis, N., & Cambourne, B. (eds.) (2011 [2003]) *Food tourism around the world*. Abingdon: Routledge.

Howard, P., Thompson, I., & Waterton, E. (eds.) (2013) *The Routledge companion to landscape studies*. New York: Routledge.
Jones, M.O., Giuliano, B.B., & Krell, R. (1983) *Foodways and eating habits: Directions for research*. Los Angeles: California Folklore Society.
Kirchenblatt-Gimblett, B. (2003) Forward. In: Long, L.M. (ed.), *Culinary tourism*. Kentucky: The University Press of Kentucky, pp. xi–xv.
Koolhaas, R. (2015) *The countryside*. Harvard GSD. www.youtube.com/watch?v=shVxB6wRHo0 (accessed 6 May 2017).
Las Vegas (2016) *How virtual tourism will help enhance real-world travel*. Mashable Australia. http://mashable.com/2016/04/22/vr-tour-travel-brandspeak/?utm_cid=p-LV-2-tw-bc#Tc9VJ80WBgqt (accessed 6 May 2017).
Locatelli, G. (2006) *Made in Italy: Food and stories*. London: Fourth Estate.
Long, L. (2013) Food in tourism studies. In: Albala, K. (ed.), *Routledge international handbook of food studies*. New York: Routledge, pp. 342–351.
Long, L.M. (ed.) (2003) *Culinary tourism*. Lexington, KY: The University Press of Kentucky.
Long, L.M. (2012) Culinary tourism. In: Pilcher, J.M. (ed.), *The Oxford handbook of food history*. Oxford: Oxford University Press, pp. 389–406.
MacDonald, J.M., & McBride, W.D. (2009) The transformation of U.S. livestock agriculture: Scale, efficiency, and risks. *Economic Information Bulletin*. No. (EIB-43) (January).
Michelin Tyre Company Ltd (1913) *Michelin guide to the British Isles*. Iliffe & Sons Ltd. https://archive.org/details/michelinguidetob00michrich (accessed 6 May 2017).
Montero, C.G. (2015) Tourism, cuisine and the consumption of culture in the Carribbean. In: Helstosky, C. (ed.), *The Routledge history of food*. New York: Routledge, pp. 291–312.
Murcott, A., Belasco, W., & Jackson, P. (eds.) (2013) *The handbook of food research*. London: Bloomsbury Academic.
Parmar, P. (2015) How culinary tourism is becoming a growing trend in travel. *Huffington Post*.
Pilcher, J.M. (ed.) (2012) *The Oxford handbook of food history*. Oxford: Oxford University Press.
PIRSA (2016) *Agriculture food and wine in South Australia*. Adelaide, Primary Industries and Regions SA.
Powell, J. (2005) *Julie and Julia: My year of cooking dangerously*. New York: Little Brown & Company.
Roudometof, V. (2016) *Glocalization: A critical introduction*. London and New York: Routledge.
Routledge (2017a) *Routledge studies of gastronomy, food and drink*. Taylor and Francis Group. www.routledge.com/tourism/series/RSGFD (accessed 6 May 2017).
Routledge (2017b) *Series in landscape*. Taylor and Francis Group. www.routledge.com/landscape/series (accessed 6 May 2017).
Routledge (2017c) *Series in tourism*. Taylor and Francis Group. www.routledge.com/tourism/series/SCSP40 (accessed 6 May 2017).
Samuel, L.R. (2001) *Brought to you by: Postwar television advertising and the American dream*. Austin, TX: University of Texas Press.
Sawaf, A., & Gabrielle, R. (2014 [2007]) *Sacred commerce: A blueprint for a new humanity*. Ojai, CA: EQ Enterprises.
Seager, J. (1995) *The state of the environment atlas*. London: Penguin Books.
Sitwell, W. (2013) *A history of food in 100 recipes*. New York: Little, Brown and Company.
Slow Food (2015) *About us*. Slow Food. www.slowfood.com/about-us/ (accessed 6 May 2017).
Smith, M., & Richards, G. (eds.) (2013) *The Routledge handbook of cultural tourism*. New York: Routledge.
Smith, S. (2015) Forward. In: J. Stanley & L. Stanley (eds.) *Food tourism: A practical marketing guide*, eds J. Stanley & L. Stanley. Wallingford, CABI, pp. xiii–xv.
Stanley, J., & Stanley, L. (2015) *Food tourism: A practical marketing guide*. Wallingford, CABI.
Sweet, R. (2015) *Cities and the grand tour: The British in Italy, c.1690–1820*. Cambridge: Cambridge University Press.
Taback, H., & Ramanan, R. (2013) *Environmental ethics and sustainability: A casebook for environmental professionals*. Boca Raton, London and New York: CRC Press.
Taylor & Francis (1985–) *Food and foodways: Explorations in the history and culture of human nourishment*. Philadelphia, Taylor & Francis (Routledge).
Timothy, D.J. (2001) *Cultural heritage and tourism: An introduction*. Bristol, New York and Ontario: Channel View Publications.
Tourism Australia (2016a) *Factsheets and media releases*. Australian Government. www.tourism.australia.com/campaigns/factsheets.aspx (accessed 6 May 2017).

Tourism Australia (2016b) *Gourmet tourism is growing.* Australian Government. www.tourism.australia.com/documents/Food_and_Wine_Fact_Sheet.pdf (accessed 6 May 2017).

Tourism Australia (2016c) *There's nothing like Australia campaign: Food and wine.* Australian Government. www.tourism.australia.com/campaigns/Food-Wine.aspx (accessed 6 May 2017).

UNWTO (2012) *Global report on food tourism, AM reports*, vol. 4. Jordan, P. (ed.) Madrid.

Urry, J., & Larsen, J. (2005 [1990]) *The tourist gaze.* London: Sage.

WFTA (2013) *About the association.* World Food Travel Association. www.asta.org/files/MainSite/Quicktime/TargetingtheCulinaryTravelerWorldFoodTravelAssociation.pdf (Accessed 6 May 2017).

Wheeler, T. (1973) *Across Asia on the cheap: A complete guide to making the overland trip.* Footscray, Lonely Planet.

Wight, C. (2008) Reengineering authenticity: Tourism encounters with cuisine in rural great Britain. In: Rubin, L.C. (ed.), *Food for thought: Essays on eating and culture.* Jefferson, NC: McFarland and Company, pp. 153–165.

Willingham, R. (2014) *Authenticity: The head, heart, and soul of selling.* New York: Prentice Hall Press.

Wilson, B. (2013) The pleasures of reading recipes. *The New Yorker.* Available from: www.newyorker.com/books/page-turner/the-pleasures-of-reading-recipes (accessed 6 May 2017).

Wolf, E. (2016) *Food tourism: The latest fad or here to stay?* WFTA. Available from: http://foodtravelnews.org/food-tourism-the-latest-fad-or-here-to-stay/ (accessed 6 May 2017).

WTTC (2016) *Tourism as a driver of peace.* World Travel and Tourism Council. www.wttc.org/research/other-research/tourism-as-a-driver-of-peace/ (accessed 6 May 2017).

36

Terroir

A socially constructed subterranean landscape gone global

Zachary Nowak

The French word *terroir* has been rendered into English as a "the taste of place," and posits "a relevant and possibly vital connection" between a specific place – the landscape – and the agricultural products that come from it (Trubek, 2008: xiii). The core concept of terroir is that every place has a certain combination of variables – soil composition, rainfall, annual light, and climate are among the most often cited – that, mediated by local traditions, knowledge, and modes of production, give food from that locale a distinctive taste. The landscape of a place is broadly conceived to include not only terrestrial variables but also subterranean realities, atmospheric conditions, and even production methods. While the concept of terroir was invented for European agriculture landscapes, it has been transported across the world by globalisation, dispersed to both strong economies as well as relatively weak sites of agricultural production. Terroir is supposed to imprint itself on the taste profiles of wines, cheeses, sausages, as well as maple syrup, pizza, sake, coffee, tequila, beer, and even fruit. This chapter will review the literature on terroir and focus on the intimate legal connection between the idea of terroir and the reality of the protected denomination of origin (PDO) production zones that supposedly contain it. As Susan Terrio has suggested, crafting products has always required crafting history (1996). With this in mind, this chapter will use specific examples such as parmesan cheese to explore terroir as a social construction that was pieced together over time. It will also examine how and why it was constructed and what this construction has to do with economics, the creation of value, and the maintenance of social boundaries. I argue that terroir has a long intellectual history and tortured logic, but nonetheless strongly influences actual material outcomes like the maintenance of certain landscapes and their monopoly use by certain, select actors.[1]

One of the principal places in which the concept of terroir is said to be found are the physical landscapes in which it is supposedly embedded, since the history of terroir is the history of the linkage between name, place, and taste. Initially, limited monopolies granted by governments to particular producers to use certain names were not recognition of unique tastes, but rather anti-fraud measures. The earliest European example[2] is a 1411 edict of Charles VI, decreeing that only cheese made in Roquefort could bear that name. Even the word terroir has had an evolution: Rengenier Rittersma has shown that in the twelfth century it referred simply to a particular soil's suitability for agriculture, and only three centuries later was it first associated with the flavor of food from a particular area. In the fifteenth century, though, "un goût de terroir" (a taste of terroir) had negative connotations: instead of referring to a well-made wine, it instead was used to describe a wine that tasted "rural," as if it were made by inexpert producers (2011: 4). Though linked to food after the fifteenth century, the specific meaning of the word terroir and its connotations oscillated in

France. Robert Parker's recent intellectual history of terroir reveals that the relative power of Paris vis-à-vis the provinces, as well as ideas of linguistic purity, influenced whether the taste of a particular place was to be celebrated or viewed with suspicion (2015).[3]

Despite these medieval precedents, the birthplace of the modern concept of terroir – that raw materials from certain precisely delimited areas processed in traditional ways give rise to high-quality foods whose tastes are unreproducible elsewhere – is Bordeaux, and terroir's coming of age was in 1855. Emperor Napoleon III requested that French wine merchants create a classification system of Bordeaux wines for the 1855 Exposition Universelle in Paris. The modernity of the concept was not simply in its insistence on the tight link between taste and geography, but in its pretensions to objectivity. The rank ordering was perfectly correlated to the reputation of the house: taste was presumed to be coincident with class. This sort of classification (and the extension of legal limited monopolies to certain producers) was extended nationwide in the 1930s with the creation of the *appellation d'origine contrôlée* (AOC) system. This system extended on the Bordeaux classifications from wines to potentially any food or beverage, confirming the political-legal linkage between taste and place (of production). Other European countries followed, although it took until 1992 to create a supra-national E.U.-wide system of classification. This law, and subsequent addenda and extensions, make the economic motivation for terroir clear. In addition to highlighting the growing demand of consumers for agricultural products with an identifiable origin, the regulation states that "the promotion of products having certain characteristics could be of considerable benefit to the rural economy, in particular to less-favoured or remote areas, by improving the incomes of farmers and by retaining the rural population in these areas" (European Economic Community, 1992: 1). While the law does gesture towards the necessity of gastronomic characteristics that are "essentially or exclusively due to a particular geographical environment with its inherent natural and human factors, and the production, processing and preparation of which take place in the defined geographical area," it makes no mention of how to evaluate these characteristics or need to demonstrate them empirically (ibid.: 2).

One of the most important debates regarding terroir concerns its boundaries: where it starts and stops, who gets to decide those limits, and who gets to employ it to commodify and add value. Terroir is an effect that anyone can taste (or fail to taste), but because it becomes "real" through the force of law, it is an intensely political process. As Kolleen Guy has shown, the determination of the boundaries of the zone of production for Champagne in the nineteenth century had little to do with soil samples and rain gauges, and much to do with old administrative boundaries, accusations of monarchical tendencies, and the power of large producers (Guy, 2003). Indeed, most terroir boundaries (the boundaries of the PDO zones) are simply administrative boundaries: the limits of administrative control of towns (such as Colonnata lard), parts of regions (the *département* of Champagne), or entire provinces (Balsamic vinegar of Modena). The lack of a process in the E.U. law cited earlier for evaluating a food's "specific quality, reputation or other characteristics attributable to that geographical origin" (European Economic Community, 1992: 2) of food – in other words, its terroir – is a significant and likely intentional oversight, as terroir is difficult to bound, and difficult to sense. While the legal boundaries established by PDO designations are quite precise (one can produce Colonnata lard in Colonnatta, but not just over the city limits in the neighbouring towns which have also produced similar spiced lard for centuries), the geophysical variables that they supposedly describe do not tend to fit neatly into administrative boundaries.[4] The marble that Colonnata lard is made in does not come simply from inside of the town of Colonnata, but from the quarries in nearby towns as well.[5] Soils – often described in terroir literature as uniform inside PDO areas – are actually quite variable, a fact often elided or obscured by producers. The larger the PDO zone of production is, the more heterogeneity there will be inside its boundaries.[6]

Parmesan cheese provides an excellent example of the complicated pasts that sometimes are forgotten in discourse about a particular food's terroir.[7] The early history of Parmesan cheese is obscure but Benedictine monasteries probably began exploiting cows for their milk around the year 1000. Before that time it seems that cattle had been draft animals, and only sheep and goats had been milked. These monks laid claim not only over the spiritual domain, but also over the terrestrial landscape: they directed the reclamation of much swampy area in the valley, but also controlled much of the territory around the cities of Parma and Reggio Emilia. The lands they owned were leased to tenants, and the monks began at some point to use their tenants' milk to make an excellent cheese. By the year 1349, "Parmigiano" was famous enough for Boccaccio to refer to it in the *Decameron*; it is fitting that in his imaginary land of plenty (ever a peasant's dream) an entire mountain was made of parmesan cheese. Boccaccio describes cooks in this fantastical world sitting atop of the mountain of Parmesan cheese, cooking macaroni in capon broth and rolling them down the hill when they were done. Parmesan appears in this fable born of a collective imaginary precisely because the monks' tenants rarely ate the cheese, though they produced the milk for it. Italian food historians Alberto Capatti and Massimo Montanari have described at length the ironic position of the peasants: they were the ultimate source of the raw materials in the countryside, but the wealth derived from the processing flowed to the urban elites. The foods were made in the countryside but took the name of the urban centres where they were exchanged: *Parmigiano* is simply the adjectival form of the name of the city Parma (Capatti and Montanari, 2003). In the sixteenth through eighteenth centuries the main producers of Parmesan cheese were feudal lords, the large abbeys (still major landholders), and the local gentry. Of course, these elites were not "producers" in the strictest sense, but rather controlled production. For the peasant tenants who did the actual work, the most salient "taste" of Parmesan cheese in the late Middle Ages and Renaissance was bitterness: they enjoyed the fruits of their labours only infrequently.[8]

There are also complicated histories for the livestock who are the ultimate primary producers for terroir products, and these histories further complicate the "traditions" on which terroir rests. To return to the Parmesan example, the actual primary producers – the cows – are not the same as in centuries past. The tenant farmer who produced parmesan cheese between the year 1000 and 1950 needed a cow that could split its energy between milk, meat, and traction. The dominant breeds were therefore the Alpine Brown and the Reggiana Red. Today only the quantity of milk is important and production relies on black and white Holsteins. While 75% of forage must be from within the PDO zone, it is likely much different from what the cows of yesteryear ate: in the past, cows grazed on semi-wild meadows and were fed a mix of hay and straw, corn and corn stalks, legumes (other than soy), and elm leaves from trees grown in intensive perennial polycultures. They are now either confined or grazed on plant species mixed to provide for maximum milk production. In addition to fodder, the rules governing the production of Parmesan cheese allow supplements permitted by Italian law such as sodium chloride, sodium bicarbonate, dicalcium phosphate, calcium carbonate, magnesium oxide, glycerol, as well as other vitamins, all common bovine feed supplements. It goes without saying that tenant farmers in the 1650s did not use these chemicals. The regulations also specifically permit alfalfa and soybeans as fodder for the cows, both of recent introduction to bovine diets in northern Italy. Chemical analyses confirm that the milk of the older breeds is quite different from Holstein's milk in its content of the proteins and short-chain fatty acids that give the final cheese its taste (Castagnetti et al., 2000). Indeed, the author of this chapter underlines the relative significance of breed vis-à-vis the relative insignificance of soil type on the organoleptic qualities of the resulting milk. If there is a taste of place, that taste has to be conceived of in both spatial and temporal axes: as the workscape (as well as the botanical landscape) of fields have changed, so too has the taste of Parmesan cheese changed over time. The move towards freeholding, the widespread use of tractors for ploughing, and the introduction of Holsteins have all surely changed the taste of Parmesan cheese. This is the story of the urban industrialism invading the countryside, urban revolutions that change not only the organisation of rural areas, but even their biota. The result is that instead of taste continuities stretching across centuries, we are sure to find a series of terroirs, ever dynamic, each different than the decade before.

Terroir has been used, as we will see later, by more than just large producers seeking monopolies, but this was undeniably one of its first broad uses. Rachel Laudan, in her discussion of a book by Slow Food co-founder Carlo Petrini, describes that organisation's calculated use of what she calls "the French Terroir Strategy." Between the 1860s and the 1930s, the French wine industry faced not only the destructive louse

Phylloxera, but also competition from Algerian vineyards planted with French grape cultivars. The solution was to proclaim that French wine could only be made in France because of terroir. Terroir was less a theory than a marketing strategy: "developed [. . .] first to prop up the French wine industry and then to encourage culinary tourism" (2004: 138). Terroir then seems to be the stepchild of modern economic imperatives (providing semi-monopoly rents to certain agricultural producers) and outdated geographic theories (regional geography). Proponents of terroir argue that the effect is the result not simply of the environment, but also of human intervention: traditional production methods would allow terroir to emerge from one zone with varied soils or weather. This "weak" version[9] of terroir decentres the fields and re-centres the artisan producer as the primary creator of terroir. This artisan, however, is in a dialectical relationship with the landscape around her. As Tim Waterman has suggested, terroir describes "patterns of habitation and the practices of daily life that have evolved in an area as a complex interrelationship between a landscape and its inhabitants" (Waterman, 2011: 64).

Despite terroir's reliance on the word "traditional," it's clear that even these exalted foods' production has been dynamic. Indeed, anthropologists and historians of food have often suggested that "traditional foods" become "traditional" not because they remain the same, but rather are able to adapt to changing political, economic, environmental, and social landscapes. The legal codes that calcify supposedly "traditional" methods of production instead are more like a photograph of current methods, back-projected into the distant past. A return to Parmesan cheese can provide examples of the change in production techniques. The two inventions crucial to producing consistent taste in almost all high-quality food products, and the two materials least discussed, are stainless steel and food-grade plastic. These two materials isolate food from contaminants, and yet they may eliminate precisely the variables that once gave foods part of their flavour. In the case of Parmesan cheese, I wonder about ashes from the fires underneath the cauldrons, or tin from the insides of buckets in which milk was collected in, or the unlacquered wood shelves that the maturing cheese rested on.[10] There are also other changes in the way Parmesan is made that certainly affects its flavour. Today, tour guides at the various cooperatives that make up the consortium of Parmesan cheese producers emphasise that nothing must be put on the outside of the cheese, as it would inevitably affect the taste. Even the dyes used to put a type of proprietary barcode on the wheels of parmesan must be made out of casein (the principal protein in cheese). Yet in the 1800s, rust was mixed with oil and applied to the outside of the wheels. Later producers darkened the cheeses' rinds with lampblack mixed with grapeseed oil. At some point, black wax was substituted, and this was used until 1964.[11] Our modern cheese is also straw-coloured, a novelty in the over seven centuries of Parmesan cheese making. The official history of the consortium notes that the tradition of flavouring and colouring the cheese with saffron persisted at least until 1940 (Zannoni, 1999: 39). The legal codes that regulate PDO zones and invent terroir are also then reflections of the *present* – its breeds and varieties, its raw materials, its materials, its methods, and everything else that is part of the alimentary landscape, broadly conceived – not necessarily just a reflection of that PDO zone's *past*.[12] Thinking about terroir means then not simply adding together the subterranean and stratospheric landscapes to the realities on the ground: it means remembering that terroir, while supposedly a reflection of something immutable, is more likely a representation of the sum of changes in production. Terroir does not negate change: it is a reflection of it, though it iterates so slowly through time that it seems eternal and unvarying.

The commercial need to fulfil an unchanging, atemporal, eternal terroir theory may ironically lead to the intentional elimination of terroir effects by producers.[13] In other words, the need to make the taste of a certain product identical across both space (the extent of the PDO zone) but also across time (from year to year) encourages producers to intervene chemically when yearly fluctuations in sunlight or precipitation or other variables change the raw materials' chemical composition. This is most clear with wine, the product terroir was born entangled with, and with which it is still most clearly linked. Even a rapid perusal of a handbook of enology reveals that many of the organic and inorganic levels in the grape must be "adjusted" by the enologist. If the grapes are not quite ripe, there are often too many organic acids in the resulting juice, or *must*. The E.U. permits the addition of tartaric acid (up to 2.5g/L) to "buffer." If mannoproteins inhibit tartrate crystallisation, the enologist can use "an industrial preparation of β-(1–3)- and β-(1–6)-glucanases (Glucanex™)" (Ribéreau-Gayon, 2005: 18, 44). For every problem caused by too little sun, too much water, or too much of a particular mineral in the soil – that is, for "problems" caused by terroir's appearance in the must or fermented wine – there is a solution: the gum arabic treatment, the ascorbic acid treatment, the potassium ferrocyanide treatment, and calcium phytate treatment, and many others. Many of these interventions are attempts to remove geophysical variables introduced by human action. For example, Australian scientists studying soil composition and plant mineral uptake in vineyards found that irrigated water brings with it minerals, which then change soil (and therefore, plant) chemistry in the irrigated area (Green et al., 2004). The laboratory of the enologist is arguably one of the most important parts of the whole process of making wine, and the place that is least described in discourse on wine. Here again we find the urban techno-culture hidden in the rural landscape: stainless steel vats, not wood; chemical reactions, not techniques handed down by artisans.[14]

The preference for the traditional food landscape over the hidden scientific (and quite modern) one is clear from the cover of a recent book whose subtitle promises an investigation of wine from "vineyard to glass." The pictures are of carefully manicured vines, hands aged by much labour, old baskets, and a dusty wine bottle, but no bank of temperature-controlling computers, no stainless steel drums, and no plastic bottles of reagents (Black and Ulin, 2013). I would argue, then, that one of the enologists' main jobs is to defeat terroir, to smooth out the dramatic variations in the content of the grape must and fermented wine. Enologists – like their counterparts in other food industries – are essentially highly specialized organic chemists whose tools are stainless steel containers, factory-produced dried yeast, and an array of white powders. Because they don't fit the overwhelmingly rural (and indeed, bucolic) ideal that we have imposed on production of supposedly unique high-quality foods, their involvement with the making of wine is obscured. The hiding of the enologist's work is perfectly consistent with maintaining the two ideas about *making*, to which Sarah Scarlett refers in her article comparing industrial and craft production. The first common idea about making things is that factory production is denigrating, and the second that "the best things are made [...] by one person working alone in a peaceful setting using age-old techniques and natural materials" (Scarlett, 2011: 30–31). Enologists certainly learn by experience, but like other professionals their primary credential, at least for their first job, is a degree in enology. If higher quality wines are more "processed," it's possible that the terroir effect in wines is to be found in low-quality wines where the hand of the enologist is relatively light.[15] In the past, at least in Italy, wines like these were common. The very little decent wine produced was drunk by the elite; the farmers had to settle for a weak, vinegary "little wine." Nowadays Italian enologists, and those everywhere, are trained using the same textbooks. Except for a very small minority, they use the same technology. Rather than let the vagaries of soil and weather be expressed in their creations, they seek to balance out the seasonal and geographic variation to *create* the very typicity that we attribute to nature. Rengenier Rittersma has suggested that it might be in the least processed food products where we might be most likely to find terroir effects, rather than in the most highly processed cheese, charcuterie, and beverages.[16] Of course, this is true only if the weight of the terroir effect rests on environmental conditions, rather than the processing skills of the artisan who does the processing.

The landscapes of terroir are now found far beyond Bordeaux and the area around Parma. This chapter has heretofore concentrated on Europe, given its role as the place where PDO zones – and with it terroir – have been institutionalised to the greatest degree. With the globalisation of trade and the increased focus of many countries on production for an international market, there has been an export of the concept of terroir from Europe. Despite the fact that terroir insists on local variables producing distinct flavours, the expansion of the "discovery" of terroir outside of Europe is well correlated with the exportation of food-making expertise, plant breeds, and animal varieties. Global tastes – like the immense global popularity for Merlot, or even for wines made with grapes rather than other fruit – may tend to eliminate local flavours. These global tastes expressed through the global market contribute to the production of landscapes that can produce these foods: *Vitis vinifera* 'Merlot' now grows on terraces carefully prepared not for potatoes or rice, but for wine grapes. Indeed, the global hegemony of the French grape varietals might undermine any pretension the French have to terroir, already greatly weakened after the infamous 1976 "Judgement of Paris."[17] Where local tastes are reinforced, it is largely those tastes that can travel (that is, be exported without degrading the quality), be decontextualised and detached from the labour that created them, and be enjoyed anywhere by anyone. If the boundaries of the sense of taste "continue to swell ever outward to include all the human senses and numerous variables such as color and composition in the presentation of food, positive and negative associations, and so on" (Waterman, 2011: 67), it seems difficult that decontextualised food can have the same taste across the flattened global market. In many cases, structural adjustment policies are intentionally written to encourage the production of cash crops (in other words, the crops for which there is already a demand on the global market; see Chapters 9 and 21). There are literally hundreds of varieties of bananas, with radically different tastes than the Cavendish cultivar most people in the world eat: yet the Cavendish's resistance to bruising and (heretofore) to disease means that its terroir is what we taste. If terroir theory is supposed to provide small, artisanal producers with a semi-monopoly rents, too often the demand of a global market, state power, and nationalism lead to large, politically connected producers using terroir to push these small producers out of the market.[18] In addition, the ability to create PDO zones and argue for products with terroir is unsurprisingly correlated with a country's historical integration with and success in exportation to global markets (see Guthman, 2007; Mincyte, 2012).

The fact that terroir's actual chemical effects are elusive and that the idea has been used by powerful producers in powerful countries does not mean that it cannot be used by other actors. While terroir has been and continues to be a marketing device harnessed by the already-wealthy and pushed politically by stronger states to the detriment of weaker ones (see for instance Handler, 2006), it can also be the economic weapon of the weak. Marginalised groups have also used terroir both to compete in a global market and to revive their local rural communities through the production of a revived (or invented) traditional food. Whether the methods of production are indeed age-old, or plastic has been substituted of late for tin, there is always enough raw material around to fashion a narrative about hard work, tradition, and their delicious fruits. Salt producers in the Sicilian town of Trapani, for example, have combined their selectively refashioned history with their handmade product and have been successful at selling their salt for much more than it would be worth without the terroir narrative attached (and the support of Slow Food).[19] But terroir can also be the impetus for the creation of a high-quality product. Heather Paxson has shown how artisans eager to work with their local production conditions and recognising their lack of "traditional" methods have intentionally created new techniques as part of their attempt to "reverse engineer" terroir in Wisconsin cheese making (2010).

Anthropologist Cristina Papa, writing about PDO products, provides an excellent summary of terroir's contradictions:

> [the typical food product] presents itself as an oxymoron, a local product but defined by supranational parameters and markets, a product that to be preserved has to be changed, a product precious for its variability but whose protection reduces that variability, a product that originated with local "how to" but that becomes calcified by "official" knowledge, a product that is characterized as part of a community's patrimony but which at the same time obeys the rules of the market.
>
> *(2002: 161)*

For all its empirical limitations, its historical elisions, and its geographic problems, terroir is a theoretical tool that can be employed in all sorts of foodscapes across the world to use delicious tastes to provide a living for hardworking people. That said, it does not follow that we should uncritically accept terroir as an integral part of agricultural landscapes whether in Europe or in the rest of the world. If terroir is a social fact, we have to continuously investigate whose discourse creates it, by whom that social fact is wielded, and to whose economic and social benefit it gets mobilised. Despite the theory's own inherent contradictions that militate market forces against particular flavours, terroir also offers the possibility of experiencing the very particular tastes of faraway places, with all the opportunities this entails for thinking globally while acting locally.

Notes

1 Friess and Jazeel offer an important reminder that the definition of landscape varies across disciplines (2017). Terroir is constituted or imagined in various landscapes – subterranean, terrestrial, stratospheric, discursive, molecular, local, global, etc. – and this essay attempts to discuss these various landscapes of terroir.

2 It's important to note that the conceptual linkage of taste and place is by no means European; similar ideas have circulated, for example, in Vietnam (see Peters, 2011). Alix Cooper has argued that the European concepts of "native" or "indigenous" plants emerged from Europeans' encounters with other biotas in the sixteenth century. These concepts could be part of terroir's intellectual history (2007).

3 This oscillation of the value placed on local food is also not strictly European: Eric Rath has argued that the meaning of "local food" and the tastes associated with it in Japan are quite modern, a relic of state mobilisation during World War II (Rath, 2015).

4 Given the incredible variation in soil types across Burgundy, James Wilson attempted to explain a homogeneous terroir there using the subsoils and bedrock (1998). Robert Ulin, discussing the preeminence of Bordeaux, has shown this is due to "its political and economic history and from a more general process of 'invention' that disguises what is social and cultural in 'natural' attire" (1995: 519). Charles Luddington has also shown the importance of Irish merchants in creating modern Bordeaux (2016).

5 For an analysis of Slow Food's cultural politics during the creation of the Colonnata lard PDO zone, see Leitch (2003).

6 Deconinck et al. argue that lobbying by interest groups, rather than geological factors, has a pronounced effect of the size of a PDO region, and that the resulting regions may be "suboptimal" (2015).

7 Much of the history of Parmesan presented here is from the Parmesan cheese consortium's official history, by Mario Zannoni (1999).

8 This is apparently still true in some modern PDO zones. See Sarah Besky's investigation of Darjeeling tea, and her description of the tea pickers drinking only low-grade broken leaf Darjeeling (2014).

9 Here I refer not to the strength of the argument, but of the environmental effect, borrowing an analogy from the debate about the two versions, strong and weak, of the Sapir-Whorf Hypothesis about linguistic relativity. The "strong" version of terroir theory is one that is focused solely on geophysical variables. Marion Demossier has shown that there has been a shift in the Burgundy region from the "geological argument"' of terroir (emphasising the importance of the soil and climate) to the wine-grower as the mediator in the expression of terroir (Demossier, 2011). Matthew Rippon has described a similar emphasis on artisan technique-based terroir (2014).

10 I would invite the reader to think about how the small bit of ash on the bottom of a pizza cooked in a wood-fired oven affects the taste, as opposed to pizza cooked in a gas oven.

11 Indeed, this is probably the reason that the logo of the cheese has a black background.

12 The transnational, cross-cultural contributions to a supposedly local product are also clear in Luddington's discussion of Irish merchants' influence on Bordeaux wine, cited earlier, as well as stories about Sikh dairymen in Italy (Povoledo, 2011).

13 Cristina Grasseni's work is useful here: she argues that Slow Food denominations for instance, have actually limited the heterogeneity of Bitto cheese in their efforts to "protect" it. She also discusses boundary-making for PDO zones, and the effect of technological changes (Grasseni, 2011).

14 Surprisingly, though, there may be fewer chemical differences to smooth out than one would expect. Despite production differences, trained panelists in one study were unable to distinguish an AOC wine from a no-name table wine produced outside of the AOC blackline (Cadot et al., 2010; see also Priilaid, 2007). This may be the result of differences in tasters' gustatory sensitivity: van Leeuwen et al. found that the climate and soil did indeed have significant effects that could be tasted, much more than the grape cultivar (2004). There are also arguments for the effect of microbes on terroir (Felder et al., 2012).

15 The increased terroir effect could also be found in "natural" wines. See Black (2013).

16 Other candidates in which we are far more likely to find terroir effects are edible salts (whose impurities differ widely, and have a significant effect on taste) and mineral water. Not having been "processed" by humans, or the metabolisms of plants or animals, they perhaps indeed possess a distinctive terroir. Bounding these tastes, however, is still problematic. Expanding the concept of *land*scape to *air*scapes (*aerroir*) and *sea*scapes (*merroir*). For a discussion

of smog meringues and dual-origin sea salt, see Twilley (2015). For a thought-provoking spoof on the class issues of being able to taste (and afford) the terroir of water, see the video "The Timmy Brothers – Water Makers" (Riccio, 2014). There is even a call to examine terroir in architecture (Camerlenghi, 2016).

17 In 1976, a panel of French judges unknowingly gave Californian wines higher marks than French wines, validating the claims of extra-European countries to quality wine production. See Taber (2006).

18 For an example from tequila production in Mexico, see Bowen and Gaytán (2012) as well as Bowen (2015) for a lengthy discussion of the political economy of PDO zones.

19 Slow Food's "help" has been critiqued as elitist and romantic. See West and Domingos (2012) and the Leitch article cited in note 5.

Further reading

Besky, S. (2014) The labor of terroir and the terroir of labor: Geographical indication and Darjeeling tea plantations. *Agriculture and Human Values*, 31(1), 83–96.

Black, R.E. (2013) Vino naturale: Tensions between nature and technology in the glass. In: Black, R.E., & Ulin, R.C. (eds.), *Wine and culture: Vineyard to glass*. London: Bloomsbury, pp. 279–294.

Black, R., & Ulin, R.C. (eds.) (2013) *Wine and culture: Vineyard to glass*. London: Bloomsbury.

Bowen, S. (2015) *Divided spirits: Tequila, mezcal, and the politics of production*. Los Angeles and Berkeley, CA: University of California Press.

Bowen, S., & Gaytán, M.S. (2012) The paradox of protection: National identity, global commodity chains, and the tequila industry. *Social Problems*, 59(1), 70–93.

Cadot, Y., Caillé, S., Samson, A., Barbeau, G., & Cheynier, V. (2010) Sensory dimension of wine typicality related to a terroir by quantitative descriptive analysis, just about right analysis and typicality assessment. *Analytica Chimica Acta*, 660(1–2), 53–62.

Camerlenghi, N. (2016) Terroir and architecture. In: Martin-McAuliffe, S.L. (ed.), *Food and architecture: At the table*. New York: Bloomsbury Academic, 23–41.

Capatti, A., & Montanari, M. (2003) *Italian cuisine: A cultural history*, O'Healy, A. (trans.). New York: Columbia University Press.

Castagnetti, G.B., Bertolini, L., & Guéguén, S. (2000) Aggiornatmento sulla conoscenza delle principali caratteristiche chimico-fisiche e tecnologie del latte di alcune razze locali (Reggiana e Modenese) nella produzione del parmigiano reggiano. *Atti della società italiana di buiatria*, 32, 251–260.

Cooper, A. (2007) *Inventing the indigenous: Local knowledge and natural history in early modern Europe*. Cambridge: Cambridge University Press.

Deconinck, K., Huysmans, M., & Swinnen, J.F.M. (2015). The political economy of geographical indications." SSRN Scholarly Paper ID 2671764. Rochester, NY: Social Science Research Network. https://papers.ssrn.com/abstract=2671764.

Demossier, M. (2011) Beyond terroir: Territorial construction, hegemonic discourses, and French wine culture. *Journal of the Royal Anthropological Institute*, 17(4), 685–705.

European Economic Community (1992) Council regulation (EEC) No. 2081/92 of 14 July 1992 on the protection of geographical indications and designations of Origin for agricultural products and foodstuffs. *Official Journal of the European Communities*, 208, 1–8.

Felder, D., Burns, D., & Chang, D. (2012) Defining microbial terroir: The use of native fungi for the study of traditional fermentative processes. *International Journal of Gastronomy and Food Science*, 1(1), 64–69.

Friess, D.A., & Jazeel, T. (2017) Unlearning 'Landscape.' *Annals of the American Association of Geographers*, 107(1), 14–21.

Grasseni, C. (2011) Re-inventing food: Alpine cheese in the age of global heritage. *Anthropology of Food*, (8), May.

Green, G.P., Bestland, E.A., & Stewart Walker, G. (2004) Distinguishing sources of base cations in irrigated and natural soils: Evidence from strontium isotopes. *Biogeochemistry*, 68(2), 199–225.

Guthman, J. (2007) The Polanyian way? Voluntary food labels as neoliberal governance. *Antipode*, 39(3), 456–478.

Guy, K. (2003) *When champagne became French: Wine and the making of a national identity*. Baltimore, MD: Johns Hopkins University Press.

Handler, M. (2006) The WTO geographical indications dispute. *The Modern Law Review*, 69(1), 70–80.

Leitch, A. (2003) Slow food and the politics of pork fat: Italian food and European identity. *Ethnos*, 68(4), 437–462.

Luddington, Chad. 2016. *Quiet revolutionaries: Irish merchants and the creation of Bordeaux grands crus wines*. Presented at the Dublin Gastronomy Symposium, Dublin, June 1.

Mincyte, D. (2012) The geopolitics of difference: Geographical indications and biocultural otherness in the New Europe. *Rachel Carson Center Perspectives,* 9, 47–51.

Papa, C. (2002) Il prodotto tipico come ossimoro: Il caso dell'olio extravergine di oliva umbro. In: Siniscalchi, V. (ed.), *Frammenti di economie. Ricerche di antropologia economica in Italia*. Cosenza, Italy: Luigi Pellegrini Editore, pp. 159–191.

Parker, T. (2015) *Tasting French terroir: The history of an idea*. Los Angeles and Berkeley, CA: University of California Press.

Paxson, H. (2010) Locating value in Artisan cheese: Reverse engineering terroir for new-world landscapes. *American Anthropologist*, 112(3), 444–457.

Peters, E.J. (2011) *Appetites and aspirations in Vietnam: Food and drink in the long nineteenth century*. Lanham, MD: AltaMira Press.

Priilaid, D.A. (2007) The placebo of place: Terroir effects in the blind and sighted quality assessments of South African varietal wines. *Journal of Wine Research*, 18(2), 87–106.

Laudan, R. (2004) Slow food: The French terroir strategy, and culinary modernism. An essay review of Carlos Petrini, Trans. William McCuaig. *Food, Culture, and Society*, 7(2), 134–144.

Rath, E.C. (2015) The invention of local food. In: Farrer, J. (ed.), *The globalization of Asian cuisines*. Springer. pp. 145–164.

Rittersma, R. (2011) Terroir, Oder Die Verklärung Des Lokalen. Sensorieller Fakt Oder Gastronomisches Dogma? Ein Denkanstoß/Eine Kritische Bestandsaufnahme. *Epikur: Journal Für Gastrosophie*, 1, 1–8.

Ribéreau-Gayon, P. (2005) *Handbook of enology*, 2nd ed., 2 vols. Hoboken, NJ: Wiley.

Riccio, P. (2014) *The Timmy Brothers – water makers*. New York: Silvercup Studios. Available from: https://vimeo.com/131422396.

Rippon, M.J. (2014) What is the geography of geographical indications? Place, production methods and protected food names: What is the geography of geographical indications? *Area*, 46(2), 154–162.

Scarlett, S.F. (2011) The craft of industrial patternmaking. *The Journal of Modern Craft*, 4(1), 27–48.

Taber, G.M. (2006) *Judgment of Paris: California vs. France and the historic 1976 Paris tasting that revolutionized wine*. New York: Scribner.

Terrio, S.J. (1996) Crafting grand cru chocolates in contemporary France. *American Anthropologist*, 98(1), 67–79.

Trubek, A. (2008) *The taste of place: A cultural journey into terroir*. Los Angeles and Berkeley, CA: University of California Press.

Twiley, N. (2015) Smog meringues. In: *Edible geography – thinking through food*. May 30. www.ediblegeography.com/smog-meringues/.

Ulin, R.C. (1995) Invention and representation as cultural capital: Southwest French winegrowing history. *American Anthropologist*, 97(3), 519–527.

Van Leeuwen, C., Friant, P., Chone, X., Tregoat, O., Koundouras, S., & Dubourdieu, D. (2004) Influence of climate, soil, and cultivar on terroir. *American Journal of Enology and Viticulture*, 55(3), 207–217.

Waterman, T. (2011) The flavor of the place: Eating and drinking in payottenland. In: Strong, J. (ed.), *Educated tastes: Food, drink, and connoisseur culture*. Lincoln: University of Nebraska Press, 58–80.

West, H.G., & Domingos, N. (2012) Gourmandizing poverty food: The serpa cheese slow food presidium. *Journal of Agrarian Change*, 12(1), 120–143.

Wilson, J.E. (1998) *Terroir: The role of geology, climate, and culture in the making of French wines*. Berkeley, CA: Wine Appreciation Guild.

Zannoni, M. (1999) *Il parmigiano-reggiano nella storia*. Parma, Italy: Silva.

37

Using the senses to write food culture and landscape

Nina Mukerjee Furstenau

Introduction: what is sensory writing?

This chapter focuses on writing that reveals nuances of food and landscape through the lens of the senses to produce engaging and effective communication. The role of food and landscape – defined as the geographical dimensions of human experience and the character of place – in just about everything becomes obvious when you settle on this one thought: the world, and human connection to it, is sensory. The sensing body of sight, sound, taste, smell and touch, and mind, "common to humans across the world today is shared with humans in the past from the moment anatomically modern *Homo sapiens* first emerged as a distinct species" (Day, 2013: 6). This sensory collective understanding provides the opportunity for writers to create an immediate coherence and connection with readers, no matter who or where they are, across boundaries of gender, cultural history, and place.

Entrikin (1991: 6) argues that the geographer's "conceptual fusion of space and experience gives areas of the earth's surface a 'wholeness' or an 'individuality'" which, if the idea of place is context, "incorporates both existential qualities of our experience of place and also our sense of place as a natural 'object' in the world." This dualistic quality of place has been at the centre of geography as a chorological science that addresses the relationship of people to their environment (Entrikin, 1991). Writing that uses the connections between people and environment – specifically place description through the lens of the senses – can bring readers into clearly communicated story.

Humans are described as sentient beings and this points quite literally to having sense perception (Ackerman, 1991). "The senses don't just make sense of life in bold or subtle acts of clarity, they tear reality apart into vibrant morsels and reassemble them into meaningful pattern" (Ackerman, 1991: xvii). Thus, sensory writing, with its ability pull those "vibrant morsels" together, communicates using shared sense and bodily knowledge. In the same way that foods can be an encounter with culture, sensory writing bridges differences of place and experience.

Readers are literally wired for sensory story. A brain-imaging study reported in *Psychological Science* reveals that the regions of the brain that process the sights, sounds, tastes, and movement of real life are activated when we are engrossed in a compelling narrative (Speer et al., 2009). The fact that humans process the sensory world and take in language in the same brain region, makes sensory storytelling instantly resonant. In *Wired for Story*, Cron says, "That's what accounts for the vivid mental images and the visceral reactions we feel when we can't stop reading, even though it's past midnight and we have to be up at dawn" (2012:

4). Since the writer's primary tool for communication is words, specific word choice greatly affects how and if writing resonates with readers. To be sure, the craft of stringing words into sentences with rhythm, correct punctuation, grammar and spelling, and so forth, play a significant role in clarity and communication. But, choosing sensory words to tease out connections between foodways – defined as the eating habits and culinary practices of a people, region, or historical period – food history, place, and taste, create deep communication (Speer et al., 2009).

First consider that each sense makes judgements all the time. Sounds, such as the crunch of a taco shell, or the sizzle of meat on a barbeque, can create a chain reaction of anticipation to salivation, as well as create the perception of enhanced flavour or repulsion for a vegetarian or vegan,

> The consumption of food and drink are among the most multisensory of our perceptual experiences. In fact, the evaluation of foodstuffs is not only influenced by the unified oral sensation (or Gestalt) of taste and smell in the mouth but also by what the foods look, feel (i.e., oral texture, temperature, viscosity, etc.) and sound like when we eat or drink them.
>
> *(Zampini and Spence, 2010: 3, 57)*

Texture and sound in chemosensory perception can mean that food crispness, for instance, is highly related with food pleasantness (Zampini and Spence, 2010). When writing using specific sensory descriptions, readers fill in story details beyond what's on the page.

In another example, deep berry colours instantly fashion the idea of marvellous fruity flavour for most human eaters. Conversely, washed out colours, such as yellow-green on decaying lettuce, conjures a different judgement. Colour has such power that many studies show that people frequently fail to perceive the flavour of a sweet solution correctly when its colour is inappropriate (Zampini and Spence, 2010). Human mouths taste the flavour of foods, but also detect texture and temperature, both causing sensate reactions. When people were asked to rate food, a crisp texture was one of the descriptors most strongly correlated with a food's perceived pleasantness (Vickers, 1983). A writer that pauses a moment to describe the sensate perception of colour, the texture of food in the mouth and its temperature, conveys a host of reactions in readers that make the story more than the sum of its parts.

Next, consider that humans bring the mind along to the table. Many times, a mother's beloved recipe is held up as perfection against every version ever after, and taste emerges out of emotion and memory as much as what is being currently eaten. Food was our first frustration, to be sure, but also our first reward. In some cultures,

> When you've hurt yourself, a kiss and cookie – either helps, both work miracles – will make it well. We learn as we go through life that the kiss and the sweet can substitute for each other in all kinds of situations. [. . .] Cookies, candy, ice cream, cake: these are for comfort and for being good and for finishing our vegetables. They are the least we deserve.
>
> *(Crumpacker, 2007: 11)*

Food as a reward for frustration or to soothe hurt is not always available at subsistence levels of farming and living. Yet, even in less agriculturally abundant areas of the world what food is available is often presented in the same vein. Writers that use the force of words to hook readers at a *gut* level, pun intended, reach across memory and time.

Then, reflect that food and landscape are inextricably linked (Potteiger, 2013). In *What Are People For*, Berry says, "eating is an agricultural act," and cannot be separated from the land that sustains plants and animals (2010: 146). For humans, food is more than sustenance, food holds memory, desire, reward from frustration, and link to place. Food can represent how people live and who they are. Food holds story. When this connection is lost, we are duped, a fatality of modern land use, an eater, "who no longer knows or imagines the connections between eating and the land, and who is therefore necessarily passive and uncritical – in short, a victim" (Berry, 2010: 146). Writing that connects to the sensory world combats this alienation from landscape.

Quite literally, food is bound to place. Historically, what sustained humans and animals alike emerged directly from what the surrounding land would support. Eating locally and seasonally are now both vanishing, due to industrial agriculture and transportation of foodstuffs across the globe, and reviving due to concerns over questionable quality and integrity of those food systems. However, place-based diets are how humankind survived and developed through history. Land use is rooted deeply in human connection (Lippard, 1997). After agriculture changed society into group settlements, humans stepped up the game – food evolved beyond sustenance and connected to family, memory, spirituality, and love (Flandrin and Montanari, 2013). These ties and more are what stirs a mother to make the same recipe for her children as an ancestor did not only miles, but also years, away in memory. Writing that takes readers to the moment of pulling grandma's casserole out of the oven, to the heat that rushes out and fogs eyeglasses, to aroma that fills the nose and tweaks memory – that writing takes readers to place as no other.

In fact, this landscape of journey – internal or external, present or past – is what makes up the perceived human world, and is one of the oldest forms of story (see *The Odyssey* and its hero's journey home, *The Iliad*'s depiction of glory, wrath, homecoming, and fate, the efforts of Rama to rescue Sita in the *Ramayana*, and the more current stories of J.R.R. Tolkien, John Cheever's "The Swimmer," and many more). Passage, from physical place to physical place, from childhood to adulthood, from one state of mind to another state of mind, has always been of interest to humankind. To effectively write about landscape, its qualities and the people and food it sustains, entails "seeing" it closely and understanding that the geographical dimensions of human experience and the character of place is compelling to readers (Lippard, 1997). When you consider the need for food, and the root of that food in landscape, it is no wonder that combining landscape and food in writing creates powerful connection with readers. Paying particular attention to the senses in writing about food and landscape creates resonance because every living being on the planet experiences the world, the centrality of landscape and food to life, in much the same way.

This is why approximately four million viewers tune in every week – for ten years running – to watch the Food Network. This is why over one million people subscribe to magazines such as *Bon Appétit*, and nearly one million to *Food and Wine* (Alliance, 2016). Why mailboxes bulge with other savoury journals from *Saveur* to *Gastronomica*, and why films such as *Ratatouille*, *Like Water for Chocolate*, *Babette's Feast*, *Big Night*, and so many more, endure. Recipe collections and cookbooks, one of the largest categories of book publishing, are getting a run for the money via computer screens that have almost limitless offerings of food blogs such as Serious Eats, Butter and Brioche, and many more, and foodie posts such as those on Storia.me/Taste-Full and foodieunderground.com. From the world of literature, consider that one line by Proust in *Remembrance of Things Past*, about petite madeleines, opened 900-plus pages of prose and an entire world: "No sooner had the warm liquid mixed with the crumbs touched my palate than a shudder ran through me and I stopped, intent upon the extraordinary thing that was happening to me" (1981: 48–51). The simple act of eating is weighted with complex cultural meanings and is sometimes even the locus of class and political allegiance.

There is a story told by Sponsler (2001) in "Eating Lessons" of Irish kings who travelled to England in 1394–95 to pledge submission to England's King Richard. A squire was to educate them in the customs of the English, refashioning their behaviour from its uncouth state. This squire tolerated their behaviour while dining for three days, noting with disapproval that they allowed their minstrels and servants to sit at the same table with them, eat from their plates, and drink from their goblets. On the fourth day, the squire,

> had the tables in the hall rearranged and laid in the "correct manner," seating the four kings at the high table, the minstrels at a separate table, and the servants at yet another table. This reconfiguration angered the Irish kings, who refused to eat on the grounds that such arrangements were contrary to

> the customs with which they had been brought up. [The squire] replied that their previous eating arrangement was not reasonable and that they would have to abandon it and adopt the English style, since that was what King Richard wished.
>
> *(2001: 1–22)*

Perceiving that a change in table manners was a necessary accessory to political submission, the Irish kings complied, replicating within the dining hall the larger battles being fought outside as the Irish were being forced under English rule. The link between eating, table manners, and subjection in the Irish kings' story underscores the importance of eating as a cultural practice. Manners ratify social divisions. Eating is central. Close to home, the dining table is the court.

The power of table manners and rituals around food preparation reaches into the current day. The common act of sitting down to eat puts a mundane, everyday face to cultural identity where powerful currents of identity and nation are literally on the table (Gilbert, 2014). In *Culinary Imagination*, a study of food as memory, myth, language, and image, issues of food and hunger are always political as a well as poetical (Gilbert, 2014). Writers that describe the sensory event of eating can craft work that unveils what food reveals and conceals about us all.

Crafting stories that resonate

With so many options to engage readers, what deserves attention? Hunger is universal, and perhaps this explains why food is an enduring literary subject, crossing boundaries of time and nation, class and ethnicity. To hook readers into seeing all the richness of food and landscape, journalists and others wishing to communicate well, craft stories with clarity, conciseness, good word choices, rhythm, and style. But, loading stories with precise descriptions is not enough. In *The Writer's Journey: Mythic Structure for Writers*, Vogler says,

> I realized that the good stories were affecting the organs of my body in various ways, and the really good ones were stimulating more than one organ. An effective story grabs your gut, tightens your throat, makes your heart race and your lungs pump, brings tears to your eyes or an explosion of laughter to your lips.
>
> *(Vogler, 2007: x)*

That reaction, right there, is directly tied to sensory writing.

Tools of sensory writing

Tool 1: specificity

There is power in place and in the language of landscape. In an interview, *The Politics of Place*, Terry Tempest Williams says,

> if we are not rooted deeply in place . . . if we don't know the names of things, if we don't know pronghorn antelope, if we don't know black-tail jackrabbit, if we don't know sage, pinyon, juniper, then I think we are living a life without specificity, and then our lives become abstractions. Then we enter a place of true desolation.
>
> *(London, 1995: 22)*

Specificity, knowing the names of things, creating a sense of intimacy with details and words well-placed, is connection. This, more than anything else, tells a reader "you are here." There may as well be a large arrow sign pointing to place. This is writing with authority, and it creates the feeling that the audience is in good hands.

Tool 2: engaging the sense of taste

Sight, sound, smell, texture, taste, and the mind, if you can conceive of your mind as a sense organ, are all tools that can set a brush fire to writing and communication. Jerome K. Jerome, in *Three Men in a Boat (To Say Nothing of the Dog)*, shows us how taste can affect mood in a way that is both humorous and apt: "It cast a gloom over the boat, there being no mustard. We ate our beef in silence" (Jerome, 2004: 104). The trigger of food creates numerous sensate interactions in the body that are commonly understood, and by using the senses readers are brought along for the ride. As Jerome demonstrates, in many ways, the mood of a piece is affected by the reaction of the senses. In another example, Graham writes,

> The other day, I turned a corner while walking my dog in town and suddenly, finally, knew the *bon mot* for the taste of the late winter air in northern New York. It tasted *tarnished*. I'd been worrying the language for weeks, and then there the word appeared dull on my tongue, a little bitter. The light, the wind, the dampness; these all tasted faintly oxidized, like a dented old bugle my grandfather had played with.
>
> *(Graham, 2012)*

Taste and the mind (memory) pack a punch for readers in Graham's prose.

Tool 3: sight

But visual clues also hit straight at the heart of a story. In "The Gumbo Chronicles," Jacobsen writes,

> In the predawn steam of a Louisiana night, I stood in a yard surrounded by catfish heads. The headlights of dualies towing fiberglass crabbing boats swept into the yard and illuminated wooden pallets stacked six feet high, holding tens of thousands of fish heads; eyes, whiskers, stringy stuff coming out the back.
>
> *(2012: 1)*

In this example, sensory language engages *sight* as well *texture*.

Tool 4: smell

It just might be that nothing is more memorable than smell, as Ackerman says in *A Natural History of the Senses*. Yet, she says, "it's almost impossible to describe it to someone who hasn't smelled it" (1991: xvii). Here, in "Road Trip to Chinatown: Tasting Your Heritage One Bite at a Time," Tai uses sensory writing to bridge that gap,

> All Chinatowns smell the same. The year is 1975; I am fourteen years old; and I could be in Los Angeles, Chicago, or New York. It doesn't matter. The smell is the same – a mix of ginger, shiitake mushrooms, durian husks, tofu, fish, and rice. It is earthy, palpable, and old country-ish. Cabbage leaves glued flat to wet sidewalks, overripe papaya, deep-fried pork.
>
> *(2008: 22)*

In this prose, it is hard to miss the slippery feel of "cabbage leaves glued flat to wet sidewalks," and the aroma that might lift as you walk by, even if the reader has not visited a Chinatown in the U.S. As McQuaid says, "aromatic images become delicate portraits of experience, etched into the nervous system as they're routed to parts of the brain involved in memory (the hippocampus) and decision making (the orbitofrontal cortex). In other words, smell literally links past and present" (2015: 93). On the page, writers link directly with readers by evoking aroma.

Tool 5: texture

Skin is what makes contact with the world. Its two-layered basic structure is how we navigate through air and space and know the sensation of touch. Recognition of this is ingrained in cultural systems of language. "Language is steeped in metaphors of touch. We call our emotions feelings, and we care most deeply when something 'touches' us. Problems can be thorny, ticklish, sticky, or need to be handled with kid gloves" (Ackerman, 1991: 70). In "Coming Home Again," Lee makes the scene come alive for readers with this,

> For *kalbi*, . . . She scored it diagonally, then sifted sugar into the crevices with her pinched fingers, gently rubbing in the crystals. . . . She wiped her hands clean of the marinade, except for her little finger, which she would flick with her tongue from time to time, because she knew that the flavor of a good dish developed not at once but in stages.
>
> *(1995: 164)*

The sensation of flicking her tongue on her little finger not only evokes texture, but a visual image that creates immediate connection with audience.

Tool 6: sound

Ackerman (1991) states that almost everything on Earth can make sound, the onrushing and withdrawing wave of molecules that begins with movement. Humans find comfort, excitement, and energy from the sense of hearing. "A campfire wouldn't be as exciting if it were silent" (Ackerman, 1991: 176). By using sensory writing to evoke hearing, authors connect at a level of vibration that perhaps underscores our world. In "True Detective", Harrington says, "Doves coo, a horseshoe clangs, quails skitter across the rain gutter. The clock radio on the night table whirs and vibrates" (1996: 22–42). The word *skitter* creates the pulse of the sentence and connects somewhere at the level of the sternum. Hard to miss.

Tool 7: mind

In *The Language of Baklava*, Abu-Jaber uses a food scene of trying to butcher a lamb to show how the internal landscape of memory no longer fit the changed external landscape of her family,

> "We thought we could still do it," Bud said. "But we couldn't." Making shish kabob always reminds the brothers of who they used to be – the heat, the spices, the preparation for cooking, and the rituals for eating were all the same as when they were children, eating at their parent's big table. But trying to kill the lamb showed them: They were no longer who they thought they were.
>
> *(Abu-Jaber, 2007: 19)*

The internal landscape of memory, during the "feather light springtime in Jordan [. . .] when they were children and their parents were still alive and nobody knew anything about the bright grocery stores of America or the way meat appeared, bloodless, gleaming with cellophane, stacked in cold rows," was gone, and the new external landscape where there were no nearby spring lambs, nor air scented with freshly grilled meat, and the brothers no longer "stood between these two events, birth and food" in sacred transformation (Abu-Jaber, 2007: 17). Both internal and external landscape was forever changed for the Abu-Jaber family, and that changed everything.

Three ingredients for food story

Writers that leverage what's on the plate in their narratives reveal much more than taste and flavour combinations. Social, political, economic, gender roles, and more are revealed in nuanced, effective ways. "Food is perhaps the most distinctive expression of an ethnic group, culture, or, in modern times, a nation" (Sonnenfeld, 2013: xvi). Then, too, conviviality is a large part of human eating behaviour,

> What differentiates human alimentary behavior from that of the animals? Culinary techniques, dietary restrictions, and religious rituals associated with eating are certainly part of the answer, but human dining is also believed to be a convivial social function.
>
> *(Flandrin and Montanari, 2013: 18)*

And, as James Boswell says, "My definition of man is a cooking animal. The beasts have memory, judgement and the faculties and passions of our minds in a certain degree; but no beast is a cook" (Boswell et al., 1852: 16). Given the idea that food reveals personal and cultural choices, descriptions of what is eaten, how it is prepared, who cooks and how people eat, and more, it carries story, atmosphere, and is a learned examination of the sense of place. Recognising that, in addition to taste and satiation, food is a lodestar for readers in three critical ways – as a personal identifier, as a gauge of cultural values, and as a catalyst throughout history (Albala, 2013) – will add depth and dimension to writing. This use of the senses as tools to reveal the nuanced meanings behind food and eating lifts writing out of the ordinary.

As a *personal identifier*, food reveals personal choices, values, ambitions, and where the eater fits in a society. "Food is central to our sense of identity," and though the notion of identity is somewhat problematic, Fischler says, it remains central "in that any given human individual is constructed, biologically, psychologically and socially by the food he/she choses to incorporate" (Fischler, 1988: 277). While this excludes eaters at sustenance levels who eat what's available to survive, for much of the world, people use food to express who they are and thus food choice is a kind of message, means of communication, just as clothes and cars can sometimes be. Writing can capitalise on this and show, rather than tell, the situation of the characters and their story as revealed through personal food choices.

Food reveals *cultural values* in that how and what people eat says a great deal about society, as in the example of the Irish Kings. "The way any given human group eats helps it assert is diversity, hierarchy and organisation, and at the same time, both its oneness and the otherness of whoever eats differently" (Fischler, 1988: 277). Writers that leverage this fact on the page, offer multi-layered prose that enhances communication with readers.

Food was and remains a major *catalyst throughout history*. The accidental discovery of new world was due to quest for foods. Population surges throughout human history – one after Columbus's travels to the New World eventually carried the sweet potato to China in 1594, or possibly earlier (O'Brien, 1972: 346) can be attributed to new sources of calories to support more people. Even modern warfare is due to advances in food – Napoleon's army marching on its stomach, for example (Albala, 2013). Many of the mass migrations through history, of course, were prompted because of food, and the landscape that would no longer support food production, from prehistoric times to the Irish potato famine. Writing that leverages taste, flavour, and what is on the plate, conveys much more than what's for dinner.

Humans have good reason to respond to landscape and food. Every time we sit down to eat, a host of images, tastes, and ideas about the food in front of us and sense of place rise to greet us. We are not just filling our bellies but our minds. Landscape and food story resonate deeply. We become rapt at food media – especially at luscious images of food, anything from a hamburger to savoury vegetarian fare

– because it is how our brains were wired for survival. The human spirit settles in nature, in connection with the land and soils that have sustained generations with not only consumable foods, but imaginative landscapes. And, because we are human, we carry memory to everything we eat and every landscape we tread. Internal landscape marks how we react in the world: which foods taste good to us, how we prepare it, produce it, and accept its arrival on our plates. Look down at the table, the plate of food there shows story – who settled and influenced the flavours of region, and more – that writers can leverage for stronger communication. In the time it takes to lick your lips, all sorts of boundaries are met, sometimes crossed, often shared. Because food can be seen as a direct link to place, to land, to sustainable health, to the agricultural hands it takes to feed us, writing that hooks readers at a gut level – sensory and of the body – is deeply resonant. Writing with food in mind, using sensory language, can lead readers on a journey to the fields of harvest, the fruit ripening on trees, and the cultures that grew those foods because their piece of earth could do so. Communication sharpens with writing details that connect to the story of the sensate world.

References

Abu-Jaber, D. (2007) *The language of baklava.* New York: Anchor, pp. 17–19.

Ackerman, D. (1991) *A natural history of the senses.* New York: Vintage, p. xvii, pp. 67–70, 175–178.

Albala, K. (2013) *Food: A cultural culinary history.* Audio Recording, Great Courses. www.thegreatcourses.com/courses/food-a-cultural-culinary-history.html.

Alliance for Audited Media (2016) https://abcas3.auditedmedia.com/MICenter/Home/Index?s=9d8ac1c0-31c5-4e27-9757-9c7715f90952#30 (accessed December 2016).

Berry, W. (2010) *What are people for? Essays.* Berkeley, CA: Counterpoint Press, p. 146.

Boswell, J., Johnson, S., & Carruthers, R. (1852) *The journal of a tour to the hebrides: With Samuel Johnson, LL.D* (Vol. 1). Office of the national illustrated Library.

Cron, L. (2012) *Wired for story: The writer's guide to using brain science to hook readers from the very first sentence.* New York: Ten Speed Press, p. 4.

Crumpacker, B. (2007) *The sex life of food: When body and soul meet to eat.* New York: Macmillan, p. 11.

Day, J. (ed.) (2013) *Making senses of the past.* Charleston, IL: Southern Illinois University, pp. 5–6.

Entrikin, J.N. (1991) *The betweenness of place: Towards a geography of modernity.* Baltimore, MD: Johns Hopkins University Press, p. 6.

Fischler, C. (1988) Food, self and identity. *Information (International Social Science Council)*, 27(2), 275–292.

Flandrin, J., & Montanari, M. (2013) *Food: A culinary history.* New York: Columbia University Press, p. 18.

Gilbert, S.M. (2014) *The culinary imagination: From myth to modernity.* New York: WW Norton & Company, p. xvi.

Graham, P.W. (2012) Sweet spot. *Alimentum: The Literature of Food*, (13). Available from: www.alimentumjournal.com/current-issue/

Harrington, W. (1996) *At the heart of it: Ordinary people, extraordinary lives.* Columbia, MO: University of Missouri Press, pp. 22–42.

Jacobson, R. (2012) *Outside online.* The Gumbo Chronicles, 1. Retrieved from www.outsideonline.com/1897631/gumbo-chronicles (accessed 9 March 2012).

Jerome, J.K. (2004) *Three men in a boat: To say nothing of the dog!: With an introduction and notes by Jeremy Lewis* (print p. 104). Penguin Classics. Literature Online. London: Penguin.

Lee, C.R. (1995) Coming home again. *New Yorker*, October 16, p. 164.

Lippard, L.R. (1997) *The lure of the local: Senses of place in a multicentered society.* New York: New Press.

London, S. (1995) The politics of place: An interview with Terry Tempest Williams. [*Radio broadcast transcript], Insight & Outlook.* Santa Barbara, CA, p. 22.

McQuaid, J. (2015) *Tasty: The art and science of what we eat.* New York: Scribner, p. 93.

O'Brien, P. (1972), The Sweet Potato: Its origin and dispersal. *American Anthropologist*, New Series, 74(3), 346. www.jstor.org/stable/671520.

Potteiger, M. (2013) Eating places food systems, narratives, networks, and spaces. *Landscape Journal*, 32(2), 261–275.

Proust, M. (1981) *Remembrance of things past.* New York: Random House, pp. 48–51.

Sonnenfeld, A. (2013) Preface. In: Flandrin, J., & Montanari, M. (eds.), *Food: A culinary history*. New York: Columbia University Press, p. xvi.

Speer, N., Reynolds, J., Swallow, K., & Zacks, J. (2009) Reading stories activates neural representations of visual and motor experiences. *Psychological Science*, 20(8), 989–999. Available from: DOI: https://doi.org/10.1111/j.1467-9280.2009.02397.x.

Sponsler, C. (2001) Eating Lessons: Lydgate's "dietary" and consumer conduct. In: Ashley, K., & Clark, R. (eds.), *Medieval conduct*. Minneapolis, MN: University of Minnesota Press, pp. 1–22. Available from: www.jstor.org/stable/10.5749/j.cttsjxn.5.

Tai, E.M. (2008) Roadtrip to Chinatown: Tasting your heritage, one bite at a time. *Gastronomica*, 8(2), 22–25.

Vickers, Z.M. (1983) Pleasantness of food sounds. *Journal of Food Science*, 48, 783–786.

Vogler, C. (2007) *The writer's journey – mythic structure for writers*. Saline, MI: McNaughton & Gunn, p. x.

Zampini, M., & Spence, C. (2010) Assessing the role of sound in the perception of food and drink. *Chemosensory Perception*, 3, 57. Available from: DOI: http://10.1007/s12078-010-9064-2.

Further reading

Aggleton, J.P., & Mishkin, M. (1986) The amygdala: Sensory gateway to the emotions. *Emotion: Theory, Research and Experience*, 3, 281–299.

Barber, D. (2014) *The third plate: Field notes on the future of food*. New York: Penguin.

Barthes, R. (1997a) Toward a psychosociology of contemporary food consumption. *Food and Culture: A Reader*, 2, 28–35.

Barthes, R. (1977b). *Elements of semiology*. London: Palgrave Macmillan.

Bourdieu, P. (1984) *Distinction: A social critique of the judgement of taste*. Cambridge, MA: Harvard University Press.

de Certeau, M., & Mayol, P. (1988) *The practice of everyday life: Living and cooking*, Vol. 2. Minneapolis, MN: Univerisity of Minnesota Press.

Fisher, M.F.K. (2004) *The art of eating*. Hoboken, NJ: Wiley.

Guixe, M., & Knolke, O. (2010) *Food designing*. Corraini Edizioni.

Harris, M. (ed.) (1987) *Food and evolution: Toward a theory of human food habits, foodways, a historical overview and theoretical prolegomenon*. Philadelphia, PA: Temple University Press, pp. 153–183.

Hughes, H. (ed.) (2017) *Best food writing 2017*. Philadelphia, PA: Da Capo Press/Perseus Books.

Korsmeyer, C. (2002) *Making sense of taste: Food and philosophy*. Ithaca, NY: Cornell University Press.

Kurlandsky, M. (2009) *Foods of a younger land*. New York: Riverhead/Penguin Group.

McQuaid, J. (2015) *Tasty: The art and science of what we eat*. New York: Scribner.

Momaday, N.S. (1969) *The way to Rainy Mountain*. Albuquerque, New Mexico: University of New Mexico Press.

Nabhan, G.P. (1993) *Songbirds, truffles, and wolves: An American naturalist in Italy*. New York: Penguin.

O'Connell, L.H. (2014) *The American plate: A culinary history in 100 bites*. Sourcebooks, Naperville, IL.

PhysOr.com (2009) Readers build vivid mental simulations of narrative situations, brain scans suggest, January 6, 2009, www.physorg.com/print152210728.html.

Sereni, C. (1987) *Casalinghitudine* (Vol. 388). Einaudi.

Standage, T. (2009) *An edible history of humanity*. New York: Bloomsbury.

Stuckley, B. (2012) *Taste what you're missing: The passionate eater's guide to why good food tastes Good*. New York: Free Press.

Wenk, G.L. (2014) *Your brain on food: How chemicals control your thoughts and feelings*. Oxford: Oxford University Press.

38

Queer space and productive landscapes

Andrea Bosio

Introduction

> It's funny. We all sit around mindlessly slagging off that vile stink-hole of a city. But in its own strange way, it takes care of us. I don't know if that ugly wall of suburbia's been put there to stop them getting in, or us getting out.
>
> (*The Adventure of Priscilla, Queen of the Desert*, 1994)

The above quotation, taken from the 1994 Australian movie *The Adventure of Priscilla, Queen of the Desert*, which narrates the journey of three drag queens in the Australian outback, introduces the questions upon which the present chapter attempts to reflect: considering the urban nature of gay meccas, can a 'queer space' exist outside of the city? How do gay and lesbian individuals relate to agricultural landscapes and its productive activities? What is the contribution of architects, landscape architects, and planners toward their inclusion – or exclusion – within those contexts?

Growing up as a city kid in Milan, I have always experienced weekends and holidays in my grandparents' countryside house as an escape from the 'big smoke'. Only when I grew up and I became conscious of my gay identity did I start to raise some questions about where I, as part of a sexual minority, would fit in the world. I was still considering, though, my urban setting to be the only context in which I could feel free to express myself, mainly by engaging in leisure activities in places where I could meet people similar to me. When other relatives of my age living in my grandparents' countryside village started to timidly come out, or at least to show homosexual tendencies, I started to compare my experience of sexual liberation to theirs, wondering not only how gay people would live freely in a less-urban setting, but mostly how they would engage potentially in agricultural activities, within a context still anchored to heteronormative traditions and values.

Years later, while conducting my PhD over the relationship between food production and consumption and the historical duality city/countryside, I collaborated with a number of Italian wine producers, farmers, and restaurateurs for fieldwork research. Subconsciously, I started to project myself, as a gay person, within the lives of those people and to ask myself, within their subtly assumed heteronormativity, where I could put myself? Would my research have been different if I had been more open about my sexuality? Would that have affected my findings and my arguments over the socio-cultural implications of food production on the construction of the agricultural landscape? While these questions are destined to remain unanswered, I attempt with this chapter to find out whether there is a place for discussions on sexuality and space within the theoretical research in the fields of architecture, landscape architecture, and urban planning. Moreover, this chapter attempts to provide an introductory investigation on the still-underdeveloped research field regarding the implications of queer sexuality within the life, the construction, and the transformation of agricultural landscapes.

In spite of a growing literature on the subject, little if no interest is offered by contemporary architecture, landscape architecture, and planning on the relationship between the contemporary changes happening in urban and rural contexts, and subjects and communities recognising themselves as other than heterosexual. In *Queers in Space*, this attitude is argued by queer theorist Brent Ingram to be so pervasive in the western mind-set that even "'progressive' urbanists, such as Jane Jacobs, were still talking publicly about 'pervert parks' [referring to the phenomenon of gay cruising in public open spaces] only a few short years before the Stonewall riots" (Ingram, 1999: 96, 97).[1] This highlights a moral judgement over the fruition of space in ways other than the ones codified by normative society. Even more problematic is the dismissive attitude toward the rural context and the life of their gay and lesbian inhabitants that Herring, in Another Country, argues to be conducted by "urban-oriented queer politics and metrobased queer academics [who] often agree with populist forms of anti-ruralism" (Herring, 2010: 5). This double problematic is argued here to become harmful when architects, landscape architects, and planners attempt to formulate theoretical and physical models for the rural environment, reiterating modalities that are indicated by scholars like Ingram and Herring to contribute to the physical exclusion of sexual minorities (Herring, 2010: 5). Thus, the chapter acknowledges the concept of queer – instead of gay – space. Differently from the notion of gay space, which is argued to play within normativity (Detamore, 2013: 86), queer space is here argued to play a significant role in allowing a redefinition of the notion of landscape in less objective and disciplinary ways, in contrast with a perspective which is linked to the "'search for certainty rather than a vehicle for individual subjectivity' and [is] tied to an often gendered and masculinist 'exercise of power over place'" (Cosgrove, 1985: 45, in Ingram et al., 1999: 57).[2]

While the notion of gay space appears in fact to be linked to urbanity and to activities of consumption – in the materialization, for instance, of gay neighbourhoods and gay clubs – the chapter argues the notion of queer space distances itself from urban/rural binaries, allowing the conception of non-heterosexual life and activities even in non-urban environments. In doing that, it evidences first how the emergence of the notion of gays and lesbians as urban subjects leads to the biased definition of queer rurality as the negative and inferior reality of its urban counterpart. How then these notions would influence the way architects and landscape architects conceive their practical and theoretical works on urban and non-urban contexts? How would they influence, perhaps, contemporary experimentations over the introduction of productive landscapes within large cities? In the attempt to answer these questions, the chapter evidences a fundamental problematic over the perception of agriculture as a tool applicable to the architecture of the city, which recognises a historical subordination of the rural landscape over the urban one.

Furthermore, in considering the contemporary emergence of mobile and flexible work/leisure patterns moving indistinctly across the physical spaces or urbanity and rurality (Bonora, 2015: 60; Behringer, 2013: 226–244), this chapter observes models which propose new ways of living either ruralised urban or urbanised rural environments, highlighting the absence of references to different ways of interpreting notions such as the one of family, couple, or community, and observing a neutral and ambiguous position on sociological compositions of the population of these new landscapes. This couples with a general biased heteronormative perspective of the surveyed literature on food studies, whether it analyses the history of agricultural practices or it studies contemporary food habits. Upon these observations, the chapter concludes by opening to some potential lines of intervention and discussion. The anti-urban narratives of Edelman and Herring and the queer readings of Foucault and Mieli pose the historical issue of nature and normativity, regarding to heterosexuality being recognised as the norm in productive environments. In light of this, the chapter advances the need to restructure the notions of nature and productive landscape in a queer perspective, suggesting the reading of Latour for a potential reframing of these notions outside of biased politicised discourses.

Defining urban and rural queer space

The chapter opens with a preliminary investigation on the relationship between non-heterosexual subjects and space, their environment of action, and the places of liberation and inclusion. The 1990s produced a vast literature on the subject, of which the chapter gives only a fractional account. From seminal books such as *Modern Homosexualities* (1992), *Sexuality and Space* (1992), *Queers in Space* (1997), and *Queer Space: Architecture and Same-Sex Desire* (1997), gay and lesbian subjects emerge to be predominantly attached to the urban landscape.

In "Bodies-Cities" (1992), Grosz's analysis of the form of gay neighbourhoods provides the opportunity to discuss the attitude which seems, within the surveyed literature, to characterise gay urban life: consumption, understood in terms of an act of freedom, and territorialisation (Grosz, 1992: 244). In "Pleasure and Danger", Bell stresses sexual citizenships are to some extent constituted through consumption, following a late-capitalist economy, and therefore tend to recognise the most visible part of the community, the shoppers, as the largest part of the gay and lesbian community (Bell, 1995: 141). Here, references to the figure of the nineteenth century French flâneur – described first by Benjamin in his 1972 *Gesammelte Schriften*, vol. 5, *Das Passagen-Werk* on the French arcades – and to a notion of citizenship which is directly understood through the physical form of the city, highlight a historical resilience of the conflation of spatial forms with moral and ethical judgement, with consumption being only the latest modality of operating such a judgement (Tester, 1994: 4).[3] The contemporary gay figure locates itself in between acceptance within the city as a consumer and isolation as part of a morally judged sexual minority. In the western history of homosexual liberation the Stonewall movement is seen as the starting point of the construction of the gay commercial ethos: in "Sex and Caring Among Men", Adam understands the urban gay figure from this period onward being "non-apologetic about his sexuality, self-assertive, highly consumerist and not at all revolutionary, though prepared to demonstrate for gay rights" (Adam, 1992: 176); Polchin, in "Have Something to Wear", asserts that the urban landscape of Christopher Street in New York – where the Stonewall riots took place and since then become a place of gay pilgrimage – "participates in the city's larger social exchanges between production and consumption [. . .] where queer sexuality can be viewed, displayed, expressed, and consumed" (Polchin, 1997: 384).

In these sources the city then appears as a landscape in which sexual minorities can easily achieve visibility and be active in public life, in light of the multiplicity of real or imaginary networks and connections of social activities and relations. The particular focus on the environment of the city during the 1990s might have ties with the AIDS epidemic, which in those years was affecting large cities across the globe. The urban dimension of the AIDS crisis reflected in a politicised activism which made the construction of urban gay narratives a necessity in order to build a sense of history and a prospect of future (Freeman, 2010: xv). The image of the city as the place where sexual liberation happens more freely through ludic activities of consumption has survived up to the present days. While Abraham re-affirms a clear and direct association between homosexuality and urban consumption through the figure of the flâneur (Abraham, 2011: 586), Kitchin, in "Sexing the City: The Sexual Production of Non-Heterosexual Space in Belfast, Manchester and San Francisco" proposes the notion of urban gay spaces as "contested sites, situated in a web of complex power geometries" (Kitchin, 2002: 206). In this sense, the contested character and the complexity within this notion of gay space could help to break the monolithic urban narrative of metronormativity, which "masks the multiplicity of queer lives in places and renders particular forms of political action, especially those based in rural areas, invisible" (Wang, 2014: 94).[4] In contrast with the sources dating from the 1990s and with their scholarly trail, the chapter highlights, indeed, the presence of a later literature, mainly of American origin and mostly included in specific journals such as *The Journal of Homosexuality* and *Thirdscape*, which began to investigate rural sexuality. Here, while lesbian narratives become more visible in comparison to earlier literature, they are still distinct from their gay counterparts. More strikingly for the scope of the chapter, this kind of literature takes the distance from the subject of the city by highlighting a clear-cut distinction between more dense, unhealthy urban settings and greener, healthier, and freer environments, even in cases of suburban and peri-urban fringes (see Chapters 17, 30 and 35).

Gay literature appears to be dominated by attitudes to consume the landscape in order to achieve freedom and to be liberated by urban stereotypes. In "Men in the Valley", Kirkey and Forsyth argue a reason for gay men to populate rural fringes of cities appears to be linked to the emergence of the nature movement during the 1970s, in which gay communes were founded with the belief that returning to an apparently more salubrious life, by harvesting organic food, would have freed them (Kirkey and Forsyth, 2001: 428, 429). This movement is linked to the queer faeries movement, a radical experiment which aimed to subvert the common stereotypes attached to masculine and feminine roles, although it progressively became more and more male dominated and more and more attached to the urban imaginary of idyllic rural retreats in which men could be free to express themselves (Cerno, 2010; Herring, 2010: 10, 64). It gave, in other words, room for the construction of the imaginary of the countryside escape, which is reprised today in the image of out-of-the-city gay meccas. In "Does Place of Residence Matter?", Wienke and Hill argue, through the analysis of the American fishing village of Provincetown in Massachusetts (which hosts the annual Mardi Gras and is characterised by established gay tourism) that the rural is often perceived by gay people as a recreational space (Wienke and Hill, 2013: 1260).

Lesbian literature has instead different recurring themes. In "Nestled into Niches", Anahita observes the American 'Landdyke' movement which, as part of the larger ecofeminist movement, relocates lesbian communities to rural areas. Here, rurality is conceived in terms of natural settings, and actions of ecological protection toward the environment are understood as a way through which achieving personal and communal liberation from a male-oriented and patriarchal society (Anahita, 2009: 719, 720). In spite of being a radical feminist movement in the 1970s, Wypler, in "Queering the Farm" (2015), argues for the Landdyke movement to be absorbed into the current movement of women in agriculture, with the feminist radical spirit being lost along with the visibility of lesbian farmers. The thin ground of lesbian farmers is also evident in "The (Mis)Translation of Masculine Femininity in Rural Space", where Sullivan analyses the visibility issues of lesbian women living in rural Northern Ontario due to the constructed masculine image of the general female rural population. Arguing this to destroy the butch imagery, the author identifies indoor places of consumption as the only spaces, within a "landscape emptied of gay people" (Sullivan, 2009: 3), in which lesbian subjects are able to territorialise their identity. However, contrary to the narratives attached to the experience of gay farmers, the lives of lesbians living in the countryside appear to be relatively more free, due to the invisibility of the woman within hegemonic logics of power of male domination. An important aspect touched by the surveyed literature is the relationship between individuality and the collectivity, or, in other words, the way rural gays and lesbians manage to live their own personal domesticity within rural communities. In Smith and Holt's "Lesbian Migrants in the 'Gentrified Valley' and 'Other' Geographies of Rural Gentrification", the small reality of lesbian households in Yorkshire rurality appears to be perceived as an idyllic retreat distant from the chaotic urban environment of the city (Smith and Holt, 2005: 317, 318).

The aspect of isolation and distance in queer rurality is, however, debated. While Ingram, in "Marginality and The Landscape or Erotic Alien(n)ations" describes distance as an issue and travelling a traumatic experience (Ingram, 1997a: 27), Rotenberk, in "Raise the Flag High" (2015), documents a plurality of different narratives relating to gays and lesbians living in rural areas, relocating the elements of distance and isolation in terms of growing up in the countryside or moving to it. In this sense, Rotenberk highlights the happiness in isolation of rural queers to be tied to the ability to form a network of peers. Moreover, studies published in Flammang's *The Taste for Civilization* and in *Food and Culture: A Reader* (2013) on urban domesticity, suggest that same-sex couples tend to adopt heterosexual gendered roles when engaged in food households, posing problematics either of female/male domination (Flammang, 2009: 42, 43), or of practical work load (Carrington, 2013, in Counihan and Van Esterik, 2013: 199–209), due to the influence of the dominating external culture. Possibly, the idyllic retreat in the countryside could then have positive effects on domesticity (especially couple domesticity), as it could have beneficial effects on the construction of gender roles outside of the model imposed by the collectivity.

However, in spite of the progressively positive narratives of the social queer rural geography, a number of assumedly problematic issues emerge from the surveyed literature when the physical and social relationship between the urban and the rural is conceived in clear-cut distinctions. First, as pointed out by Gorman-Murray in *Sexuality, Rurality and Geography*, due to assumed environmental and social aspects of rural areas such as the extreme isolation, the very poor accessibility to media, and the limited social connection, "rural to urban displacement is normalised in accounts of queer migration" (Gorman-Murray, 2013: 100). Second, the agricultural rural environment becomes commodified and gains the symbolical role of physical and moral object of freedom, with activities pertaining to production, including agriculture, being under-considered. Consequently, the visibility zone for queer subjects within rural landscapes gets associated with either urban consumption or with the fruition of nature in utilitarian terms; and thus the activities of fruition to gay and lesbian rural subjects are compared with an obviously more rich and vibrant reality of the urbanity of larger cities. Consequently, in light of rurality being conceived physically as opposed, and culturally as inferior, to urbanity, an interest is observed in researching the seeking of a gay and lesbian rural identity but little interest in the maintenance of such an identity through established rural practices.

Queer theorist Herring argues, in his anti-urbanist theory *Another Country*, that assuming the city as the natural environment for queer subjects means not only to frame in a general reductive perspective "queer kids of any gender, class, race or region" (Herring, 2010: 216), but also to provide an arguable "distant reference point for non-urban identified queers" (Herring, 2010: 224). In *Sexuality, Rurality, and Geography*, Browne and McGlynn advocate for a redefinition of the notion of rural "understanding this through imaginaries of gender/sexual difference and tolerance of othering, rather than 'objective' measures" (Browne and McGlynn, 2013: 36–37, 39). The physical opposition of rurality to urbanity is, however, a perspective that becomes obsolete when considered before the evolution of social interaction and communication. The advent of the internet, and especially of the web 2.0, provides also another instrument for the destruction of the assumed isolation and irremediable need for queer rural folks to move to the city. The internet provides an impalpable communication and information platform, differently from books and physical publications which "made reading a potentially dangerous act if you were reading the wrong thing" (Kennicott, 2014). The dating platform Muddy Matches, described as "the Tinder for farmers" (Tyzak, 2014), is an example: it utilises, due to the greater distance of rural areas and the impossibility to connect immediately as in cities, an algorithm based on events shared instead of the geographical location of the users.

Already in 1997 Aaron Betsky was arguing that queer space was disappearing from large cities due to the advent of internet virtual places which are " indistinguishable from those of their heterosexual peers" (Betsky, 1997: 14) and which are now pervasive. In this sense, the homogenization of the gay experience through internet apps has made city and countryside closer. However, while cities are experiencing a decrease of gay venue and commercial spaces which weakens their gay identity and threatens the existence of gay neighbourhood altogether, rural places are benefiting from the broadening of inclusive attitudes towards gay and lesbians, also promoted by the media (Meko, 2007).[5] The shift from earlier, dramatic documentaries such as *Fighting in Southwest Louisiana: Gay Life in Rural America* (1992), to more recent, positive ones including *Farm Family: in Search of Gay Life in Rural America* (2004) and *Forbidden: Undocumented and Queer in Rural America* (2016) testifies to the opening of rurality not only to living sexuality freely but also to living sexuality freely in light of being integrated within working communities, including agricultural ones.

This progressive opening up of the countryside to global media has led to the inclusion of sexualities other than heterosexuality even in uncharted territories, such as dating reality shows, romantic comedies, and soap operas, with, for example the Italian version of *Farmer Wants a Wife* including a gay farmer seeking a boyfriend (Grasso, 2016). This recent phenomenon, which destabilises the stereotyped image of the heteronormative countryside is both influenced by, and is influencing a progressively greater visibility of rural queers. While projects of activism and liberation born within queer rural communities of farmers, such as the documentary *Out Here* (2013) or historical journals such as *Country Women*, are giving the strength to other rural gay and lesbian individuals to be visible, influence coming from the outside, such as movies or even reality shows portraying rural farmers, is contributing to building a new image of rural queerness (Underhill, 2015). Social landscapes that rural queers were imagining by reading novels or were planning to join by accessing urban journals are now actually realisable. The question now is: can disciplines such as architecture, landscape architecture, and planning incorporate this vision into their theoretical and practical activity? Is there already a space for such a conception? And if not, is there a possibility to build it?

Agriculture, buildings, and gender

In order to investigate how architects, landscape architects, and planners understand the concepts of rural and urban in inclusive terms for subjects who do not recognise themselves as heterosexual, the chapter argues the need to question the formulation of the notion of landscape itself. In spite of semantic differences between countries (particularly between northern Europe and the Mediterranean cultures) in regards to the importance given either to human perception of nature-made aspects or to the human-made transformation upon nature, the notion of landscape serves in Europe (and the western world) to define a physical and transformative human presence within the natural environment. Mather notes that "man may be a creature of the city, but land is still his habitat, and he values its scenic beauty as well as its fruit" (Mather, 1986: 15).[6] Historically, the value that the inhabitants of the city see in the countryside shapes the prevailing sense of the city's dominance over the rural, and the choices as to how to use and administer the land beyond the city walls.[7] Today, it can be argued this relationship between the urban and the rural to be at its historical peak, as "more than even before, it can be said that the earth's surface is urbanised to some degree" (Soja and Kanai, 2007: 62), with the city emerging as the ground of problems such as overpopulation, pollution, soil consumption, and energy wasting (Chapter 16).

In order to address those issues, contemporary architecture, landscape architecture, and planning have elaborated models and theories that attempt to integrate agricultural elements within the city (see Chapters 12, 26 and 27). Within the theories in support of those practical experimentations, there has been a sensible shift in the past decade from the conception of agriculture as a mere design tool for architects and designers to more holistic notions of landscape in both physical and social terms. Examples such as Steiner's *The Living Landscape: An Ecological Approach to Landscape Planning* (1991) or Tutino Vercelloni's *Costruire per l'agricoltura* (Building for agriculture; 1996) still conceive the productive landscape as an element to be planned, built, and controlled. In contrast, movements such as landscape urbanism, illustrated in Waldheim's manifesto-book *The Landscape Urbanism Reader* (2006), and its subsequent evolution theorised in Mostafavi's *Ecological Urbanism* (2010), have shifted toward a notion of landscape which embeds and encompasses architecture and urbanism. Such a shift proves then to invalidate the twentieth-century ideological conception that assumes nature to be defined by what grows from the soil, which constitutes a preconception of a neat distinction between a consumerist urban environment and a productive agricultural one.

Current experimentations also attempt to integrate new contemporary working habits and customs, such as the mobility and externalisation of large corporations, or simply the possibility to work from home, with architectural systems characterised by the inclusion of natural elements. In *Campagnes Urbaines*, French agronomist Donadieu conceives "hybrid countrysides, which are made of farmers who love the city and are in need of its inhabitants [. . .] and in which the farmers-entrepreneurs will capitalize their knowledge and their income [. . .] for other farmers" (Donadieu, 1998: 26). Through design projects mixing office jobs, urban residential living, and agricultural activities, Cibic explores, in *Rethinking Happiness*, post-fordist ways of living by theorising the notion of community to be linked to the construction of a "lovely place [. . .] where people can meet together and do things together, where children can play in a safe environment, where there is greenery" (Cibic, 2010: 21). A problem here emerges in the absence of any clarification of who might inhabit those theorised and designed architectures and landscapes, in spite of those models theorising post-fordist landscapes characterised by new conceptions of work and leisure and new ways of inhabiting urban and rural environments. This absence inhibits in fact to make any considerations over the formation of new potential types of communities for gender roles, sexuality, and type of community habits and customs. Consequently, for the interest of the chapter, it is thus impossible to tell whether queer subjects are included within the future imagined by contemporary architects.

A further problem for queer subjects appears to be coming from the literature and the theory in food studies, which indirectly influences the conception of architectural, landscape architecture, and planning models. This specific literature, while attempting to address either the history of agriculture or cultural and social habits of food production and consumerist practices or their mythicised symbolisms, is argued to be tending inevitably to fall into an idealised and heteronormative binary conception of the world.[8] In practical terms, it assumes, for instance, gender-based labour distinctions to be attached to general historical notions such as the human passage from gatherers to hunters as universal and taken for granted values which are still valid today. Examples of such a literature include Bevilacqua's *Genius Loci*, in which agriculture and nature are respectively conceived as female and male, with their union being attributed to the myth of the balance between humans and nature (Bevilacqua, 2010: 24, 31). In *Food: The Key Concepts*, the figure of the woman is attached to images such as the ones of the nurturer, the gatherer, the cook, as the author claims either that "preparing and serving food has long given women power," or that "even when women are restricted to feeding their own immediate family, they find leverage through cooking," or perhaps that "through their foodwork they may hold families together," and "may also tame men" (Belasco, 2008: 41, 42). In *The Anthropology of Food and Body*, food is portrayed as an instrument of power and liberation in a feminist perspective, albeit the woman is always pictured in a biological binary opposition with the man. Curiously, the only references to homosexuality relate here to the eating disorder and the obsession for thinness, and to homosexual tribal practices of males dominating nature (Counihan, 1999: 66, 218). In *Il riposo della polpetta* (The Resting of the Meatball), Montanari, while analysing the male-oriented rite of the barbecue, argues "every culture sometimes feels nostalgia for this past, more mythical than real; every culture wishes to complete itself within the nature from which it presumably detached. However, even this 'nature' is a cultural construction" (Montanari, 2009: 155).

The risk here is represented by these cultural constructions to be potentially taken as ideological assumptions. It is the case, for example, of the literature around the Slow Food philosophy, which attempts to define a new agricultural system and a new type of gastronomy. Within a type of literature addressing sociopolitical issues as crucial stances for the construction of such a fairer and more sustainable agricultural reality, official and unofficial publications such as Petrini's *Buono, pulito e giusto* (Good, Clean and Fair, 2005), Honoré's *In Praise of Slow* (2004), and publications on sustainable agriculture mentioning the Slow Food movement such as Calori's *Coltivare la citta* (Cultivating the City, 2009) avoid addressing sexual diversity as a value by referring either to the ambiguous term 'family', or to the generic terms 'producers' and 'consumers'. Other publications are instead more explicitly gender and sexually biased; *Territorio zero* envisions a technological and socially sustainable short-supply-chain-oriented agricultural system in which male strength is used for manual labour and the "female innate intellect" is used for the conception and maintenance of the system itself (Di Cori Modigliani, 2013: 61). This is only a small percentage of the existing literature on food studies, and certainly I do not want to insinuate, by pointing out the surveyed readings, that all the literature on food studies takes a heteronormative and gender-biased attitude. I have aimed, though, to point out that it exists as a fraction of such a literature which postulates, justifies, and propagates within academic and non-academic environments sexual and gender-biased arguments that, due to the heteronormative attitude of contemporary western society, are not counterargued. Furthermore, I have sought to question whether and how much this biased perspective influences other disciplinary grounds such as the ones of architecture, landscape architecture, and planning.

Conclusion

This chapter starts by questioning the identity of the definition of queer space, through a research on literature specifically focusing on spatial implications of sexual identities. In doing that, it highlights the perceived urban attitude of gay and lesbian subjects, and the emergence of a link between sexual visibility and urban consumption. This link, the chapter argues, is also present in rural landscapes, with consequent exclusion of productive activities from the sphere of sexual liberation of queer subjects. The chapter then, in the second section, considers the contemporary architectural experimentations over the inclusion of productive landscapes within the urban environment of western cities. In doing that, it evidences how a shift toward more holistic theories merging landscape stances with architectural and urban views helps to break the assumedly imbalanced perception of the notions of urban and rural. It however questions the absence, in the observed architecture theoretical and practical models, of considerations over the potential inhabitants of the new imagined landscapes and systems, thanks also to the influence of a general heteronormative and gender-biased perspective of publications on food studies and landscape history, which act as a support to those architectural theories.

In conclusion, in reprising this last aspect in a queer perspective, a critique comes from Edelman, who, in *No Future* (2004) claims the monocultural and universal prospect of future given by an heteronormative society gives all other non-conforming individuals no hope, and thus contributes to create an over-consideration on the consumption of the present, rather than on the construction of the future. A similar critical discussion on the moral and gender-based perspective attached to production is already traceable in two philosophers from the 1970s. Foucault, in *The History of Sexuality*, repositions the relationship between power and sexuality within the advent of capitalism, arguing its transforms from relationship based on repression to one based on production, "an integral part of the bourgeois order" (Foucault, 1976: 5). Mieli, in *Elementi di critica omosessuale* (Element of Homosexual Critique), instead refers to the process of imposition of heterosexuality as the norm for the socialisation of production and for the formal submission of labour to the Marxist notion of capital, by talking of "educastration" (Mieli, 1977: 13–16).[9]

Edelman's criticism induces the chapter to advance the need to restructure not only the notion of rural, as claimed before, but the entire notion of nature. In leaving the problem open to further research, the chapter suggests the reading of Latour's *Politics of Nature* (2004) as a starting point. Latour highlights the properly western attitude, traceable from the ancient Greeks, of pairing the conception of politics and the notion of nature with the configuration of the latter emerging from a subjective artificial interpretation that is used as a sociopolitical instrument (Latour, 2004: 28, 43). Finally, in highlighting the heavy reliance of the present research on a corpus of documents of almost solely American origin, the chapter opens up to further narratives and to future theories, coming from different countries, and pertaining to different cultures, and different landscapes.

Notes

1 The Stonewall Riots were a series of spontaneous violent protests on June 28, 1969, at the Stonewall Inn, Greenwich Village, New York City. Acted by the local gay community over the police, who arrested, beat, and violently abused the gay population in the US in the 1960s, the demonstration of protesting against police officers without having the fear of being arrested symbolised an act of sexual liberation, and since then has been considered the single most important event leading to the construction of an American first, and then international, gay liberation movement. For further discussion on the Stonewall Riots, see Martin Duberman, *Stonewall* (Penguin Books: London, 1993) and Nicholas Edsall, *Toward Stonewall: Homosexuality and Society in the Modern Western World* (University of Virginia Press: Charlottesville, 2003).

2 For a definition of sexual minority, the chapter refers to Butler: "To be part of a sexual minority means, most emphatically, that we are also dependent on the protection of public and private spaces, on legal sanctions that protect us from violence, on safeguards of various institutional kinds against unwanted aggression imposed upon us, and the violent actions sometimes instigated" (Butler, 2004: 20).

3 For further discussion on flâneur, see Walter Benjamin, *The Arcades Project*, translated and edited by Rolf Tiedermann (Cambridge, MA: Harvard University Press, 1982). Originally published as *Gesammelte Schriften*, vol. 5, *Das Passagen-Werk* (Frankfurt-am-Main: Suhrkamp, 1972).

4 For a definition of metronormativity, see Halberstam, J. (2005) *In a Queer Time and Place: Transgender Bodies, Subcultural Lives*, New York: New York University Press, 36–37.

5 For a description of the ecological, political, and socio-economic history of urban gay neighbourhoods, see Ghaziani, A. (2015) "The queer metropolis," in Delamater, J. and Plante, R. F. (eds.) *The Handbook of the Sociology of Sexuality*, New York: Springer, 305–330.

6 The notion of landscape itself varies from place to place. The chapter quotes Venturi Ferriolo's interpretations, claiming the term landscape is included in the geographical term environment, as it is characterized by a human action toward the territory. Venturi Ferriolo, *Etiche del paesaggio* (2002): 134.

7 This argument can be traced back in Aristotle's idea of city, which is conceived as the place where science and technology was created in leisure rather than by means of physical work, and thus separates with city walls the virtuous from the outside savages. For more discussion on Aristotle's idea of city, see Swanson, J.A. and Corbin C.D. (2009) *Reader's Guides: Aristotle's Politics*, London: Continuum International Publishing; Chan, M.D. (2006) *Aristotle and Hamilton on Commerce and Statesmanship*, Columbia: University of Missouri Press.

8 For further readings on a mythical and poetical perspective of the landscape and agriculture, see Venturi Ferriolo, *Etiche del paesaggio* (2002); Schama, *Landscape and Memory* (1997).

9 Normativity proves to be an issue both for heterosexual and homosexual reality. Brown suggests homonormativity needs to go outside of the boundaries created by neoliberalism, as it has generated a notion of gay and lesbian space of action shaped by commodity and subjected to for-profit businesses serving the desires of gay consumers (Brown, 2012: 7). Bernini, in *Apocalissi Queer* (2013), criticizes the combination of Edelman's no-future attitude with the capitalistic type of economy based on consumption typical of urban queer communities, as acting as a counter-flywheel on discussions over the normativisation of realities other than that traditional heteronormative disciplinarity.

Further reading

Abraham, J. (2011) The homosexuality of cities. In: Bridge, G., & Watson, S. (eds.), *The new Blackwell companion to the city*. Oxford: Wiley-Blackwell.

Adam, B.D. (1992) Sex and caring among men: Impacts of AIDS on gay people. In: Plummer, K. (ed.), *Modern homosexualities: Fragment of lesbians and gay experience*. London: Routledge.

Anahita, S. (2009) Nestled into niches: Prefigurative communities on lesbian land. *Journal of Homosexuality*, 56(6), 719–737.

Behringer, W. (2013) *Storia culturale del clima: dall'Era glaciale al Riscaldamento globale*, trans. from *Kulturgeschichte des klimas*. originally published in 2009. Gravellona Toce: Bollati Boringhieri.

Belasco, W. (2008) *Food: The key concepts*. Oxford and New York: Berg.

Bell, D. (1995) Pleasure and danger: The paradoxical spaces of sexual citizenship. *Political Geography*, 14(2), 139–153.

Benjamin, W. (1982 [1972]) *The arcades project*, Tiedermann, R. (trans., & ed.). Cambridge, MA: Harvard University Press.

Bernini, L. (2013) *Apocalissi Queer: Elementi di teoria antisociale*. Pisa: Edizioni ETS.

Betsky, A. (1997) *Queer space: Architecture and same-sex desire*. New York: William Morrow.

Bevilacqua, F. (2010) *Genius Loci: Il dio dei luoghi perduti*. Soveria Mannelli: Rubbettino.

Bonora, P. (2015) *Fermiamo il consumo di suolo: Il territorio tra speculazione, incuria e degrado*. Bologna: Il mulino.

Brown, G. (2012) Homonormativity: A metropolitan concept that denigrates 'Ordinary' gay lives. *Journal of Homosexuality*, 59(7), 1065–1072.

Browne, K., & McGlynn, N. (2013) Rural lesbian, gay, bisexual, and trans equality. In: Gorman-Murray, B.P., & Bryant, L. (eds.), *Sexuality, rurality, and geography*. Lanham, MD: Lexington Books.

Burdett, R., & Sudjic, D. (2007) *The endless city: The urban age project by the London School of economics and Deutsche Bank's Alfred Herrhausen society*. London: Phaidon.

Butler, J. (2004) *Undoing gender*. London: Routledge.

Calori, A. (2009) *Coltivare la città: Giro del mondo in dieci progetti di filiera corta.* Milan: Terredimezzo editore.

Carrington, C. (2013) Feeding lesbigay families. In: Caunihan, C., & Van Esterik, P. (eds.), *Food and culture: A reader*, 3rd ed. New York: Routledge.

Caunihan, C., & Van Esterik, P. (eds.) (2013) *Food and culture: A reader*, 3rd ed. New York: Routledge.

Cerno, T. (2010) Nel bosco degli elfi gay. *L'Espresso*, October 10, 2010. http://espresso.repubblica.it/visioni/societa/2010/10/18/news/nel-bosco-degli-elfi-gay-1.24885.

Chan, M.D. (2006) *Aristotle and Hamilton on commerce and statesmanship.* Columbia: University of Missouri Press.

Cibic, A. (2010) *Rethinking happiness.* Milan: Corrine Edizioni.

Cosgrove, D. (1985) Prospect, perspective and the evolution of the landscape idea. *Transactions of the Institute of British Geographers, New Series*, 10(1), 45–62.

Counihan, C. (1999) *The anthropology of food and body: Gender, meaning, and power.* New York: Routledge.

Counihan, C., & Van Esterik, P. (2013) *Food and culture: A reader*, 3rd ed. New York: Routledge.

De Santoli, L., & Consoli, A. (2013) *Territorio Zero.* Rome: Minimum Fax.

Detamore, M. (2013) Queering the hollow. In: Gorman-Murray, B.P., & Bryant, L. (eds.), *Sexuality, rurality, and geography.* Lanham, MD: Lexington Books.

Di Cori Modigliani, S. (2013) La narrative esistenziale di Territorio Zero. In: De Santoli, L., & Consoli, A. (eds.), *Territorio zero.* Rome: Minimum Fax.

Donadieu, P. (1998) *Campagnes urbaines.* Versailles: Acted Sud.

Duberman, M. (1993) *Stonewall.* London: Penguin Books.

Edelman, L. (2004) *No future: Queer theory and the death drive.* Durham, NC and London: Duke University Press.

Edsall, N. (2003) *Toward Stonewall: Homosexuality and society in the modern western world.* Charlottesville, VA: University of Virginia Press.

Flammang, J.A. (2009) *The taste for civilization: Food, politics, and civil society.* Urbana, Chicago, IL and Springfield: University of Illinois Press.

Foucault, M. (1978 [1976]) *The history of sexuality, Volume I: An introduction*, Hurley, R. (trans.). New York: Pantheon Books.

Freeman, E. (2010) *Time binds: Queer temporalities, queer histories.* Durham, NC: Duke University Press.

Ghaziani, A. (2015) The queer metropolis. In: Delamater, J., & Plante, R.F. (eds.), *The handbook of the sociology of sexuality.* New York: Springer, pp. 305–330.

Gorman-Murray, B.P. (2013) Documenting lesbian and gay lives in rural Australia. In: Gorman-Murray, B.P., & Bryant, L. (eds.), *Sexuality, rurality, and geography.* Lanham, MD: Lexington Books.

Gorman-Murray, B.P., & Bryant, L. (2013) *Sexuality, rurality, and geography.* Lanham, MD: Lexington Books.

Grasso, A. (2016) Donne e gay nel nuovo romanzo dei contadini innamorati. *Corriere della sera*, April 28, 2016. www.corriere.it/spettacoli/16_aprile_28/donne-gay-nuovo-ro . . . o-contadini-innamorati-4efd438c-0ca4-11e6-a4db-66a817deaada.shtml.

Grosz, E. (1992) Bodies-cities. In: Colomina, B. (ed.), *Sexuality and space.* New York: Princeton Papers on Architecture.

Halberstam, J. (2005) *In a queer time and place: Transgender bodies, subcultural lives.* New York: New York University Press.

Herring, S. (2010) *Another country: Queer anti-urbanism.* New York: New York University Press.

Honoré, C. (2004) *In praise of slow: How a worldwide movement is challenging the cult of speed.* London: Orion Books.

Ingram, G.B. (1997a) Marginality and the landscapes of erotic alien(n)ations. In: Ingram, G.B., Bouthilette, A.M., & Retter, Y. (eds.), *Queers in space: Communities, public spaces, sites of resistance.* Seattle: Bay Press, pp. 27–52.

Ingram, G.B. (1997b) Open space as strategic queer sites. In: Ingram, G.B., Bouthilette, A.M., & Retter, Y. (eds.), *Queers in space: Communities, public spaces, sites of resistance.* Seattle: Bay Press, pp. 95–125.

Ingram, G.B. (1999) *Queer geographies: Diverging theories of homoerotic space for urban policy, planning and design.* Guest Lecture at the University of Amsterdam, Department of Sociology.

Ingram, G.B., Bouthillette, A.M., & Retter, Y. (1997a) *Queers in space: Communities, public spaces, sites of resistance.* Seattle: Bay Press.

Ingram, G.B., Bouthillette, A.M., & Retter, Y. (1997b) Narratives of place: Subjective and collective. In: Ingram, G.B., Bouthilette, A.M., & Retter, Y. (eds.), *Queers in space: communities, public spaces, sites of resistance.* Seattle: Bay Press, 55–60.

Ingram, G.B., Bouthillette, A.M., & Retter, Y. (1997c) Strategies for (re)constructing queer communities. In: Ingram, G.B., Bouthilette, A.M., & Retter, Y. (eds.), *Queers in space: Communities, public spaces, sites of resistance.* Seattle: Bay Press, 447–457.

Kennicott, P. (2014) Smuggler: A memoir of gay male literature. *VQR*, Fall 2014. www.vqronline.org/memoir-articles/2014/10/smuggler.

Kirkey, K., & Forsyth, A. (2001) Men in the valley: Gay male life on the suburban-rural fringe. *Journal of Rural Studies*, 17, 421–441.

Kitchin, R. (2002) Sexing the city: The sexual production of non-heterosexual space in Belfast, Manchester and San Francisco. *City*, 6(2), 205–218.

Latour, B. (2004) *Politics of nature: How to bring the sciences into democracy*, Porter, C. (trans.). Cambridge MA: Harvard University Press.

Mather, A.S. (1986) *Land use*. Essex: Longman Scientific & Technical.

McCarthy, L. (2000) Poppies in a wheat field. *Journal of Homosexuality*, 39(1), 75–94.

Meko, J. (2007) We're all concerned about the loss of gay neighborhoods. *The Fogcity Journal*, July 16. Available from: www.fogcityjounral.com/news_in_brief/jm_soma_070716.shtml.

Mieli, M. (2002 [1977]) *Elementi di critica omosessuale*. Milan: Feltrinelli.

Montanari, M. (2009) *Il riposo della polpetta, e altre storie intorno al cibo*. Bari: Laterza.

Mostafavi, M., & Doherty, G. (2010) *Ecological urbanism*. Zurich: Lars Muller Publishers.

Petrini, C. (2005) *Buono, pulito e giusto: Principi di nuova gastronomia*. Turin: Einaudi.

Polchin, J. (1997) Having something to wear: The landscape of identity on Christopher street. In: Brent Ingram, G., Anne-Marie Bouthilette, A.M., & Retter, Y. (eds.), *Queers in space: Communities, public spaces, sites of resistance*. Seattle: Bay Press.

Rotenberk, L. (2015) Raise the flag high: Queer farming in rural America. *Modern Farmer*, January 22, 2015. http://modernfarmer.com/2015/01/raise-flag-high-queer-farming-rural-america/.

Schama, S. (1997) *Landscape and memory*. London: Vintage Books.

Smith, D., & Holt, L. (2005) Lesbian migrants in the gentrified valley and 'Other' geographies of rural gentrification. *Journal of Rural Studies*, 21, 313–322.

Soja, E.W., & Kanai, M. (2007) The urbanization of the world. In: Burdett, R., & Sudjic, D. (eds.), *The endless city: The urban age project by the London school of economics and Deutsche bank's Alfred Herrhausen society*. London: Phaidon.

Steiner, F. (1991) *The living landscape: An ecological approach to landscape planning*. New York: McGraw Hill.

Sullivan, R.E. (2009) The (mis)translation of masculine femininity in rural space: (Re)reading 'Queer' women in Northern Ontario, Canada. *Thirdscape: A Journal of Feminist Theory & Culture*, 8(2). Available from: http://journals.sfu.ca/thirdspace/index.php/journal/article/viewArticle/sullivan/247.

Swanson, J.A., & Corbin, C.D. (2009) *Reader's guides: Aristotle's politics*. London: Continuum International Publishing.

Tester, K. (1994) *The Flâneur*. London: Routledge.

Tutino Vercelloni, I. (1996) *Costruire per l'agricoltura: Storia, sperimentazioni, ipotesi*. Milan: Skira.

Tyzak, A. (2014) Tinder for farmers? New app for those seeking a muddy match. *The Telegraph*, September 21, 2014. www.telegraph.co.uk/women/sex/online-dating/11104840/Tinder-for-farmers-New-app-for-those-seeking-a-muddy-match.html.

Underhill, V. (2015) Say hello to the queer farmers of America. *Bitchmedia*, April 9, 2015. https://bitchmedia.org/post/say-hello-to-the-queer-farmers-of-america.

Venturi Ferriolo, M. (2002) *Etiche del paesaggio: Il progetto del mondo umano*. Rome: Editori Riuniti.

Waldheim, C. (2006) *The landscape urbanism reader*. New York: Princeton Architectural Press.

Wang, S. (2014) Encountering metronormativity: Geographies of queer visibility in Central New York. *Wagadu*, 12, 91–124.

Wienke, C., & Hill, G.J. (2013) Does place of residence matter? Rural-urban differences and the wellbeing of gay men and lesbians. *Journal of Homosexuality*, 60(9), 1256–1279.

Wypler, J. (2015) Queering the farm. *Our Lives Magazine*. http://ourlivesmadison.com/article/queering-the-farm/.

39

The cultural and spiritual aspects of growing edible plants

Testing for meaningfulness in Leeds, UK

Ann Light and Christina Welch

Introduction

Grown Edible Meaningful (GEM) was a small-scale project funded by the UK's Arts and Humanities Research Council (AHRC) as part of its Connected Communities programme (Connected Communities, 2009). The landscape of the GEM project is one of the most ethnically and religiously diverse parts of inner-city Britain: a densely packed district of the northern English city of Leeds. Leeds is home to at least 140 different ethnic groups, and over 170 different languages are spoken by school pupils across the city (Leeds, 2011a). High numbers of Hindus, Jews, Muslims, and Sikhs live in Leeds (ibid.), which also gives home to many Christian faith communities, including a concentration of new evangelical churches with roots in African Pentecostalism (Callaghan, 2015; Gipton and Harehills, QS210EW (2011); Leeds City Council, 2011a and 2011b). Leeds' population as recorded in the 2011 UK census stood at 751,500 residents, with the electoral ward of Gipton and Harehills, where the project took place, composed of a population of 27,078 (Gipton and Harehills, 2011b).

The specific focus for the GEM project, however, was on the small gardens and pots on windowsills that make up the growing spaces for many of the residents of Harehills, a part of East Leeds. Harehills is an ethnically varied and religiously diverse area of the city, and includes a large number of immigrants and asylum-seekers (Gipton and Harehills, 2011a). It has relatively high levels of deprivation (Gipton and Harehills, 2011b), with government statistics placing Harehills in the most deprived 20% of the country (Democracy Leeds, 2010).

The GEM project's data gathering – about whether people found meaning in growing edible food – was conducted through (1) in-depth interviews with five participants following a food-growing intervention with about 50 Harehills residents, and (2) collecting information and stories from a food festival organized to allow the 150 or so participants to meet, talk, and share food with others interested in local food-growing initiatives. The ethnic and cultural diversity of the area contributed a strong sense of diaspora to the project, both of people and of plants, but what marked the study as unusual was the analysis of spiritual factors evident in the stories gathered from participants (Light and Welch, 2014).

Although what was found cannot be considered representative of any of the different ethnicities and faiths encountered in the study, the meaningfulness of grown edible plants across the ethnicities and faiths was pertinent. Some stories of food growing referenced beliefs and traditions associated with spiritual lifeways, while others revealed the almost psychospiritual transformative effect of working with the soil to bring food into being (Taylor, 2009). Notable also was the passion with which people talked about land and plants left behind in other countries and how a connection to the ancestral home and its flora travelled with them, pointing to the value of reflection on growing food in areas with recent immigration.

Schroeder, an environmental psychologist, has long noted the importance of a connection with the natural world for wellbeing and wholeness, examining the perceptual connection between the environment and ourselves as sensory human-beings (1990). Although largely explored in wilder settings, Schroeder's work highlights the human body's response to plants and trees; the 'felt sense' he discovered incorporates not only an understanding of place, but also with a perception of spirituality. Works by Wilson (1983), Kellert and Wilson (1993), Ingold (2000), Edensor (2010), and Edensor and Jayne (2012) also explore this relationship. Verain et al. (2012) comment from a social science perspective on the association between the natural world and spirituality: in this case, a connection between high organic food consumption and a person's affiliation with the spiritual. Anthropologists, theologians, and religious studies scholars have documented a connection between food and faith, especially in regard to preparation and consumption, though production of food is also religiously significant (Douglas, 2003; Wirzba, 2011; Harvey, 2013). Food growing has also been associated with a reaction against an overly consumerist society (Bauman in Rojek, 2004) and with health and economic issues (Church et al., 2015). With food so central to human life, the importance of it to all factors of existence is, perhaps, unsurprising and the findings of the GEM project related to all these studies, although can be framed particularly by the postcolonial discourse on diaspora (Cook and Harrison, 2003; Wilson, 2016). In many of the accounts, participants noted an ecological sense of connection, which acknowledged the environment but extended it as time, place, and a sense of identity met in the production of grown foodstuffs.

Food in diaspora

The importance of bringing the natural world, especially edible plants and fruit-bearing trees, into refugee camps is well documented by the Lemon Tree Trust, who use urban agriculture and greening initiatives to help transform the land and lives of those affected by forced displacement (Hunt and Tomkins, 2016). Their work reveals the significance to migrants and refugees, even those in transit, of growing food and flowers, but stops short of drawing conclusions about any spiritual as well as physical benefits of the project. The principal participants in the GEM project were more settled, being a diverse mix of first- and second-generation immigrants residing in a densely built part of a major English city.

Existing research has demonstrated the links between food and a sense of home for those in diaspora. Temple (1999) explored Polish diasporic women's perception of identity noting the importance of Polish food to their sense of ethnicity. Morgan et al. (2005: 96) examined migrant culture in Western Sydney and found that "by growing and cooking traditional foods [disaporic peoples] reconnect with their homeland". Brown et al. (2010) found that international postgraduate students in England used food from their country of origin as a form of comfort, while Tookes (2015) noted that foods perceived as traditional were central in adjusting to a new life in a new country for Barbadians. Other research has focussed on food growing for wellbeing. Graham and Connell's study of Greek and Vietnamese migrants in a Sydney suburb ascertained that growing plants from their homeland gave a sense of place, providing surroundings that were reassuring and allowing them to retain traditional foodways (2006). Moulin-Doos' study of intercultural gardening (where migrants from different origins garden together) found that sharing knowledge and experiences of agriculture and foodways from different places round the world resulted in respectful interactions between differing groups of diasporic peoples with widely divergent backgrounds (2013). Central to this was the production of food (ibid.).

Mazumdar and Mazumdar (2009) have explored how the growing of significant plants is used by Hindu immigrants to Southern California to sacralise their home and garden spaces. Their 2012 study extended this research into other immigrants groups into the same area and found that migrants generally developed ecospaces, or landscapes, reminiscent of home, which allowed them to continue practicing sacred rituals and remember past places and people. Although, surprisingly little has been written about the relationship between the spiritual aspects of food growing and migrancy, Cone and Myhre (2000) have noted there is a spiritual aspect to community-supported food growing in the US.

Where the relationship between food growing and diasporic living has been explored, it has tended to be in countries where the local climate allows for a range of grown produce that resembles that of home. Both Graham and Connell's (2006) and Mazumdar and Mazumdar's (2009; 2012) studies, noted earlier, explore the significance of plants that are grown in both the new and ancestral home locations.

A move to cooler climate, such as from Bangladesh to Britain, brings with it major problems when attempting to grow fruits and vegetables that require more heat or sun than found in the UK. Datta's work on Bangladeshi immigrants in London (2009) touches on some of these issues, noting that, for these migrants, the garden is typically a productive, rather than leisure, space, used to produce food used in everyday cooking. This resonates with the findings of the GEM project, where Fahima, a second-generation Bangladeshi woman, spoke of her father's determination to grow coriander in the front garden of his Leeds terraced house. The herb was unavailable in England when he arrived with his family in the 1980s, yet was central to their traditional foodways and provided a connection with 'back home' (Light and Welch, 2014). She related how Asian people frequently knocked on their door, asking to take some. The importance of 'back home' for Fahima can be seen in how she has struggled to keep alive a Bangladeshi lemon tree that allows her to flavour traditional dishes and connect sensually with her ancestral roots.

A further issue that disaporic food growers often face is the lack of access to growing spaces, for Britain is a country where land occupancy is dense and many homes have no, or very little, outside area (for instance, the back-to-back houses and flats of Leeds). In the GEM study, accounts were given by women from a network of Afghan refugees. They reported to the GEM researchers about what they had cultivated before leaving Afghanistan, and related their grief at no longer having the means to grow traditional produce. Mazumdar and Mazumdar (2012) also found that, for immigrants without a garden from homelands where food and flower growing was the norm, the lack of access to a plot of land was a deeply felt sadness.

The importance of traditional food and flora for migrants cannot be underestimated. They allow a connection with home and thus provide a sense of the familiar in strange surroundings, positively affecting wellbeing. And they can be central to religious practice (also see Chapter 28). Sered's sociological study of Middle-Eastern Jewish women found that food was a crucial "medium for passing on divine blessing" (1988: 132) and traditional ingredients were vital for continuing ethnic identity and religious belief. Further, both Graham and Connell (2006: 384) and Mazumdar and Mazumdar (2009: 262, 2012: 260) observe that, for Buddhist and Hindu families, home-grown flowers are hugely important in shrine decoration and thus central to ritual observance. Nonetheless, the spiritual aspect of food and flora grown by migrants is largely a marginalised subject (Gabaccia, 1998; Diner, 2001; Jennings et al., 2014), frequently only noted in participant comments but rarely explored in any critical depth. Refugees cited in "Transforming Land, Transforming Lives", for example, speak of gardening as being 'in their blood' and 'living within me' (Hunt and Tomkins, 2016). Some even state that they rescued seeds before they fled (ibid.). Yet the holistic implications of this are absent from the report on how urban agriculture can help transform refugee camps and rebuild displaced lives.

The GEM project explicitly sought to explore the connections between food growing, culture, environment, and the spiritual in more depth.

Talking to people about what they grow

The GEM project followed a growing season, starting its active phase in April 2013 and culminating that September in a food festival; the time frame being significant for reconnecting urban households with the idea of seasonal food and its production. The overall aim of the project was to see if growing food would encourage participants to consider how to eat and live more sustainably with in-depth interviews with key participants, and data gathered at the food festival providing material for extensive analysis. It was not a controlled study of attitudes, but a piece of action research (Kindon et al., 2010; Reason and Bradbury, 2013), slotted into people's everyday lives, where actual behaviour could be observed and discussed. The researchers acted principally as enablers, partly to learn what people were thinking, and partly to take a view on the role of reflection on practice, as an adjunct to growing food, in promoting environmental behaviour. To this end, the researchers worked with existing community organisations and their agenda of events to introduce food cultivation locally. This made the selection of participants within the specified area of East Leeds quite arbitrary, but provided introductions to a range of people from Harehills already interested in growing food. The researchers followed up the growing intervention with individual interviews to provide a bank of rich data, and organised a food-growing festival that brought a wider set of people together to talk about the possible meaningfulness of growing edible produce. The festival signalled the end of the intervention for the participants, and provided an opportunity for them to share their experiences and stories with each other and the wider community. Attendees at the festival were interviewed about their perceptions of, and connections with grown food, and workshops at the event used probes in the form of images of edible plants to garner further data (Gaver et al., 1999).

Indeed, all the events of the project could be said to act as probes. This approach can be seen in both the workshop interventions and the festival and follows design methodology on probing (e.g. ibid.). The research team adopted open-ended contextual interviews (Beyer and Holzblatt, 1998) to avoid asking people directly about their behaviour and attitudes (and in this case, their relationship to food, the environment and nature in general). On the whole, direct questioning results in less informative material (ibid.), so the team favoured eliciting information on practices by encouraging reflection after action and creating an evocative situation in which to talk.

People

GEM was run by a combination of academics and community partners. In total about 150 people, across a range of communities and cultures, were involved in the project, which was run by the authors working closely with two Leeds-based partner community organisations: Leeds Love It Share It, and Space 2.

Leeds Love It Share It Community Interest Company (*LLISI*) is a community organisation that works on projects that support the citizens of Leeds in tackling the problems of climate change at a grassroots neighbourhood level. LLISI was interested to gain a deeper understanding of how and why people engage with food-growing projects, particularly as many food-growing projects in Leeds have struggled to maintain the engagement needed to take people through the full growing and harvesting process. LLISI provided a coordinator for GEM.

Space2 is a community arts organisation that promotes arts projects to change lives in some of Leeds' most challenged neighbourhoods. Space2 delivered workshops on growing food with GEM. Space2 also brought in two groups for the growing season of 2013:

- *Markham Avenue*: a residents group initiated by Fahima, a Bangladeshi Muslim who hosted workshops in her own house and garden. Several family members live on the same street and the group comprised these and neighbours from a number of ethnic and religious groups. People from this group planted up small containers with edible produce, and took them away with them to tend in their gardens.
- An *Afghan refugee group*, making a transition from a formal support group run by the Refugee Forum, to a participant-led group run by Bahar Maryam, a local resident who was setting up an organisation to support women and children from Afghanistan, her country of origin. This group used a small patio area at a local community centre as a communal growing location. Here people could leave the container plants they had potted up, making a community container garden.

Methods

The researchers invited collaborations with people from the Harehills district of East Leeds using the partner organisations existing contacts, and made new connections with churches and community centres through social and print media. This meant that members of several communities with a diverse range of cultural and religious backgrounds participated in the project's workshops and interviews. Light and Welch gathered data from these workshops and interviews and analysed the material.

Three types of growers emerged as participants, loosely characterised as those:

- Who had left gardens and vegetable plots in other countries and come to Leeds without anywhere to grow anything;
- For whom tending plants was something they learned from their parents or who found out later that they had green fingers; and/or
- Who wanted to learn skills of cultivation, either for the first time or because they were trying to grow new plants.

Thus, there was already investment in growing food by virtue of the selection process, but it varied from long-term to interrupted to being a new interest. Participants were anonymised in all descriptions of the workshops and resulting accounts, except where consent for publication was received. In the booklet produced (Light and Welch, 2014, Figure 39.1) both group leaders, Fahima and Bahar Maryam, are featured by name, with a narrative.

Information gathering was approached in three ways: workshops, food festival, and interviews, as detailed next.

Community food-growing intervention workshops

The food-growing intervention workshops comprised the two community groups, led by Space2. Each group took part in four growing workshops and then met for a meal to share their harvest in the autumn. The plants and seeds made available to the groups were chosen as practical-to-grow, robust food sources, and types of food significant to the individual or family participating. Each session was attended by the coordinator from LLISI who focused primarily on troubleshooting any practical growing issues raised by the groups.

Figure 39.1 Cover and page 7 of the booklet that the authors provided to participants as a record of the research.

Source: Light and Welch, 2014

A food-growing festival

In September 2013, a festival at a local community hall was set up by the LLISI coordinator. Local people were invited to take part in workshops, tell stories about plant growing and food production, share in a meal, and plant and take away seeds. Food and pictures of things that can be grown were used as prompts (probes) for stories to do with food and other growing activity. Additionally, there was a display of hand-drawn allotment plants to which people could pin paper leaves with their comments about growing.

Many of the two food-growing groups attended, with significant numbers of other local people from many different communities, totalling about 150 people, as noted. In fact, Space2, who have worked in Leeds extensively, commented: "There was something really important about having [the festival]. It's probably been the most diverse event we've ever had" (pers. comm.).

As well as interviewing during the afternoon, there were two story circles used to elicit narratives of growing, common experiences and emotional responses to growing food. These made use of pictures of fruit and vegetables and also the presence of a healthy and a dying pot of sage to help overcome language barriers. In the event, one circle made up entirely of Afghan women from the Refugee group and their children (apart from Light as facilitator) was seeded in English but conducted in Dari and later translated. The other circle was in English and brought together a wide variety of participants attending the festival to talk about food growing and what it has meant to them.

Interviews

Participants in the intervention workshops and at the festival were invited to share their thoughts in more depth. These were semi-structured interviews of up to an hour's length; the project conducted five such interviews.

What we learned

Methodologically, using the platform of food-growing worked well to engage people in wider discussions about lifestyle, values, and beliefs, even if they only had a basic interest in plants. The workshops, however, were not conducive to in-depth data gathering. Rather, they acted as a precursor to involving the groups in the festival and building trust, although abundant data was gathered from interviewing the groups' leaders Fahima and Bahar Maryam (extracts from their interviews follow).

POEM: I GROW FOOD FOR COLOURS

This poem was written by Michelle Scally Clarke in September 2013 after attending the Food Growing Festival, based on the experiences she heard there.

I GROW FOOD FOR COLOURS
THE CYCLE OF LIFE
WHEN ONE PLANT DIES A DIFFERENT FOOD ARRIVES
AND THRIVES
AS I SOW AND SEED, AND I'M ON MY KNEES
MY HERITAGE STANDS WITH PRIDE
DIG FOR VICTORY
A POLE OF LAND
GIVEN TO EVERY WOMAN AND MAN
FROM GROUND TO HAND
MY GRANDFATHER
GREW CELERY AND CUCUMBER
FOR CHRISTMAS DINNER
HE ALSO KEPT CHICKENS
AND ONE YEAR WE ATE SARAH

9

BEING WITH MY STEP FATHER
PLANTING SEEDS
AS A KID, I HAVE A MEMORY
SITTING UNDER THE ALLEYS ARCHES OF RUNNER BEANS
ALWAYS HAVING SOMETHING TASTY TO EAT
BLACKBERRIES AND ELDERBERRY WINE
CORIANDER, SPINACH, LETTUCE, RADISH, CRESS, CHIVES
IT ENRICHES MY SOUL, BODY AND MIND
AND PUT ME IN TOUCH WITH MOTHER NATURE
WE GROW, RHUBARB, PEAS, STRAWBERRIES AND LEEKS
BUT THE LEEKS FLOWERED
SO WE LEFT THEM FOR THE BEES
AS THEIR TREAT
I GROW FOOD TO SHOW MY CHILDREN
THAT NATURE IS PERFECT IN IT'S DIFFERENT SHAPES AND SIZES

THAT PERFECT DOES NOT COME FROM THE SUPERMARKET
I DRAW THE FOOD, THE WHOLE CYCLE
OF HOW WE ARE LIVING AND FOR A SOURCE OF IMAGERY
I ENJOY BEING OUT IN THE GARDEN.
THE CATHEDRAL OF LEAVES, IT'S MY PLACE OF PEACE
MY SANCTUARY.
I EAT IT, SHARE IT, GIVE IT AWAY
IT LOOKS GOOD BUT TASTES BETTER
I LIKE SINGING TO MY PLANT
A BIT OF A SHOW-TUNE
THEY LOVE A BIT OF COUNTRY OR DOLLY PARTON
THE JOY OF GROWING, SOIL AND WEATHER
THE CYCLE OF LIFE
WHEN ONE PLANT DIES ANOTHER FOOD THRIVES.

Michelle Scally Clarke, 2013

Michelle Scally Clarke reads her poem.

10

Figure 39.2 A poem written by Michelle Scally Clarke from ideas gathered at the festival.

By contrast, the festival was a very rich source of material. Fun activities, free food, and a central location were factors in the success. It was set up as a public event with a broad invitation to people who grow food to come and share their experiences. The event was held on a Saturday afternoon in a community hall. Activities for children allowed parents to participate, while the presence of a poet gathering material for a poem added interest (Figure 39.2). Story circles were successful at getting mixed groups of local people to share their feelings and anecdotes about food growing. The use of visual stimuli and discussion of the dying herb helped prompt and promote engagement, especially for the non-English speakers.

Thematically, the aggregated data revealed that, for many participants, growing food connects them to the wider environment. We used discourse analysis to make interpretations and discoveries in our data (e.g. Potter and Wetherell, 1987), such as recurring themes and types of association. Here we present some top-level existential themes, touched on by many participants, followed by particular excerpts showing more detailed emotional and spiritual responses to the work of growing.

Reflections across accounts about what is learnt about life from growing food included the following thoughts:

- Growing, picking, and eating emphasises the rhythms of life and death.
- Failure with plants (and crops) is a regular part of gardening, and is a good introduction to 'letting things go'.
- When growing, one is able to control some things, but there is also the force of life itself, which pushes the shoots up . . . or not.
- There are seasons to growing, running through a person's life as well as the year, and related to people's access to space and time as lives change from childhood to retirement and beyond.

After (multiple) researcher analyses of recurring themes and clustering of statements, the accounts of food growing were separated into those that:

- Stressed environmental factors,
- Referenced nature and spirituality,
- Spoke of a sense of place,
- As well as comments about practicalities for growing and eating (which we do not include here).

There was some overlap in the dominant theme of each account, particularly between the spiritual and place responses, perhaps reflecting the numbers of migrants in our sample, for whom both religion and migration had a direct relationship with grown food. Environmental concerns tended to be a more discrete category (comments were not co-located with other themes or appeared in different accounts). That said, it was a small sample of a very varied group so we regard our classification as preliminary, noting that, collectively, the themes echo the findings of, amongst others, Wilson (1983), Schroeder (2008), Ingold (2000), Taylor (2009), and Edensor (2010).

Examples of comments from the GEM project participants relating to environment included these, excerpted from Light et al. (2014):

- "I am more aware of the changes in seasons, and at times the weather changes resulting to climate change."
- "Growing things changes your attitude to the weather, accepting the rain."
- "We are expecting to eat so many foods out of season we have lost connection to our own lands and what we can grow."
- "Growing food gives kids a practical understanding of environmental green issues. You value water more if you have to collect it for your garden."

Among some, there was explicit awareness of the politics of food production (though not in all groups), for instance one participant at the festival notes that, "I cannot grow food without considering ecological issues . . . the diversity of approaches in growing is massive and a political issue" (Light et al., 2014).

A second group of comments spoke to the spiritual aspects of working the land:

- "Nature is a living thing; when we damage or abuse nature we do the same to ourselves . . . people have been disconnected from nature and society disconnects you from food and reconnecting is a basic fundamental thing."
- "I love growing because it reconnects me to the soil and I feel at peace when I'm working with the soil." (Light et al., 2014)

In a variant of this, one person said: "Food growing is calming, it gives me space to think and time to talk to others" whilst another said: "I gain peace of mind from growing my own food". Yet, what was striking in the findings was the absence of discourses about cheapness or health as such, aside from healthiness of gardening. When people were asked why they grew food, they spoke with love about their plants or regret about the lack of opportunity to grow them. Examples of motivation for food growing included: "There is a satisfaction from seeing things grow, especially from seed"; "It is satisfying to pick produce and eat, or preserve it and give it or sell to people in your neighbourhood"; "I started growing food through the influence of my neighbour who has a nice garden . . . I am happy to see the results and the fruits I got"; "Food growing makes me happy, being in the garden, watering, planting, weeding"; and "Food tastes so much better when you know you have grown it yourself. What a sense of pride!" (Light et al., 2014).

A related set of comments were more explicitly religious in nature, for instance:

- "There is a sort of faith connection, in terms of thinking about the earth and protection of the earth."
- "Buddhism teaches us not to waste anything, any food on our plate, so we eat the left overs the next day."
- "There's a group of eco-friendly church members; growing things is well supported as it's a community-led initiative." (Light et al., 2014).

One narrative was explicit about a faith connection:

> There was a robin and it started picking about and looking at me and then it flew back under the bush but it flew back and kept coming a bit nearer and I felt really loved. I just felt, because I believe in God, that God was showing me that, even though there are really tough things going on in my life, He loves me.
>
> *(Light et al., 2014)*

Here the self-awareness of the act of faith is telling, since Britain has one of the highest ratios of agnostics and non-believers in the world (Noack, 2015).

Overall, the emotional attachment participants expressed about land (and the food grown there), and any spiritual connectedness with it, was significant, and resonated with the findings of Cone and Myhre in their study of community-supported agriculture in the US (2000). Further, they once more echoed the importance of a connection with the natural world as expressed by scholars including Wilson (1983), Schroeder (2008), Ingold (2000), Taylor (2009), and Edensor (2010).

Two accounts of spirituality, place, and growing

The organizers of the two groups were interviewed at greater length than most of the other group members. It is here that we can see themes of spirituality, place, and growing in some detail.

One interviewee was Fahima, introduced earlier, who has strong Bangladeshi roots. She lives with her husband and children, is religiously observant and is very active in food-growing projects as well as growing plants that hold significance for, her and her family. Talking of her planting, she refers to family and homeland (Light and Welch, 2014):

> I started off growing sunflowers a good few years ago. My mum was always into growing things. A while back, my sister-in-law bought some lime tree seeds from Bangladesh. She grew four trees from the seeds and gave me one. She gave one to mum and to my brother, my brother's survived, but mine and my mum's died. So on my last trip there, on the morning we were leaving, I dug up a small tree from the family garden and brought it home. I brought two guava trees back too. They remind me of Bangladesh and my relatives there.
>
> *(Light and Welch, 2014)*

The GEM food-growing season for Fahima included Ramadan, the Muslim month of obligatory fasting (*sawm*) for all physically capable adults that commemorates the first revelation of the Qur'an to the Prophet Mohammed. This proved to be a significant time for her gardening activities. Her reflection on cultivation during this period captures the intersection between religious observance and practical growing practices:

> At the crack of dawn, you'd have your morning food, and then you'd do your prayer, and then you could water the plants because the sun was coming out and you'd water the plants just as they were ready for a feed. At the break of fast, what we'd do is come in and have some fruit, then go out into the garden, do the break of fast prayer, come back in and have our dinner. Then, while the kids are doing the dishes, I'd water the plants . . . it just worked so well and, honest to God, the garden looked better during Ramadan than on normal times because it got its nutrition at the right times.
>
> *(Light and Welch, 2014)*

The other group leader was Bahar Maryam. She is originally from Afghanistan but now lives in Leeds with her husband and children. She grows a number of plants from her homeland, and mentors other women in growing skills. She describes the migrant experience thus:

> Sometimes you miss your own food. When you live in a country with different weather, climate, vegetables, it can be difficult to find your own. So it's good to have one thing or two things that remind me about my country. And also it's a different feeling when you are having your own fruit and vegetables in your own garden and it's organic, fresh . . . they're growing, so it's everything natural.
>
> *(Light and Welch, 2014)*

This has significance for how she sees relations with her children and their understanding of Afghan culture:

> They need to know about their background and their culture . . . their fruit . . . knowing what they're called. It is actually important because their parents are Afghan British and if somebody asks them something they should be prepared to answer. [This tree is a] *Senjed* . . . S-e-n-j-e-d. This grows in our country . . . we dry them then we use them in the winter for our New Year. And also it's really good like a medicine for tummy ache.
>
> It's 8 years old. Someone sent it from Afghanistan, from our relatives, just to use for New Year, so we just decided: let's try to plant it, if we can grow them in this country we can use the fruit for every New Year. We put the seeds in the pot. One day we realised that it's growing so I've kept it until now.
>
> In the New Year, when we cook food, food or fruit anything, we use this tradition, to find 7 fruit or foods beginning with the same letter [the letter S], so it's *senjed*, one of them, we put *sabze* like grass,

> this is a symbol of spring, and garlic which is [called] *sir*. We put fishes; we put *sib* which is an apple, *sabzi* which is spinach. *Sabzi Challow* is white rice with the spinach. *Sirka* which is vinegar, so we're making like 7 special foods or fruit and put on the table so people can eat it – we call it *Haft Sin* or 7 S's.
>
> *(Light and Welch, 2014)*

What is immediately apparent in both accounts is the transnational aspect of these planting and growing activities – the seeds are from overseas, the gift of the extended family. This goes beyond the symbolic relationship that the plants have with transported social, cultural, and religious rituals. All these aspects are interwoven, since the acts of importing and growing, and the memories of the plants from elsewhere are distinct, yet linked to the statement of identity and/or faith that these rituals perform.

Further accounts of place and time

In considering transplanting, the group leaders are far from alone, as these three extracts from festival participants show:

- "I bring back seeds from Africa and am growing ginger, turmeric, sweet potatoes."
- "Immigrant communities have enriched our food – when they arrived they started to grow things that weren't here before."
- "Growing food reminds me of home. I had a big garden and used to share it with my neighbours. I live in a flat now and miss growing flowers most of all." (Light et al., 2014).

Transplanting not only happens in space, but time. Many of the participants' accounts dwelt on people the plants evoked. Some were friends or lost loved ones: "A friend gave me a bay tree last year for my birthday and I feel quite protective of it . . . plants have connections with people, and that's really nice"; "I tried moving some plants from my parents' garden when they died. The plants didn't actually come across and that allowed me to sob again at their loss, but in a whole different way" (Light et al., 2014).

Childhood featured in some stories: "Peas remind me when I was little, but the peas never made it to the pot as we would eat them from the plants"; "It brings back memories of childhood. Dad had an allotment . . . it was a men's space mostly", or "I was thinking about being the person who is selling my father's house, which was the family home for all my life, with memories of gardening and growing veg and it's quite sad to let go. As a memory I am taking an apple tree . . . ", showing that many food growers are born of gardening parents (Light et al., 2014).

Yet some stories were of how love, even reverence, developed without family precedent:

> In 2000, in South Africa, I was working alongside African women growing things and I knew nothing about it, but I did know how to dig. So I used to dig and then they used to plant things and they grew some of the best vegetables. They made the garden, built upon sand, using compost year after year. This garden was their first land or place that actually belonged to them . . . it was one of the most sacred places of my life: their gardens.
>
> *(Light et al., 2014)*

Discussion

As hoped, the GEM project showed that growing food offers a practical shared reference point as a way into discussions about the environment and beliefs. It demonstrated that talking primarily about plants serves to level ethnic and religious differences and leads well into cultural topics. Indeed, one has to question how far the environmental reflections gathered during the project are an intrinsic part of the act of growing, or part of the act of reflecting on growing, brought about by the research. If, as it seemed, the latter, then cultural events to reflect on growing are particularly valuable.

It was apparent from the project findings that growing (and memories of it) grounded people in both the local and the global. Growing food appeared as a special kind of *making* that drew together connections of past and future, living and dying, trusting and controlling, and working with forces outside of one's own control; natural (and metaphysical) forces. Using reflection and storytelling, the project drew attention to these connections and the seasonal, cyclical nature of growing in people's lives. In particular, we detected a period when people stopped growing, often at a point of unsettledness between leaving their parent's home and making their own. Many stories harked back to the person who had introduced the participant to growing and revealed the significance of childhood to many active growers. This was also linked with narratives of ecology and wonder, religious belief and festival, such as celebrations of Ramadan and the Afghan New Year.

Because of the extreme cultural diversity of the workshops and food festival, the project did not gather representative data from any one religious group. Instead, the researchers spoke to many people with a connection to other growing contexts, either of their youth or of another country. This included a diaspora of immigrant families who have lost access to the land, their land of origin and the cultures implicit in their growing practices. The researchers also met many who now have nowhere to grow food after an earlier experience of food growing. Their accounts held wistfulness for the work of growing and for plants that might not grow well in Leeds' climate and soil.

It would seem that for food growers, regardless of place of origin, the significance of edible produce rarely lay in economic factors, but was located in flora that provided the link with wellbeing, land, place, or ancestral home, and thus food growing gave surroundings a sense of the familiar and secure.

Noting this impact, a key reflection from the project concerned current marketing of growing your own.

> I think the subsistence argument is a bit of a red herring in terms of encouraging people to garden ... even if you do it really cheaply with all recycled stuff, you pull out your two little stringy carrots when you could have got a pound for 40p in the greengrocers.
>
> *(Space2, pers. comm.)*

> There's a motivation here for getting more people involved in sustainable activities that will benefit their health and well-being and the environment [but] it's about looking at some of those more complex reasons as to why people do gardening.
>
> *(LLISI, pers. comm.)*

Further reading

Beyer, H., & Holtzblatt, K. (1998) *Contextual design: Defining customer-centered systems*. San Francisco, CA: Morgan Kauffan.

Brightwell, M. (2012) *A taste of home: Food, identity and belonging among Brazilians in London*. PhD thesis, Royal Holloway, University of London, UK: https://pure.royalholloway.ac.uk/portal/files/4699582/A_Taste_of_Home_Thesis_Brightwell_2012.pdf.

Brown, L., Edwards, J., & Hartwell, H. (2010) A taste of the unfamiliar: Understanding the meanings attached to food by international postgraduate students in England, *Appetite*, 202–207.

Callaghan, J. (2015) *Changing landscapes: Gipton and Harehill (Leeds): A superdiverse inner city ward. Working Papers in Translanguaging and Translation (WP. 7)*. www.birmingham.ac.uk/generic/tlang/index.aspx.

Church, A., Mitchel, R., Ravenscroft, N., & Stapleton, L.M. (2015) '"Growing Your Own": A multi-level modelling approach to understanding personal food growing trends and motivations in Europe. *Ecological Economics*, 110, 71–80.

Cone, C., & Myhre, A. (2000) Community-supported agriculture: A sustainable alternative to industrial agriculture? *Human Organization*, 59(2), 187–197.

Connected Communities. (2009-) https://connected-communities.org.

Cook, I., & Harrison, M. (2003) Cross over food: Re-materializing postcolonial geographies. *Transactions of the Institute of British Geographers*, 28(3), 296–317.

Datta, A. (2009) Making space for Muslims; housing Bangladeshi families in East London. In: Phillips, R. (ed.), *Muslim spaces of hope: Geographies of possibility in Britain and the West*. London: Zed Books, pp. 120–136.

Democracy Leeds. (2010) *Depravation*. http://democracy.leeds.gov.uk/documents/s62979/appendix.

Diner, H.R. (2001) *Hungering for America: Italian, Irish and Jewish foodways in the age of migration*. Cambridge, MA: Harvard University Press.

Douglas, M. (2003) *Food in the social order: Studies of food and festivities in three American communities*. London: Routledge.

Edensor, T. (ed.) (2010) *Geographies of rhythm: Nature, place, motilities and bodies*. Farnham, Surrey: Ashgate.

Edensor, T., & Jaynes, M. (eds.) (2012) *Urban theory beyond the West: A world of cities*. London: Routledge.

Gabaccia, D. (1998) *We are what we eat: Ethnic food and the making of Americans*, Cambridge, MA: Harvard University Press.

Gaver, W.W., Dunne, A., & Pacenti, E. (1999) Cultural probes. *Interactions*, 6(1), 21–29.

Gipton and Harehills. (2011a). *Gipton and Harehills demographics*. Leeds, England. http://gipton-and-harehills.localstats.co.uk/census-demographics/england/yorkshire-and-the-humber/leeds/gipton-and-harehills.

Gipton and Harehills. (2011b). *A summary of the results of the 2011 census*. http://westyorkshireobservatory.org/resource/view?resourceld=2771.

Gipton and Harehills QS210EW (2011) *Religion detailed (quick Stats)* www.ukcensusdata.com/gipton-and-harehills-e05001424#sthash.ceejBzSw.dpbs.

Graham, S., & Connell, J. (2006) Nurturing relationships: The gardens of Greek and Vietnamese migrants in Marrickville, Sydney. *Australian Geographer*, 37(3), 375–393.

Harvey, G. (2013) *Food, sex and strangers: Understanding religion as everyday life*. Abingdon, Oxford: Acumen.

Hunt, S., & Tomkins, M. (2016) *Transforming land, transforming lives: Greening innovation and urban agriculture in the context of forced displacement*. http://lemontreetrust.org/wp-content/uploads/2016/05/LTT_Brochure.pdf.

Ingold, T. (2000) *Perceptions of the environment: Essays in livelihood, dwelling and skill*. Abingdon, Oxon: Routledge.

Jenkins, W. (ed.) (2010) *Berkshire encyclopaedia of sustainability: The spirit of sustainability* Great Barrington, MA: Berkshire.

Jennings, H.M., Thompson, J.L., Merrell, J., Bogin, B., & Henrich, M. (2014) Food, home and health: The meanings of food amongst Bengali women in London. *Journal of Ethnobiology and Ethnomedicine*, 10(44). Available from: www.ncbi.nlm.nih.gov/pmc/articles/PMC4035805/

Kellert, S.R., & Wilson, E.O. (eds.) (1993) *The biophilia hypothesis*. Washington, DC: Island Press.

Kindon, S., Pain, R., & Kesby, M. (eds.) (2010) *Participatory action research approaches and methods; connceting people, participation and place*. London: Routledge.

Leeds City Council (2011a). *Faith communities in leeds by electoral ward*. Office of National Statistics. https://observatory.leeds.gov.uk/resource/view?resourceId=3462.

Leeds City Council (2011b) *Leeds census 2011 – another update from the CDWs*. www.touchstonesupport.org.uk/2013/02/leeds-census-2011-another-update-from-the-cdws/.

Light, A., & Welch, C. (2014) *Sprouting bonds: How growing our food connects us to places, land, families and cultures*: http://bookleteer.com/book.html?id=3023#page/1/mode/2up.

Light, A., Welch, C., & Hill, K. (2014) *GEM (Grown, Edible, Meaningful) report for the AHRC, AHRC*. https://connected-communities.org/index.php/project/gem-grown-edible-meaningful/.

Mazumdar, S., & Mazumdar, S. (2009) Hindu space in Southern California. *Journal of Environmental Psychology*, 29(2), 256–266.

Mazumdar, S., & Mazumdar, S. (2012) Immigrant home gardens: Places of religion, culture, ecology, and family. *Landscape and Urban Planning*, 105, 258–265.

Morgan, G., Rocha, C., & Poynting, S. (2005) Grafting cultures: Longing and belonging in immigrants' gardens and backyards in Fairfield. *Journal of Intercultural Studies*, 26, 93–105.

Moulin-Doos, C. (2013) Intercultural gardens: The use of space by migrants and the practice of respect. *Journal of Urban Affairs* (special edition 00), 1–11.

Noack, R. (2015) These are the world's least religious countries. *Washington Post*. 14 April 2015. www.washingtonpost.com/news/worldviews/wp/2015/04/14/map-these-are-the-worlds-least-religious-countries/.

Potter, J., & Wetherell, M. (1987) *Discourse and social psychology: Beyond attitudes and behaviour*. Thousand Oaks, CA: Sage.

Reason, P., & Bradbury, H. (eds.) (2013) *Sage handbook of action research: Participative inquiry and practice*. London: Sage.

Research Council UK. (2017) *GEM (Grown, Edible, Meaningful): Herbal probes for community driven environmental change: Outcomes*. http://gtr.rcuk.ac.uk/projects?ref=AH/K003976/1.

Rojek, C. (2004) The consumerist syndrome in contemporary society: An interview with Zygmunt Bauman. *Journal of Consumer Culture*, 4(3), 291–312.

Schroeder, H.W. (2008) The felt sense of natural environments. *Proceedings*, Twenty-first Annual Conference of the Environmental Design Research Association, Champaign-Urbana: Illinois, 63–72.www.focusing.org/folio/Vol21No12008/06_TheFeltSenseTRIB.pdf.

Schroeder, H.W. (2012) Giving voice to the experiental value of natural environments. *The Humanistic Psychologist*, 40, 136–152.

Sered, S.S. (1988) Food and holiness: Cooking as a sacred act amongst Middle-Eastern Jewish Women. *Anthropological Quarterly*, 61(3), 129–139.

Taylor, B. (2009) *Dark green religion: Nature spirituality and the plenetary future*. Berkeley, CA: University of California Press.

Temple, B. (1999) Diaspora, diaspora space and Polish women. *Women's Studies International Forum*, 22(1), 17–14.

Tookes, J.S. (2015) "The food represents": Barbadian foodways in diaspora. *Appetite*, 90, 65–73.

Verain, M.C.D., Bartels, J., Daigevos, S.J., Sijtsema, S.J., Onwezen, M.C., & Antonides, G. (2012) Segments of sustainable food: A literature review. *International Journal of Consumer Studies*, 36(2), 123–132.

Wilson, E.O. (1983) *Biophilia*. Cambridge, MA: Harvard University Press.

Wilson, M. (ed.) (2016) *Postcolonialism, indigeneity and struggles for food sovereignty*. London: Routledge.

Wirzba, N. (2011) *Food and faith: A theology of eating*. New York: Cambridge University Press.

40

utopia landscape food utopia

Jody Beck

In the introduction to her book, *Utopia as Method*, Ruth Levitas claims that a utopian method, an imaginative reconstruction of society, relevant to our contemporary world would:

- Provide a critical tool for exposing the limitations of current policy discourses about economic growth and ecological sustainability;
- Facilitate genuine holistic thinking about possible futures, combined with reflexivity, provisionality and democratic engagement with the principles and practices of those futures; and
- Require us to think about our conceptions of human needs and human flourishing in those possible futures (Levitas, 2013: xi).

This provides a crucial framework for considering the future of food and the landscapes associated with its production at this time when humanity faces twin epidemics of malnutrition and obesity, relies on a food system which wastes many of the food calories produced, and promotes agricultural production which is polluting our air, water, and soil. Some key statistics outline problems with our global food system. While we have made progress in addressing hunger, there were still 795 million undernourished people in 2015. (FAO et al., 2015) At the same time, obesity continues to be a concern. In 2014, there were more than 1.9 billion overweight adults, 600 million of whom were obese. There were also 41 million children under the age of five who were overweight or obese (WHO, 2016). One in four calories produced to feed people is not eaten by a person, and if food waste was its own country, it would be the third largest greenhouse gas emitter in the world behind only China and the United States (Lipinisky, 2015; Chapter 23). Thirty-four million hectares of agricultural land are impacted by salinisation related to agricultural practices, and nitrate contamination of groundwater has been reported as a major concern in Europe, the United States, and South and East Asia (Mateo-Sagasta and Burke, n.d.; also see Chapter 16). In addition to these material problems which are integral to our current food system, the imbalance of political and economic power that frames the global food system is yet another motivation for turning to a utopian method for considering the future of food – assuming that equity and self-determination matter (Chapters 24, 25, 31, 32 and 33). The foundational logic of the current food system is clearly inadequate relative to any conception of human needs and human flourishing as indicated by the basic indicators of obesity and hunger just described. In order to resolve the problems in our food system, we need to fundamentally re-imagine the conceptual foundations of our food system, and therefore the very conceptual foundations of our society.

The trajectories of our current food system indicate that the only possible future of food relies, in fact, on some level of radical change. (Kunstler, 2005; Roberts, 2009; Ladner, 2011; Stock et al., 2015; Weis, 2007) As the saying goes, how can we not be reasonable and demand the impossible?

In his book *The Real Cost of Cheap Food*, Michael Carolan gives an important outline of the negative impacts that our contemporary food system has on our health, society, and environment (Carolan, 2013). Another excellent book for understanding the problems with our food system is *Food Systems Failure: The Global Food Crisis and the Future of Agriculture* (Rosin et al., 2014). Not only does the book explain the problems at hand, but it also shows how these problems result from this neoliberal organisation of the global food system (Harvey, 2005; Springer, 2016; Otero et al., 2013; McMichael, 2009). The book concludes with a call to imagine a food utopia in opposition to the neoliberal food agenda. The authors of *Food Utopias* (Stock et al., 2015), picking up where *Food Systems Failure* left off, draw on Sargent (2010) and propose that food utopias are useful tools in three different modalities – as productive critiques, as optimistic and engaged experimentations and as discursive processes. They agree with Levitas (2013) that utopia ought to be used as a verb, a way of attacking a problem – not as a noun, a proposal with embedded finality. It is important to note that the authors of *Food Utopias* describe their project as a political proposal that calls for a multiplicity of utopian food stories which, en masse, can begin to make the food system more just and sustainable while at the same time preserving if not expanding freedom.

In the same way that utopian thinking is an ideal method for considering the future of food given the radical changes required, landscape – as a method – is an ideal way to work at utopian thinking. Landscapes unavoidably intertwine the human and non-human worlds and the environmental, social, cultural, political, and economic characteristics of landscapes are inseparable from each other (Waldheim, 2016; McHarg and Steiner, 1998; Groth and Bressi, 1997; Jackson and Horowitz, 2000; Meinig, 1979). To imagine a landscape is to imagine this entire complex. Landscape as a method (see a similar view in Nassauer, 2012) posits complete futures, and the best landscape thinking takes a holistic approach to all of the previously described aspects of the world around us and their interconnections (Hester, 2006; Mostafavi, 2016; Zeunert, 2017). One can see this method at work in the attempts to address rapidly changing technological and social conditions in the early twentieth-century work of practitioners such as John Nolen and the earlier suburban planning work of Frederick Law Olmsted (Beck, 2013). Charles Waldheim claims that the very origin of modern landscape architecture as a profession in the early twentieth century was set in the aspiration to provide a professional and "progressive response to the social and environmental challenges of rapid urbanization in the second half of the nineteenth century" (Waldheim, 2016: 160). Certainly the work and writings of Ian McHarg propose that landscape be used as a method which would support a utopian projection of a better future (McHarg, 1992). One can see this trajectory of understanding landscape as a method in more recent theory as well. In his introduction to the influential text *Recovering Landscape*, James Corner wrote in a utopian vein that:

> landscape may still embrace naturalistic and phenomenological experience but its full efficacy is extended to that of a synthetic and strategic art form, one that aligns diverse and competing forces (social constituencies, political desires, ecological processes, program demands, etc.) into newly liberating and interactive alliances.
>
> *(Corner, 1999: 2)*

Anyone looking for a wider consideration of this possibility of understanding landscape as a method which supports utopian work would do well to start with Part 2: Design Process in *Theory in Landscape Architecture* edited by Simon Swaffield (Swaffield, 2002). More recently, Joan Iverson Nassauer, in her chapter "Landscape as Method and Medium for the Ecological Design of Cities" in *Resilience and Ecology in Urban Design*, claims that landscape as method "pragmatically employs the imaginative artifice of design to produce potential innovations that anticipate the future" (Nassauer, 2012: 83). To connect this understanding of landscape back to food and reinforce the motivation for utopia as the mediating concept between them, as Carolan argues for an understanding of the embodied politics of food (Carolan, 2016), this chapter is grounded in the proposal that landscape relations are the material realities of these embodied politics.

If a utopian method is a good way to explore the problem of the future of food seriously, then landscape as a method is a good way to bring that utopian thought to ground. The sequence of thought from food to utopia to landscape returns neatly to utopia since our food is intimately connected to land and the production and distribution of food is integral to landscapes in their richest, most holistic sense (Kloppenburg et al., 1996; Ackerman-Leist, 2013). Using landscape as a methodology to motivate and organise utopia as a method is a powerful way to produce an imaginative reconstruction of society which can act as a guiding principle for action in the present.

Indeed, the relationship between food, utopia, and landscape is long and fruitful. Thomas More's work *Utopia* is a touchstone for any discussion of utopian thought, though by no means the first expression of humanity's utopian impulse in support of seeking a better way of living (More, 1964). Both the motivation for More's work as well as the content of his imagined society are closely related to his critique of landscape and food system concerns at his time. The recounting of Hythloday's travels to the nation of Utopia is offered as a counter-example to the many problems with laws in England and in support of More's argument that stealing should not be a capital offence. More argued that the root cause of theft was poverty, and the root cause of poverty at the time was the enclosure of the commons for wool production and cattle fattening by which the powerful would 'stop the course of agriculture', destroy towns and villages and turn people to vagrancy. More set the social problem of poverty at the feet of capital accumulation and those who used their economic and political power to do so by controlling land at the expense of livelihoods and daily nutritional needs of the less powerful members of their society. Similar critiques are at the core of such movements as Food Sovereignty and Food Justice (see Chapters 31 and 32). More also claimed that the change in land use resulting from the primacy of capital accumulation not only left many people with no way to grow food or generate income, it also increased the cost of food itself by reducing supply. We can see a contemporary parallel in the impact that biofuel production has had on food prices (Donohue and Cunningham, 2009).

Not only was *Utopia* written as a critical response to a failure in the food system and the restructuring of agrarian land and life in ways that did not give priority to livelihoods and nutrition, the organisation of society around landscapes of food production and distribution was central to More's 'imaginative restructuring of society'. In Utopia, the size of towns was related to the countryside which provided food for their inhabitants. This parallels current concepts of food and ecological footprints (Rees, 1992; Khan and Hanjra, 2009; WWF, 2016). This allowed those inhabitants proximate access to the agricultural land for two pragmatic reasons. All Utopians rotated through the occupation of agriculture for a number of years and the extra labour needed for the annual harvest was provided by town dwellers – there were no semi-vagrant agricultural labourers in Utopia. The residential structure of the Utopian towns themselves was organised around the halls in which the citizens shared meals. This understanding of food as one of several social resource systems is shared with many other Utopian projects including, for example, two other utopian novels which had wide readership – Bellamy's *Looking Backward* (1926) and Morris' *News from Nowhere* (1891). The production of food was also the foundation of several utopian proposals that employed landscape as a method; Frank Lloyd Wright's Broadacre City, Ebenezer Howard's Garden City, and Ludwig Hilberseimer's New Regional Pattern are foremost among them (Waldheim, 2016: ch. 7). These proposals are relevant in our time because they indicate potential answers to many of the problems we now face. In all three cases, the control and use of land for the common good is a beginning point for their work and each used the design and arrangement of landscapes of food production as a central philosophical component of their proposal – though they had decidedly different assessments of the appropriate relationship between individuals and society.

Ebenezer Howard published his first book *To-morrow: A Peaceful Path to Real Reform* in 1898 and thoroughly revised the text, though not the basic premise, for the 1902 publication titled *Garden Cities of To-morrow* (Howard, 1902 and 2010). His proposed response to the growth of urban populations and the loss of rural population, which was a concern of many at the time, was a reduction in the size of cities and their integration back into the agricultural landscape. All land in Howard's proposal was held in common, and his proposed combination of taxes and rent captured the value of land for the common good rather than the enrichment of speculators and landlords. According to Howard's calculations, the rent from agricultural lands would have generated a significant proportion of the funding which made the entire project feasible while still allowing agriculturalists to be profitable. This settlement pattern would have integrated food production into civic life in a number of meaningful ways (Viljoen, 2005; Hall and Ward, 1998). First, it would have given producers ready access to markets for fresh vegetables as well as staple crops such as wheat and corn. In fact, a central motivation for Howard's proposal was that it would favour local agriculturalists over producers in distant lands by significantly reducing transportation costs for the former – thereby encouraging people to remain in agricultural professions and provide fresh food for those living in cities. Howard described the flow of vegetables and other foodstuffs from the rural belt around his garden cities into the cities as only one part of a resource cycle. The other half of that cycle was the flow of sewage in the opposite direction to be used as fertiliser to enrich the soil, further reinforcing the connections between the urban and the rural land.

Additionally, Howard proposed a belt of industrial land between the city centre and the agricultural land. Drawing on ideas put forward by Kropotkin (1968), Howard proposed that people could work part-time in industrial facilities when not engaged in agricultural production. This imagination of a future of industrial work which kept people directly involved in the production of food was picked up by the Resettlement Authority in the United States, though it did not find fruitful application (Beck, 2013). While the concept of the garden city spawned an international movement and much emulation, its foundational integration of landscapes of agricultural production and urban form has not been applied in any meaningful way (Hall and Ward, 1998; Buder, 1990). At least in part, this is due to the fact that the relationship between the city and rural land that Howard projected relied on revolutionary changes to both social and economic structures intimately connected to food production and land ownership. Many of the goals of movements and proposals for alternative forms of food production such as food sovereignty and locavorism (Fitzgerald, 2016; Chapters 31 and 33) would be achieved through the realisation of Howard's proposal that close geographic and economic ties between agricultural production and city dwelling/industrial production ought to form the basis for civilisation.

Landscapes of food production were even more politically central to Frank Lloyd Wright's proposal for Broadacre City (1932–35). In fact, the politics of food served as the primary principle around which he organised the inhabitation of landscapes and his proposal for urban form. Wright's concept developed over several decades and was published in three books and several articles in addition to being exhibited as a large model in the Rockefeller Center in 1935 (Wright, 1951, 1995, 1970). The last of these books, *The Living City* (1970) was the most developed statement of his concept. In Wright's analysis, rent was the foundational evil of modern society and left people practically enslaved and therefore politically impotent. Rent was integral to modern society, in his analysis, precisely because people were not independent and relied on wages for sustenance. Wright's ultimate goal was an urban form that would foster a genuine democracy and a true form of capitalism based on autonomy and citizen engagement through the potential of household food production and family self-sufficiency. The connections that Wright drew between household food production and the possibility for a genuine democracy cannot be overstated. Democracy, Wright proclaimed, was "civilization of man and ground – really organic agronomy" (Wright, 1970: 25). While there was a place for farming, forestry, and other forms of agriculture in Broadacre City, the central concept in his proposal was that every family have enough land to feed themselves should they need or want to – resulting in a fine-grained productive landscape – and allowing them to opt out of wage relationships should they so desire. Wright's connection of food production to political power is echoed in many of the proposals for alternative forms of food production today. The similarities between Wright's early twentieth-century critique and proposal and that of contemporary critiques of the food system which are focused on food justice is striking. Ron Finley's critique of the role gardening can play in mitigating the injustices of power in South Los Angeles, for example, is perhaps one of the most politically eloquent associations between growing food at a small scale and resistance to domination and oppression within a social structure (Finley, 2013).

Ludwig Hilberseimer, a German architect who emigrated to the United States in 1938, wrote about planning decentralising cities on the basis of an organic order. His most complete statement regarding an ideal form of cities and society was published as *The New Regional Pattern* (1949). His analysis of historical precedents as well as problems contemporary to his work ties the control of land and food production to the economic and social health of society. He points to two sources of both social and economic problems. First, he lists the process of removing people from land as agricultural operations expanded in size in order to solidify economic and class power. He draws on examples ranging across history from the upper classes of Imperial Rome, to the capitalists of England that More railed against, to the large farming corporations of the United States. Second, he describes the growing separation of agricultural production from industrial work. The ideal historical settlement pattern, for Hilberseimer, was the medieval town which not only balanced industrial work and agricultural production on a small scale but, in his analysis, was integrated with the productivity of the local countryside. In his view, the Industrial Revolution altered this arrangement and destroyed well balanced regions in order to concentrate the production of raw materials in some locations and industrial activity and trade in others – thereby separating people from agricultural landscapes by concentrating them in cities as a pool of labour and producing disastrous consequences for agricultural and urban landscapes as well as the people who lived in them. While Hilberseimer clearly tied the health of society and the genuine freedom and happiness of people to this urban organization, he did not set out to design towns. Rather, he began with the region which he defined as a landscape – in the full complexity of the term as used earlier in describing landscape as methodology – that can largely sustain itself (Hilberseimer, 1949: 88–92). These regions are based on the capacity for industrial work and agricultural work to be integrated – in much the same way that was proposed in *Utopia*, the Garden City, and Broadacre City – with an emphasis on a balanced economy tied to the cultivation of landscapes for food and fibre and the diversification of not only production but also farm sizes. Like Howard and Wright, Hilberseimer proposed a decentralised structure of settlement in which cities are not distinct from agricultural landscapes but integrated with them. The vast contemporary and popular urban homesteading and local food literature illustrates the current attractiveness of Hilberseimer's desire to re-integrate food production into modern urban life (Markham, 2014; Kraft, 2011; Caswell and Siskin, 2012; Allen, 2012; Gayeton, 2014).

Howard, Hilberseimer, and Wright all grounded their utopian ideas on proposals regarding how the production of food – in the breadth of its social, economic, and environmental parameters – would support ideal forms of social relations. Each worked from proposed social goals back to the primacy of food production and then from the production of food outward to a proposal for material forms of inhabitation. In this way, each of them used landscape as a method to generate the imagination of how their utopian proposal could manifest on solid earth. These proposals cannot be dismissed for their totality as utopian impulses because, on a close reading, each individual built into their proposal an expectation that the realisation of their project would vary with time, geography, and the desires of the individuals living in their imagined landscapes (Howard, 1902: 22; Wright, 1970: 120–121; Hilberseimer: 90–92).

Instead of developing a food system which prioritised food production and the associated landscapes as a politically enabling category in the way that Howard, Wright, and Hilberseimer proposed, modern global society has developed a food system which prioritises profit from food. This results in the political disenfranchisement of people from their food and the landscapes associated with its production (Vallianatos, 2006: ch. 5; Weis, 2007; Albritton, 2009; Nestle, 2002). Howard, Wright, and Hilberseimer saw the concentration of people into urban settings as distinct from rural food production as a source of social problems, and each sought to re-integrate urban inhabitation and engagement of agricultural landscapes through systems of urban and peri-urban agriculture. In all three cases, the physicality of the proposal for a re-imagined landscape was inseparable from an imaginative reconstruction of society itself. The trend toward concentration and separation they were resisting is apparent in the contemporary ownership of land and the landscapes generated by food production in the same way that it was clear to Thomas More as he was writing *Utopia*. It is telling that Hilberseimer's analysis, made in the middle of the twentieth century, parallels and includes analysis that aligns with More's analysis from the early sixteenth century. The consolidation of agricultural land in industrialised countries has continued along that trajectory of consolidation seen in early sixteenth century Britain (Lowder et al., 2016). In the United States for example, the number of farms has decreased by approximately two-thirds while the farm size has more than doubled since 1935 (USDA, 2016). One of the critiques of the Green Revolution is that it disadvantaged small producers and concentrated wealth, even if it did not reduce the number of farms, in the developing world (Shiva, 1991; Bowonder, 1979; Beck, 1995; Chapters 9, 10, 12 and 16).

The reality of political and material disenfranchisement inherent in the development of our food system, however, does not imply that there has not been a robust discourse regarding other possible futures. On the contrary, there has been a broad history of thinking about the future of food. In his book *Meals to Come*, Warren Belasco offers a detailed and well-constructed history of the conceptualisation of the future of food from the late eighteenth century to the contemporary world and in his conclusion offers a hint at why we have ended up with the food system we have (Belasco, 2006). Belasco defines this field of thought as a debate between those who believe that scientific, agricultural, and political advances can keep pace with population growth and always provide for the needs, if not the wants, of humanity (Fedoroff et al., 2010) and those who feel that the exponential growth of human population and our demands on the natural environment will ultimately overtake our capacity to produce enough food and lead to famines and political unrest if radical and globally scaled changes are not made (FAO, 2017; Chapter 16). Belasco begins this narrative in the 1790s. On one side of the debate he places William Godwin and the Marquis de Condorcet who held a cornucopian belief in the essentially unlimited capacity of mental production. On the other side he places Thomas Malthus with his pivotal work, *An Essay on the Principle of Population*, who believed quite the opposite (Malthus, 1976). Belasco describes the trajectories of this debate over the last centuries with great clarity and then groups the many imaginations of food in the future of humanity into three different cornucopian futures: the classical future, the modernist future, and the recombinant future. In the classical future, institutions and ways of life – including food and foodways – continue much as they have been in the past except for steady improvements made by an ongoing process of development, largely based on the workings of theoretically free market forces. Food in the modernist future, by contrast, relies largely on discontinuities – radical new technologies, shockingly new foods, and a trend toward minimalism and exactness which is in contrast to the profusion and plenty promised by the markets of the classical future. These two ideologies about the future of food – classicism and modernism – have largely defined thought about the future of the food system and attempts at remediating the problems of the present. Both the modernist and the classical future present an imaginative reconstitution of society.

However, it is Belasco's last future – the recombinant future – which best defines both the present state of our food system as well as its current trajectory. The recombinant future is not organised around an imaginative reconstruction of society. In it, the interplay between the motivational forces of the classical and the modernist future allow marketing, with an absence of pretence to address the needs of future in an ethical way, to drive the food system. In the recombinant future as defined by Belasco, changes to the underlying structure of food production as radical as those imagined in the modernist future do take place – they are just papered over by marketing and food engineering which together maintain the appearance of continuity with just enough novelty for brand differentiation. The radicality of these unseen changes is not in the service of the good of all, but rather in the service of profit.

While both the modernist and the classical futures defined by Belasco focus on long-term outcomes, the recombinant future allows the near-horizon outcome of capital accumulation to dominate precisely because it prioritises the immediacy of consumption and therefore food choices detached from any associated ethical issues (Chapter 24). It is this tension between the apparent continuity of the consumption experience (which ultimately relies on the discontinuities of food production methods) and the resistance to these underlying discontinuities which underpins the current debate about the future of food taking place between advocates for conventional and alternative forms of agriculture (Chapters 10, 12, 16, 31, 32 and 33). This resistance takes place in the broadest socio-environmental terms and is often founded upon a care about and for landscapes. These discontinuities include a wide range of materials and processes from textured vegetable protein to genetically modified salmon to organic apples shipped from New Zealand to the United States out of season. At a more foundational and broadly philosophical scale however, monocropping and the selection of seed for maximisation of yield under optimal conditions instead of traditional seed saving which prioritises resilience under a variety of climatic possibilities may be the most significant discontinuity with the history of human agriculture (Clawson, 1985; Altieri, 1987) Together, the combined effect of these discontinuities is the conceptual disassociation of diet from landscape, and therefore the practical disassociation of people from their most intimate and critical relationship to landscapes.

As the recombinant future continues to play out and our food system is driven by food engineering and marketing in the service of profit, the discourse on the future of food has not slackened. Fraser et al. (2016) identify four distinct themes in their recent literature review. These include (1) increasing the capacity to grow more food through the development and application of biotechnological and other technical solutions; (2) correcting the current inequity in food distribution through changes to diets as well as revision of global power dynamics which leave many outside the market; (3) increasing local food sovereignty in opposition to the control of the food system by international organizations, corporations, and trade agreements; and (4) addressing market failures through proposing policies and regulations that would remove perverse incentives which generate negative outcomes with regard to human health, environmental health, and social justice. Using both utopia and landscape as methods of inquiry, the question to ask relative to each theme is whether it enriches and enlivens the full range of interrelated social and environmental systems associated with the production of food and takes a meaningful stance on issues of social justice.

While Fraser et al. (2016) present the arguments supporting and challenging each of these positions and do not evaluate the content of any of these themes, they do posit that any resolution of food system issues needs to include components of all these approaches. In addition to laying out the incredible breadth of literature that addresses problems with our food system, this review is particularly valuable for another aspect of its conclusion. The authors conclude that the largest barrier to resolving the problems with our food system is, in fact, political, and requires something like a mediated modelling process able to combine bottom-up participation and top-down advice from experts. If one, however, considers the widespread resistance to the biotechnology integral to the first theme and the parallel and politically resistive motivations underlying the three subsequent themes, one can see that much of the contemporary debate over food is grounded precisely in the grass-roots mistrust of the experts and the corporations who have been dictating the material, financial and regulatory frames of the current food system (see Chapters 9, 12, 25, 31 and 32). While the authors conclude with a hope that these two political logics can find a way to get along, all evidence presented by the authors in their literature review is that they are mutually exclusive and will continue to vie for the future of our food system.

The central reason that it seems unlikely that the two positions will find a middle ground is that they are approaching the food problem from opposing utopian impulses. The first category of proposals for the future of the food system – using biotechnology and other technical solutions to produce more food – is focused on maximising productivity by employing capital to develop and implement technological solutions while maintaining if not supporting the political and economic status quo. This world is the current political world, just technologically made more abundant with all concentration of political power that it implies (Collins, 2011). The correlate to increasing technology and the use of capital is the reduction in labour needs (Chapters 9, 12 and 16). Just as in More's analysis of the fate of people in the English countryside in the late eighteenth century, this focus on increasing the capital-generating capacity of land – often required just to pay for the technology itself – reduces the number of people who can make their living from the land and increases the number of people pushed into urban environments where they become dependent on wages for their livelihoods. This is exactly the situation that Howard and Wright challenged and which Hilberseimer was reacting against.

Additionally, while some technology, such as Monsanto's trademarked Crop Production System, allows a more fine-grained analysis of soil through the use of costly GIS technology-enabled farm equipment, most increased use of technology actually reduces the requirement to know the land intimately and supports instead the broad brush of monoculture. This monocultural method of farming clearly maps onto the abstraction of capitalism as it produces commodity crops which can be traded as an industrial feedstock sight-unseen, based on the barest description of quality and qualities. This technological approach fails to address the complexity of the food system in a holistic way and, as its opponents point out directly or by implication, therefore misses large sections and perhaps even the core of the problems with our food system as it exists (Carson, 2002; Warner, 2007; Estabrook, 2012). This blindness, I suggest here, is neither genuine nor benign. Rather, it is strategically necessary in order to shift some of the cost of the business of producing food to society in the forms of social and environmental externalities, thereby increasing profit (Pretty et al., 2001; Spittler et al., 2011).

It is precisely the lack of political change and the externalising of impacts implicit in increasing the role of technology and therefore capital in the production of food which forms the discontented heart and the utopian vision of the last three themes identified within the literature – addressing inequity, increasing local and community control of food, and addressing market failures by removing perverse incentives. These three themes focus directly not on the end of producing food itself but rather on significant changes to the social, political, and economic means which frame the future of food as a social and cultural, not simply a material, category. These last three categories of response all focus on the means by which food is produced and distributed rather than assuming that landscapes are simply a growing medium (as does the technological proposal). They take the holistic integration of environmental, cultural, and economic aspects of landscape and its contribution to food as a foundational concept. They seek closed, integrated resource chains that do not externalise the negative aspects of food production but rather bring accountability and responsibility into the food system. Fraser et al. (2016) offer a wide-ranging bibliography of texts related to these themes.

Within the literature which proposes solutions to our current food problems, there is a subset of books from within the fields of planning and landscape architecture that have addressed the role alternative forms of agriculture can play in re-imagining society and which are discussed later. While many books cross the lines between the following categories and there are categories such as homesteading that are valuable but not included here, within the framework of utopian and landscape method it makes sense to consider these works in three different categories. First, some volumes take on the broadly philosophical problems of our food system and discuss the political or social framework for making significant changes within our inhabited landscape. Second, there are books that approach the problems with our food system and the possibility of alternatives through the lens of regional and metropolitan planning and therefore remain at the level of policy. Lastly, a significant number of books present folios of projects which – taken together or independently – offer already realised landscapes or particular landscape proposals from which to imagine more complete imaginative reconstructions of society.

Three books stand out in the first category: *Civic Agriculture: Reconnecting Farm, Food and Community* by Thomas A. Lyson (2004); *Rebuilding the Foodshed: How to Create Local, Sustainable and Secure Food Systems* by Philip Ackerman-Leist (2013); and *Public Produce: Cultivating Our Parks, Plazas and Streets for Healthier Cities* by Darrin Nordahl (2012). Lyson defines civic agriculture, in contrast to the industrialised and globalised food system, as a collection of agriculture and food practices that not only produce better food but also create jobs and strengthen community identities. He claims that a "new social blueprint for agriculture" must come from below – from the grass-roots practices of building community relationships around food (Lyson, 2004: 103). Ackerman-Leist begins with a discussion of the meaning of local, and then discusses the drivers involved in rebuilding the food system (Ackerman-Leist, 2013). One of the most fruitful and poignant discussions in his book is the role that energy currently plays in food production and, by implication, the role that a thorough examination of energy ought to play in imagining new forms of food systems which would have benefits in terms of his proposed set of values – health, equity, and environment. Nordahl takes a look at the space of cities and civility and sees a way to systematically integrate food production into our public spaces and therefore into our civic lives (Nordahl, 2012). He points, as many do, to the success of the victory garden programs of the past and proposes that we ought to make a similar peace-time effort at a social scale to engage some of our most easily accessible land, public land in cities.

Outstanding texts in the second category, those that consider the problem as one of planning in the broadest sense are *The Urban Food Revolution: Changing the Way We Feed Cities* by Peter Ladner (2012); *Second Nature Urban Agriculture: Designing Productive Cities* edited by Andre Viljoen and Katrin Bohn (2014); and *Designing Urban Agriculture: A Complete Guide to the Planning, Design, Construction, Maintenance and Management of Edible Landscapes* by April Philips (2013). Ladner's book is valuable because it asks some of the hard questions about local food and the real value that it offers (Ladner, 2012). He proposes that the resolution of our food problems needs to come through a combination of individual choice, community engagement, and government action. In addition, his work stands out because of his emphasis on the preservation of rural land in combination with a proposal for alternative forms of agriculture and food production. *Second Nature Urban Agriculture* (Viljoen and Bohn, 2014) is a sequel to the book *Continuous Productive Urban Landscapes* (Viljoen, 2005). It offers a theoretical basis for the imagination of alternative food systems embedded in alternative urban forms and offers actions that can be taken at the level of the city, backed up by proposed and completed urban projects. While Philips' book, *Designing Urban Agriculture*, is perhaps not the exhaustive practical guide to urban agriculture the title promises, it is yet a highly valuable book that frames chapters about ecology, planning strategies, form, systems integration, lifecycle operations, and outreach and community around examples of urban agricultural installations and pragmatic items such as a soil assessment checklist (Philips, 2013).

Finally, *Farming the City: Food as a Tool for Today's Urbanization* edited by Miazzo and Minkjan (2013), *Urban Farms* by Rich and Benson (2014), and *Carrot City: Creating Places for Urban Agriculture* by Gorgolewski, Komisar, and Nasr (2011) all provide a broad coverage of examples of experimentation in alternative food system components ranging from production to distribution and from digital information sharing to soil building. As the subtitle suggests, *Farming the City* examines food as a tool for urbanisation and presents cases that support this understanding of food (Miazzo, 2013). It is important to point out that the book presents urbanisation not in terms of traffic and population densities but in terms of economic relationships and an urban society comprised of citizens, artists, and activists. *Urban Farms* offers less writing and a less theoretical basis than the other volumes in this list, but its sixteen examples are well photographed and therefore the book offers a valuable tool in communicating the vision of food system components integrated with spaces of inhabitation and how significantly such a system could support an imaginative reconstruction of society (Rich and Benson, 2014). Lastly, *Carrot City* offers significant write-ups about projects that have not been realised as well as those that have (Gorgolewski, 2011). This juxtaposition of the real and the more numerous proposed projects is valuable in that it gives the reader an easy way to move from existing alternatives to the current food system to landscape and utopian thinking at a broad scale.

All of these texts propose imaginative reconstructions of society, all of them are utopian in that they propose a better way for living through a holistic re-integration of life, landscape, and food. They operate at many different scales and with different particular goals – both projected and realised. In some way, whether stated or not, they all challenge the priorities underlying the current neoliberal structure of property ownership and the distribution of benefits of the use of land through the mechanism of profit, and in particular through the extraction of food from the land as one means to that end. While all of these efforts at revising the food system to be more socially just, more ecologically sound and more politically enabling are positive and productive, we are, however, missing visions at the scale of Howard, Wright, and Hilberseimer that directly tie the radical political change the world needs directly to food production and distribution. We are missing visions that see political change and food system change as inseparable and that fully engage landscape as method in order to visualise an alternative, perhaps utopian, future which contrasts with the present. To rephrase a utopian call to arms, how can we not be reasonable and demand impossibly grand visions of a better world? And in our demanding, should we not look to the theories and practices of landscape architecture in which one finds the ability to be both truly and practically utopian?

Further reading

Ackerman-Leist, P. (2013) *Rebuilding the foodshed how to create local, sustainable, and secure food systems*. Santa Rosa, CA: Post Carbon Institute.

Albritton, R. (2009) *Let them eat junk: How capitalism creates hunger and obesity*. New York: Pluto Press.

Allen, W. (2012) *The good food revolution: Growing healthy food, people and communities*. New York: Gotham Books.

Altieri, M.A., & Merrick, L.C. (1987) In situ conservation of crop genetic resources through maintenance of traditional farming systems. *Economic Botany*, 41(1), January–March, 86–96.

Beck, J. (2013) *John Nolen and the metropolitan landscape*. USA and Canada: Routledge.

Beck, T. (1995) The green revolution and poverty in India. *Applied Geography*, 15(2). 161–181.

Belasco, W.J. (2006) *Meals to come: A history of the future of food*. Berkeley, CA: University of California Press.

Bellamy, E. (1926) *Looking backward: 2000 to 1887*. Boston, MA and New York: Houghton Mifflin Company.

Bowonder, B. (1979) Impact analysis of the green revolution in India. *Technological Forecasting and Social Change*, 15(4), 297–313.

Buder, S. (1990) *Visionaries and planners: The Garden City movement and the modern community*. New York: Oxford University Press.

Carolan, M. (2013) *The real cost of cheap food*. Florence: Taylor and Francis.

Carolan, M. (2016) *Embodied food politics*. London: Routledge.

Carson, R. (2002) *Silent spring*. New York: First Mariner Books.

Caswell, D., & Siskin, D. (2012) *The little house in the suburbs: Backyard farming and home skills for self-sufficient living*. Cincinnati: Betterway Homes.

Clawson, D.L. (1985) Harvest security and intraspecific diversity in traditional tropical agriculture. *Economic Botany*, 39(1), January–March, 56–67.

Collins, K. (2001) Statement of Keith Collins, Chief Economist, U.S. Department of Agriculture before the U.S. Senate Committee on Appropriations Subcommittee on Agriculture, Rural Development and Related Agencies, May 17, 2011. [on-line]. www.usda.gov/oce/newsroom/archives/testimony/2005-1997files/051701co.html (accessed 2 May 2017).

Corner, J. (1999) *Recovering landscape: Essays in contemporary landscape architecture*. New York: Princeton Architectural Press.

Donohue, M., & Cunningham, D.L. (2009) Effects of grain and oilseed prices on the costs of US poultry production. *Journal of Applied Poultry Research*, 18(2), 325–337. Available from: DOI: https://doi.org/10.3382/japr.2008-00134.

Estabrook, B. (2012) *Tomatoland: How modern industrial agriculture destroyed our most alluring fruit*. Kansas City: Andrews McMeel Publishing.

FAO. (2017) *The future of food and agriculture – trends and challenges*. Rome: FAO. Available from: www.fao.org/3/a-i6583e.pdf [Accessed 5 April 2017].

FAO, IFAD and WFP. (2015) *The state of food insecurity in the world 2015. Meeting the 2015 international hunger targets: Taking stock of uneven progress*. Rome: FAO.

Fedoroff, N.V., Battisti, D.S., Beachy, R.N., Cooper, P.J.M., Fischhoff, D.A., Hodges, C.N., Knauf, V.C., Lobell, D., Mazur, B.J., Molden, D., Reynolds, M.P., Ronald, P.C., Rosegrant, M.W., Sanchez, P.A., Vonshak, A., & Zhu, J.-K. (2010) Radically rethinking agriculture for the 21st century. *Science*, 327(5967), 833–834.

Finley, R. (2013) A guerilla gardener in South Central LA. *TEDtalks*. www.ted.com/talks/ron_finley_a_guerilla_gardener_in_south_central_la?language=en (accessed 16 April 2016).

Fitzgerald, K. (2016) Thinking globally, acting locally: Locavorism and humanist sociology. *Humanity and Society*, 40(1), 3–21.

Fraser, Evan et al. (2016) Biotechnology or organic? Extensive or intensive? Global or local? A critical review of the potential pathways to resolve the global food crisis. *Trends in Food Science and Technology*, 48, 78–87.

Gayeton, D. (2014) *Local: The new face of food and farming in America*. New York: Harper Design.

Gorgolewski, M., Komisar, J., & Nasr, J. (2011) *Carrot city: Creating places for urban agriculture*. New York: The Monacelli Press.

Groth, P., & Bressi, T. (1997) *Understanding ordinary landscapes*. New Haven, CT: Yale University Press.

Gumpert, D.E. (2013) *Life, liberty, and the pursuit of food rights: The escalating battle over who decides what we eat*. White River Junction: Chelsea Green Publishing Company.

Hall, P., & Ward, C. (1998) *Sociable cities: The legacy of Ebenezer Howard*. Chichester: John Wiley and Sons.

Harvey, D. (2005) *A brief history of neoliberalism*. Oxford: Oxford University Press.

Harvey, D. (2008) *Spaces of hope*. Berkeley, CA: University of California Press.

Hester, R. (2006) *Design for ecological democracy*. Cambridge, MA: MIT Press.

Hilberseimer, L. (1949) *The new regional pattern: Industries and gardens, workshops and farms*. Chicago. Paul Theobald.

Howard, E. (2010) *To-morrow: A peaceful path to real reform*. Cambridge: Cambridge University Press.

Howard, E., Mumford, L., & Osborn, F.J. (1902) *Garden cities of to-morrow (being the second edition of "To-morrow: A Peaceful Path to Real Reform")*. London: Swan Sonnenschein and Co., Ltd. Accessed via babel.hathitrust.org October 29, 2015, permanent link, http://hdl.handle.net/2027/hvd.32044029579976.

Jackson, J., & Horowitz, H. (2000) *Landscape in sight: Looking at America*. New Haven, CT: Yale University Press.

Khan, S., & Hanjra, M. (2009) Footprints of water and energy inputs in food production – global perspectives. *Food Policy*, 34(2), 130–140.

Kloppenberg, J., Hendrickson, J., & Stevenson, G. (1996) Coming in to the foodshed. In: *Rooted in the land*. New Haven, CT: Yale University Press, pp. 113–123.

Kraft, E. (2011) *The complete idiot's guide to urban homesteading*. New York: Alpha Books.

Kropotkin, P. (1968) *Fields, factories and workshops: Or industry combined with agriculture and brain work with manual work*. New York: Greenwood press.

Kunstlerr, J. (2005) *The long emergency: Surviving the converging catastrophes of the twenty-first century*. New York: Atlantic Monthly Press.

Ladner, P. (2011) *Urban food revolution: Changing the way we feed cities*. Gabriola Island: New Society Publ.

Levitas, R. (2013) *Utopia as Method: The imaginary reconstruction of society*. New York: Palgrave Macmillan.

Lipinisky, B. (2015) What's food loss and waste got to do with sustainable development? A lot, actually. published by World Resources Institute on September 23, 2015. www.wri.org/blog/2015/09/what%E2%80%99s-food-loss-and-waste-got-do-sustainable-development-lot-actually (accessed 19 April 2016).

Lowder, S., Skoet, J., & Raney, T. (2016) The number, size, and distribution of farms, smallholder farms and family farms worldwide. *World Development*, 87, 16–29. Rome, Italy: Food and Agriclture Organization of the United Nations.

Lyson, T.A. (2004) *Civic agriculture: Reconnecting farm, food, and community*. Medford, MA: Tufts.

Malthus, T.R. (1976) *Essay on the principle of population*. New York: Norton.

Markham, B. (2014) *The mini-farming Bible: The complete guide to self-sufficiency on ¼ acre*. New York: Skyhorse Publishing.

Mateo-Sagasta, J., & Burke, J. (nd) *Agriculture and water quality interactions: A global overview*. SOLAW Background Thematic Report – TR08. Food and Agriculture Organization.

McHarg, I.L. (1992) *Design with nature*. New York: John Wiley and Sons, Ltd.

McHarg, I., & Steiner, F. (1998) *To heal the Earth: Selected writings of Ian L. McHarg*. Washington, DC: Island Press.

McMichael, P. (2009) A food regime geneology. *The Journal of Peasant Studies*, 31(1), 139–169. Available from: DOI: 10.1080/03066150902820354.

Meinig, D. (ed.) (1979) *The interpretation of ordinary landscapes: Geographical essays*. New York: Oxford University Press.

Miazzo, F., & Minkjan, M. (2013) *Farming the city: Food as a tool for today's urbanization*. Amsterdam: Valiz.

More, T. (1964) *Utopia*. New Haven, CT: Yale University Press.

Morris, W. (1891) *News from nowhere*. London: Reeves and Turner. Permanent link: Accessed via babel.hathitrust.org January 18, 2017, permanent link, https://hdl.handle.net/2027/dul1.ark:/13960/t7cr6hc61.

Mostafavi, M. (2016) *Ecological urbanism*. Zurich: Lars Muller.

Nassauer, J.I. (2012) Landscape as method and medium for the ecological design of cities. *Future City Resilience in Ecology and Urban Design*, 79–98. doi:10.1007/978-94-007-5341-9_5.

Nestle, M. (2002) *Food politics: How the food industry influences nutrition and health*. Berkeley, CA: University of California Press.

Nordahl, D. (2012) *Public produce the new urban agriculture*. Washington, DC: Island Press.

Otero, G., Pechlaner, G., & Gurcan, E. (2013) The political economy of "Food Security" and trade: Uneven and combined dependency. *Rural Sociology*, 78(3), 263–289. Available from: DOI: 10.1111/ruso.12011.

Patel, R. (2012) *Stuffed and starved: The hidden battle for the world food system*. Brooklyn, NY: Melville House.

Philips, A. (2013) *Designing urban agriculture: A complete guide to the planning, design, construction, maintenance, and management of edible landscapes*. Hoboken, NJ: Wiley.

Pretty, J., Brett, C., Gee, D., Hine, R., Mason, C., Morison, J., Rayment, M., Van der Bijl, G., & Dobbs, T. (2001) Policy challenges and priorities for internalizing the externalities of modern agriculture. *Journal of Environmental Planning and Management*, 44(2), 263–283.

Rees, E. (1992) Ecological footprints and appropriated carrying capacity: What urban economics leaves out. *Environment and Urbanization*, 4(2), 121–130.

Rich, S.C., & Benson, M. (2014) *Urban farms*. New York: Abrams.

Roberts, P. (2009) *The end of food: The coming crisis in the world food industry*. London: Bloomsbury.

Rosin, C.J., Stock, P., & Campbell, H. (2014) *Food systems failure: The global food crisis and the future of agriculture*. London: Routledge.

Sargent, L.T. (2010) *Utopianism: A very short introduction*. Oxford: Oxford University Press.

Sen, A. (1982) The food problem: Theory and policy. *Third World Quarterly*, 4(3), 447–459. Available from: DOI: 10.1080/01436598208419641.

Shiva, V. (1991) *The violence of the green revolution*. London: Zed Books Ltd.

Shiva, V. (2016) *Stolen harvest: The hijacking of the global food supply*. Lexington, KY: University Press.

Spittler, J., Ross, R., & Block, W. (2011) The economic impact of agricultural subsidies in the United States. *The Journal of Social, Political and Economic Studies*, 36(3), Fall, 301–317.

Springer, S. (2016) *The discourse of neoliberalism: An anatomy of a powerful idea*. London: Roman and Littlefield.

Steel, C. (2013) *Hungry city: How food shapes our lives*. London: Vintage Books.

Stock, Corolan, & Rosin (eds.) (2015) *Food utopias: Reimagining citizenship, ethics and community*. Abingdon, Oxon and New York: Routledge.

Swaffield, S. (2002) *Theory in landscape architecture: A reader*. Philadelphia, PA: University of Pennsylvania.

USDA Economic Research Service (2016) *The number of farms has leveled off at about 2.1 million*. www.ers.usda.gov/data-products/chart-gallery/gallery/chart-detail/?chartId=58268 (accessed 1 May 2017).

Vallianatos, E. (2006) *This land is their land: How corporate farms threaten the world*. Monroe, ME: Common Courage Press.

Viljoen, A. (ed.) (2005) *Continuous productive urban landscapes: Designing urban agriculture for sustainable cities*. Great Britain: Architectural Press.

Viljoen, A., & Bohn, K. (2014) *Second nature urban agriculture: Designing productive cities: Ten years on from the continuous productive urban landscape (CPUL City) concept*. London: Routledge.

Waldheim, C. (2016) *Landscape as urbanism: A general theory*. Princeton, NJ: Princeton University Press.

Warner, K. (2007) *Agroecology in action: Extending alternative agriculture through social networks*. Cambridge, MA: The MIT Press.

Weis, T. (2007) *The global food economy: The battle for the future of farming*. London: Zed Books.

Winne, M. (2010) *Closing the food gap: Resetting the table in the land of plenty*. Boston, MA: Beacon Press.

World Health Organization. (2016) *Obesity and overweight factsheet, updated June 2016*. www.who.int/mediacentre/factsheets/fs311/en/ (accessed 30 April 2017).

Wright, F.L. (1951) *When democracy builds*. Chicago: University of Chicago Press.

Wright, F.L. (1970) *The living city*. New York: New American Library.

Wright, F.L. (1995) *Broadacre city*. Tucson, AZ: University of Arizona Press.

WWF (2016) *Living planet report 2016. Risk and resilience in a new era*. Gland: WWF International.

Zeunert, J. (2017) *Landscape architecture and environmental sustainability: Creating positive change through design*. London: Bloomsbury Academic.

Index

Note: Information in figures is indicated by page numbers in *italics*. Information in tables is indicated by page numbers in **bold.**

FSC
www.fsc.org
MIX
Papier | Fördert
gute Waldnutzung
FSC® C083411

Kolibri 360 GmbH
Ferdinand-Jühlke-Straße 7
99095 Erfurt, Deutschland
produktsicherheit@kolibri360.de